The Theory of the Moiré Phenomenon Volume II: Aperiodic Layers

Computational Imaging and Vision

Volume 34

The Theory of the Moiré Phenomenon

Volume II: Aperiodic Layers

by

Isaac Amidror

Peripheral Systems Laboratory,
Ecole Polytechnique Fédérale de Lausanne (EPFL), Switzerland

 Springer

A C.I.P. Catalogue record for this book is available from the Library of Congress.

ISBN-13 978-90-481-7373-0
ISBN-10 1-4020-5458-0 (e-book)
ISBN-13 978-1-4020-5458-7 (e-book)

Published by Springer,
P.O. Box 17, 3300 AA Dordrecht, The Netherlands.

www.springer.com

Printed on acid-free paper

To my parents

How can it be that mathematics, being after all a product of human thought which is independent of experience, is so admirably appropriate to the objects of reality?

Albert Einstein [Einstein02 p. 209]

Front cover image: A star-shaped Glass pattern in the superposition of two slightly transformed copies of an aperiodic dot screen. See Problem 3-18 at the end of Chapter 3.

Back cover images: Radial (or circular) dot trajectories in the superposition of two copies of an aperiodic dot screen, where one of the two superposed layers has been slightly scaled (or slightly rotated). See Figs. 2.1(e) and 2.1(c), respectively.

Contents

Preface xiii

1. Introduction 1

   1.1   The moiré effect between aperiodic structures 1
   1.2   Brief historical background and main applications 4
   1.3   The scope of the present book 5
   1.4   Overview of the following chapters 7
   1.5   About the exercises and the moiré demonstration samples 9

2. Background and basic notions 11

   2.1   Introduction 11
   2.2   Periodic, repetitive and aperiodic layers 15
   2.3   Superposition of aperiodic layers 19
        2.3.1   Glass patterns and correlation 19
        2.3.2   Stable vs. singular moiré-free superpositions 24
        2.3.3   Macrostructures and microstructures in the superposition 26
   2.4   The element distribution in the original layers and its influence 28
   2.5   Multilayer superpositions 30
        Problems 32

3. Glass patterns and fixed loci 47

   3.1   Introduction 47
   3.2   The fixed point theorem 48
   3.3   Behaviour of Glass patterns and periodic moirés under affine mappings . . . 51
        3.3.1   Behaviour under layer rotations 51
        3.3.2   Behaviour under layer scalings 53
        3.3.3   Behaviour under layer shifts 54
        3.3.4   Behaviour under a general affine transformation 59
   3.4   Behaviour of Glass patterns under general layer transformations 63
        3.4.1   Examples with non-linear layer mappings 65
   3.5   Mappings in both layers; mutual fixed loci 72
   3.6   Synthesis of fixed loci in the superposition 83
   3.7   Almost fixed points 87
        Problems 96

4. Microstructures: dot trajectories and their morphology 105

4.1   Introduction 105
4.2   Morphology of the microstructures; dot trajectories 106
4.3   Dot trajectories as solution curves of a system of differential equations 107
4.4   Dot trajectories as a vector field 109
       4.4.1   The curve equations of the dot trajectories 114
4.5   The dot trajectories when both layers undergo transformations 124
4.6   Synthesis of dot trajectories 134
4.7   Dot trajectories in periodic and in repetitive cases 137
4.8   The microstructures under different superposition rules 139
4.9   The visual interpretation of microstructures 140
       Problems 148

5. Moiré phenomena between periodic or aperiodic screens 157

5.1   Introduction 157
5.2   Brief review: moiré patterns, Glass patterns and dot trajectories 158
5.3   A few detailed examples to illustrate the formal results 161
5.4   Invariance properties of moiré patterns, Glass patterns and dot trajectories . . 175
       Problems 181

6. Glass patterns in the superposition of aperiodic line gratings 187

6.1   Introduction 187
6.2   Glass patterns in the superposition of straight line gratings 188
       6.2.1   Superposition of 1D vs. 2D aperiodic layers 189
       6.2.2   Superposition of periodic vs. aperiodic line gratings 192
6.3   Mathematical derivations: generalization of the indicial equations method . . 193
6.4   Examples of Glass patterns in the superposition of curved line gratings 196
6.5   The effect of adding constraints to the original layers 208
6.6   A first step towards the intensity profiles of Glass and moiré patterns 214
       Problems 221

7. Quantitative analysis and synthesis of Glass patterns 225

7.1   Introduction 225
7.2   Brief review of the Fourier approach in the periodic case 226
       7.2.1   Spectra of periodic and aperiodic layers 228
       7.2.2   Moiré effects in the superposition of periodic gratings 230
       7.2.3   Moiré effects in the superposition of periodic dot screens 233
       7.2.4   Shape of the intensity profile of the moiré pattern 236
       7.2.5   Orientation and size of the moiré cells 238

7.3   Intensity profile of Glass patterns in the superposition of aperiodic
        gratings 238
        7.3.1   Superposition of correlated gratings 240
        7.3.2   Superposition of uncorrelated gratings 246
7.4   Intensity profile of Glass patterns in the superposition of aperiodic
        screens 248
        7.4.1   Superposition of correlated screens 249
        7.4.2   Shape of the intensity profile of the Glass pattern 252
        7.4.3   Orientation and size of the Glass pattern 253
        7.4.4   Cases with several fixed points or with continuous fixed lines 254
        7.4.5   Superposition of uncorrelated screens 258
        7.4.6   Discussion 258
7.5   Higher order moirés 259
7.6   Intermediate, partly aperiodic cases 260
7.7   Intermediate, partly correlated cases 262
7.8   Glass patterns and cross correlation 264
        Problems 270

Appendices

A. Fixed point theorems for first- and second-order polynomial mappings . . . 281

A.1   Introduction 281
A.2   The fixed point theorem for linear or affine mappings 281
A.3   The fixed point theorem for second-order polynomial mappings 284
A.4   Mutual fixed points between two mappings; application to the
        moiré theory 288

B. The various interpretations of a 2D transformation 289

B.1   Introduction 289
B.2   Interpretation as two surfaces over the plane or as two sets of
        level lines 289
B.3   Interpretation as a mapping from the plane into itself 290
B.4   Interpretation as a domain transformation $r(\mathbf{g}(x,y))$ 294
B.5   Interpretation as a coordinate change 295
B.6   Interpretation as a 2D vector field 298
B.7   Relationship between the different representations of $\mathbf{g}(x,y)$ 301
B.8   Remark on the local reflection of a 2D transformation 306

C. The Jacobian of a 2D transformation and its significance 309

C.1   Introduction 309
C.2   Geometric interpretation of the Jacobian 309
C.3   Properties of the transformation $\mathbf{g}(x,y)$ that can be deduced from
        its Jacobian 311
C.4   The local orientation properties of a transformation $\mathbf{g}(x,y)$ 319
C.5   Other properties of $\mathbf{g}(x,y)$ that can be deduced from its Jacobian matrix 322

D. Direct and inverse spatial transformations 327
D.1   Introduction 327
D.2   Background and basic notions 327
D.3   A deeper look into the domain and range planes of the
        mapping $(u,v) = \mathbf{g}(x,y)$ 332
D.4   2D transformations and their inverse 336
        D.4.1   The image of the standard Cartesian grid under the
                 transformations $\mathbf{g}$ and $\mathbf{g}^{-1}$ 337
        D.4.2   The image of a general curve under the
                 transformations $\mathbf{g}$ and $\mathbf{g}^{-1}$ 340
D.5   The active and passive interpretations of a transformation 343
D.6   Domain and range transformations of a function 347
        D.6.1   The 1D case 349
        D.6.2   The 2D case 351
        D.6.3   The effect of transformation $\mathbf{g}$ on objects and on their
                 characteristic functions 355
D.7   The relative point of view: object deformations
        vs. coordinate deformations 356
D.8   Examples 357
D.9   Other possible sources of confusion 386
        D.9.1   Forward and backward mapping algorithms in digital imaging . . . 386
        D.9.2   Pre-multiplication and post-multiplication based notations 389
D.10   Implications to the moiré theory: issues related to the figures 390
D.11   Fixed points of a superposition in terms of direct or inverse
        transformations 399
        D.11.1   Fixed points when only one layer is transformed 399
        D.11.2   Fixed points when both layers undergo transformations 401
D.12   Useful approximations 405

E. Convolution and cross correlation 411
E.1   Introduction 411
E.2   Convolution 411

E.3   Cross correlation   413
E.4   Extension to more general cases   415
E.5   The Fourier transform of convolution and of cross correlation   416
E.6   Methods for quantifying the correlation; similarity measures   418

F. The Fourier treatment of random images and of their superpositions   421
F.1   Introduction   421
F.2   Stochastic processes and their power spectra   421
F.3   Possible stochastic modelizations of random screens and gratings   425
        F.3.1   Point processes   425
        F.3.2   Shot noise   426
        F.3.3   Random fields   429
F.4   Stochastic modelization of layer superpositions   430
F.5   Evaluation of the stochastic vs. deterministic approaches for
        our application   430

G. Integral transforms   433
G.1   Introduction   433
G.2   Fourier decomposition of periodic and aperiodic structures   433
G.3   Generalized Fourier decomposition of geometrically transformed
        structures   434
G.4   Integral transforms and their kernels   435
G.5   The use of generalized Fourier transforms in the moiré theory   439

H. Miscellaneous issues and derivations   443
H.1   Classification of the dot trajectories   443
        H.1.1   Classification of the dot trajectories in the linear case   444
        H.1.2   Classification of the dot trajectories in the non-linear case   447
H.2   The connection between the vector fields $\overline{\mathbf{h}}_1(x,y)$ and $\overline{\mathbf{h}}_2(x,y)$ in Sec. 4.5 . . .   451
H.3   Hybrid (1,-1)-moiré effects whose moiré bands have 2D
        intensity profiles   452

I. Glossary of the main terms   457
I.1   About the glossary   457
I.2   Terms in the image domain   457
I.3   Terms in the spectral domain   461
I.4   Terms related to moiré   463
I.5   Terms related to light and colour   466
I.6   Miscellaneous terms   467

List of notations and symbols 473

List of abbreviations 475

References 477

Index 485

Preface

Since *The Theory of the Moiré Phenomenon* was published it became the main reference book in its field. It provided for the first time a complete, unified and coherent theoretical approach for the explanation of the moiré phenomenon, starting from the basics of the theory, but also going in depth into more advanced research results. However, it is clear that a single book cannnot cover the full breadth of such a vast subject, and indeed, this original volume admittently concentrated on only some aspects of the moiré theory, while other interesting topics had to be left out.

Perhaps the most important area that remained beyond the scope of the original book consists of the moiré effects that occur between correlated random or aperiodic structures. These moiré effects are known as Glass patterns, after Leon Glass who described them in the late 1960s. However, this branch of the moiré theory remained for many years less widely known and less understood than its periodic or repetitive counterpart: *Less widely known* because moiré effects between aperiodic or random structures are less frequently encountered in everyday's life, and *less understood* because these effects did not easily lend themselves to the same mathematical methods that so nicely explained the classical moiré effects between periodic or repetitive structures. Only recently has it been shown that in spite of their very different appearance and properties, moiré patterns between periodic or repetitive structures and Glass patterns between aperiodic or random structures are, in fact, particular cases of the same basic phenomenon, and all of them satisfy the same fundamental rules. These new research results have accumulated into a considerable volume that could no longer fit into an additional chapter in the original book. Rather, it became evident that this new material merits a new volume of its own, that should be entirely devoted to the aperiodic case.

And indeed, the study of the moiré phenomena which occur between random or aperiodic structures is not less interesting (and certainly not less surprising and fascinating) than the study of their periodic or repetitive counterparts. In a way, the diversity of the phenomena which occur in aperiodic superpositions is even more surprising than that of periodic cases, due to the unique interplay between the macroscopic and microscopic structures that appear in the superposition, and their very different properties. And yet, it turns out that using a suitable approach, earlier results from the periodic case can be extended to englobe the aperiodic case, too, although such an extension could *a-priori* seem hopeless. This applies to the main mathematical approaches already known from the periodic case, including the indicial equations method and the Fourier approach, both of which have been successfully extended to the aperiodic case. It should be noted, however, that just as in the periodic case, the indicial equations method as well as the Fourier approach are only applicable to the study of the macrostructures (the macroscopic phenomena in the superposition), but not to the study of the microstructures. And indeed, as it is clearly shown in the present volume, the study of the microstructures requires a different arsenal of mathematical tools. (Note that this is also true in the periodic case, and indeed, in the first volume, too, the microstructures had to be discussed in a separate chapter, Chapter 8.)

The present book is intended to be a complementary, yet stand-alone companion to the original volume. Just like the first volume, it provides a full general purpose and application-independent exposition of the subject. It leads the reader through the various phenomena which occur in the superposition of correlated aperiodic layers, both in the image and in the spectral domain. And just like its predecessor, this volume favours a pictorial, intuitive approach which is supported by mathematics, and it includes a large number of figures and illustrative examples, some of which are visually striking and even spectacular.

Although this book has no pretentions to cover the entire moiré theory between aperiodic or random layers, it lays the foundations of this theory and gives for the first time an extended, comprehensive introduction to the subject. But it also goes in many aspects beyond the introductory level, and treats in depth some of the most interesting research results that have been recently obtained. Among other subjects, this book gives the full mathematical explanation of Glass patterns, describes their various properties, and explores their behaviour when the original layers undergo geometric transformations. It explains in detail the nature and the mathematical properties of the dot trajectories that often accompany Glass patterns in aperiodic screen superpositions, and it also investigates the one-dimensional counterparts of the classical Glass patterns that are generated in the superposition of correlated aperiodic line gratings. Most importantly, this book also provides the full Fourier-based approach, which allows us to explore Glass patterns and their intensity levels *quantitatively* and not only *qualitatively*. And finally, last but not the least, this book shows throughout the entire text and the accompanying figures the similarities as well as the differences between the moiré patterns which are obtained in periodic and in aperiodic cases, and it presents the theoretic results in a unified and coherent manner that clearly shows the relationship between the periodic and the aperiodic cases.

As already mentioned, this book is intended to be independent, as far as possible, from its predecessor. For this reason some fundamental notions from the original volume that are necessary for the comprehention of the new material are shortly reviewed here in Chapter 2, and some figures and results from the original volume are also reminded (or slightly adapted) in later chapters. Nevertheless, in order to avoid excessive repetitions, reference is occasionally made to some specific points in the original volume. On the other hand, it is interesting to note that in many ways the present volume also sheds new light on topics that were already explained in its predecessor, for example, by showing how these topics can be interpreted within a larger, unified framework that covers both periodic and aperiodic cases. And indeed, the wider point of view offered by the present volume often deepens our understanding of classical results from the periodic moiré theory.

This book is intended for students, scientists and engineers wishing to widen their knowledge of the moiré effect and its relationship with random or aperiodic structures. In particular, it will be very useful for people interested in the various moiré applications and in moiré-based technologies. The reader will find in this book not only a theoretical explanation of the moiré phenomena in question, but also practical recipes accompanied by many examples for the synthesis of aperiodic layers that give in their superposition moiré patterns having desired shapes, intensity profiles or microstructures. The prerequisite mathematical background is limited to an elementary familiarity with calculus and with the Fourier theory. But even occasional readers with no mathematical background will certainly enjoy the beauty of the effects illustrated throughout this book, and — it is our hope — may be tempted to learn more about their nature and their properties.

The material in this book is based on the author's personal research at the EPFL (*Ecole Polytechnique Fédérale de Lausanne*). This work would have never been possible without the support and the excellent research environment provided by the EPFL. The author wishes to express his gratitude to Prof. Roger D. Hersch, the head of the Peripheral Systems Laboratory of the EPFL, for his encouragement throughout the different stages of this project. Many thanks are also due to the publishers for their continued helpfulness and availability throughout the publishing cycle.

Chapter 1

Introduction

1.1 The moiré effect between aperiodic structures

It is a well-known fact that the superposition of periodic layers may give rise to new periodic structures that do not exist in any of the original layers (see, for example, Fig. 1.1(a)). These structures, known as *moiré effects,* have been thoroughly studied over the years, and their mathematical foundations are today fully understood. The same is also true for moiré effects between repetitive layers, i.e. between geometric transformations of periodic layers (see, for example, Fig. 1.1(b)). Moiré phenomena which occur between periodic or repetitive layers have been extensively explained and illustrated in *Vol. I* of the present work.

It turns out, however, that moiré effects may be also generated in the superposition of aperiodic or random layers. These effects look very different from the moiré effects which occur in the superposition of periodic or repetitive layers, and they often consist of a single structure resembling a top-viewed funnel or a distant galaxy in the night sky (see, for example, Fig. 1.2(a)). This phenomenon is known in literature as a *Glass pattern,* after Leon Glass who described it in the late 1960s [Glass69, Glass73].

And yet, moiré effects that occur between random or aperiodic layers are less widely known to the large public than their periodic or repetitive counterparts, probably because they are less frequently encountered in everyday's life. Everybody has surely seen, in one occasion or another, moiré patterns which appear in the intersection of repetitive structures (for example, between the railings on both sides of a bridge, between two far picket fences, or even at home, between the folds of a nylon curtain). On the other hand, intersections of aperiodic structures such as random line gratings or random dot screens are certainly less likely to be encountered.

But there also exists a deeper reason why Glass patterns are less frequently observed than their periodic or repetitive counterparts. While moiré effects readily appear in almost any superposition of periodic or repetitive layers — in fact, it is often much more difficult to find such layers that *do not* generate moiré effects in their superposition — aperiodic structures are much more "capricious", and their moiré effects only manifest themselves when some strict conditions are satisfied. The main condition is that the two superposed structures must be at least partially correlated; as it is clearly shown in Fig. 1.2(b), two uncorrelated structures will not give a moiré effect in their superposition. But even if the two aperiodic layers in question are almost or even fully identical, no moiré effects will be seen in their superposition if the relative angle and the relative displacement between the two layers are not appropriate. Furthermore, moiré effects between aperiodic layers do not

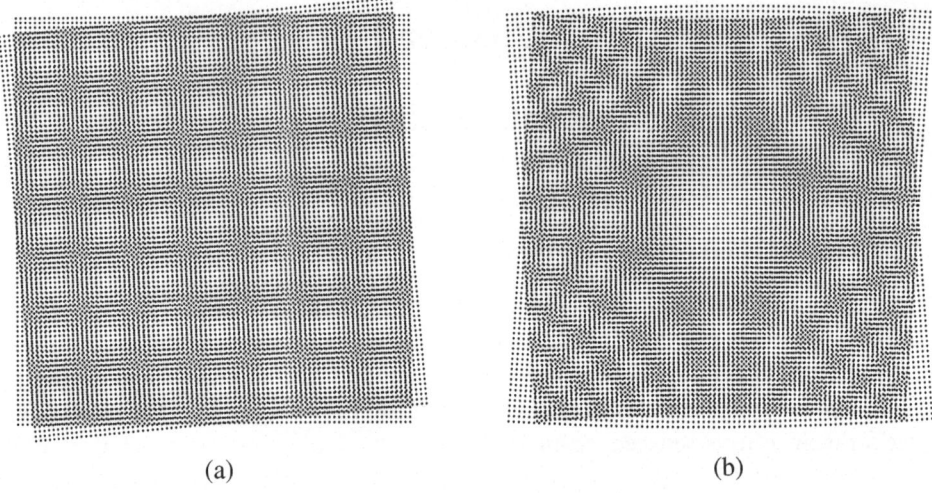

(a)   (b)

Figure 1.1: (a) A moiré effect in the superposition of two periodic layers.
(b) A moiré effect in the superposition of two repetitive layers.

extend throughout the superposed area, as their periodic counterparts do, and they are usually visible only in a small part of the superposition. The slightest shift between the two layers may cause this moiré effect to move far away from the center, so that within the region of interest (i.e. inside the frame of the figure) no moiré effect will be visible.

These reasons explain, indeed, why Glass patterns are less frequently encountered than their periodic or repetitive counterparts. Probably, the most likely situation in everyday's life to observe a moiré pattern between aperiodic structures occurs when making identical transparent photocopies of a document or of a patterned image. If two of the resulting transparencies are placed on top of each other with a small angle difference, and their mutual displacements are not too large, a circular Glass pattern may become visible in their superposition, as shown in Figs. 1.2(a) or 1.3. This is, indeed, the setting that led to the discovery of this phenomenon, as described by Leon Glass in [Glass02].

But although the phenomena related to the superposition of aperiodic layers are not new, they remained over the years much less understood than their periodic or repetitive counterparts, and it was often believed that their systematic analysis was not possible.[1] One probable reason is that these phenomena did not easily lend themselves to the same mathematical methods that so nicely explained the classical moiré effects between periodic or repetitive layers (geometric considerations, indicial equations, Fourier series developments, etc.). It is therefore our aim in the present book to "demystify" these

---

[1] See, for example, [Wesner74 p. 1710]: "Moirés and other unwanted patterns ... can occur with random screens just as they do with regular screens. With regular screens these can be analyzed and dealt with in a systematic manner, but with random screens this is not possible".

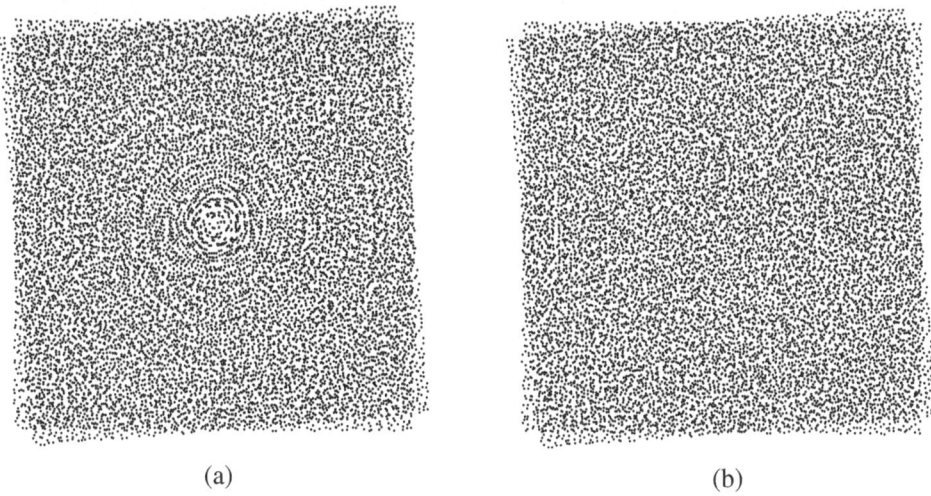

(a)   (b)

Figure 1.2: (a) A Glass pattern in the superposition of two identical aperiodic layers that are slightly rotated on top of each other. (b) The superposition of two uncorrelated aperiodic layers does not generate a Glass pattern.

phenomena, and to give them the right and just place they deserve within the framework of a general, unified moiré theory that englobes both periodic and aperiodic cases. The main difficulty here is that the mathematical methods used to analyze the moiré phenomena between periodic or repetitive layers highly depend on the periodic or repetitive nature of the original layers, properties which obviously do not exist in aperiodic or random layers. And yet, it turns out that in spite of their different visual appearance, moirés between periodic, repetitive or aperiodic layers are in fact particular cases of the same basic phenomenon, and they all follow the same fundamental rules which explain what happens in the superposition of any layers, periodic or not. Our task will be, therefore, twofold: to see how the moiré theory can be generalized to the aperiodic case, too, and, on the other hand, to find which parts of the theory remain proper to periodic or repetitive cases, or, possibly, to aperiodic cases alone.

In fact, as we can see when observing Figs. 1.1–1.2 under a magnifying glass, all the different types of moiré effects — periodic, repetitive and aperiodic — occur due to a similar interaction mechanism between the overlaid structures. More precisely, all of these phenomena result from the geometric distribution of dark and bright areas in the superposition: Areas where dark elements of the original structures fall on top of each other contain more open spaces, and therefore they appear brighter than areas in which dark elements fall between each other. But in spite of their common origins, moiré effects which occur in aperiodic cases look very different from their periodic or repetitive counterparts. The study of these differences and their explanation will be one of our main concerns in the chapters which follow.

And indeed, throughout this volume we put a special emphasis on the comparison between Glass and moiré patterns. As far as possible we provide for each case being discussed an illustrative figure showing side by side the aperiodic superposition and its periodic or repetitive counterpart. This facilitates the visual comparison of the Glass and moiré patterns in question, and helps to clearly understand the similarities and the differences between them.

1.2 Brief historical background and main applications

Since they have been described in [Glass69] and [Glass73], Glass patterns have aroused a large research interest, particularly in fields related to the human visual system, its modelization and its understanding. In addition to Glass' own work, which has been surveyed retrospectively in [Glass02], numerous other contributions were published in this field; see, for example, [Dakin97], [Wilson98] and [Cardinal03], to mention just a few.

Glass patterns have also found application in various other fields. Because of their extreme sensitivity to the slightest displacements in the overlaid structures, Glass patterns have been used in applications such as high-precision registration of superposed layers [Dey91], or the optical alignment of images [Schuette97]. Other applications of Glass patterns include the determination of the axis of rotation between two moving parts [Walker80a]; the comparison of two or more patterns to determine whether they are essentially the same [Prokoski99]; the identification of a given pattern within an image [*ibid.*]; and document authentication and anti-counterfeiting [Amidror04a]. Glass patterns can be used also as a versatile, qualitative "visual equation solver" for illustrating the solution curves (trajectories) of a system of two differential equations [Glass73 p. 361], and for visualizing the flow lines of a vector field [Glass02 pp. 39–40]. As we will see later, they can be also used for finding qualitatively the fixed points of a transformation, for showing the solutions of an equation $f(x,y) = 0$ or the solutions of a system of two such equations (see Problems 3-10 to 3-12 in Chapter 3), and even for visualizing the curve family defined by the indefinite integral $\int f(x)dx + const.$ (see Problem 4-10 in Chapter 4). Various other applications are provided in the problem sections at the end of each of the following chapters. And finally, last but not the least, Glass patterns have been used also in art [Huck03], and even just for fun [Walker80].

Correlations between random patterns and their displaced or distorted counterparts — albeit not Glass patterns *per se* — have been also used in speckle metrology for the detection and measurement of slight displacements or deformations [Gåsvik95, Svanbro04]. And in a completely different context, the superposition of aperiodic layers has also been investigated in the fields of halftoning and colour printing, with the aim of improving colour printing algorithms that are based on the superposition of aperiodic dot screens [Kang99]. The main concern in this case is preventing, as far as possible, the appearance of disturbing artifacts in the original aperiodic dot screens and in their

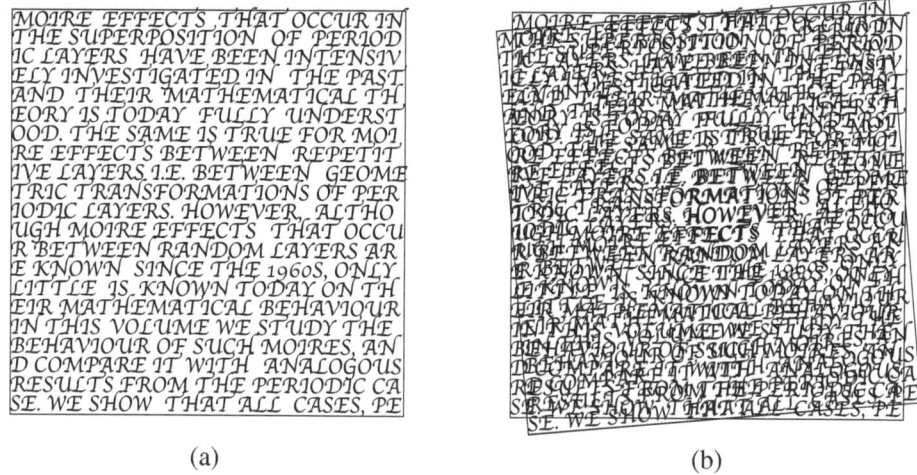

(a)   (b)

Figure 1.3: (a) An arbitrary aperiodic structure. (b) The superposition of two identical copies of this aperiodic structure with a small angle difference gives a moiré effect in the form of a Glass pattern.

superpositions, and the design of screens that would minimize such artifacts [Lau01, Lau02, Lau02a]. Naturally, in this application one is not interested in studying the properties of Glass patterns between correlated aperiodic layers, but, on the contrary, in ways to eliminate or at least to minimize any visible effects in the superposition.

However, in spite of these many diverse applications, the mathematical understanding of moiré patterns between aperiodic structures, as well as their properties and their relationship with classical moiré effects between periodic or repetitive structures, did not progress significantly over the years. Only recently has it been shown that in spite of their very different appearance and properties, moiré patterns between periodic or repetitive structures and Glass patterns between aperiodic or random structures are, in fact, particular cases of the same basic phenomenon, and all of them satisfy the same fundamental rules [Amidror03a, Amidror03b, Amidror03c]. These and other new research results form, indeed, the basis for the present volume.

1.3 The scope of the present book

Our main goal in this volume is to present the theory of the moiré phenomenon between aperiodic or random layers, and to show how the classical moiré theory can be extended to cover phenomena which occur between such layers, too. This book provides the full mathematical explanation of Glass patterns, describes their various properties, and explores their behaviour when the original layers undergo geometric transformations.

Among other topics, it explains in detail the nature and the mathematical properties of the dot trajectories that often accompany Glass patterns in aperiodic screen superpositions, and it also investigates the one-dimensional counterparts of the classical Glass patterns that occur in the superposition of correlated aperiodic line gratings. Finally, following our experience in *Vol. I* of the present work, we also provide here the Fourier-based approach, which allows us to explore Glass patterns and their intensity levels not only *qualitatively* but also *quantitatively*. In addition to the analysis of the various phenomena being discussed, we also show how to synthesize them. Thus, for each of these phenomena we also provide practical recipes accompanied by illustrated examples that show how one can synthesize the desired effect by properly designing the superposed layers.

It is clear, however, that not all of the interesting subjects concerning aperiodic layers and their moiré phenomena could be included in the present book. In the following list we enumerate the main points which have remained beyond the scope of our work.

- First of all, we limit ourselves here to the analysis of moiré effects in the superposition of static layers. Temporal moirés or other subjects that are related to the speed or the kinematics of moving layers are not directly addressed in the present book, although they can be considered as natural extensions of the theory presented here.

- We do not consider here effects such as light scattering, light diffraction, or any other physical questions concerning the nature of light (coherent / incoherent) and its influence on the moiré. In particular, we will always assume that the spacing between the individual elements in each layer is coarse enough for diffraction effects to be ignored.

- We do not consider here, either, the discrete nature of grating or screen elements which are produced on digital devices such as laser printers, high-resolution filmsetters, etc., and the influence of this discrete nature on the moiré. The jagged aspect of discrete lines or dots is considered here as a real-world constraint, and we try to avoid it (or at least to reduce its influence) by producing our samples on appropriate devices with high enough resolutions (normally, at least 600 dots per inch).

- We suppose here that the different layers are superposed *in contact*, for example by overprinting, and we ignore the possible effects of the distance between the layers on the resulting moiré patterns, such as parallax-related phenomena or the Talbot effect.

- We do not take into account statistical considerations such as the distribution or the probability density of our aperiodic layers (random dot screens, etc.) and of their superpositions. In fact, as explained in Sec. F.5 of Appendix F, we consider all our images as being *deterministic*, and we always treat them as such.

- We do not consider here phenomena such as the nebulous or worm-like microstructure artifacts which may occur in random screens or in the superposition of *uncorrelated* random screens. Such artifacts are widely known in the field of halftoning, where superposed aperiodic screens are used for colour printing [Kang99, Lau01]. While the research of such phenomena is certainly not less interesting than the study of dot trajectories in the superposition of correlated layers, it requires a different mathematical approach that is more closely related to the statistical distribution of the elements in the original aperiodic layers. But such considerations remain beyond the scope of our present work.

- We also intentionally content ourselves here with a simplified model of the human visual system, and we avoid going any further into the complex questions related to the modelization of the human visual system and its performance in an inhomogeneous environment (like the perception of ordered shapes within a random dot screen superposition). Nor shall we dwell here on the applications of Glass patterns in the research of the human visual system, a subject that has been largely covered in the existing literature (see Sec. 1.2 above), but which remains beyond our main field of interest here — the mathematical study of Glass patterns and of their various properties. More details about human vision and its modelization can be found, for example, in [Wandell95] and [Daly92].

- Finally, we generally prefer a pictorial, intuitive approach supported by mathematics over a rigorous mathematical treatment. In many cases we give informal demonstrations rather than formal proofs, or defer detailed derivations to an appendix.

It should be noted that just like *Vol. I* of this work, this book has not been written with a specific application in mind. In fact, our principal aim is to present the theoretical aspects of the moiré phenomenon in a general, application-independent way. And yet, for the benefit of the interested readers, we have included among the problems at the end of each chapter some of the main applications of the theory being discussed, along with additional references. This should give the reader a general idea about some of the applications that Glass and moiré patterns have found in various fields.

1.4 Overview of the following chapters

Chapter 2 lays the foundations for the rest of this book, and reviews the basic notions and terminology that will be needed in the sequel. It presents the various macro- and microstructure phenomena that may appear in superpositions of aperiodic layers, discusses the influence of the element distribution within the original layers on the superposition, and shows what happens in cases involving multilayer superpositions. Some applications of the various phenomena which may occur in the superposition of aperiodic layers are reviewed in the problems at the end of the chapter.

Chapter 3 presents the fixed point theorem and shows its fundamental role in the explanation of Glass patterns and their behaviour under layer transformations. It shows how we can determine mathematically the geometric locus of the resulting Glass pattern given the transformation undergone by each of the two individual layers. And finally, it shows how our theoretical approach can be used not only for the *analysis* of Glass patterns, but also for the *synthesis* of Glass patterns, namely, for the design of layer transformations that, when applied to the original layers, generate in the superposition Glass patterns having any desired geometric locus in the plane.

In Chapter 4 we concentrate on the dot trajectories that may occur in the superposition of aperiodic layers and on their mathematical explanation. We start our discussion with an overview of dot trajectories and the mathematical tools that have been proposed for their analysis. Based on this mathematical understanding we proceed to the *synthesis* of dot trajectories, and we show how we can design aperiodic screens that give in their superposition any desired dot trajectories. We also explain why dot trajectories are only visible between aperiodic layers but not between periodic layers. And finally, we discuss the microstructures which appear in cases based on other superposition rules, and we briefly explain why they look different from the familiar dot trajectories of the standard multiplicative superposition rule (black dots overprinted on black dots).

Then, in Chapter 5 we investigate the mathematical relationship between the phenomena that occur in the superposition of periodic or aperiodic dot screens under the same geometric layer transformations. We explain the similarities as well as the striking differences that exist between these effects, and we also show what happens in in-between cases, where the superposed layers are intermediate between periodic and aperiodic.

In Chapter 6 we study Glass patterns which occur in the superposition of aperiodic line gratings, either straight or curved. We compare these Glass patterns with their 2D counterparts, and explain the similarities and the differences between them. On the other hand, we compare their behaviour with that of the moiré patterns which are obtained in superpositions of periodic or repetitive line gratings. We also provide the mathematical derivations of the curve shapes of the Glass patterns in the 1D case, and show how they are related to the 2D case and to the periodic case.

Finally, in Chapter 7 we investigate the layer superpositions and the resulting Glass patterns using a Fourier-based approach, which is a direct extension of the theory that governs the superposition of periodic layers. This powerful mathematical approach allows us to analyze, both qualitatively and quantitatively, the surprising intensity profile of the Glass patterns in the superposition of aperiodic screens with any given dot shapes. Furthermore, it also leads us to a method for synthesizing Glass patterns having any desired shapes and intensity profiles. As an additional bonus, this approach also explains some other intriguing questions, such as why higher order moirés cannot exist in aperiodic cases, while they clearly do exist in periodic or repetitive cases.

The main body of the book is accompanied by several appendices:

Appendix A provides a deeper insight into the fixed point theorem that has been presented in Chapter 3. It explains in detail two of the simplest but most important special cases of this theorem, one for linear or affine mappings, and the other for second-order polynomial mappings. The importance of these particular cases is twofold: On the one hand, these cases cover many of the layer transformations being used in the figures and in the examples throughout this book; but on the other hand, these cases also serve as simple illustrations of the general case, where more complex mappings may appear. A further generalization to the case of mutual fixed points between two mappings is also provided.

In Appendix B we review the various possible interpretations of the general 2D transformation, $(u,v) = \mathbf{g}(x,y)$. Each such transformation can be interpreted in several different yet completely equivalent ways, for example, as two surfaces over the plane, as two sets of level lines, as a mapping from the plane into itself, as a coordinate change, or as a 2D vector field. Each of these interpretations turns out to be useful in some other facets of our moiré theory. But because all of these interpretations are mathematically equivalent, we are always free to choose among them the one that best suits the circumstances.

Appendic C provides the geometric interpretation of the Jacobian of a 2D transformation $(u,v) = \mathbf{g}(x,y)$, and reviews the main properties of $\mathbf{g}(x,y)$ that can be deduced from its Jacobian.

In Appendix D we shed more light on the use and on the behaviour of direct and inverse 2D transformations, and explain the main sources of confusion that may occur in the handling of such transformations as domain or as range transformations. We also discuss other related issues, such as the implementation of such transformations by forward or backward mapping algorithms in computer applications, and we show the various implications of all these results to the moiré theory.

Appendix E is devoted to the main properties of convolution and cross correlation, both of which are largely used in the moiré theory.

In Apprndix F we evaluate the stochastic and the deterministic approaches in the context of the aperiodic moiré theory. We first provide a short overview of the stochastic approach, and then we explain why we have chosen to use the deterministic rather than the stochastic approach.

In Appendix G we show how one can generalize Fourier decompositions to the case of geometrically transformed layers by using appropriate integral transforms.

Appendix H is devoted to miscellaneous issues and derivations that we preferred, for different reasons, not to include in the main text of our work.

And finally, in Appendix I we provide a glossary of the most important terms that have been used in the present book.

1.5 About the exercises and the moiré demonstration samples

Keeping the same conventions as in *Vol. I* of this work, we provide at the end of each chapter a section containing a number of problems and exercises. Many of these problems are not merely routine exercises, but really intriguing and sometimes even challenging problems. Their aim is not only to aid the assimilation of the material covered by the chapter, but also to develop new insights beyond it. As already mentioned, these problems also include examples of real-world applications of the theory discussed in the chapter, along with references to existing publications on these applications (books, scientific papers, patents, etc.). We therefore highly encourage readers to dedicate some time for reviewing these exercises.

Since moiré effects are best appreciated by a hands-on experience, some of the key figures of this book have been also provided in the form of *PostScript*® programs [Adobe90], which can be printed on transparencies using any standard desktop laser printer. These PostScript programs and the instructions for using them can be found in the Internet site of this book, at the address:

`http://lspwww.epfl.ch/books/moire/`

This site has been significantly updated and extended, and it includes now many demonstration samples for aperiodic cases, too, in addition to the periodic and repetitive cases that already accompanied *Vol. I*. By printing these samples the reader will obtain a kit of transparencies offering a vivid illustration of the various moiré effects and their dynamic behaviour in the superposition. This demonstration set will allow the interested reader to make his own experiments by varying different parameters (angles, scaling factors, geometric transformations, etc.) in order to better understand their effects on the resulting moirés. This will not only be a valuable aid for the understanding of the material, but certainly also a source of amusement and fun.

*   *   *

Finally, a word about our notations. Throughout this book we adopt the following notational conventions:

Sec. 3.2	—	Section 2 of Chapter 3.
Sec. A.2	—	Section 2 of Appendix A.
Fig. 3.2	—	Figure 2 of Chapter 3.
Fig. A.2	—	Figure 2 of Appendix A.
(3.2)	—	Equation or formula 2 of Chapter 3.
(A.2)	—	Equation or formula 2 of Appendix A.

Similar conventions are also used for enumerating tables, examples, propositions, remarks, etc.; for instance, Example 3.2 is the second example of Chapter 3.

Whenever reference is made to the first volume of this work, we use the abbreviation "*Vol. I*". When referring to a section, a figure or an equation in *Vol. I*, we simply add the prefix "*I*" to the specified number; for example, Sec. *I*.3.2 refers to Sec. 2 of Chapter 3 in *Vol. I*, Fig. *I*.3.2 means Fig. 2 of Chapter 3 in *Vol. I*, and Eq. (*I*.A.2) refers to the second equation in Appendix A of *Vol. I*.

Chapter 2

Background and basic notions

2.1 Introduction

The various phenomena which occur in the superposition of periodic layers or in the superposition of repetitive layers (namely, geometrically transformed periodic layers) are today fully understood. It is well known that such superpositions may give rise to periodic (or, respectively, repetitive) moiré patterns, and various mathematical tools, both in the image domain and in the Fourier spectral domain, have been devised to explain these phenomena, qualitatively as well as quantitatively. These phenomena and their mathematical explanations have been treated at length in *Vol. I*.

In the present volume we try to widen our scope of interest, and to see what happens in the superposition of aperiodic layers such as random or pseudo-random screens. In fact, such cases are not really new. For example, it is well known in the field of colour printing that by overprinting random dot screens instead of periodic dot screens one can obviate the generation of moiré effects in the superposition (see, for example, Sec. 1.4.3 in [Kipphan01]). On the other hand, it is equally well known that the superposition of aperiodic screens may give, in certain circumstances, typical aperiodic moiré effects, known as *Glass patterns*, whose visual appearance seems quite different from that of classical moirés between periodic or repetitive layers (see Figs. 2.1–2.2).

But as already mentioned in Chapter 1, although the phenomena related to the superposition of aperiodic layers are not new, they remained over the years much less understood than their periodic or repetitive counterparts. It is therefore our aim in the present volume to investigate these phenomena. In particular, it would be interesting to see if there exists a general approach which can provide a unified framework for the explanation of all cases, periodic or not. And indeed, it turns out that in spite of their very different visual appearance, moirés between periodic, repetitive or aperiodic layers are in fact particular cases of the same phenomenon, and they all follow the same fundamental rules. However, as clearly shown in *Vol. I*, the analysis of moiré phenomena between periodic or repetitive layers highly depends on the periodic or repetitive nature of the original layers, properties which obviously do not exist in aperiodic or random layers. Our main task will be, therefore, to see how the moiré theory can be generalized to the aperiodic case, too, and which parts of the theory must remain proper to periodic or repetitive cases only (or, possibly, to aperiodic cases only).

The present chapter lays the foundations for the rest of this book, and it reviews the basic notions and terminology that will be needed in the sequel. We start in Sec. 2.2 with a short review of the main background notions. Some of these notions have already been

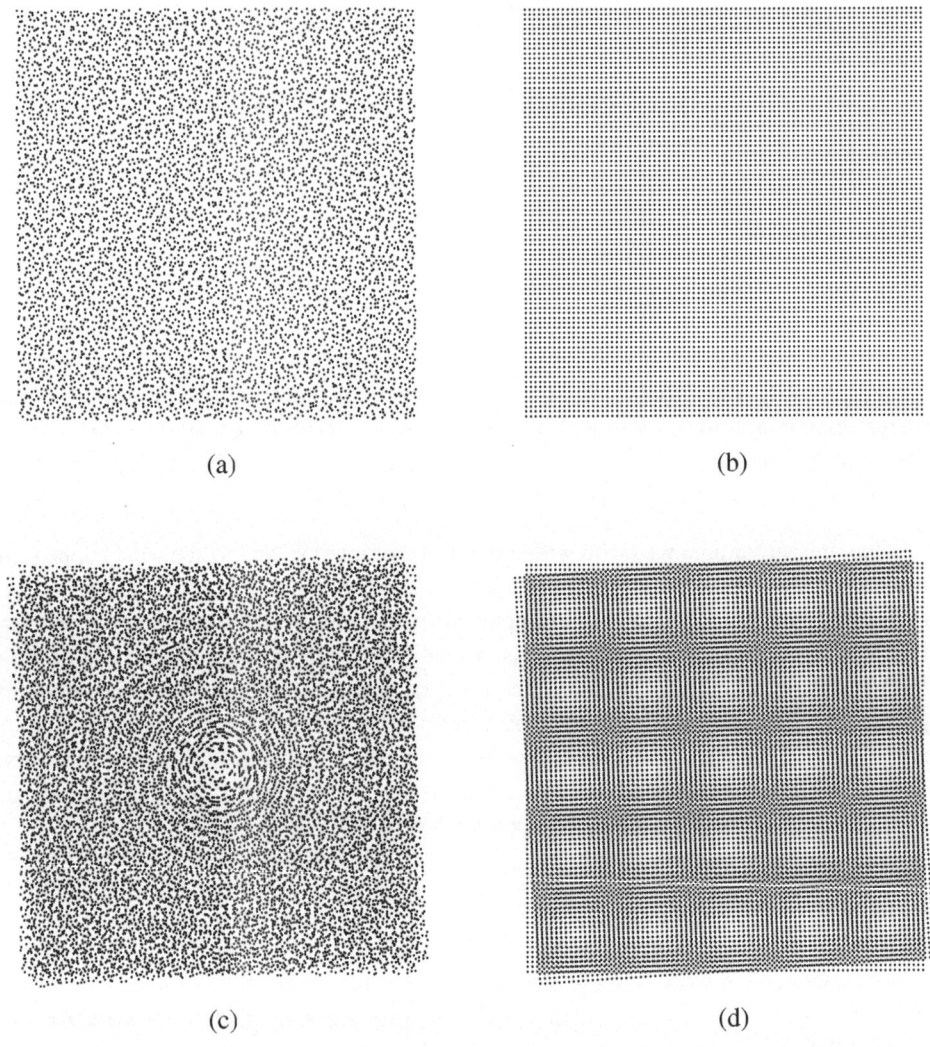

(a)   (b)

(c)   (d)

Figure 2.1: Glass patterns between aperiodic dot screens, and their moiré
counterparts between periodic dot screens. The aperiodic dot
screen (a) and its periodic counterpart (b) are used for
generating all the superpositions shown in Figs. 2.1–2.5.
(c) The superposition of two identical copies of aperiodic dot
screen (a) with a small angle difference gives a Glass pattern
about the center of rotation. Note its typical microstructure
consisting of concentric circular dot trajectories. (d) When the
superposed layers are periodic, a Glass pattern is still generated
about the center of rotation, but due to the periodicity of the
layers, this pattern is periodically repeated throughout the
superposition, thus generating a periodic moiré pattern.

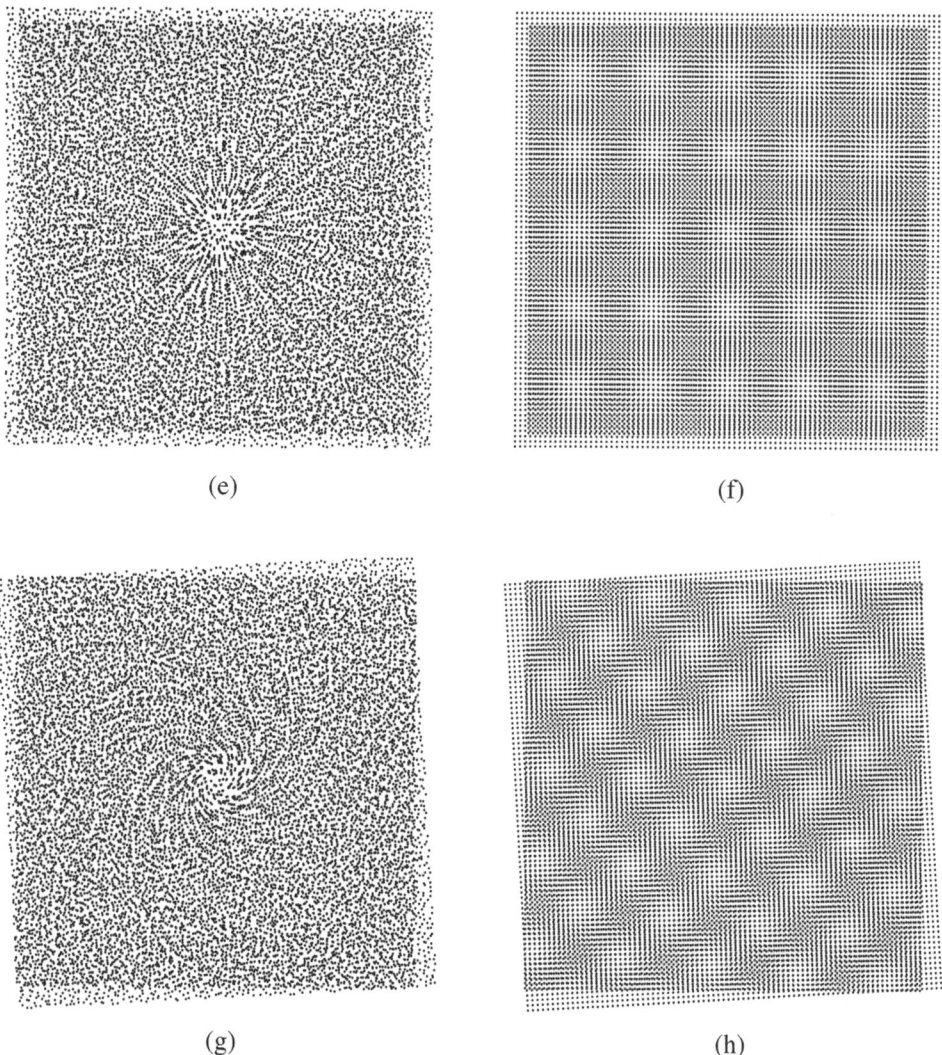

(e)

(f)

(g)

(h)

Figure 2.1: (*continued.*) (e),(f) Same as (c),(d), but with a small scaling difference rather than a small angle difference between the superposed layers. In this case the microstructure consists of radial dot trajectories. (g),(h) Same as (e),(f), but with both a small angle and a small scaling difference between the superposed layers. In this case the microstructure consists of spiral dot trajectories.

introduced in *Vol. I*, so readers who are already familiar with them may rapidly skim over this introductory material and return to it later, whenever the need may arise. In Sec. 2.3 we present the various macro- and microstructures that may appear in superpositions of aperiodic layers. Then, in Sec. 2.4 we discuss the influence of the element distribution within the original layers on the superposition, and finally in Sec. 2.5 we proceed to cases

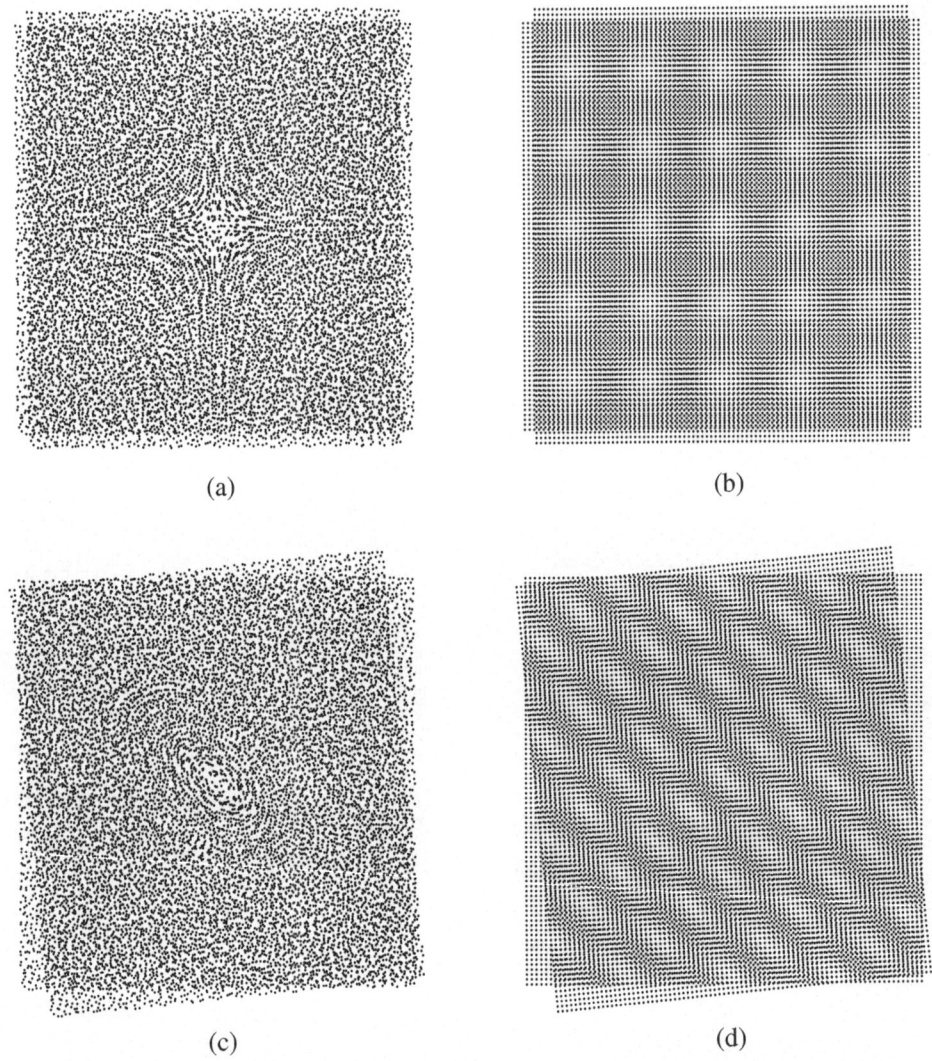

(a)   (b)

(c)   (d)

Figure 2.2: (a),(b) Same as Fig. 2.1 (e),(f), but here the scaled layer undergoes
unequal scaling rates in the x and y directions: horizontal shrinking
and vertical expansion. Note the hyperbolic dot trajectories in the
superposition. (c),(d) Same as (a),(b), but with a small angle
difference, too, between the superposed layers. In this case the
microstructure consists of elliptic dot trajectories.

involving multilayer superpositions. Some applications of the various phenomena which
occur in superpositions of aperiodic layers are described in the problems at the end of the
chapter. Note that the figures in this chapter are mainly given to illustrate the various basic
phenomena, but they will be revisited and more deeply explained in the following chapters.

2.2 Periodic, repetitive and aperiodic layers

Throughout this book we will be concerned with bidimensional (2D) structures in the continuous x,y plane, that we call *images* or *layers*, and with their superpositions. We therefore start by explaining these notions and their main properties, both in the image domain and in the Fourier, spectral domain.[1] A layer (or image) is the most general term we use to cover "anything" in the image domain. It can be periodic or not, continuous or binary, etc. And yet, we still need to make some basic assumptions concerning our images. These assumptions and their implications are reviewed in the following paragraphs.

First of all, we limit ourselves here to monochrome, black and white images.[2] This means, as already explained in Sec. *I.2.2*, that each of our images can be represented by a *reflectance* function $r(x,y)$, which assigns to any point (x,y) of the image a value between 0 and 1 representing its light reflectance: 0 for black (i.e., no reflected light), 1 for white (i.e., full light reflectance), and intermediate values for in-between shades. In the case of transparencies, the reflectance function is replaced by a *transmittance* function defined in a similar way.

A superposition of such images can be obtained in several ways, for example by overprinting or by laying printed transparencies on top of each other. Since the superposition of black and any other shade always gives black, this suggests a *multiplicative* model for the superposition of monochrome images, just as in the periodic case (see Sec. *I.2.2*). Thus, when m monochrome images, periodic or not, are superposed, the reflectance (or transmittance) of the resulting image is the *product* of the reflectance (or transmittance) functions of the individual images:

$$r(x,y) = r_1(x,y) \cdot \ldots \cdot r_m(x,y) \qquad (2.1)$$

Therefore, according to the convolution theorem [Bracewell86 p. 244], if we denote the Fourier transform of each function in Eq. (2.1) by the respective capital letter and the 2D convolution by $**$, the spectrum of the superposition is given by:

$$R(u,v) = R_1(u,v) ** \ldots ** R_m(u,v) \qquad (2.2)$$

Remark 2.1: It should be noted, however, just as in the periodic case, that the multiplicative model is not the only possible superposition rule, and in other situations different superposition rules can be appropriate. For example, when images are superposed by making multiple exposures on a positive photographic film (assuming that we do not exceed the linear part of the film's response), intensities at each point are summed up, which implies an *additive* rule of superposition. In another example, when

---

[1] As shown in *Vol. I*, Fourier considerations are a fundamental tool in the study of moiré effects between periodic layers. But since in the present volume the Fourier theory is not used before chapter 7, we do not give here a full introductory discussion on the use of the Fourier approach, as we did in Chapter 2 of *Vol. I*. Instead, we provide the needed background in Chapter 7, and here we only give some introductory remarks. See also Problems 2-1–2-13 and Figs. 2.10–2.13 at the end of the chapter.

[2] The extension of our discussion to the polychromatic case is rather straightforward, and it can be done in the same manner as in the periodic case (see Chapter 9 in *Vol. I*).

images are superposed by making multiple exposures on a negative photographic film (again, assuming a linear response) an *inverse additive* rule can be appropriate. More exotic superposition rules (involving, for example, various Boolean operations etc.) can be artificially generated by computer, even if they do not correspond to any physical reality. Note that different superposition rules in the image domain will have different spectrum composition rules in the spectral domain, which are determined by properties of the Fourier transform. For example, in the case of the additive superposition rule, where Eq. (2.1) is replaced by:

$$r(x,y) = r_1(x,y) + \dots + r_m(x,y) \tag{2.3}$$

the spectrum of the superposition is no longer the spectrum-convolution given by Eq. (2.2), but rather the *sum* of the individual spectra:

$$R(u,v) = R_1(u,v) + \dots + R_m(u,v) \tag{2.4}$$

In cases where no explicit rule is known that relates the individual spectra $R_i(u,v)$ to the spectrum $R(u,v)$ of the superposition, $R(u,v)$ can still be found directly by applying the Fourier transform to $r(x,y)$. ∎

Let us now explain what we mean by periodic and by aperiodic or random layers. A 1D function $f(x)$ is said to be *periodic* if there exists a non-zero number p such that for any $x \in \mathbb{R}$, $f(x + p) = f(x)$. Similarly, a 2D layer $r(x,y)$ is said to be periodic if there exists a non-zero vector $\mathbf{p} = (p_1, p_2)$ such that for any $(x,y) \in \mathbb{R}^2$, $r(x + p_1, y + p_2) = r(x,y)$. If there exist two independent vectors having this property, $r(x,y)$ is said to be 2-fold periodic. A layer $r(x,y)$ is said to be *aperiodic* if it is not periodic.[3] For example, the image of a human portrait or a natural landscape is aperiodic. As a second example, a random dot screen consisting of randomly positioned black dots is also aperiodic. Note, however, that this random dot screen may also be regarded as a *stochastic layer*, from a more statistical point of view, if we consider the screen in question as just one possible realization of a stochastic process having some given statistical distribution. In the context of random dot screens (or more generally, random scatter; see Chapter 17 in [Bracewell95]) the terms aperiodic layer, stochastic layer and random layer are often used interchangeably.

From the spectral point of view, the Fourier transforms of periodic layers and of their superpositions are *purely impulsive*, while the Fourier transform of an aperiodic layer is basically *continuous* (for example: the Fourier transform of a unit cube is a 2D sinc function; see [Bracewell95 pp. 150–151]). But when the structure of the aperiodic layer is very complex, as in the case of a random dot screen, its Fourier spectrum becomes very jumpy or noisy and admits a typical *diffuse* appearance (see [Bracewell95 pp. 586–590;

---

[3] More precisely, a layer is said to be aperiodic if it is not *almost-periodic*, which also implies that it is not periodic (see Fig. *I*.B.3 in Appendix *I*.B, and the definitions in Sec. *I*.D.2 of Appendix *I*.D). Note, however, that although repetitive layers formally fall within the scope of this definition of aperiodic layers, we prefer to exclude them from our present definition, because they are still structurally ordered, and they have already been investigated in Chapters 10 and 11 of *Vol. I*. We therefore adopt the definition saying that a layer is aperiodic if it is neither periodic (or almost-periodic) nor a geometrically transformed version thereof.

600–601]). Further properties of the Fourier transform of aperiodic or random images are discussed in Problems 2-1 to 2-10 and in Appendix F.

It should be noted that there exist also intermediate cases between pure periodic and pure aperiodic cases, for example: randomly perturbed periodic layers, where a certain percent of random behaviour is being added to an originally periodic structure. As one would expect, the Fourier spectrum of such layers is intermediate between the spectra of periodic and aperiodic cases: it consists of slightly blurred impulses corresponding to the frequencies in question plus a diffuse background that is typical to random cases. Such intermediate cases will be discussed in Sec. 7.6.

Another important notion that is largely used throughout this book is that of *correlation* between two images. Intuitively, two images are well correlated if they are similar to each other, and they are non-correlated if they are totally different. For example, two identical copies of the same image are fully correlated, but gradually adding random noise to one or both of them will reduce their correlation, until they finally become non-correlated. Thus, the degree of correlation gives an indication to the similarity between the images.

Two images are said to be *globally correlated* if they highly agree with each other throughout their entire domain of definition. For example, if $r_1(x,y)$ and $r_2(x,y)$ are two identical copies of the same aperiodic image (one or both of which may have been subject to some amplitude deformations or some additive random noise), then they are globally correlated. However, the image $r_1(x,y)$ is no longer globally correlated with the shifted image $r_2(x-x_0, y-y_0)$ or with the stretched image $r_2(ax, ay)$.

Similarly, two images are said to be *locally correlated* (or to be well correlated in a certain location) if they highly agree with each other in the specified location. As its name indicates, this is a local property, so that two given images can be highly correlated in some areas but not at all correlated in other areas. As we will see below, local correlation between two superposed aperiodic layers is the reason for the appearance of a Glass pattern in that area of the superposition. For example, the Glass patterns shown in Fig. 2.1(c) consist of a brighter area in the center because within this area the two superposed layers are well correlated (the dots of both layers almost coincide), while in the remaining areas of the figure the correlation between the two layers is low.

Another related term is that of *cross correlation* between two images (or functions). The cross correlation between two functions $f(x,y)$ and $g(x,y)$ is a third function $c_{f,g}(x,y)$ that indicates the relative amount of agreement between $f(x,y)$ and $g(x,y)$ for all possible degrees of misalignments (shifts). At each point (x,y) the value of the function $c_{f,g}(x,y)$ is defined as the volume under the product of f and g after g has been shifted by x,y. For example, if $r_1(x,y)$ and $r_2(x,y)$ are defined as above, then their cross correlation function consists of a high peak about the origin, and low values everywhere else. The reason is that r_1 and r_2 are globally correlated when the shift between them is $(x,y) = (0,0)$, but for any other shift they are no longer globally correlated. In general, points (x,y) in which the value of the cross correlation function is high indicate displacements of x,y between the two images in which the volume under the product of the two entire layers is high. It should be noted, however, that the cross correlation can detect displacements x,y in which the two *entire* images are *globally* well correlated, but it cannot detect isolated zones of *local* correlation between the two images, since the contribution of such local zones is relatively small, and it may be buried and lost within the global volume under the product of the entire images. The formal definition of cross correlation as well as its main mathematical properties are reviewed in Appendix E, and they should be consulted whenever the need arises.

Next, let us review the notions of *macrostructure* and *microstructure*. It is well known that when periodic layers (line gratings, dot screens, etc.) are superposed, new structures of two distinct levels may appear in the superposition, which do not exist in any of the original layers: macrostructures and microstructures (see, for example, Chapter 8 in *Vol. I*). The macrostructures, also known as *moiré patterns*, are much coarser than the detail of the original layers, and they are clearly visible even when observed from a distance. The microstructures, on the contrary, are almost as small as the periods of the original layers (typically, just 2–5 times larger), and therefore they are visible only when one is examining the superposition from a close distance or through a magnifying glass. These tiny structures are also called *rosettes* owing to the various flower-like shapes they often form in the superposition of periodic dot screens. Macrostructure and microstructure phenomena may also become visible in the superposition of aperiodic or random layers. Here, too, the macrostructures (Glass patterns) can be clearly seen when observed from a distance, whereas the microstructure dot arrangements (such as "dot trajectories", "worms", etc; see Sec. 2.3.3 below) are visible only when examining the superposition from a close distance.

Macrostructures and microstructures may coexist in the same superposition. In fact, we will see in Sec. 2.3.3 that macrostructures are the macroscopic consequence of local variations in the microstructures of the superposition, just as in the periodic case. But like in the periodic case, microstructures may exist even when no macrostructures are generated (see, for instance, Figs. 2.3(e),(g)).

Remark 2.2: The distinction between macrostructures and microstructures may seem at first quite artificial and subjective. But in fact, macrostructures and microstructures are simply two different facets of the same physical reality in the layer superposition. They only represent two different scales at which we consider the same phenomenon: In the macroscopic scale we consider the global, average behaviour of the phenomenon (much like the rules of classical physics), whereas in the microscopic scale we study the behaviour of the same phenomenon from the point of view of the individual screen elements and the interactions between them (much like the study of the same physical rules through the behaviour of molecules and atoms).[4] ∎

---

[4] The study of a given phenomenon at different scales is commonplace in physics and in science in general. A more detailed discussion on this subject can be found, for example, in [Poston78 p. 217], where the different scales of modeling physical phenomena are explained in the context of the flow of fluids.

Since the microstructure behaviour is most clearly perceived in the superposition of layers consisting of black dots on a white background, we will usually illustrate our discussions with such cases. The study of the general case will be deferred until Chapter 7, where we will see the influence of the shapes and intensity levels of the dots in the individual layers on the intensity profile of the resulting Glass patterns.

Finally, for the sake of simplicity, we will generally concentrate on layers having a uniform distribution of their microstructure elements (and hence a constant mean gray level), although our results hold also in more complex structures such as halftone gradations, halftoned images with varying gray levels, etc.

2.3 Superposition of aperiodic layers

While the superposition of two similar periodic layers generates moiré effects that are themselves periodic, the superposition of two similar aperiodic layers generates an aperiodic moiré effect, that is known as a Glass pattern (compare, for example, Fig. 2.1(c) and 2.1(d) or Figs. 2.1(e) and 2.1(f)). This moiré pattern is typically concentrated about a certain point in the superposition, and it gradually fades out and disappears farther away from this point. Depending on whether it was obtained by rotation of one of the superposed layers, by a scaling transformation, or by a combination of the two, it gives rise to an intriguing ordering of the microstructure elements in the superposition in "dot trajectories" having a circular, radial or spiral shape, as shown in Figs. 2.1(c), (e) and (g) [Glass73]. Other layer transformations may give rise to Glass patterns having hyperbolic, elliptic or other geometrically shaped dot trajectories, as shown for example in Figs. 2.2(a),(c) [Glass73]. However, unlike periodic moirés, Glass patterns do not reappear when we invert or rotate one of the superposed layers by 180° (see Figs. 2.6(a),(b)).

2.3.1 Glass patterns and correlation

As already pointed out by Glass, this phenomenon occurs thanks to the local correlation between the structures of the two superposed layers; in fact, the strength of the effect can be used as a visual indication of the degree of correlation between the two layers at each point of the superposition.[5] Thus, when two identical layers having the same arbitrary structure are slightly rotated on top of each other (see Fig. 2.1(c)), a visible Glass pattern is generated about the center of rotation, indicating the high correlation between the two layers in this area. Within the center of this Glass pattern the corresponding elements from the two layers fall almost exactly on top of each other, but slightly away from the center they fall just next to each other, generating circular trajectories of point pairs. Farther away from the center the correlation between the two layers becomes smaller and smaller, and the elements from both layers fall in an arbitrary, non-correlated manner; in

---

[5] Note that for the time being we use the term "correlation" rather intuitively. We defer the formal mathematical discussion on correlation and its quantitative values to Chapter 7, where it naturally arises in the context of Fourier analysis.

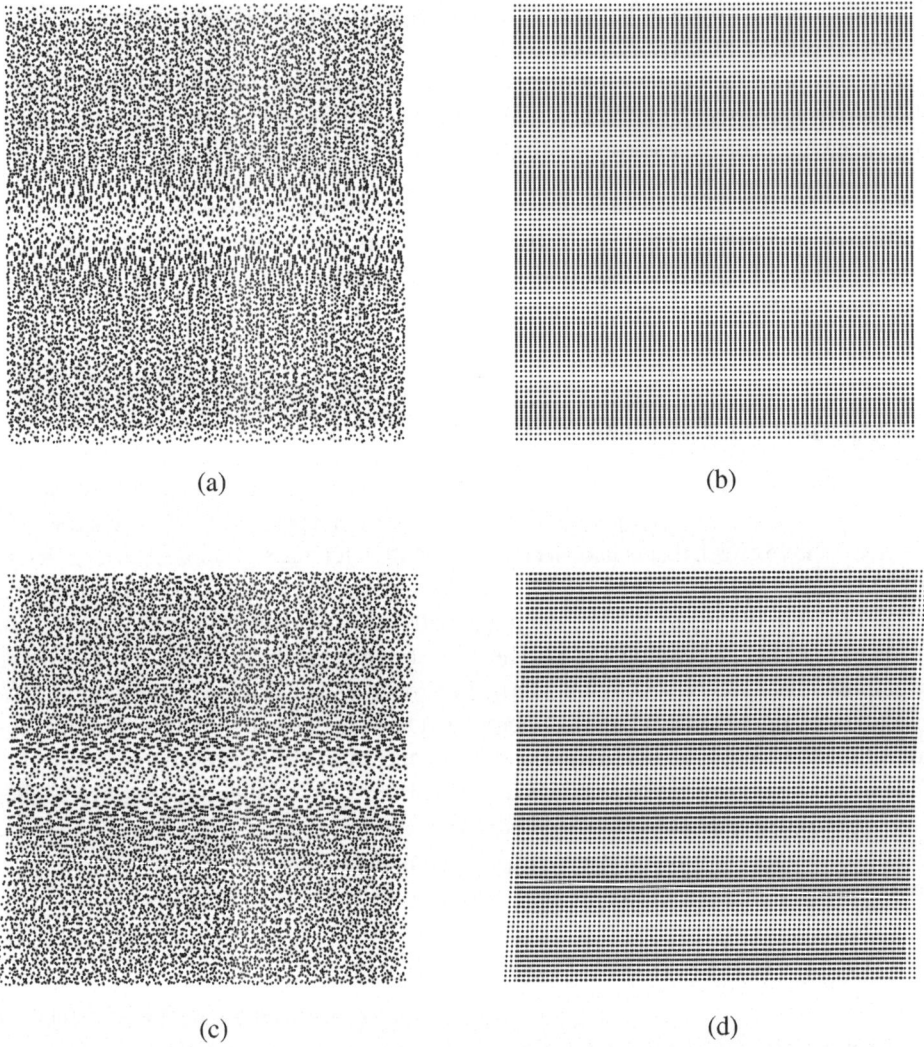

(a)   (b)

(c)   (d)

Figure 2.3: (a),(b) The superposition of two identical copies of the aperiodic dot screen of Fig. 2.1(a) (or of the periodic dot screen of Fig. 2.1(b)), where one of the layers undergoes a slight vertical expansion. In this case a linear Glass pattern is generated along the x axis, and its microstructure consists of straight dot trajectories in the vertical direction. (c),(d) Same as (a),(b), but with a slight horizontal shear insteasd of the vertical scaling. Here, again, a linear Glass pattern is generated along the x axis, but this time the microstructure consists of horizontal dot trajectories parallel to the Glass pattern.

this area the Glass pattern is no longer visible. This explains why the Glass pattern gradually decays and disappears as we go away from its center. It is clear, however, that

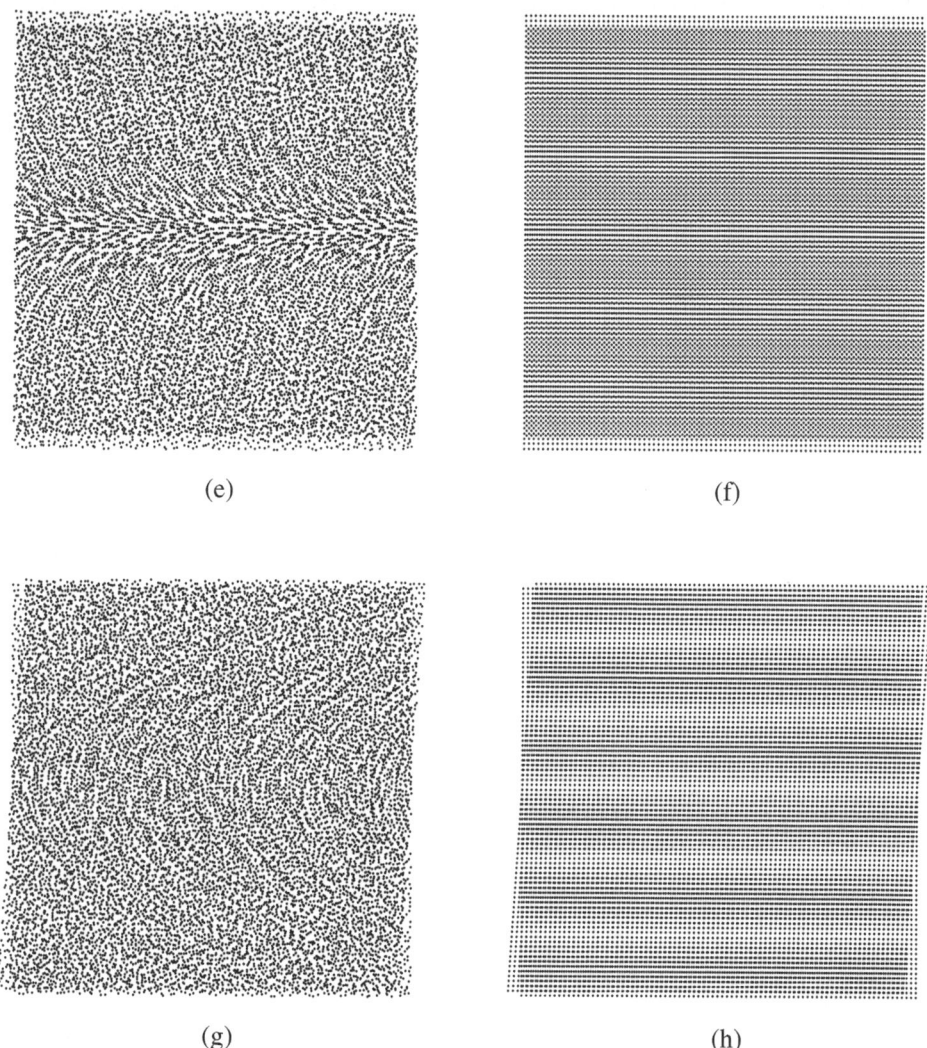

(e)

(f)

(g)

(h)

Figure 2.3: (*continued.*) (e),(f) Same as (a),(b), but here the unscaled layer is slightly shifted to the right. (g),(h) Same as (c),(d), but here the untransformed layer is slightly shifted downward.

when the two superposed layers are not at all correlated, no Glass pattern appears in the superposition (this is, indeed, what happens when one of the aperiodic transparencies is rotated by 180° on top of its identical copy, as shown in Fig. 2.6(a)). In intermediate cases, where the two superposed layers are only partially correlated (for example, when one layer is a copy of the other with a certain amount of random noise being added), the Glass pattern becomes weaker and less perceptible, depending on the degree of the correlation

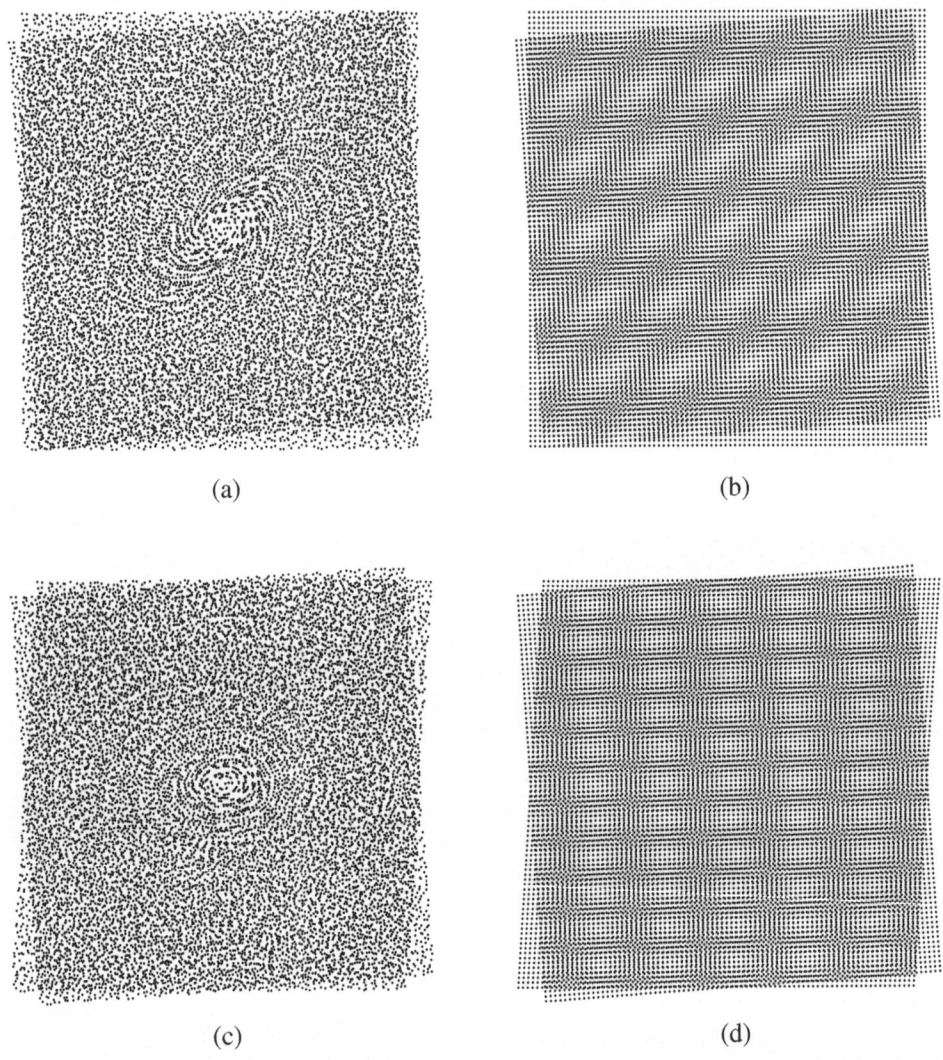

(a)   (b)

(c)   (d)

Figure 2.4: Same as in Figs. 2.3(a)–(d), but with a small rotation in one of the
layers. As we can clearly see in (a) and (c), the linear Glass patterns
of Figs. 2.3(a),(c) turn into a spiral or elliptic Glass pattern around
the origin.

which still remains between the superposed layers (see Sec. 7.7). These facts are
succinctly formulated by the following general result, that will be further discussed in Sec.
7.8:

Proposition 2.1: A layer superposition gives rise to Glass patterns *iff* there exists some
degree of correlation between the superposed layers. ■

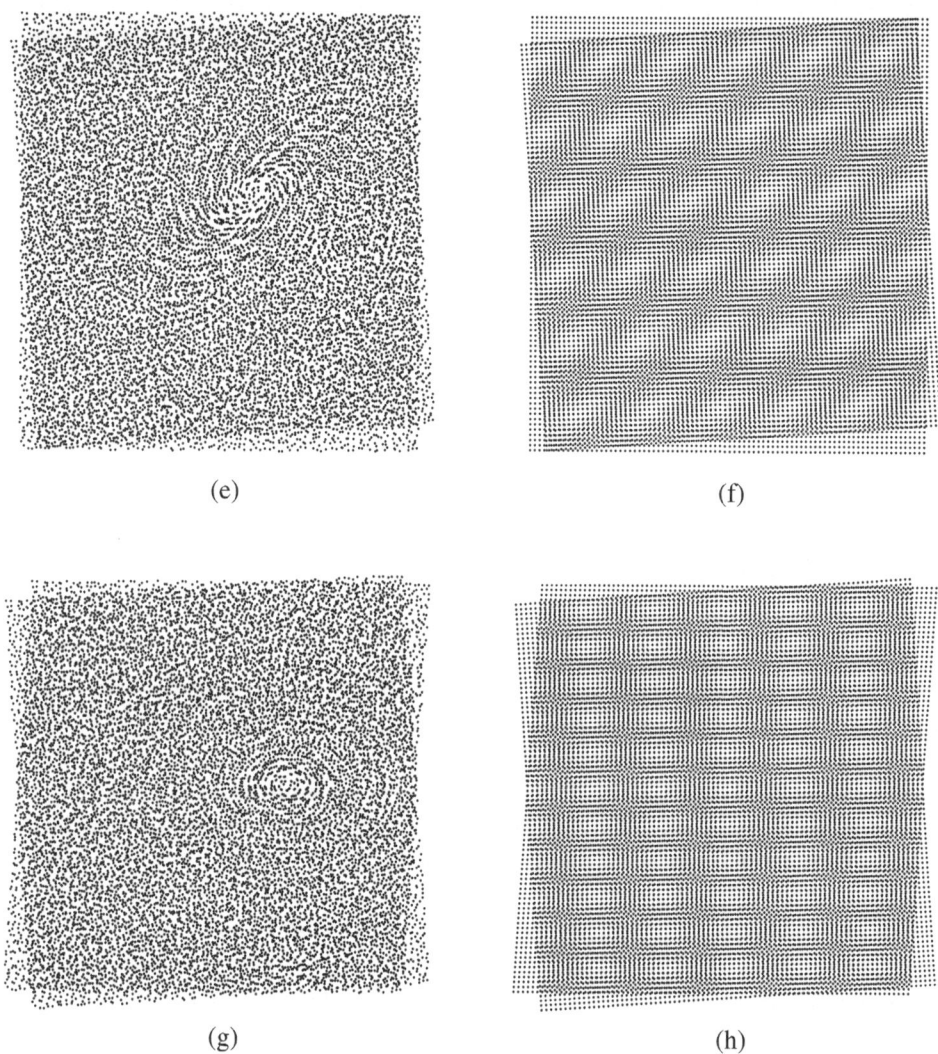

(e)   (f)

(g)   (h)

Figure 2.4: (*continued.*) Same as in Figs. 2.3(e)–(h), but with a small rotation in one of the two layers. As we can see in (e) and (g), the singular moiré-free superpositions of Fig. 2.3(e),(g) turn into clearly visible spiral or elliptic Glass patterns, which are similar to those obtained in (a) and (c), but slightly shifted away from the origin.

In fact, as we will see in Sec. 2.3.3, the bright and dark areas which form the Glass pattern occur due to variations in the local correlation between the superposed layers. The particular case in which the correlation is constant throughout the superposition will be explained in Sec. 2.3.2 below.

It should be noted that although this formulation of Proposition 2.1 uses the term "Glass patterns", it is in fact completely general, and it covers both periodic and aperiodic cases. As we will see later, periodic cases are, indeed, particular cases of a general layer superposition, and in spite of their apparently different look, they still satisfy the same fundamental rules as any other layers. But because of their additional internal structure, periodic cases also satisfy several additional specific rules (that are expressed in terms of periods or frequencies, as described by the classical periodic moiré theory), rules that are not valid for general aperiodic cases.

Remark 2.3: Throughout this book we adopt the terminological convention under which the macroscopic phenomena that are generated in the superposition of originally periodic layers are called "moiré patterns", while their counterparts in the superposition of aperiodic layers are called "Glass patterns". The microscopic phenomena that appear in the aperiodic case are called "dot trajectories". However, it is also customary to lump all of these different phenomena under the common generic term "moiré patterns". Although this usage is not consistent with our usual convention, it may still be practical when no risk of confusion may exist (see, for example, the title of this book). ■

2.3.2 Stable vs. singular moiré-free superpositions

Just as in the periodic case (see Sec. 2.9 in *Vol. I*), we can distinguish in the general case, too, between two different types of moiré-free superpositions. Suppose that two identical layers, periodic or not, are superposed exactly on top of one another (or with a small constant lateral shift, as shown in Figs. 2.5(a),(b)), possibly with some fixed percent of random noise being added throughout. In this case the correlation between the layers remains constant throughout the superposition, and no macro moiré effects are visible. Upon first observation, this situation resembles the moiré-free case that occurs when the two superposed layers are completely independent of each other and have no correlation at all, as in Figs. 2.6(a),(b). However, a significant difference exists between these two types of moiré-free superposition: In the first case, the moiré (or Glass) pattern does exist, but it is not visible because it is infinitely big; such a moiré-free superposition is very unstable, since any slight deviation in the angle or in the scaling of any of the superposed layers may cause the moiré to "come back from infinity" and to become highly visible. This can be clearly seen by comparing Figs. 2.5(a),(b) with Figs. 2.5(c),(d). Just as in the periodic case, this situation is called a *singular* moiré-free superposition. On the other hand, moiré-free superpositions where the superposed layers are completely independent of each other are *stable* moiré-free superpositions: in such cases no moiré (or Glass) patterns become visible even when angle or scale deviations occur between the individual layers. This kind of situation is, indeed, intended when people say that "the superposition of random screens does not generate moiré effects"; such phrases are often heard in the context of random screen halftoning, for example in colour printing.

Another example of a singular moiré-free superposition is shown in Fig. 2.3(g). As we can easily see, in spite of the visible microstructure no perceptible Glass patterns exist in

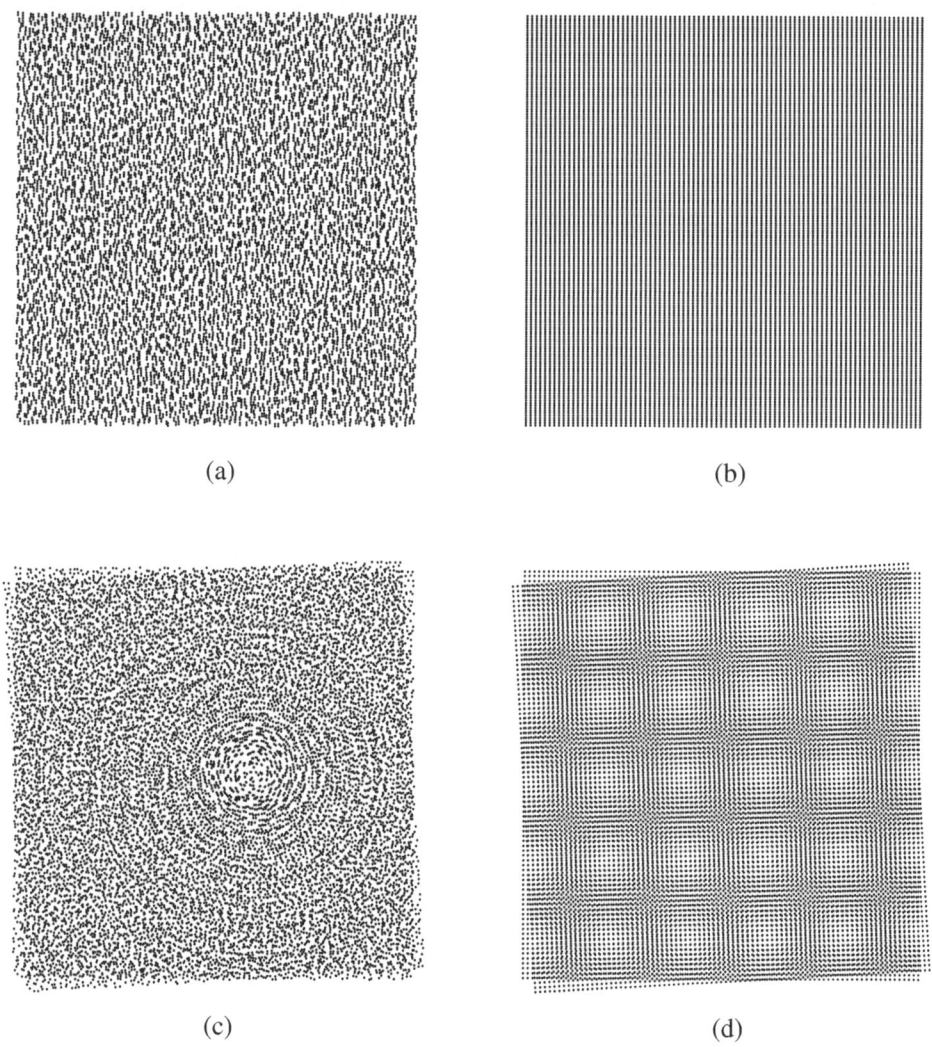

(a)

(b)

(c)

(d)

Figure 2.5: The simplest case of singular moiré-free superpositions: (a),(b) The superposition of two identical copies of the aperiodic dot screen of Fig. 2.1(a) (or of the periodic dot screen of Fig. 2.1(b)) with a small vertical shift. (c),(d) A small rotation in one of the two layers in (a) or (b) is sufficient to cause the reappearance of highly visible Glass (or moiré) patterns. Note that these patterns are simply shifted versions of their counterparts in Figs. 2.1(c),(d); this is explained in Sec. 3.3.3.

this superposition. But the slightest angle deviation in any of the two layers immediately causes the reappearance of a highly visible Glass pattern, as shown in Fig. 2.4(g).

Finally, it is interesting to note that periodic structures are much less liable to form moiré-free configurations than aperiodic structures. This is true for both singular and stable moiré-free superpositions. For example, the periodic counterpart of the singular moiré-free superposition of Fig. 2.3(g), shown in Fig. 2.3(h), as well as the periodic counterparts of the stable moiré-free superpositions of Figs. 2.6(a),(b), are *not* moiré-free superpositions. The converse, however, is always true: The aperiodic counterpart of a moiré-free superposition between periodic layers is always moiré-free, too.

2.3.3 Macrostructures and microstructures in the superposition

As we can see, the explanation of Glass patterns in Sec. 2.3.1 is based on an observation of the individual elements of the original layers and their behaviour in the superposition. We say, therefore, that this explanation is based on the *microstructure*. To obtain the point of view of the *macrostructure*, we have to look at the layers and their superposition from a greater distance, where the individual elements of the layers are no longer discerned by the eye and what we see is only a gray-level average of the microstructure in each area of the superposition. From the point of view of the macrostructure, the center of the Glass pattern consists of a brighter gray level than areas farther away (due to the partial overlapping of the microstructure elements of the two layers in this area); farther away, the macroscopic gray level is darker (because elements from the two layers are more likely to fall side by side, thus increasing the covering rate and the macroscopic gray level). This means that the Glass pattern is not just an optical illusion, and it corresponds, indeed, to the physical reality. In fact, just as in the periodic case (see Proposition 8.1 in *Vol. I*), moiré or Glass patterns are simply the macroscopic interpretation of the variations in the microstructures throughout the superposition.

On the other hand, the ordering of the microstructure elements within a Glass pattern into circular, radial or other "dot trajectories" is no longer visible from far away (try to observe the Glass patterns in Fig. 2.1 from a distance of 3–4 meters, where the individual elements of the layers are no longer discerned by the eye, and only the global gray level variations can be perceived). The dot trajectories are not part of the macrostructure description, and they belong to the microstructure of the superposition, just like rosettes in the periodic case. And indeed, from the point of view of the macrostructure there is no difference between the gray levels that are obtained when the neighboring elements in the superposition are located on circular trajectories or on radial trajectories: What counts in both cases is the resulting mean coverage rate, which determines the overall gray level, and not the specific geometric arrangement of the dots.

What happens, now, in terms of the macro- and microstructures, in singular and in stable moiré-free superpositions? In *singular* cases, where the correlation is constant or nearly constant (but non-zero), a consistent pattern of microstructures (dot trajectories) may be observed throughout the superposition; but this pattern corresponds to a constant or nearly constant gray level throughout, meaning that no macrostructures (Glass patterns) are visible. Such cases are illustrated in Figs. 2.3(e), 2.3(g) and 2.5(a); by observing these

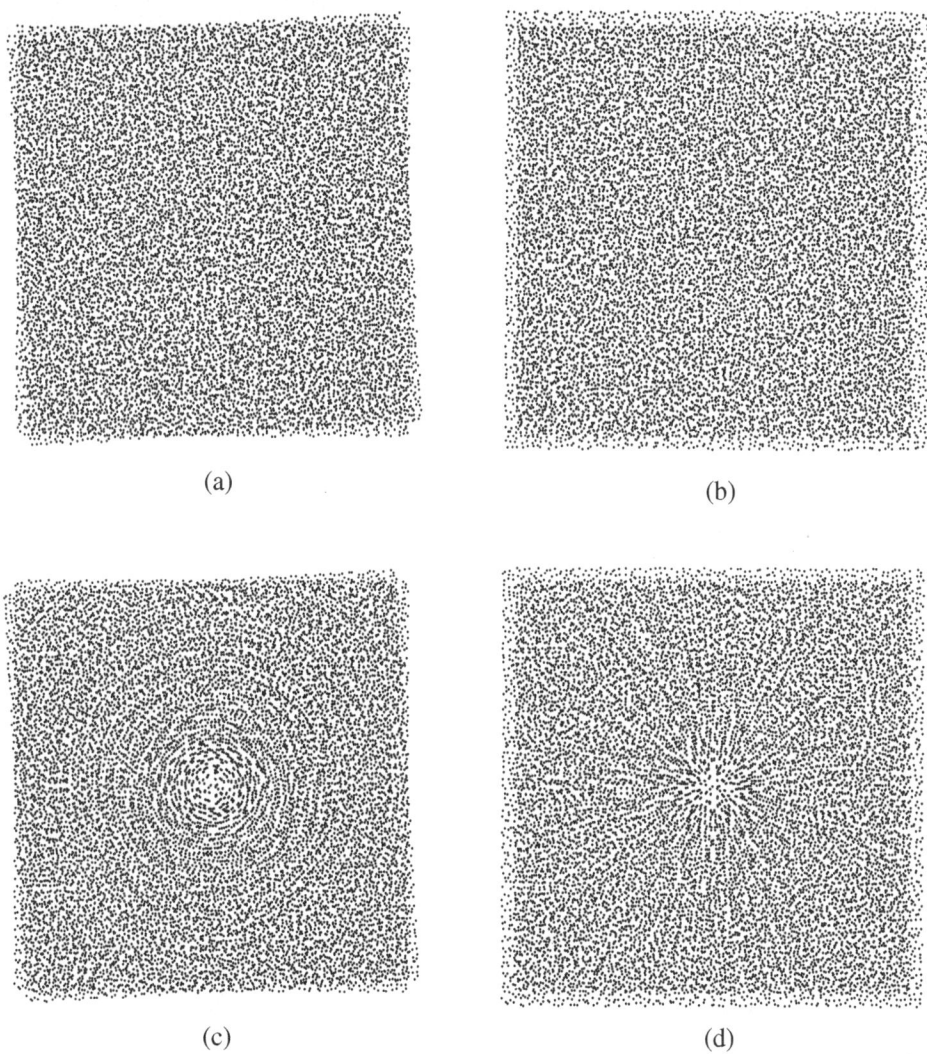

Figure 2.6: (a),(b) Same as (c) and (e) in Fig. 2.1, but with one of the layers being rotated by 180°; unlike in the periodic case, no Glass patterns are generated here. (c),(d) Same as (c) and (e) in Fig. 2.1, but using another aperiodic screen instead of the screen shown in Fig. 2.1(a) (the same screen for both layers). Note that the Glass patterns and the dot trajectories remain the same as in Figs. 2.1(c) and (e), although the individual dot locations have changed.

figures from a distance of 3–4 meters it can be confirmed that they contain no visible gray-level variations, unlike Figs. 2.3(a) and 2.3(c) or Figs 2.1–2.2. Proceeding now to *stable* moiré-free cases, i.e. superpositions of *uncorrelated* random screens, it is clear that they do not generate macrostructures, either, but the difference is situated in the microstructures: in this case, no consistent microstructure patterns (dot trajectories) can be observed in the superposition. This does not yet mean that the microstructure in stable cases is fully smooth and uniform, and in fact, it still may contain various random dot alignments such as nebulous or worm-like structures or artifacts [Lau01], just as the original aperiodic layers themselves. But unlike in singular cases, these microstructure artifacts are irregular and nonuniform, and they cannot be considered as dot trajectories. Although these artifacts do not influence the macroscopic overall gray level of the superposition, they still may be more or less conspicuous when viewed from a close distance, depending on the statistical nature or distribution of the points in the original screens (fully random, "blue noise" [Ulichney88], "green noise" [Lau98, Lau02], etc.).

2.4 The element distribution in the original layers and its influence

As explained at the end of Sec. 2.2, in order to increase the visibility of the various phenomena that occur in the superposition, we illustrate them using layers that consist of distinct microstructure elements such as dots, lines, etc. For example, our layers may consist of black dots on a white background, as shown in Figs. 2.1(a),(b).

In the case of aperiodic layers, the individual elements (in our present example, the black dots) can be distributed within the original layer in various different ways. In a *random scatter* [Bracewell95 pp. 561–563] the individual elements are scattered in the plane in a fully random manner; but the elements can be also scattered according to any other spatial distributions. For example, in order to avoid the disgraceful blotchiness due to the clustering and avoidance effects of a purely random scatter [Bracewell95 pp. 561–562], one may use instead a *pseudo-random scatter*, that gives an apparently random arrangement that looks more pleasing and uniform than purely random scatter [Bracewell95 pp. 593–595]. Such a pseudo-random layer may be constructed on the basis of a periodic dot screen, by perturbing each original dot location (x,y) by a small random displacement $(\Delta x, \Delta y)$ where Δx and Δy are selected randomly within a predefined range, say, $-0.5T \leq \Delta x, \Delta y \leq 0.5T$, T being the period of the originally periodic dot screen. Such pseudo-random dot screens look "cleaner" and much more uniform than purely random dot screens (compare the pseudo-random dot screen of Fig. 2.1(a) with the purely random dot screen shown in Fig. 2.7(a)). For this reason we have generated the aperiodic figures throughout this book using a pseudo-random element distribution. But one could also generate the layers using any other distribution rule, such as a 2D normal distribution, resulting in a non-uniform random scatter [Bracewell95 pp. 599–601].

It may be asked, therefore, whether the element distribution within the original aperiodic layers has any influence on the resulting phenomena that are observed in the layer superposition. It turns out that the artifacts of *statistical* origin, such as the various nebulous or worm-like dot arrangements in the microstructure of moiré-free superpositions, do vary depending on the specific element distribution being used in the individual layers [Lau98, Lau02]. However, other superposition phenomena, such as the

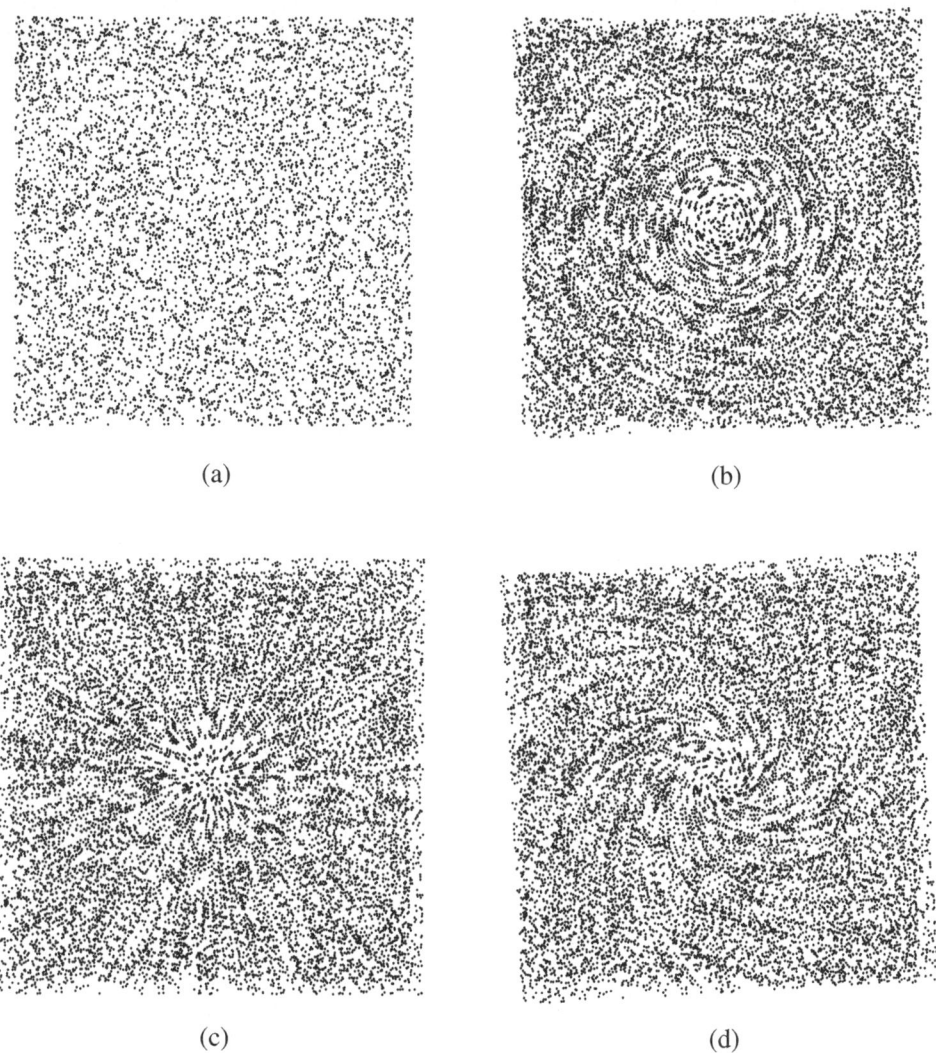

(a)   (b)

(c)   (d)

Figure 2.7: Glass patterns that are generated between purely random dot screens.
(a) The purely random dot screen that is used for generating all the
superpositions shown in this figure. Note the visible blotchiness due
to the clustering and avoidance effects of the purely random dot
distribution, as opposed to the more uniform look of the pseudo-
random dot distribution used in Fig. 2.1(a). (b) The superposition of
two identical copies of dot screen (a) with a small angle difference;
compare with Fig. 2.1(c). (c) Same as (b), but with a small scaling
difference rather than a small angle difference between the superposed
layers; compare with Fig. 2.1(e). (d) Same as (c), but with both a small
angle and a small scaling difference between the superposed layers;
compare with Fig. 2.1(g). Note that the Glass patterns as well as the
dot trajectories remain the same as in Fig. 2.1, although the individual
dot locations in the two figures are different.

Glass patterns and the microstructure dot trajectories, are not affected by the specific random distribution being used in the individual layers (as long as they remain aperiodic).[6] For example, Figs. 2.6(c),(d) show the same superpositions as Figs. 2.1(c),(e) using a different aperiodic screen. As we can clearly see by comparing these figures, the resulting Glass patterns and dot trajectories remain invariant, in spite of the different individual dot locations. A more radical example appears in Fig. 2.7, which shows the same superpositions as in the left-hand column of Fig. 2.1, using this time the purely random dot screen of Fig. 2.7(a) instead of the pseudo-random dot screen of Fig. 2.1(a). As we can clearly see by comparing Figs. 2.1 and 2.7, the resulting Glass patterns and dot trajectories remain unchanged — although the microstructure dot locations are obviously different in both cases. The reason is that these phenomena depend on the *correlation* between the superposed layers, but not on the specific dot distributions within the individual layers. This allows us, indeed, to use throughout this book (with the only exception of Figs. 2.7 and 7.18(b),(d)) pseudo-random dot screens instead of purely random ones, without affecting the superposition phenomena.[7]

It is important to note, however, that the element shapes and sizes in the original layers do influence the resulting phenomena in the superposition. We will return to this question in more detail in Chapter 7.

2.5 Multilayer superpositions

So far we have only mentioned phenomena that occur in the superposition of two aperiodic layers. What happens, now, when three or more aperiodic layers are superposed? As we may guess, each layer pair whose two layers are sufficiently correlated will generate in the superposition a Glass pattern of its own, but uncorrelated pairs will not generate Glass patterns. For example, if in a given superposition of three aperiodic layers all of the three layers are mutually correlated, three different Glass patterns will be generated in the superposition (see Fig. 2.8). Depending on their locations in the superposition these Glass patterns may be partially or even fully overlapping, but because Glass patterns are not periodic structures, they do not interfere with each other and do not create new, multilayer effects as moiré patterns do in the superposition of periodic layers. Thus, in the aperiodic case there are no equivalents to multilayer moirés that are truly generated by more than two layers, such as the three-layer moirés shown in Figs. 2.8(h) and 3.5 of *Vol. I*. (Note that in our Fig. 2.8 each of the Glass patterns is only generated by *two* layers, and it does not disappear when the third layer is removed; this is clearly shown

---

[6] Obviously, the use of a non-uniform dot distribution will cause a non-uniform distribution of gray levels in the original layers, and hence also in their superposition. But the new phenomena that occur due to the superposition will still have the same shapes as in the case of uniform dot distribution.

[7] Note, however, that the use of pseudo-random layers instead of purely random layers *does* influence the superposition phenomena if it introduces some residual periodicity. As we will see in Sec. 7.6 (see also Figs. 5.3–5.4), the introduction of some residual periodicity into the aperiodic layers may cause in the superposition the appearance of darker bands that surround the bright center of the Glass pattern. See also Problem 7-19.

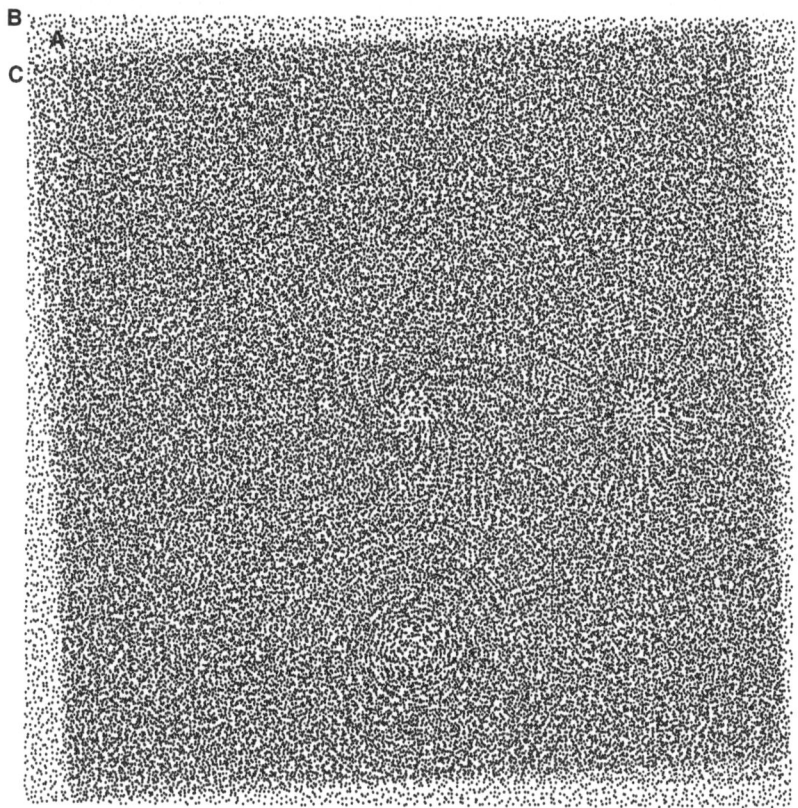

Figure 2.8: A superposition of three aperiodic layers A, B and C, where layer
B is a slightly scaled-up copy of layer A, and layer C is a slightly
rotated copy of layer A. Layer A has been slightly shifted to the
right, in order to move the two Glass patterns it generates away
from the origin. Because each of the three layer pairs in this
superposition consists of two mutually correlated layers, three
Glass patterns are generated: Layers A and B generate a radial
Glass pattern (like in Fig. 2.1(e)); layers A and C generate a
circular Glass pattern (like in Fig. 2.1(c)); and layers B and C
generate a spiral Glass pattern (like in Fig. 2.1(g)). Note,
however, that each of these Glass patterns is slightly masked by a
third aperiodic layer that does not contribute to its generation.
The three unmasked two-layer Glass patterns are clearly shown
in Fig. 2.9(b)–(d).

in Fig. 2.9). For this reason we can limit our discussion in the rest of this book to the
study of Glass patterns between two layers, without any loss of generality. For example,
the locations of the Glass patterns in a three-layer superposition are determined by the
three two-layer superpositions involved (see Problems 3-27 to 3-29 in Chapter 3).

PROBLEMS

2-1. *The Fourier spectrum of random images.* The Fourier theory is very well adapted to the study of random structures (see, for example, [Bracewell86 Chapters 15–16], [Champeney73 Chapter 6] or [Coulon84 Chapters 5–6]). However, unlike in deterministic images, in the case of random images all the phase information is lost, and one can only deal with power spectra (see, for example, [Castelman79 pp. 199–201] and [Champeney73 pp. 79–80]). And yet, given a random image such as the random dot screen shown in Fig. 2.1(a), one can still consider it just like any other image and apply to it the Fourier transform. The resulting spectrum can be expressed in terms of its real part and its imaginary part, and hence also in terms of the amplitude spectrum (or its square, the power spectrum) and the phase spectrum. But the phase spectrum of an image contains precisely all of its phase information. How do you explain this contradiction? *Hint*: As explained in Sec. F.2 of Appendix F, a *random process* (or a *stochastic process*) can be viewed as an infinite set of signals (or images) that represent different instances of the same random process (much like different series of results obtained by rolling the same dice). It is clear, therefore, that in a stochastic process all the phase information is lost, and one can only speak of the power spectrum of the given process. However, any *particular instance* of the stochastic process is, in itself, a deterministic signal (or image), and as such, it does have its own phase information, and one may consider its Fourier transform (including its phase spectrum) just like any other image. It should be remembered, however, that because the structure of such an image is very complex, its Fourier spectrum is very jumpy or noisy, and it often admits a typical *diffuse* appearance [Bracewell95 pp. 586–590, 600–601]; see also the following problem.

2-2. *The Fourier spectrum of a random dot screen.* Fig. 2.10 shows, step by step, how one can understand the Fourier spectrum of a random dot screen. Explain this figure, and using it explain the diffuse nature of the spectra of random screens. *Hint*: Part (a) of the figure shows a single white square dot on a black background; this image is expressed mathematically by the function $d(x,y) = \text{rect}(x/\tau, y/\tau)$ where τ is the width of the square dot. The Fourier transform (spectrum) of this function is given by $D(u,v) = \tau^2 \text{sinc}(\tau u)\text{sinc}(\tau v)$ [Bracewell95 p. 150]. Part (b) of the figure shows the inverse video of the image in part (a), namely, a black dot on a white background. This image is expressed by $1 - d(x,y)$, and therefore its Fourier transform is $\mathcal{F}[1 - d(x,y)] = \mathcal{F}[1] - \mathcal{F}[d(x,y)] = \delta(u,v) - D(u,v)$, where $\delta(u,v)$ is an impulse at the origin (note that this impulse is clearly visible in the center of the spectrum in Fig. 2.10(b)). Part (c) of the figure shows a shifted copy of the black dot. Because this image is not symmetric about the origin, its spectrum is no longer purely real-valued, and it consists of both real and imaginary parts. According to the 2D shift theorem [Bracewell95 p. 156], if $R(u,v)$ is the spectrum of a given image $r(x,y)$ then the spectrum of the shifted image $r(x - a, y - b)$ is given by $R_{a,b}(u,v) = \mathcal{F}[r(x - a, y - b)] = e^{-i2\pi(ua+vb)} \cdot R(u,v)$. Its real and imaginary parts, $\text{Re}[R_{a,b}(u,v)] = \cos 2\pi(ua + vb)R(u,v)$ and $\text{Im}[R_{a,b}(u,v)] = -\sin 2\pi(ua + vb)R(u,v)$, explain, indeed, the modulated look of the spectrum in the figure. Part (d) of the figure shows a superposition of the black dot of part (c) with another shifted black dot, and the resulting spectrum. Note the doubly modulated look of this spectrum, where each of the two modulations is contributed by one of the shifted dots. Finally, part (e) of the figure shows a random dot screen consisting of several shifted copies of the black dot shown in part (b). Here, too, each of the

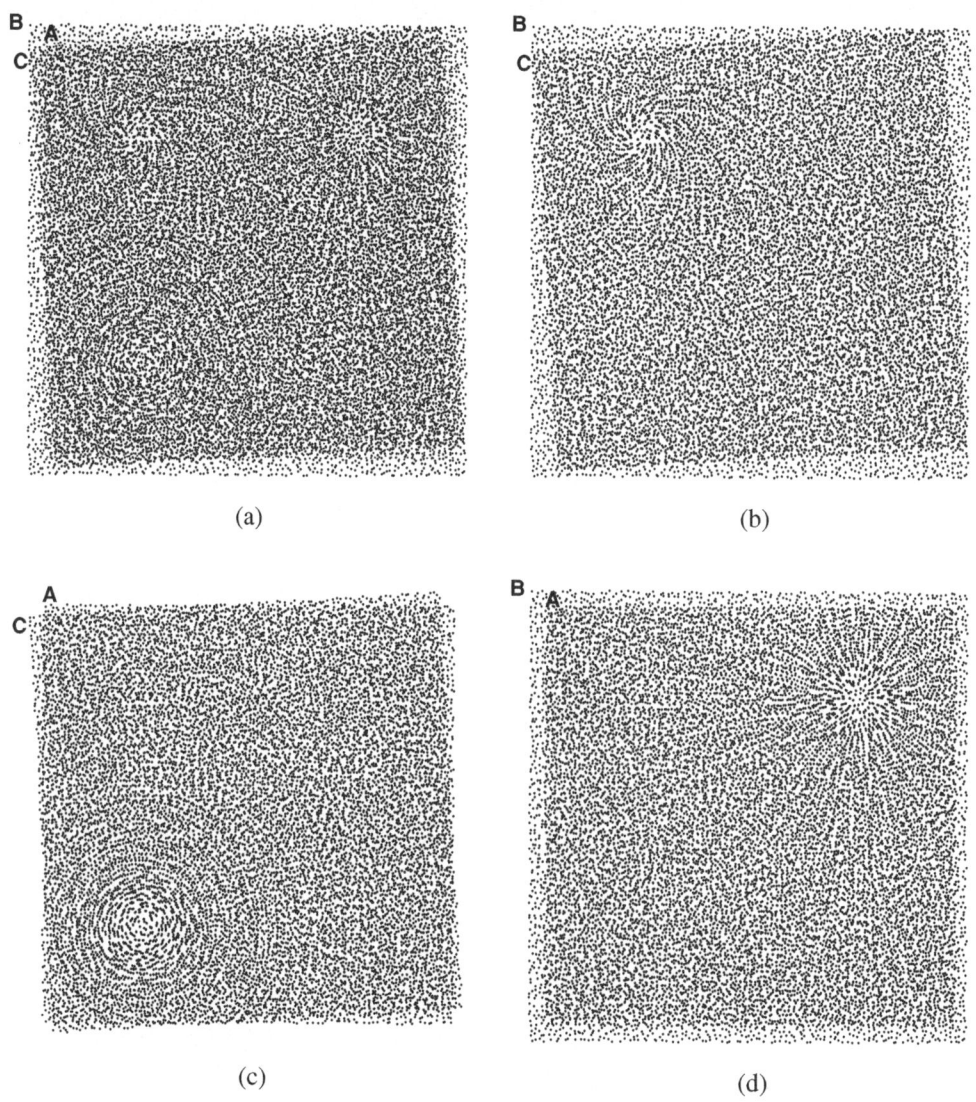

(a)

(b)

(c)

(d)

Figure 2.9: (a) The same three-layer superposition as in Fig. 2.8, showing only the lower right-hand part of the figure where the three Glass patterns are located. (b) Same as in (a), but without layer A. (c) Same as in (a), but without layer B. (b) Same as in (a), but without layer C.

superposed dots contributes its own modulation to the overall spectrum; this results in the typical diffuse nature of this spectrum.

$r(x,y)$ $\qquad\qquad$ $\mathrm{Re}[R(u,v)]$ $\qquad\qquad$ $\mathrm{Im}[R(u,v)]$

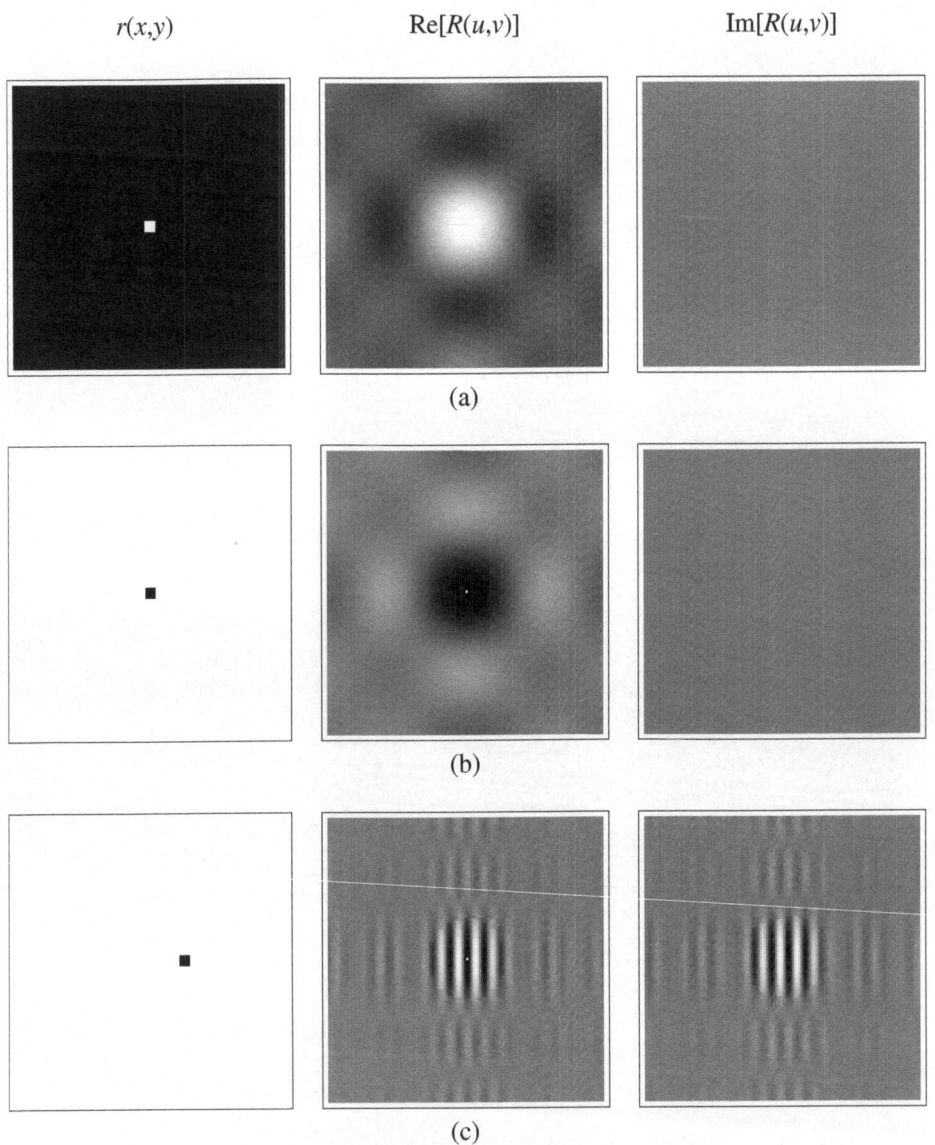

Figure 2.10: Explanation of the Fourier spectrum of a random dot screen. For each image $r(x,y)$, $\mathrm{Re}[R(u,v)]$ and $\mathrm{Im}[R(u,v)]$ show the real and the imaginary parts of the spectrum $R(u,v)$ as obtained on computer by 2D DFT. (a) A single, centered square white dot on a black background and its spectrum. (b) A single, centered square black dot on a white background and its spectrum. (c) A shifted copy of the black dot, and its spectrum.

$r(x,y)$ $\quad\quad\quad$ $\mathrm{Re}[R(u,v)]$ $\quad\quad\quad$ $\mathrm{Im}[R(u,v)]$

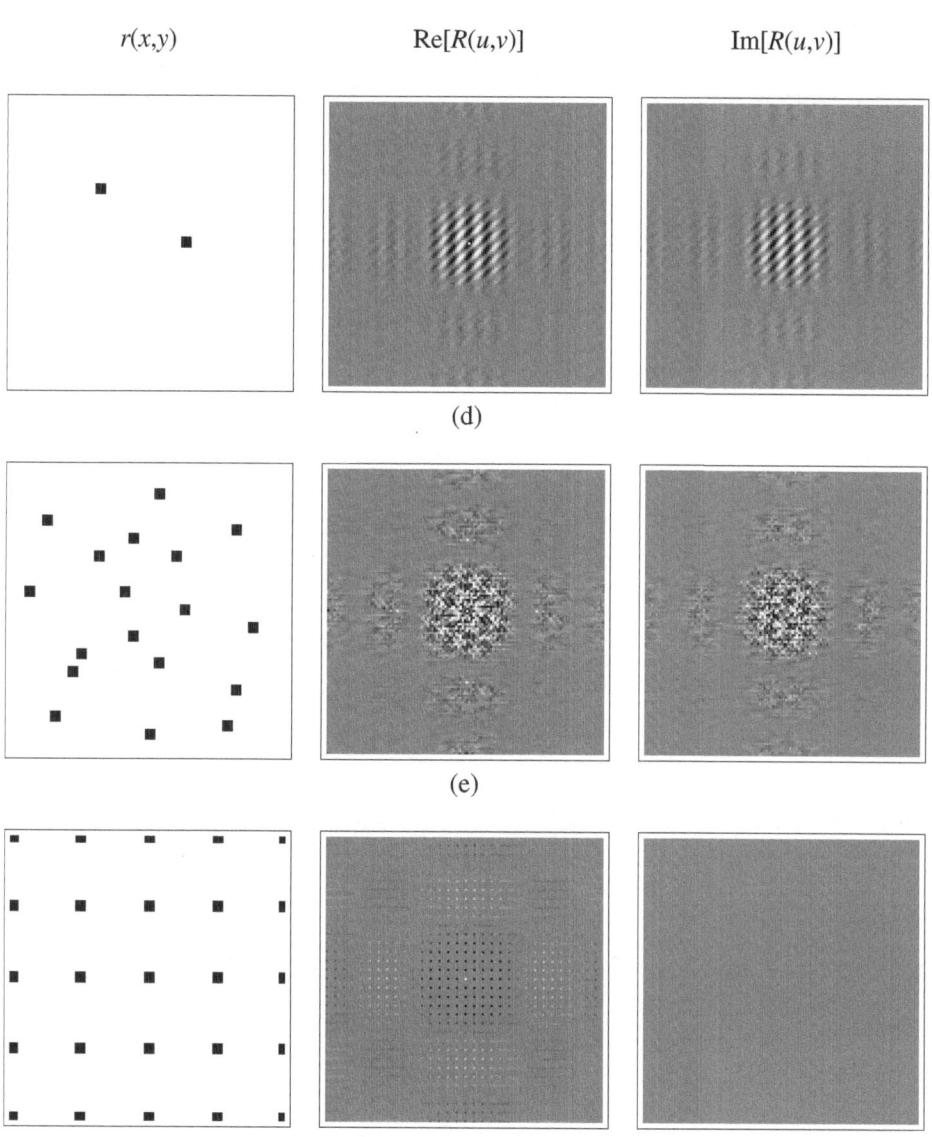

Figure 2.10: (*continued.*) (d) A superposition of the black dot shown in (c) with another shifted black dot, and the resulting spectrum. (e) A random dot screen consisting of several shifted copies of the black dot shown in (b), and its spectrum. For the sake of comparison, (f) shows a periodic counterpart of the screen shown in (e), along with its purely impulsive spectrum (see also Fig. 2.12 in *Vol. I*). The background gray level in all of the spectra represents the value zero, while white represents positive values and black represents negative values. In the image domain, however, black represents zero and white represents one, as in our usual convention (see Sec. 2.2).

r(x,y)   Abs[R(u,v)]   Arg[R(u,v)]

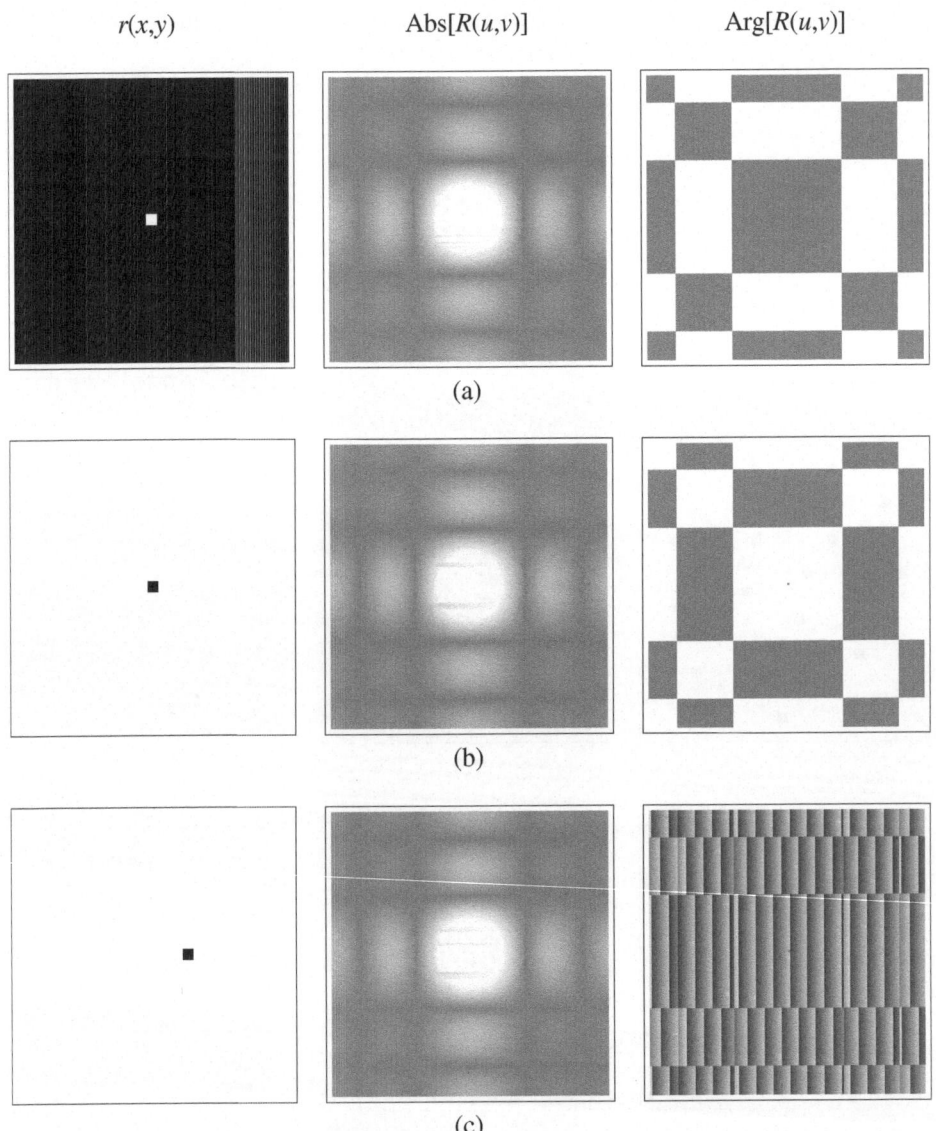

(a)

(b)

(c)

Figure 2.11: Same as Fig. 2.10, but here each of the spectra is represented by its magnitude Abs[$R(u,v)$] and its phase Arg[$R(u,v)$] rather than by its real part Re[$R(u,v)$] and its imaginary part Im[$R(u,v)$]. (a) A single, centered square white dot on a black background and its spectrum. (b) A single, centered square black dot on a white background and its spectrum. (c) A shifted copy of the black dot, and its spectrum.

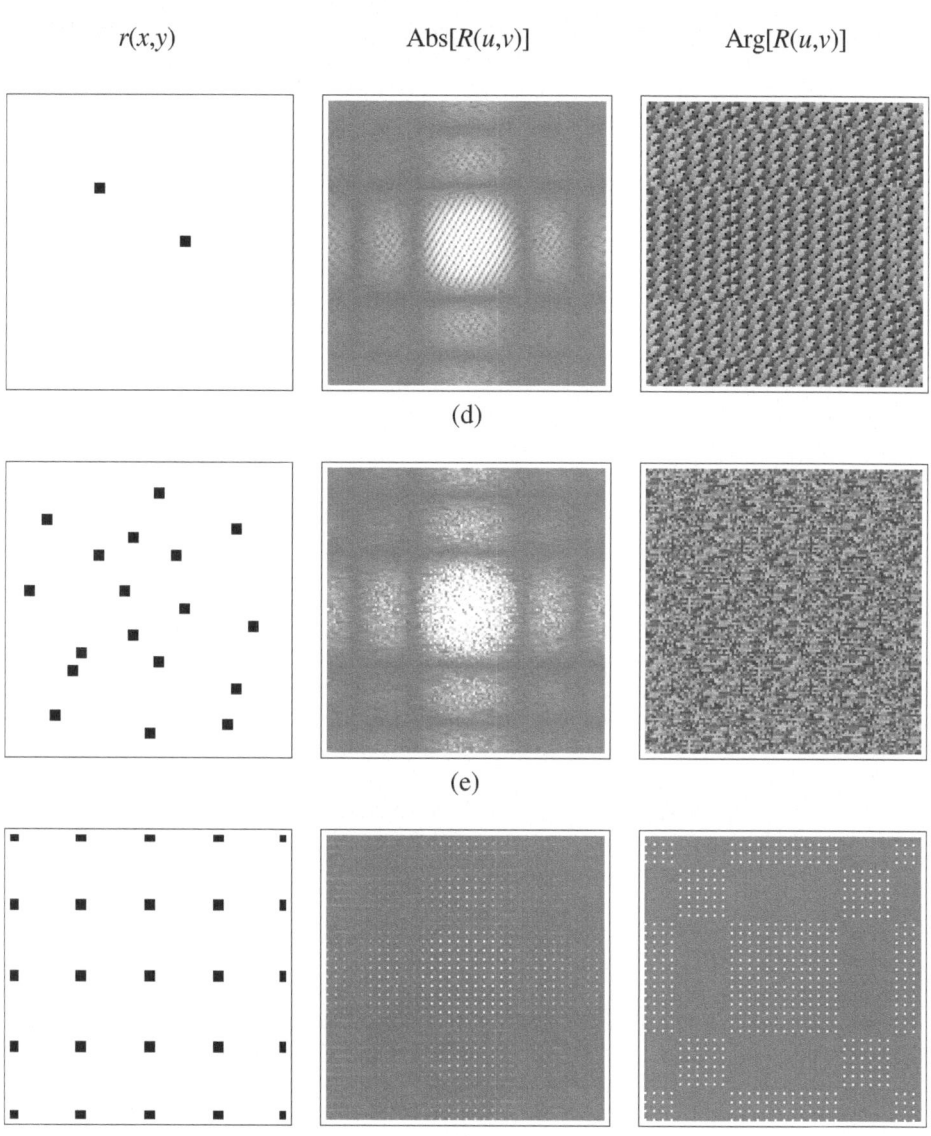

Figure 2.11: (*continued.*) (d) A superposition of the black dot shown in (c) with another shifted black dot, and the resulting spectrum. (e) A random dot screen consisting of several shifted copies of the black dot shown in (b), and its spectrum. For the sake of comparison, (f) shows a periodic counterpart of the screen shown in (e), along with its purely impulsive spectrum. The gray level conventions for the image domain and for the spectral domain are the same as in Fig. 2.10. Note that the magnitude of the spectrum is non-negative, while the phase values vary between $-\pi$ and π.

2-3. *The Fourier spectrum of a random dot screen (continued).* What would have changed in Fig. 2.10 if parts (c)–(e) were drawn with white dots on a black background, as in part (a), and the inverse-video operation (the transition to black dots on a white background) were applied only then?

2-4. *The Fourier spectrum of a random dot screen (continued).* Being a complex-valued entity, the Fourier transform $F(u,v)$ of a function $f(x,y)$ can be presented in two equivalent ways: either in terms of its real part $\mathrm{Re}[F(u,v)]$ and its imaginary part $\mathrm{Im}[F(u,v)]$, where $F(u,v) = \mathrm{Re}[F(u,v)] + i\mathrm{Im}[F(u,v)]$, or, using the polar notation, in terms of its amplitude spectrum $\mathrm{Abs}[F(u,v)]$ and its phase spectrum $\mathrm{Arg}[F(u,v)]$, where $F(u,v) = \mathrm{Abs}[F(u,v)] \cdot e^{i\,\mathrm{Arg}[F(u,v)]}$. Fig. 2.11 shows the same images as Fig. 2.10, but this time the respective spectra are represented by their magnitude $\mathrm{Abs}[F(u,v)]$ and their phase $\mathrm{Arg}[F(u,v)]$ rather then by their real part $\mathrm{Re}[F(u,v)]$ and their imaginary part $\mathrm{Im}[F(u,v)]$. Explain the connection between the two representations of each of the spectra in Figs. 2.10 and 2.11.

2-5. *The Fourier spectrum of a random dot screen: influence of the dot shape.* What would have changed in the spectra of Fig. 2.10 if the individual screen dots were circular rather than square? *Hint:* In this case, part (a) of the figure would consist of a *circular* white dot on a black background; this image is expressed mathematically by the function $d(x,y) = \mathrm{rect}(r/(2a))$ where $r = \sqrt{x^2 + y^2}$ and a is the radius of the dot. The Fourier transform of this function is given by $D(u,v) = aJ_1(2\pi aq)/q$ where $q = \sqrt{u^2 + v^2}$ and J_1 is the first-order Bessel function of the first kind [Bracewell95 p. 354]. This spectrum is similar to the spectrum of a square white dot, $D(u,v) = \tau^2 \mathrm{sinc}(\tau u)\,\mathrm{sinc}(\tau v)$ (see Fig. 2.10(a)), except that it has a full circular symmetry. Consequently, the new spectra in all parts of the figure would be similar to their counterparts in the existing figure, except that each of them would have a circular nature.

2-6. What would have changed in the spectrum of Fig. 2.10(e) if the dot screens were periodic? This is illustrated in Fig. 2.10(f); a similar case is also shown in Fig. 2.10 of *Vol. I,* which explains the spectrum of a periodic screen consisting of *white* square dots. What would you expect to see in the spectrum if the dot screen were intermediate between fully periodic and fully random? How does the spectrum of the pseudo-random dot screen of Fig. 2.1(a) differ from the spectrum of the purely random dot screen of Fig. 2.7(a)? See also Problem 2-9 below.

2-7. Can you suggest a general rule that intuitively explains the spectrum of a dot screen that consists of identical elements? *Hint:* It is helpful to consider such a dot screen as a convolution of a single, centered screen element $d(x,y)$ and an impulse nailbed whose individual impulses indicate the screen's element locations. Using the convolution theorem it is clear that the spectrum of this screen is the product of the spectrum of the single element and the spectrum of the impulse nailbed. Now, the Fourier transform of a single, centered screen element generally gives a widespread, continuous spectrum (such as the 2D sinc function $D(u,v) = \tau^2 \mathrm{sinc}(\tau u)\,\mathrm{sinc}(\tau v)$ in the case of a square dot, or its circular counterpart $D(u,v) = aJ_1(2\pi aq)/q$ with $q = \sqrt{u^2 + v^2}$ in the case of a circular dot). On the other hand, the Fourier transform of an impulse nailbed gives a highly structured spectrum (for example, an impulse nailbed in the periodic case, or a diffuse spectrum in the aperiodic case). It follows, therefore, that in the product of these two spectra, the spectrum of the single, centered screen element plays the role of a continuous "envelope" that modulates the dense, fine structure of the other spectrum. Thus, the global shape (the "envelope") of the spectrum of our screen is determined by the Fourier transform of *a single screen element* that is centered at the origin, while the precise internal structure of this global envelope is determined by the arrangement (the locations) of the screen's elements. For example, if the screen elements are arranged periodically this envelope modulates a nailbed of equally spaced impulses (as shown in Fig. 2.10(f); see also Fig. 2.12 in *Vol. I*), and if the screen dot locations are fully random the envelope modulates a continuous diffuse-looking spectrum (as shown in Fig. 2.10(e)). It should be noted, however, that although this rule is true when the dot screen is either purely periodic or purely random, as shown by Eqs. (F.6) and (F.10) in Appendix F, it does not necessarily hold in random dot screens in which the dot locations are correlated or otherwise disturbed. As clearly indicated by Eqs. (F.9) and (F.11) in Appendix F, the power spectrum in such cases is not only determined by the individual dot shape, but also by the distribution (or probability density) of the dot locations. For example, random dot screens in which the distances between neighbouring dots are highly correlated, with a characteristic nearest neighbour distance of r, tend to give a ring-like power spectrum where the mean radius of the ring is $1/r$ (see, for example, the figures in [Yellott82] and [Yellot83], Plates 15–16 in [Harburn75], or Fig. 10.25 in [Glassner95 p. 433]).

2-8. To illustrate the fact that the envelope shape of the spectrum of a dot screen is not always determined only by the shape of the spectrum of the individual dot, a student suggests the following experiment: Starting from Figs. 2.10(e) and 2.10(f), multiply their spectra by a ring-like band-pass filter (having the value 1 between the radiuses r_1 and r_2 and 0 everywhere else). The diffuse or impulsive ring-like spectra thus obtained have a ring-like envelope which *does not* originate from the shape of the spectrum of the individual dot. What do you think of this experiment? What do you expect to see in the image domain when applying an inverse Fourier transform to these ring-like spectra? *Hint*: Indeed, the ring-like envelope of the spectrum in such cases does not originate from the shape of the spectrum of the individual dot. However, this brute-force approach for generating a spectrum whose envelope shape is not determined by the spectrum of a single dot misses its goal, since the structures in the image domain that belong to these spectra are no longer dot screens, but rather continuous images (that resemble blurred dot screens). Note that it *is* possible to construct dot screens having a ring-like spectrum envelope that is not determined by the spectrum of an isolated screen dot, as indeed shown in the references given at the end of the previous problem, but the construction of such dot screens requires more thought than the brute-force method suggested here. In any case, dot screens having such ring-like spectra are certainly not purely random nor purely periodic, since in both of these cases the screen's spectrum is only modulated by the individual dot's spectrum (see Eqs. (F.6) and (F.10) in Appendix F).

2-9. The Fourier spectrum of a fully periodic structure such as a periodic dot screen or a periodic line grating is purely impulsive (see, for example, Fig. 2.10(f), or Fig. 2.12 in *Vol. I*). Suppose now that we slightly perturb the periodicity by adding some random noise to the locations of the periodic elements (this operation is often called "jittering"). How would this influence the spectrum of the structure? What would you expect to see in the spectrum when you gradually increase this random noise? *Hint*: See Eq. (F.11) in Appendix F.

2-10. *The influence of the structure of a dot screen on its spectrum.* In order to get a deeper insight into the influence of the structure of a dot screen on its Fourier spectrum it is highly recommended that you make your own Fourier experiments on various dot screens, using any 2D Fourier transform program or software package. For example, you may try to see how the spectrum is influenced by the individual dot shape, by the dot size, by the dot distances, by the dot arrangement, or by any other parameter. In order to develop a better intuition it may be also helpful to have a look at the drawings in any book on Fourier transforms, for example in [Bracewell95 Chapter 4],

$r(x,y)$ Re[$R(u,v)$]   Im[$R(u,v)$]

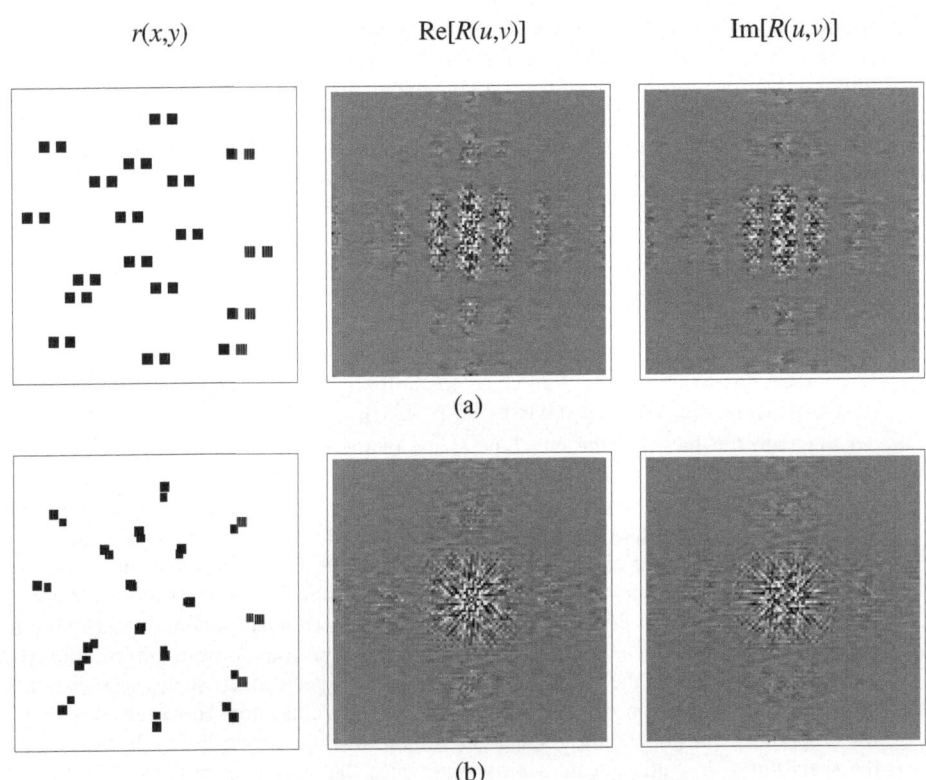

(a)

(b)

Figure 2.12: (a) The superposition of two identical copies of the random dot screen of Fig. 2.10(e) that have been slightly shifted from each other in opposite directions, and the corresponding spectrum. Note the visible cosinusoidal modulation of the spectrum, which clearly indicates the size and the direction of the layer displacement. (b) The superposition of two identical copies of the random dot screen of Fig. 2.10(e), one of which was slightly scaled by a factor $s \approx 1$. The spectrum of this superposition is not as easy to interpret as that of part (a).

[Bracewell86 pp. 246–247 and 411–426] or [Champeney73 pp. 20–39]. Another excellent source of visual information is *The Atlas of Optical Transforms* [Harburn75], which provides an extensive set of various 2D shapes (including various periodic and random dot screens) and their 2D power spectra; apart from being highly instructive, it also offers several examples having surprising and sometimes even quite spectacular spectra.

2-11. *The spectrum of the superposition of random dot screens.* Fig. 2.12(a) shows the superposition of two identical random dot screens which have been slightly displaced in opposite directions along the x axis. Explain this spectrum. *Hint*: Let $r(x,y)$ be a random screen consisting of white dots on a black background, and let $R(u,v)$ be its spectrum. The superposition of two copies of this dot screen that have been shifted in opposite directions is given by the product $r(x+a, y)r(x-a, y)$, but if there is no

r(x,y) Abs[R(u,v)] Arg[R(u,v)]

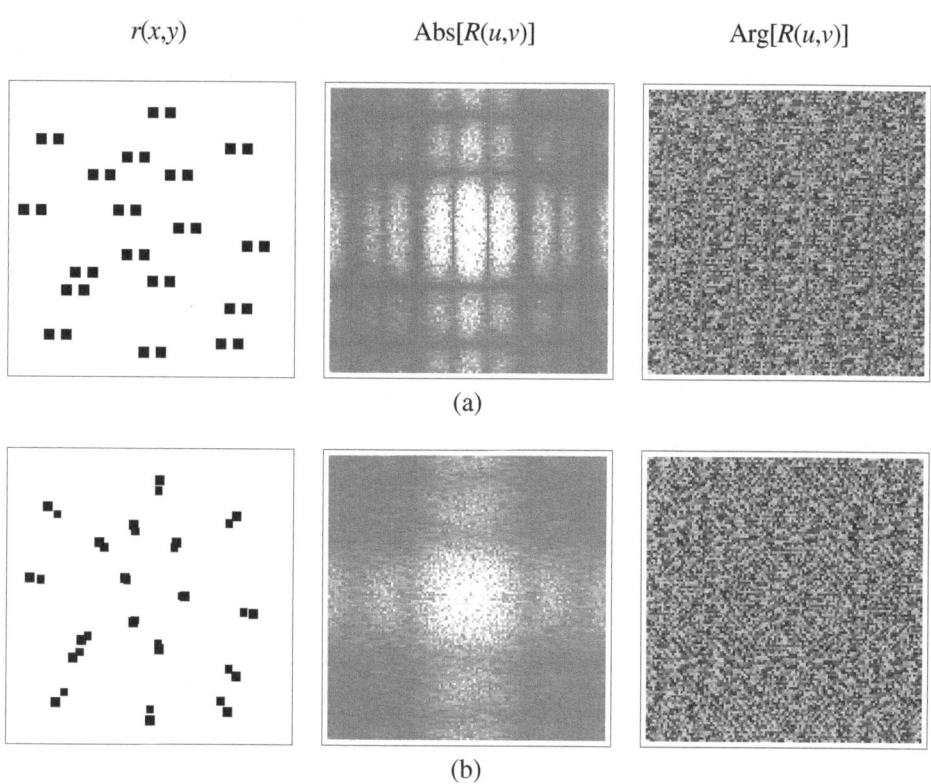

(a)

(b)

Figure 2.13: Same as Fig. 2.12, but here each of the spectra is represented by its magnitude Abs[R(u,v)] and its phase Arg[R(u,v)] rather than by its real part Re[R(u,v)] and its imaginary part Im[R(u,v)].

overlapping between the dots of the two shifted screens, their superposition can be also expressed by the sum $r(x+a, y) + r(x-a, y)$. According to the modulation theorem [Bracewell86 pp. 108–109, 244], the spectrum of this sum is given by $2R(u,v)\cos(2\pi a)$. This multiplication by $\cos(2\pi a)$ explains, indeed, the modulated nature of this spectrum. However, the screen superposition in our figure is the *negative* of this superposition, since it consists of black dots on a white background; it is expressed, therefore, by $(1 - [r(x+a, y) + r(x-a, y)])$, and its spectrum is $\delta(u,v) - 2R(u,v)\cos(2\pi a)$, where $\delta(u,v)$ is an impulse at the origin. Note that the background gray level in the spectrum represents the value zero, while white represents positive values and black represents negative values. Thus, one period of the modulating cosine corresponds to *two* vertical bands in the spectrum. The frequency a of the spectral modulation (i.e. the reciprocal value of the period of the cosine) equals half of the distance $2a$ between the two shifted layers.

2-12. *The spectrum of the superposition of random dot screens (continued).* Fig. 2.13(a) shows the same screen superposition as Fig. 2.12(a), but this time the respective spectra are represented by their magnitude Abs[F(u,v)] and their phase Arg[F(u,v)] rather then by their real part Re[F(u,v)] and their imaginary part Im[F(u,v)]. Explain the connection between the two representations of the respective spectra in Figs. 2.12 and 2.13. Note that the displacement between the two layers is inversely proportional to the fringe spacing in the spectrum, and the direction of displacement is perpendicular to the fringes.

2-13. Fig. 2.12(a) shows the spectrum of the superposition of two slightly shifted copies of a random dot screen. As shown in the previous problems, this spectrum provides useful information about the displacement between the two copies of the random screen and about its orientation. What would you expect to see in the spectrum of the superposition of two copies of a random screen that slightly differ in their *orientation* or in their *scaling* rather than in their *displacement*? What useful information can you extract from these spectra? *Hint*: The spectrum of a rotated screen is simply a rotated version of the original spectrum, and the spectrum of a spatially scaled screen is basically an inversely scaled version of the original spectrum; the exact relation is given by the 2D similarity theorem: $\mathcal{F}[f(ax,by)] = \frac{1}{|ab|}F(u/a,v/b)$ [Bracewell86 p. 244]. However, the spectrum of the superposition in the cases of rotation or scaling is more difficult to interpret than in the case of displacement, and its visual inspection does not provide easily perceptible cues on the superposition as in the case of displacement.

2-14. *Laser speckle*. The invention of the laser created great anticipation among users of optics because it appeared to be the answer to many illumination problems: Indeed, lasers provide a beam of light that is intense, collimated, monochromatic and coherent. However, disappointment soon followed, as it turned out that visual or photographic images of objects illuminated by a laser were covered with a random grainy structure that severely limited the effective resolution [Cloud95, Sec. 18.1]. This phenomenon, known as *laser speckle*, is caused by the interference of coherent light waves that are randomly scattered by a rough surface; a more detailed explanation of this phenomenon and its various properties can be found, for example, in [Cloud95, Chapter 18]. For some years laser speckle was considered a nuisance that caused the laser to be less useful than expected, but as understanding of this phenomenon was developed, it evolved from a problem into the basis of new precision measurement technologies. Can you think of possible applications of laser speckle to the detection and measurement of displacements or deformations? Some of these applications are briefly described in the following problems. More details can be found, for example, in [Erf78], [Cloud95] and [Gåsvik95].

2-15. *Detection and measurement of slight in-plane displacements or deformations*. Speckle metrology is a valuable alternative to the classical moiré-based methods in metrology. Instead of using periodic gratings, speckle techniques make use of the random pattern of dark and bright spots (known as *speckle*) that is formed when a diffusely reflecting object is illuminated by coherent laser light. But if a random pattern already exists on the object's surface naturally (or if it can be applied thereto), there is no need to use coherent laser light to generate speckle, and the object can be simply illuminated by incoherent light. In either case, when the object is slightly moved or deformed, the observed speckles on its surface also move accordingly. Because the random speckle pattern is a unique characteristic of the microstructure of the specific surface area, the displacements or deformations undergone by the speckle can be used to accurately detect and measure in-plane motions or distortions of the object. Before computers were used to record images, a photographic plate was exposed by two views of the speckle, one *before* and one *after* the object was displaced or deformed. The resulting doubly-exposed plate, known as a *specklegram*, was then illuminated by a laser beam, giving on a viewing screen placed behind the specklegram a circular halo of speckle with a clearly visible pattern of linear interference fringes [Erf78 p. 60]. (These fringes are often called by abuse of language *Young's fringes*, although in reality they do not originate from Young's double-slit experiment but from a specklegram consisting of pairs of random speckles, which can be roughly regarded as a cluster of many randomly distributed copies of a double slit [Bahuguna88].) The spacing and orientation of these fringes provide information about the global displacement between the two exposures, and hence about the displacement undergone by the object. This technique is known as *speckle photography*. Nowadays computers are being used instead of specklegrams, and the images are recorded by a CCD detector and are stored on separate frames in the computer. Because the speckle images before and after the object displacement are on different frames, it is possible to use a cross-correlation algorithm to determine the displacement. Moreover, unlike speckle photography, which only provides global (or averaged) information on the entire frame, cross-correlation can be performed locally, by using a small moving window, and the result is a full 2D *displacement field*, i.e. a vector field that indicates at each point the local displacement of the speckle (including the magnitude and the direction, but not the sign of the displacement). This allows to obtain a detailed view of the displacement or deformation undergone by the object on a point-by-point basis (rather than just global, averaged information on the entire frame, as in the older method). This technique is known as *speckle correlation*. More details on these and other speckle-based techniques in metrology can be found, for example, in [Erf78], [Cloud95] and [Svanbro04].

(a) How are these techniques related to the Fourier spectrum of two superposed copies of a random dot screen, such as in Fig. 2.13(a)? Can you use the information provided by this Fourier spectrum for the detection and measurement of slight displacements or deformations?

(b) How do you interpret the information provided by the fringes in a specklegram? *Hint*: The average displacement of the object is inversely proportional to the fringe spacing within the specklegram, and the direction of displacement is perpendicular to the fringes.

(c) How can the techniques described above be improved for measuring out-of-plane deformations, too?

(d) All techniques which are based on two images (before and after the deformation) suffer from a sign ambiguity: they provide accurate information on the magnitude and on the direction of the displacement, but not the sign of the displacement. Is there a way to eliminate this sign ambiguity?

2-16. *Speckle interferometry*. While in speckle correlation the *position* of the speckle before and after deformation is recorded, speckle interferometry studies the *phase change* of the speckles. As opposed to classical moiré interferometry where the information is obtained from a fine grating that is affixed to the surface of the specimen, in speckle interferometry optically rough surfaces are used, and it is precisely their speckle that provides the desired information. The motion or deformation of the object's surface introduces a change in the phase of the individual speckles. This change is extracted either optically, by studying the interference fringes generated between the coherent light reflected from the speckles and a coherent light reference, or electronically. In the latter case (known as *electronic speckle pattern interferometry* or *ESPI*) the phase change of the speckle is extracted by a computer by subtracting the speckle image after the distortion from the speckle image before the distortion, and displaying the absolute value of the difference (to avoid negative values). This gives on the display interference fringes consisting of black lines that cover the surface of the object. These are lines of constant deformation, i.e. lines which connect points on the object's surface that underwent an equal amount of deformation or rigid body motion. This fringe pattern gives us accurate information on the displacement or deformation undergone by the object. More details on this method and its different variants can be found, for example, in [Cloud95]. Note, however, that if the displacement or the deformation are too large, the fringes become too dense and may vanish due to speckle decorrelation [Cloud95 Sec. 18.7; Svanbro04 p. 17]. Speckle interferometry is therefore only suited for measuring very small deformations that are smaller than a single speckle diameter, while speckle correlation is best suited for measuring larger deformations. Thus, speckle correlation and speckle interferometry complement each other. Can you think of methods for combining them in order to improve the performance of the system? Such methods have been studied in [Svanbro04]. An important advantage of speckle interferometry over classical moiré interferometry is that it does not require the application of a grating to the surface of the specimen, and thus combersome sample preparation is obviated. It should be noted, however, that just like moiré interferometry, speckle interferometry suffers from a very high sensitivity to disturbances in the environment, and it may require high mechanical stability and vibration-isolated optical benches.

2-17. *Stereo matching.* Stereo vision is one of the various possible approaches for measuring the distance to an object. In a typical stereo imaging system two identical cameras are placed side by side and take simultaneously a left and a right image of the same scene. The distance to a point P in the scene can be found from the distance between the cameras and the difference between the coordinates of the left and right images of the point P. In order to detect the distance one has to find, therefore, corresponding points in the left and right images. In a conventional implementation this is done by comparing a region from the left image with a moving window in the right image, that is shifted horizontally until the region with the highest correlation is found and selected as the match. However, in an alternative implementation, the matching points can be found by rotating one of the images. Because the two images differ only by a small translation and are otherwise almost identical, rotating one of the two gives in their superposition a visible Glass pattern which surrounds the fixed point (see Chapter 3). By locating the fixed point we obtain the disparity between the left image and the right image at the location of the fixed point. Given the distance between the two cameras, this allows us to calculate the distance to the object P in the scene whose image is located at the fixed point. This technique is described in more detail in [PoChe´95]. What are the main advantages and drawbacks of this technique?

2-18. *Avoiding sampling moirés by random sampling.* All continuous signals or images that are to be processed numerically must be first sampled. Sampling is the process of recording the values of the original continuous signal (or image) at a certain discrete set of locations; see, for example, [Bracewell86 Chapter 10] or [Bracewell95 Chapter 7]. In classical signal processing the sampling of signals or images is done at regular spacings, i.e. on a periodic lattice. The main advantage of regular sampling is that it is very easy to perform, and its theoretic and practical aspects are well understood and widely described in the literature. But regular sampling has also a significant drawback: if the signal or image being sampled contains periodic or structured detail, highly visible aliasing (or moiré) effects may appear in the sampled image due to the interaction between the image frequencies and the sampling frequency. Such sampling artifacts can be avoided by using *random sampling* (also known as *stochastic sampling*), i.e. by sampling the original signal on a random lattice. This kind of sampling is explained and illustrated in Chapters 9–10 of [Glassner95]. But it turns out that although random sampling completely eliminates the risk of aliasing (or moiré) artifacts in the sampled signal, it introduces instead unstructured, irregular noise artifacts (see Fig. 9.3 in [Glassner95 p. 374]). Whether this is an advantage or not depends on the intended use of the sampled signal. If it is an image to be viewed by a person, the noisy version is often superior because the human visual system is much more sensitive to structured aliasing (or moiré) artifacts than to unstructured noise. Thus a noisy picture can look better than one that is aliased, though neither is more accurate than the other in terms of its quality of match to the underlying original signal. Can you think of situations in which a regular sampling method could be better than random scanning? See Sec. 9.1 in [Glassner95].

2-19. *Random scanning.* Random scanning would be the best way to avoid the appearance of moiré effects when scanning images. Several hardware devices and software algorithms have been devised for this end over the years (see, for example, [Hardy48], [Ahumada83]). However, because of various implementation difficulties this method is rarely used. Explain the practical problems in the implementation of this method; can you think of ways to overcome them?

2-20. *Avoiding superposition moirés in colour printing.* Most existing printing devices are bilevel, meaning that they are only capable of printing solid ink or leaving the paper unprinted, but they are unable to produce intermediate ink levels. This is also the case in most colour printing devices, where each of the primary printing colours (usually, cyan, magenta, yellow and black, or in short, CMYK) is only bilevel. This problem is solved in the printing world by using the *halftoning* technique (see Sec. 3.2 in *Vol. I*): The original continuous-tone image (or each of its CMYK colour planes, in case of a colour image) is broken into tiny dots whose size varies depending on the tone level. When printed, this gives to the eye (looking from a normal viewing distance) an illusion of a full range of intermediate tone levels, although in reality the printing device is only bilevel. Traditionally, high- and medium-end printing devices use a halftoning method that is based on periodic dot screens: Each of the colour planes of the original continuous-tone colour image is transformed into a regular screen of equidistant dots, where the size of the screen dots varies according to the image tone level, but the frequency and the angle of the screen remain fixed. The resulting dot screens of the different colour planes are then printed on top of each other at standard predefined orientations to produce the final colour image. However, this superposition of periodic dot screens may be the source of a serious problem: If particular care is not being taken to perfectly adjust the periods and the angles of the individual superposed screens, objectionable moiré patterns may be generated in the printed image (see Sec. 3.2 in *Vol. I*). One of the most effective ways to avoid the risk of such moiré effects in colour printing is by using random dot screens rather than periodic ones (obviously, this is only true if the random screens being used are mutually uncorrelated). What are the main advantages and the main drawbacks of using such random dot screens in colour printing? *Hint*: The main advantage of using random screens is that they completely solve the moiré problems in the superposition of the halftone screens of the different primary colours. Moreover, the superposition of random screens is no longer sensitive to small angle or scaling deviations like the traditional screen combination (which is a singular superposition, and hence an unstable moiré-free state). And most importantly, the use of random screens completely removes the limitation of using up to 4 primary colours, which was imposed in the traditional method by the fact that moiré-free combinations of more than 4 periodic screens are very difficult to find (see Chapter 3 in *Vol. I*). And indeed, when the number of screens to be superposed exceeds 4 (for example, when using non-standard or special-purpose inks), the use of random screens becomes the only practical way for avoiding superposition moirés. On the other hand, arguments against random screening include their graininess and their higher sensitivity to dot gain. A more detailed comparison between periodic and random screen halftoning can be found in [Widmer92], [Rodriguez94], [Schläpfer94] and [Sharma03 pp. 398–399]. Although random-screen halftoning is technically more complicated than periodic halftoning, several such techniques are already available in colour printing (see illustrations in [Kipphan01 Sec. 1.4.3]). It is not impossible that when such techniques become sufficiently mature, they will render the traditional screening methods obsolete, and the moiré problems between superposed screens will belong to the past. More information on random halftoning (often called *stochastic screening* or *frequency modulation screening*) can be found, for example, in [Kang99 Sec. 1.5], [Kipphan01 Sec. 1.4.3] and [Sharma03 Secs. 6.2, 6.3 and 6.8].

2-21. *Optimization of random halftone screens based on their spectral properties.* The spectral properties of random halftone screens have been thoroughly studied in the literature, with the aim of improving the visual quality of such halftone screens in the printing world (see, for example, [Allebach76], [Ulichney87], [Lau02a], [Lau03] and many of the references therein). Although a detailed discussion on the various types of halftone screens and their spectral analysis remains beyond the scope of the present book, can you see how spectral analysis can help in improving the visual image quality in black and white printing and in colour printing?

2-22. *Comparison of patterns.* Can you devise a method for the comparison of two or more 2D patterns to determine whether they are essentially the same? This subject is treated in [Prokoski99], where Glass patterns are called "flash correlation artifacts" (or in short, "FCAs").

2-23. *Identifying the presence of a given pattern within an image.* How can the technique of the previous problem be used for identifying the presence of a given pattern within an image? Can you think of useful applications? See [Prokoski99].

2-24. *Detection of residual periodicities.* One possible way to detect the existence of residual periodicities in a given image is by applying to it the Fourier transform (or FFT), and searching for peaks (blurred impulses) in the spectrum. However, this solution is not always practical. Propose a visual solution to this problem which is based on the superposition of a transparency on top of the given image. Supposing that the image contains some residual periodicities, what do you expect to see when you superpose a line grating of a given frequency on top of the image? See, for example, the figure in [Oster63 p. 62]. What happens when you rotate the grating? And what happens if you use instead a dot screen? Can you think of a more sophisticated structure for the transparency, that would allow the simultaneous detection of a wider range of hidden frequencies in the image?

2-25. *Moiré effects between periodic gratings and random screens.* The superposition of a periodic grating (or a periodic dot screen) and a random dot screen may give low-frequency nebulous, irregular structures that are sometimes considered as random moiré effects. Can you explain these effects using spectral considerations? A reasoning of this kind has been used in [Williams92] to explain sampling and aliasing phenomena that may sometimes be perceived due to the structure of the retina in the eye.

Chapter 3

Glass patterns and fixed loci

3.1 Introduction

The concept of a *fixed point* or a *fixed locus* plays a central role in the understanding of Glass patterns. A transformation is said to have a fixed point (or a fixed locus) if there exists a point (or locus) that is not affected by the transformation and remains in its original location after the transformation has been applied.

The fixed point theorem is a fundamental result in mathematical topology, that was first established in 1910 by the Dutch mathematician L. E. J. Brouwer [EncMath88 Vol. 1 p. 482]. It roughly says that any continuous transformtion from a non-empty, convex, compact subset of $\mathbb{R}^n$ into itself has at least one fixed point. This theorem is more popularly called the "crumpled paper theorem" [Dey91 p. 2], owing to one of its quite surprising consequences: Given two identical sheets of paper, if one of the sheets is crumpled and placed on top of the other, then there must be at least one point on the crumpled sheet that is directly above the corresponding point on the bottom sheet. (Another surprising consequence of this theorem is that when a map of the country is laid down on the floor, there exists at least one point of the map that is located exactly on top of the real point it represents.)

But although this mathematical result dates back to the beginning of the 20th century, it seems that the fundamental connection between Glass patterns and the fixed point theorem was only explicitly reported in the mid 1990s, in the context of an application to stereo matching [Pochec95].

In the present chapter we will see the two-fold relationship between Glass patterns and the fixed point theorem: On the one hand, this theorem provides the basic explanation to Glass patterns and their behaviour; but on the other hand, Glass patterns can also serve as an excellent graphical tool for visually illustrating the theorem.

We start our discussion in Sec. 3.2 with a short presentation of the fixed point theorem and its variants that are relevant to our needs. Based on this background we explain in Sec. 3.3 the behaviour of Glass patterns under affine layer transformations (for example, rotations, scalings, translations and their combinations). Then, in Sec. 3.4, we discuss the behaviour of Glass patterns under general, non-linear layer transformations. In Sec. 3.5 we extend the discussion to cases in which both of the superposed layers undergo transformations. And finally, in Sec. 3.6 we show that our theoretical approach can be used not only for the *analysis* of Glass patterns, but also for the *synthesis* of Glass patterns, namely, the design of layer transformations that generate in the superposition Glass patterns having any desired geometric locus in the plane.

We terminate this chapter with an interesting generalization of fixed points (or fixed loci): It turns out that macroscopic Glass patterns can be also generated in limit cases where, strictly speaking, there exist no fixed points, but only *almost* fixed points (i.e. points with only almost-perfect coincidence between the two layers). This generalization is explained and illustrated at the end of the chapter, in Sec. 3.7.

3.2 The fixed point theorem

There exist many different variants of the fixed point theorem, each being adapted to some particular needs or applications. The basic 1D version of the theorem (see, for example, [Weisstein99 p. 653]) says that any continuous function $g(x)$ that maps the domain $D = [a,b]$ onto itself, $g: [a,b] \rightarrow [a,b]$, has at least one fixed point in $[a,b]$ (namely, a point $x_F \in [a,b]$ that is mapped by $g(x)$ to itself: $g(x_F) = x_F$). This is clearly illustrated in Fig. 3.1(a).

This fundamental theorem can be easily generalized to higher dimensions, although in such cases it can no longer be graphically illustrated as in Fig. 3.1(a). For example, a 2D version of the fixed point theorem states that any continuous mapping $\mathbf{g}(x,y)$ that maps the disk $D = \{(x,y) \mid x^2 + y^2 \leq r\}$ into itself has at least one fixed point in D, namely, a point $(x_F,y_F) \in D$ that is mapped by $\mathbf{g}(x,y)$ to itself: $\mathbf{g}(x_F,y_F) = (x_F,y_F)$ [Weisstein99 p. 176].

It is interesting to note, however, that the fixed point theorem is not generally valid for infinite domains D such as $D = \mathbb{R}$, or, in the 2D case, $D = \mathbb{R}^2$ (the entire x,y plane). In such cases the theorem still holds for many functions, but there exist other functions for which the theorem fails. This is illustrated, for the 1D case, in Fig. 3.1(b): Although any function of the type $g(x) = x + c$ (with $c \neq 0$) is continuous and fully maps $\mathbb{R}$ onto itself, these functions do not have any fixed point $x_F \in \mathbb{R}$ such that $g(x_F) = x_F$ (unless we admit that parallel lines meet at infinity, in which case we may say that $x_F = \infty$ is a fixed point).[1] However, other continuous functions that map $\mathbb{R}$ onto itself, such as $g(x) = x^3$, do have fixed points, since they do cross the diagonal $y = x$ at least at one point x_F. A similar situation exists also in the 2D case: while many continuous mappings $\mathbf{g}(x,y)$ from $\mathbb{R}^2$ onto itself, such as scalings or rotations, have a fixed point, other mappings, such as translations: $\mathbf{g}(x,y) = (x-x_0,y-y_0)$, do not have fixed points (again, unless we consider infinity as a fixed point). However, a first important result for our needs may be formulated as follows:

The affine fixed point theorem: All affine mappings from $\mathbb{R}^2$ onto itself, $\mathbf{g}(x,y) = (a_1x + b_1y + x_0, a_2x + b_2y + y_0)$, that are non-degenerate, have a single fixed point.   ∎

This theorem asserts that mappings such as rotations, scalings, etc. as well as their combinations have, indeed, a fixed point; this also includes all of their combinations with

---

[1] Note that the function $g(x) = x + c$, $c \neq 0$, is not a valid counter-example for the fixed point theorem with $D = [a,b]$, simply because it does not map D onto itself.

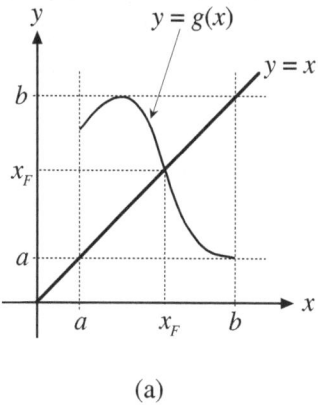
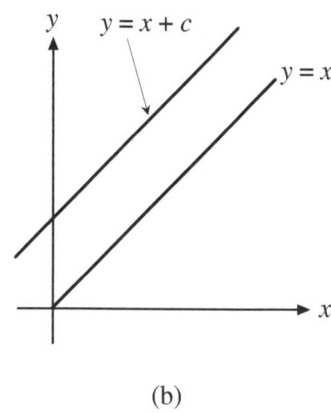

(a) (b)

Figure 3.1: (a) Illustration of the fixed point theorem in the 1D case. Any continuous function $y = g(x)$ that maps a domain $D = [a,b]$ onto itself crosses the diagonal $y = x$ within the domain $[a,b]$ at least once. At each such point x_F we have, therefore, $g(x_F) = x_F$. Moreover, due to the continuity of the function g, for any point x_G within a near neighborhood of x_F we have: $g(x_G) \approx x_G$. (b) The fixed point theorem is not generally valid when D is the full range of $\mathbb{R}$. This can be illustrated by any continuous function of the type $g(x) = x + c$, $c \neq 0$. Although these functions map $\mathbb{R}$ onto itself, they are parallel to the diagonal $y = x$, and hence they never cross it at any finite value $x_F \in \mathbb{R}$, meaning that for no point $x_F \in \mathbb{R}$ we have $g(x_F) = x_F$.

translations, but pure translations are excluded. This theorem is explained and illustrated in Sec. A.2 of Appendix A, where the precise meaning of the term "non-degenerate" is also given. A first generalization of this theorem that covers all affine mappings over $\mathbb{R}^2$, degenerate or not, is given at the end of Sec. A.2, and a further generalization to polynomial mappings of order 2 is provided in Sec. A.3.

Let us see now how the fixed point theorem can help us formalize our subject of interest, the superposition of similar structures, periodic or not. This mathematical formalization will allow us to deduce important facts on Glass patterns and their behaviour in the superposition. Suppose we are given a layer $r_1(x,y)$ consisting of an arbitrary structure as explained in Sec. 2.2. We generate a second, slightly modified layer $r_2(x,y)$ by applying to $r_1(x,y)$ a continuous mapping (coordinate transformation) $\mathbf{g}(x,y)$ that maps the x,y plane $\mathbb{R}^2$ onto itself. For example, $r_2(x,y)$ could be a slightly rotated version of $r_1(x,y)$. We now superpose the two layers $r_1(x,y)$ and $r_2(x,y)$, for example by overprinting, or by laying their transparencies on top of each other. As explained in Sec. 2.2, the superposition thus obtained is represented mathematically by the product:

$$r(x,y) = r_1(x,y)\, r_2(x,y) \tag{3.1}$$

Suppose that the continuous mapping $\mathbf{g}(x,y)$ has a fixed point (x_F,y_F). This means that at the point (x_F,y_F) we have $r_2(x_F,y_F) = r_1(\mathbf{g}(x_F,y_F)) = r_1(x_F,y_F)$, so that the point (x_F,y_F,z_F) belonging to the surface $z = r_1(x,y)$ remains unchanged after applying the mapping $\mathbf{g}(x,y)$. For example, if it was a black point, it remains a black point in $r_2(x,y)$, and if it was a white point, it remains a white point in $r_2(x,y)$. Furthermore, in the neighbourhood of this fixed point, any point (x_G,y_G,z_G) of $r_1(x,y)$ has only been slightly displaced in $r_2(x,y)$. How does this affect the superposition of Eq. (3.1)?

Clearly, the superposition $r(x,y)$ is darker than each of the individual layers, since it becomes black wherever any of the superposed layers is black. However, the mean gray level of the superposition remains brighter in a close neighbourhood about the fixed point, since in this area the black dots of $r_2(x,y)$ fall almost exactly on top of their original counterparts in $r_1(x,y)$, so that the superposition is only slightly darker than $r_1(x,y)$. But as we go farther from the fixed point (x_F,y_F), the mean gray level of the superposition stabilizes at a darker level, since the correlation between the dots of $r_2(x,y)$ and the dots of $r_1(x,y)$ gradually decreases, and the black points of $r_2(x,y)$ fall more often between black points of $r_1(x,y)$, leaving less white area in the superposition. This explains, indeed, why a Glass pattern appears in the superposition about the fixed point, as shown in Figs. 2.1 and 2.2. If the transformation $\mathbf{g}(x,y)$ has a full locus of fixed points (i.e. a *fixed locus*), a Glass pattern is generated along this entire locus. For example, Fig. 2.3(a) shows a linear Glass pattern that is generated about the *fixed line* of the transformation $\mathbf{g}(x,y) = (x, ay)$.

In fact, we will see later that similar Glass patterns can be generated even in cases where the transformation does not have a real fixed point (or points) but only an *almost* fixed point (or points), where $\mathbf{g}(x_F,y_F)$ is not fully identical to (x_F,y_F) but only very close to it. Although in such cases there is no perfect coincidence between the two superposed layers, the elements of both layers around such points still fall very close to each other, while farther away the correlation gradually decreases. This generates a visible Glass pattern whose center is just slightly darker than in the case of perfect coincidence. We will return to almost fixed points later, in Sec. 3.7, but for the time being we will continue our discussion using cases having real fixed points, because these are the simplest study cases for the understanding of Glass patterns and their behaviour in the layer superposition.

Now, how do we explain the difference between the phenomena which occur in the superposition of periodic or aperiodic layers? If the dots of $r_1(x,y)$ (and hence the dots of $r_2(x,y)$) are randomly distributed, then far away from the fixed point (x_F,y_F) there will no longer be any correlation between the points of the two layers, and the resulting gray level in the superposition will remain constant as we go farther from (x_F,y_F). However, if $r_1(x,y)$ is a periodic structure, such as a periodic dot screen, then as we go farther from the fixed point (x_F,y_F) the mean gray level will periodically become darker and brighter, because zones of in-phase superposition, where elements of the two layers fall on top of one another, repeatedly alternate with zones of counter-phase superposition, where elements of the two layers fall between each other (compare Figs. 2.1(c) and (d)). It is interesting to note that in the superposition of partly random layers, such as periodic dot screens with a certain degree of randomness being added, the resulting Glass patterns have, indeed, an intermediate look: Depending on the case, they still may have about the center oscillations between darker and brighter areas, but since the correlation between the layers decreases with the distance, these oscillations gradually fade away and disappear as we go farther from the center of the Glass pattern. We will return to this point in more detail in Sec. 7.6.

The correspondence between Glass patterns and their periodic moiré counterparts will be further illustrated in the next section.

3.3 Behaviour of Glass patterns and periodic moirés under affine mappings

Having understood the mathematical meaning of Glass patterns,[2] let us try to see their behaviour when any of the superposed layers undergoes a transformation such as rotation, scaling, translation, etc. Moreover, since the behaviour of periodic moirés under such transformations is already fully known from the classical moiré theory, it would be interesting to compare the behaviour of both cases, periodic and aperiodic, and to see if they follow the same mathematical rules.

Remark 3.1: In order to study the behaviour of a Glass pattern, we must, of course, make sure that a Glass pattern is indeed generated in our layer superposition. Therefore, we have to superpose layers that are sufficiently correlated. The easiest way of doing so is to assume full correlation, i.e. that the superposed layers, periodic or not, be fully identical before the application of the layer mappings in question.[3] And indeed, in the following discussions, until Chapter 7, we will implicitly make this assumption. Note that this assumption does not cause a loss of generality, since in cases where the original layers are only partly correlated (for example, due to the presence of some random noise), the Glass patterns may look somewhat weaker or different, but their behaviour under layer mappings remains the same (see Sec. 7.7). ■

3.3.1 Behaviour under layer rotations

The simplest nontrivial layer transformation consists of a rotation of one of the superposed layers; this case has the practical advantage of being very easy to experiment by manipulating superposed transparencies. Suppose we have two identical transparencies consisting of the same arbitrary dot pattern, periodic or not. We superpose the two transparencies precisely on top of each other, and while keeping the first transparency fixed, we slightly rotate the other one by a small angle α about the origin, so that a Glass pattern becomes visible about the fixed point at the center of rotation (the origin). As we

---

[2] In fact, we have so far concentrated on the topological point of view; other mathematical aspects of Glass patterns, such as their morphological properties and their intensity profiles, will be discussed in later chapters.

[3] In the case of random layers this implies, of course, the use of the same random number generator with the same seed for the generation of both layers.

have already seen, the center of the Glass pattern is brighter than areas further away, due to the partial overlapping of the black elements of both layers about the fixed point. This behaviour at the center is common to both periodic and aperiodic cases, and indeed, the macroscopic difference between them becomes apparent only farther away from the fixed point: In an *aperiodic* case, the mean gray level of the superposition gradually stabilizes at a certain darker level (see Fig. 2.1(c)), because farther from the fixed point the correlation between the two layers becomes negligible. But in a *periodic* case (see Fig. 2.1(d)), as we go away from the fixed point the gray level becomes alternately darker and brighter. This periodic oscillation in the gray level occurs because zones of in-phase superposition, where elements of the two periodic layers fall on top of one another, repeatedly alternate with zones of counter-phase superposition, where elements of the same two layers fall between each other.

As we can see, the Glass pattern which is generated about the fixed point in a periodic case is periodically repeated throughout the superposition, forming the bright areas of the periodic moiré pattern. From this point of view, the period tiles of a periodic moiré pattern are simply duplicates of the main Glass pattern which is generated about the fixed point, and the period length of the moiré corresponds to the distance between these duplicates.[4] This does not mean, of course, that our rotation transformation $\mathbf{g}(x,y)$ has more fixed points when the two superposed layers are periodic than when the layers are aperiodic: obviously, in both cases $\mathbf{g}(x,y)$ has exactly one fixed point. But when the two superposed layers are periodic, we also have infinitely many points of coincidence between the two superposed layers, where the two layers happen to coincide because of the periodicity in their internal structure. But these points of coincidence are not fixed points of the underlying mapping $\mathbf{g}(x,y)$.

It can be said, therefore, that in a periodic case the glass pattern that is generated in the superposition is periodic and extends throughout the entire plane. Note, however, that we will usually prefer to use a different convention, according to which in the case of a periodic moiré the term Glass pattern does not refer to the entire periodic pattern, but only to the central moiré period which is located about the fixed point. All the other periods will be considered as duplicates of this Glass pattern, and the entire periodic pattern will be called, as usual, a moiré pattern.

This leads us to the following general remark:

Remark 3.2: In a superposition of periodic layers one of which undergoes a non-degenerate affine mapping $\mathbf{g}(x,y)$, the fixed point of $\mathbf{g}(x,y)$ determines the *main* periodic tile of the moiré (that we call here the Glass pattern), while all the other periodic tiles are only duplicates which exist due to the periodicity of the superposed layers. This observation is,

---

[4] It is important to note, however, that these duplicates are *not* necessarily identical in their microstructure. The periodicity of the moiré concerns only its macrostructure, namely, the moiré intensity profile (the variation in the mean gray level that is observed from such a distance that the microstructure detail of the original layers is no longer discerned by the eye). In other words, although the microstructure in the superposition of two periodic layers is not always periodic, the intensity profile of the isolated moiré is, indeed, periodic (see Sec. 6.3 in *Vol. I*).

indeed, quite surprising, since so far, during the study of periodic moirés (for example, in *Vol. I*), there was no reason to suspect that one of the infinitely many periods of the periodic moiré was more fundamental or "authentic" than the others, and all of them were considered as being fully equivalent. ■

Returning to the case of layer rotations, a short experimentation with the superposed layers shows that in spite of all the differences between the moiré patterns in periodic and aperiodic superpositions, their fundamental behaviour under layer rotations remains basically the same: In both cases, when the angle α departs from 0, the Glass pattern (respectively, the periodic tile of the moiré) becomes smaller and smaller until it completely disappears; and conversely, as the angle α tends to 0, the Glass pattern (respectively, the periodic tile of the moiré) becomes bigger and bigger, until when α reaches 0 we obtain a singular superposition with an infinitely large moiré, which is no longer visible.

This similarity in the behaviour of moiré effects between periodic or aperiodic layers is not merely a particularity of the case involving layer rotations. As we will see below, it turns out that this behaviour is, in fact, much more general.

3.3.2 Behaviour under layer scalings

A similar effect occurs also in the case of a scaling transformation. Note, however, that in this case the visual study of the effect by using superposed transparencies is not as easy as in the case of rotation, because it is not possible to manually stretch or shrink transparencies. For experimenting with this case one needs, therefore, to prepare in advance a set of reduced or enlarged copies of the original layer (for example, zoomed photocopies). A better solution would be to make simulations on a computer screen, since this would allow a continuous observation of the superposition as the scaling rate of one of the layers is gradually being varied.

Suppose we have two identical layers consisting of the same arbitrary dot pattern, periodic or not. We superpose the two layers precisely on top of each other, and while keeping the first layer fixed, we slightly scale the other one (see Fig. 2.1(e)). Like in the case of rotations, this gives in the superposition a fixed point at the origin. And indeed, once again, a Glass pattern becomes visible about the fixed point, whose center is brighter than areas farther away. This happens, just as in the case of rotation, due to the partial overlapping of the black elements of both layers about the fixed point. But although the microstructure obtained in this case is different than in the case of layer rotations (it consists of *radial* rather than *circular* dot trajectories; compare Figs. 2.1(e) and (c)), the macroscopic properties of the Glass pattern remain the same.

Let us now compare the behaviour of the superposition around the fixed point in periodic and in aperiodic cases. Once again, it turns out that the macroscopic difference between periodic and aperiodic cases becomes apparent only farther away from the fixed point: As we can see in Fig. 2.1(e) that illustrates the aperiodic case, when we go farther away from the fixed point, the mean gray level of the superposition stabilizes at a certain darker level. But in the corresponding periodic case, shown in Fig. 2.1(f), when we go farther away from the fixed point, the gray level becomes alternately darker and brighter, and it continues oscillating repeatedly as the elements of the two layers periodically fall on top of each other (in phase) or between each other (in counter phase). Thus, we may say, just as in the case of rotation, that while in the aperiodic case there exists only one Glass pattern, which is located about the fixed point, in the periodic case, the Glass pattern which is generated about the fixed point is periodically repeated throughout the superposition, forming the bright areas of the periodic moiré pattern.

But once again, in spite of the significant difference between the Glass patterns in periodic and aperiodic superpositions, their fundamental behaviour under layer scalings remains basically the same: In both cases, when the scaling factor s gradually departs from 1, the Glass pattern (respectively, the periodic tile of the moiré) becomes smaller and smaller; and conversely, as the scaling factor s tends to 1, the Glass pattern (respectively, the periodic tile of the moiré) becomes bigger and bigger, until when s reaches 1 we obtain a singular superposition with an infinitely big moiré, which is no longer visible. It should be mentioned, however, that while in the periodic case new higher-order moirés may occur around $s = 2$, 3, or $s = \frac{1}{2}, \frac{1}{3}$, etc., in the purely aperiodic case no higher order moirés exist, since at such scaling rates there is no correlation between the superposed layers (for instance, a random screen $r(x,y)$ is not correlated with $r(2x,2y)$). We will return to this point in more detail in Sec. 7.5.

It would be tempting to ask at this point why the dot trajectories, which are so conspicuous in the aperiodic cases, are not visible in the corresponding periodic cases (compare, for example, Figs. 2.1(c) and (d), or Figs. 2.1(e) and (f)). We will defer this interesting question until Sec. 4.7, where we will be better placed to see the answer.

3.3.3 Behaviour under layer shifts

So far we have studied the effects of rotations and scalings, both of which have a fixed point at the origin. It would be interesting to see now the effects of layer shifts on this fixed point and on the Glass pattern which surrounds it in the superposition.

As we have seen in Sec. 2.3.2, pure translations (layer shifts) do not generate Glass patterns (see, for example, Fig. 2.5(a)); rather, they give a singular moiré-free superposition. And indeed, transformations consisting of pure translations, $g(x,y) = (x–a, y–b)$, have no fixed points at all. Therefore, in order to study the effect of layer shifts on Glass patterns, we start with a layer superposition that generates a Glass pattern, like in the cases of rotation or scaling, and we study the effect of layer shifts on the already existing Glass pattern.

Suppose we have two identical transparencies that are superposed with a small angle difference or a small scaling difference, so that a visible Glass pattern is generated about the fixed point at the origin. What happens to this Glass pattern when we laterally translate one of the transparencies with respect to the other? In the case of periodic moirés the answer is already well known: the moiré pattern is simply translated (by a much larger distance than the original layer shift), without undergoing any other modifications. As explained in Sec. 7.6 of *Vol. I*, the extent and the direction of this translation are determined by the extent and the direction of the shifts in the original layers. But since the superposition of periodic layers is a particular case of the superposition of any general layers, it would be reasonable to expect that the behaviour of a periodic moiré under layer shifts should be a particular case of the behaviour of a Glass pattern under the same layer shifts. And indeed, as described in detail below, a simple experimentation with two superposed transparencies shows that exactly the same results are obtained in both periodic and aperiodic cases.

Suppose that we generate a Glass pattern (respectively, a periodic moiré pattern) about the center by rotating the second transparency by a small angle α counterclockwise, and that we slightly shift the first transparency (the unrotated layer) in a given direction.[5] As shown in Fig. 3.2 for aperiodic layers and in Fig. 3.3 (or Fig. 7.6 of *Vol. I*) for periodic layers, the resulting effect in both cases is a much larger shift of the Glass pattern (respectively, the periodic moiré), in a direction which is basically perpendicular to the shift of the first transparency:

- When the first transparency is slightly shifted to the right, the Glass (or moiré) pattern largely moves downward;

- When the first transparency is slightly shifted to the left, the Glass (or moiré) pattern largely moves upward;

- When the first transparency is slightly shifted upward, the Glass (or moiré) pattern largely moves to the right;

- And when the first transparency is slightly shifted downward, the Glass (or moiré) pattern largely moves to the left.

The identical qualitative behaviour of periodic and aperiodic cases further confirms our assumption that both cases are, indeed, two different facets of the same phenomenon. But if our assumption is correct, the behaviour of both cases must be identical quantitatively, too. Since the quantitative behaviour of the periodic moiré under layer shifts is already well known (see Sec. 7.6 in *Vol. I*), we will try now to determine quantitatively the behaviour of the aperiodic case (i.e. the shift of the Glass pattern), in order to see if we obtain the same results.

In order to do so, let us try to locate the fixed point (i.e. the center of the Glass pattern) when the second layer is rotated by angle α, and the first, unrotated layer is shifted laterally in the original x and y directions by (x_0, y_0). This is also equivalent to rotating the second layer by angle α, and then shifting it along the original x and y directions by

---

[5] We choose this layer configuration in order to remain compatible with the figures and the examples given for the periodic case in Chapter 7 of *Vol. I*.

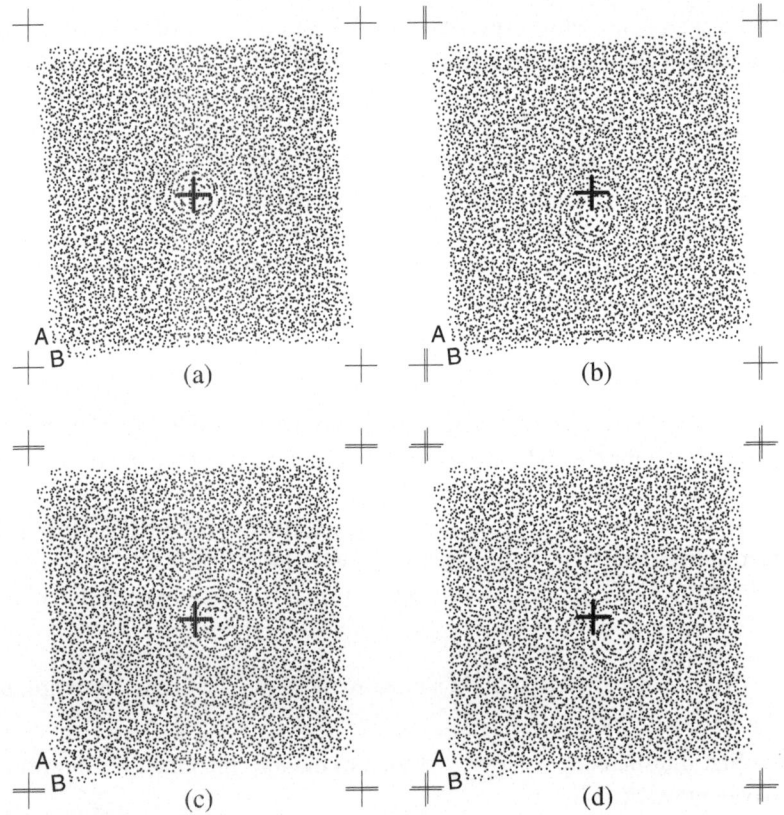

Figure 3.2: A Glass pattern between two identical aperiodic dot screens that are superposed
with a small angle difference α, and its behaviour under layer shifts. The origin of
each image is indicated by a cross. (a) Both layers and the resulting Glass pattern
are centered on the origin, in their initial position. (b) Layer A is shifted by x_0 to
the right; consequently, the Glass pattern is shifted downward. (c) Layer A is
shifted by y_0 upward; consequently, the Glass pattern is shifted to the right.
(d) Layer A is shifted by x_0 to the right and by y_0 upward; consequently, the Glass
pattern is shifted downward and to the right. The small shifts of x_0 and y_0 in the
layer A can be best perceived in the hairline crosses that surround each case.

$-(x_0,y_0)$, while the first layer remains fixed. The mapping $\mathbf{g}(x,y)$ is given, therefore, by:[6]

$$x' = x\cos\alpha + y\sin\alpha + x_0$$
$$y' = -x\sin\alpha + y\cos\alpha + y_0 \tag{3.2}$$

---

[6] Note that unless otherwise mentioned we always assume here, just as in Chapters 10 and 11 of *Vol. I*,
that $\mathbf{g}(x,y)$ is applied to the original dot screen as a *domain* transformation, so that it is considered in
fact as an *inverse* transformation (see Footnote 1 in Sec. 10.2 of *Vol. I*). A more detailed explanation is
provided in Appendix D; see also Problems 5-7, 5-8 and 7-16.

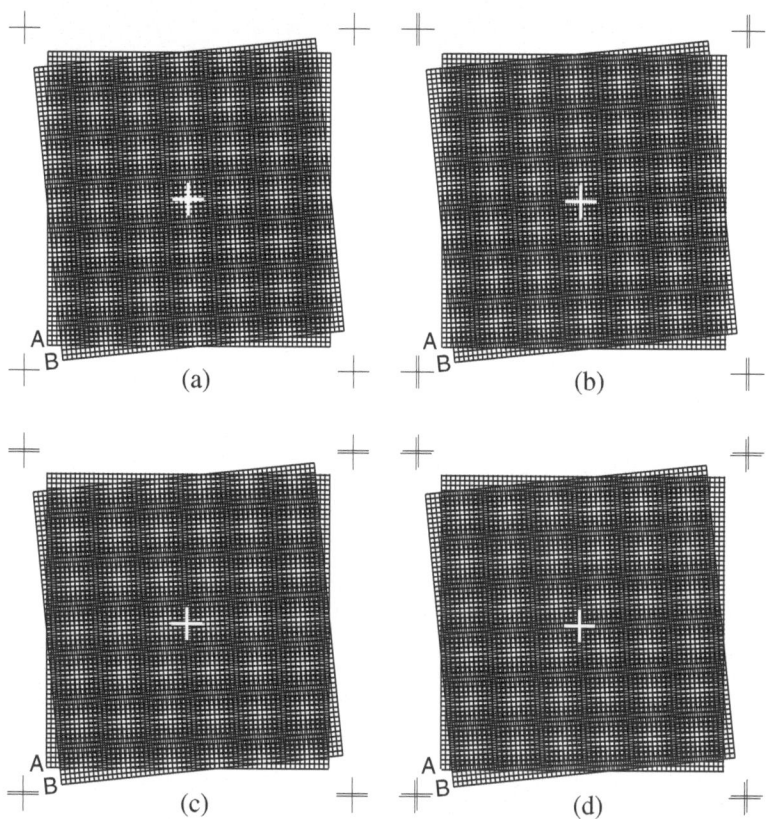

Figure 3.3: A periodic moiré between two identical periodic grids of period T that are superposed with a small angle difference α, and its behaviour under layer shifts. The origin of each image is indicated by a cross. (a) Both grids and the resulting moiré are centered on the origin, in their initial position. (b) Grid A is shifted by 1/2 period (i.e. by $T/2$) to the right; consequently, the moiré is shifted 1/2 moiré period downward. (c) Grid A is shifted by 1/2 period (i.e. by $T/2$) upward; consequently, the moiré is shifted 1/2 moiré period to the right. (d) Grid A is shifted by 1/2 period to the right and 1/2 period upward; consequently, the moiré is shifted 1/2 moiré period downward and 1/2 moiré period to the right. Note that the angle difference α between the superposed layers as well as the layer shifts are the same as in Fig. 3.2: $\alpha = 6°$, $x_0 = T/2$, $y_0 = T/2$. And indeed, the resulting shift of the moiré pattern is identical to the resulting shift of the Glass pattern in Fig. 3.2 (although in the aperiodic case it cannot be expressed in terms of moiré periods).

where x,y are the coordinates *before* applying the rotation and the shift, and x',y' are the coordinates *after* the mapping. Our problem of finding the fixed point of $\mathbf{g}(x,y)$ consists of finding when (x',y') equals (x,y). This happens, of course, at the points (x,y) where:

$$x = x\cos\alpha + y\sin\alpha + x_0$$
$$y = -x\sin\alpha + y\cos\alpha + y_0$$

(3.3)

which gives us the following linear set of equations for x and y:

$$x(1 - \cos\alpha) - y\sin\alpha = x_0$$
$$x\sin\alpha + y(1 - \cos\alpha) = y_0$$

(3.4)

It can be easily shown (for example, using Cramer's rule) that the solution of this set of equations is given by:

$$x = \frac{x_0(1 - \cos\alpha) + y_0\sin\alpha}{2(1 - \cos\alpha)}$$
$$y = \frac{-x_0\sin\alpha + y_0(1 - \cos\alpha)}{2(1 - \cos\alpha)}$$

(3.5)

or in matrix form:

$$\begin{pmatrix} x \\ y \end{pmatrix} = \frac{1}{2} \begin{pmatrix} 1 & \frac{\sin\alpha}{1 - \cos\alpha} \\ -\frac{\sin\alpha}{1 - \cos\alpha} & 1 \end{pmatrix} \begin{pmatrix} x_0 \\ y_0 \end{pmatrix}$$
$$= \frac{1}{2} \begin{pmatrix} 1 & \cot(\alpha/2) \\ -\cot(\alpha/2) & 1 \end{pmatrix} \begin{pmatrix} x_0 \\ y_0 \end{pmatrix}$$

(3.6)

Note that this solution (fixed point) exists and is unique whenever $1 - \cos\alpha \neq 0$, i.e. whenever $\alpha \neq 0$.

Eq. (3.6) means that the fixed point (x,y) is simply a linear transformation of the lateral shift (x_0,y_0) undergone by the original layer. The matrix of this transformation has the form $\frac{1}{2}\begin{pmatrix} 1 & -b \\ b & 1 \end{pmatrix}$ with $b = -\cot(\alpha/2)$. This matrix is a particular case of the similarity matrix $\begin{pmatrix} a & -b \\ b & a \end{pmatrix}$. As we know from linear algebra (see, for example, [Lay03 p. 339]), every similarity matrix corresponds to a linear transformation that is composed of:

- a rotation by angle θ, where: $\cos\theta = \dfrac{a}{\sqrt{a^2 + b^2}}$
- and a scaling by: $s = \sqrt{a^2 + b^2}$

By inserting here $a = \frac{1}{2}$ and $b = -\frac{1}{2}\cot(\alpha/2)$ (taking into account the factor $\frac{1}{2}$ before the matrix) and using the identities $1/\sqrt{1 + \cot^2 x} = \sin x$ [Spiegel68 p. 15] and $\sin x = \cos(x - \frac{\pi}{2})$ we see that the location (x,y) of the fixed point is obtained from the layer shift (x_0,y_0) by:[7]

- a rotation by angle θ, where: $\cos\theta = \dfrac{1}{\sqrt{1 + \cot^2(\alpha/2)}} = \sin\dfrac{\alpha}{2} = \cos(\dfrac{\alpha}{2} - \dfrac{\pi}{2})$
- namely: $\theta = \dfrac{\alpha}{2} - \dfrac{\pi}{2}$

(3.7)

---

[7] Note that the same result can be also obtained without relying on the properties of a similarity matrix, as shown in [Amidror03a].

- and a scaling by:
$$s = \tfrac{1}{2}\sqrt{1 + \cot^2(\alpha/2)} = \frac{1}{2\sin(\alpha/2)} \tag{3.8}$$

And indeed, as we have expected, this quantitative result fully corresponds to the moiré shift obtained in the case of a periodic moiré:

(a) As predicted by proposition 7.2 in *Vol. I*, the moiré in the periodic case is shifted in its own direction, which is, as shown in Fig. 7.6 (or Fig. 4.8 in *Vol. I*), exactly $\frac{\alpha}{2} - \frac{\pi}{2}$.

(b) Furthermore, the extent of the shift of the periodic moiré is given, in terms of periods, by Eqs. (*I*.7.26) and (*I*.2.10), namely:

$$b_M = T_M(\phi_1 - \phi_2) = \frac{T}{2\sin(\alpha/2)}(\phi_1 - \phi_2) \tag{3.9}$$

where b_M is the resulting shift of the moiré, T_M is the period of the moiré, T is the period of the original layers, and ϕ_1 and ϕ_2 are the shifts of the original layers in terms of periods T. Noting that in our case $\phi_2 = 0$ (the second layer is not shifted), we see that for a shift of d in the first layer (i.e., $\phi_1 = d/T$ periods), the extent of the resulting shift of the moiré is:

$$b_M = \frac{1}{2\sin(\alpha/2)} d \tag{3.10}$$

Hence, the shift of the periodic moiré is obtained by scaling up the shift d of the first layer by the factor $s = \frac{1}{2\sin(\alpha/2)}$, exactly as predicted by Eq. (3.8) above according to fixed point considerations.

Note that this result fully explains our observations at the beginning of the present section (see Fig. 3.2). For example, consider the Glass pattern that is obtained at the origin by rotating one of the two identical aperiodic dot screens by angle α. When the other layer is shifted to the right or to the left, the Glass pattern moves forward or backward along the line emanating from the origin at the angle $\frac{\alpha}{2} - \frac{\pi}{2}$, which is precisely the bisector of angle α measured from the negative side of the y axis (see Fig. 7.6, or Fig. 4.8 in *Vol. I*).

We see, therefore, that the resulting moiré shifts in the periodic case and in the aperiodic case are, indeed, identical, and both are explained by a shift of the fixed point. Note, however, that in the periodic case, the same result can also be obtained in terms of periods, frequencies, Fourier series developments, etc., as it was done, indeed, in Chapter 7 of *Vol. I*. But in the general, aperiodic case such period-based considerations are no longer applicable, and we therefore must revert to our more general analysis in terms of the fixed point theorem, which gives us the same results in the more general terms of *distances* rather than in terms of *periods*.

3.3.4 Behaviour under a general affine transformation

In the most general affine case, when the transformation $\mathbf{g}(x,y)$ is given by:

$$\begin{aligned} x' &= a_1 x + b_1 y + x_0 \\ y' &= a_2 x + b_2 y + y_0 \end{aligned} \tag{3.11}$$

the fixed point is given by the set of equations:

$$(1 - a_1)x \quad - \quad b_1y = x_0$$
$$-a_2x + (1 - b_2)y = y_0$$

(3.12)

whose solution is:

$$x = \frac{(1 - b_2)x_0 + b_1y_0}{1 - a_1 - b_2 + a_1b_2 - a_2b_1}$$
$$y = \frac{a_2x_0 + (1 - a_1)y_0}{1 - a_1 - b_2 + a_1b_2 - a_2b_1}$$

(3.13)

or in matrix form:

$$\begin{pmatrix} x \\ y \end{pmatrix} = \frac{1}{1 - a_1 - b_2 + a_1b_2 - a_2b_1} \begin{pmatrix} 1 - b_2 & b_1 \\ a_2 & 1 - a_1 \end{pmatrix} \begin{pmatrix} x_0 \\ y_0 \end{pmatrix}$$

(3.14)

This means that even in the general case where the layer transformation is given by the affine mapping $g(x,y)$ of Eq. (3.11), the location (x,y) of the fixed point is still a linear transformation of the shift (x_0,y_0) undergone by the original layer.

It should be noted, however, that this solution (i.e., the fixed point) exists and is unique *iff* the determinant of the homogenous equations in Eqs. (3.12), i.e. the denominator of Eq. (3.14), is nonzero:

$$1 - a_1 - b_2 + a_1b_2 - a_2b_1 \neq 0$$

(3.15)

If this condition is not satisfied, then either there exist no solutions (fixed points) at all (this happens, for example, when $g(x,y)$ is a pure translation), or there exists an infinity of solutions. The two examples which follow illustrate such situations.

Example 3.1: A fixed line (i.e. a full line of fixed points) along the entire x axis:

Consider a superposition of two identical aperiodic layers one of which undergoes a vertical scaling, as shown in Fig. 2.3(a). In this case the transformation $g(x,y)$ undergone by the distorted layer is given by:

$$x' = x$$
$$y' = sy$$

so that we have in Eq. (3.11) $a_1 = 1$, $b_1 = 0$, $a_2 = 0$ and $b_2 = s$; it follows, therefore, that in this case Eq. (3.15) equals zero. What does this mean in terms of the fixed points? As usual, the fixed locus of this mapping consists of all the points for which (x',y') equals (x,y), namely:

$$x = x$$
$$y = sy$$

The solution of this set of equations is $y = 0$ for all x; this means that the fixed locus of our mapping does not consist of a single point, but rather of the entire x axis. And indeed, as we can clearly see in Fig. 2.3(a), a linear Glass pattern is generated all along the x axis.

Now, what happens if we shift the transformed layer horizontally by x_0? In this case the layer transformation $\mathbf{g}(x,y)$ becomes:

$$x' = x - x_0$$

$$y' = sy$$

As we can see, the coefficients a_1, b_1, a_2 and b_2 remain here unchanged, meaning that Eq. (3.15) still equals zero. However, the fixed locus does not remain here unchanged. The fixed points of this mapping consist of all the points (x,y) that satisfy:

$$x = x - x_0$$

$$y = sy$$

but here, because of the shift x_0, the first equation is inconsistent, and hence no points (x,y) can satisfy the system. This means that in this case there are no fixed points at all. And indeed, as shown in Fig. 2.3(e), after the application of the shift the macroscopic Glass pattern no longer exists in the superposition, and what we see in this case is just the dot trajectories of the microstructure. (Recall from Sec. 2.3 that a Glass pattern between two transformed copies of the same original layer consists of macroscopic gray level variations — usually a bright center where the elements from both layers fall almost exactly on top of each other, that is surrounded by a darker area where the layer elements are less correlated and fall arbitrarily between each other.) The absence of a Glass pattern in this case can be easily verified by comparing Figs. 2.3(a) and 2.3(e) while observing them from a distance of 3–4 meters, where the microstructure elements of the layers are no longer discerned by the eye, and only the macrostructures remain visible. ∎

Example 3.2: Another case with a fixed line along the entire x axis:

Consider a superposition of two identical aperiodic layers one of which undergoes a horizontal shear transformation, as shown in Fig. 2.3(c). In this case the transformation $\mathbf{g}(x,y)$ undergone by the distorted layer is given by:

$$x' = x - ay$$

$$y' = y$$

so that we have in Eq. (3.11) $a_1 = 1$, $b_1 = -a$, $a_2 = 0$ and $b_2 = 1$; it follows, therefore, that Eq. (3.15) equals zero. What does this mean in terms of the fixed points? As usual, the fixed locus of this mapping consists of all the points for which (x',y') equals (x,y), namely:

$$x = x - ay$$

$$y = y$$

Once again, the solution of this set of equations is $y = 0$ for all x, meaning that the fixed locus of our mapping consists of infinitely many points, the entire x axis. And indeed, as we can clearly see in Fig. 2.3(c), a linear Glass pattern is generated all along the x axis.

Now, what happens if we shift the transformed layer vertically by y_0? In this case the layer transformation $\mathbf{g}(x,y)$ becomes:

$$x' = x - ay$$

$$y' = y - y_0$$

As we can see, the coefficients a_1, b_1, a_2 and b_2 remain here unchanged, so that Eq. (3.15) still equals zero. However, the fixed locus does not remain here unchanged. The fixed points of this mapping consist of all the points (x,y) that satisfy:

$$x = x - ay$$

$$y = y - y_0$$

but here, because of the shift y_0, the second equation is inconsistent, and hence no points (x,y) can satisfy the system. This means that in this case there are no fixed points at all. And indeed, as shown in Fig. 2.3(g), no Glass pattern (i.e. macroscopic gray level variation) exists in this superposition after having applied the shift. This can be seen by comparing Figs. 2.3(c) and 2.3(g) while observing them from a distance of 3–4 meters, where the microstructure elements of the layers are no longer discerned by the eye, and only the macrostructures remain visible. ■

These two examples illustrate, indeed, the fact that when Eq. (3.15) equals zero the affine transformation (3.11) does not have a single fixed point, but either infinitely many fixed points, or no fixed points at all. This subject is further explained in Appendix A, Sec. A.2.

Finally, it is interesting to note that expressions (3.11)–(3.14) above can be written in a more compact way using matrix notation. If we write Eqs. (3.11) using u_1, v_1, u_2, v_2 instead of a_1, b_1, a_2, b_2 we can rewrite the general affine transformation $\mathbf{g}(\mathbf{x})$ as follows:

$$\mathbf{x}' = \mathbf{g}(\mathbf{x}) = \begin{pmatrix} u_1 x + v_1 y + x_0 \\ u_2 x + v_2 y + y_0 \end{pmatrix} = F\mathbf{x} + \mathbf{x}_0$$

where $F = \begin{pmatrix} u_1 & v_1 \\ u_2 & v_2 \end{pmatrix} = \begin{pmatrix} \mathbf{f}_1 \\ \mathbf{f}_2 \end{pmatrix}$, $\mathbf{x} = \begin{pmatrix} x \\ y \end{pmatrix}$ and $\mathbf{x}_0 = \begin{pmatrix} x_0 \\ y_0 \end{pmatrix}$. This notation clarifies, indeed, the connection with the case of periodic screens, where $\mathbf{f}_1$ and $\mathbf{f}_2$ can be interpreted as the two vector frequencies of the periodic screen in question (see also the remark following Proposition 10.4 in *Vol. I*). Using this matrix notation the fixed point of the affine transformation $\mathbf{g}(\mathbf{x})$ is given by the equation:

$$\mathbf{x} = F\mathbf{x} + \mathbf{x}_0$$

namely: $(I - F)\mathbf{x} = \mathbf{x}_0$

(where I is the identity matrix, $\left(\begin{smallmatrix} 1 & 0 \\ 0 & 1 \end{smallmatrix}\right)$). The solution of this equation is given by:

$$\mathbf{x} = (\mathbf{I} - \mathbf{F})^{-1}\mathbf{x}_0$$

which is the implicit matrix form of Eq. (3.14). The advantage of this notation is not only in its compact form, but also in clearly showing the interpretation of this result in the periodic case, as mentioned above.

What does all this mean in terms of Glass patterns in the layer superposition? If we superpose two aperiodic layers one of which is a slightly transformed copy of the other under an affine transformation $\mathbf{g}(x,y)$, a Glass pattern is generated about the fixed locus (fixed point or fixed line) of the transformation, as shown, indeed, in Figs. 2.1–2.2. But if the transformation has no fixed points, as in the case of pure translation, no Glass pattern is generated in the superposition.

Having understood the behaviour of Glass and moiré patterns under linear and affine layer transformations, we proceed in the next section to the more general case of non-linear layer transformations.

3.4 Behaviour of Glass patterns under general layer transformations

Similar considerations also prevail in more general layer transformations, such as second-order polynomial transformations, logarithmic transformations, trigonometric transformations, etc. However, finding the fixed points in such cases may require more complex calculations, using either analytic or numeric methods. And indeed, as we will see later, Glass patterns may be used as a practical tool for visually locating fixed points.

Note that in the most general case the fixed points may be far more versatile than in a linear or affine transformation: A general mapping may have, depending on the case, no fixed points at all, one or more isolated fixed points, or even one or more straight or curved lines consisting of fixed points. Consequently, in the most general case Glass patterns may be generated simultaneously about one or several fixed points, along one or several straight or curved lines, or not be generated at all. Some simple examples illustrating the behaviour of Glass and moiré patterns under non-linear layer transformations are presented later in this section.

The superposition of *repetitive* layers (i.e. geometrically transformed periodic layers) can be considered as a particular case of a superposition under general layer transformations, in which the original layers, before undergoing the transformations, are periodic. But once again, the fact that the original layers are periodic does not influence the behaviour of the fixed points; it only adds some new structural information, which causes the resulting moiré effects to be repetitive.

And indeed, it turns out that the following universal rule holds for periodic, repetitive as well as aperiodic layers, and explains the fundamental behavior of their moiré effects:

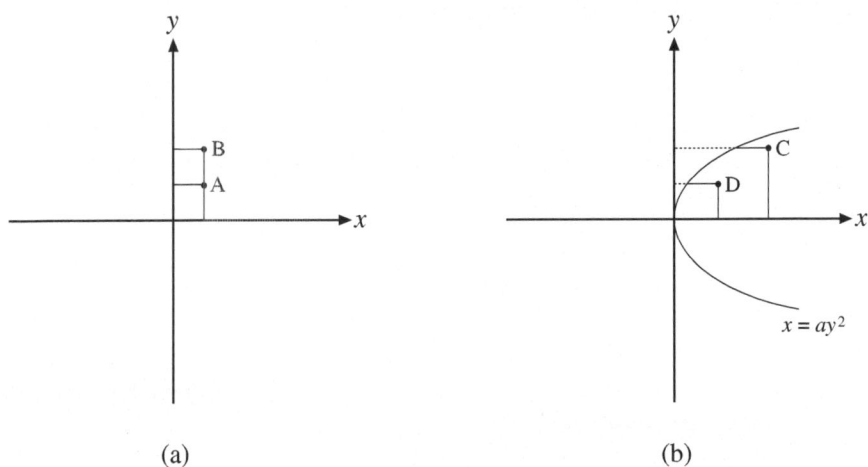

(a)   (b)

Figure 3.4: Schematic view of the parabolic mapping that we use in Figs. 3.5–3.15. This mapping moves each point (x,y) of the original layer horizontally by a distance of ay^2, which depends on the square of the y coordinate. Thus, the original x axis remains unchanged, while the original y axis is transformed into the parabola $x = ay^2$. Points A and B are transformed, respectively, into points C and D on the parabola $x = ay^2 + c$. Note that this mapping is, in fact, a non-linear horizontal shear transformation. As explained in Appendix D (see Sec. D.6 and Example D.6), this mapping can be expressed mathematically either as a *direct* transformation $(x,y) \mapsto (x + ay^2, y)$, or as a *domain* transformation $(x',y') = (x - ay^2, y)$, which is the inverse of $(x',y') = (x + ay^2, y)$. Thus, although our parabolic mapping maps each point (x,y) to $(x + ay^2, y)$, the parabolically transformed version of a layer such as $p(x',y') = \cos(2\pi f x')$ is given by $r(x,y) = p(x - ay^2, y) = \cos(2\pi f[x - ay^2])$; see Fig. D.16(g) in Appendix D.

Proposition 3.1: The behaviour of a moiré or Glass pattern under rotations, scalings and shifts of the individual layers (or more generally, under any layer mappings, linear or not), is determined by the fixed points of the mappings in question and their properties.[8] ■

As we can see, our two universal rules, Propositions 2.1 and 3.1, determine the conditions for the generation of all moiré effects, as well as the behaviour of these moiré effects under rotations, scalings, shifts, or any other layer transformations.

This provides, indeed, a unifying insight into the basic properties of all types of moiré effects: moirés between aperiodic, periodic or repetitive layers. In Chapter 7 we will see that this unifying insight can be extended even further, using the Fourier theory, so as to cover quantitatively the intensity profiles of the Glass (or moiré) patterns.

---

[8] We will see later, in Sec. 3.7, that Glass patterns can be also generated about *almost* fixed points, where the dots from both layers *almost* coincide. In such cases the behaviour of the Glass patterns is determined by the almost fixed points of the mappings. We can therefore further generalize Proposition 3.1 as follows: The behaviour of the moiré under any layer mappings is determined by the fixed or almost fixed points of the mappings and by their properties.

3.4.1 Examples with non-linear layer mappings

As we have just seen, Proposition 3.1 states that the behaviour of both Glass and moiré patterns under any layer mappings is determined by the fixed points of the mappings and their properties. Having so far illustrated this rule for linear and affine mappings alone, we wish to illustrate it now for non-linear layer mappings, too. This will also give us a deeper insight into the relationship between aperiodic and repetitive cases (namely, between superpositions of aperiodic layers and superpositions of periodic layers where the respective layers have undergone the same non-linear mappings). As already mentioned above, when a non-linear mapping is used there may exist more possible configurations for the fixed points than in the case of a linear or affine transformation. While in a linear or affine transformation there may exist either one fixed point, or a full line of fixed points, or no fixed points at all (ignoring the trivial case of the identity mapping, in which all the points of the plane are fixed points), in a general mapping there may also exist cases with several isolated fixed points, with several straight fixed lines, or even with one or several curved fixed lines. In order to illustrate these interesting cases, that are inexistent in linear or affine mappings, we will use here a simple family of non-linear mappings of the second order, namely: parabolic mappings. These mappings transform every vertical coordinate line $x' = c$ into a horizontally oriented parabola $x = ay^2 + c$.[9] In order to apply such a parabolic mapping to our original dot screen (periodic or not), we simply have to horizontally move each point (x,y) of the original screen by a distance which depends on the square of its y coordinate:[10]

$$\begin{pmatrix} x' \\ y' \end{pmatrix} = \begin{pmatrix} x - ay^2 \\ y \end{pmatrix}$$ (3.16)

The coefficient a determines the "bending rate" of the resulting parabolas.[11] Such a mapping (coordinate transformation) is schematically illustrated in Fig. 3.4. Note, however, that for the sake of clarity we have drawn Fig. 3.4 with a relatively large value of a; in reality we will use much smaller values of a, as shown in Figs. 3.5(a),(b), in order not to fully destroy the correlation between the superposed layers. These simple second-order mappings offer us three main advantages: They clearly illustrate the main fixed point configurations which occur in non-linear mappings; their mathematical handling is still tractable; and finally, the results for the corresponding repetitive cases (i.e. where the original, untransformed layers are periodic) are already known (see Sections 10.7.3–10.7.4 in *Vol. I*), which enables us to compare the results obtained for periodic and aperiodic layers. For the sake of completeness, a full discussion on the fixed points in a general mapping of the second order is provided in Sec. A.3 of Appendix A.

---

[9] We choose this convention (a horizontally oriented parabola) in order to remain compatible with the figures and the examples given in Secs. 10.7.3–10.7.4 of *Vol. I*.

[10] Note that although each point (x,y) is moved by this mapping to $(x + ay^2, y)$, Eq. (3.16) is defined with a *minus* sign, so that the straight coordinate lines $x' = c$ be transformed into the desired curved coordinate lines, the parabolas $x - ay^2 = c$ (i.e. $x = ay^2 + c$). In other words, we must always consider here the *inverse* transformation, since we are applying it to the given layer $r(x',y')$ as a domain transformation: $r(x - ay^2, y)$ (see Appendix D and in particular Sec. D.6, Example D.6 and Fig. D.16).

[11] Note that the term *curvature* is defined in mathematics in a different way (see [Courant88 p. 86]).

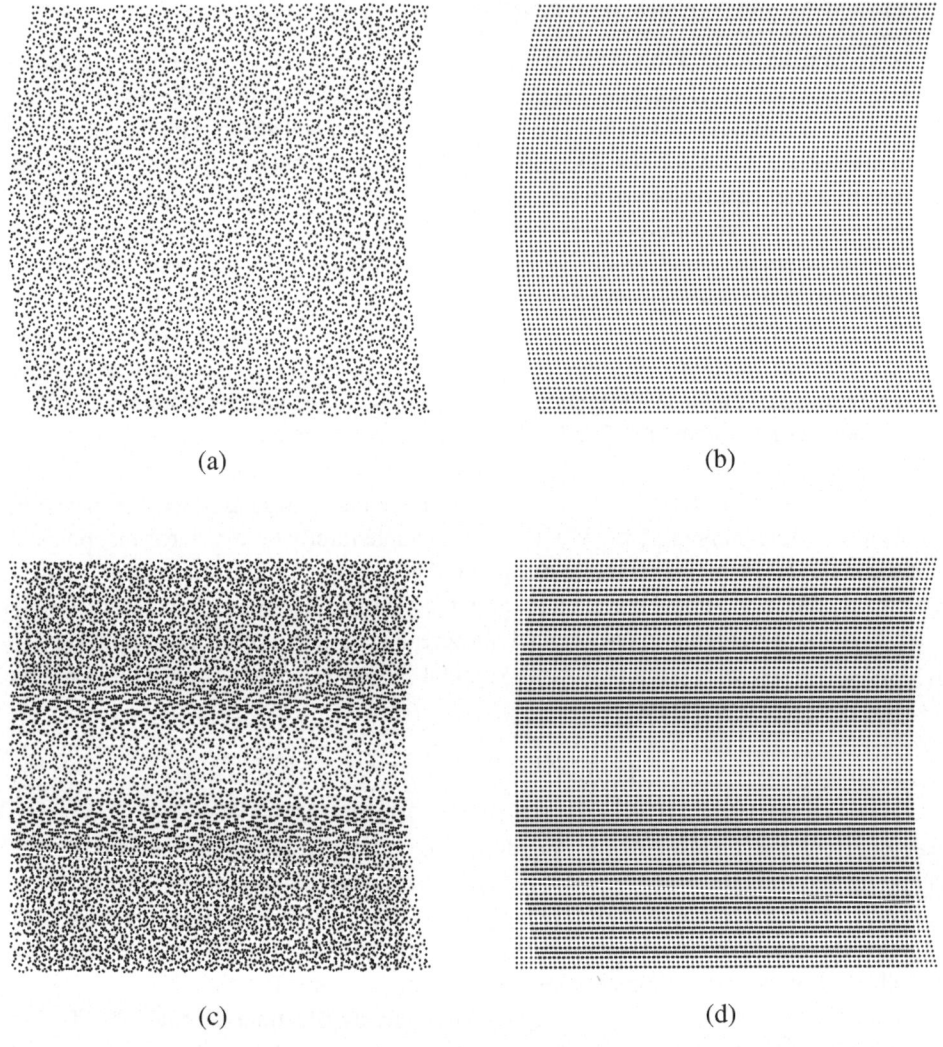

(a)   (b)

(c)   (d)

Figure 3.5: (a) The aperiodic dot screen of Fig. 2.1(a) after having undergone
the parabolic transformation $\mathbf{g}(x,y) = (x - ay^2, y)$. (b) The periodic
dot screen of Fig. 2.1(b) after having undergone the same
parabolic transformation. (c) The superposition of two identical
aperiodic dot screens, one of which has undergone the parabolic
transformation $\mathbf{g}(x,y)$. Since this transformation does not involve
layer shifts, the two layers clearly coincide along the x axis.
(d) The superposition of two identical periodic dot screens, one of
which has undergone the same parabolic transformation $\mathbf{g}(x,y)$. In
both (c) and (d) a linear Glass pattern is generated precisely at the
same location.

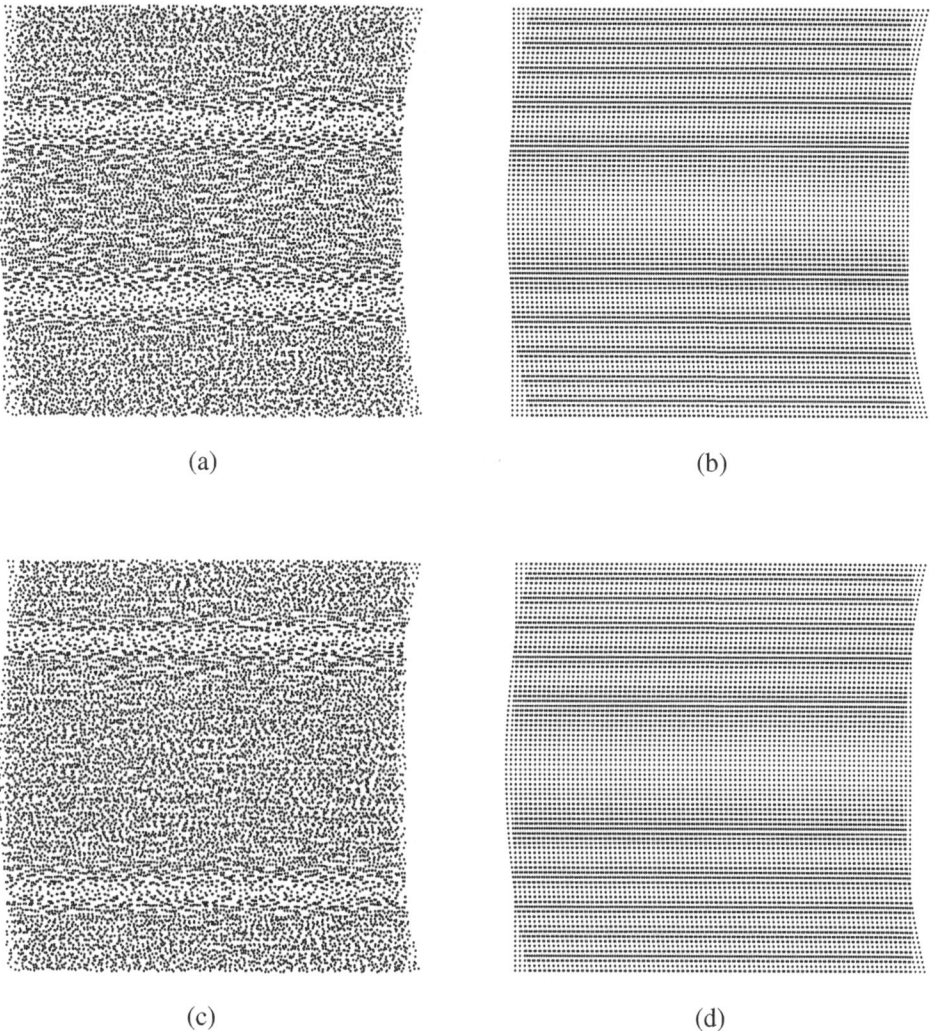

(a)

(b)

(c)

(d)

Figure 3.6: (a),(b) The superposition of two identical aperiodic (respectively, periodic) dot screens, one of which has undergone the parabolic transformation $\mathbf{g}(x,y) = (x - ay^2, y)$, while the untransformed screen has been slightly shifted by $x_0 = T$ to the right (T equals one period of the periodic dot screen of (b)). In both (a) and (b) a pair of linear Glass patterns is generated precisely at the same locations. (c),(d) Same as in (a),(b) but with a horizontal shift of $x_0 = 2T$.

Example 3.3: Parabolic mapping with translation:

Let us superpose two identical untransformed layers on top of each other, and apply to one of them the parabolic mapping $\mathbf{g}(x,y) = (x - ay^2, y)$. This is shown in Fig. 3.5(c) for an aperiodic case (namely, where the two originally identical dot screens, before applying the

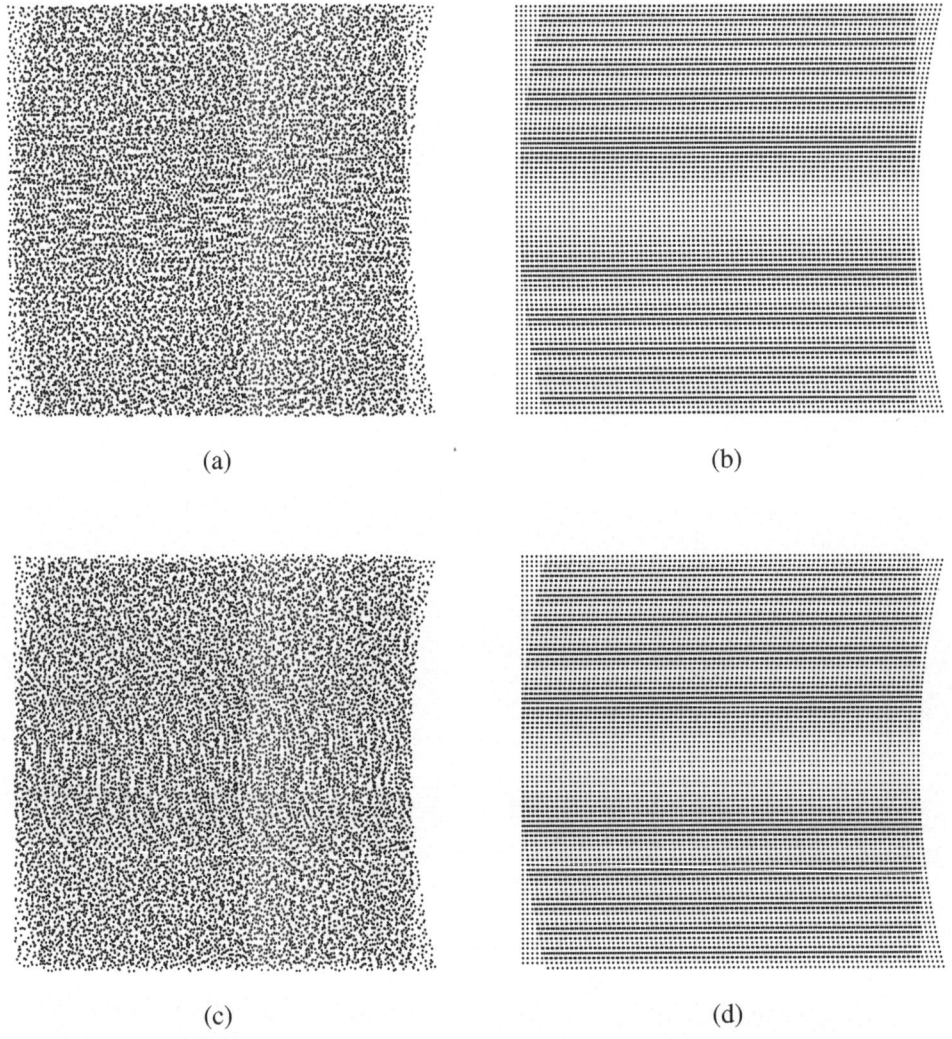

(a)   (b)

(c)   (d)

Figure 3.7: (a),(b) Same as in Figs. 3.6(a),(b) but with a slight layer shift of $x_0 = -T$.
(c),(d) Same as in Figs. 3.6(a),(b), but with a slight vertical shift of $y_0 = T$
upward.

transformation, were aperiodic), and in Fig. 3.5(d) for a repetitive case (namely, where the
two originally identical dot screens, before applying the transformation, were periodic).
The mapping in question has infinitely many fixed points, which together form the entire x
axis. This can be easily demonstrated as follows:

Our non-linear mapping $\mathbf{g}(x,y)$ is given by the system of equations:

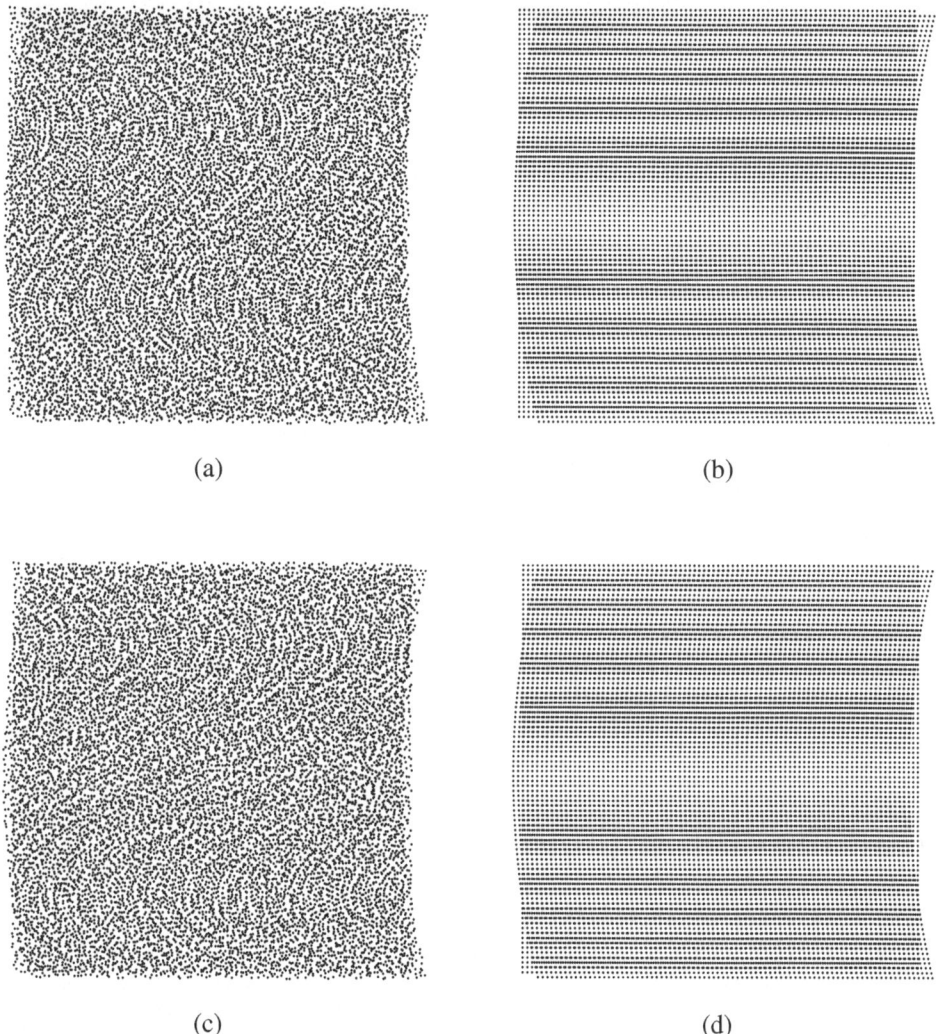

(a)   (b)

(c)   (d)

Figure 3.8: (a),(b) Same as in Figs. 3.6(a),(b) but with a slight layer shift of $(x_0,y_0) =$ (T,T). (c),(d) Same as in Figs. 3.6(a),(b), but with a slight layer shift of $(x_0,y_0) = (2T,T)$.

$$x' = x - ay^2$$
$$y' = y \tag{3.17}$$

Clearly, the fixed points of this mapping are those points for which (x',y') equals (x,y), namely:

$$x = x - ay^2$$
$$y = y \tag{3.18}$$

And indeed, the solution of this simple set of equations is $y = 0$ for all x, meaning that the fixed points of mapping (3.17) consist of the entire x axis.

Consequently, a linear Glass pattern is generated in the layer superposition along the x axis; this Glass pattern is clearly visible in Fig. 3.5(c). As we can see in the figure, the two superposed layers are fully correlated along the x axis, but as we go farther away from the x axis the correlation between the layers gradually decreases, and the Glass pattern fades away. Note, however, that in the corresponding repetitive case (Fig. 3.5(d)) a moiré effect still exists even beyond the range of the Glass pattern. This happens due to the additional ordering which exists in the internal structure of repetitive layers: thanks to their repetitivity, the superposed layers also have infinitely many points of coincidence which are not fixed points of the underlying mapping $\mathbf{g}(x,y)$.

Now, suppose that we shift one of the dot screens (say, the untransformed one) horizontally by x_0. In this case our transformation becomes:

$$x' - x_0 = x - ay^2$$
$$y' = y \tag{3.19}$$

and the set of equations determining its fixed points is:

$$x - x_0 = x - ay^2$$
$$y = y \tag{3.20}$$

The solution of this set of equations is clearly:

$$y = \pm\sqrt{x_0/a} \tag{3.21}$$

for all x. However, we should distinguish here between two possible cases: If the untransformed dot screen is shifted *to the right*, meaning that $x_0 > 0$, or more precisely: if x_0 and a have the same sign, then the fixed points form two parallel horizontal lines whose vertical distances from the x axis are given by $\pm\sqrt{x_0/a}$. This is shown in Figs. 3.6(a),(c), where the resulting Glass patterns are, indeed, clearly visible. As we can see in these figures, when we gradually shift the untransformed screen to the right, the two parallel fixed-point lines simultaneously move away from the x axis; and inversely, as x_0 tends to zero, the two parallel fixed-point lines move closer and closer to the x axis, until they coincide with it when $x_0 = 0$ (Fig. 3.5(c)). But if the untransformed screen is shifted *to the left*, meaning that $x_0 < 0$, or more precisely: if x_0 and a have opposite signs, then no real fixed points may exist, and indeed, as shown in Fig. 3.7(a), the Glass pattern disappears. It is interesting to note, however, that in the corresponding repetitive case (Fig. 3.7(b)) the moiré effect is still clearly visible. This happens, again, due to the additional ordering which exists in the internal structure of the repetitive layers: as we have already seen, thanks to their repetitivity, the superposed layers also have infinitely many points of coincidence which are not fixed points of the underlying mapping $\mathbf{g}(x,y)$.

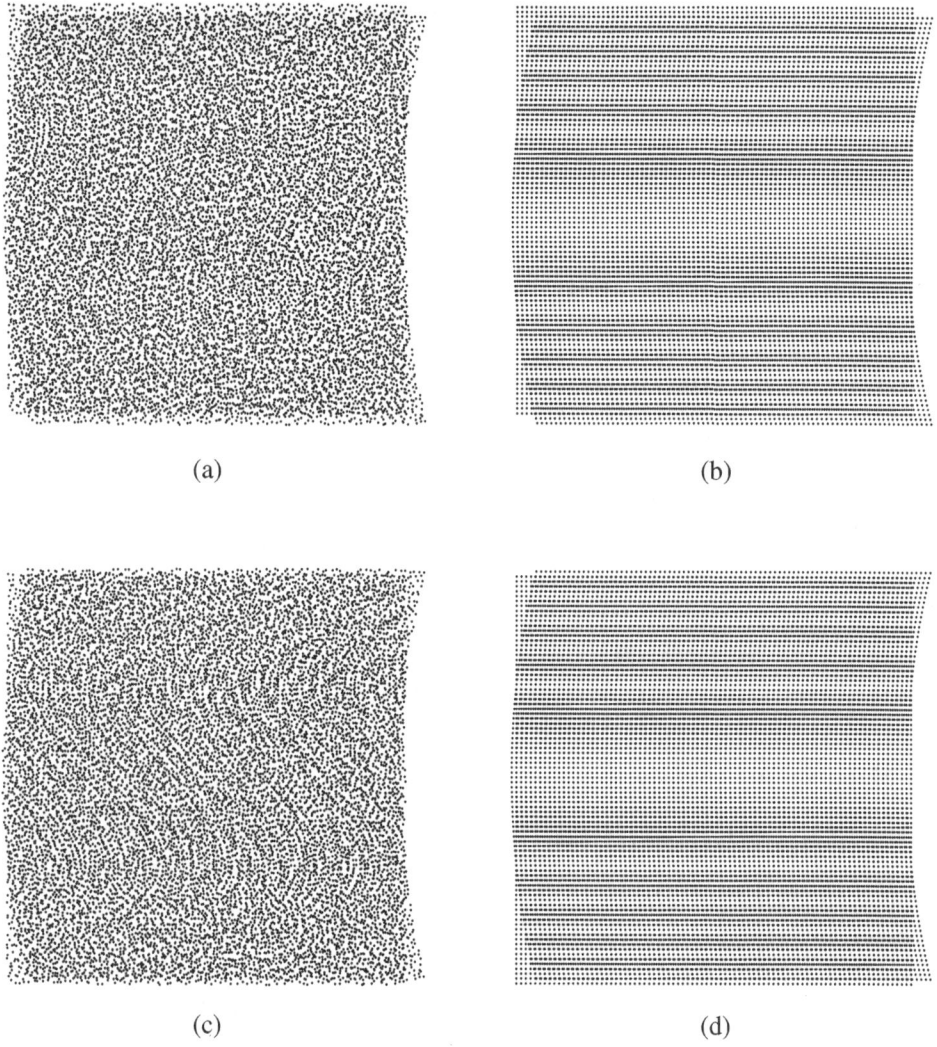

Figure 3.9: (a),(b) Same as in Figs. 3.6(a),(b), but with a slight layer shift of $(x_0,y_0) =$ $(T,2T)$. (c),(d) Same as in Figs. 3.6(a),(b), but with a slight layer shift of $(x_0,y_0) = (T,-T)$.

What happens now if we shift the untransformed dot screen vertically? Let us start again from the initial superposition of the two layers origin on origin (Fig. 3.5(c)), this time gradually applying a vertical layer shift. In this case the transformation becomes:

$$x' = x - ay^2$$
$$y' - y_0 = y$$

(3.22)

and its fixed points are given by the system of equations:

$$x = x - ay^2$$

$$y - y_0 = y$$

(3.23)

However, since the second equation in system (3.23) is inconsistent, no points (x,y) solve this set of equations. And indeed, as shown in Fig. 3.7(c), no fixed points, and hence no Glass patterns, exist in the superposition (see Sec. 2.3.3). Glass patterns do not exist, either, when we apply both a horizontal layer shift x_0 and a vertical layer shift y_0 (see Figs. 3.8 and 3.9(a),(b)). It is interesting to note, however, the microstructures (dot trajectories) which are generated in such cases due to the layer shift of (x_0,y_0): As shown in Figs. 3.8 and 3.9(a), when $y_0 > 0$ this microstructure consists of inversed S-like dot trajectories, whose curve extrema form two parallel lines equally spaced above and below the x axis. These parallel lines are located at the same geometric locations as the fixed-point lines that are generated when we only have a pure horizontal shift of $(x_0,0)$; this can be clearly seen by comparing Figs. 3.6(a) and 3.8(a), or Figs. 3.6(c) and 3.8(c). Finally, when the vertical shift is applied in the negative direction (see Fig. 3.9(c)), the orientation of the S-like dot trajectories is inverted, but their extrema still remain at the same locations. These microstructure phenomena will be explained later (see Example 4.10 in Chapter 4 and Example 5.4 in Chapter 5). ∎

3.5 Mappings in both layers; mutual fixed loci

Suppose now that both of the superposed layers have been transformed, the first layer by mapping $g_1(x,y)$ and the second layer by mapping $g_2(x,y)$. What are the mutual fixed points between the two mappings?[12] In fact, we have already encountered such cases, for instance, in Example 3.3 above, where one of the layers has been transformed and the other (untransformed) layer has been shifted. In such cases we no longer have a transformed layer and a reference layer, but rather two transformed layers. What we are looking for in such cases is simply the points (x,y) in the plane which are mapped by both of the mappings $g_1(x,y)$, $g_2(x,y)$ to the same destination points: $g_1(x,y) = (x',y')$, $g_2(x,y) = (x',y')$. In other words, we are looking for the points (x,y) for which $g_1(x,y) = g_2(x,y)$, which are the zeros of the mapping:[13]

$$g_M(x,y) = g_1(x,y) - g_2(x,y) = (0,0)$$

(3.24)

Let us now consider a few such examples.

---

[12] We use the term *mutual fixed point* of g_1 and g_2 to designate a point (x_F,y_F) for which $g_1(x_F,y_F) = g_2(x_F,y_F)$. Similarly, a *mutual fixed locus* of g_1 and g_2 is a locus in the plane that consists of all the mutual fixed points of g_1 and g_2. Note that the term *common fixed point* of g_1 and g_2 is already used in the mathematical literature for a point (x_F,y_F) that satisfies $g_1(x_F,y_F) = (x_F,y_F) = g_2(x_F,y_F)$, but this definition is too restrictive for our needs.

[13] Once again, $g_1(x,y)$, $g_2(x,y)$ and $g_M(x,y)$ are interpreted here as inverse transformations, as explained in Appendix D, since they are used as *domain* transformations.

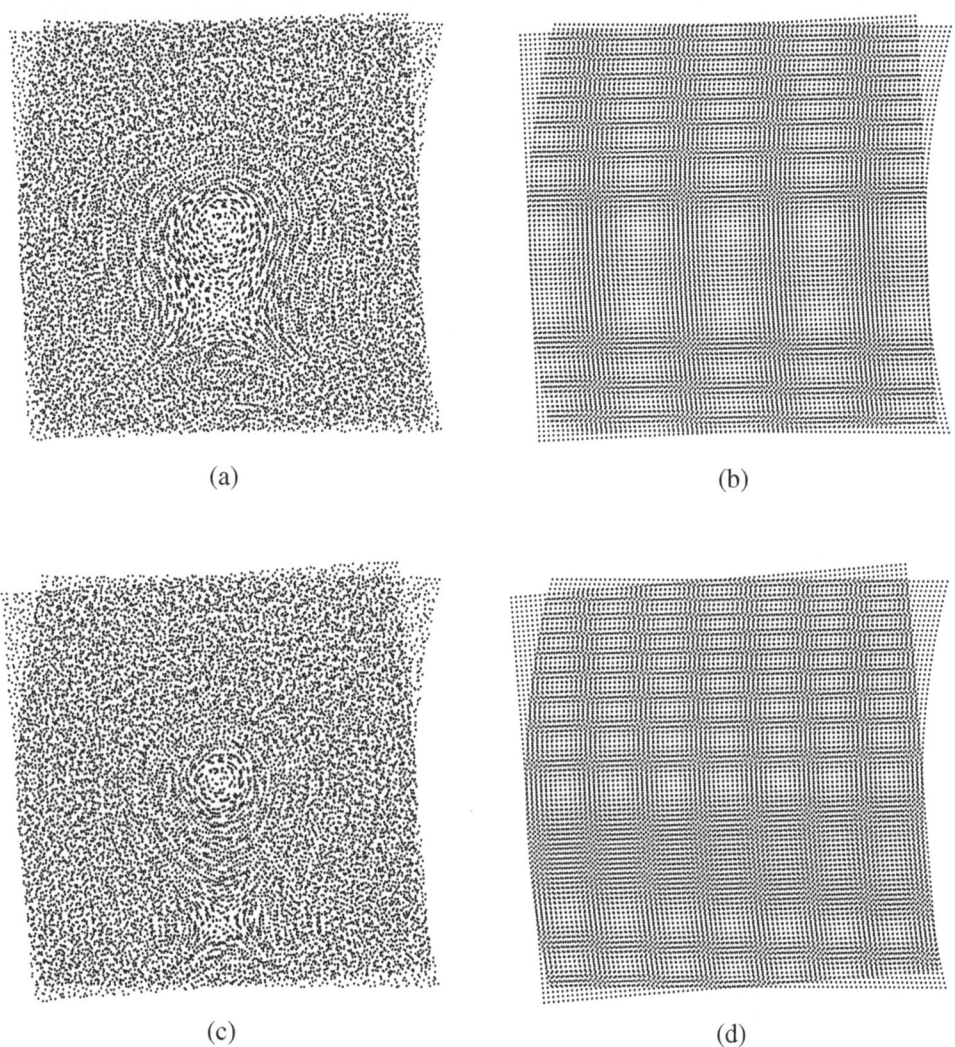

(a)

(b)

(c)

(d)

Figure 3.10: (a),(b) The superposition of two identical aperiodic (respectively, periodic) dot screens, one of which has undergone the parabolic transformation $\mathbf{g}(x,y) = (x - ay^2, y)$, while the untransformed layer has been slightly rotated by angle $\alpha = 3°$. In both (a) and (b) two Glass patterns are generated precisely at the same locations (fixed points): one about the origin and the other below the origin. (c),(d) Same as in (a),(b) but with $\alpha = 5°$.

Example 3.4: Parabolic mapping with rotation:

Suppose that we apply to one of the two layers the same parabolic transformation as in Example 3.3, while the other layer is rotated by a small angle α. As we can see in Fig. 3.10, this generates in the superposition two Glass patterns: An elliptic Glass pattern

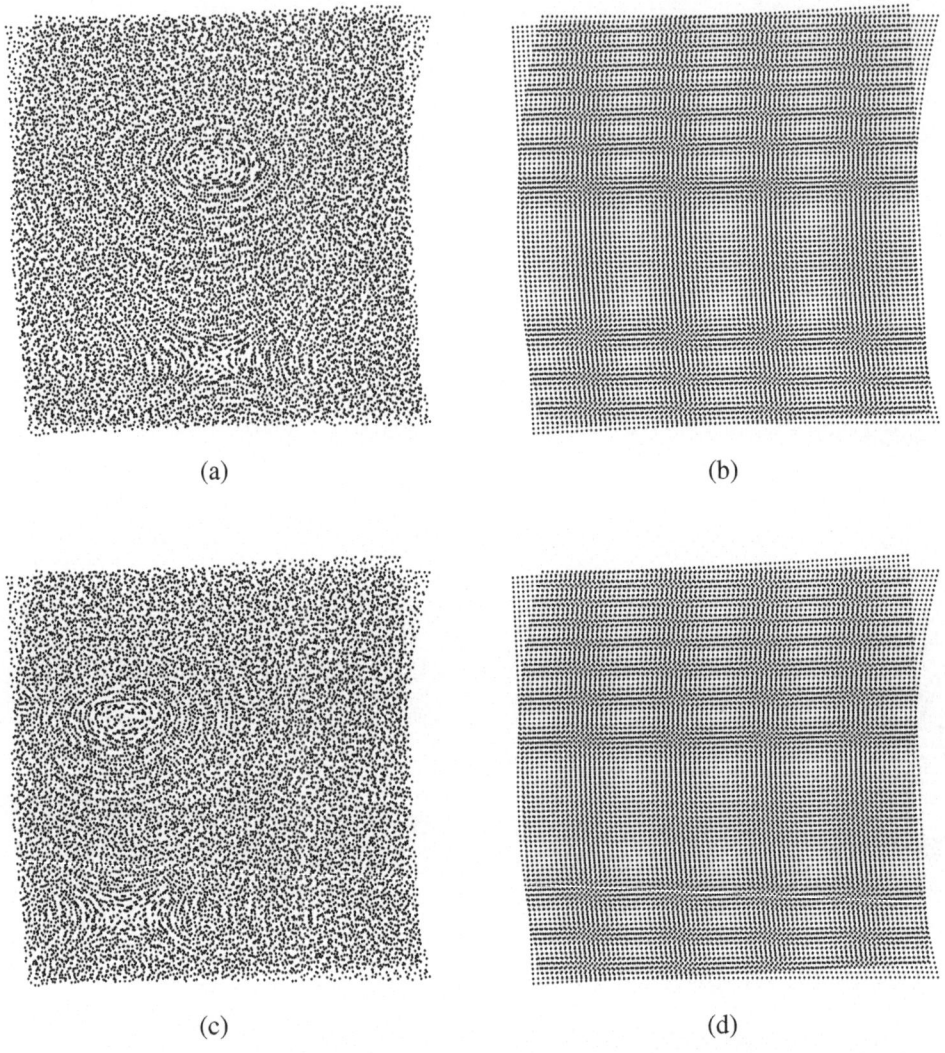

(a)   (b)

(c)   (d)

Figure 3.11: (a),(b) Same as in Figs. 3.10(a),(b) but here the rotated layer has also undergone a slight horizontal shift of $x_0 = T$ to the right. (c),(d) Same as in Figs. 3.10(a),(b) but here the rotated layer has also undergone a slight shift of $(x_0, y_0) = (T, T)$.

which is centered about the origin, and a hyperbolic Glass pattern which is vertically displaced from the origin by a distance depending on the angle α. When α increases, the hyperbolic Glass pattern moves away from the origin, and when α tends to zero, the hyperbolic Glass pattern moves closer to the origin, thus forming together with its elliptic counterpart a curious hybrid Glass pattern having elliptic trajectories in one side and hyperbolic trajectories in the other side. When α reaches 0 we return to the situation of

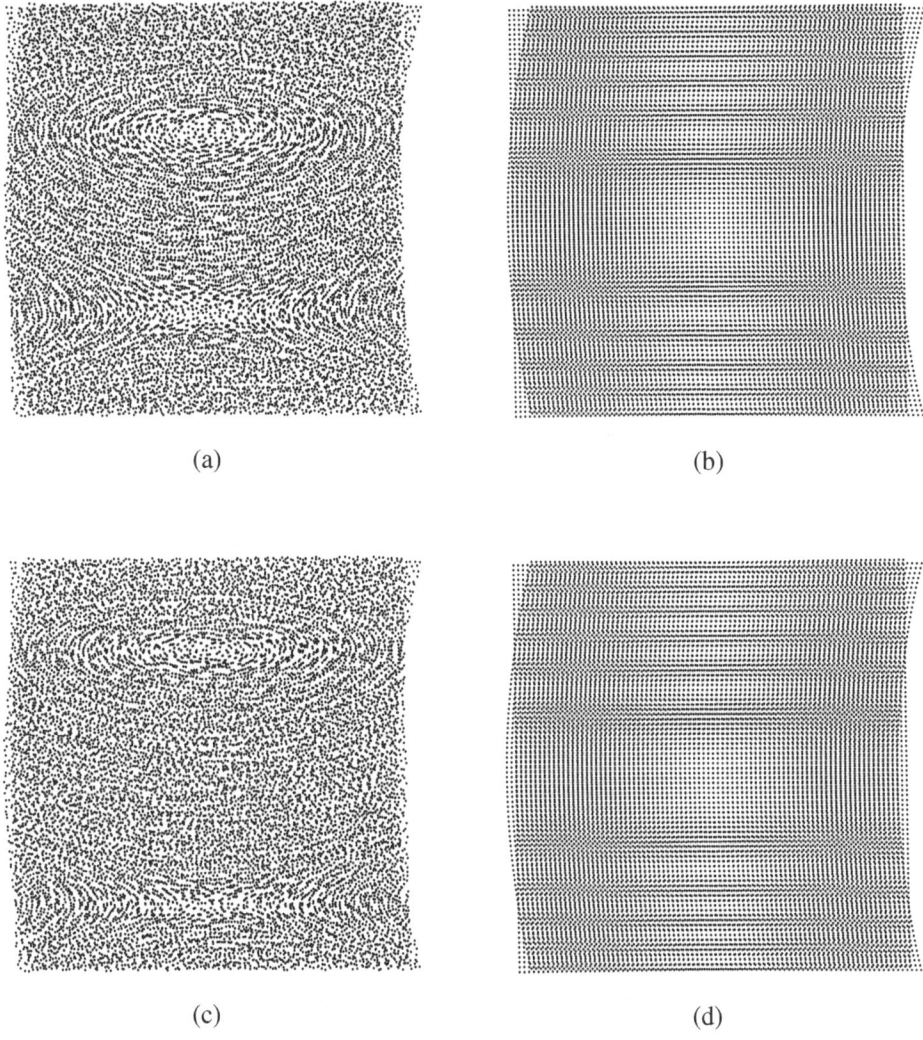

(a)   (b)

(c)   (d)

Figure 3.12: (a),(b) Same as in Figs. 3.11(a),(b) but with $\alpha = 1°$. (c),(d) Same as
(a),(b) but with a horizontal shift of $x_0 = 2T$. Compare with Fig.
3.6(a)–(d), where $\alpha = 0°$.

Fig. 3.5(c), namely, the elliptic and hyperbolic trajectories become completely flat, and we
obtain a horizontal linear Glass pattern that is centered along the x axis.

Let us analyze this case mathematically. The transformations $\mathbf{g}_1(x,y)$ and $\mathbf{g}_2(x,y)$
undergone by our two layers are given here by:

$$
\begin{aligned}
x' &= x\cos\alpha + y\sin\alpha & x' &= x - ay^2 \\
y' &= -x\sin\alpha + y\cos\alpha & y' &= y
\end{aligned}
\tag{3.25}
$$

The mutual fixed points of these transformations are given by the system of equations:

$$x\cos\alpha + y\sin\alpha = x - ay^2$$
$$-x\sin\alpha + y\cos\alpha = y$$

(3.26)

Using the trigonometric identity $\dfrac{\sin\alpha}{1-\cos\alpha} = \cot\dfrac{\alpha}{2}$ we obtain from the second equation of this system:

$$y = -x\cot\frac{\alpha}{2}$$

(3.27)

and by inserting this into the first equation of system (3.26) and simplifying the resulting expression, we obtain:

$$a\cot^2\frac{\alpha}{2}x^2 + (\cos\alpha - \cot\frac{\alpha}{2}\sin\alpha - 1)x = 0$$

Some further simplifications using the identity $\cot\dfrac{\alpha}{2} = \dfrac{1+\cos\alpha}{\sin\alpha}$ show that:

$$(\cos\alpha - \cot\frac{\alpha}{2}\sin\alpha - 1) = -2$$

and hence we finally obtain:

$$a\cot^2\frac{\alpha}{2}x^2 - 2x = 0$$

This second order equation in x has two solutions, $x = 0$ and $x = \dfrac{2}{a}\tan^2\dfrac{\alpha}{2}$. The respective y values obtained from Eq. (3.27) are $y = 0$ and $y = -\dfrac{2}{a}\tan\dfrac{\alpha}{2}$. The set of equations (3.26) has, therefore, two solutions, which are the mutual fixed points of our mappings (3.25):

$$(x,y) = (0,0) \qquad \text{and} \qquad (x,y) = (\frac{2}{a}\tan^2\frac{\alpha}{2}, -\frac{2}{a}\tan\frac{\alpha}{2})$$

(3.28)

This means that our mappings have one mutual fixed point at the origin, and a second mutual fixed point whose location depends on the angle α; note that since for small values of α we have $\tan^2\dfrac{\alpha}{2} \ll \tan\dfrac{\alpha}{2}$, the second fixed point is located almost vertically below (or above) the origin. This fully confirms, indeed, our observations based on Fig. 3.10.

What happens now if, in addition to the rotation, we allow also layer shifts of (x_0,y_0)? As we can see in Fig. 3.11(a), shifting the rotated layer horizontally by $x_0 > 0$ will cause the two mutual fixed points and their Glass patterns to move away from each other vertically; and conversely, as x_0 tends to zero the two mutual fixed points approach each other. If we allow x_0 to become negative, the two fixed points will further approach each other, until at a certain negative value of x_0 both of them coincide and fuse into a single point. If we further decrease x_0, the fixed points completely disappear. On the other hand, shifting the rotated layer vertically by y_0 will only cause the fixed points to move horizontally to the left or to the right, depending on the sign of y_0 (compare Figs. 3.11(a) and 3.11(c)). It is also interesting to see the influence of angle α on the resulting Glass patterns in the presence of a constant layer shift. As we can see by comparing Figs. 3.12 and 3.6, when $\alpha \to 0$ we return to the situation of Fig. 3.6, namely: the elliptic and hyperbolic trajectories become completely flat, and we obtain a pair of horizontal linear Glass patterns located at equal distances of $\pm\sqrt{x_0/a}$ above and below the x axis.

As we can see, the mutual fixed points of our layer mappings $\mathbf{g}_1(x,y)$ and $\mathbf{g}_2(x,y)$ are clearly visible, both in the aperiodic and in the repetitive cases, thanks to the Glass patterns which are generated around them in the layer superposition (compare, for example, Figs. 3.10(c) and 3.10(d)). However, in the corresponding repetitive case (Fig. 3.10(d)) a moiré effect still exists even beyond the range of the Glass patterns. As we have already seen, this happens due to the additional ordering which exists in the internal structure of repetitive layers: thanks to their repetitivity, the superposed layers also have infinitely many points of coincidence which are not mutual fixed points of the underlying mappings $\mathbf{g}_1(x,y)$ and $\mathbf{g}_2(x,y)$. We will return to this interesting result in more detail in Sec. 6.5. ∎

Example 3.5: Parabolic mapping with scaling:

Suppose that we apply to one of the two layers the same parabolic transformation as in the previous examples, while the other layer is slightly scaled by a scaling factor of $s \approx 1$. As we can see in Figs. 3.13(a), a single fixed point appears at the origin; note, however, the interesting parabolic dot-trajectories which surround the Glass pattern. Now, if we shift the scaled layer by x_0 horizontally (see Figs. 3.13(c),(d)), the fixed point simply moves to the right or to the left, depending on the sign of x_0. And if we also shift the scaled layer vertically by y_0 (see Fig. 3.14(a),(b)), our fixed point is displaced along a parabolic curve which is a function of x_0 and y_0^2. This can be easily demonstrated as follows:

Our layer mappings $\mathbf{g}_1(x,y)$ and $\mathbf{g}_2(x,y)$ are given in this case by:

$$\begin{aligned} x' &= sx - x_0 & x' &= x - ay^2 \\ y' &= sy - y_0 & y' &= y \end{aligned} \tag{3.29}$$

The mutual fixed points between these mappings are given by the system of equations:

$$\begin{aligned} sx - x_0 &= x - ay^2 \\ sy - y_0 &= y \end{aligned} \tag{3.30}$$

From the second equation of this system we obtain $y = \dfrac{1}{s-1}y_0$. And by inserting this into the first equation of the system and simplifying the resulting expression we get:

$$x = \frac{1}{s-1}x_0 + \frac{a}{(s-1)^3}y_0^2$$

System (3.30) has, therefore, a single solution, which is the single mutual fixed point of our mappings (3.29):

$$(x,y) = \left(\frac{1}{s-1}x_0 + \frac{a}{(s-1)^3}y_0^2, \frac{1}{s-1}y_0\right) \tag{3.31}$$

And indeed, as we already observed above, the location of this fixed point is a function of x_0 and of y_0^2.

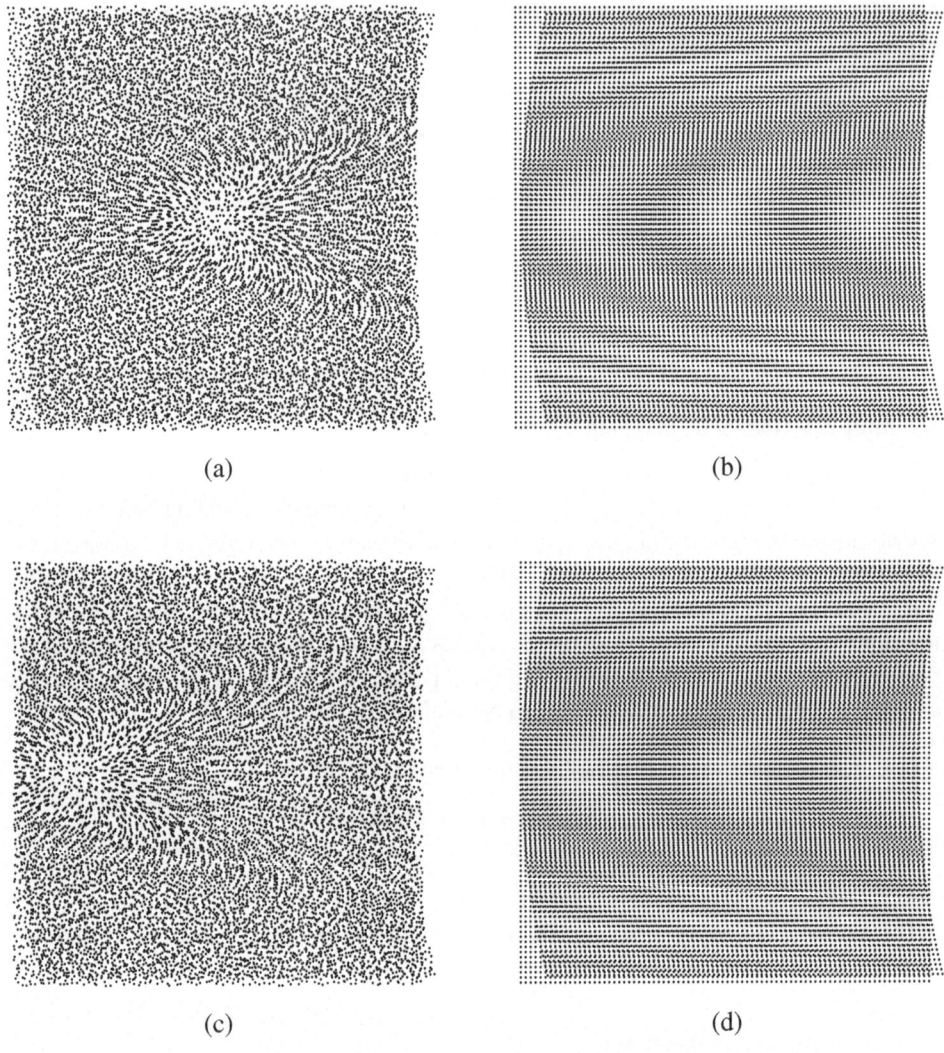

(a)   (b)

(c)   (d)

Figure 3.13: (a),(b) The superposition of two identical aperiodic (respectively, periodic) dot screens, one of which has undergone the parabolic transformation $\mathbf{g}(x,y) = (x - ay^2, y)$, while the untransformed layer has been slightly scaled up. (c),(d) Same as in (a),(b) but where the scaled layer has been also slightly shifted by $x_0 = T$ to the right.

What happens now if we apply both scaling and rotation? As we can see in Figs. 3.14(c),(d), the result is similar to the case of rotation without scaling (Figs. 3.10(c),(d)), namely, an elliptic Glass pattern centered at the origin, plus a hyperbolic Glass pattern which is displaced from the origin. Note, however, that the dot trajectories in this case have a spiral form, which is typical to cases involving both rotation and scaling, as we have already seen earlier (see Fig. 2.1(g)).

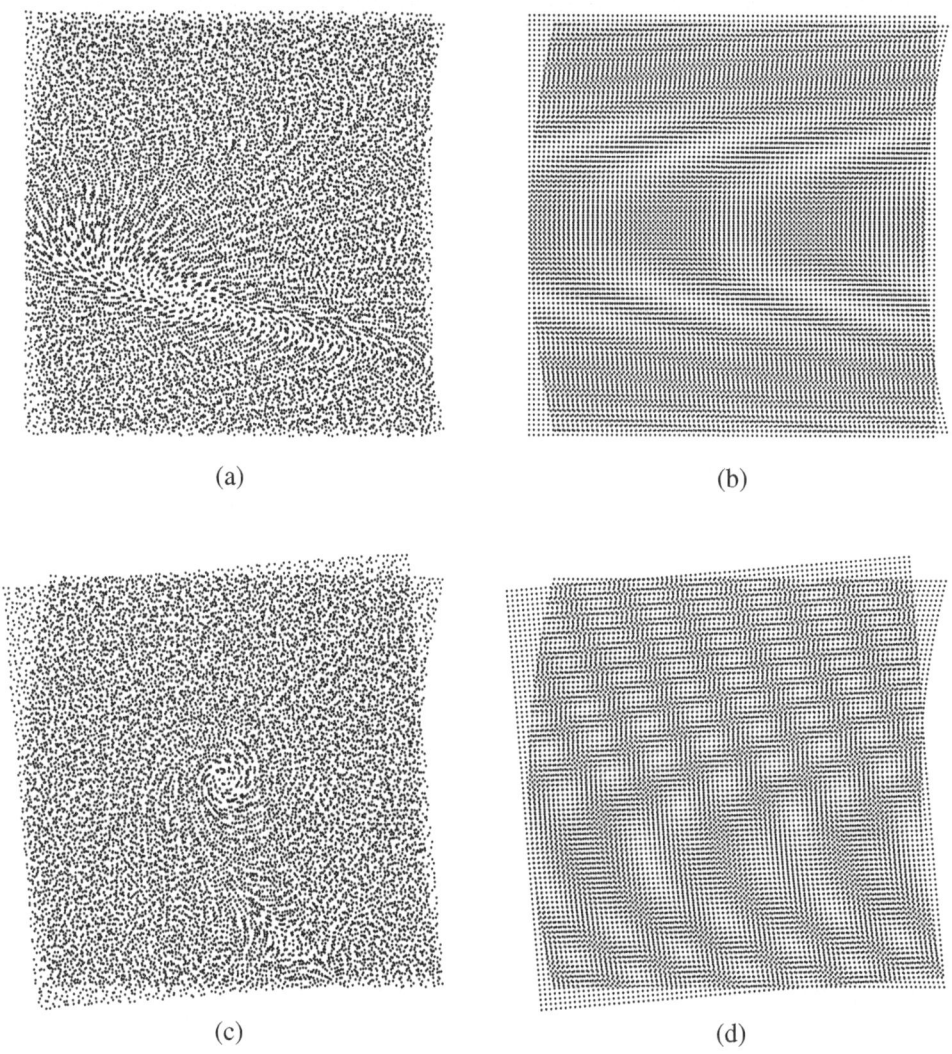

(a)   (b)

(c)   (d)

Figure 3.14: (a),(b) Same as in Figs. 3.13(a),(b) but where the scaled layer has been also slightly shifted by $(x_0,y_0) = (T,T/2)$. (c),(d) Same as in Figs. 3.13(a),(b) but with a slight rotation by angle $\alpha = 5°$. Note the two Glass patterns which are generated in both (c) and (d) precisely in the same locations.

Once again, we can clearly see in Figs. 3.13 and 3.14 that a Glass pattern is generated, both in the aperiodic and in the repetitive cases, about each of the fixed points. But in the repetitive cases the moiré effect contains, in addition, infinitely many transformed repetitions of these Glass patterns. ∎

Example 3.6: An example with four fixed points:

In all the examples we have seen so far there existed either 0, 1 or 2 fixed points, or one or two full lines of fixed points (if the second-order mapping in question was degenerate). However, as shown in Sec. A.3 of Appendix A, a non-degenerate second order mapping may have up to 4 isolated fixed points. Let us see now an example with 4 mutual fixed points.

One possible way to obtain such a case is by superposing two parabolic dot screens. For example, we may start with two identical dot screens, and apply to one of them our horizontal parabolic mapping (3.16), while to the other screen we apply a similar parabolic mapping that operates in the vertical direction.[14] If we superpose the two resulting layers origin on origin, we will observe in the superposition a single Glass pattern centered about the origin. But if we now introduce a small horizontal layer shift of $-x_0$ in the horizontal parabolic layer and a small vertical layer shift of $-y_0$ in the vertical parabolic layer, where $x_0, y_0 > 0$, we will obtain, indeed, four different Glass patterns in the superposition, as shown in Fig. 3.15(a): two elliptic Glass patterns on one diagonal, and two hyperbolic Glass patterns on the other diagonal.

To derive the locations of the fixed points mathematically, suppose that the first layer has been transformed by the vertical parabolic mapping $\mathbf{g}_1(x,y)$:

$$x' = x$$
$$y' = y - ax^2 \tag{3.32}$$

while the second layer has been transformed by the horizontal mapping $\mathbf{g}_2(x,y)$ given by:

$$x' = x - ay^2$$
$$y' = y \tag{3.33}$$

The mutual fixed points between these two mappings are given, therefore, by:

$$x - ay^2 = x$$
$$y = y - ax^2 \tag{3.34}$$

Obviously, there exists exactly one point which satisfies both equations, namely: $(x,y) = (0,0)$. However, if we apply a shift of $-y_0$ to the first layer and a shift of $-x_0$ to the second layer, where $x_0, y_0 > 0$, our mappings $\mathbf{g}_1(x,y)$ and $\mathbf{g}_2(x,y)$ become:

$$x' = x \qquad\qquad\qquad x' = x + x_0 - ay^2$$
$$\text{and:} \tag{3.35}$$
$$y' = y + y_0 - ax^2 \qquad\qquad y' = y$$

---

[14] It is important to note that in aperiodic cases we cannot take for the vertically transformed layer a copy of the horizontally transformed layer that is rotated by 90°, since such a rotation destroys the correlation between the two originally identical layers.

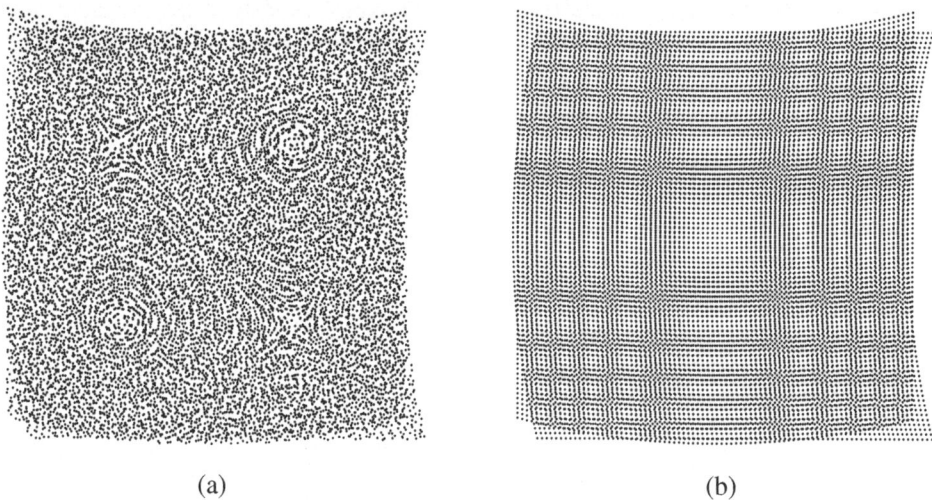

(a) (b)

Figure 3.15: (a) An example with 4 mutual fixed points: The superposition of two identical aperiodic dot screens, one of which has undergone a vertical parabolic transformation plus a slight vertical shift of $y_0 = -T$, while the other has undergone a horizontal parabolic transformation plus a slight horizontal shift of $x_0 = -T$. Note the two circular and the two hyperbolic Glass patterns which are clearly visible about the fixed points in the superposition. (b) The superposition of two identical *periodic* dot screens, which have undergone exactly the same transformations as in (a). This example is further explained in Example 5.3 of Chapter 5. The reason that two of the Glass patterns in (a) are circular while the others are hyperbolic is explained in Example H.4 of Appendix H.

and the mutual fixed points between them are given by:

$$x + x_0 - ay^2 = x$$
$$y = y + y_0 - ax^2$$

(3.36)

which means, after simplification:

$$ay^2 = x_0$$
$$ax^2 = y_0$$

(3.37)

The solutions of this system are:

$$x = \pm\sqrt{y_0/a}, \quad y = \pm\sqrt{x_0/a}$$

(3.38)

This means, indeed, as we can see in Fig. 3.15(a), that the mappings $\mathbf{g}_1(x,y)$ and $\mathbf{g}_2(x,y)$ have in this case 4 mutual fixed points:

$$(x,y) = (\pm\sqrt{y_0/a}, \pm\sqrt{x_0/a})$$

(3.39)

This assumes, of course, that $x_0 > 0$ and $y_0 > 0$, or, more precisely, that x_0, y_0 and a have the same sign. If this assumption does not hold, then there may exist no fixed points at all (if x_0, y_0 have the same sign but the sign of a is opposite), or there may only exist two fixed points (if x_0 and y_0 have opposite signs). ∎

We may conclude our present discussion with the following general result:

Proposition 3.2: When we superpose two identical aperiodic layers which have been slightly transformed by the layer mappings $\mathbf{g}_1(x,y)$ and $\mathbf{g}_2(x,y)$, respectively, a Glass pattern becomes visible in the superposition about each of the mutual fixed loci of the two mappings, i.e. about the points (x,y) for which $\mathbf{g}_M(x,y) = \mathbf{g}_1(x,y) - \mathbf{g}_2(x,y) = (0,0)$. ∎

A few remarks are, however, in order:

1. If only one of the layers (say, the first one) has been transformed, then we have $\mathbf{g}_1(x,y) = \mathbf{g}(x,y)$, $\mathbf{g}_2(x,y) = (x,y)$, and the mutual fixed loci of the two mappings reduce to the fixed loci of the transformation $\mathbf{g}(x,y)$ itself, as in Secs. 3.2–3.4. To see this, note that in this case the points which satisfy $\mathbf{g}_M(x,y) = \mathbf{g}_1(x,y) - \mathbf{g}_2(x,y) = (0,0)$ are precisely the points that satisfy $\mathbf{g}(x,y) = (x,y)$.

2. If the superposed layers are more than just *slightly* transformed with respect to each other, the correlation between them may become too low even very close to the fixed loci of the transformation. Although in such cases Glass patterns are still theoretically generated about the fixed loci, in practice these Glass patterns are no longer visible. For example, a rotation by a large angle such as 90° or 180° clearly has a fixed point in the center of rotation, but in practice no Glass pattern can be perceived about it in the superposition (see, for example, Fig. 2.6(a)). Another example, featuring a full fixed line, is shown in Fig. 3.22 and discussed in Problem 3-13.

3. More generally, it is clear that the fixed loci of a layer transformation $\mathbf{g}(x,y)$ (or the mutual fixed loci of the layer transformations $\mathbf{g}_1(x,y)$ and $\mathbf{g}_2(x,y)$) are independent of whether or not the superposed layers are correlated. However, according to Proposition 2.1, Glass patterns are only generated about these fixed loci if there exists some degree of correlation between the superposed layers.

4. If the original layers are *periodic* rather than aperiodic, then infinitely many additional replicas of the Glass patterns are generated in the superposition; these replicas are not located about fixed loci of the transformation.

5. As we have already seen in Sec. 3.2, there exist layer transformations having no fixed loci at all. But it turns out that Glass patterns may still exist in such cases, if the layer superposition has *almost fixed loci* rather than fixed loci. We will return to this point in Sec. 3.7.

6. The layer mappings $\mathbf{g}_1(x,y)$ and $\mathbf{g}_2(x,y)$ as well as the mapping $\mathbf{g}_M(x,y)$ are understood here as *domain* transformations. A more detailed explanation of this subject is provided in Sec. D.11 of Appendix D.

3.6 Synthesis of fixed loci in the superposition

In the previous examples we saw how the Glass patterns which appear in the superposition are related to the fixed loci of the layer mappings, and how we can find mathematically the fixed loci of the given layer mappings. But because the fixed loci determine the basic shape of the resulting Glass patterns in the layer superposition, it may be interesting to see if we can also go the other way around, and design a layer mapping that will produce a fixed locus (and hence a Glass pattern) of any desired shape.

In fact, all that we need to do for this end is to find a mapping (or equivalently, as explained in Sec. B.6 of Appendix B, a vector field) $\mathbf{k}(x,y)$ whose zeros are precisely the points (x,y) which form the desired locus. Equivalently, if we regard $\mathbf{k}(x,y)$ as being composed of two scalar functions, $\mathbf{k}(x,y) = (k_1(x,y), k_2(x,y))$ (see Sec. B.1 in Appendix B), then all we need is to find a system of two equations:

$$k_1(x,y) = 0$$
$$k_2(x,y) = 0 \tag{3.40}$$

whose solutions (x,y) form the desired locus in the x,y plane. Once we have found such a pair of equations (or the mapping $\mathbf{k}(x,y)$ whose components are $k_1(x,y)$ and $k_2(x,y)$), we simply choose two layer mappings $\mathbf{g}_1(x,y)$ and $\mathbf{g}_2(x,y)$ whose difference gives $\mathbf{k}(x,y)$. For example, we may choose:

$$\mathbf{g}_1(x,y) = (x,y) + \mathbf{k}(x,y)$$
$$\mathbf{g}_2(x,y) = (x,y) \tag{3.41}$$

or:

$$\mathbf{g}_1(x,y) = (x,y) + \tfrac{1}{2}\mathbf{k}(x,y)$$
$$\mathbf{g}_2(x,y) = (x,y) - \tfrac{1}{2}\mathbf{k}(x,y) \tag{3.42}$$

Now, if we apply the mappings $\mathbf{g}_1(x,y)$ and $\mathbf{g}_2(x,y)$, respectively, to two identical aperiodic layers (such as random dot screens) that were initially superposed exactly on top of each other, the transformed layers will have in the superposition precisely our desired fixed locus. The reason is that the mutual fixed locus $\mathbf{g}_1(x,y) - \mathbf{g}_2(x,y) = (0,0)$ (see Eq. (3.24)) consists exactly of the points (x,y) that satisfy $\mathbf{k}(x,y) = (0,0)$. And indeed, this method can be also used for visually illustrating the solutions of the given system of equations (3.40) (see Problem 3-11).[15]

Remark 3.3: It is interesting to note that the fixed locus remains practically unchanged whether the mappings $\mathbf{g}_1(x,y)$ and $\mathbf{g}_2(x,y)$ are applied to the given layers as domain transformations or as direct transformations. The reasons and the significance of this point are explained in Sec. D.11 of Appendix D. See also Problem 3-22. ∎

---

[15] Note that it is often best to use $\varepsilon\mathbf{k}(x,y)$ rather than $\mathbf{k}(x,y)$, where ε is a small positive fraction, in order not to lose the visual effect due to a loss of correlation between the two superposed layers.

Let us illustrate the synthesis of fixed loci and of their corresponding Glass patterns with a few examples.

Example 3.7: Suppose we wish to design a layer superposition whose mutual fixed locus consists of a given curve $y = f(x)$, say, a top-opened parabola $y = ax^2$.

For this end, we have to find a transformation $\mathbf{k}(x,y)$ (or equivalently, a pair of equations $k_1(x,y)$ and $k_2(x,y)$) whose zeros form the curve $y = ax^2$. One possible way of doing so consists of finding a surface $z = k_1(x,y)$ whose zero level line coincides with the curve $y = ax^2$. Such a surface is given, of course, by $z = y - ax^2$. We can choose, therefore, the pair of equations:

$$k_1(x,y) = y - ax^2$$
$$k_2(x,y) = 0 \qquad\qquad (3.43)$$

which corresponds to the transformation $\mathbf{k}(x,y) = (y - ax^2, 0)$. In this case we obtain, according to (3.41):[15]

$$\mathbf{g}_1(x,y) = (x,y) + (y - ax^2, 0) = (x + y - ax^2, y)$$
$$\mathbf{g}_2(x,y) = (x,y) \qquad\qquad (3.44)$$

These two transformations have, indeed, the desired mutual fixed locus, and when they are applied to two identical aperiodic screens they give in the superposition a curvilinear Glass pattern that lies along the curve $y = ax^2$ (see Fig. 3.16(a)).

Note, however, that this solution is by no means unique. For example, we could also choose the pair of equations:

$$k_1(x,y) = 0$$
$$k_2(x,y) = y - ax^2 \qquad\qquad (3.45)$$

which corresponds to the transformation $\mathbf{k}(x,y) = (0, y - ax^2)$. In this case we obtain, according to (3.41):[15]

$$\mathbf{g}_1(x,y) = (x,y) + (0, y - ax^2) = (x, 2y - ax^2)$$
$$\mathbf{g}_2(x,y) = (x,y) \qquad\qquad (3.46)$$

The transformation $\mathbf{g}_1(x,y)$ of this pair is clearly different from that of the pair (3.44); note, in particular, that the transformation $\mathbf{g}_1(x,y)$ in (3.44) is a non-linear variant of horizontal shearing, while the transformation $\mathbf{g}_1(x,y)$ in (3.46) is a non-linear variant of vertical scaling. And yet, both transformation pairs (3.44) and (3.46) give the same mutual fixed locus, which consists of the curve $y = ax^2$. The Glass patterns that are generated by these two transformation pairs are shown, respectively, in Figs. 3.16(a) and 3.16(c). Note that although the macroscopic Glass patterns in both cases are identical, the microstructure in each of the superpositions is different. ∎

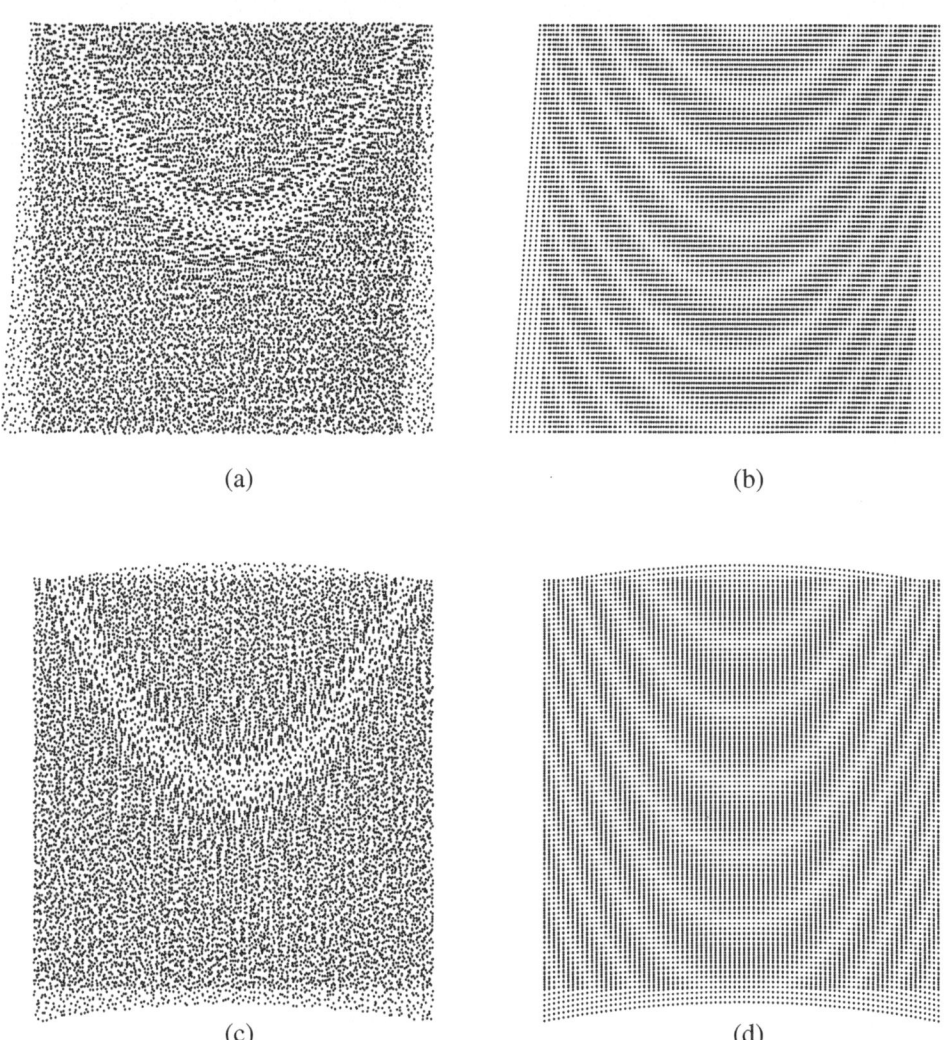

Figure 3.16: (a) The Glass pattern generated by applying the layer mappings (3.44) to two identical aperiodic dot screens. (c) The Glass pattern generated by applying the layer mappings (3.46) to the same identical aperiodic dot screens. (b) and (d) are the periodic counterparts of (a) and (c). Note that in (a) and (b) the transformation $\mathbf{g}_1(x,y)$ that is applied to the first layer is a non-linear variant of horizontal shearing, while in (c) and (d) the transformation $\mathbf{g}_1(x,y)$ is a non-linear variant of vertical scaling. In both cases the resulting macroscopic Glass patterns lie along the same curve $y = ax^2$, although the microstructures surrounding them are clearly different.

Example 3.8: As a second example, suppose we wish to design a layer superposition whose mutual fixed locus consists of a family of equispaced concentric circles around the origin. This locus is most easily expressed in terms of polar coordinates as the set of all the points (r,θ) that satisfy the equation $\sin r = 0$. How can we find a transformation $\mathbf{k}(x,y)$ whose solutions are precisely given by $\sin r = 0$? We start by seeking such a transformation in terms of the polar coordinates r,θ, and once we have found one we will convert it back into the Cartesian coordinates x,y.

Consider the transformation $\mathbf{k}(r,\theta)$ whose two components are given by:

$$k_1(r,\theta) = r\cos\theta$$
$$k_2(r,\theta) = r\sin\theta$$

(3.47)

Clearly, the system (3.47) has its solutions at the points where both $r\cos\theta$ and $r\sin\theta$ are zero. But since $\cos\theta$ and $\sin\theta$ are never zero at the same angle θ, it turns out that the transformation $\mathbf{k}(r,\theta)$ has a zero only where $r = 0$, i.e. at the origin. In order to obtain a new transformation with more interesting zeros, we may replace r in (3.47) with any desired function $z = f(r)$, or even $z = f(r,\theta)$. The resulting transformation will have its zeros wherever $f(r,\theta) = 0$. For example, if we take $f(r,\theta) = \sin r$ we obtain the system:

$$k_1(r,\theta) = \sin r\cos\theta$$
$$k_2(r,\theta) = \sin r\sin\theta$$

(3.48)

whose zeros occur, exactly as we desired, wherever $\sin r = 0$.[16]

Now, in order to convert this transformation back to Cartesian coordinates we substitute $r = \sqrt{x^2 + y^2}$, $\cos\theta = \dfrac{x}{\sqrt{x^2 + y^2}}$ and $\sin\theta = \dfrac{y}{\sqrt{x^2 + y^2}}$, which gives us:

$$k_1(x,y) = \frac{x}{\sqrt{x^2 + y^2}}\sin\sqrt{x^2 + y^2}$$
$$k_2(x,y) = \frac{y}{\sqrt{x^2 + y^2}}\sin\sqrt{x^2 + y^2}$$

(3.49)

Clearly, the solutions of the system of equations $k_1(x,y) = 0$, $k_2(x,y) = 0$ give our desired fixed locus in terms of the Cartesian coordinates x,y. Therefore, we can take $k_1(x,y)$ and $k_2(x,y)$ as the two components of our desired transformation $\mathbf{k}(x,y)$.

Now, having found $\mathbf{k}(x,y)$, we can choose, for example, according to (3.41):[17]

$$\mathbf{g}_1(x,y) = (x,y) + \mathbf{k}(x,y)$$

$$\mathbf{g}_2(x,y) = (x,y)$$

---

[16] Note that when viewed over the x,y plane, both $z = \sin r\cos\theta$ and $z = \sin r\sin\theta$ are surfaces having a radial profile of $\sin r$, so both of them intersect the x,y plane on all concentric circles around the origin whose radius r satisfies $\sin r = 0$. In addition, $z = \sin r\cos\theta$ intersects the plane along the radial lines whose angles θ satisfy $\cos\theta = 0$, and $z = \sin r\sin\theta$ intersects the plane along the radial lines whose angles θ satisfy $\sin\theta = 0$; but these radial lines are not commom zeros to both surfaces, and therefore they are not zeros of the system (3.48).

[17] Once again, it may be better to replace here $\mathbf{k}(x,y)$ by $\varepsilon\mathbf{k}(x,y)$, where ε is a small positive fraction, in order not to lose the correlation between the two superposed layers.

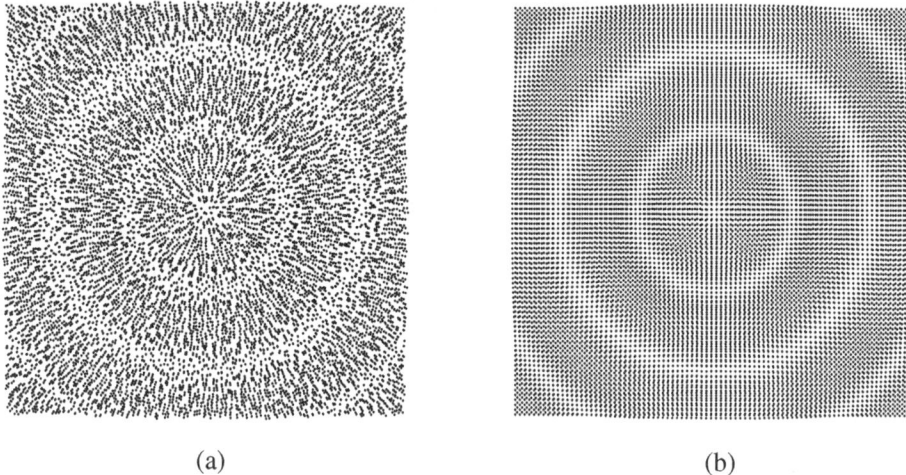

(a)   (b)

Figure 3.17: (a) An example with a fixed locus consisting of an isolated point at the origin and a family of equispaced concentric circles surrounding it. This example has been synthesized by applying appropriate layer transformations to two identical aperiodic dot screens, as explained in Example 3.8. (b) The superposition of two identical periodic dot screens, which have undergone the same transformations as in (a).

If we apply the mappings $\mathbf{g}_1(x,y)$ and $\mathbf{g}_2(x,y)$ to our two initially identical layers, then the resulting fixed points between them will include all the points for which $\mathbf{k}(x,y) = (0,0)$. This locus consists of all the points (x,y) where $\sin\sqrt{x^2 + y^2} = 0$, namely, the origin $(0,0)$ plus a family of equispaced concentric circles around it (see Fig. 3.17(a)). ■

This example illustrates, indeed, a rather complex fixed locus which consists of an isolated point at the origin plus an infinite number of curved lines surrounding it. Other interesting examples, including an example that generates the figure on the front cover of this book, are presented in Problems 3-16–3-20.

3.7 Almost fixed points

We have seen throughout this chapter the central role that fixed points (or fixed loci) play in the understanding of Glass patterns. However, as we have already hinted in the beginning of this chapter, Glass patterns can be sometimes generated even in cases where fixed points do not exist. This may happen when the layer transformations do not have real mutual fixed points, but only *almost fixed points*, i.e. points where $\mathbf{g}_1(x_F,y_F)$ is not fully identical to $\mathbf{g}_2(x_F,y_F)$ but only very close to it. If several such points exist, they form together an *almost fixed locus*. Although in almost fixed points there is no perfect coincidence between the two superposed layers, the elements of both layers around such points still fall very close to each other, while farther away the correlation gradually decreases. This generates about the almost fixed point (or locus) a visible Glass pattern whose center is just slightly darker than in the case of perfect coincidence.

As far as Glass patterns are concerned, almost fixed points are a straightforward generalization of fixed points. The reason we have concentrated so far on cases having real fixed points (or loci) is purely didactic, because these are the simplest study cases for the understanding of Glass patterns. But once we have understood the nature and the main properties of Glass patterns involving fixed points, we may now complete our discussion and extend it to the case of Glass patterns involving almost fixed points.[18] We will do so with the help of the following examples.

Example 3.9: A single almost fixed point at the origin:

Suppose we apply to one of the superposed layers a scaling transformation, while the other layer remains unchanged. Therefore, the two layer transformations being used are:

$$\mathbf{g}_1(x,y) = (sx,sy) = (s\,r\cos\theta,\, s\,r\sin\theta)$$

$$\mathbf{g}_2(x,y) = (x,y) = (r\cos\theta,\, r\sin\theta)$$

The reason for using here the notation involving polar coordinates will become clear shortly. These transformations have a single mutual fixed point (a *real* fixed point) at the origin, the only point where $\mathbf{g}_1(x,y) - \mathbf{g}_2(x,y) = (0,0)$. And indeed, as we can see in Fig. 3.18(a), when $\mathbf{g}_1(x,y)$ and $\mathbf{g}_2(x,y)$ are applied to two identical aperiodic dot screens, they generate in the superposition a clearly visible Glass pattern about this fixed point. Note that in the following we will assume that $s > 1$; for the case of $0 < s < 1$ see Problem 3-24.

Now, what happens if we replace sr in $\mathbf{g}_1(x,y)$ by $s(r + \delta)$, where δ is a small positive constant? In this case the transformation $\mathbf{g}_1(x,y)$ becomes a *non-linear* radial scaling, and the fixed points occur where $\mathbf{g}_1(x,y) - \mathbf{g}_2(x,y) = ([s(r + \delta) - r]\cos\theta,\ [s(r + \delta) - r]\sin\theta) = (0,0)$. This happens where $f(r) = s(r + \delta) - r = 0$, i.e. where $r = -s\delta/(s - 1)$. But because $s > 1$, $f(r)$ has no zeros for any $r \geq 0$. It follows, therefore, that the layer transformations $\mathbf{g}_1(x,y)$ and $\mathbf{g}_2(x,y)$ have no mutual fixed points; and yet, because δ is small, it is clear that around the origin the difference $\mathbf{g}_1(x,y) - \mathbf{g}_2(x,y)$ closely approaches $(0,0)$. This means that the origin is an almost fixed point of $\mathbf{g}_1$ and $\mathbf{g}_2$. And indeed, as we can clearly see in Fig. 3.18, when δ differs from zero just slightly a Glass pattern is still visible around the origin, because the elements from both layers still fall around the origin very close to each other. But as we gradually increase δ, the macroscopic Glass pattern becomes less and less visible, and finally it completely disappears. It is interesting to note, however, that the dot trajectories around the Glass pattern survive longer than the macroscopic Glass pattern itself: As we can clearly see in Fig. 3.18, the dot trajectories only disappear at a higher value of δ. This reminds us of our earlier observations about the behaviour of the

---

[18] We could have called such Glass patterns "almost Glass patterns", but in fact the distinction between "real" and "almost" Glass patterns is not quite necessary, and we will continue to call them simply "Glass patterns".

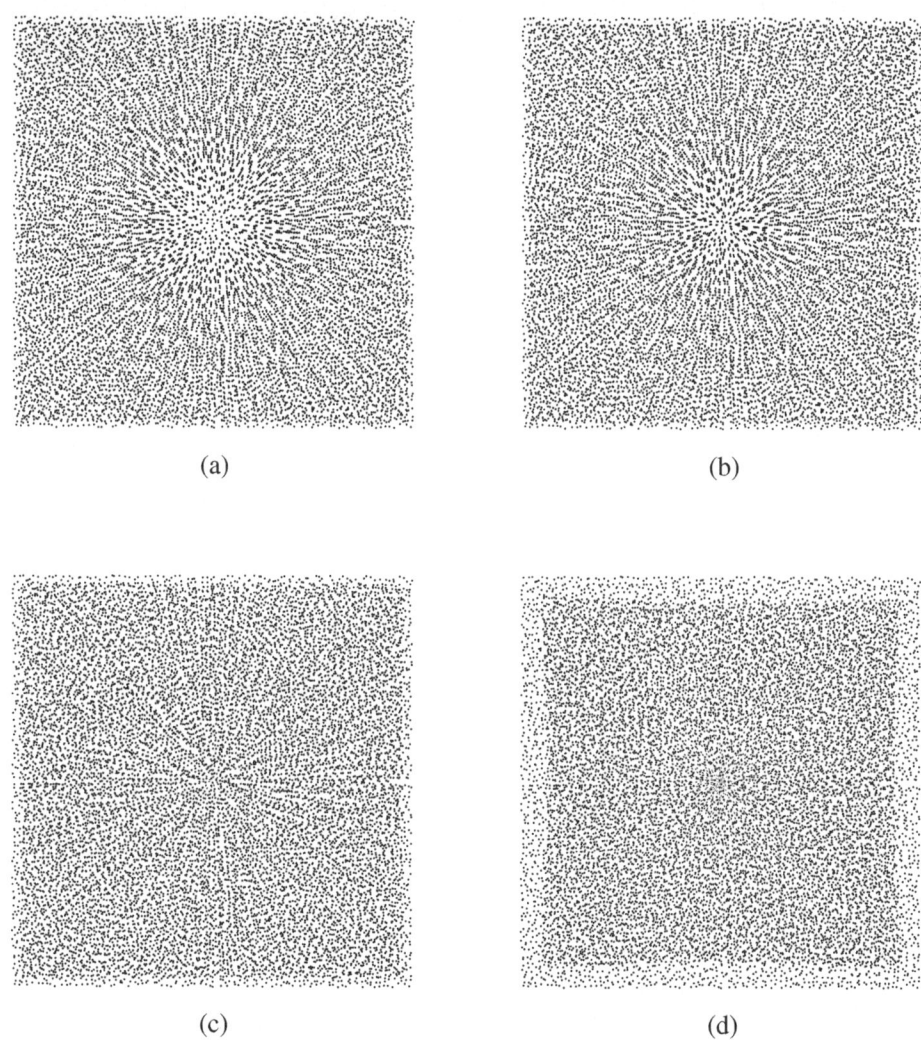

(a)   (b)

(c)   (d)

Figure 3.18: A superposition with an almost fixed point at the origin (see Example 3.9). In all parts of this figure one of the superposed layers undergoes a weak radial scaling $(s(r + \delta)\cos\theta, s(r + \delta)\sin\theta)$: in (a) $\delta = 0$, while in (b)–(d) the value of δ gradually increases. As shown in (a), when $\delta = 0$ the superposed layers have a real fixed point at the origin, and a visible Glass pattern appears about it. When δ slightly increases the fixed point is destroyed and turns into an *almost* fixed point. And yet, as shown in (b), as long as δ is sufficiently small, this almost fixed point is still surrounded by a clearly visible Glass pattern. When we further increase δ the macroscopic Glass pattern becomes less and less visible, and finally it completely disappears. As shown in (c), the dot trajectories survive longer than the macroscopic Glass pattern, and they only disappear at a higher value of δ, as shown in (d).

microstructure around a real fixed point in the superposition (see, for example, Fig. 2.1(c)): Here, too, the dot trajectories fade out and disappear farther away from the fixed point than the macroscopic gray level differences. This means, indeed, that the dot trajectories provide a better and more sensitive indication of correlation than the macroscopic gray level variations. ■

Example 3.10: An almost fixed line along the x axis:

As a second example, we return to the case already described in Example 3.2 above. Consider the superposition of two identical aperiodic layers, where one of the layers, say, the first one, undergoes a slight horizontal shear transformation. As shown in Fig. 3.19(a) (or 2.3(c)), this gives a horizontal fixed line along the x axis, which is surrounded by a clearly visible linear Glass pattern. For the sake of convenience, we repeat here the mathematical explanation given in Example 3.2, using a slightly different notation.

The transformations undergone by the two layers in this case are given by:

$$\mathbf{g}_1(x,y) = (x - ay, y)$$

$$\mathbf{g}_2(x,y) = (x, y)$$

The locus of their mutual fixed points is given by the set of all the points (x,y) that satisfy $\mathbf{g}_1(x,y) = \mathbf{g}_2(x,y)$; these points are the solutions of the system of equations:

$$x = x - ay$$

$$y = y$$

and they consist, indeed, of the line $y = 0$, i.e. the entire x axis.

Now, what happens if we slightly shift one of the layers (say, the unsheared one) by y_0 downward? This clearly destroys the horizontal fixed line, since after having applied this shift no points of the two layers remain in full coincidence. To see this mathematically, we note that our layer transformations become after the shift:

$$\mathbf{g}_1(x,y) = (x - ay, y)$$

$$\mathbf{g}_2(x,y) = (x, y + y_0)$$

The locus of their mutual fixed points is given, therefore, by the points (x,y) that satisfy:

$$x = x - ay$$

$$y = y + y_0$$

but in this case, because of the shift y_0, the second equation is inconsistent, and hence no points (x,y) can satisfy the system. This means, indeed, that in this case there exist no fixed points at all. And yet, as we can see in Fig. 3.19(b), if the vertical shift y_0 is small enough, then along the x axis the elements from both layers still fall very close to each other, while

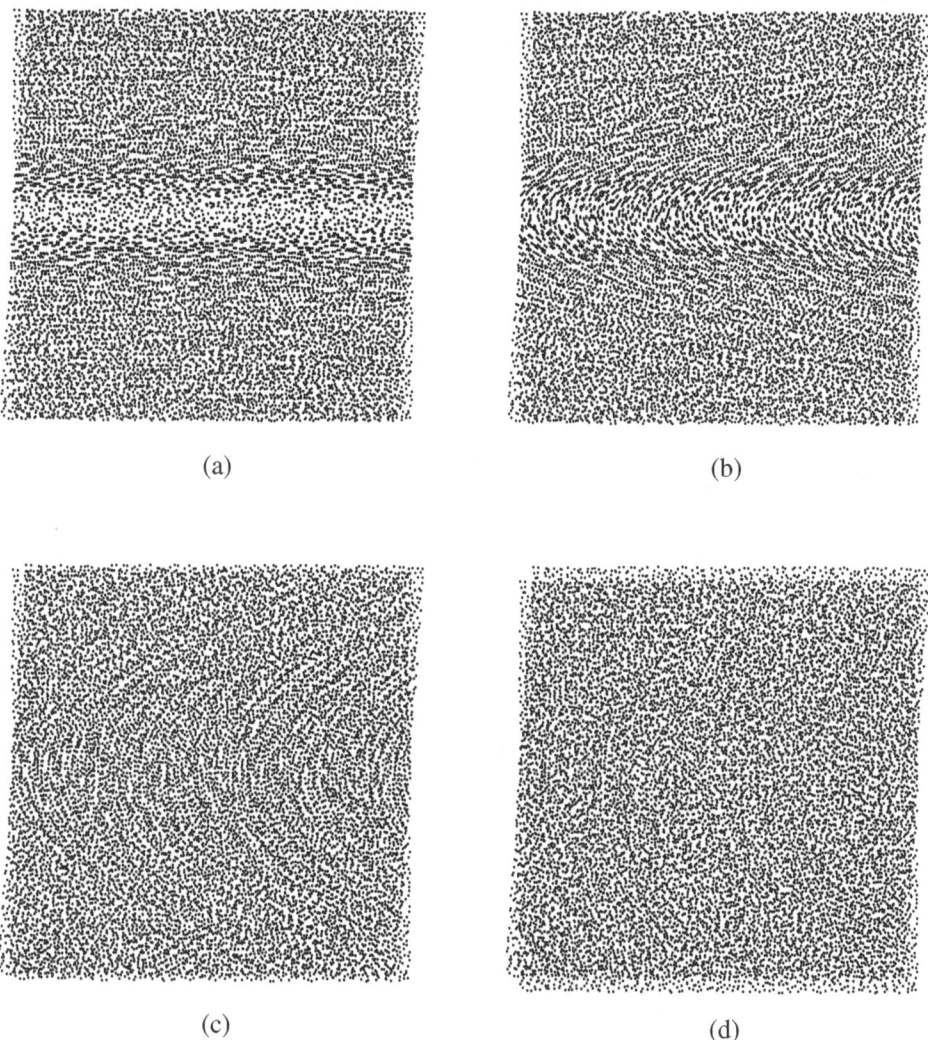

(a)

(b)

(c)

(d)

Figure 3.19: A superposition with an almost fixed line along the x axis (see Example 3.10). In all parts of this figure one of the superposed layers has undergone a slight horizontal shear transformation, $\mathbf{g}_1(x,y) = (x - ay, y)$, while the other layer is shifted down by y_0, with $y_0 = 0$ in (a), and with gradually increasing values of y_0 in (b)–(d). As shown in (a), when $y_0 = 0$ the superposed layers have a fixed line along the x axis, and a visible linear Glass pattern appears along it. When y_0 slightly increases the fixed line along the x axis is destroyed and turns into an *almost* fixed line. And yet, as shown in (b), as long as the shift y_0 is sufficiently small, this almost fixed line is still surrounded by a clearly visible Glass pattern. But when we gradually increase y_0 the macroscopic Glass pattern becomes less and less visible, and finally it completely disappears, as shown in (c). Note that the dot trajectories survive longer than the macroscopic Glass pattern, and they only disappear at a higher value of y_0. See also Problem 4-10.

farther away from the x axis the correlation between the two layers gradually decreases. Thus, although the x axis is no longer a fixed line, it still remains an *almost* fixed line; and indeed, as we can see in Fig. 3.19(b), it is still surrounded by a visible linear Glass pattern. How can we see this mathematically?

Although in this case no points (x,y) satisfy $\mathbf{g}_1(x,y) = \mathbf{g}_2(x,y)$, i.e. $\mathbf{g}_1(x,y) - \mathbf{g}_2(x,y) = (0,0)$, we can still find those points for which the difference $\mathbf{k}(x,y) = \mathbf{g}_1(x,y) - \mathbf{g}_2(x,y)$ is *almost* zero, or, more precisely, the points (x,y) for which we obtain the minimum of this difference.

Hence, if we denote the components of the difference $\mathbf{k}(x,y)$ by $k_1(x,y)$ and $k_2(x,y)$:

$$\mathbf{k}(x,y) = (k_1(x,y), k_2(x,y))$$

then what we are looking for is the loci of the minima of the function $k(x,y)$:

$$k(x,y) = \sqrt{k_1(x,y)^2 + k_2(x,y)^2}$$

that gives the length of the vector $\mathbf{k}(x,y)$ (or, in other words, the distance of $\mathbf{k}(x,y)$ from $(0,0)$). Alternatively, for the sake of simplicity, we may prefer to find the loci of the minima of its squared version, $n(x,y)$:

$$n(x,y) = k_1(x,y)^2 + k_2(x,y)^2 \tag{3.50}$$

In our present example we have:

$$\mathbf{k}(x,y) = \mathbf{g}_1(x,y) - \mathbf{g}_2(x,y) = (-ay, -y_0)$$

so that $k_1(x,y) = -ay$ and $k_2(x,y) = -y_0$, and therefore the function $n(x,y)$ is given by:

$$n(x,y) = a^2y^2 + y_0^2$$

And indeed, it is clear that this surface gets its minima along the line $y = 0$, the x axis.

Thus, although our transformations $\mathbf{g}_1$ and $\mathbf{g}_2$ do not have any mutual fixed points, they still have an *almost* fixed locus which is situated along the entire x axis. When the shift y_0 is reduced to zero, these minima become real zeros of $\mathbf{k}(x,y)$, and we obtain along the x axis a real fixed line rather than an almost fixed line. But when the shift y_0 departs from zero and starts getting bigger, the high correlation between the two layers along the x axis gradually drops, and the linear Glass pattern gradually fades out until at a certain value of y_0 it is no longer perceptible (see Fig. 3.19).

It is interesting to note, however, that microstructure dot trajectories may still be visible in the superposition even when the macroscopic Glass pattern (the macroscopic gray level variation) is no longer perceptible, and they only disappear at higher values of the shift y_0 (see Fig. 3.19). The parabolic shape of the dot trajectories in the present example will be explained in Chapter 4 (see Example 4.9). ■

Example 3.11: A pair of almost fixed lines running in parallel to the x axis:

We consider now the interesting case of Fig. 3.8(a) (the inversed S-like dot trajectories), which is reproduced here, for the sake of convenience, in Fig. 3.20. As we have seen in Example 3.3, the layer transformations undergone by the two layers in this case consist of a parabolic transformation in the first layer, and a shift of (x_0, y_0) in the other layer:

$$\mathbf{g}_1(x,y) = (x - ay^2, y)$$

$$\mathbf{g}_2(x,y) = (x - x_0, y - y_0)$$

so that the difference $\mathbf{k}(x,y)$ is:

$$\mathbf{k}(x,y) = \mathbf{g}_1(x,y) - \mathbf{g}_2(x,y) = (x_0 - ay^2, y_0)$$

As we did in Example 3.3, let us first consider the case in which $y_0 = 0$. In this case the mutual fixed points of $\mathbf{g}_1$ and $\mathbf{g}_2$ are given by the system of equations (3.20):

$$x - x_0 = x - ay^2$$

$$y = y$$

whose solutions consist of all the points (x,y) that satisfy:

$$y = \pm\sqrt{x_0/a}$$

(see Eq. (3.21)). Assuming that x_0 and a have the same sign, this means that the fixed locus consists here of two parallel horizontal lines whose vertical distances from the x axis are given by $\pm\sqrt{x_0/a}$. And indeed, as shown in Fig. 3.20(a), two linear Glass patterns are clearly visible along these horizontal lines.

Now, suppose that we add to the shifted layer a slight vertical shift of y_0. This clearly destroys the two horizontal fixed lines, since when $y_0 \neq 0$ the mutual fixed points of $\mathbf{g}_1$ and $\mathbf{g}_2$ are given by the system of equations:

$$x - x_0 = x - ay^2$$

$$y - y_0 = y$$

but this system has no solutions because its second equation is inconsistent. And yet, as we can see in Fig. 3.20(b), if the vertical shift y_0 is small enough, then along the two horizontal lines $y = \pm\sqrt{x_0/a}$ the elements of both layers still fall very close to each other, while farther away from these lines the correlation between the two layers gradually decreases. Thus, although these two lines are no longer fixed lines, they still remain *almost* fixed lines; and indeed, as we can see in Fig. 3.20(b), they are still surrounded by visible linear Glass patterns.

Thus, although in this case no points (x,y) satisfy $\mathbf{g}_1(x,y) = \mathbf{g}_2(x,y)$, i.e. $\mathbf{g}_1(x,y) - \mathbf{g}_2(x,y) = (0,0)$, we can still find those points for which the difference $\mathbf{k}(x,y) = \mathbf{g}_1(x,y) - \mathbf{g}_2(x,y)$ is *almost* zero, or, more precisely, the points (x,y) for which we obtain the minimum of this difference.

To do so, we consider again the function $n(x,y)$ (Eq. (3.50)). In our present case $n(x,y)$ is given by:

$$n(x,y) = (x_0 - ay^2)^2 + y_0^2$$

$$= a^2y^4 - 2ax_0y^2 + x_0^2 + y_0^2$$

Clearly, the minima of $n(x,y)$ are obtained where its partial derivatives:

$$\frac{\partial}{\partial x}n(x,y) = 0$$

$$\frac{\partial}{\partial y}n(x,y) = 4a^2y^3 - 4ax_0y$$

are both zero, i.e. where:

$$4a^2y^3 - 4ax_0y = 0$$

This happens either for $y = 0$ or for $y = \pm\sqrt{x_0/a}$. A quick verification shows that $y = 0$ is a local maximum, while $y = \pm\sqrt{x_0/a}$ are, indeed, local minima of $n(x,y)$. This means that although our layer superposition does not have any fixed loci, it does have *almost* fixed loci which are situated along the two parallel horizontal lines $y = \pm\sqrt{x_0/a}$. If the vertical shift of y_0 in the first layer is reduced to zero, these minima become real zeros of $\mathbf{k}(x,y)$, and hence they become real fixed lines rather than almost fixed lines. This can be clearly seen in the layer superpositions by comparing Figs. 3.20(b) and 3.20(a). Note that here, too, the macroscopic Glass patterns are only visible for values of y_0 that are very close to 0; but the dot trajectories surrounding the almost fixed lines survive longer, and they still remain visible even in higher values of y_0 (see Fig. 3.20(c)). The inversed S-like shape of these dot trajectories will be explained in Chapter 4 (see Example 4.10). ■

Example 3.12: No almost fixed points at all:

As a last example, consider the superposition of two identical aperiodic layers one of which, say, the second one, undergoes a small vertical shift of y_0 (see Fig. 2.5(a)). In this case we have:

$$\mathbf{k}(x,y) = (x,y) - (x,y-y_0) = (0,y_0)$$

Here the difference $\mathbf{k}(x,y)$ has a constant value of $(0,y_0)$ throughout the plane, so that it never becomes zero, and moreover, it doesn't have any minima. Therefore, in this case there exist no fixed points and no almost fixed points, so that no Glass patterns (macroscopic gray level variations) are generated in the superposition. And yet, as long as the shift y_0 is small enough, vertical dot trajectories are clearly visible throughout the superposition (see Fig. 2.5(a)). This shows, once again, that dot trajectories can exist in the superposition even when no macroscopic Glass patterns are generated. In fact, this case is a singular moiré-free superposition, as explained in Sec. 2.3.2. ■

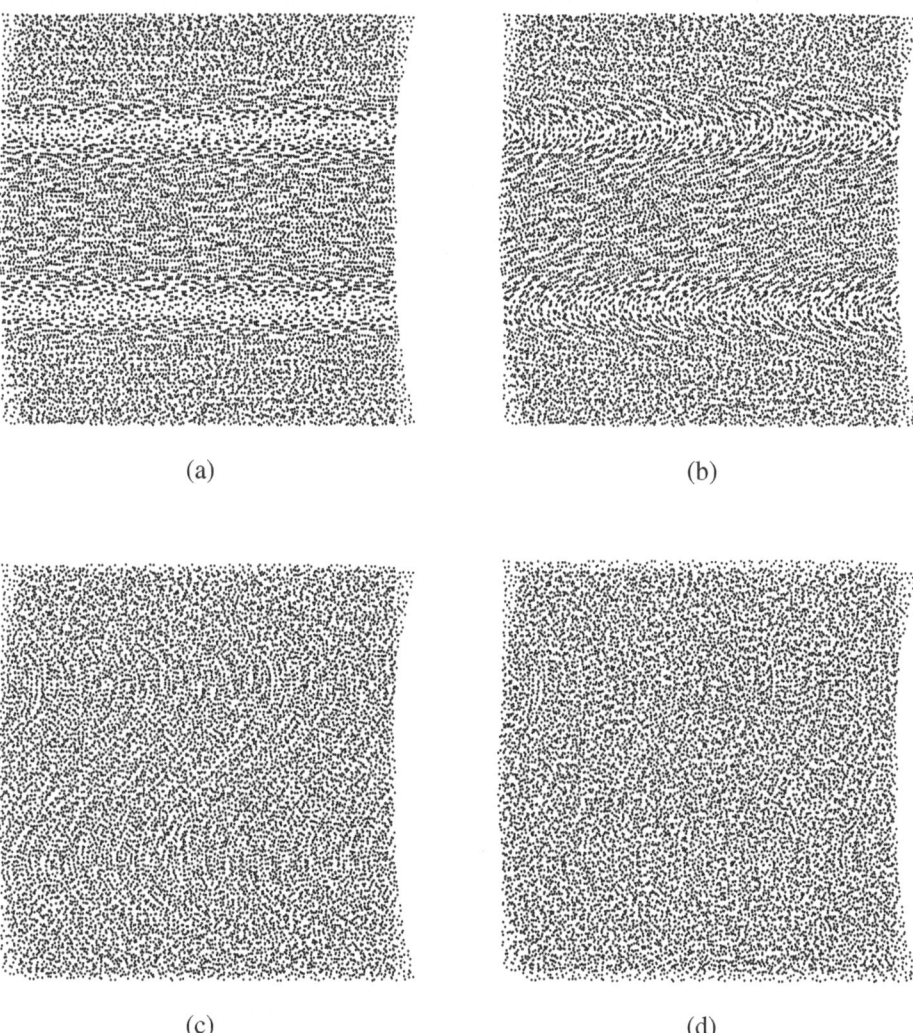

(a)   (b)

(c)   (d)

Figure 3.20: A superposition with a pair of almost fixed lines (see Example 3.11). In all parts of this figure one of the superposed layers has undergone a weak parabolic transformation, $\mathbf{g}_1(x,y) = (x - ay^2, y)$, while the other layer is shifted by (x_0, y_0), with $y_0 = 0$ in (a), and with gradually increasing values of y_0 in (b)–(d). The value of x_0 remains constant in all cases. As shown in (a), when $y_0 = 0$ the superposed layers have a pair of fixed lines running parallel to the x axis, and visible linear Glass patterns appear along them. When y_0 slightly increases the fixed lines are destroyed and they turn into *almost* fixed lines. And yet, as shown in (b), as long as the shift y_0 is sufficiently small, these almost fixed lines are still surrounded by clearly visible Glass patterns. But when we gradually increase y_0 the macroscopic Glass patterns become less and less visible, and finally they completely disappear, as shown in (c). Note that the dot trajectories survive longer than the macroscopic Glass patterns, and they only disappear at a higher value of y_0. See also Problem 4-10.

Finally, it should be noted that turning fixed loci into almost fixed loci is just one possible way of decorrelating two superposed layers (i.e., reducing the correlation between them). Another possible way of decorrelation consists of adding to one or both of the original layers an increasing rate of random noise (see Fig. 3.21). Interestingly, in all of these cases the microstructures (dot trajectories) resist better to the decorrelation than the macrostructures, and hence they provide a better indication to the existence of correlation between the superposed layers than the macrostructures.

PROBLEMS

3-1. *Behaviour of Glass patterns under layer shifts.* Suppose we are given two identical random dot screens that are superposed on top of each other in full coincidence, dot on dot. We rotate the first screen by angle $\alpha/2$ and the other screen by angle $-\alpha/2$, and obtain a Glass pattern at the origin. Now, we shift the second layer forward or backward along the x axis. Show that as a result the Glass pattern will move downward or upward, respectively, along the y axis. What happens to the Glass pattern when we shift the second layer forward or backward along the y axis, or along the main diagonal $y = x$? How will the Glass pattern move in each of these cases if the first layer is also scaled up by the factor 1.1?

3-2. What would you expect to see in Figs. 3.2 and 3.3 if layer B were a *scaled-up* version of layer A rather than a *rotated* version of layer A? And if it were a *scaled-down* version of layer A?

3-3. *Determination of the axis of rotation between two moving parts.* Can you think of a method for the determination of the axis of rotation between two moving parts using Glass patterns? Such a method, which was reportedly used in dental surgery for the correction of certain types of deformity, is described in [Walker80a].

3-4. *Alignment of similar images.* It is often desired to align two slightly different images of the same subject (for example, images obtained from the same patient at different times or with different imaging techniques, a satellite and an aerial view of the same area, or two images that have been taken from slightly different angles or using different photographic equipment). Can you think of a method based on Glass patterns for aligning such images? *Hint:* Such a method has been provided in [Schuette97]: By superposing the two original images with a small rotation one obtains a Glass pattern whose center indicates a common point in both images. Then, by slightly shifting one of the superposed layers, one obtains a Glass pattern around another point that is common to both images. The two images can be aligned after having found in them (at least) two different common points. If one of the two original images is non-linearly transformed with respect to the other, this method may be used to find a certain number of common reference points between the two images, that can be used later as control points for scattered data interpolation (see, for example, [Amidror02a]).

3-5. *High-precision registration.* How can you use two identical random screens for the high-precision registration of two objects? How can you improve this method by using instead a random screen and its negative image? (see, for example, [Dey91]).

3-6. *Comparing the scale of two copies of an image.* How can you use Glass patterns to determine if two copies of an image are of the same scale, or if one of them has been slightly scaled in one direction or in two directions?

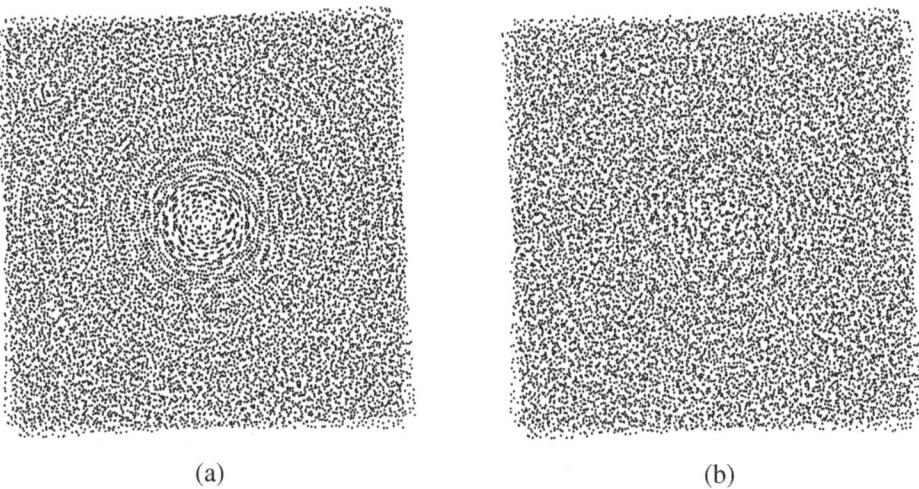

(a) (b)

Figure 3.21: Microstructures (dot trajectories) resist better to decorrelation than the corresponding macrostructures (macroscopic gray level variations). (a) The superposition of two identical aperiodic dot screens with a small angle difference. (b) The same superposition after the addition of some random noise to the dot locations in one of the superposed layers. Note that the circular dot trajectories are still visible in (b), although the macroscopic gray level variations have almost fully disappeared. A further increase in the random noise level will finally destroy the dot trajectories, too.

3-7. *Detecting small displacements.* How can you use Glass patterns to detect small displacements between two objects? Can you also determine the direction of the displacement?

3-8. *Latent images.* Techniques for the generation and the detection of latent images (as used for document authentication) are usually based on the superposition of periodic structures (see, for example, Problems 7-8 to 7-17 at the end of Chapter 7 in *Vol. I*). But similar techniques have been also devised for use in the superposition of aperiodic, random structures [Hines75; McGrew95]. What are the main differences between latent images that are produced by periodic or by random structures? *Hint:* What happens in each case when the detecting layer is slightly shifted or rotated?

3-9. *Glass patterns as a limit case of moiré patterns.* Looking at Figs. 2.1(c),(d) a student suggests to explain the Glass pattern in (c) as a limit case of the periodic moiré effect in (d) which is obtained when the period T of the two superposed layers tends to infinity: According to Eq. (2.10) in *Vol. I*, $T_M = \dfrac{T}{2\left|\sin\left(\alpha/2\right)\right|}$, it follows that when $T \to \infty$ and the angle α is kept fixed, the period T_M of the moiré tends to infinity. What do you think of this explanation?

3-10. *Visual illustration of the fixed loci of a transformation.* The precise calculation of the fixed loci of non-linear transformations may be quite complicated or even intractable, depending on the complexity of the transformations involved. Can you devise a method for visually illustrating the fixed loci of a transformation, based on the superposition of aperiodic dot screens? What are its limitations?

3-11. *A method for visualizing the solutions of a system of equations.* Can you devise a method for visually illustrating the solutions of a system of two equations $k_1(x,y) = 0$, $k_2(x,y) = 0$, namely, $\mathbf{k}(x,y) = (0,0)$? *Hint*: Apply the layer transformations (3.41) or (3.42) to two identical aperiodic dot screens, and observe the fixed points (the centers of the Glass patterns) in the superposition. What are the limitations of this method? Can it be used for any given transformation, or only for weak transformations?

3-12. *A method for visualizing the solutions of an equation* $f(x,y) = 0$. Can you devise a method for visually illustrating the zeros of the equation $f(x,y)$, i.e. the curve $f(x,y) = 0$? *Hint*: Take in the previous problem $\mathbf{k}(x,y) = (f(x,y),0)$. See also Example 3.7.

3-13. *A fixed line with no visible Glass pattern in the superposition of aperiodic screens.* Fig. 3.22(a) shows the superposition of two identical aperiodic dot screens, one of which has been flipped about the y axis and placed face down on top of the other layer. Fig. 3.22(c) shows a similar case with an additional rotation of $\alpha = 5°$ between the superposed layers. Since in these cases no correlation exists between the two superposed layers, no Glass patterns are generated in the superposition. And yet, a close inspection of the figures clearly reveals a vertical line of coincidence passing through the center, along which some symmetrical artifacts can be clearly distinguished in the microstructure.

(a) How do you explain this phenomenon? (*Hint:* in both cases the superposition has a full line of fixed points, passing through the center; see also Sec. A.2 of Appendix A, point (2)).

(b) Find the angle of this fixed line as a function of the rotation angle α. Note that this fixed line remains equally visible for all rotation angles α, while a real Glass pattern disappears when the rotation angle α increases. How do you explain this difference?

(c) Parts (b) and (d) of the figure show the behaviour of the superposition in the corresponding periodic cases. How do you explain the difference between the aperiodic and the periodic cases?

(d) What do you expect to see if we also shift one of the layers on top of the other?

3-14. *A hybrid case with both a Glass pattern and a 1-fold periodic moiré.* Fig. 3.23(a) shows an aperiodic image consisting of an arbitrary printed text. When an identical copy of this image is superposed on top of the first one with a small angle difference, a circular Glass pattern becomes clearly visible in the superposition (Fig. 3.23(b)). If we continue rotating the second layer, the Glass pattern becomes smaller and weaker, but new vertical periodic moiré bands start to show up (Fig. 3.23(c)), whose period becomes smaller and smaller as we continue the rotation.

(a) How do you explain this phenomenon? *Hint:* The original superposed layers are not really 2D-randomly distributed structures, since they consist of horizontal *text lines* which are separated by fixed inter-line spaces. Viewed from a distance, the text lines in the two layers may be seen as horizontal line gratings that generate in the superposition their own periodic moiré bands. In fact, this is simply a hybrid case, in which both periodic and aperiodic moirés are visible simultaneously. Can we learn from Fig. 3.23(c) anything about the radius of a Glass pattern, given that we already know how to calculate quantitatively the period of the vertical moiré bands, using Eq. (2.10) of *Vol. I* (namely, $T_M = \dfrac{T}{2\,|\sin(\alpha/2)|}$)?

(b) Suppose now that we place one of the aperiodic layers face down on top of the other. In this case, the Glass pattern completely disappears, but the vertical moiré bands remain clearly visible (see Fig. 3.23(d) or Fig. 2 in [Garavaglia01]). How do you explain this result?

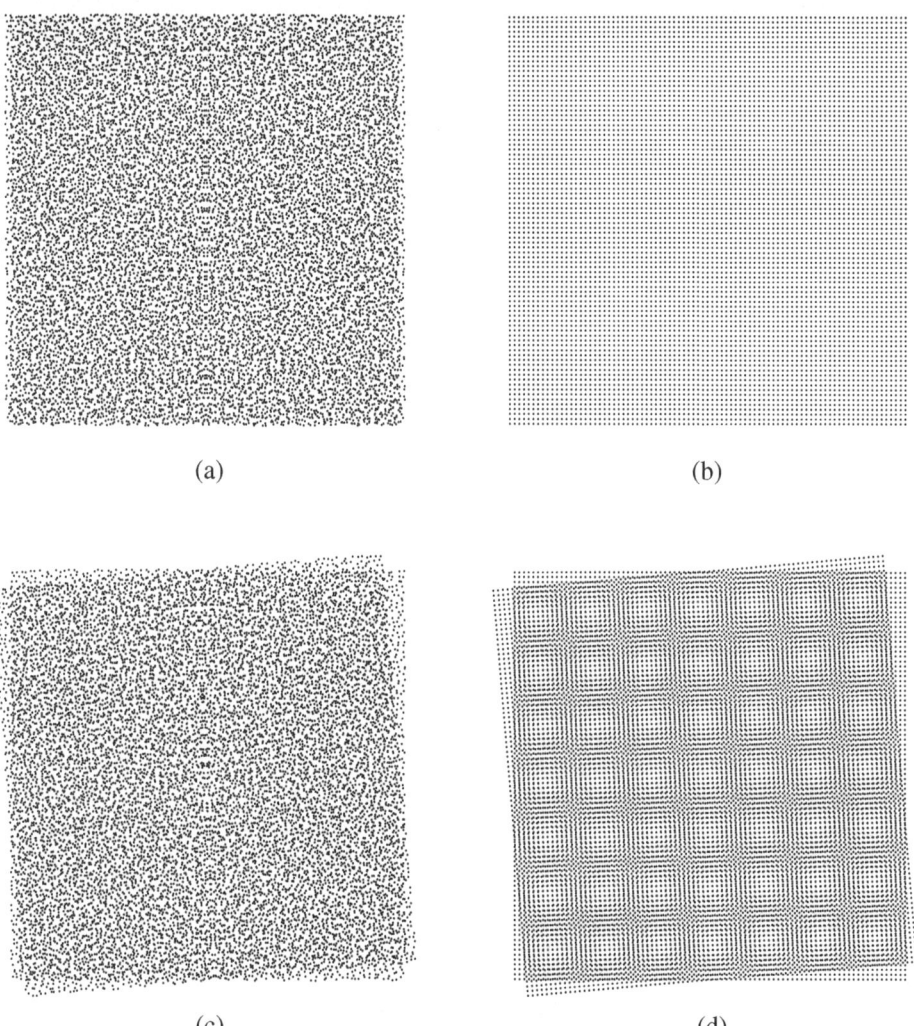

Figure 3.22: (a) The superposition of two identical aperiodic dot screens, one of which has been flipped about the y axis. This layer transformation corresponds, in fact, to the linear mapping $\mathbf{g}(x,y) = (-x,y)$, i.e. mirror imaging of the layer about the y axis. The y axis itself is clearly a fixed line of this transformation. Note, however, that the horizontal dot trajectories that appear along this fixed line are only visible in the layer superposition very close to the y axis, because farther away the correlation between the two layers immediately vanishes. (b) The counterpart of (a) with periodic layers; note that in this case the dots of the two layers coincide. (c) The same superposition as in (a), but with an additional rotation of $\alpha = 5°$ between the superposed layers. (d) The counterpart of (c) with periodic layers. Note the perfect symmetry of the microstructures about the fixed line in both periodic and aperiodic cases.

(c) What would you expect to see in cases (a) and (b) if the letters were spatially distributed in a random way, instead of being ordered in horizontal lines?

3-15. What do you expect to see in Figs. 3.10–3.12 when the layer rotation is applied to the opposite direction (i.e. when angle α is negative)?

3-16. *Synthesis of a layer superposition having a predefined fixed locus.* Design layer transformations $\mathbf{g}_1(x,y)$ and $\mathbf{g}_2(x,y)$ that will produce in the superposition of two initially identical random screens a fixed locus consisting of two concentric circles around the origin. *Hint*: Consider the fourth-order parabola defined by $z = x^2(x^2 - 2)$. This W-shaped function has a local maximum at $x = 0$ (where $z = 0$) and two local minima at $x = \pm 1$ (where $z = -1$), and its solutions (the points where $z = 0$) are $x = 0$ and $x = \pm 2$. Therefore, the revolution surface $z = r^2(r^2 - 2)$ defined by this function (where $r = \sqrt{x^2 + y^2}$ is the radius around the origin) intersects the x,y plane at $r = 0$ (the origin) and along the perimeter of the circle defined by $r = 2$. Now, if we raise this revolution surface by a constant $0 < z_0 < 1$, for example $z_0 = 0.5$, it will intersect the x,y plane along the perimeter of two concentric circles. We repeat therefore the same procedure as in Example 3.8, but this time using in Eq. (3.48) the radial profile $r^2(r^2 - 2) + 0.5$ instead of $\sin r$. What happens to the fixed locus (and to the Glass pattern in the layer superposition) when we gradually vary the value of the constant z_0 between 0 and 1? And what happens below and above this range?

3-17. *Synthesis of a layer superposition having a predefined fixed locus (continued).* Design layer transformations $\mathbf{g}_1(x,y)$ and $\mathbf{g}_2(x,y)$ that will produce in the superposition of two initially identical random screens a fixed locus consisting of the origin plus a family of concentric heart-like curves around it. *Hint*: Construct a surface that intersects the x,y plane along the perimeter of a heart-like shape such as a cardioid. For example, think of a top-opened conic surface whose level lines are cardioids. The cardioid is defined in polar coordinates by the equation [Bronshtein97 p. 78]: $r = c(1 + \cos\theta)$ where c is a positive constant. (This is, in fact, a simple generalization of the polar equation of the circle, $r = c$, where the radius length varies with the angle θ so as to follow the perimeter of the heart; try to see how the original radius c is modified by the factor $(1 + \cos\theta)$ at different angles θ around the origin.) Consequently, you need to construct a top-opened conic surface whose level line at any level z is given by $r = z(1 + \cos\theta)$. The equation $z = f(r,\theta)$ of this surface is given, therefore, by $z = r/(1 + \cos\theta)$. Now, in order to have this surface intersect the x,y plane along a cardioid, you need to lower it by some constant z_0: $z = r/(1 + \cos\theta) - z_0$. But if you wish to obtain a more complex surface that intersects the x,y plane on a family of concentric cardioids, you may consider the surface: $z = \sin(r/(1 + \cos\theta))$. All that remains to do now is to repeat the same procedure as in Example 3.8, but this time using in Eq. (3.48) the angle-dependent radial profile $\sin(r/(1 + \cos\theta))$ instead of $\sin r$. Note, however, that all of the cardioids thus obtained have a commom singular point at the origin (since the slope of the surface $z = r/(1 + \cos\theta)$ is vertical at $\theta = 180°$). This can be avoided by using instead of the cardioid a more general curve known as *limaçon*, whose polar equation is [Bronshtein97 p. 78]: $r = c(a + \cos\theta)$. If we choose, for example, $a = 1.2$, this curve is still close to a cardioid, and yet it avoids the singularity at the origin when $\theta = 180°$.

3-18. *Synthesis of a layer superposition having a predefined fixed locus (continued).* Design layer transformations $\mathbf{g}_1(x,y)$ and $\mathbf{g}_2(x,y)$ that will produce in the superposition of two initially identical random screens a fixed locus consisting of a star-like curve that surrounds the origin, as shown in the figure on the front cover of this book. *Hint*: In this case, you may consider a top-opened conic surface having star-like level lines,

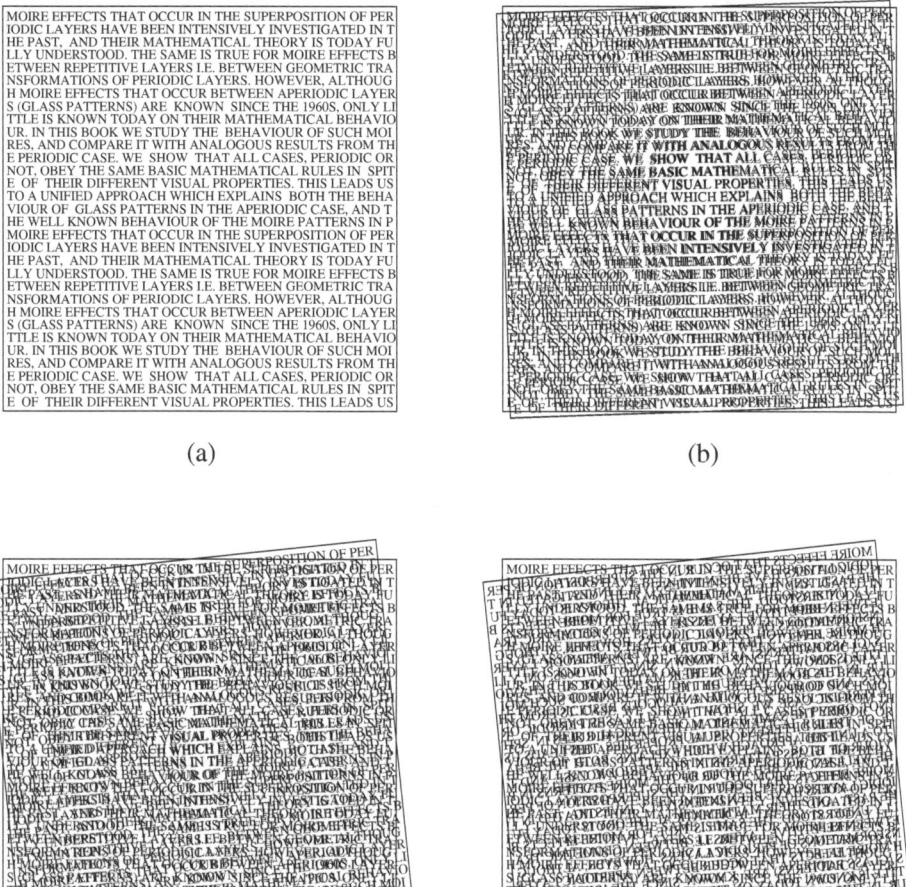

Figure 3.23: (a) An aperiodic structure consisting of lines of text. (b) The superposition of two identical copies of this structure with a small angle
difference generates a circular Glass pattern around the center of
rotation. (c) When the angle difference between the two layers increases
the Glass pattern becomes smaller and weaker, but new vertical moiré
bands become visible in the superposition. (d) If we keep the same
angle difference as in (c) but place one of the two identical layers face
down on top of the other, the Glass pattern completely disappears —
but the vertical moiré bands remain clearly visible. See Problem 3-14.

such as $z = r(1 + 0.5\cos5\theta)$, or, possibly, $z = r/(1 + 0.5\cos5\theta)$, which gives a slightly
different star. You may adjust the orientation of the star by replacing cos by sin or by
–sin, as seems suitable. In order to have this surface intersect the x,y plane along a star,

you need to lower it by some constant z_0: $z = r(1 + 0.5\cos 5\theta) - z_0$. But if you wish to obtain a more complex surface that intersects the x,y plane on a family of concentric stars, you may consider a surface such as: $z = \sin(r(1 + 0.5\cos 5\theta))$.

3-19. *Synthesis of a layer superposition having a predefined fixed locus (continued)*. The surface $z = \sin(r(c + \cos\theta))$ intersects the x,y plane on a family of curves whose shape depends on the parameter c: a family of straight lines if $c = 0$; a family of hyperbolas if $0 < c < 1$; a family of parabolas if $c = 1$; and a family of ellipses if $c > 1$. Generate the corresponding layer superpositions, experiment with them, and explain.

3-20. *Synthesis of a layer superposition having a predefined fixed locus (continued)*.
(a) Design layer transformations $\mathbf{g}_1(x,y)$ and $\mathbf{g}_2(x,y)$ that produce in the superposition of two initially identical random screens a Glass pattern consisting of periodic horizontal bands, as shown in Fig. 3.24(a). *Hint*: You may try the transformations $\mathbf{g}_1(x,y) = (x, y + a\cos(2\pi fy))$ and $\mathbf{g}_2(x,y) = (x,y)$, where $\mathbf{g}_1(x,y)$ is a non-linear variant of vertical scaling. Another possibility (not shown in the figure) consists of using $\mathbf{g}_1(x,y) = (x + a\cos(2\pi fy), y)$ and $\mathbf{g}_2(x,y) = (x,y)$, where $\mathbf{g}_1(x,y)$ is a non-linear variant of horizontal shearing. The linear counterparts of these two transformations are shown in Figs. 2.3(a),(b) and 2.3(c),(d), respectively.

(b) Do you see any contradiction in the fact that two *aperiodic* layers generate in their superposition a *periodic* Glass pattern? Explain.

(c) Design layer transformations $\mathbf{g}_1(x,y)$ and $\mathbf{g}_2(x,y)$ that give in the superposition a Glass pattern that is periodic along both main directions, as shown in Fig. 3.24(c).

(d) Part (d) of Fig. 3.24 is obtained by applying the same layer transformations $\mathbf{g}_1(x,y)$ and $\mathbf{g}_2(x,y)$ as in part (c) to two identical *periodic* dot screens. Compare under magnification the microstructures that are generated in different areas of Fig. 3.24(c) with their counterparts in Fig. 3.24(d). What are the main similarities and the main differences between the microstructures in the aperiodic case and in the periodic case?

3-21. In most of the examples we have seen so far the locus of the Glass pattern in the aperiodic case significantly differs from the locus of the moiré effect in the corresponding periodic (or repetitive) case. And yet, in some cases such as in Figs. 3.17 and 3.24, the Glass and moiré patterns have exactly the same loci. Can you characterize the cases in which this phenomenon occurs?

3-22. *Designing the transformation $\mathbf{k}(x,y)$ as a vector field*. In some cases the design of a transformation $\mathbf{k}(x,y)$ whose zeros form a given locus may be simplified by regarding $\mathbf{k}(x,y)$ as a *vector field* (as explained in Sec. B.6 of Appendix B, any transformation $(u,v) = \mathbf{g}(x,y)$ can be also interpreted as a vector field). Let us consider, for instance, the case already discussed in Example 3.8. Regarded as a vector field, transformation (3.47) assigns to each point (r,θ) in the x,y plane the radial vector $(r\cos\theta, r\sin\theta)$ whose length is r and whose direction is θ. Clearly, the zeros of this vector field are located where the vector length is zero, i.e. at the origin. In order to obtain a new vector field with more interesting zeros, we can modulate the length of its vectors by replacing r with any desired function $z = f(r)$, or even $z = f(r,\theta)$. The vectors of the new vector field remain radial, but their length is $f(r,\theta)$, which means that the new vector field has its zeros wherever $f(r,\theta) = 0$. This implies that we regard $\mathbf{k}(r,\theta)$ as a vector field; however, a vector field is by definition a *direct* transformation, whereas in principle all parts of Eq. (3.24) are considered as domain transformations (see point 6 at the end of Sec. 3.5). Should this fact preclude the design of $\mathbf{k}(x,y)$ as a vector field? *Hint*: As explained in Sec. D.11 of Appendix D, the fixed locus of a transformation $\mathbf{g}(x,y)$ is identical to that of its inverse, $[\mathbf{g}(x,y)]^{-1}$. Furthermore, the mutual fixed locus of the transformations $\mathbf{g}_1(x,y)$ and $\mathbf{g}_2(x,y)$ is almost identical to that of their inverse transformations. This

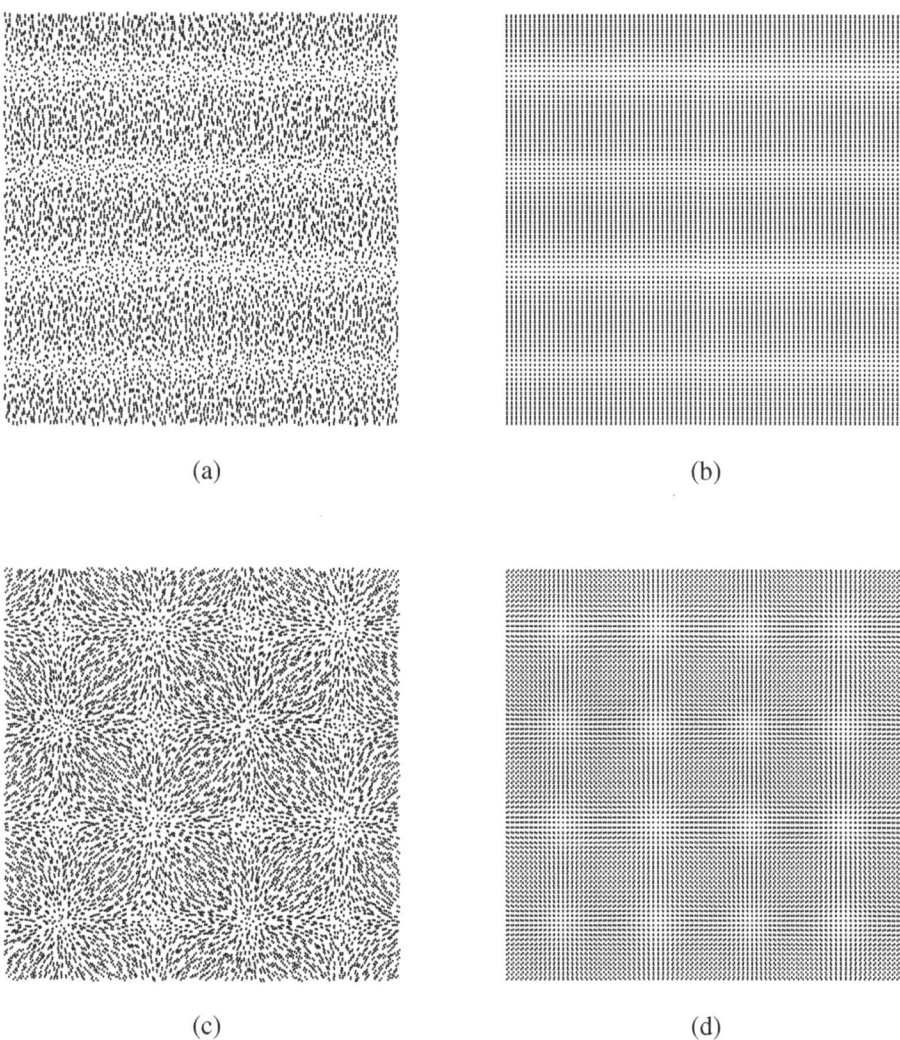

(a)   (b)

(c)   (d)

Figure 3.24: (a) The superposition of two originally identical aperiodic dot screens
that have undergone layer transformations $g_1(x,y)$ and $g_2(x,y)$ may give
rise to a periodic Glass pattern that consists of equidistant horizontal
bands. (b) The application of the same transformations $g_1(x,y)$, $g_2(x,y)$ to
two originally periodic dot screens gives a moiré pattern whose locus is
the same as in (a). (c),(d) The two-fold periodic counterparts of (a),(b).
See Problems 3-20 and 3-21.

means that the transformations $g_1(x,y)$ and $g_2(x,y)$ of Eqs. (3.41) and (3.42) (or their
inverse transformations) can be applied to our superposed layers either as domain or as
direct transformations, giving the same (or practically the same) fixed loci in the
superposition. This also allows us to construct the desired transformation $k(x,y)$ either
as a domain or as a direct transformation: in the first case $g_1(x,y)$ and $g_2(x,y)$ of Eqs.

(3.41) and (3.42) will be domain transformations, and in the second case they will be direct transformations. This also allows us to design transformation $\mathbf{k}(x,y)$ as a vector field whenever we feel it easier or more natural, as could be the case, for example, in Problems 3-16–3-20.

3-23. *Almost fixed points.* Example 3.9 shows how we can obtain in the superposition of two aperiodic layers an *almost* fixed point that is still surrounded by a visible Glass pattern. This is done by slightly modifying the layer transformations of an original case having a real fixed point in such a way that the fixed point is lost, but because the perturbation is so small we still remain with an *almost* fixed point. In Example 3.9 the layer transformation being used is a non-linear radial scaling. Is it possible to obtain a similar almost fixed point if we restrict ourselves to *linear* radial scaling transformations, namely, $\mathbf{g}(x,y) = (sx,sy)$?

3-24. What do you expect to see in Example 3.9 (Fig. 3.18) when $0 < s < 1$? *Hint*: In this case $f(r)$ does have a zero at $r = -s\delta/(s-1) > 0$; this gives us a circular fixed locus.

3-25. *Almost fixed line.* Consider the superposition of two identical aperiodic layers where one of the layers, say, the first one, undergoes a slight vertical scaling: $\mathbf{g}_1(x,y) = (x, sy)$, while the other layer remains unchanged: $\mathbf{g}_2(x,y) = (x,y)$. As shown in Fig. 2.3(a), this gives a horizontal fixed line along the x axis, which is surrounded by a clearly visible horizontal linear Glass pattern. What happens if we slightly shift one of the layers (say, the unscaled one) by x_0 horizontally? This can be clearly seen by comparing Figs. 2.3(a) and 2.3(c). Explain your observation mathematically, as we have done in Example 3.10.

3-26. *Pseudo Glass patterns?* In some cases, a non-linear transformation that is applied to a given layer may "chase away" all the elements of the transformed layer from a certain region, creating a "void" within the transformed layer. Inside this region the layer superposition only contains elements of the other layer, and the gray level of the superposition is brighter. But unlike in a Glass pattern, where the bright center is due to a zone of high correlation between the two layers, in the present case the visible bright area is due to the "void" in the transformed layer. Can you think of layer transformations that would produce such an effect?

3-27. *Glass patterns in three-layer superpositions.* Explain the locations of the three Glass patterns in Fig. 2.8. Why does the slight shift of layer A to the *right* cause the Glass pattern between layers A and B to move to the *right*, while the Glass pattern between layers A and C moves *downward*? *Hint*: See Secs. 3.3.3–3.3.4; compare also with formulas (7.26), (2.10) and (2.11) in *Vol. I* for the case of periodic layers.

3-28. *Glass patterns in three-layer superpositions (continued).* When superposing three identical aperiodic dot screens that are only rotated and shifted with respect to each other (without any scalings), we obtain three circular Glass patterns whose centers are located, unlike in Fig. 2.8, along a straight line. How do you explain this observation?

3-29. *Glass patterns in three-layer superpositions (continued).* Eq. (3.14) determines the location of a Glass pattern that is generated between two identical aperiodic layers, one of which has undergone a non-degenerate affine transformation (i.e. an affine transformation that satisfies conditions (3.15)). How can you predict the location of each of the three Glass patterns in a three layer superposition such as in Fig. 2.8 using Eq. (3.14)? Can you find a way to determine the required layer shifts that should be applied to layers A, B and C in order that the three Glass patterns appear in some predefined locations (x_a,y_a), (x_b,y_b) and (x_c,y_c)? *Hint*: You will have to solve a simultaneous set of equations consisting of one instance of Eq. (3.14) for each of the three layer pairs: A and B, A and C, and B and C.

Chapter 4

Microstructures: dot trajectories and their morphology

4.1 Introduction

Perhaps the most striking property of moiré effects between aperiodic screens, which clearly distinguishes them from their periodic or repetitive counterparts, is the appearance in the superposition of intriguing microstructure dot alignments, also known as "dot trajectories". These dot trajectories may have various geometric shapes, depending on the transformations undergone by the superposed layers. In the case of simple linear transformations such as layer rotations, layer scalings, etc. the resulting dot trajectories are rather simple (circular, radial, spiral, elliptic, hyperbolic, linear, etc.); see Figs. 2.1–2.3. But when the layer transformations are more complex, the resulting dot trajectories may have more interesting and sometimes even quite spectacular shapes. And yet, if the same layer transformations are applied to *periodic* layers, no dot trajectories are visible in the superposition, and the resulting moiré effects look completely different, as we can clearly see in the right hand side of each of the figure pairs throughout this book. What is the reason for this surprising difference?

This intriguing question is, indeed, one of our main subjects of interest in the present book. But before we can answer this question, we first need to develop an appropriate mathematical approach for the study of the microstructures, that will allow us to analyze the dot trajectories and to explain their morphology and their various other properties. This will be, indeed, our main task in the present chapter. As an additional bonus, this mathematical approach will also allow us, later in this chapter, to *synthesize* screen superpositions having dot trajectories of any desired geometric shapes.

The present chapter is structured as follows: We start our discussion in Secs. 4.2–4.5 with an overview of dot trajectories and the different mathematical tools that have been proposed for their analysis. Based on this mathematical understanding we proceed in Sec. 4.6 to the *synthesis* of dot trajectories, and we show how we can design aperiodic screens that give in their superposition any desired dot trajectories. This result can be also used as a simple tool for visually illustrating the trajectories (field lines) of 2D vector fields and the solution curves of systems of two differential equations in the plane. In Sec. 4.7 we explain why dot trajectories are visible between aperiodic layers but not between periodic layers. Finally, in Sections 4.8–4.9 we discuss the microstructures which appear in cases based on other superposition rules, and we briefly explain why they look different from the familiar dot trajectories that appear when using the standard multiplicative super-position rule (black dots overprinted on black dots).

4.2 Morphology of the microstructures; dot trajectories

The surprising geometric diversity of the microstructure shapes in superpositions of aperiodic screens that undergo different geometric transformations was first reported in [Glass73]. And indeed, as shown in many of the figures throughout the present book, fixed points in the superposition of aperiodic screens are often surrounded by microstructure dot alignments having various typical shapes, such as circles, ellipses, hyperbolas, spirals, etc., depending on the case. These microstructure dot alignments are called *dot trajectories*. For example, a fixed point which occurs due to the rotation of one of the superposed layers is typically surrounded by circular dot trajectories (Fig. 2.1(c)), while a fixed point due to a uniform scaling transformation is typically surrounded by radial dot trajectories (Fig. 2.1(e)). Spiral dot trajectories, on their part, are typical of intermediate cases involving both rotation and scaling (Fig. 2.1(g)), while hyperbolic dot trajectories typically occur around the fixed point in scaling transformations involving expansion along one direction and contraction along a different direction (Fig. 2.2(a)). In terms of the analysis we have done so far the Glass patterns obtained in all of these cases are equivalent, because they are all generated around a fixed point at the origin, and macroscopically they all consist of a brighter area at the center, which gradually fades out as we go away from the center, due to the decreasing correlation between the layers. And yet, each of these Glass patterns clearly differs from the others in terms of the dot trajectories. Furthermore, it turns out that dot trajectories may appear even in superpositions where no fixed points exist (see, for example, Figs. 2.3(e),(h), 3.8 and 3.9). This large versatility in the dot trajectories and in their shapes certainly deserves investigation. It is therefore our aim here to study the dot trajectories, their morphology and their various properties.

One important observation that can be done immediately is that the shapes of the dot trajectories depend on the transformations undergone by the original, aperiodic layers, but they do not depend on the specific dot distribution of these aperiodic layers. This can be clearly illustrated by comparing Figs. 2.1(c),(e) with Figs. 2.6(c),(d) and Figs. 2.7(b),(c): As we can see, the shapes of the dot trajectories obtained in these cases are, indeed, *equivalent* — although they need not be *identical* in terms of the individual dot locations. This fundamental fact implies that our investigation of the dot trajectories in the superposition of aperiodic layers should be focused on the study of the transformations that are applied to the superposed layers rather than on the study of the layers' dot distributions.[1]

Suppose, to begin with, that we superpose two identical aperiodic screens exactly on top of each other, and that we apply to one of the superposed screens a transformation $\overline{g}(x,y)$ given componentwise by:

---

[1] The influence of the dot shapes and of the degree of periodicity within the original layers will be studied later, in Chapter 7. The role of the superposition rule that is used between the two layers is discussed at the end of the present chapter, in Secs. 4.8–4.9; for the time being we assume here the classical multiplicative superposition rule (see Sec. 2.2).

$$x' = \bar{g}_1(x,y)$$
$$y' = \bar{g}_2(x,y) \tag{4.1}$$

(Note that x' and y' do not represent here derivatives, but simply a different coordinate system in the plane. The reason for using the barred notations $\bar{g}$, $\bar{g}_1$ and $\bar{g}_2$ will become clear in the following.) As a simple example, we may think of the linear transformation:

$$x' = a_1x + b_1y$$
$$y' = a_2x + b_2y \tag{4.2}$$

The dot trajectories that are generated in these screen superpositions are physically made of pairs of dots, one dot from each layer. Each such pair in the superposition represents, in fact, two successive locations of the same dot, namely, the dot's location *before* and *after* the layer transformation $\bar{g}(x,y)$ has been applied. Let us try to see how these dot trajectories can be analyzed mathematically.

4.3 Dot trajectories as solution curves of a system of differential equations

A first attempt to interpret such dot trajectories has already been presented in [Glass73]. According to this approach, we consider the finite transformation $\bar{g}(x,y)$ as an iteration of an infinitesimal transformation. This reasoning leads one from Eqs. (4.1) to a pair of first-order differential equations that correspond to our infinitesimal transformation:

$$\frac{d}{dt}x(t) = \bar{g}_1(x(t),y(t))$$
$$\frac{d}{dt}y(t) = \bar{g}_2(x(t),y(t)) \tag{4.3}$$

For example, in the particular case of linear transformation (4.2) we get:

$$\frac{d}{dt}x(t) = a_1x(t) + b_1y(t)$$
$$\frac{d}{dt}y(t) = a_2x(t) + b_2y(t) \tag{4.4}$$

The solution of system (4.3) consists of a family of curves in the x,y plane, whose members differ from each other by some constants c [Kreyszig93 180–186]. The parametric representation of each of these curves is $(x(t),y(t))$, where the parameter t may be thought of as time. Each of these solution curves is called a *trajectory* since it traces out the evolution of the curve as t is being varied. (Note that we reserve the term *dot trajectories* for the dot alignments that occur in random screen superpositions. When confusion may arise we will call the trajectories of the differential equations *solution curves*.)

Proceeding with the approach of [Glass73], the dot trajectories observed in the superposition of aperiodic dot screens simply represent the trajectories of this underlying pair of differential equations.[2] And indeed, this explanation agrees with experimental evidence in various cases.

However, a deeper examination of the question shows that this reasoning is not always true. This is clearly demonstrated by the following examples.

Example 4.1: Consider the identity transformation $\overline{\mathbf{g}}(x,y) = (x,y)$, which is a particular case of (4.2) with $a_1 = b_2 = 1$ and $a_2 = b_1 = 0$. Obviously, in this case the two layers perfectly coincide on top of each other, and no dot trajectories are generated. However, the solution of the corresponding system of differential equations (4.4) with $a_1 = b_2 = 1$ and $a_2 = b_1 = 0$ gives the family of straight lines $y = cx$ for any constant c, and its solution curves (trajectories) consist of radial lines emanating from the origin (see [Kreyszig93 p. 168]). While such radial dot trajectories can be expected in the layer superposition when the mapping (4.2) being applied to one of the screens is a scaling by $s \neq 1$ (see Fig. 2.1(e)), it is clear that in our present case, where $s = 1$, no dot trajectories will appear in the superposition. ∎

Example 4.2: Suppose that we apply to one of the two superposed layers a transformation $\overline{\mathbf{g}}(x,y)$ consisting of rotation by a small angle α, as illustrated in Fig. 2.1(c). This linear transformation is clearly a particular case of (4.2), with $a_1 = \cos\alpha$, $b_1 = -\sin\alpha$, $a_2 = \sin\alpha$ and $b_2 = \cos\alpha$. According to Fig. 2.1(c) we would expect the solution curves of the differential equations (4.4) to consist of circular trajectories around the center of rotation. However, a short verification shows that this is not always the case. Although for $\alpha = 90°$, where $a_1 = b_2 = 0$, $a_2 = 1$ and $b_1 = -1$, the differential equations do have circular trajectories as expected, it turns out that for $\alpha = 0°$ (the identity transformation, where $a_1 = b_2 = 1$ and $a_2 = b_1 = 0$) the solution curves of the differential equations consist of a family of radial lines emanating from the origin. Furthermore, for all intermediate rotation angles $0° < \alpha < 90°$ the solution curves consist of a family of spiral trajectories (see, for example, [Birkhoff89 p. 144]); as $\alpha \to 90°$ these spirals gradually approach circular trajectories, but as $\alpha \to 0°$, i.e. for small rotation angles, the spiral trajectories straighten out and gradually approach radial straight lines. This result does not agree with our observations based on Fig. 2.1(c). ∎

Example 4.3: As a third example, suppose that we apply to one of the superposed layers the linear mapping $\overline{\mathbf{g}}(x,y) = (-x,y)$ which is obtained from (4.2) when $a_1 = -1$, $b_2 = 1$ and $a_2 = b_1 = 0$. This mapping corresponds to a horizontal flipping of the layer in question about the y axis. In this case, the solution of the corresponding system of differential equations (4.4) with $a_1 = -1$, $b_2 = 1$ and $a_2 = b_1 = 0$ gives the family of hyperbolas $y = c/x$ for any constant c, and its trajectories consist of hyperbolic curves around the origin (see [Kreyszig93 pp. 169–170]). However, once again, such trajectories cannot be expected in our layer superposition when we horizontally flip one of the screens about the y axis (see

---

[2] For more details on the interpretation of such systems of differential equations and on the significance of their trajectories see Sec. B.6 in Appendix B. In essence, these trajectories (solution curves) correspond to the field lines that are observed when the mapping $\overline{\mathbf{g}}(x,y)$ is represented as a 2D vector field that assigns to each point (x,y) in the x,y plane the vector $\overline{\mathbf{g}}(x,y)$.

Fig. 3.22(a)). Furthermore, while our flipping transformation has a full line of fixed points along the y axis, the corresponding system of differential equations (4.4) only has in our case a single fixed point at the origin. ∎

These three examples show that the system of differential equations (4.4) does not always correspond to the dot trajectories and to the fixed points that appear in our layer superposition when the mapping (4.2) is applied to one of the superposed layers. The reasons for this failure as well as ways to overcome it will become clear shortly.

4.4 Dot trajectories as a vector field

A better understanding of this problem can be obtained by considering the situation from another point of view. As explained in Appendix B, any 2D transformation $\bar{g}(x,y)$ can be also interpreted as a *vector field* that assigns to each point (x,y) in the x,y plane the vector $\bar{g}(x,y)$. This vector field can be illustrated graphically by drawing starting from each point (x,y) an arrow having the length and the orientation of the vector $\bar{g}(x,y)$. This interpretation has the important advantage of clearly showing in a visual way the effect of the transformation $\bar{g}(x,y)$ on any point of the x,y plane. It should be noted that the system of differential equations (4.3) is simply a different representation of the same vector field $\bar{g}(x,y)$, since its solution curves $(x(t),y(t))$ express the field lines (trajectories) of the vector field $\bar{g}(x,y)$ (see Sec. B.6 in Appendix B).

Remark 4.1: Before we proceed with our discussion, let us see now the reason for using here the barred notation $\bar{g}$ rather than simply $\mathbf{g}$. As explained in Sec. D.6 of Appendix D, every given transformation $\mathbf{g}(x,y)$ can be used in two different ways: either as a direct transformation, or as an inverse transformation. Consider, for example, the transformation $\mathbf{g}(x,y) = (2x,2y)$. Clearly, this transformation maps each point (x,y) to the new location $(2x,2y)$, symbolically: $(x,y) \mapsto (2x,2y)$, and thus it expands the original layer by two. This is, indeed, the interpretation of $\mathbf{g}(x,y)$ as a *direct* transformation. However, when the same transformation $\mathbf{g}(x,y)$ is used as a *domain* transformation, i.e. when it acts on the original layer $r(x',y')$ to give $r(2x,2y)$, its effect is inversed: $r(2x,2y)$ is a two-fold shrinked version of $r(x',y')$, while the two-fold expansion of $r(x',y')$ is expressed by $r(x/2,y/2)$, i.e. by using the *inverse* transformation $\mathbf{g}^{-1}(x,y) = (x/2,y/2)$. This inversion effect of domain transformations is explained in detail in Appendix D (see Sections D.6 and D.10).

It should be noted that in the classical moiré theory between periodic or repetitive layers one always uses the interpretation of the layer transformations $\mathbf{g}(x,y)$ as *domain* (and hence, *inverse*) transformations. This is the case, for example, when we study the moiré effects that are obtained when we apply the domain transformations $\mathbf{g}_1(x,y)$ and $\mathbf{g}_2(x,y)$ to the original periodic layers $p_1(x',y')$ and $p_2(x',y')$, giving the distorted layers $p_1(\mathbf{g}_1(x,y))$ and $p_2(\mathbf{g}_2(x,y))$ (see, for example, Propositions *I*.10.2 and *I*.10.5).[3] The same is also true in the

---

[3] Exceptionally, direct rather than inverse transformations are being used in *Vol. I* in Eq. (*I*.3.1) and in its vector form (see Sec. 3.4.1 in *Vol. I*).

case of aperiodic layers, as we have seen throughout Chapter 3 (see, for example, Footnote 6 in Sec. 3.3 or the 6th remark following Proposition 3.2). However, when it comes to dot trajectories, the transformation $g(x,y)$ that is applied to a dot screen is understood as an operation that moves each point (x,y) of the original dot screen to its new destination under g, namely: $(x,y) \mapsto g(x,y)$. This means that in the context of dot trajectories we are interested in the effect of g as a *direct* transformation.[4] However, this double meaning of the notation $g(x,y)$ may cause confusion if one cannot be certain which of the two possible interpretations is intended. This risk of confusion could exist in particular in the figure legends throughout this book, because the same figures are used to illustrate all sections, including those dealing with dot trajectories. To take a concrete example, if we say that one of the layers has undergone the transformation $g(x,y) = (2x,2y)$, do we mean that the layer has been shrinked by applying $g(x,y) = (2x,2y)$ as a domain transformation, or do we mean that it has been expanded by applying $g(x,y) = (2x,2y)$ as a direct transformation, i.e. $(x,y) \mapsto (2x,2y)$? For reasons of consistency with the classical moiré theory, we prefer as a rule to use throughout our work, including in the figure legends, the convention taking $g(x,y)$ as a *domain* transformation. Hence, $g(x,y) = (2x,2y)$ should be understood as the transformation that shrinks the layer $r(x',y')$ into $r(2x,2y)$ rather than as a two-fold expansion $(x,y) \mapsto (2x,2y)$. And yet, as we have just seen, it turns out that in the context of dot trajectories we have to deal with *direct* transformations.

Therefore, in order to avoid any possible confusion, we must use a distinctive notation that clarifies which of the two meanings of $g(x,y)$ is intended in each case. For this end, we introduce here the following convention: Whenever we wish to use $g(x,y)$ as a *direct* transformation, we mark it by an upper bar. Thus, $\bar{g}(x,y) = (2x,2y)$ should be understood as the direct transformation $(x,y) \mapsto (2x,2y)$ that expands the plane by factor 2, while $g(x,y) = (2x,2y)$ represents, as usual, the domain transformation that shrinks the original layer $r(x',y')$ into its two-fold reduced copy $r(2x,2y)$. Hence, a twofold layer magnification will be routinely expressed, including in our figure legends, by $g(x,y) = (x/2,y/2)$, giving $r(g(x,y)) = r(x/2,y/2)$. But in the context of dot trajectories the very same layer magnification must be expressed by $\bar{g}(x,y) = (2x,2y)$, as a *direct* transformation.[5] A few concrete examples of the use of g and $\bar{g}$ will be given in Sec. 5.3, based on the material presented in Chapters 3 and 4. ■

We return now to our superposition of two identical aperiodic screens, one of which has undergone the direct transformation $(x,y) \mapsto \bar{g}(x,y)$. Remember that the dot trajectories in our superposition of two aperiodic dot screens consist of pairs of dots, which represent the location of a screen dot *before* and *after* the layer transformation $\bar{g}(x,y)$ has been applied. These dot pairs can be represented, therefore, as a vector field, which assigns to

---

[4] Note, in particular, that the vector field interpretation of a transformation $g(x,y)$ is based on the effect of g as a direct transformation. This is explained in detail in Sec. B.6 of Appendix B.

[5] In fact, because $\bar{g}$ and g are the inverse of each other, and because the notation g is already used in the classical moiré theory (including *Vol. I*) for the domain transformation, we could have used here g^{-1} rather than $\bar{g}$ to denote the direct transformation. However, in order to avoid the awkward situation in which the *direct* transformation is denoted by g^{-1} while the *inverse* (domain) transformation is denoted by g, we have chosen to use here for the direct transformations the lighter and more neutral notation $\bar{g}$.

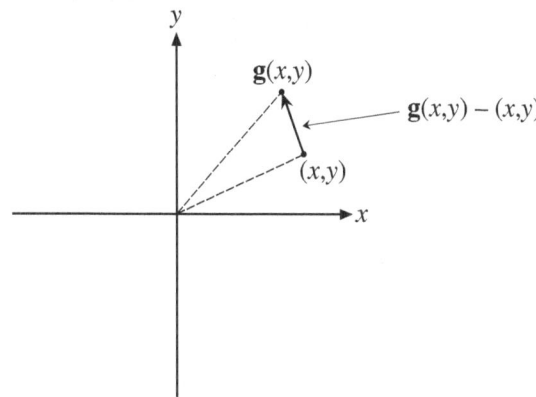

Figure 4.1: Point (x,y) in the x,y plane and its image $\bar{\mathbf{g}}(x,y)$ under the transformation $\bar{\mathbf{g}}: \mathbb{R}^2 \to \mathbb{R}^2$. The vector connecting the original point (x,y) to its destination $\mathbf{g}(x,y)$ under transformation $\bar{\mathbf{g}}$ is given by $\bar{\mathbf{g}}(x,y) - (x,y)$.

each point $(x,y) \in \mathbb{R}^2$ a vector that connects (x,y) to its new location $\bar{\mathbf{g}}(x,y) \in \mathbb{R}^2$ under the transformation $\bar{\mathbf{g}}$ [Glass02]. It is important to note, however, that the vector field of the transformation $\bar{\mathbf{g}}(x,y)$ itself *does not* have this property; that is, the vector it assigns to (x,y) does not connect (x,y) to its destination $\bar{\mathbf{g}}(x,y)$, but rather to the point $(x,y) + \bar{\mathbf{g}}(x,y)$. For instance, if we consider the identity transformation $\bar{\mathbf{g}}(x,y) = (x,y)$ (see Example 4.1), it is clear that in this case the vector attached to each point (x,y) is the vector (x,y) itself, which points, therefore, to the point $(2x,2y)$ and not to the destination point under $\bar{\mathbf{g}}$, which is the point (x,y).

Therefore, in order to obtain a vector field that correctly represents our dot trajectories, we must consider, instead of the transformation $\bar{\mathbf{g}}(x,y)$ itself (the transformation that has been applied to one of the superposed layers), the *relative* transformation between the two layers, which is given by:

$$\bar{\mathbf{h}}(x,y) = \bar{\mathbf{g}}(x,y) - (x,y) \tag{4.5}$$

And indeed, if we draw the vector field representation of this transformation, we obtain exactly what we desired: The vector field of $\bar{\mathbf{h}}(x,y)$ assigns to each point (x,y) the vector $\bar{\mathbf{g}}(x,y) - (x,y)$ which connects the original point (x,y) to its destination under the layer transformation $\bar{\mathbf{g}}$, the point $\bar{\mathbf{g}}(x,y)$ (see Fig. 4.1).[6]

It should be noted, however, that the dot trajectories in our layer superposition can be represented by either of the vector fields $\bar{\mathbf{h}}(x,y)$ or $-\bar{\mathbf{h}}(x,y)$. This is because the two dots that compose each dot pair in the layer superposition are identical, so that the dot pairs

---

[6] Note that for any points (a,b) and (c,d) in the plane, when the tail of the vector $(a,b) - (c,d)$ is attached to the point (c,d), its head is located at the point (a,b); this means that the vector $(a,b) - (c,d)$ connects the point (c,d) to the point (a,b).

(and hence the dot trajectories in the superposition) remain unchanged when we interchange the two layers. This means that the dot trajectories in the superposition do not show the direction (the positive or negative sense) of the difference vector. In other words, the sense of the vectors of the vector field has no significance for the dot trajectories in the superposition. The above results can be therefore summarized as follows:

Proposition 4.1: Suppose that we are given two identical aperiodic dot screens that are superposed on top of each other in full coincidence, dot on dot. When we apply to one of the layers a transformation $\overline{g}(x,y)$, we obtain in the superposition dot trajectories that correspond to the vector field $\overline{h}(x,y) = \overline{g}(x,y) - (x,y)$ (or, equivalently, to the vector field $\overline{h}(x,y) = (x,y) - \overline{g}(x,y)$). ∎

Remark 4.2: It should be understood that although dot trajectories are theoretically generated in the superposition for any layer transformation $\overline{g}(x,y)$, they can only be visible (and correspond to the vector field $\overline{h}(x,y)$) if the layer transformation $\overline{g}(x,y)$ is not too "violent". Otherwise, the correlation between the layers is strongly reduced, and the visual effect in the layer superposition may be lost and no longer correspond to the vector field. Note also that the vector field $\overline{h}(x,y)$ matches the dot trajectories only in areas of the superposition where the corresponding dots of both layers remain close to each other (i.e. in areas where the arrow length in $\overline{h}(x,y)$ are small). In areas where the dots get farther apart (and the arrow lengths in $\overline{h}(x,y)$ increase) the correlation between the layers is reduced, and the dot trajectories disappear and no longer correspond to $\overline{h}(x,y)$ (compare, for example, Figs. 4.2(a) and 2.1(c)). ∎

Let us now see a few examples to illustrate how this approach explains the dot trajectories obtained in our aperiodic screen superpositions. We start with the three cases that caused us trouble in Examples 4.1 – 4.3 above.

Example 4.4: Suppose that one of the superposed screens, say, the first one, is scaled by $s > 0$. In order not to lose the correlation between the two screens we assume that $s \approx 1$. This layer transformation is expressed here by $\overline{g}(x,y) = (sx,sy)$ (although we could also say, for reasons of compatibility with the notations used in the classical moiré theory and in *Vol. I*, that the scaled layer has undergone the domain transformation $g(x,y) = (x/s,y/s)$). The relative transformation between the two layers is given in this case by:

$$\overline{h}(x,y) = (sx,sy) - (x,y) = (s-1)(x,y)$$

Regarding this linear transformation as a vector field, we see that it assigns to any point (x,y) in the x,y plane a vector $\overline{h}(x,y) = (s-1)(x,y)$. Clearly, if $s > 1$ then $(s-1) > 0$ and the vector field consists of radial trajectories emanating from the origin (see Fig. 4.2(b)), whereas if $0 < s < 1$ the vector field consists of radial trajectories pointing to the origin (Fig. 4.2(d)). The origin itself is a fixed point of $\overline{g}(x,y)$, and hence a zero of the vector field $\overline{h}(x,y)$. And indeed, this fully agrees with the radial dot trajectories that we observe in the screen superposition. Note, however, that the dot trajectories in the screen superposition (Fig. 2.1(e)) do not show the *direction* along the trajectories (which is indicated by the arrowheads in the vector fields of Figs. 4.2(b) or 4.2(d)).

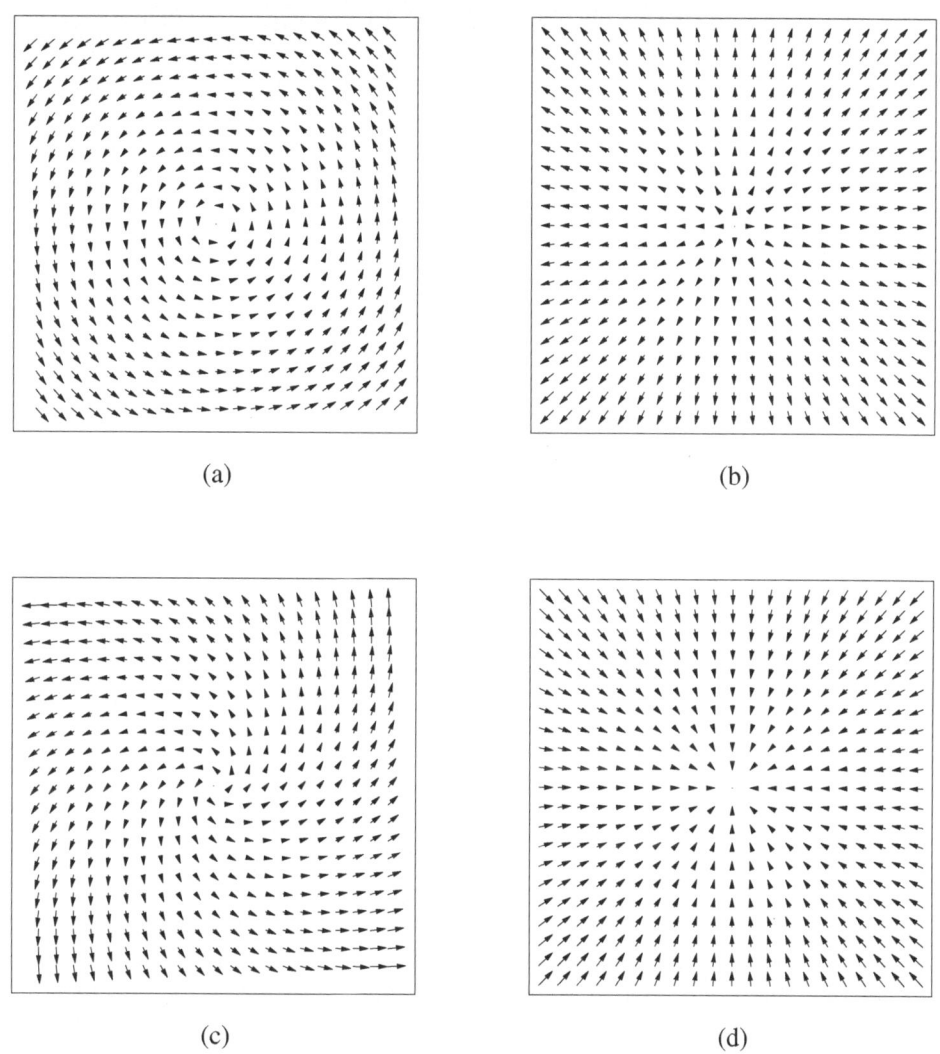

(a)

(b)

(c)

(d)

Figure 4.2: Vector field representation of the relative transformation $\overline{\mathbf{h}}(x,y) = \overline{\mathbf{g}}(x,y) - (x,y)$ where $\overline{\mathbf{g}}(x,y)$ (the transformation undergone by one of the layers) is: (a) a small rotation; (b) a small expansion; (c) both a small rotation and a small expansion; (d) a small shrinking. Compare with Figs. 2.1(c), (e) and (g).

Finally, let us examine the case of $s = 1$, which caused us trouble in Example 4.1 above. In this case (scaling by 1) our vector field $\overline{\mathbf{h}}(x,y)$ assigns to each point (x,y) in the x,y plane the zero vector; hence, this transformation has no fixed points and no trajectories — exactly as we would expect when the two identical screens fully coincide on top of each other. Thus, by considering the vector field of the relative layer transformation $\overline{\mathbf{h}}(x,y)$ we have overcome our troubles from Example 4.1. ■

Example 4.5: Let us consider the linear mapping $\overline{g}(x,y)$ which corresponds to a rotation of the first layer about the origin by a small angle α. As we have seen in Example 4.2 this linear mapping is given by $\overline{g}(x,y) = (x\cos\alpha - y\sin\alpha, x\sin\alpha + y\cos\alpha)$. Here, too, the vector field $\overline{h}(x,y) = \overline{g}(x,y) - (x,y)$ assigns to each point (x,y) in the x,y plane a vector that connects it to its new location after the rotation. This vector field consists of a circular arrangement of arrows (or dot pairs) about the origin, as we can see in Fig. 4.2(a). This agrees, indeed, with the dot trajectories that are generated in the screen superposition, as shown in Fig. 2.1(c). ∎

Example 4.6: Consider now the linear mapping discussed in Example 4.3 above, which corresponds to horizontally flipping the second layer about the y axis. This layer mapping is given by $\overline{g}(x,y) = (-x,y)$, and therefore the relative transformation between the two superposed layers is given in this case by:

$$\overline{h}(x,y) = (x,y) - (-x,y) = (2x,0)$$

Considering this degenerate linear transformation as a vector field, we see that it assigns to any point (x,y) in the x,y plane a vector $(2x,0)$. Clearly, this vector field consists of horizontal trajectories emanating from the y axis, while the y axis itself is a fixed line of this transformation. This agrees, indeed, with what we observe in the screen superposition (see Fig. 3.22(a)). Note, however, that the horizontal trajectories of this vector field are only visible in the superposition near the y axis, because farther away there is no longer any correlation left between the two superposed layers. ∎

These three examples show clearly that by considering the vector field of the relative layer transformation $\overline{h}(x,y)$ we have overcome our troubles from Examples 4.1 – 4.3. To further illustrate our mathematical interpretation of the dot trajectories in the superposition as a vector field, Figs. 4.2–4.6 provide a graphical representation of the vector field $\overline{h}(x,y)$ for some of the most interesting superpositions that we have seen so far.[7] These figures include both linear and non-linear layer transformations, and they illustrate dot trajectories that occur around one or more fixed points (or fixed lines), as well as dot trajectories that are generated when no fixed points exist in the superposition. Note the agreement between the dot trajectories in each of the superpositions and the arrows showing the dot displacements in the corresponding vector field $\overline{h}(x,y)$.

4.4.1 The curve equations of the dot trajectories

Although the vector field $\overline{h}(x,y)$ pictures the graphical shapes of the dot trajectories qualitatively, it does not yet provide the explicit formulas of these curves. For example, looking at Fig. 2.3(g) and Fig. 4.4(a) we get the visual impression that in this case the dot trajectories resemble right-opened parabolas; but because we do not yet have the precise mathematical expression of the curves, we cannot be certain whether they are really parabolas or only curves that resemble parabolas.

---

[7] In fact, some of these figures illustrate cases in which *both* layers have undergone transformations; such cases are discussed in detail in Sec. 4.5.

Is there a way to find the explicit mathematical expression of these curves? As explained in Sec. B.6 of Appendix B, it turns out that the trajectories (field lines) of a vector field $\overline{g}(x,y)$ are expressed mathematically by the family of solution curves of the system of differential equations:

$$\frac{d}{dt}x(t) = \overline{g}_1(x(t),y(t))$$

$$\frac{d}{dt}y(t) = \overline{g}_2(x(t),y(t))$$

where $\overline{g}_1(x,y)$ and $\overline{g}_2(x,y)$ are the two Cartesian components of the vector field $\overline{g}(x,y)$. Thus, we see that the system of differential equations (4.3) is simply a different representation of the transformation $\overline{g}(x,y)$, since the solution curves of Eqs. (4.3) express the field lines of the vector field $\overline{g}(x,y)$. This explains, indeed, why the system of differential equations (4.3) fails to represent our dot trajectories, as we have seen in Sec. 4.3: The reason is that Eqs. (4.3) express the field lines of the vector field $\overline{g}(x,y)$, while the dot trajectories in our layer superposition are represented by the relative vector field $\overline{h}(x,y) = \overline{g}(x,y) - (x,y)$ and not by $\overline{g}(x,y)$ itself. But if we replace $\overline{g}(x,y)$ in Eqs. (4.3) by $\overline{h}(x,y)$, we get a new, modified pair of differential equations:

$$\frac{d}{dt}x(t) = \overline{h}_1(x(t),y(t))$$

$$\frac{d}{dt}y(t) = \overline{h}_2(x(t),y(t)) \tag{4.6}$$

which provides, indeed, the mathematical expression of the field lines of the relative vector field $\overline{h}(x,y)$. It can be expected, therefore, that by using Eqs. (4.6) instead of Eqs. (4.3) we can also "cure" the differential-equations approach of Sec. 4.3 and make it work properly.

Let us examine a few examples to see how well this modified system of differential equations predicts the shapes of the dot trajectories in the superposition.

Example 4.7: Consider the superposition of two identical aperiodic screens one of which undergoes the scaling transformation $\overline{g}(x,y) = (sx,sy)$ (see Example 4.4 and Fig. 2.1(e)). In this case the relative transformation between the two layers is:

$$\overline{h}(x,y) = (sx,sy) - (x,y) = (s-1)(x,y)$$

or componentwise:

$$\overline{h}_1(x,y) = (s-1)x$$

$$\overline{h}_2(x,y) = (s-1)y$$

The dot trajectories of our layer superposition are given, therefore, by the field lines of the vector field $\overline{h}(x,y)$ (see Fig. 4.2(b)), which are expressed, as we have just seen, by the trajectories (solution curves) of the modified pair of differential equations (4.6):

$$\frac{d}{dt}x(t) = (s-1)x(t)$$

$$\frac{d}{dt}y(t) = (s-1)y(t)$$

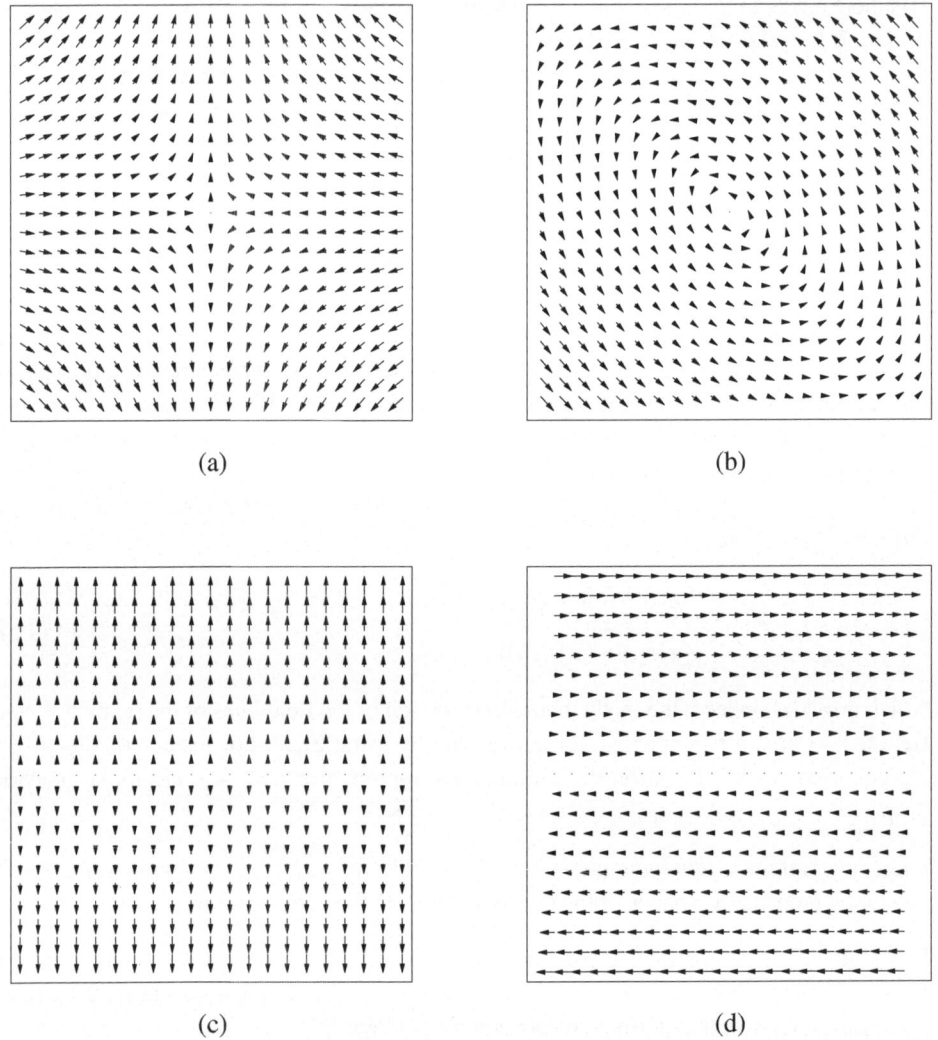

(a)   (b)

(c)   (d)

Figure 4.3: Vector field representation of the relative transformation $\overline{\mathbf{h}}(x,y) = \overline{\mathbf{g}}(x,y) - (x,y)$ where $\overline{\mathbf{g}}(x,y)$ (the transformation undergone by one of the layers) is: (a) a small horizontal shrinking and a small vertical expansion; (b) same as in (a) but with a small rotation, too; (c) a small vertical expansion; (d) a small horizontal shear. Compare with Figs. 2.2(a),(c) and 2.3(a),(c).

The solution curves of this system of differential equations consist of the family of radial straight lines $y = cx$ passing through the origin (see [Kreyszig93 p. 168]). This is, indeed, the explicit expression of the radial field lines of the vector field $\overline{\mathbf{h}}(x,y)$ (see Fig. 4.2(b)), and it perfectly agrees with the dot trajectories obtained in our layer superposition (Fig. 2.1(e)). ∎

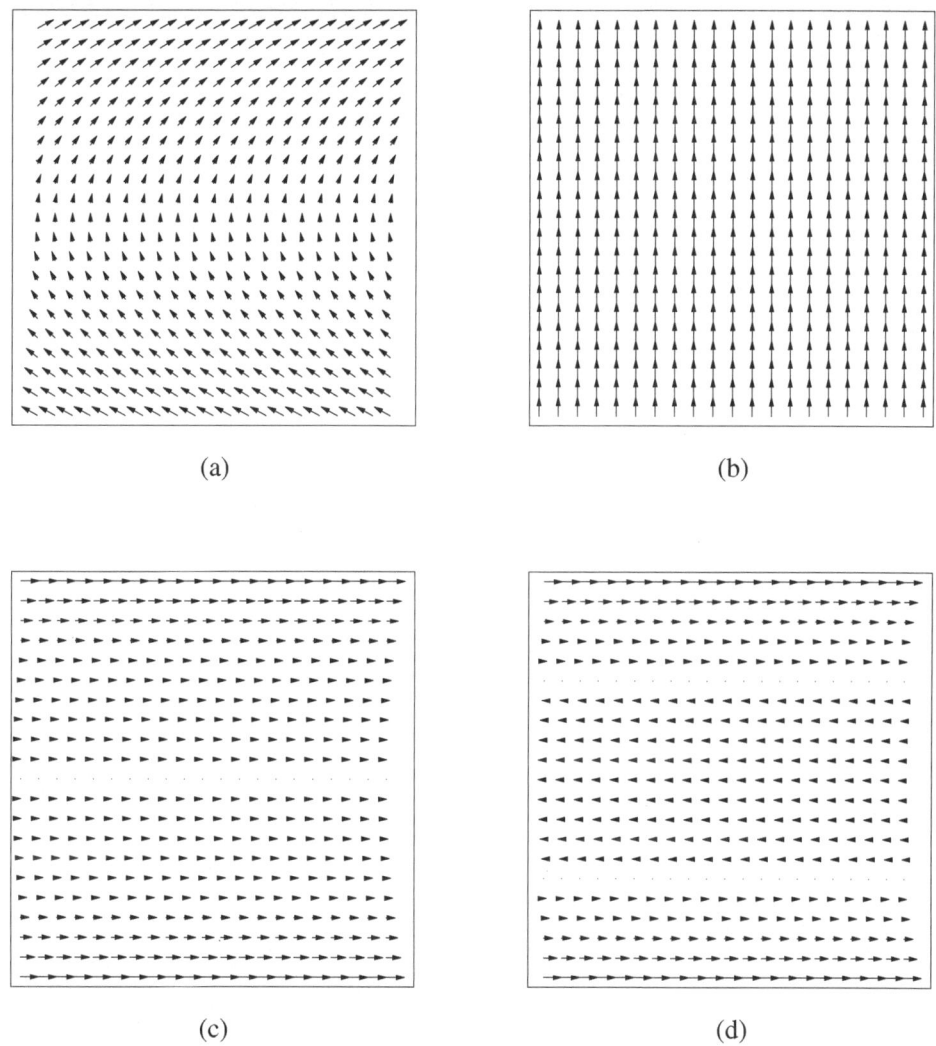

(a)

(b)

(c)

(d)

Figure 4.4: Vector field representation of the relative transformation $\overline{\mathbf{h}}(x,y) = \overline{\mathbf{g}}_1(x,y) - \overline{\mathbf{g}}_2(x,y)$ where: (a) $\overline{\mathbf{g}}_1(x,y)$ is a slight horizontal shear and $\overline{\mathbf{g}}_2(x,y)$ is a small vertical shift; (b) $\overline{\mathbf{g}}_1(x,y)$ is a small vertical shift and $\overline{\mathbf{g}}_2(x,y)$ is the identity transformation; (c) $\overline{\mathbf{g}}_1(x,y)$ is a slight parabolic transformation and $\overline{\mathbf{g}}_2(x,y)$ is the identity transformation; (d) $\overline{\mathbf{g}}_1(x,y)$ is a slight parabolic transformation and $\overline{\mathbf{g}}_2(x,y)$ is a small horizontal shift of $x_0 = T$ to the right. Compare with Figs. 2.3(g), 2.5(a), 3.5(c) and 3.6(a).

Example 4.8: Consider now the superposition of two identical aperiodic screens one of which undergoes a slight horizontal scaling of $s_x = 1 - \varepsilon$ and a slight vertical scaling of $s_y = 1 + \varepsilon$, where ε is a small positive fraction (see Fig. 2.2(a)). This inhomogeneous

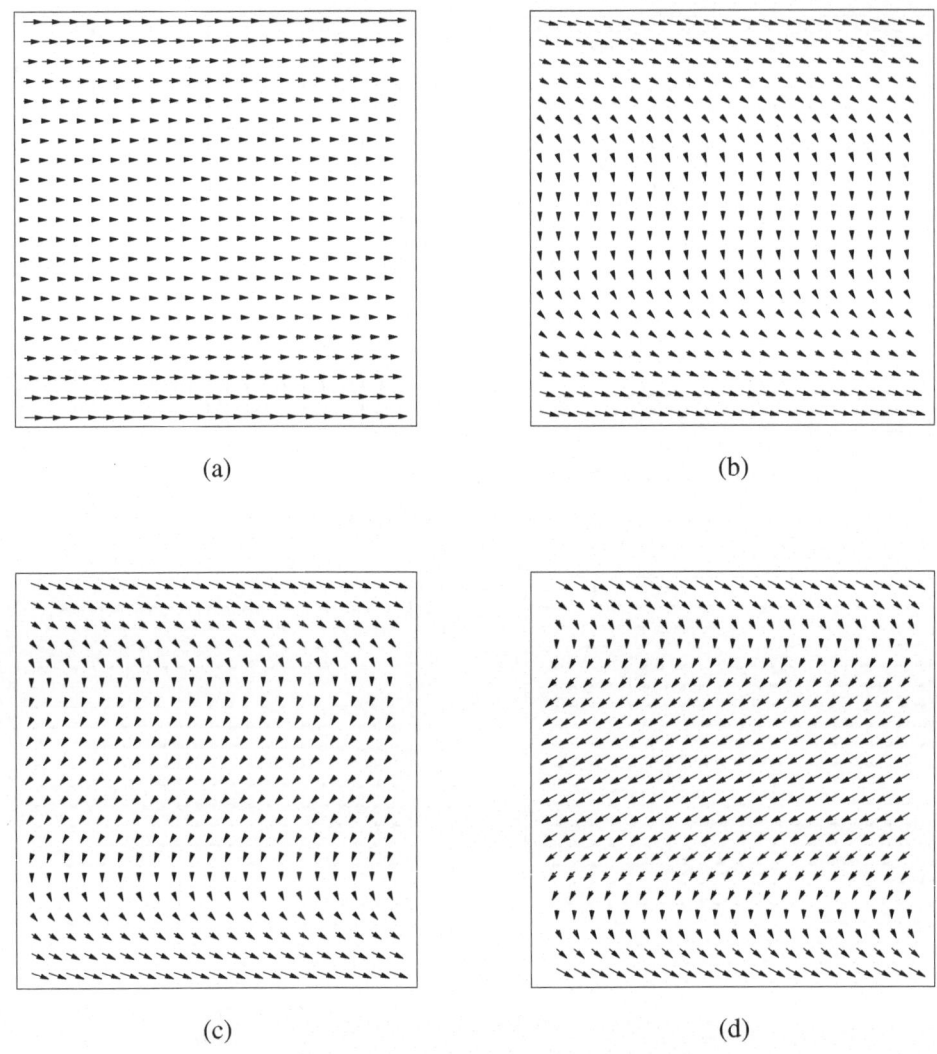

(a)   (b)

(c)   (d)

Figure 4.5: Vector field representation of the relative transformation $\overline{\mathbf{h}}(x,y) =$ $\overline{\mathbf{g}}_1(x,y) - \overline{\mathbf{g}}_2(x,y)$ where: (a) $\overline{\mathbf{g}}_1(x,y)$ is a slight parabolic transformation and $\overline{\mathbf{g}}_2(x,y)$ is a small horizontal shift of $x_0 = -T$; (b) same as in (a) but with a small vertical shift of $y_0 = T$ upward; (c) same as in (a) but with a small shift of $(x_0,y_0) = (T,T)$; (d) same as in (a) but with a small shift of $(x_0,y_0) = (2T,T)$. Compare with Figs. 3.7(a),(c) and 3.8(a),(c).

scaling transformation is expressed by $\overline{\mathbf{g}}(x,y) = ((1-\varepsilon)x, (1+\varepsilon)y)$. In this case the relative transformation between the two layers is:

$$\overline{\mathbf{h}}(x,y) = ((1-\varepsilon)x, (1+\varepsilon)y) - (x,y) = (-\varepsilon x, \varepsilon y)$$

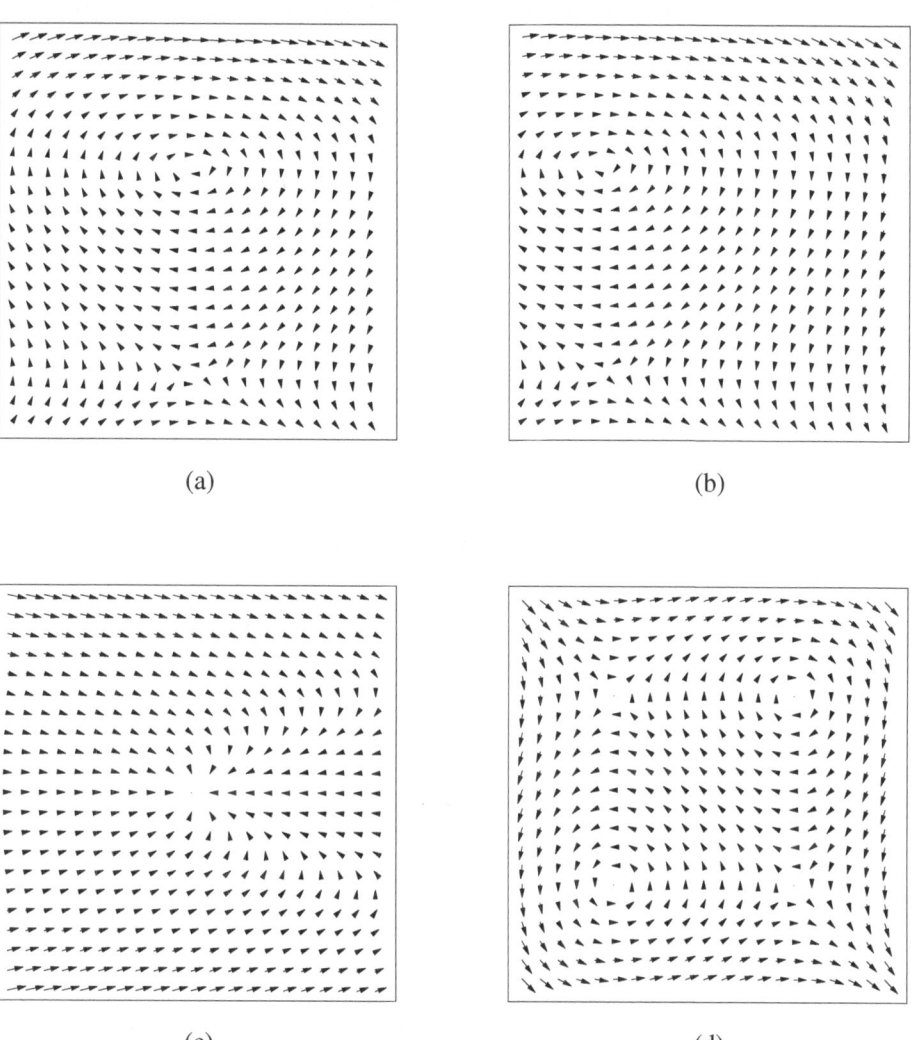

(a)

(b)

(c)

(d)

Figure 4.6: Vector field representation of the relative transformation $\overline{h}(x,y) = \overline{g}_1(x,y) - \overline{g}_2(x,y)$ where: (a) $\overline{g}_1(x,y)$ is a slight parabolic transformation, and $\overline{g}_2(x,y)$ is the transformation consisting of a small rotation of $3°$ and a small horizontal shift of $x_0 = T$ to the right; (b) same as in (a) but with a small shift of $(x_0,y_0) = (T,T)$; (c) $\overline{g}_1(x,y)$ is a slight parabolic transformation, and $\overline{g}_2(x,y)$ is a small scaling up; (d) $\overline{g}_1(x,y)$ is a slight vertical parabolic transformation with a small vertical shift of $y_0 = -T$, and $\overline{g}_2(x,y)$ is a slight horizontal parabolic transformation with a small horizontal shift of $x_0 = -T$. Compare with Figs. 3.11(a), 3.11(c), 3.13(a) and 3.15(a).

or componentwise:

$$\bar{h}_1(x,y) = -\varepsilon x$$

$$\bar{h}_2(x,y) = \varepsilon y$$

The dot trajectories of our layer superposition are given, therefore, by the field lines of the vector field $\bar{\mathbf{h}}(x,y)$ (see Fig. 4.3(a)), and they are expressed by the trajectories (solution curves) of the modified pair of differential equations (4.6):

$$\frac{d}{dt}x(t) = -\varepsilon x(t)$$

$$\frac{d}{dt}y(t) = \varepsilon y(t)$$

The solution curves of this system of differential equations consist of the family of hyperbolas $y = c/x$ (see [Kreyszig93 p. 170]). This is, indeed, the explicit expression of the hyperbolic field lines of the vector field $\bar{\mathbf{h}}(x,y)$ (see Fig. 4.3(a)), and it perfectly agrees with the dot trajectories obtained in our layer superposition (Fig. 2.2(a)). ■

We proceed now with two further examples which illustrate how Eqs. (4.6) explain the shapes of the dot trajectories in two of the more intriguing cases that we have met in the previous chapters: the parabolic dot trajectories of Fig. 2.3(g), and the inversed S-like dot trajectories of Fig. 3.8(a).

Example 4.9: Consider the superposition shown in Fig. 2.3(g). In this case the two originally identical layers have undergone, respectively, a slight horizontal shear transformation, and a small shift of y_0 downward. The *direct* transformations that have been applied to the two layers are given therefore by:

$$\bar{\mathbf{g}}_1(x,y) = (x + ay, y)$$

$$\bar{\mathbf{g}}_2(x,y) = (x, y - y_0)$$

Equivalently, we can also shift the sheared layer in the inverse direction, and leave the second layer unshifted. In this case the transformed layer becomes:

$$\bar{\mathbf{g}}(x,y) = (x + ay, y + y_0)$$

while the other layer remains in its original state. Therefore, the relative transformation between the two layers is:[8]

$$\bar{\mathbf{h}}(x,y) = (x + ay, y + y_0) - (x,y) = (ay, y_0)$$

This transformation is illustrated as a vector field in Fig. 4.4(a). Now, because the two components of this transformation are $\bar{h}_1(x,y) = ay$ and $\bar{h}_2(x,y) = y_0$, Eqs. (4.6) become:

---

[8] Note that although this consideration seems straightforward, its formal justification is only given in Sec. 4.5 (see Eq. (4.8) there).

$$\frac{d}{dt}x(t) = ay(t)$$

$$\frac{d}{dt}y(t) = y_0$$

Although this system of differential equations is non-homogeneous (due to the non-zero constant y_0; see [Kreyszig93 Sec. 4.6]), its solution is almost trivial. From the second equation of the system we obtain by simple integration:

$$y(t) = y_0 t + c_2$$

and therefore, by substitution in the first equation, we obtain:

$$\frac{d}{dt}x(t) = ay_0 t + ac_2$$

This gives by integration:

$$x(t) = \tfrac{1}{2} a y_0 t^2 + ac_2 t + c_1$$

Now, comparing $x(t)$ and $y(t)$ we immediately see that $x(t)$ has the general form of $[y(t)]^2$. To find the precise relationship between them, we note that:

$$[y(t)]^2 = y_0^2 t^2 + 2y_0 c_2 t + c_2^2$$

and therefore we have:

$$x(t) = \frac{a}{2y_0}[y(t)]^2 + const.$$

or, in other words, by eliminating the parameter t and expressing y as a function of x:

$$y = \sqrt{\frac{2y_0}{a}x} + const.$$

This means that the trajectories (solution curves) of our system of differential equations are right-opened parabolas. This is, indeed, the explicit expression of the parabolic field lines of the vector field $\overline{\mathbf{h}}(x,y)$ (see Fig. 4.4(a)), and it perfectly agrees with the dot trajectories that we see in Fig. 2.3(g). Furthermore, this result indicates that as the shift y_0 is increased the parabolas become flat, while as $y_0 \to 0$ the slope of the parabolas tends to zero, until their two branches become horizontal lines when $y_0 = 0$. This is confirmed, indeed, by Fig. 3.19. Note that when a and y_0 have opposite signs the dot trajectories turn into left-opened parabolas. ■

Example 4.10: Consider now the superposition shown in Fig. 3.8(a), with the inversed S-like dot trajectories. As we have seen in Examples 3.3 and 3.11, the transformations which have been applied in this case to the two originally identical layers consist of a weak parabolic transformation in one of the layers and a shift of (x_0,y_0) in the other:

$$\mathbf{g}_1(x,y) = (x - ay^2, y)$$

$$\mathbf{g}_2(x,y) = (x - x_0, y - y_0)$$

Note, however, that these transformations are given in Examples 3.3 and 3.9 as *domain* transformations, while here, in accordance with Remark 4.1, we need to express them as *direct* transformations. We therefore have:

$$\bar{\mathbf{g}}_1(x,y) = [\mathbf{g}_1(x,y)]^{-1} = (x + ay^2, y)$$

$$\bar{\mathbf{g}}_2(x,y) = [\mathbf{g}_2(x,y)]^{-1} = (x + x_0, y + y_0)$$

Equivalently, if we prefer to modify only one of the layers, we can apply a shift in the inverse direction to the layer that underwent the parabolic transformation, and leave the other layer unshifted. In this case the transformed layer becomes:

$$\bar{\mathbf{g}}(x,y) = (x + ay^2 - x_0, y - y_0)$$

while the other layer remains in its original state. Therefore, the relative transformation between the two layers is:[7]

$$\bar{\mathbf{h}}(x,y) = (x + ay^2 - x_0, y - y_0) - (x,y) = (ay^2 - x_0, -y_0)$$

This transformation is illustrated as a vector field in Fig. 4.5(c). Now, because the two components of this transformation are $\bar{h}_1(x,y) = ay$ and $\bar{h}_2(x,y) = y_0$, Eqs. (4.6) become:

$$\frac{d}{dt}x(t) = a[y(t)]^2 - x_0$$

$$\frac{d}{dt}y(t) = -y_0$$

Just like in the previous example, this system of differential equations is non-homogeneous, due to the non-zero constants x_0 and y_0, but its solution is quite simple. From the second equation of the system we obtain by integration:

$$y(t) = -y_0 t + c_2$$

and therefore, by substitution in the first equation, we obtain:

$$\frac{d}{dt}x(t) = a(-y_0 t + c_2)^2 - x_0$$

$$= ay_0^2 t^2 - 2ay_0 c_2 t + ac_2^2 - x_0$$

which gives by integration:

$$x(t) = \frac{ay_0^2}{3}t^3 - ay_0 c_2 t^2 + ac_2^2 t - x_0 t + c_1$$

Now, comparing $x(t)$ and $y(t)$ we see that $x(t)$ has the general form of $[y(t)]^3$. To find the precise relationship between them, we note that raising $y(t)$ to the third power gives:

$$[y(t)]^3 = -y_0^3 t^3 + 3c_2 y_0^2 t^2 - 3c_2^2 y_0 t + c_2^3$$

and hence:

$$-\frac{a}{3y_0}[y(t)]^3 = \frac{ay_0^2}{3}t^3 - ac_2 y_0 t^2 + ac_2^2 t - \frac{a}{3y_0}c_2^3$$

Therefore we have:

$$x(t) = -\frac{a}{3y_0}[y(t)]^3 - x_0 t + const.$$

and noting from $y(t) = -y_0 t + c_2$ that $x_0 t$ is, in fact, $-\frac{x_0}{y_0} y(t) + const.$:

$$x(t) = -\frac{a}{3y_0}[y(t)]^3 + \frac{x_0}{y_0} y(t) + const.$$

Therefore, by eliminating the parameter t we obtain:

$$x = -\frac{a}{3y_0}y^3 + \frac{x_0}{y_0}y + const.$$

This means that the trajectories (solution curves) of our system of differential equations are third-order parabolas. This is, indeed, the explicit expression of the field lines of the vector field $\overline{h}(x,y)$ (see Fig. 4.5(c)), and it perfectly agrees with the dot trajectories that we see in Fig. 3.8(a) and explains their intriguing inversed S-like shapes. Furthermore, this result explains why the third-order parabolas become flat as y_0 is increased (compare Figs. 3.8(a) and 3.9(a)), and why their extrema move away from the x axis as x_0 is increased (compare Figs. 3.8(a) and 3.8(c)). This last result follows from the fact that the extrema of these third-order parabolas are attained where $\frac{d}{dy}x = -\frac{a}{y_0}y^2 + \frac{x_0}{y_0} = 0$, namely, along the horizontal lines $y = \pm\sqrt{x_0/a}$. Incidentally, these horizontal lines are identical to those of Eq. (3.21), which express the fixed lines in the case where $y_0 = 0$ (see Example 3.3).   ■

As we can see, in all of the examples above the trajectories of the modified pair of differential equations (4.6) perfectly correspond to the dot trajectories in the layer superposition. However, it turns out that in some particular cases there still exist minor discrepancies: When the layer transformation $\overline{g}(x,y)$ is a rotation by angle α, the solution curves of the system of differential equations (4.6) consist of spirals that converge to the center of rotation (see [Birkhoff89 p. 144]), while the dot trajectories we observe in the screen superposition (Fig. 2.1(c)) as well as the arrow alignment of the vector field $\overline{h}(x,y)$ (Fig. 4.2(a)) consist of circles surrounding the center of rotation. How can we explain this discrepancy?

It turns out that the arrow representation of the vector field $\overline{h}(x,y)$ fully corresponds to our dot trajectories, because each of the arrows connects the location of the same screen dot in the transformed and in the untransformed layers. But the solution curves (trajectories) of the modified system of differential equations (4.6) do not always correspond to the motion of a screen dot under the given layer transformation: Although the solution curves of this system of differential equations express mathematically the field lines of the relative layer transformation $\overline{h}(x,y)$, these field lines are not always precisely what we are looking for but only a close approximation. This point is discussed further in Problem 4-15. It should be mentioned, however, that since we in any way must restrict ourselves to weak layer transformations (in order not to destroy the correlation in the superposition), this discrepancy — whenever it occurs — is rather marginal and it can be neglected for all practical needs.

4.5 The dot trajectories when both layers undergo transformations

What happens now to the dot trajectories if we take one step further and allow both of the superposed layers to be transformed, one by a mapping $\overline{\mathbf{g}}_1(x,y)$ and the other by a mapping $\overline{\mathbf{g}}_2(x,y)$ (see Sec. 3.5 above)? As a straightforward generalization of Eq. (4.5), one would expect the dot trajectories in this case to be represented by the vector field of the relative transformation between the two layers, namely:

$$\overline{\mathbf{h}}(x,y) = \overline{\mathbf{g}}_1(x,y) - \overline{\mathbf{g}}_2(x,y) \tag{4.7}$$

Note, however, that in this case the dot pairs which make up the dot trajectories in the superposition no longer represent a dot's location before and after the layer transformation has been applied, but rather the new locations of the same original dot under the transformation $\overline{\mathbf{g}}_1(x,y)$ and under the transformation $\overline{\mathbf{g}}_2(x,y)$. And indeed, unlike the vector field (4.5), which perfectly corresponds to the dot trajectories that are obtained when *one* of the superposed layers is transformed, it turns out that in cases where *both* layers are transformed the vector field (4.7) only provides an *approximation* to the dot trajectories that are generated in the superposition.[9] This fact can be clearly illustrated as follows:

Consider the vector field (4.7), $\overline{\mathbf{h}}(x,y) = \overline{\mathbf{g}}_1(x,y) - \overline{\mathbf{g}}_2(x,y)$. Obviously, if we replace the transformations $\overline{\mathbf{g}}_1(x,y)$ and $\overline{\mathbf{g}}_2(x,y)$ in this vector field by the transformations $\overline{\mathbf{g}}_1(x,y) + \overline{\mathbf{f}}(x,y)$ and $\overline{\mathbf{g}}_2(x,y) + \overline{\mathbf{f}}(x,y)$, respectively, where $\overline{\mathbf{f}}(x,y)$ stands for any arbitrary transformation, the resulting vector field remains unchanged, because:

$$(\overline{\mathbf{g}}_1(x,y) + \overline{\mathbf{f}}(x,y)) - (\overline{\mathbf{g}}_2(x,y) + \overline{\mathbf{f}}(x,y)) = \overline{\mathbf{g}}_1(x,y) - \overline{\mathbf{g}}_2(x,y) = \overline{\mathbf{h}}(x,y)$$

However, it turns out that if we apply to our original layers the transformations $\overline{\mathbf{g}}'_1(x,y) = \overline{\mathbf{g}}_1(x,y) + \overline{\mathbf{f}}(x,y)$ and $\overline{\mathbf{g}}'_2(x,y) = \overline{\mathbf{g}}_2(x,y) + \overline{\mathbf{f}}(x,y)$ rather than $\overline{\mathbf{g}}_1(x,y)$ and $\overline{\mathbf{g}}_2(x,y)$, the shapes of the dot trajectories in the superposition will generally be altered (although this does not necessarily happen in all cases). This is clearly demonstrated in Fig. 4.7, where the straight dot trajectories shown in (a) turn into slightly curved dot trajectories in (c) due to the addition of the same transformation $\overline{\mathbf{f}}(x,y) = (0, ax^2)$ to both $\overline{\mathbf{g}}_1(x,y)$ and $\overline{\mathbf{g}}_2(x,y)$. Another indication to the fact that the vector field (4.7) is just a close approximation to our dot trajectories is discussed in Problem 4-18.

If so, what is the generalization of vector field (4.5) that accurately represents the dot trajectories in superpositions where both layers undergo transformations? A first hint to the answer is obtained from the following observation:

Suppose we are given two identical aperiodic dot screens that are superposed on top of each other in full coincidence, dot on dot. Clearly, if we apply to both layers the same

---

[9] Note that the discrepancy mentioned at the end of the last section, where only one of the layers is transformed, does not affect the representation of the dot trajectories by the vector field $\overline{\mathbf{h}}(x,y) = \overline{\mathbf{g}}(x,y) - (x,y)$ but only their representation by the differential equations (4.6). However, in the present case, where both layers are transformed, we are facing a more serious discrepancy that affects the vector field $\overline{\mathbf{h}}(x,y) = \overline{\mathbf{g}}_1(x,y) - \overline{\mathbf{g}}_2(x,y)$ itself, and consequently Eqs. (4.6) are affected, too.

transformation $\bar{\mathbf{f}}(x,y)$, we still remain with two identical aperiodic screens that are superposed in full coincidence. Therefore, if we now apply transformations $\bar{\mathbf{g}}_1(x,y)$ and $\bar{\mathbf{g}}_2(x,y)$ to the two transformed layers, the resulting dot trajectories will have the same shapes as the dot trajectories that would be obtained by applying $\bar{\mathbf{g}}_1(x,y)$ and $\bar{\mathbf{g}}_2(x,y)$ to the original, untransformed layers. This is clearly illustrated in Fig. 4.8, where the dot trajectories obtained in (a) and (c) have equivalent shapes, although the individual dot locations may have changed. In other words, we have the following result:

Proposition 4.2: The dot trajectories obtained by applying the transformations $\bar{\mathbf{g}}_1(x,y)$ and $\bar{\mathbf{g}}_2(x,y)$ to two identical aperiodic dot screens are *equivalent* to the dot trajectories that are obtained by applying to the same original dot screens the transformations $\bar{\mathbf{g}}'_1(x,y) = \bar{\mathbf{g}}_1(\bar{\mathbf{f}}(x,y))$ and $\bar{\mathbf{g}}'_2(x,y) = \bar{\mathbf{g}}_2(\bar{\mathbf{f}}(x,y))$, where $\bar{\mathbf{f}}(x,y)$ is any arbitrary transformation. ∎

It is interesting to note, however, that this invariance under transformation composition is only true for the shapes of the dot trajectories, but it has no parallel for moiré effects in the periodic case, which do vary under such compositions (see Figs. 4.10(b),(d)). We will return to this point in more detail in Sec. 5.4 of Chapter 5.

Now, since $\bar{\mathbf{f}}(x,y)$ stands here for any arbitrary transformation, it is clear that Proposition 4.2 remains also true in the particular case where $\bar{\mathbf{f}}$ is the inverse of the transformation $\bar{\mathbf{g}}_2(x,y)$, namely, $\bar{\mathbf{f}}(x,y) = \mathbf{g}_2(x,y)$. This means that the dot trajectories obtained by applying the transformations $\bar{\mathbf{g}}_1(x,y)$ and $\bar{\mathbf{g}}_2(x,y)$ to our original screens are equivalent to the dot trajectories that are obtained by applying to our original screens the transformations $\bar{\mathbf{g}}'_1(x,y) = \bar{\mathbf{g}}_1(\mathbf{g}_2(x,y))$ and $\bar{\mathbf{g}}'_2(x,y) = (x,y)$, where $\mathbf{g}_2$ is the inverse of the direct transformation $\bar{\mathbf{g}}_2$. Now, this last superposition has the particularity that only one of its two layers has been transformed. Therefore, by virtue of Proposition 4.1, we see that the vector field which accurately represents the dot trajectories of this superposition, and hence also the dot trajectories obtained by applying the transformations $\bar{\mathbf{g}}_1(x,y)$ and $\bar{\mathbf{g}}_2(x,y)$ to our original screens, is given by:

$$\bar{\mathbf{h}}_1(x,y) = \bar{\mathbf{g}}_1(\mathbf{g}_2(x,y)) - (x,y) \tag{4.8}$$

(or equivalently, by $-\bar{\mathbf{h}}_1(x,y)$).

Fig. 4.11 shows in parts (a) and (b) the vector fields (4.7) and (4.8) for the superposition of Fig. 4.7(c). Similarly, Fig. 4.12 shows in parts (a) and (b) the vector fields (4.7) and (4.8) for the superposition of Fig. 4.8(c). As we can see, in both figures the vector fields (4.8) shown in part (b) perfectly corresponds to the dot trajectories, while the vector field (4.7) shown in part (a) provides only an approximation.

Note, however, that using a similar reasoning we could also take $\bar{\mathbf{f}}$ in Proposition 4.2 to be the inverse of the transformation $\bar{\mathbf{g}}_1(x,y)$, namely $\bar{\mathbf{f}}(x,y) = \mathbf{g}_1(x,y)$. This means that the dot trajectories obtained by applying $\bar{\mathbf{g}}_1(x,y)$ and $\bar{\mathbf{g}}_2(x,y)$ to our original screens are also equivalent to the dot trajectories that are obtained by applying to our original screens the transformations $\bar{\mathbf{g}}'_1(x,y) = (x,y)$ and $\bar{\mathbf{g}}'_2(x,y) = \bar{\mathbf{g}}_2(\mathbf{g}_1(x,y))$, where $\mathbf{g}_1$ is the inverse of the

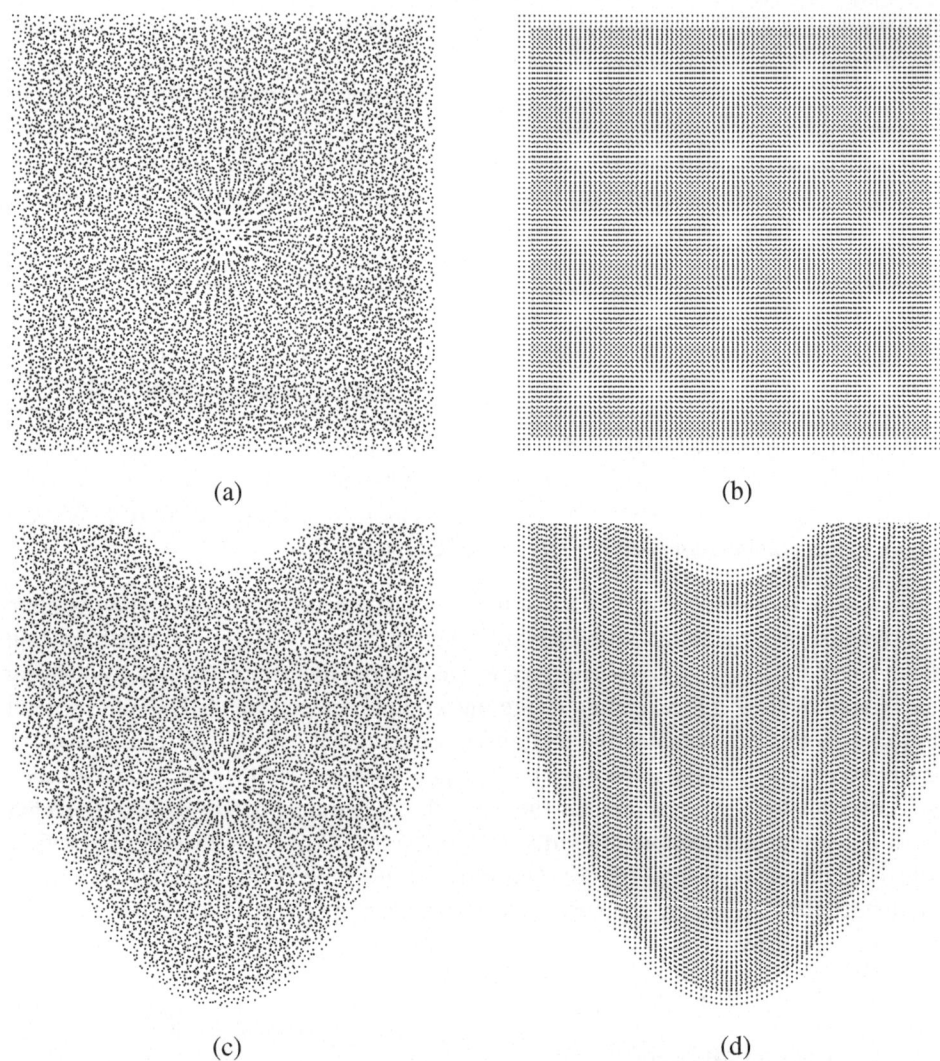

(a)   (b)

(c)   (d)

Figure 4.7: The dot trajectories obtained in the superposition of aperiodic screens are *not* invariant under the addition of a transformation $\bar{\mathbf{f}}(x,y)$ to the two direct transformations $\bar{\mathbf{g}}_1(x,y)$ and $\bar{\mathbf{g}}_2(x,y)$ undergone by the original layers. The same is also true for the $(1,-1)$-moirés in the periodic case, although the $(1,-1)$-moirés *are* invariant under additions in terms of *domain* transformations, as shown in Fig. 5.5. (a) Two identical random screens that have undergone the direct transformations $\bar{\mathbf{g}}_1(x,y) = (sx,sy)$ and $\bar{\mathbf{g}}_2(x,y) = (x,y)$. (b) The periodic counterpart of (a). (c) The same original random screens after having undergone, instead, the transformations $\bar{\mathbf{g}}'_1(x,y) = \bar{\mathbf{g}}_1(x,y) + \bar{\mathbf{f}}(x,y)$ and $\bar{\mathbf{g}}'_2(x,y) = \bar{\mathbf{g}}_2(x,y) + \bar{\mathbf{f}}(x,y)$, where $\bar{\mathbf{f}}(x,y) = (0, ax^2)$. Note the clear difference between the dot trajectories in (c) and (a). (d) The counterpart of (c) based on periodic screens. The moiré effects in (b) and (d) are not identical, since in terms of the domain transformations we have in (b) $\mathbf{g}_1(x,y) = \bar{\mathbf{g}}_1^{-1}(x,y) = (x/s,y/s)$ and $\mathbf{g}_2(x,y) = (x,y)$, while in (d) we have $\mathbf{g}'_1(x,y) = [\bar{\mathbf{g}}'_1(x,y)]^{-1} = (x/s, y/s - (a/s^3)x^2)$ and $\mathbf{g}'_2(x,y) = [\bar{\mathbf{g}}'_2(x,y)]^{-1} = (x, y - ax^2)$ which do not satisfy $\mathbf{g}'_1 - \mathbf{g}'_2 = \mathbf{g}_1 - \mathbf{g}_2$.

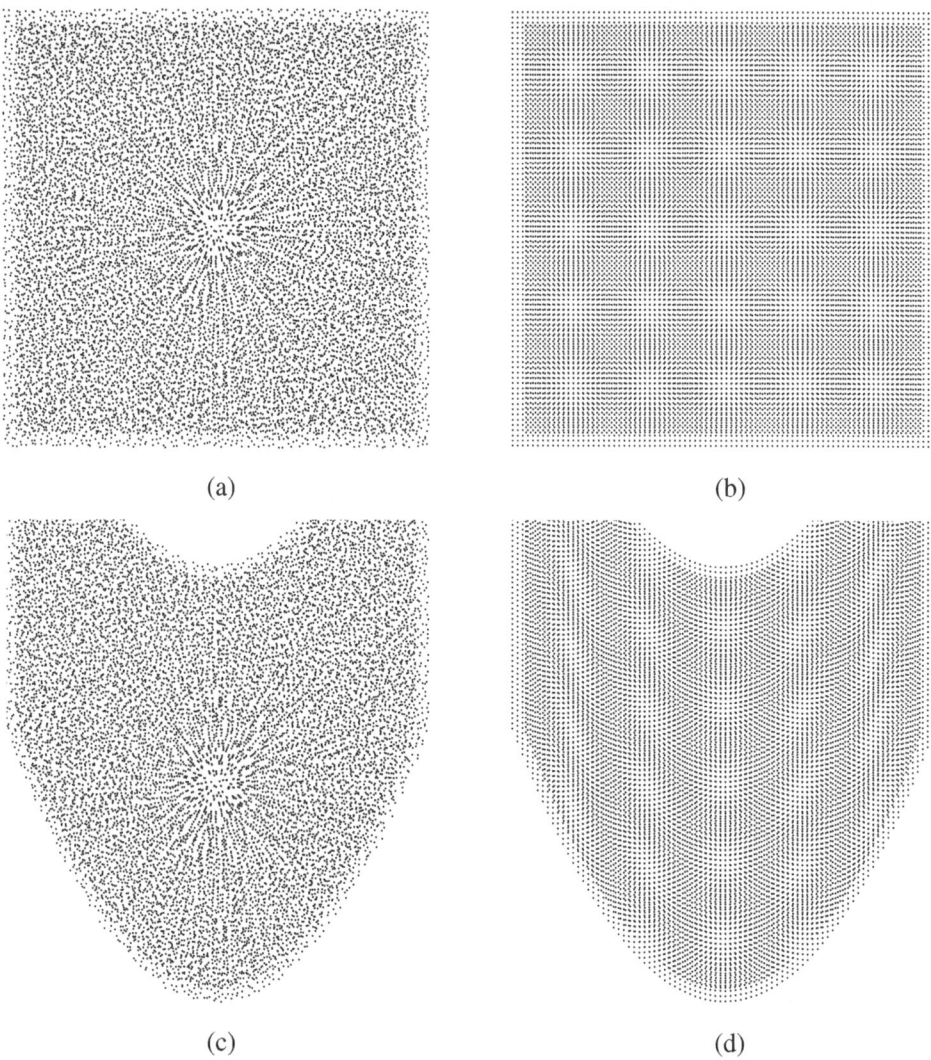

(a)   (b)

(c)   (d)

Figure 4.8: The dot trajectories obtained in the superposition of aperiodic screens are invariant under composition: Applying $\overline{\mathbf{g}}_1(x,y)$ and $\overline{\mathbf{g}}_2(x,y)$ to two aperiodic layers gives the same dot trajectories as the application of $\overline{\mathbf{g}}_1(x,y)$ and $\overline{\mathbf{g}}_2(x,y)$ to the same original layers after they have been transformed by $\mathbf{f}(x,y)$. (a) The same superposition as in Fig. 4.7(a): Two identical random screens that have undergone the direct transformations $\overline{\mathbf{g}}_1(x,y) = (sx,sy)$ and $\overline{\mathbf{g}}_2(x,y) = (x,y)$. (b) The periodic counterpart of (a). (c) The same original random screens after having undergone the transformations $\overline{\mathbf{g}}'_1(x,y) = \overline{\mathbf{g}}_1(\mathbf{f}(x,y))$ and $\overline{\mathbf{g}}'_2(x,y) = \overline{\mathbf{g}}_2(\mathbf{f}(x,y))$ instead of $\overline{\mathbf{g}}_1(x,y)$ and $\overline{\mathbf{g}}_1(x,y)$, where $\mathbf{f}(x,y) = (x, y+ax^2)$. The composite transformations that have been applied are, therefore, $\overline{\mathbf{g}}'_1(x,y) = (sx, sy+sax^2)$ and $\overline{\mathbf{g}}'_2(x,y) = (x, y+ax^2)$. Unlike in Fig. 4.7, the dot trajectories in (c) and (a) are equivalent. (d) The counterpart of (c) based on periodic screens.

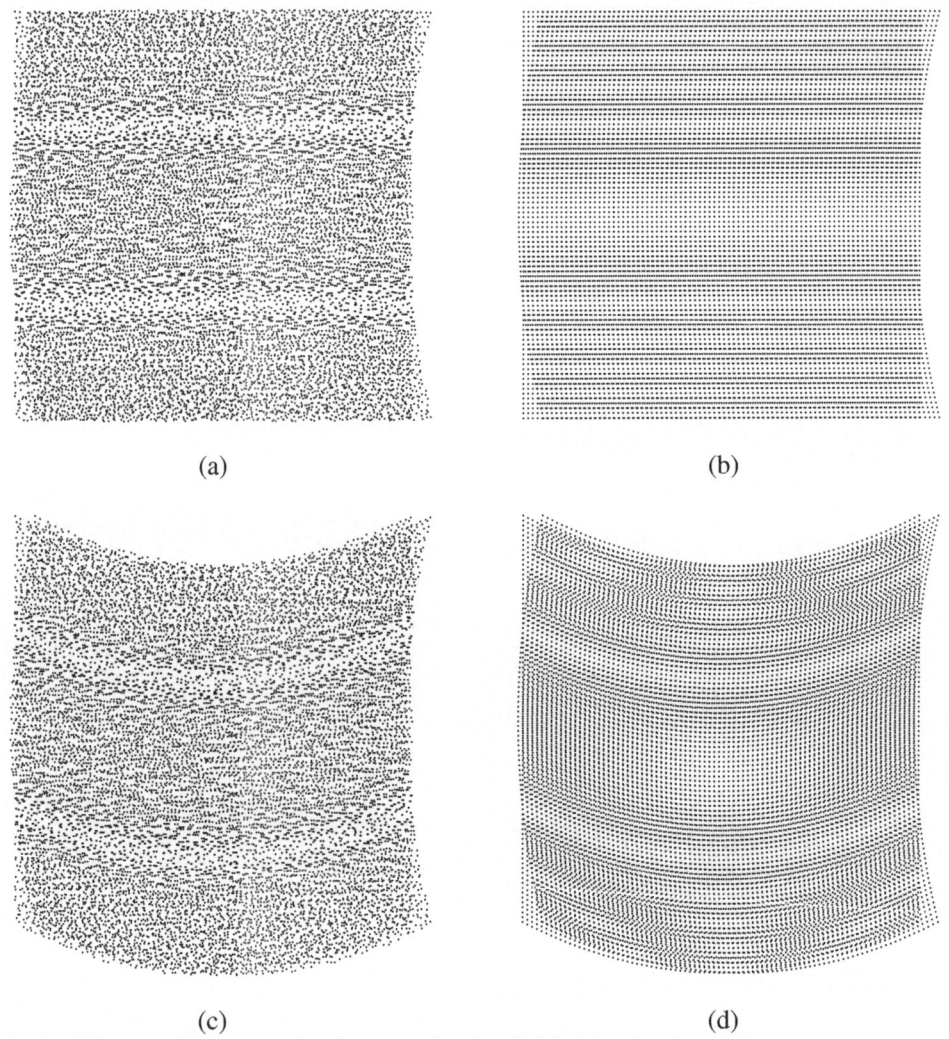

(a)   (b)

(c)   (d)

Figure 4.9: Moiré and Glass patterns are *not* invariant under the addition of a transformation
$\bar{\mathbf{f}}(x,y)$ to the two direct transformations $\bar{\mathbf{g}}_1(x,y)$ and $\bar{\mathbf{g}}_2(x,y)$ undergone by the
original layers (although they *are* invariant under additions in terms of *domain*
transformations, as shown in Fig. 5.7). Note that although in this particular case
the dot trajectories do remain invariant, this is not generally true (see Fig. 4.7 and
Problem 4-16). (a) The Glass pattern obtained by applying the direct trans-
formations $\bar{\mathbf{g}}_1(x,y) = (x+ay^2, y)$ and $\bar{\mathbf{g}}_2(x,y) = (x+x_0, y)$ to two identical random
screens. (b) the full (1,-1)-moiré obtained by applying the same direct trans-
formations to two identical periodic screens. (c) The results obtained by applying
to the same original random screens the direct transformations $\bar{\mathbf{g}}_1(x,y) + \bar{\mathbf{f}}(x,y)$ and
$\bar{\mathbf{g}}_2(x,y) + \bar{\mathbf{f}}(x,y)$, where $\bar{\mathbf{f}}(x,y) = (0, bx^2)$, $b > a$. Note that the Glass patterns have
been distorted, although the dot trajectories remain unchanged. (d) The periodic
counterpart of (c); note that the macroscopic shape of the (1,-1)-moiré bands has
been distorted, along with the dot alignments in the microstructure.

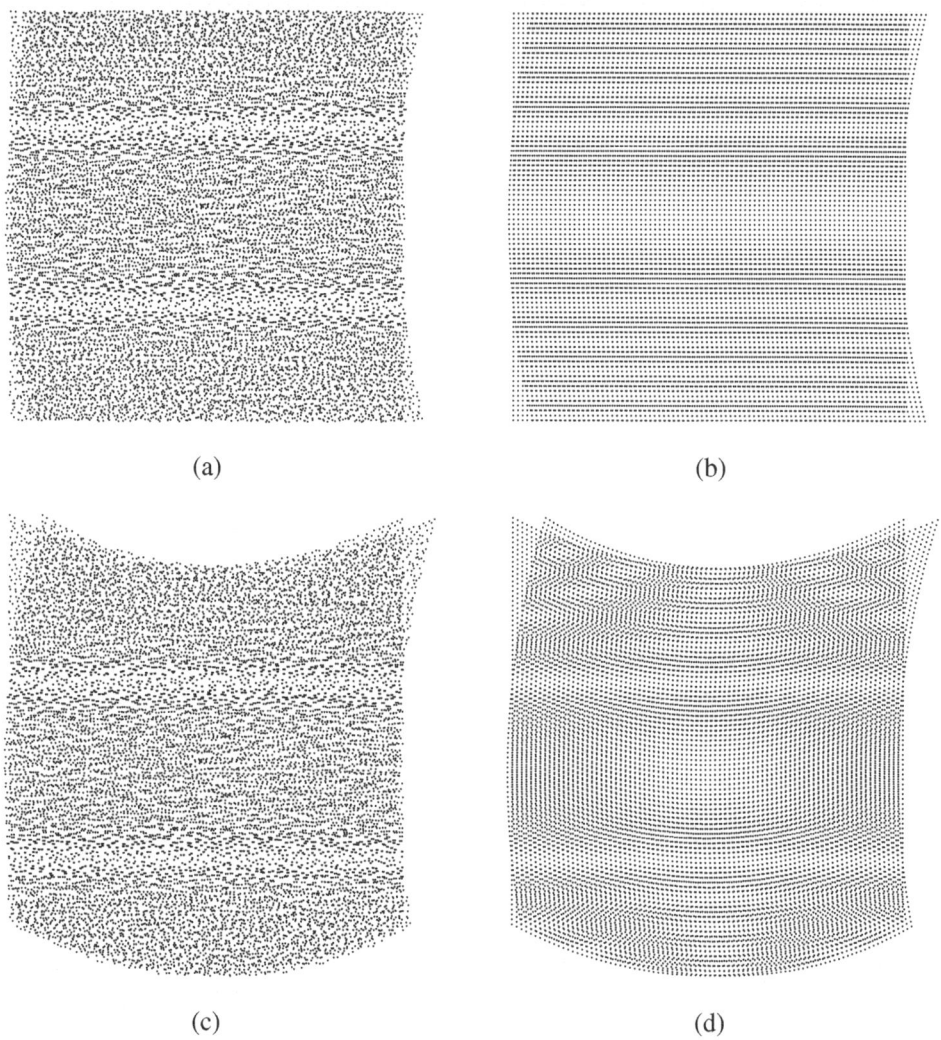

(a)

(b)

(c)

(d)

Figure 4.10: Dot trajectories and Glass patterns are invariant under composition: Applying $\bar{\mathbf{g}}_1(x,y)$ and $\bar{\mathbf{g}}_2(x,y)$ to two aperiodic layers gives the same dot trajectories as the application of $\bar{\mathbf{g}}_1(x,y)$ and $\bar{\mathbf{g}}_2(x,y)$ to the same original layers after they have been transformed by $\mathbf{f}(x,y)$. (a) Two identical random screens that have undergone the direct transformations $\bar{\mathbf{g}}_1(x,y) = (x+ay^2, y)$ and $\bar{\mathbf{g}}_2(x,y) = (x+x_0, y)$. (b) The periodic counterpart of (a). (c) The same original random screens after having undergone the direct mappings $\bar{\mathbf{g}}'_1(x,y) = \bar{\mathbf{g}}_1(\mathbf{f}(x,y))$ and $\bar{\mathbf{g}}'_2(x,y) = \bar{\mathbf{g}}_2(\mathbf{f}(x,y))$ instead of $\bar{\mathbf{g}}_1(x,y)$ and $\bar{\mathbf{g}}_2(x,y)$, where $\mathbf{f}(x,y) = (x, y+bx^2)$, $b > a$. The composite transformations that have been applied are, therefore, $\bar{\mathbf{g}}'_1(x,y) = (x+a[y+bx^2]^2, y+bx^2)$ and $\bar{\mathbf{g}}'_2(x,y) = (x+x_0, y+bx^2)$. (d) The counterpart of (c) based on periodic screens; note that only the real Glass patterns (i.e. the moiré bands that are generated around the two fixed lines) remain invariant, while all the other bands of the (1,-1)-moiré have been distorted. The dot alignments in the microstructure of (d) follow the same distortions as the original layers.

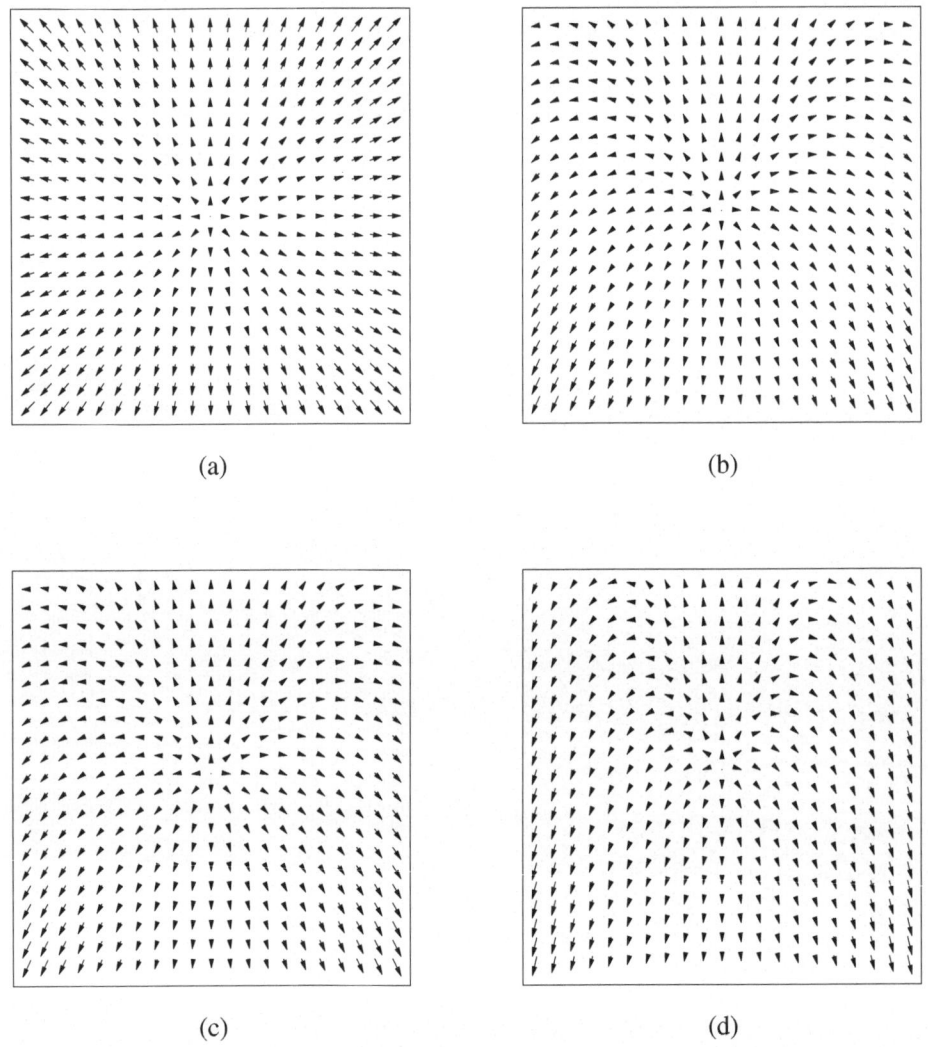

(a)   (b)

(c)   (d)

Figure 4.11: The vector fields obtained by Eqs. (4.7), (4.8) and (4.9) for the super-position of Fig. 4.7(c). (a) The vector field $\mathbf{h}(x,y) = \overline{\mathbf{g}}'_1(x,y) - \overline{\mathbf{g}}'_2(x,y)$, where $\overline{\mathbf{g}}'_1(x,y) = \overline{\mathbf{g}}_1(x,y) + \mathbf{f}(x,y)$ and $\overline{\mathbf{g}}'_2(x,y) = \overline{\mathbf{g}}_2(x,y) + \mathbf{f}(x,y)$, does not represent the dot trajectories of Fig. 4.7(c) accurately because the layer transformations $\overline{\mathbf{g}}'_1(x,y)$ and $\overline{\mathbf{g}}'_2(x,y)$ are not sufficiently weak (due to the addition of $\mathbf{f}(x,y) = (0, ax^2)$). (b) The vector field $\mathbf{h}_1(x,y) = \overline{\mathbf{g}}'_1(\mathbf{g}'_2(x,y)) - (x,y)$ of Eq. (4.8) represents correctly the dot trajectories of Fig. 4.7(c). (c) The vector field $\mathbf{h}_2(x,y) = (x,y) - \overline{\mathbf{g}}'_2(\mathbf{g}'_1(x,y))$ of Eq. (4.9), too, represents correctly the dot trajectories of Fig. 4.7(c). (d) The difference between the vector fields $\mathbf{h}_1(x,y)$ and $\mathbf{h}_2(x,y)$ is not zero, meaning that they are not identical. The connection between them is explained in Sec. H.2 of Appendix H.

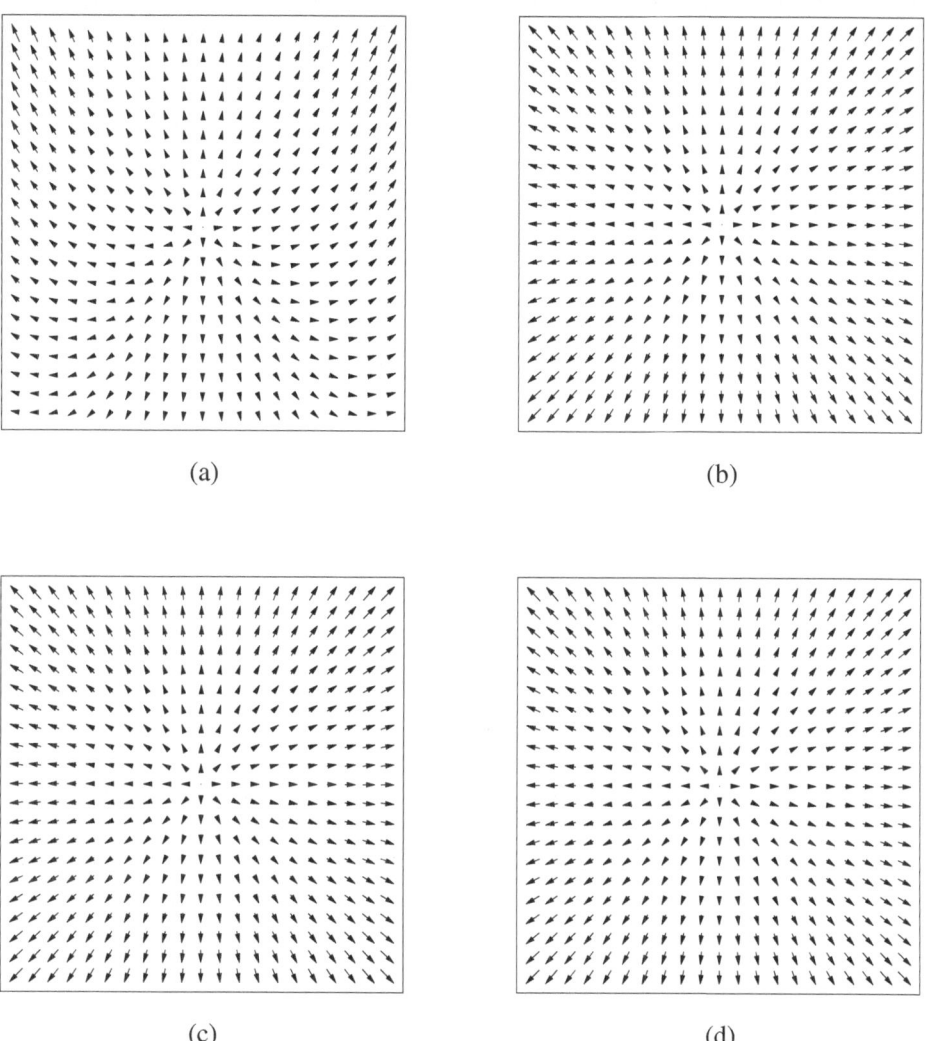

(a)   (b)

(c)   (d)

Figure 4.12: The vector fields obtained by Eqs. (4.7), (4.8) and (4.9) for the super-position of Fig. 4.8(c). (a) The vector field $\mathbf{h}(x,y) = \overline{\mathbf{g}}'_1(x,y) - \overline{\mathbf{g}}'_2(x,y)$, where $\overline{\mathbf{g}}'_1(x,y) = \overline{\mathbf{g}}_1(\overline{\mathbf{f}}(x,y))$ and $\overline{\mathbf{g}}'_2(x,y) = \overline{\mathbf{g}}_2(\overline{\mathbf{f}}(x,y))$, does not represent the dot trajectories of Fig. 4.8(c) accurately because the layer transformations $\overline{\mathbf{g}}'_1(x,y)$ and $\overline{\mathbf{g}}'_2(x,y)$ are not sufficiently weak (due to the composition with $\overline{\mathbf{f}}(x,y) = (x, y + ax^2)$). (b) The vector field $\overline{\mathbf{h}}_1(x,y) = \overline{\mathbf{g}}'_1(\mathbf{g}'_2(x,y)) - (x,y)$ of Eq. (4.8) represents correctly the dot trajectories of Fig. 4.8(c). (c) The vector field $\overline{\mathbf{h}}_2(x,y) = (x,y) - \overline{\mathbf{g}}'_2(\mathbf{g}'_1(x,y))$ of Eq. (4.9), too, represents correctly the dot trajectories of Fig. 4.8(c). (d) The difference between the vector fields $\overline{\mathbf{h}}_1(x,y)$ and $\overline{\mathbf{h}}_2(x,y)$ is not zero, meaning that they are not identical. The connection between them is explained in Sec. H.2 of Appendix H.

direct transformation $\overline{g}_1$. In this case we obtain, by virtue of Proposition 4.1, that the precise mathematical representation of our dot trajectories is provided by the vector field:

$$\overline{h}_2(x,y) = (x,y) - \overline{g}_2(g_1(x,y)) \tag{4.9}$$

(or equivalently, by the vector field $-\overline{h}_2(x,y)$). This is illustrated in Fig. 4.11(c), which shows the vector field (4.9) that corresponds to the superposition of Fig. 4.7(c). Similarly, Fig. 4.12(c) shows the vector field (4.9) that corresponds to the superposition of Fig. 4.8(c). In both cases this vector field correctly represents the dot trajectories.

The connection between the vector fields (4.8) and (4.9) is explained in detail in Sec. H.2 of Appendix H.

It should be noted, finally, as shown in Proposition D.11 in Appendix D, that when $\overline{g}_1(x,y)$ and $\overline{g}_2(x,y)$ are weak transformations, vector field (4.7) closely approximates the accurate vector fields (4.8) and (4.9). This explains, indeed, why vector field (4.7) gives a good approximation to the dot trajectories in cases where both of the layers are being transformed, provided that both layer transformations are rather weak.

We can therefore present our results so far by the following generalization of Proposition 4.1:

Proposition 4.3: Suppose that we are given two identical aperiodic dot screens that are superposed on top of each other in full coincidence, dot on dot. When we apply to the two layers the transformations $\overline{g}_1(x,y)$ and $\overline{g}_2(x,y)$, respectively, we obtain in the superposition dot trajectories that correspond to the vector fields $\overline{h}_1(x,y) = \overline{g}_1(g_2(x,y)) - (x,y)$ or $\overline{h}_2(x,y) = (x,y) - \overline{g}_2(g_1(x,y))$ (or equivalently, to the vector fields $-\overline{h}_1(x,y)$ or $-\overline{h}_2(x,y)$), where g_1 and g_2 are the inverse of the direct transformations $\overline{g}_1$ and $\overline{g}_2$. Furthermore, if $\overline{g}_1$ and $\overline{g}_2$ are weak transformations, then the vector field $\overline{h}(x,y) = \overline{g}_1(x,y) - \overline{g}_2(x,y)$ (or $-\overline{h}(x,y)$) provides a close approximation to the dot trajectories in the superposition. ■

It should be noted, however, that although the vector field $\overline{h}(x,y) = \overline{g}_1(x,y) - \overline{g}_2(x,y)$ only *approximates* our dot trajectories, it often turns out to be more practical to use than the accurate vector fields $\overline{h}_1(x,y) = \overline{g}_1(g_2(x,y)) - (x,y)$ or $\overline{h}_2(x,y) = (x,y) - \overline{g}_2(g_1(x,y))$. The reason is that the explicit forms of the vector fields $\overline{h}_1(x,y)$ and $\overline{h}_2(x,y)$ may be quite complex, because they include mapping compositions; furthermore, in many cases they may even not be available, since we do not always have the explicit forms of both of the direct and inverse transformations required. Therefore, in cases where the use of the precise vector fields $\overline{h}_1(x,y)$ and $\overline{h}_2(x,y)$ is not practical, we will often use instead the approximated vector field $\overline{h}(x,y) = \overline{g}_1(x,y) - \overline{g}_2(x,y)$. Of course, this approximation is only valid under the assumption that both $\overline{g}_1$ and $\overline{g}_2$ are weak transformations. But this assumption is fully justified in our case: As we have seen in Remark 4.2, although dot trajectories are theoretically generated in the superposition in all cases, they can only be clearly visible if the layer transformations $\overline{g}_1(x,y)$ and $\overline{g}_2(x,y)$ are not too "violent". Otherwise, the correlation between the superposed layers is strongly reduced, and the visual effect in the layer superposition may be lost and no longer correspond to the vector field.

Remark 4.3: As mentioned earlier in Remark 4.1, it is often preferable to express the transformation undergone by a layer $r(x,y)$ as a domain transformation $r(\mathbf{g}(x,y))$ rather than as a direct transformation $(x,y) \mapsto \overline{\mathbf{g}}(x,y)$ (for example: $r(x/2,y/2)$ rather than $(x,y) \mapsto (2x,2y)$). If one prefers to use the domain transformations $\mathbf{g}_1$ and $\mathbf{g}_2$ rather than their direct equivalents $\overline{\mathbf{g}}_1$ and $\overline{\mathbf{g}}_2$,[10] then, in view of Proposition D.10 in Appendix D, the approximated vector field (4.7) can be restated using the approximation:

$$\overline{\mathbf{h}}(x,y) \approx \mathbf{g}_2(x,y) - \mathbf{g}_1(x,y) \tag{4.10}$$

Once again, this approximation only holds if $\mathbf{g}_1$ and $\mathbf{g}_2$ are weak transformations; but as we have just seen, in our case we are anyway limited to weak layer transformations in order not to lose the visual effect in the superposition. Under these circumstances we have, indeed (see Proposition D.11 in Appendix D):

$$\overline{\mathbf{g}}_1(\mathbf{g}_2(x,y)) - (x,y) \approx \mathbf{g}_2(x,y) - \mathbf{g}_1(x,y) \approx \overline{\mathbf{g}}_1(x,y) - \overline{\mathbf{g}}_2(x,y)$$

$$(x,y) - \overline{\mathbf{g}}_2(\mathbf{g}_1(x,y)) \approx \mathbf{g}_2(x,y) - \mathbf{g}_1(x,y) \approx \overline{\mathbf{g}}_1(x,y) - \overline{\mathbf{g}}_2(x,y)$$

Similarly, it is also possible to restate Proposition 4.1 in terms of the domain transformation $\mathbf{g}$ rather than in terms of the direct transformation $\overline{\mathbf{g}}$. ∎

Finally, as we already know from Sec. 3.7, dot trajectories may appear even in aperiodic layer superpositions in which no fixed points exist (see, for example, Figs. 2.3(e),(g), 2.5(a), 3.8(a),(c) and 3.9(a),(c)).[11] Although in such cases there exist no points of perfect coincidence between the two layers, in some areas of the superposition the dots of both layers fall very close to each other, meaning that in these areas the two layers are still correlated. And indeed, dot trajectories are clearly visible in such layer superpositions wherever the correlation between the layers is sufficiently high. These dot trajectories are, indeed, represented by the field lines of the corresponding vector field $\overline{\mathbf{h}}(x,y)$, as shown, for example, in Figs. 4.4(a), 4.4(b) and 4.5(c) (which correspond, respectively, to the superpositions of Figs. 2.3(g), 2.5(a) and 3.8(a)). Let us have a closer look at these three examples.

In the case of Fig. 2.5(a) (a small vertical shift in one of the layers) the vector field $\overline{\mathbf{h}}(x,y)$ has a constant value of $(0,y_0)$ throughout the plane. Therefore, all the arrows of the vector field have the same length, meaning that the distance between the dots in each of the dot pairs which form the dot trajectories is constant throughout the layer superposition, and no Glass patterns (macroscopic gray level variations) are generated. In fact, as we have seen in Sec. 2.3.2, this case is a singular moiré-free superposition.

---

[10] For example, for reasons of compatibility with Chapters 10 and 11 of *Vol. I*, or if the available layer transformation software is based on a backward mapping algorithm (see Sec. D.9.1 in Appendix D).

[11] Note that the cases shown in these figures have really no fixed points — in contrast to other cases where fixed points do exist but they are not visible because they are located outside the image frame.

In the case of Fig. 2.3(g) the vector field is given by $\bar{\mathbf{h}}(x,y) = (ay,y_0)$ (see Example 4.9). This vector field has clearly no zeros, since its second component, y_0, is constant and non-zero throughout the plane. However, unlike in the previous example, in the present case the arrows of the vector field $\bar{\mathbf{h}}(x,y)$ do not have a constant length (i.e. the distance between the dots in each of the dot pairs forming the dot trajectories is not constant). The points (x,y) in which this length is minimal correspond to the almost fixed points of the superposed layers (see also Problem 4-18(d)).

Similarly, in the case of Fig. 3.8(a) the vector field is given by $\bar{\mathbf{h}}(x,y) = (ay^2-x_0, -y_0)$ (see Example 4.10). This vector field, too, has no zeros, since its second component, $-y_0$, is constant and non-zero throughout the plane. But once again, the arrow lengths in $\bar{\mathbf{h}}(x,y)$ are not constant, and the points (x,y) in which this length is minimal correspond to the almost fixed points of the superposed layers.

To conclude this section, Figs. 4.2–4.6 above illustrate, for some of the most interesting cases that we have seen so far, the mathematical interpretation of the superposition's dot trajectories as a vector field. In cases where only one of the layers is transformed the vector field being used is (4.5), and in cases where both layers undergo transformations the vector field is approximated by (4.7) or by (4.10). Note the agreement between the dot trajectories in each of the superpositions and the arrows showing the dot displacements in the corresponding vector field. It should be always remembered, however, that this agreement only concerns areas in the superposition in which the correlation between the two superposed layers is sufficiently high, meaning that the corresponding dots of both layers remain close to each other. In areas where the dots go farther apart from each other (and hence the arrow lengths in the vector field increase) the correlation between the layers is reduced, and the dot trajectories are no longer visible. Note, however, that in such areas field lines still do exist in the vector field — in fact, the vectors in these areas are even longer, since the distance between the corresponding dots in the two layers is bigger. Thus, a visual agreement between the dot trajectories and the vector field is only possible in areas where the correlation between the superposed layers is sufficiently high (meaning that the vectors in the vector field are not too long). This fact can be clearly observed when comparing the vector fields of Figs. 4.2(a), 4.2(b) and 4.5(a) with the respective layer superpositions in Figs. 2.1(c), 2.1(e) and 3.7(a).

4.6 Synthesis of dot trajectories

Having succeeded in analyzing the dot trajectories that occur in the superposition of aperiodic screens, it may be asked now whether we can also *synthesize* such effects. In other words, can we synthesize, for any given vector field $\bar{\mathbf{h}}(x,y)$, two aperiodic dot screens that generate in their superposition dot trajectories corresponding to $\bar{\mathbf{h}}(x,y)$?

Suppose that we are given a vector field $\bar{\mathbf{h}}(x,y)$ or, equivalently, a system of differential equations $\frac{d}{dt}x(t) = \bar{h}_1(x(t),y(t))$, $\frac{d}{dt}y(t) = \bar{h}_2(x(t),y(t))$. We wish to synthesize an aperiodic screen superposition whose dot trajectories visually illustrate our given vector field (or system of differential equations).

We start by superposing two identical aperiodic dot screens on top of each other in full coincidence. As we already know from Proposition 4.1, if we apply a transformation $\bar{g}(x,y)$ to one of the layers, we obtain in the superposition the dot trajectories that correspond to $\bar{h}(x,y) = \bar{g}(x,y) - (x,y)$ (or equivalently to $-\bar{h}(x,y) = (x,y) - \bar{g}(x,y)$). Therefore, in order to obtain in the superposition dot trajectories which correspond to $\bar{h}(x,y)$, we may apply to one of the superposed layers the transformation:

$$\bar{g}(x,y) = (x,y) + \bar{h}(x,y) \tag{4.11}$$

(or equivalently, $\bar{g}(x,y) = (x,y) - \bar{h}(x,y)$), while the other layer remains unchanged.

Similarly, it follows from Proposition 4.3 that by applying a transformation $\bar{g}_1(x,y)$ to one of the superposed layers and a transformation $\bar{g}_2(x,y)$ to the other layer, we obtain in the superposition the dot trajectories that correspond to the vector field $\bar{h}_1(x,y) = \bar{g}_1(\bar{g}_2(x,y)) - (x,y)$ or to the vector field $\bar{h}_2(x,y) = (x,y) - \bar{g}_2(\bar{g}_1(x,y))$. Furthermore, from the second part of the same proposition it follows that if $\bar{g}_1$ and $\bar{g}_2$ are weak transformations then the resulting dot trajectories are closely approximated by the vector field $\bar{h}(x,y) = \bar{g}_1(x,y) - \bar{g}_2(x,y)$. The first part of Proposition 4.3 is not practical to use here because of the transformation compositions that are involved. But the approximation provided by the second part of proposition 4.3 gives us, indeed, an alternative way for synthesizing dot trajectories that correspond well to $\bar{h}(x,y)$, by applying to the two superposed layers transformations $\bar{g}_1(x,y)$ and $\bar{g}_2(x,y)$ whose difference gives $\bar{h}(x,y)$. For example, we may distribute the deformation $\bar{h}(x,y)$ in equal parts between the two layers as follows:

$$\bar{g}_1(x,y) = (x,y) + \tfrac{1}{2}\bar{h}(x,y)$$
$$\bar{g}_2(x,y) = (x,y) - \tfrac{1}{2}\bar{h}(x,y) \tag{4.12}$$

which gives, indeed, $\bar{g}_1(x,y) - \bar{g}_2(x,y) = \bar{h}(x,y)$. Note that all the examples which follow have been generated by distributing $\bar{h}(x,y)$ in equal parts between the two superposed layers; the reasons for this choice are explained in Sec. D.10 of Appendix D. It should be remembered, however, that the dot trajectories generated by Eq. (4.12) are based on the vector field $\bar{h}(x,y) = \bar{g}_1(x,y) - \bar{g}_2(x,y)$, which only provides a good approximation to the dot trajectories. A perfect correspondence can be obtained by using Eq. (4.11), i.e. by deforming only one of the two layers.

Finally, if we prefer to use the domain transformations g_1 and g_2 rather than their direct equivalents $\bar{g}_1$ and $\bar{g}_2$ (see Remark 4.3 above), we can use here approximation (4.10), $\bar{h}(x,y) \approx g_2(x,y) - g_1(x,y)$, where:

$$g_1(x,y) = (x,y) + \tfrac{1}{2}\bar{h}(x,y)$$
$$g_2(x,y) = (x,y) - \tfrac{1}{2}\bar{h}(x,y)$$

Having understood the main principles, let us now illustrate the synthesis of dot trajectories by the following examples. Further examples are provided in the problems at the end of this chapter.

Example 4.11: Suppose we wish to synthesize an aperiodic screen superposition whose dot trajectories visually illustrate the vector field $\overline{\mathbf{h}}(x,y)$ defined by:

$$\overline{\mathbf{h}}(x,y) = (2xy,\ y^2 - x^2)$$

(see Fig. 4.14(a)). As we have just seen, this can be done, for example, by applying to the two original layers the following transformations $\overline{\mathbf{g}}_1(x,y)$ and $\overline{\mathbf{g}}_2(x,y)$, which distribute the deformation $\overline{\mathbf{h}}(x,y)$ in equal parts between the two layers:

$$\overline{\mathbf{g}}_1(x,y) = (x,y) + \tfrac{1}{2}\overline{\mathbf{h}}(x,y) = (x + xy,\ y + \tfrac{1}{2}y^2 - \tfrac{1}{2}x^2)$$

$$\overline{\mathbf{g}}_2(x,y) = (x,y) - \tfrac{1}{2}\overline{\mathbf{h}}(x,y) = (x - xy,\ y - \tfrac{1}{2}y^2 + \tfrac{1}{2}x^2)$$

And indeed, if we apply the transformations $\overline{\mathbf{g}}_1(x,y)$ and $\overline{\mathbf{g}}_2(x,y)$, respectively, to the two superposed layers, we obtain in the superposition dot trajectories that correspond to the given vector field $\overline{\mathbf{h}}(x,y)$. This is clearly illustrated in Fig. 4.13(a). In order to improve the visual effect, the layer transformations $\overline{\mathbf{g}}_1(x,y)$ and $\overline{\mathbf{g}}_2(x,y)$ used in Fig. 4.13(a) have been softened by replacing $\overline{\mathbf{h}}(x,y)$ in both equations by $\varepsilon\overline{\mathbf{h}}(x,y)$, where ε is a small positive fraction; this guarantees that the correlation between the two superposed layers remains strong enough to give clearly visible dot trajectories.

Note that the moiré effect obtained under the same layer transformations in the case of periodic screens looks completely different (Fig. 4.13(b)). This point will be explained in detail in Chapter 5. ∎

Example 4.12: As a second example, let us choose a system of two differential equations that is well known in physics, such as the non-linear system describing the behaviour of a free undamped pendulum (see, for example, [Kreyszig93 pp. 181–182] or [Tabor89 pp. 11–17]). The system of differential equations is given in this case by:

$$\tfrac{d}{dt}x(t) = y(t)$$

$$\tfrac{d}{dt}y(t) = -k\sin(x(t))$$

where k is a positive constant. The solution curves $(x(t),y(t))$ of this system are illustrated in Fig. 4.14(c). As explained in Sec. B.6 of Appendix B, the solution curves of this system are the field lines of the vector field $\overline{\mathbf{h}}(x,y) = (y, -k\sin x)$ (see Fig. 4.14(b)). We therefore consider the two layer transformations:

$$\overline{\mathbf{g}}_1(x,y) = (x,y) + \tfrac{1}{2}\overline{\mathbf{h}}(x,y) = (x + \tfrac{1}{2}y,\ y - \tfrac{1}{2}k\sin x)$$

$$\overline{\mathbf{g}}_2(x,y) = (x,y) - \tfrac{1}{2}\overline{\mathbf{h}}(x,y) = (x - \tfrac{1}{2}y,\ y + \tfrac{1}{2}k\sin x)$$

And indeed, as shown in Fig. 4.13(c), if we apply these transformations to our superposed aperiodic screens, we obtain in the superposition dot trajectories that correspond to Figs. 4.14(b) and (c).

Finally, note that just as in the previous example, the moiré effect obtained under the same layer transformations in the case of periodic screens looks completely different (Fig. 4.13(d)). This point will be explained in Chapter 5. ∎

These examples illustrate, indeed, how our results can be used to synthesize aperiodic screen superpositions having any desired dot trajectories. This not only confirms our theoretic results, but it also gives us an interesting application for the visualization of 2D vector fields (or the solution curves of systems of differential equations in two variables). It should be repeated, however, that the dot trajectories can be clearly visible in the superposition only when the transformations that we apply to the two layers are not too "violent"; otherwise the correlation between the layers is strongly reduced, and the visual effect is lost.

4.7 Dot trajectories in periodic and in repetitive cases

We now return to the intriguing question that we have already mentioned earlier: Why do we observe dot trajectories only in aperiodic superpositions and not in their corresponding periodic or repetitive superpositions (compare, for example, Fig. 3.10(c) with Fig. 3.10(d))? A careful examination, under a magnifying glass, of the microstructure in a periodic or a repetitive superposition (for example, in Fig. 3.10(d)) will reveal that similar circular, elliptical or hyperbolic arrangements of dot pairs do exist about the fixed points in such cases, too. However, in periodic or repetitive cases the dot locations are strongly constrained by an imposed ordering, which limits the degrees of freedom in the microstructure, and makes the dot trajectories much less conspicuous than in aperiodic superpositions. Furthermore, this imposed ordering also generates in the superposition a highly visible moiré structure whose presence obscures the arrangements of the correlated dot pairs in the microstructure. In aperiodic cases the microstructure ordering is not perturbed by any other imposed structures, and therefore the dot trajectories can manifest themselves freely in the superposition.

Note, however, that the dot trajectories are less visible when the correlation between the superposed layers is lower (for example, when a certain amount of random noise is added to the original layers), or when each of the superposed layers consists of elements of a different shape (for example, "1"-shaped elements in one layer and tiny pinholes in the other layer, as we will see in Chapter 7).

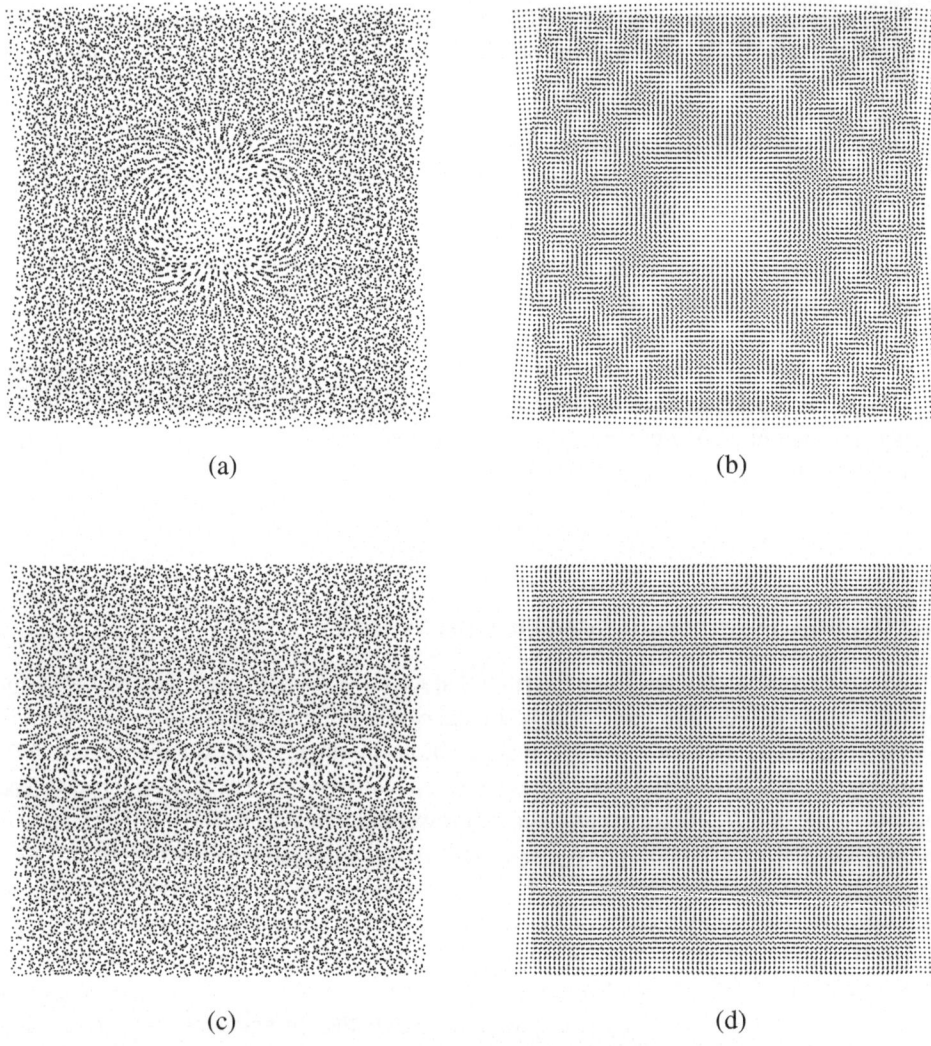

(a)   (b)

(c)   (d)

Figure 4.13: Examples of synthesized aperiodic layers that give in their
superposition dot trajectories having a given desired shape:
(a) Dot trajectories that show the flow lines of the vector field
$\mathbf{h}(x,y) = (2xy, y^2-x^2)$ (see Example 4.11 and Fig. 4.14(a)).
(c) Dot trajectories that illustrate the non-linear system of
differential equations describing the behaviour of a free
undamped pendulum (see Example 4.12 and Figs. 4.14(b),(c)).
(b) and (d) show, respectively, the moiré effects obtained when
the same layer transformations are applied to periodic layers.
The transformations applied to the original layers are $\overline{\mathbf{g}}_1(x,y) =
(x,y) + \frac{1}{2}\varepsilon\mathbf{h}(x,y)$ and $\overline{\mathbf{g}}_2(x,y) = (x,y) - \frac{1}{2}\varepsilon\mathbf{h}(x,y)$, where in (a) and
(b) $\mathbf{h}(x,y) = (2xy, y^2-x^2)$, and in (c) and (d) $\mathbf{h}(x,y) = (y, -k\sin x)$.

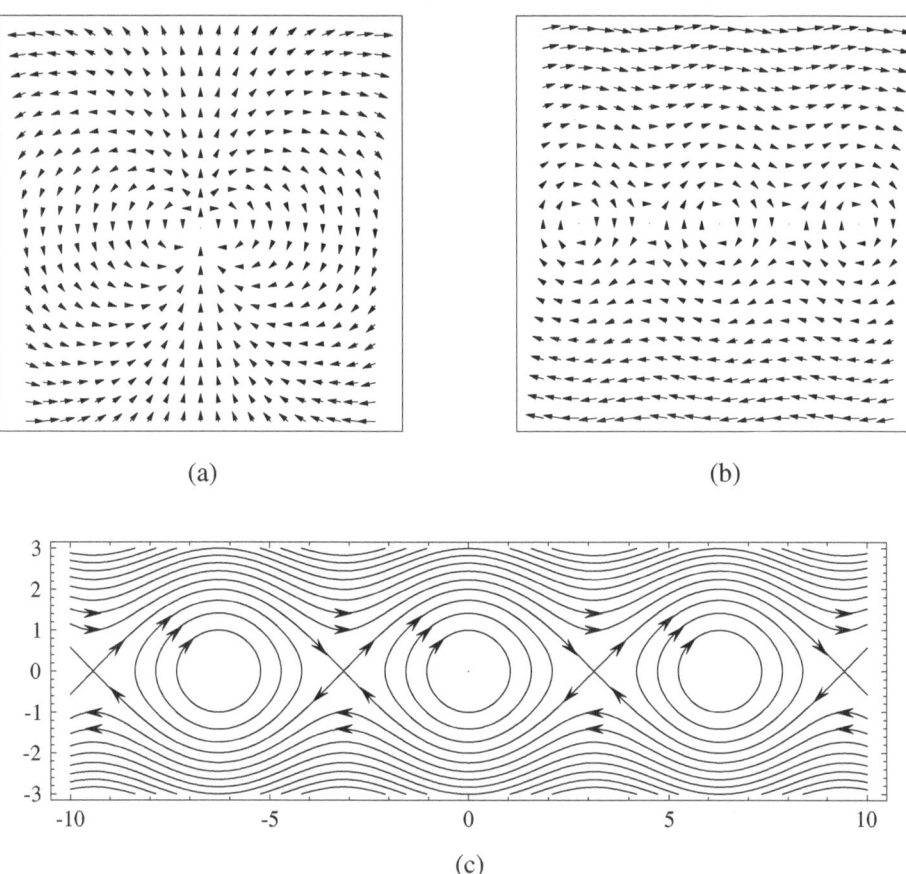

(a)   (b)

(c)

Figure 4.14: (a) Vector field representation of $\overline{\mathbf{h}}(x,y) = (2xy, y^2-x^2)$ (see Example 4.11 and Fig. 4.13(a)). (b) Vector field representation of the non-linear system of differential equations describing the behaviour of a free undamped pendulum (see Example 4.12 and Fig. 4.13(c)). (c) Solution curves (trajectories) of the system of differential equations of (b), drawn in a larger scale for clarity.

4.8 The microstructures under different superposition rules

As already mentioned in Sec. 2.2, layer superpositions are not necessarily limited to cases consisting of black elements on a white background and to the multiplicative superposition rule. For example, we may equally well take the negative of one or both of the participating layers, or replace the multiplicative superposition rule by any other rule. And indeed, it turns out that the microstructures observed in the superposition of aperiodic layers may considerably depend on the superposition rule between the layers (see Fig.

4.15), and even, to a lesser extent, on the individual dot shapes (see Problem 4-20).[12] Already in 1970, just one year after Glass' first publication on the circular microstructures observed in the superposition of two mutually rotated identical random screens [Glass69], a paper by Anstis showed that the observed microstructure may radically change, turning into a petal-like structure, if one of the superposed layers is replaced by its own negative [Anstis70 p. 1424 and Fig. 6]. This visual effect is even more pronounced if we replace the black background of the negative screen by an intermediate gray level, in order to make both black and white dots equally visible in the superposition [Glass76]. These two new superposition rules are shown in Figs. 4.15(b),(d), using the same layers as in the original, classical case that is shown in part (a) of the figure. Yet another superposition rule is shown in part (c) of the figure. As we can see, the circular dot trajectories that are so clearly perceived in the classical configuration (Fig. 4.15(a)) completely disappear in the new configurations, giving place to different petal-like microstructures, or to no apparent structures at all. It appears that circular dot trajectories are generated in the superposition of mutually rotated layers only when the contrast between the dots and the background is similar in both of the layers, while layers with opposite contrast yield the petal-like microstructure (compare Figs. 4.16(a) and 4.16(c), or their twofold magnifications shown in Figs. 4.20 and 4.21). A similar petal-like effect is also obtained in the case of opposite contrasts when the small angular mismatch between the two layers is replaced by a small scaling mismatch, although the detail of the petals in this case is different (compare Figs. 4.16(c) and 4.17(c)). However, it turns out that under other layer transformations, for example those shown in Figs. 2.2–2.3, this petal-like phenomenon is much less perceptible: Although the original dot trajectories of the classical superposition clearly disappear when we use layers with opposite element contrast, it is hard to identify what geometric shapes, if any, appear in the new microstructure (compare, for example, Figs. 4.18(a) and 4,18(c) or Figs. 4.19(a) and 4.19(c)).

In the following section we try to give a possible explanation to these phenomena; note, however, that this issue is clearly bordering on properties of the human visual system, a subject which remains beyond the scope of our discussion.

4.9 The visual interpretation of microstructures

The perception of order in disorder is a well-known property of the human visual system [Walker80]. Perhaps the best example is our identification of star constellations in the night sky, which consists of grouping together unrelated objects (stars) into "meaningful" or "familiar" shapes such as various animals or objects known from everyday's life. It seems that our brain tries to "identify" or to "give sense" to what it sees in terms of its prior experience, in a way that would be most useful to the observer.[13]

---

[12] The influence of the dot shapes on the *macrostructures* is described separately in Chapter 7.

[13] Another well-known example consists of the high priority our visual system gives to the detection of motion within a complex structure. This can be explained by the vital need of the organism to detect moving objects and to quickly tell apart a potential prey from a potential predator.

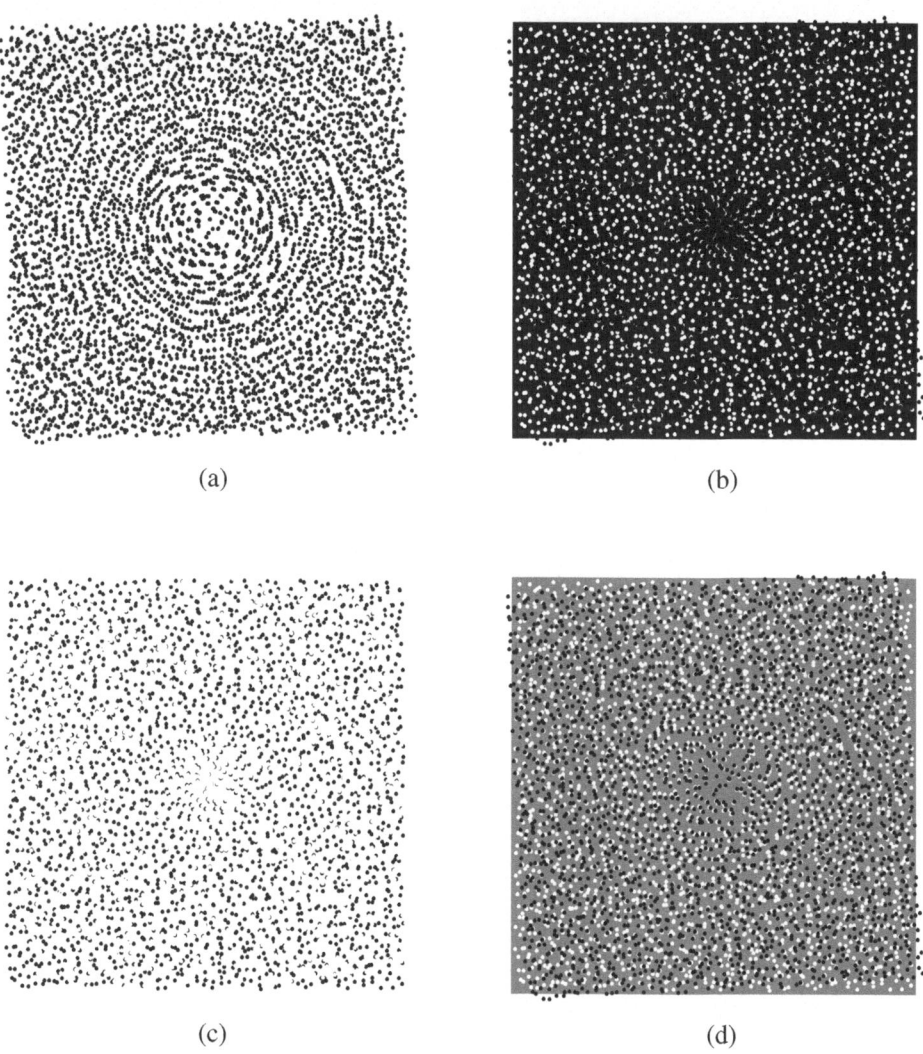

(a)   (b)

(c)   (d)

Figure 4.15: The microstructure observed in the superposition of aperiodic layers may
considerably depend on the superposition rule between the layers. This figure
shows a given layer superposition (consisting of two identical aperiodic dot
screens with a small angle difference) that is printed in (a)–(d) using four
different superposition rules: (a) The classical multiplicative superposition
rule. (b) Same as in (a), but here the unrotated layer has been replaced by its
own negative. (c) The dots of the second layer are printed in opaque white ink
covering the black dots of the first layer. (d) Same as in (b), but here the black
background of the negative layer has been replaced by an intermediate gray
level, in order to facilitate the perception of the dots of both layers. In this case
each dot pair consists of a black dot from the rotated layer and a white dot
from the first, unrotated layer (which may be fully or partially covered by a
black dot). Note the differences between the microstructures perceived in the
different superposition rules.

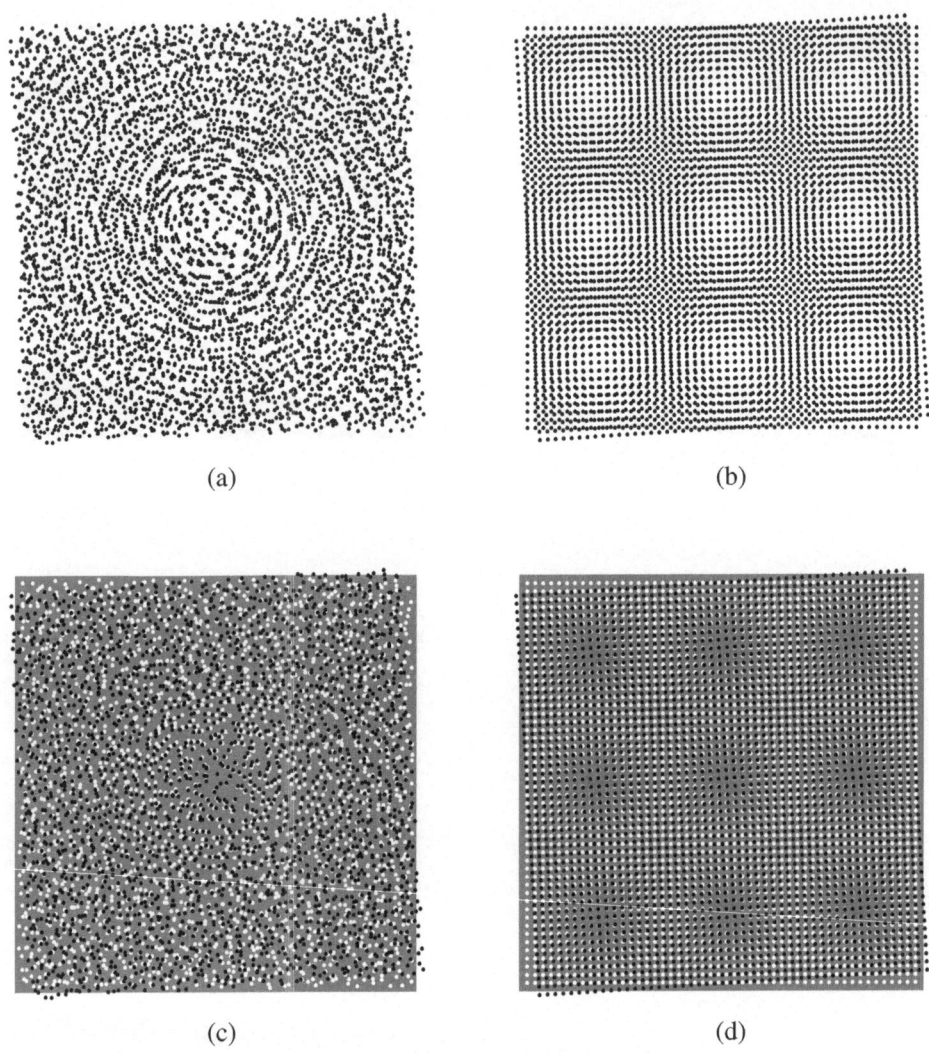

(a)   (b)

(c)   (d)

Figure 4.16: (a) The superposition of two identical aperiodic dot screens with a small angle difference. (b) The periodic counterpart of (a). (c) Same as (a), but here the untransformed layer has been replaced by an identical copy of itself made of white dots on a gray background. (d) the periodic counterpart of (c). While the microstructure in (a) consists of circular dot trajectories, the microstructure observed in (c) consists of an intriguing petal-like form.

This is certainly a mental, cognitive process, and therefore expressing it mathematically does not seem to be a realistic task. At most, mathematics may help in accounting for properties of the given structures like their degree of uniformity, granularity or randomness, or in grouping together individual elements into "clusters" according to

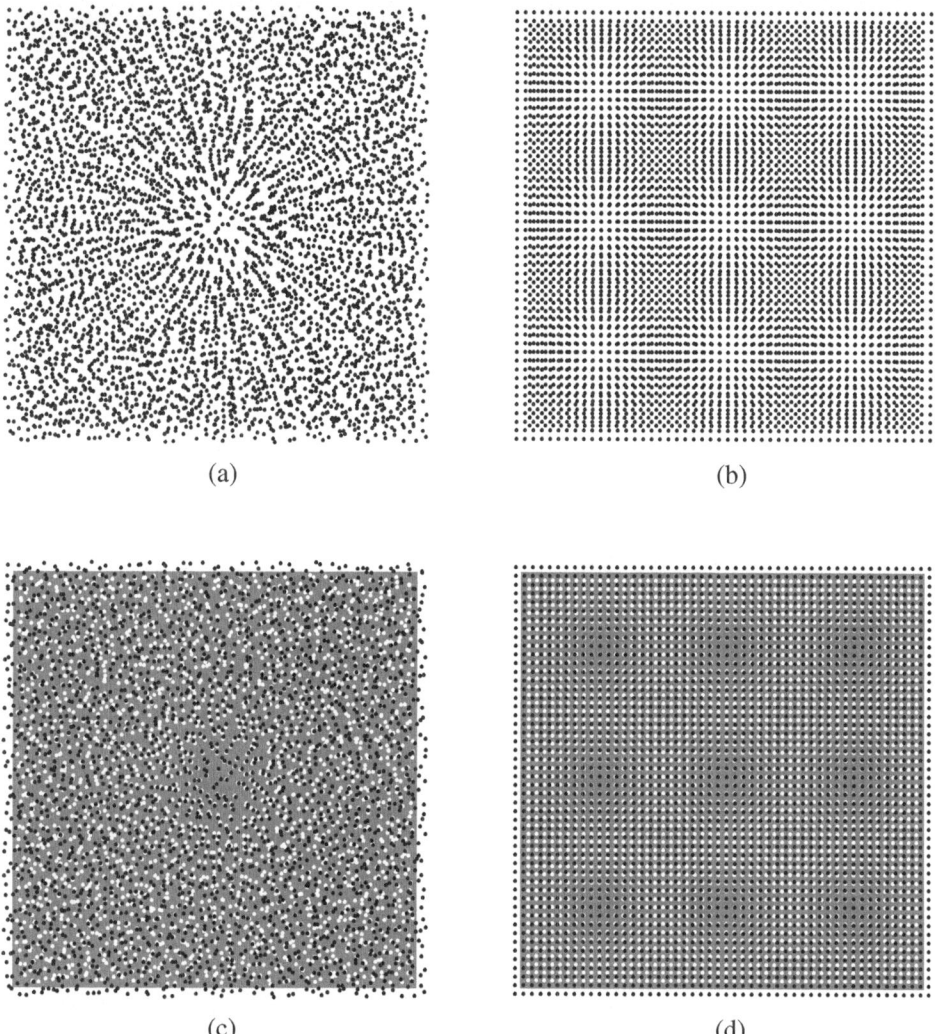

(a)   (b)

(c)   (d)

Figure 4.17: (a) The superposition of two identical aperiodic dot screens where the second layer has undergone a small scaling transformation. (b) The periodic counterpart of (a). (c) Same as (a), but here the untransformed layer has been replaced by an identical copy of itself made of white dots on a gray background. (d) the periodic counterpart of (c). While the microstructure in (a) consists of radial dot trajectories, the microstructure in (c) consists of an intriguing petal-like form (which is distinct from that of Fig. 4.16(c)).

their locations, brightness, etc. But the identification of suggestive shapes such as animals ("snakes", "worms") or flowers ("petals") in the disorder is performed by our brain, and it is based on our subjective experience, and not on objective mathematical

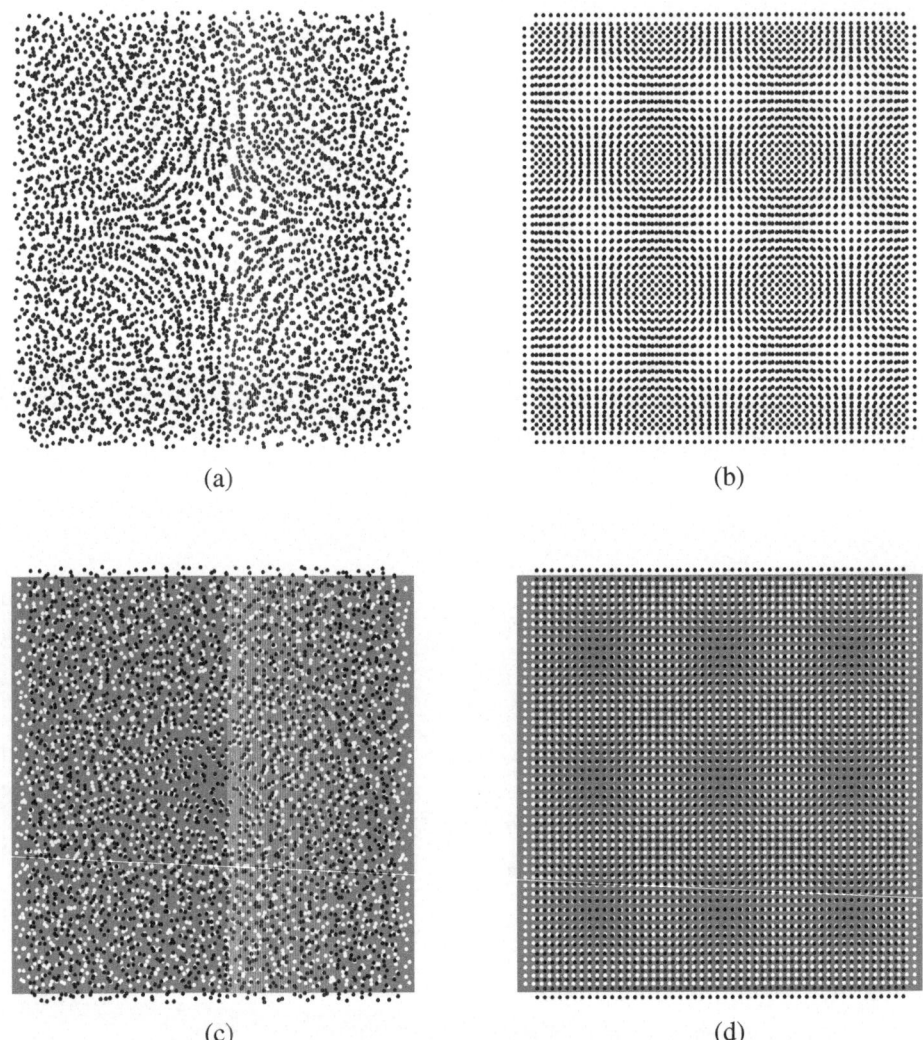

(a)   (b)

(c)   (d)

Figure 4.18: (a) The superposition of two identical aperiodic dot screens where the second layer has undergone a small inhomogeneous scaling transformation: a slight vertical expansion and a slight horizontal shrink. (b) The periodic counterpart of (a). (c) Same as (a), but here the untransformed layer has been replaced by an identical copy of itself made of white dots on a gray background. (d) the periodic counterpart of (c). Note the difference between the microstructures in (a) and in (c).

considerations. Only if a geometrical, statistical or other objective tendency is present can a mathematical model be sought in order to modelize the phenomenon in question and to predict its behaviour.

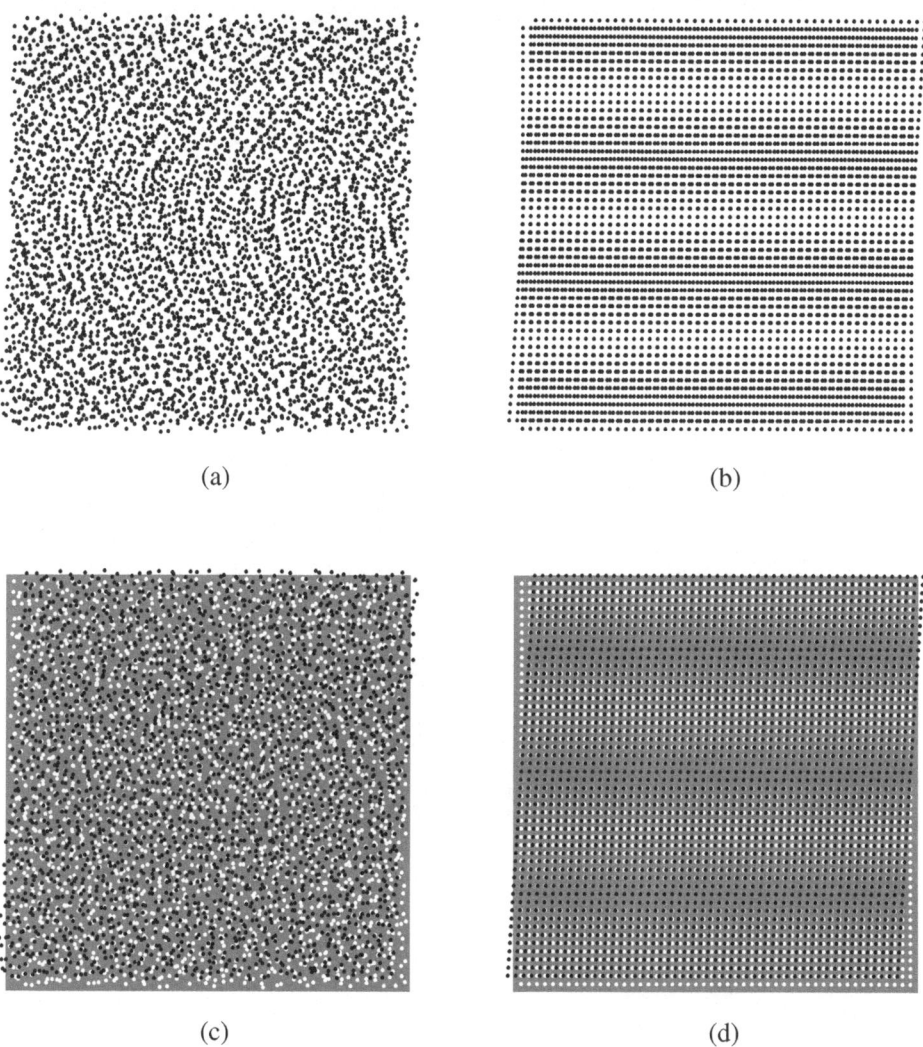

(a)   (b)

(c)   (d)

Figure 4.19: (a) The superposition of two identical aperiodic dot screens where the second layer has undergone a small horizontal shearing plus a small vertical shift. (b) The periodic counterpart of (a). (c) Same as (a), but here the untransformed layer has been replaced by an identical copy of itself made of white dots on a gray background. (d) the periodic counterpart of (c). Note the difference between the clear parabolic microstructure in (a) and the intriguing, ambiguous microstructure in (c).

To take an example, consider Fig. 2.5(a), which shows the superposition of two identical copies of a random screen with a small vertical mismatch. Here, in spite of the random nature of the structure, there obviously exists a strong statistical preference to vertical dot

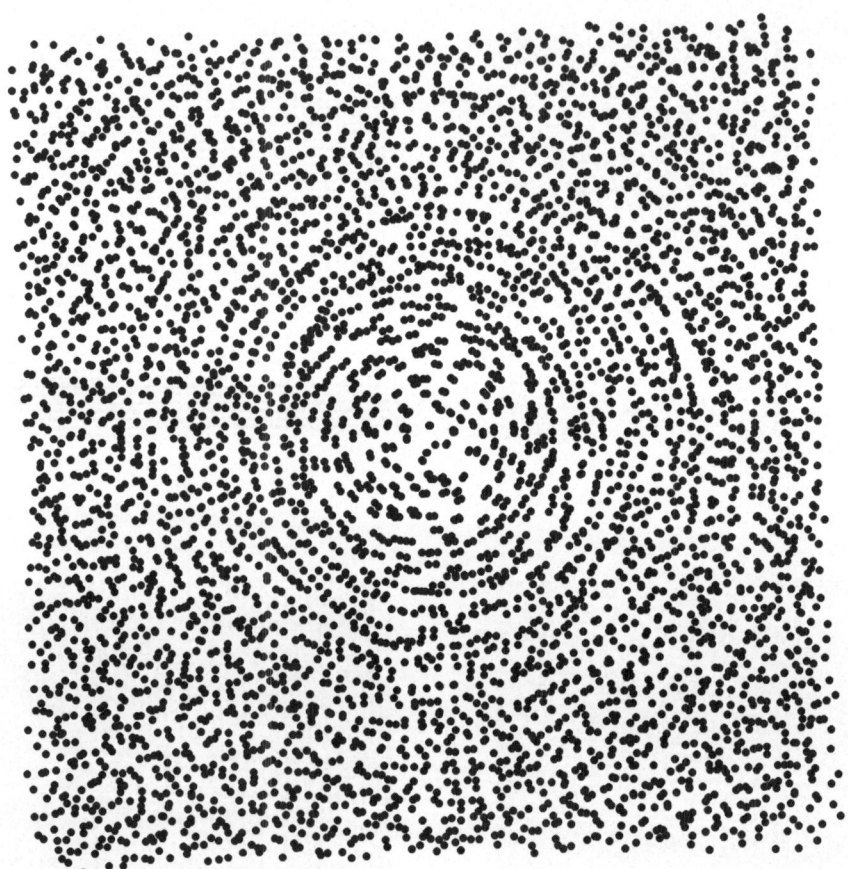

Figure 4.20: A two-fold magnification of Fig. 4.16(a) showing the detail of its microstructure.

ordering; and indeed, it is not surprising that we can easily identify in this superposition various vertical snake- or worm-like structures. The same goes to other cases, too, such as Fig. 2.1(c), which shows the superposition of two identical random screens with a small angle difference: Here, the microstructure has a strong statistical preference to circular dot ordering, because it is composed of dot pairs surrounding the center. And indeed, as we have seen in Sec. 4.4, this microstructure can be modelized mathematically as a vector field whose vectors connect between the two dots of each dot pair. Now, when we replace one of the two superposed screens by its negative, it is obvious that each of the dots in the layer superposition remains exactly in the same location; and still, our visual impression of "ordering" in the superposition is completely changed. *A-priori*, this visual impression cannot be predicted or accounted for by mathematical considerations alone; but if we can find any rules that express our subjective visual impression of the microstructure using

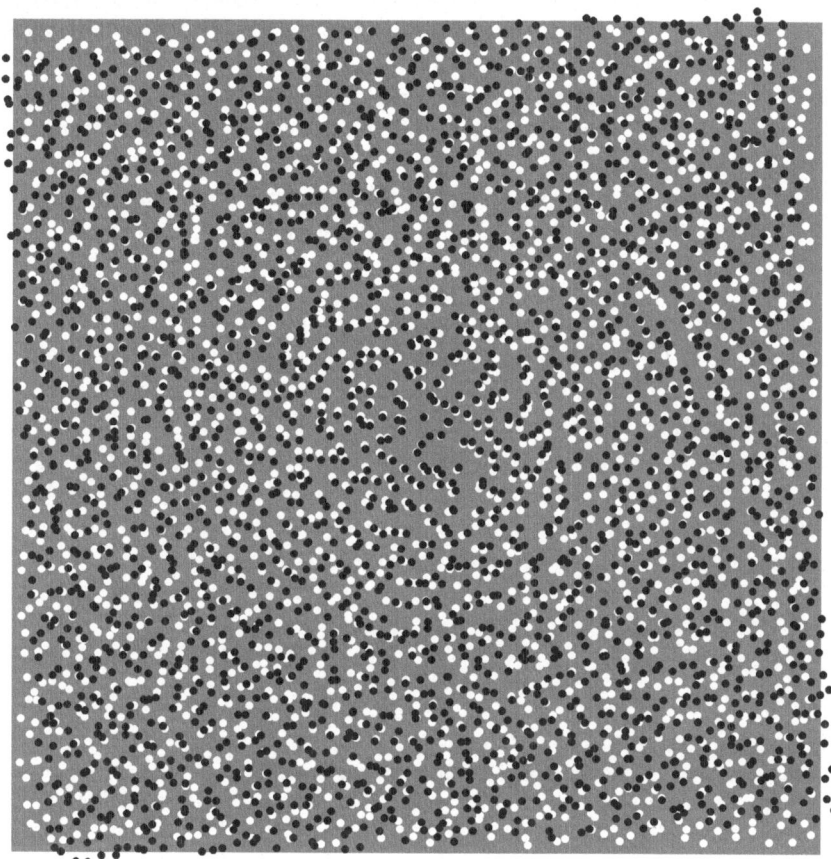

Figure 4.21: A two-fold magnification of Fig. 4.16(c) showing the detail of its microstructure.

geometrical, statistical or other terms, we may, perhaps, find a mathematical formulation that could modelize such visual phenomena *a-posteriori*, up to certain limitations, just as we have done in the case of the classical dot-trajectories. See also Problem 4-21 which may give a starting point for the understanding of the microstructures in Fig. 4.16.

The fact that our mathematical model alone cannot fully account for all the visual phenomena in question is also illustrated by the fact that dot trajectories can only be visible in the superposition if the layer transformations $\bar{\mathbf{g}}_1(x,y)$ and $\bar{\mathbf{g}}_2(x,y)$ are not too violent. For example, circular dot trajectories due to an angular mismatch can only be observed in the superposition when the rotation undergone by one of the layers is small, whereas the vector field connecting the dot pairs (before and after the transformation) exists mathematically in *all* cases, even if the transformation is strong and the dot trajectories are no longer visible to the eye. Thus, although the mathematical model remains true in all cases, the scope of its applicability to the phenomenon in question is determined by considerations that are extraneous to the mathematical model itself. Note that the same reasoning is also true in the periodic case: As shown in Chapter 3 of *Vol. I* (in the context of moiré minimization), the question of whether or not a specific moiré is visible in the superposition of periodic layers does not only depend on mathematical quantities such as its period or its amplitude, but also on other, more subjective and hard to modelize criteria, such as the background noise, the illumination, the viewing conditions, the viewer's concentration, etc.

From the point of view of vision research, phenomena such as dot trajectories demonstrate the ability of the human visual system to detect local correlations in the presence of noise, and to combine local correlations from different regions of the visual field in such a manner that a simple percept is formed [Glass69]. The difficulties in the perception of sharp, unambiguous microstructures when one of the layers is replaced by its own negative indicates that the human visual system is less efficient in detecting local correlations when the two layers consist of opposite contrasts [Glass76]. But such vision-research considerations remain beyond the scope of our discussion.

PROBLEMS

4-1. *Visual illustration of the solution curves of a system of two differential equations.* It has been suggested in [Glass73 p. 361] that the superposition of random dot screens can be used to obtain qualitative information about the trajectories (solution curves) of a system of two differential equations. Can you explain? *Hint*: remember that the dot trajectories you see in the superposition correspond to the trajectories of the system of differential equations (4.6), where $\bar{h}_1(x,y)$ and $\bar{h}_2(x,y)$ are the two components of the transformation $\bar{\mathbf{h}}(x,y) = \bar{\mathbf{g}}(x,y) - (x,y)$.

4-2. *Visual illustration of a vector field.* In a similar way, can you propose a method based on the superposition of random dot screens to illustrate visually the flow lines of a given vector field? Will this method work correctly for any vector field? (Consider, for example, the vector field induced by a rotation through angle α when $\alpha = 1°$ and when $\alpha = 90°$).

4-3. *The shape of the dot trajectories in various linear scaling transformations.* Suppose we are given two identical random or pseudo-random dot screens that are superposed on top of each other in full coincidence, dot on dot. When we apply to one of the layers a scaling transformation having the same scaling effect along the x and y axes there appear in the superposition radial dot trajectories, as shown in Fig. 2.1(e). If, however, the scaling transformation we apply to one of the layers consists of expansion along one axis and contraction along the other axis, the dot trajectories we obtain in the superposition are hyperbolic (see Fig. 2.2(a)). Yet another case is shown in Fig. 2.3(a), where the scaling transformation only affects one direction; in this case the resulting dot trajectories are parallel straight lines that extend along the scaling direction. Can you guess (or check in an appropriate aperiodic layer superposition) what will be the shape of the dot trajectories when the scaling transformation that we apply to one of the layers consists of two different expansions (or two different contractions) along the two axes? See the mathematical solution in Problem 4-11 below.

4-4. *Synthesis of dot trajectories.* The pair of linear differential equations (4.4) may have trajectories (solution curves) of various shapes, depending on the values of its coefficients a_1, b_1, a_2 and b_2. A complete classification of the different possible critical points and trajectory shapes in such linear differential equation pairs is given in Table H.1 of Appendix H (see also [Kreyszig93 pp. 176–178], [Gray97 pp. 551–567] or Sec. 1.4 of [Tabor89]). These different types of trajectories can be illustrated by means of the dot trajectories in the superposition of appropriate aperiodic layers. Some of these different types, including nodes, center points, spirals, saddle points, etc. are illustrated in Figs. 2.1–2.3. And yet, these figures do not illustrate all the different cases that appear in Table H.1. Synthesize aperiodic layer superpositions whose dot trajectories show the remaining types of trajectories (solution curves) that are not already shown in these figures (one such case appears in the previous problem). Remember that the transformations you suggest must be weak enough in order not to excessively reduce the correlation between the two superposed layers.

4-5. *Synthesis of dot trajectories (continued).* Can you synthesize a pair of aperiodic layers that give in their superposition wavy cosinusoidal dot trajectories along the main axis? More generally, can you synthesize aperiodic layers that give in their superposition dot trajectories that correspond to the family of parallel curves $y = f(x) + const.$? *Hint*: Using the same approach as in Examples 4.9 and 4.10 show that the curves $y = f(x) + const.$ are the solution curves of the system of differential equations:

$$\tfrac{d}{dt}x(t) = 1$$
$$\tfrac{d}{dt}y(t) = \tfrac{d}{dx}f(x(t))$$

This means that they are also the field lines of the vector field $\overline{\mathbf{h}}(x,y) = (1, \tfrac{d}{dx}f(x))$. Therefore, in order to obtain such dot trajectories in the superposition, you have to generate transformations $\overline{\mathbf{g}}_1(x,y)$ and $\overline{\mathbf{g}}_2(x,y)$ whose difference is $\overline{\mathbf{h}}(x,y)$, and to apply them to the two originally identical aperiodic layers. Can you obtain in a similar way dot trajectories having the shape of the parallel curves $x = f(y) + const.$ like in Figs. 3.19(c) and 3.20(c)? Note that in all of these cases the superposition does not have fixed points, since one of the components of the vector field is a non-zero constant. This means that all of these superpositions only consist of microstructures, just as in Figs. 3.19(c) and 3.20(c).

4-6. *Synthesis of dot trajectories (continued).* How can you synthesize a pair of aperiodic layers that give in their superposition dot trajectories that correspond to the level curves of the surface $z = g(x,y)$? *Hint*: According to Sec. B.7 in Appendix B, the vector field whose field lines are the level curves of the surface $z = g(x,y)$ is given by $\overline{\mathbf{h}}(x,y) = (\tfrac{\partial}{\partial y}g(x,y), -\tfrac{\partial}{\partial x}g(x,y))$. Similarly, the vector field whose field lines are the lines of maximal slope of the surface $z = g(x,y)$ is given by $\overline{\mathbf{h}}(x,y) = (\tfrac{\partial}{\partial x}g(x,y), \tfrac{\partial}{\partial y}g(x,y))$.

4-7. *A horizontal pair of linear Glass patterns with vertical dot trajectories.* Can you find the counterpart of the superposition shown in Fig. 3.6(a) having the same two linear Glass patterns, but with *vertical* rather than horizontal dot trajectories? *Hint*: Consider Figs. 2.3(a) and (c), which are generated by two different linear transformations (a vertical scaling or a horizontal shear). The superposition shown in Fig. 3.6(a) is a non-linear generalization of the linear shear of Fig. 2.3(c), that gives two linear Glass patterns (fixed lines) rather than one, but still has the same horizontal dot trajectories. Can you find a similar non-linear generalization to the vertical scaling of Fig. 2.3(a), that gives two linear Glass patterns (fixed lines) rather than one, but still has vertical dot trajectories like in Fig. 2.3(a)?

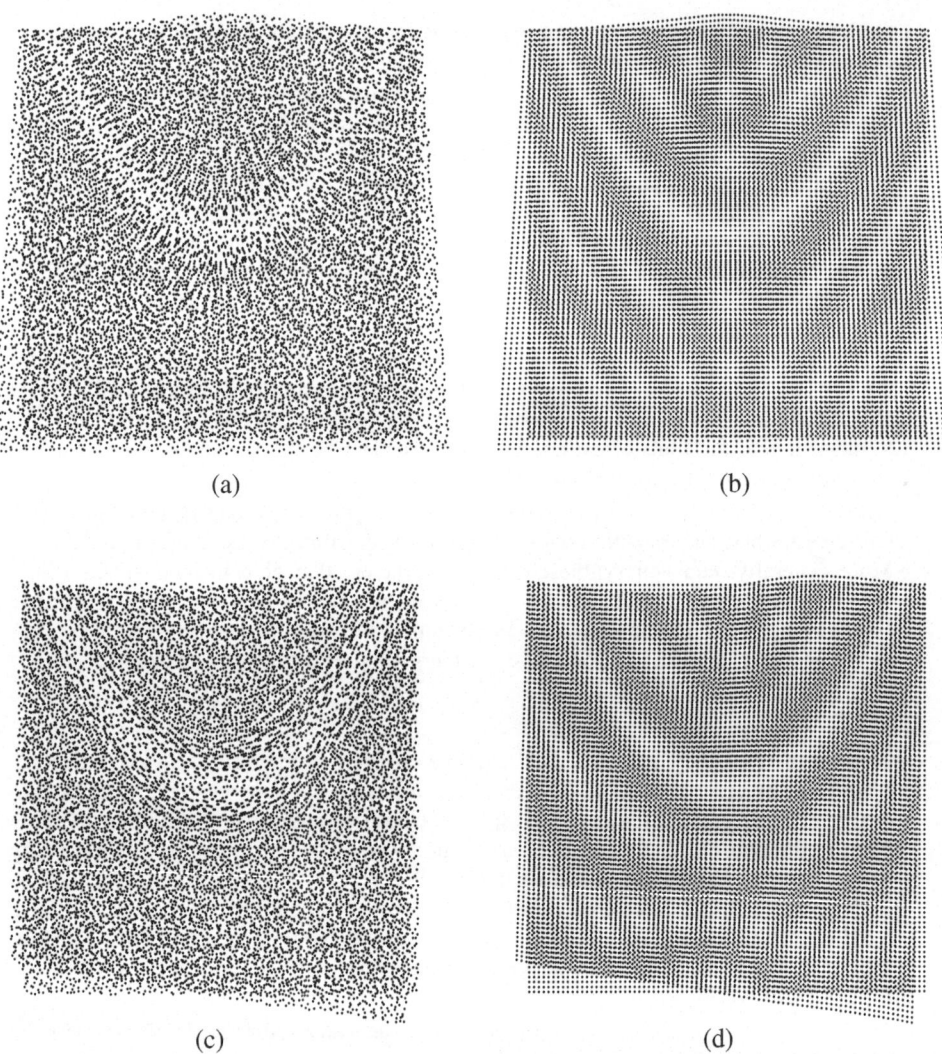

(a)   (b)

(c)   (d)

Figure 4.22: (a) A curvilinear Glass pattern lying along the curve $y = ax^2$, with dot
trajectories that are perpendicular to the curve. (c) A curvilinear Glass
pattern lying along the same parabolic curve, with dot trajectories that are
parallel to the curve. (b) and (d) are the periodic counterparts of (a) and
(c). Compare with Fig. 3.16, where the same curvilinear Glass patterns
are surrounded by straight horizontal or vertical dot trajectories; see
Problem 4-8.

4-8. *Synthesis of the dot trajectories along a given curvilinear Glass pattern.* We have seen
   in Example 3.7 and in Fig. 3.16 that there exist different ways for synthesizing a Glass
   pattern along a given curve $y = f(x)$, which only differ from each other in their
   microstructure: If we apply the layer transformations (3.41) or (3.42) taking $\mathbf{k}(x,y) =$
   $(y - f(x), 0)$ we obtain the desired $f(x)$-shaped Glass pattern surrounded by straight

horizontal dot trajectories (see Fig. 3.16(a)), whereas if we choose $\mathbf{k}(x,y) = (0, y - f(x))$ the same $f(x)$-shaped Glass pattern is surrounded by straight *vertical* dot trajectories (see Fig. 3.16(c)). More generally, by choosing $\mathbf{k}(x,y) = (a\,[y - f(x)], b\,[y - f(x)])$ with arbitrary coefficients a and b we obtain the same $f(x)$-shaped Glass pattern, but this time with straight dot trajectories that are oriented in the direction $\theta = \arctan(b/a)$. In fact, it is even better to use the normalized coefficients $a/\sqrt{a^2 + b^2}$ and $b/\sqrt{a^2 + b^2}$ rather than a and b; why? (*Hint*: This can be best understood by considering $\mathbf{k}(x,y)$ as a vector field. At any point (x,y), the vector returned by the vector field $(y - f(x), 0)$ and the vector returned by the vector field $(0, y - f(x))$ have the same lengths, and they only differ in their directions: horizontal or vertical, respectively. The vector returned by the third vector field, $(a\,[y - f(x)], b\,[y - f(x)])$, is oriented in the direction $\theta = \arctan(b/a)$, but in order for it to have the same length as in the first two vector fields, the coefficients a and b must be normalized as indicated.)

(a) Can you generalize these results to a Glass pattern that lies along the curve $f(x,y) = 0$ rather than along the curve $y = f(x)$?

(b) These results show us how to generate a curvilinear Glass pattern that lies along a given curve, and that is surrounded by straight, parallel dot trajectories that are all oriented in one fixed direction. However, it would be more interesting to obtain dot trajectories that are parallel or perpendicular to the curvilinear Glass pattern itself. How can we obtain such dot trajectories? *Hint*: Show that if we use instead of the constant coefficients a and b the variable coefficients $a(x,y) = \frac{\partial}{\partial x} f(x,y)$ and $b(x,y) = \frac{\partial}{\partial y} f(x,y)$ (or their normalized counterparts, $a(x,y)/\sqrt{a(x,y)^2 + b(x,y)^2}$ and $b(x,y)/\sqrt{a(x,y)^2 + b(x,y)^2}$, for the same reasons as above), we obtain dot trajectories that are *perpendicular* to the $f(x,y)$-shaped curvilinear Glass pattern (see Fig. 4.22(a)). Similarly, show that if we take $a(x,y) = \frac{\partial}{\partial y} f(x,y)$ and $b(x,y) = -\frac{\partial}{\partial x} f(x,y)$ (or rather their normalized counterparts, $a(x,y)/\sqrt{a(x,y)^2 + b(x,y)^2}$ and $b(x,y)/\sqrt{a(x,y)^2 + b(x,y)^2}$) we obtain dot trajectories that are *parallel* to the same Glass pattern (see Fig. 4.22(c)). See also Sec. B.7 in Appendix B.

(c) As we already know from Chapter 3, there exist infinitely many ways to generate a macroscopic Glass pattern that lies along the same given curve. The present problem shows how to choose among them the ones that satisfy some additional conditions that we impose on the desired microstructure. Can you think of any other criteria for selecting a preferred solution among the infinitely many possibilities?

4-9. *Visual illustration of the indefinite integral of a function $y = f(x)$.* Can you propose a method based on the superposition of two random dot screens for illustrating visually the curves $y = \int f(x)dx + const.$? *Hint*: As we have seen in Problem 4-5 (see also Examples 4.9 and 4.10), the field lines of the vector field $\overline{\mathbf{h}}(x,y) = (1, \frac{d}{dx} f(x))$ correspond to the curves $y = f(x) + const.$ This means, therefore, that the field lines of the vector field $\overline{\mathbf{h}}(x,y) = (1, f(x))$ correspond to the curves of the indefinite integral $y = \int f(x)dx + const.$ Hence, all that we need to do is to generate transformations $\overline{\mathbf{g}}_1(x,y)$ and $\overline{\mathbf{g}}_2(x,y)$ whose difference is $\overline{\mathbf{h}}(x,y)$, to apply them to two originally aperiodic dot screens, and to observe the dot trajectories in the resulting superposition.

4-10. Based on Problems 4-5 and 4-9, give a full explanation of the macroscopic and microscopic phenomena that occur in Figs. 3.19(a)–(d) and 3.20(a)–(d). *Hint*: In part (a) of both figures the *domain* transformations that were undergone by the two layers are $\mathbf{g}_1(x,y) = (x - f(y), y)$ and $\mathbf{g}_2(x,y) = (x,y)$, so that the difference transformation is $\mathbf{k}(x,y) = \mathbf{g}_1(x,y) - \mathbf{g}_2(x,y) = (-f(y), 0)$; therefore, we obtain in the superposition a macroscopic Glass pattern that lies along the lines $f(y) = 0$ (see Example 3.7 and Problem 3-12 in Chapter 3). The *direct* transformations that were undergone by the same two layers are $\bar{\mathbf{g}}_1(x,y) = (x + f(y), y)$ and $\bar{\mathbf{g}}_2(x,y) = (x,y)$, so that their difference is $\bar{\mathbf{h}}(x,y) = \bar{\mathbf{g}}_1(x,y) - \bar{\mathbf{g}}_2(x,y) = (f(y), 0)$; therefore the macroscopic Glass pattern is surrounded by horizontal, straight dot trajectories (see Problem 4-8 above). In part (c) of both figures the *domain* transformation that is undergone by the second layer is $\mathbf{g}_2(x,y) = (x, y+1)$, so that the difference transformation is $\mathbf{k}(x,y) = \mathbf{g}_1(x,y) - \mathbf{g}_2(x,y) = (-f(y), -1)$; therefore, as explained in Examples 3.10 and 3.11, we no longer see in the superposition the macroscopic Glass pattern but only the dot trajectories. To find the shape of these dot trajectories, note that the *direct* transformation that is undergone by the second layer is $\bar{\mathbf{g}}_2(x,y) = (x, y-1)$, so that the difference is $\bar{\mathbf{h}}(x,y) = (f(y), 1)$. This means, as we have seen in Problem 4-9, that the dot trajectories correspond to the curves $x = \int f(y)dy + const.$ Part (b) in both figures, where the difference transformation is $\bar{\mathbf{h}}(x,y) = (f(y), c)$ with $0 < c < 1$, corresponds to an intermediate case between (a) and (c); and part (d) in both figures corresponds to the difference transformation $\bar{\mathbf{h}}(x,y) = (f(y), c)$ with $c > 1$, where the dot trajectories become flatter (and less visible) as the value of c increases.

4-11. *Finding the curve equations of dot trajectories.* Consider the superposition of two identical aperiodic dot screens one of which undergoes a slight horizontal scaling of $s_1 = 1 + \varepsilon_1$ and a slight vertical scaling of $s_2 = 1 + \varepsilon_2$ where ε_1 and ε_2 are small positive fractions. Find the curve equations of the resulting dot trajectories. *Hint*: Proceeding like in Example 4.8, we have in this case $\bar{\mathbf{h}}(x,y) = ((1+\varepsilon_1)x, (1+\varepsilon_2)y) - (x,y) = (\varepsilon_1 x, \varepsilon_2 y)$. According to Sec. B.6 in Appendix B, the field lines of this vector field are the solution curves of the system of differential equations:

$$\tfrac{d}{dt}x(t) = \varepsilon_1 x(t)$$
$$\tfrac{d}{dt}y(t) = \varepsilon_2 y(t)$$

The solutions of this system are, of course, $x(t) = c_1 e^{\varepsilon_1 t}$, $y(t) = c_2 e^{\varepsilon_2 t}$ where c_1 and c_2 are arbitrary constants [Birkhoff89 p. 133], which means in explicit form $y^{\varepsilon_1} = kx^{\varepsilon_2}$ for all k. This corresponds to a family of parabolas emanating from the origin, whose order depends on the given values of ε_1 and ε_2.

4-12. What can we learn from the shape of the dot trajectories in a superposition of correlated dot screens (for example, circular, hyperbolic, radial or spiral dot trajrctories surrounding a fixed point, or S-like dot trajectories without fixed points) on the nature and on the mathematical properties of the layer transformations which generate them?

4-13. What happens to the dot trajectories of Fig. 3.13(a) when the untransformed layer is scaled up rather than scaled down? And when it is scaled up in one direction and scaled down in the other direction? Draw the vector field and the field lines (trajectories) of each of these cases.

4-14. *Direction fields and dot trajectories.* A student argues that dot trajectories in the superposition of aperiodic layers are in fact represented by *direction fields* rather than by vector fields, since the lengths of the vectors are immaterial and only their directions count. For example, all the vector fields having horizontal vectors throughout the plane (such as $\bar{\mathbf{h}}(x,y) = (x,0)$, $\bar{\mathbf{h}}(x,y) = (x^2,0)$, etc.) give equivalent dot trajectories in the superposition. What is your opinion?

4-15. *Circular dot trajectories represented by spiral solution curves?* Suppose we are given two identical aperiodic dot screens which are superposed on top of each other in full coincidence, dot on dot, and that we apply a slight geometric transformation $\bar{\mathbf{g}}(x,y)$ to one of the two layers and observe the dot trajectories in the resulting superposition. We have seen in Sec. 4.4 that the vector field of the *relative* layer transformation $\bar{\mathbf{h}}(x,y) = \bar{\mathbf{g}}(x,y) - (x,y)$ (unlike that of the *absolute* layer transformation $\bar{\mathbf{g}}(x,y)$) fully corresponds to the dot trajectories that are generated in the screen superposition. The reason is, of course, that the arrows of this vector field connect the departure and the

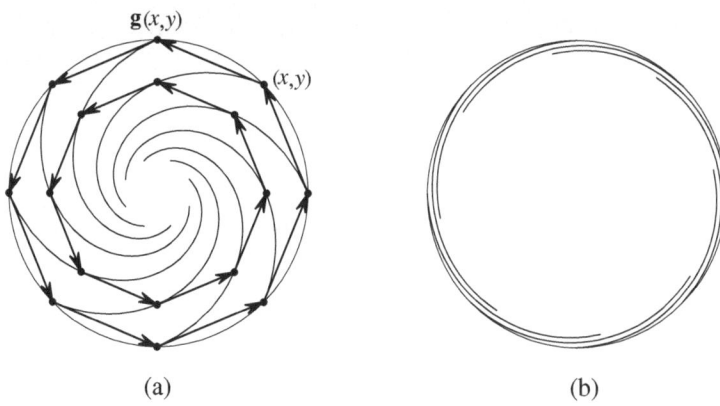

$\mathbf{g}(x,y)$

(x,y)

(a)   (b)

Figure 4.23: (a) The arrow (vector) connecting point (x,y) to its new location $\overline{\mathbf{g}}(x,y)$
after the application of the rotation transformation $\overline{\mathbf{g}}$ is not tangential to
the circle, but rather a chord. Therefore the field line emanating from
(x,y) slightly spirals inward, and it is not purely circular. (b) Clearly, the
smaller the rotation angle α, the closer the field line follows the circle.
The rotation angles used in (a) and (b) are, respectively, 45° and 5°.

destination points of each screen dot under the transformation $\overline{\mathbf{g}}(x,y)$. Now, remember
that the system of differential equations (4.3) is simply a different representation of the
vector field $\overline{\mathbf{g}}(x,y)$, since its solution curves $(x(t),y(t))$ express the field lines (trajectories)
of the vector field $\overline{\mathbf{g}}(x,y)$ (see Sec. B.6 in Appendix B). Therefore, by rewriting the
system of differential equations (4.3) for the *relative* layer transformation $\overline{\mathbf{h}}(x,y)$ rather
than for the *absolute* layer transformation $\overline{\mathbf{g}}(x,y)$, we obtain a new system of differential
equations, Eq. (4.6), whose solution curves correspond to the field lines of the vector
field $\overline{\mathbf{h}}(x,y) = \overline{\mathbf{g}}(x,y) - (x,y)$. It can be expected, therefore, that the solution curves of this
system of differential equations, unlike those of the original system (4.3), do
correspond to the dot trajectories observed in the superposition. However, it turns out
that in some particular cases the resulting solution curves do not fully agree with the
dot trajectories in the superposition: When the layer transformation $\overline{\mathbf{g}}(x,y)$ is a rotation
by angle α, the solution curves of the system of differential equations belonging to
$\overline{\mathbf{h}}(x,y) = \overline{\mathbf{g}}(x,y) - (x,y)$ consist of spirals that converge to the center of rotation (see
[Birkhoff89 p. 144]), while the dot trajectories we observe in the screen superposition
(Fig. 2.1(c)) as well as the arrow alignment of the vector field $\overline{\mathbf{h}}(x,y)$ (Fig. 4.2(a))
consist of circles surrounding the center of rotation. How can you explain this
discrepancy?

Hint: Remember that by definition, the field line at each point (x,y) of a vector field is
tangential to the arrow emanating from the same point. However, the arrow connecting
the point (x,y) to its rotated location $\overline{\mathbf{g}}(x,y)$ is not tangential to the circle's arc traced by
the rotating point; rather, it is a *chord* connecting the points (x,y) and $\overline{\mathbf{g}}(x,y)$ on the
circumference of the circle. Therefore, the field line passing through point (x,y) follows
the direction of the chord emanating from (x,y), and hence it is not tangent to the
circle, but rather oriented slightly inwards (see Fig. 4.23(a)). This explains why the
field lines in this case are, indeed, spirals that converge into the center of rotation,
although the arrows of the vector field $\overline{\mathbf{h}}(x,y)$ are arranged circularly. But such a spiral
obviously does not trace out the motion of a screen dot under the rotation
transformation $\overline{\mathbf{g}}(x,y)$, since each dot clearly moves along the circumference of the

circle. In other words, although the modified system of differential equations (4.6) that correspond to $\overline{\mathbf{h}}(x,y)$ does express the field lines of our vector field $\overline{\mathbf{h}}(x,y)$, these field lines simply do not represent the circular dot trajectories we observe in the layer superposition (Fig. 2.1(c)).[14] It should be noted, however, that when the rotation angle α is small, the spiral field lines are in fact very close to perfect circles, and the discrepancy is negligible (compare the spiral field lines in Figs. 4.23(a) and 4.23(b) which correspond, respectively, to $\alpha = 45°$ and $\alpha = 5°$). The difference between the spiral field lines and the circular dot trajectories becomes significant only for large angles α, but in these cases the visual effect of the dot trajectories is lost anyway, because the correlation between the superposed layers is strongly reduced. On the other hand, the graphic representation of the vector field $\overline{\mathbf{h}}(x,y)$ in terms of arrows connecting the departure and the destination points of each screen dot under the rotation $\overline{\mathbf{g}}(x,y)$ (Fig. 4.2(a)) remains always faithful to the dot trajectories in the superposition (as long as they are visible, as explained in the paragraph following Proposition 4.1 in Sec. 4.4).

It should be noted, more generally, that trying to follow the arrows of a vector field by hand in order to get an idea of the shape of the field lines (the solution curves) is not always as easy as it sounds, and it may be sometimes quite misleading (see Remark B.4 in Sec. B.6 of Appendix B). In our present case, for example, one may be tempted to guess from the circular nature of the vector field that by following the arrows one obtains circular solution curves, although in reality the solution curves are spirals (that just slightly deviate from perfect circles). More details on this point, as well as several illustrated examples, can be found in [Schwalbe97, pp. 39–42 and 69].

4-16. In general, dot trajectories between two aperiodic screens are not invariant under the addition of a transformation $\overline{\mathbf{f}}(x,y)$ to the transformations $\overline{\mathbf{g}}_1(x,y)$ and $\overline{\mathbf{g}}_2(x,y)$ undergone by the original layers (see Secs. 4.5 and 5.4 and Fig. 4.7). However, it turns out that in some particular cases the dot trajectories do remain invariant under such additions; this situation occurs, for example, in Fig. 4.9. Can you explain why in this particular case the dot trajectories remain invariant? *Hint*: The layer transformations being used in this case are $\overline{\mathbf{g}}_1(x,y) = (x + ay^2, y + f(x))$ and $\overline{\mathbf{g}}_2(x,y) = (x + 1, y + f(x))$. Find the explicit expression of the inverse transformation $\mathbf{g}_2(x,y) = [\overline{\mathbf{g}}_2(x,y)]^{-1}$ and then, using Eq. (4.8), show that the vector field $\overline{\mathbf{h}}_1(x,y) = \overline{\mathbf{g}}_1(\mathbf{g}_2(x,y)) - (x,y)$ that describes accurately the dot trajectories, has a zero vertical component. This means that the vector field and the dot trajectories remain horizontal under the addition of any $\overline{\mathbf{f}}(x,y) = (0, f(x))$ to both layers.

4-17. Calculate the vector fields $\overline{\mathbf{h}}(x,y)$, $\overline{\mathbf{h}}_1(x,y)$ and $\overline{\mathbf{h}}_2(x,y)$ that are provided by Eqs. (4.7), (4.8) and (4.9) for the superpositions shown in Figs. 4.7(c) and 4.8(c). Verify that in both cases the vector fields $\overline{\mathbf{h}}_1(x,y)$ and $\overline{\mathbf{h}}_2(x,y)$ represent the dot trajectories in the superposition faithfully, while the vector field $\overline{\mathbf{h}}(x,y)$ does not represent them accurately (see also Figs. 4.11 and 4.12).

4-18. *Mutual fixed points between two transformations and the mutual fixed points between their inverse transformations.* Suppose we are given two identical random dot screens $r_1(x,y)$ and $r_2(x,y)$, and that we apply to these layers the domain (inverse) transformations $\mathbf{g}_1(x,y)$ and $\mathbf{g}_2(x,y)$, respectively. The resulting layers are given, therefore, by $r_1(\mathbf{g}_1(x,y))$ and $r_2(\mathbf{g}_2(x,y))$. These two transformed layers have a mutual fixed point (x_F, y_F) if $\mathbf{g}_1(x_F, y_F) = \mathbf{g}_2(x_F, y_F)$, i.e. if $\mathbf{g}_1(x_F, y_F) - \mathbf{g}_2(x_F, y_F) = (0,0)$. On the other hand, we

---

[14] In fact, each of the circular dot trajectories can be seen as the *orbit* of a point (x_0, y_0) under the rotation transformation, i.e. as the set of all points (x,y) that can be reached by applying the rotation transformation to the point (x_0, y_0). Such orbits can be studied using techniques from Lie's theory of transformation groups [Cantwell02; Steeb96; Olver86].

can also express the transformations undergone by the two layers in terms of their influence on the location of the individual dots of the original screens, namely $(x,y) \mapsto \overline{\mathbf{g}}_1(x,y)$ and $(x,y) \mapsto \overline{\mathbf{g}}_2(x,y)$, where $\overline{\mathbf{g}}_1$ and $\overline{\mathbf{g}}_2$ are the respective *direct* transformations: $\overline{\mathbf{g}}_1(x,y) = \mathbf{g}_1^{-1}(x,y)$, $\overline{\mathbf{g}}_2(x,y) = \mathbf{g}_2^{-1}(x,y)$. The distance between the two destinations of the same point (x,y) under the transformations $\overline{\mathbf{g}}_1$ and $\overline{\mathbf{g}}_2$ is given by $\overline{\mathbf{h}}(x,y) = \overline{\mathbf{g}}_1(x,y) - \overline{\mathbf{g}}_2(x,y)$. Therefore, in each mutual fixed point, where points of the two transformed layers coincide, we would expect to have also a coincidence between the destination points $\overline{\mathbf{g}}_1(x_F,y_F)$ and $\overline{\mathbf{g}}_2(x_F,y_F)$, i.e. $\overline{\mathbf{g}}_1(x_F,y_F) - \overline{\mathbf{g}}_2(x_F,y_F) = (0,0)$. For example, we would expect the vector field shown in Fig. 4.6(d) to have its zeros $\overline{\mathbf{h}}(x,y) = \overline{\mathbf{g}}_1(x,y) - \overline{\mathbf{g}}_2(x,y) = (0,0)$ at the same points where the superposed layers in Fig. 3.15(a) have their mutual fixed points, i.e. where $\mathbf{k}(x,y) = \mathbf{g}_1(x,y) - \mathbf{g}_2(x,y) = (0,0)$. However, it turns out that the mutual fixed points of $\mathbf{g}_1$ and $\mathbf{g}_2$ are not always identical to the mutual fixed points of $\overline{\mathbf{g}}_1$ and $\overline{\mathbf{g}}_2$. Although these fixed points are identical in transformations that satisfy the identity $\mathbf{g}_1 - \mathbf{g}_2 \equiv \overline{\mathbf{g}}_2 - \overline{\mathbf{g}}_1$, in the general case we only have the approximation provided by Proposition D.8 in Appendix D: If $\mathbf{g}_1$ and $\mathbf{g}_2$ are weak transformations, then $\mathbf{g}_1(x_F,y_F) = \mathbf{g}_2(x_F,y_F)$ implies that $\overline{\mathbf{g}}_1(x_F,y_F) \approx \overline{\mathbf{g}}_2(x_F,y_F)$, and $\overline{\mathbf{g}}_1(x_F,y_F) = \overline{\mathbf{g}}_2(x_F,y_F)$ implies that $\mathbf{g}_1(x_F,y_F) \approx \mathbf{g}_2(x_F,y_F)$.

(a) Can you find a pair of transformations $\mathbf{g}_1$ and $\mathbf{g}_2$ for which the mutual fixed point of $\mathbf{g}_1$ and $\mathbf{g}_2$ is not identical to the mutual fixed point of $\overline{\mathbf{g}}_1$ and $\overline{\mathbf{g}}_2$? Consider, for example, the layer transformations $\overline{\mathbf{g}}_1(x,y) = (sx,sy)$ and $\overline{\mathbf{g}}_2(x,y) = (x + a, y + b)$ and their inverse transformations $\mathbf{g}_1(x,y) = (x/s,y/s)$ and $\mathbf{g}_2(x,y) = (x - a, y - b)$. What are the mutual fixed points between $\mathbf{g}_1$ and $\mathbf{g}_2$ and between $\overline{\mathbf{g}}_1$ and $\overline{\mathbf{g}}_2$?

(b) The mutual fixed points of the superposed layers are obtained where $\mathbf{g}_1(x,y) = \mathbf{g}_2(x,y)$, but on the other hand, in the same superposition, the stationary points in which the two destinations of an original dot coincide occur where $\overline{\mathbf{g}}_1(x,y) = \overline{\mathbf{g}}_2(x,y)$. How do you explain the fact that these two loci are not necessarily identical? *Hint*: Remember that the vector field $\overline{\mathbf{h}}(x,y) = \overline{\mathbf{g}}_1(x,y) - \overline{\mathbf{g}}_2(x,y)$ provides only an approximation to the dot trajectories, while the precise result is provided by the vector field $\overline{\mathbf{h}}_1(x,y) = \overline{\mathbf{g}}_1(\mathbf{g}_2(x,y)) - (x,y)$ (or, equivalently, by the vector field $\overline{\mathbf{h}}_2(x,y) = (x,y) - \overline{\mathbf{g}}_2(\mathbf{g}_1(x,y))$). And indeed, it is easy to see that $\mathbf{g}_1(x_F,y_F) = \mathbf{g}_2(x_F,y_F)$ occurs *iff* $\overline{\mathbf{g}}_1(\mathbf{g}_2(x_F,y_F)) = (x_F,y_F)$ and *iff* $(x_F,y_F) = \overline{\mathbf{g}}_2(\mathbf{g}_1(x_F,y_F))$, although it does not necessarily imply that $\overline{\mathbf{g}}_1(x_F,y_F) = \overline{\mathbf{g}}_2(x_F,y_F)$, too. This means that when we use the approximation $\overline{\mathbf{h}}(x,y) = \overline{\mathbf{g}}_1(x,y) - \overline{\mathbf{g}}_2(x,y)$ to represent the vector field corresponding to the dot trajectories in our layer superposition, there may exist a tiny discrepancy between the location of the fixed point of this vector field and the location of the fixed point in the superposition. But when $\overline{\mathbf{g}}_1$ and $\overline{\mathbf{g}}_2$ are weak transformations (as they must be in order not to lose the correlation between the two superposed layers and hence the dot trajectories), this discrepancy is negligible.

(c) In the case shown in Fig. 4.6(d), does the vector field $\overline{\mathbf{h}}(x,y)$ have its zeros exactly where the superposed layers of Fig. 3.15(a) have their mutual fixed points, i.e. where $\mathbf{k}(x,y) = (0,0)$? *Hint*: As we have seen in Sec. 3.5, the domain transformations undergone by the two layers in this case are given by $\mathbf{g}_1(x,y) = (x, y + y_0 - ax^2)$ and $\mathbf{g}_2(x,y) = (x + x_0 - ay^2, y)$. Find the corresponding direct transformations, $\overline{\mathbf{g}}_1(x,y)$ and $\overline{\mathbf{g}}_2(x,y)$, and check if $\overline{\mathbf{h}}(x,y) = \overline{\mathbf{g}}_1(x,y) - \overline{\mathbf{g}}_2(x,y)$ and $\mathbf{k}(x,y) = \mathbf{g}_1(x,y) - \mathbf{g}_2(x,y)$ have the same zeros.

(d) Do almost fixed points of transformations $\mathbf{g}_1(x,y)$ and $\mathbf{g}_2(x,y)$ (see Sec. 3.7) coincide with almost fixed points of the transformations $\overline{\mathbf{g}}_1(x,y)$ and $\overline{\mathbf{g}}_1(x,y)$?

4-19. Knowing that the mutual fixed points between the direct transformations $\overline{\mathbf{g}}_1$ and $\overline{\mathbf{g}}_2$ are not necessarily the same as the mutual fixed points between the domain transformations $\mathbf{g}_1$ and $\mathbf{g}_2$ (see Proposition D.8 in Appendix D), design a test showing that Glass patterns are generated about the mutual fixed points of $\mathbf{g}_1$ and $\mathbf{g}_2$ and not about the mutual fixed points of $\overline{\mathbf{g}}_1$ and $\overline{\mathbf{g}}_2$. *Hint*: Find a pair of simple layer transformations $\mathbf{g}_1$ and $\mathbf{g}_2$ for which the mutual fixed points of $\mathbf{g}_1$ and $\mathbf{g}_2$ are clearly different from those of $\overline{\mathbf{g}}_1$ and $\overline{\mathbf{g}}_2$, and show in the corresponding aperiodic screen superposition that the Glass patterns are indeed generated about the mutual fixed points of $\mathbf{g}_1$ and $\mathbf{g}_2$.

4-20. *Influence of the dot shapes on the dot trajectories.* Does the geometric form of the dot trajectories in the superposition of two aperiodic layers depend on the shapes of the individual microstructure elements (dots) in the superposed layers? For example, consider the circular dot trajectories in the superposition of two slightly rotated, aperiodic dot screens made of small black dots, as in Fig. 4.15(a). What happens in the superposition when we replace the black dots in one or both layers by other small elements, such as "|", "—", "O", "L" or any combinations thereof?

4-21. *A possible starting point for the explanation of the microstructures in Fig. 4.16.* As pointed out in [Barlow04], the microstructure in the superposition shown in Fig. 4.16(a) consists of dot pairs, while the microstructure in the superposition shown in Fig. 4.16(c) consists of "anti-pairs", i.e. dot pairs whose individual dots have opposite polarities. Assuming that Fig. 4.16(a), too, is printed on a gray background, one can say that the anti-pairs of Fig. 4.16(c) are the *annihilators* of the dot pairs of Fig. 4.16(a), since by taking the average intensity of the two figures the pairing effect disappears, and we are left with a dot screen that consists of single black dots on a gray background. In other words, the addition of the petal-like "anti-screen" of Fig. 4.16(c) undoes the screen superposition of Fig. 4.16(a), and gives back only one of the two originally superposed layers (with a gray background). How can this help us to explain the different microstructures in Figs. 4.16(a) and 4.16(c)? You may also consider Figure 8 in [Barlow04] which illustrates a more symmetric case where both of the images consist of black and white dots on a gray background. As clearly shown by this figure, the addition of the two images (the first having a circular microstructure due to the pairing effect and the second having a petal-like microstructure due to the anti-pairing effect) destroys the visible microstructures, and gives a random image that is devoid of any apparent dot trajectories. Interestingly, as shown there, the same effect occurs even if the dot locations in the two complementary images (the one with circular dot trajectories and the one with petal-like dot trajectories) do not coincide; in this case the intensity cancellation is not done on a dot-by-dot basis but only on the average.

4-22. *The microstructures under different superposition rules.* The "logical or" superposition rule gives white wherever *at least* one of the two superposed layers is white, and black wherever *both* of the superposed layers are black. It has been shown in [Garavaglia01] that when this superposition rule is used the resulting dot trajectories in the superposition of two aperiodic dot screens seem to be perpendicular to the dot trajectories that are obtained in the usual multiplicative superposition. For example, in the superposition of two identical aperiodic screens with a small angle difference the resulting dot trajectories are radial lines instead of concentric circles. How can you explain this phenomenon? *Hint*: The resulting elements in this type of superposition consist of the intersection of black dots from both layers. What would you expect to see in the intersection of neighbouring black dots that are arranged circularly about the center as in Fig. 4.16(a)? And what would you expect to see in the intersection of neighbouring black dots that are arranged along radial lines about the center as in Fig. 4.17(a)? What do you expect to see in the superposition when both layers are periodic?

Chapter 5

Moiré phenomena between periodic or aperiodic screens

5.1 Introduction

Having prepared the required groundwork in the previous chapters, we are ready now to return to some of our questions that still remain open. Perhaps the most outstanding and the most intriguing one concerns the mathematical relationship between the phenomena which occur in the superposition of periodic or aperiodic dot screens that undergo the same geometric layer transformations. As we already know, the superposition of two *periodic* dot screens that have undergone geometric transformations, linear or non-linear, gives geometrically transformed moiré effects, whose properties are explained by the classical moiré theory (see, for example, Chapters 10 and 11 in *Vol. I*). However, if the same layer transformations are applied to two correlated *aperiodic* dot screens, their superposition gives instead other phenomena, Glass patterns and dot trajectories, whose visual properties are completely different. And yet, because the same geometric transformations have been applied to the original layers in both periodic and aperiodic cases, there should exist some mathematical relationship between the phenomena that occur in their respective superpositions, in spite of their different visual properties. And indeed, it turns out that such a mathematical relationship does exist, although it is not really obvious and it may even seem surprising at first sight.

It is therefore our aim in the present chapter to investigate the mathematical relationship between the phenomena that occur in the superposition of periodic or aperiodic dot screens under the same geometric layer transformations. We will explain the similarities as well as the striking differences that exist between these phenomena, and we will also see what happens in in-between cases, where the superposed layers are intermediate between periodic and aperiodic.

We start in Sec. 5.2 with a brief review of the main results we have obtained so far concerning moiré patterns, Glass patterns and dot trajectories, in order to have them readily available whenever they will be needed in the sections which follow. In Sec. 5.3 we give a few detailed examples to illustrate the relationship between the phenomena that occur in periodic or aperiodic cases, and we formulate our findings in a more formal way. Then, in Sec. 5.4 we study the behaviour and the invariance properties of periodic and aperiodic screen superpositions under two different operations: addition of transformations and composition of transformations. Table 5.1 at the end of this chapter provides a brief comparative summary of the various phenomena which occur in superpositions of periodic or aperiodic dot screens. In Chapter 6 this comparison will be extended to the 1D case of periodic or aperiodic line gratings, too.

5.2 Brief review: moiré patterns, Glass patterns and dot trajectories

As we can see in our figures throughout the preceding chapters, for example, in Figs.
2.1–2.2 or 3.5–3.15, the application of the same layer mappings to periodic or to aperiodic
dot screens gives quite different results. If the original dot screens before applying the
mappings are periodic, as in the right hand part of each figure, the superposition of the
transformed layers may give rise to periodic or repetitive moiré patterns; but if the original
dot screens before applying the mappings are aperiodic, as in the left hand part of each
figure, the superposition of the transformed layers may give rise to Glass patterns about
the mutual fixed points (or the almost fixed points) of the transformations. Note that in
both periodic and aperiodic cases microstructure dot alignments may be also generated in
the superposition, with or without the presence of macroscopic moiré or Glass patterns.
While in the case of periodic layers these microstructures simply follow the distortions of
the original layers, in the case of aperiodic layers the microstructures may consist of dot
trajectories of more exotic shapes.

But since in all of our figures the same layer mappings have been applied to both the
periodic and the aperiodic dot screens, there should exist some mathematical relationship
between the structures which appear in the periodic and aperiodic superpositions, in spite
of the significant differences that exist between them. Having already acquired the
mathematical background for understanding both periodic and aperiodic cases, it is time
for us now to elucidate this interesting question. But before we start, let us first review the
main results that will be needed for this end.

Consider, first, the periodic case. As we already know from the classical moiré theory
for periodic layers, when two initially periodic layers having the same periodicity (but
possibly different intensity profiles) undergo weak geometric transformations, linear or
non-linear, a highly visible moiré effect may be generated in their superposition (see, for
example, the right hand side images in Figs. 2.1–2.2 or 3.5–3.15). Suppose that the
geometric transformations undergone by the originally identical layers are $(u,v) = \mathbf{g}_1(x,y)$
and $(u,v) = \mathbf{g}_2(x,y)$, respectively. Then the geometric transformation $(u,v) = \mathbf{g}_M(x,y)$ of the
resulting moiré effect with respect to the original, undistorted layers (for example, the
transformation which relates the moiré pattern of Fig. 2.1(d) with the original periodic
pattern of Fig. 2.1(b)) is given by the difference between the layer transformations $\mathbf{g}_1(x,y)$
and $\mathbf{g}_2(x,y)$ (see Proposition 10.5 in *Vol. I*):[1]

$$\mathbf{g}_M(x,y) = \mathbf{g}_1(x,y) - \mathbf{g}_2(x,y) \tag{5.1}$$

Note that the three transformations $\mathbf{g}_1(x,y)$, $\mathbf{g}_2(x,y)$ and $\mathbf{g}_M(x,y)$ are considered here, as
explained in Remark 4.1, as *domain* transformations; for example, if the original,

---

[1] In fact, the superposition of initially periodic layers may also give rise to moiré effects of higher orders,
as explained in Sec. 10.9 of *Vol. I*. But because such moiré effects have no equivalents in the aperiodic
case (see Sec. 7.5) we ignore them here and only consider the first order subtractive moiré effect given
by Eq. (5.1), i.e. the (1,-1)- or (1,0,-1,0)-moiré effects. For the sake of brevity we will refer to these
moiré effects using the generic term "first-order moirés".

undistorted periodic layers are expressed by $p_1(x,y)$ and $p_2(x,y)$, then the distorted layers after the application of the transformations are expressed by $r_1(x,y) = p_1(\mathbf{g}_1(x,y))$ and $r_2(x,y) = p_2(\mathbf{g}_2(x,y))$.

The geometric layout of the bands (or cells) of the moiré pattern (5.1) are determined by the loci of the points of coincidence between the two layers after they have undergone the domain transformations $\mathbf{g}_1(x,y)$ and $\mathbf{g}_2(x,y)$. If each of the distorted layers is regarded as an indexed family of lines (or dots), their loci of coincidence (and hence the moiré patterns in the layer superposition) form a new indexed family of magnified and distorted bands (or cells). Their equations can be deduced from the equations of the two original layers using the well-known indicial equations method for line gratings (see, for example, Sec. 11.2 in *Vol. I*), or a straightforward 2D generalization thereof. In this generalization the individual layers being superposed are no longer line gratings but rather 2D line grids (or dot screens), and therefore each of the two layers is regarded as two indexed families of line gratings, or, equivalently, as a doubly-indexed family of crossing lines (or of dots). The loci of coincidence between the two distorted layers are given in this case by the points (x,y) that satisfy:

$$\mathbf{g}_1(x,y) - \mathbf{g}_2(x,y) = (p,q) \qquad p,q \in \mathbb{Z} \tag{5.2}$$

or in componentwise notation:

$$\begin{pmatrix} g_{1,1}(x,y) \\ g_{1,2}(x,y) \end{pmatrix} - \begin{pmatrix} g_{2,1}(x,y) \\ g_{2,2}(x,y) \end{pmatrix} = \begin{pmatrix} p \\ q \end{pmatrix} \qquad p,q \in \mathbb{Z}$$

where $\mathbf{g}_1(x,y) = (g_{1,1}(x,y), g_{1,2}(x,y))$ and $\mathbf{g}_2(x,y) = (g_{2,1}(x,y), g_{2,2}(x,y))$. This corresponds, indeed, to two indexed families of moiré bands, one being indexed by p and the other being indexed by q (or, equivalently, to one doubly indexed family of moiré cells being indexed by (p,q)). The connection between Eqs. (5.1) and (5.2) is that the coordinate lines of the moiré transformation (5.1), i.e. the level lines of its components, $g_{1,1}(x,y) - g_{2,1}(x,y)$ and $g_{1,2}(x,y) - g_{2,2}(x,y)$, consist, respectively, of the curve families $g_{1,1}(x,y) - g_{2,1}(x,y) = p$ and $g_{1,2}(x,y) - g_{2,2}(x,y) = q$.[2] This is clearly illustrated in the examples in Sec. 5.3 below. Alternatively, these moiré bands can be also interpreted as the straight or curved coordinate lines that result from applying the domain transformation $\mathbf{g}_M(x,y)$ to a unit grid having the periodicity of the original, undistorted periodic layers. Note that in cases like Fig. 2.3(b),(d), Fig. 3.5(d) and Fig. 3.6(b),(d) there exists only one system of moiré bands, because the other one has an infinitely large periodicity and is not visible.

We proceed now to the aperiodic case. As shown in the left hand side of Figs. 2.1–2.2 or 3.5–3.15, if the two originally identical dot screens are aperiodic, then the application of the same geometric transformations $\mathbf{g}_1(x,y)$ and $\mathbf{g}_2(x,y)$ to the original layers may give rise in the superposition to a Glass pattern (namely, a visible zone of high correlation between

---

[2] Remember that each transformation $(u,v) = \mathbf{g}(x,y)$ consists of two scalar functions, $u = g_1(x,y)$ and $v = g_2(x,y)$, namely, $\mathbf{g}(x,y) = (g_1(x,y),g_2(x,y))$. Each of these two scalar functions can be regarded as a surface over the x,y plane, and described graphically by its set of level lines. See Appendix B for more details.

the two superposed layers) that is generated about the locus of the mutual fixed points between the two layer mappings $g_1(x,y)$ and $g_2(x,y)$. In other words, the Glass pattern is generated about the locus of the points (x,y) that satisfy $g_1(x,y) = g_2(x,y)$, namely:

$$g_1(x,y) - g_2(x,y) = (0,0) \tag{5.3}$$

(Note that here too, just as in the periodic case, the layer transformations $g_1(x,y)$ and $g_2(x,y)$ are considered, in accordance with Remark 4.1, as *domain* transformations.)

A comparison of Eqs. (5.2) and (5.3) shows, as we already know from Chapter 3, that the Glass pattern is, in fact, a subset of the moiré pattern that is obtained between originally periodic layers under the same layer transformations $g_1(x,y)$ and $g_2(x,y)$. And indeed, as we can clearly see by comparing the two images in each row of Figs. 2.1–2.2 or 3.5–3.15, the moiré pattern that is generated in the superposition of periodic layers includes, in addition to the locus of the fixed points, in which $(p,q) = (0,0)$, infinitely many repetitions (possibly distorted ones) of this locus, in which $(p,q) \neq (0,0)$. These repetitions, that only occur due to the periodicity of the original layers, are the loci of the points where the distorted periodic structures happen to coincide, but they are not mutual fixed points of the layer transformations $g_1(x,y)$ and $g_2(x,y)$.

Another characteristic feature proper to the superposition of aperiodic screens is the dot trajectories that appear in the microstructure of the superposition. As we already know, such dot trajectories may be generated in the superposition of aperiodic layers even in cases where no Glass patterns exist (see, for example, Fig. 3.8(a)). The dot trajectories are made of correlated dot pairs, one dot from each of the two superposed layers. If only one of the two originally identical layers has been transformed, each screen dot appears in the layer superposition in its original location, (x,y), and then once again in its transformed dot location. Each such dot pair in the superposition represents, therefore, two successive locations of the same dot, namely, the dot's location before and after the layer transformation has been applied. And indeed, it was shown in Sec. 4.4 that the dot trajectories in this case are described mathematically by the vector field that consists of the vectors connecting the original dot locations (x,y) to their destinations $\overline{g}(x,y)$:

$$\overline{h}(x,y) = \overline{g}(x,y) - (x,y) \tag{5.4}$$

(or by the vector field $-\overline{h}(x,y) = (x,y) - \overline{g}(x,y)$, as the order of the dots within each dot pair is irrelevant), where $\overline{g}(x,y)$ is the transformation undergone by the distorted layer. Note that unlike in Eqs. (5.1)–(5.3) above, the transformations being considered here are *direct* transformations, as explained in Remark 4.1, whence the use of the barred notation.[3]

If, however, both of the original layers have been transformed, the dot pairs which make up the dot trajectories in the superposition no longer represent a dot's location before and after a layer mapping has been applied, but rather the new destinations of the same original

---

[3] For example, if one of the layers has been magnified by a factor of 1.1, this layer mapping must be expressed in Eq. (5.4) as a *direct* transformation $\overline{g}(x,y) = (1.1x, 1.1y)$, while in Eqs. (5.1)–(5.3) it must be expressed as a *domain* (inverse) transformation, $g(x,y) = (x/1.1, y/1.1)$. See also Problem 5-7.

dot under each of the two layer mappings. In this case the dot trajectories that are generated in the superposition are given by Eq. (4.8) (or Eq. (4.9)). But as long as the layer mappings being used are sufficiently weak, we prefer to use the practical approximation to the dot trajectories that is provided in Proposition 4.3 by the vector field:

$$\overline{\mathbf{h}}(x,y) = \overline{\mathbf{g}}_1(x,y) - \overline{\mathbf{g}}_2(x,y) \tag{5.5}$$

(or, equivalently, by the vector field $-\overline{\mathbf{h}}(x,y)$). When only one of the layers is being transformed Eq. (5.5) reduces into Eq. (5.4), and the approximation turns into an accurate representation of the dot trajectories.

5.3 A few detailed examples to illustrate the formal results

Let us now consider a few examples that will help us understand in more detail the mathematical relationship between the phenomena that occur in periodic and aperiodic cases under the same layer mappings. In particular, these examples illustrate how the two alternative representations of the given layer mappings, as direct or as domain transformations, are to be used. In each of these examples we first explain the dot trajectories and the Glass patterns that occur in the aperiodic case, and then we proceed with the explanation of the moiré effects that occur in the corresponding periodic case. The formal results that are illustrated by these examples are formulated at the end of this section.

Example 5.1: Radial dot trajectories surrounding a fixed point:

Consider the screen superpositions shown in Figs. 2.1(e) and 2.1(f). In both of these superpositions one layer is a slightly scaled-up version of the other, with a magnification rate of $s = 1 + \varepsilon$ (ε being a small positive fraction). In this case the layer mappings can be expressed, therefore, by the direct transformations:

$$\overline{\mathbf{g}}_1(x,y) = ((1+\varepsilon)x, (1+\varepsilon)y)$$

$$\overline{\mathbf{g}}_2(x,y) = (x,y)$$

Alternatively, the same layer mappings can be also expressed as domain transformations $\mathbf{g}_1(x,y)$ and $\mathbf{g}_2(x,y)$ where $\mathbf{g}_1(x,y) = \overline{\mathbf{g}}_1^{-1}(x,y)$ and $\mathbf{g}_2(x,y) = \overline{\mathbf{g}}_2^{-1}(x,y)$. To find the explicit expression of $\overline{\mathbf{g}}_1^{-1}(x,y)$ we write $\overline{\mathbf{g}}_1(x,y)$ in its full componentwise form:

$$u = (1+\varepsilon)x$$

$$v = (1+\varepsilon)y$$

and solve for x and y. As we can see, the inverse of $\overline{\mathbf{g}}_1(x,y)$ is given explicitly by:

$$x = \frac{1}{1+\varepsilon}u = (1-\delta)u$$

$$y = \frac{1}{1+\varepsilon}v = (1-\delta)v$$

where $\delta = \frac{\varepsilon}{1+\varepsilon}$ is a small positive fraction (that is just slightly smaller than ε).

Hence, the inverse transformations of $\overline{\mathbf{g}}_1(x,y)$ and $\overline{\mathbf{g}}_2(x,y)$, expressed again as mappings in the x,y plane, are given by:

$$\mathbf{g}_1(x,y) = ((1-\delta)x, (1-\delta)y)$$

$$\mathbf{g}_2(x,y) = (x,y)$$

These mappings represent our given layer mappings in terms of domain transformations.

Having found the two alternative representations of our layer mappings as *direct* and as *domain* transformations, we now proceed to the explanation of the various phenomena which occur in our layer superpositions in Figs. 2.1(e) and 2.1(f).

We start with the dot trajectories that occur in the aperiodic case. As we already know, the dot trajectories are expressed by the field lines of the vector field $\overline{\mathbf{h}}(x,y)$ that is given by Eqs. (5.4) or (5.5). In our example we have, therefore:

$$\overline{\mathbf{h}}(x,y) = \overline{\mathbf{g}}_1(x,y) - (x,y) = (\varepsilon x, \varepsilon y)$$

This vector field assigns to each point (x,y) a vector equal to $(\varepsilon x, \varepsilon y)$, pointing in a radial direction away from the origin, as shown in Fig. 4.2(b). And indeed, this vector field clearly reflects the dot trajectories that surround the Glass pattern in the aperiodic superposition shown in Fig. 2.1(e). Note that this vector field gives an accurate representation of the dot trajectories, because in this case only one of the superposed layers has been transformed.

Having understood the dot trajectories of Fig. 2.1(e), we proceed to the explanation of the macroscopic Glass pattern. The Glass pattern of Fig. 2.1(e) is the zone of high correlation between the two superposed layers that is generated, according to Eq. (5.3), about the mutual fixed points of the *domain* transformations undergone by our layers. In other words, the Glass pattern is generated around the points (x,y) in which $\mathbf{g}_1(x,y) = \mathbf{g}_2(x,y)$. It is easy to see that in our example this condition is satisfied exactly at the point $(x,y) = (0,0)$, the only point where $(1-\delta)x = x$ and $(1-\delta)y = y$. This explains, indeed, why a Glass pattern is generated in our case about the origin.

Finally, how do we explain the straight moiré bands which are obtained when the same transformations $\overline{\mathbf{g}}_1(x,y)$ and $\overline{\mathbf{g}}_2(x,y)$ are applied to two identical *periodic* dot screens, as shown in Fig. 2.1(f)? The domain transformation undergone by the resulting first-order moiré is given according to Eq. (5.1) by:

$$\mathbf{g}_M(x,y) = \mathbf{g}_1(x,y) - \mathbf{g}_2(x,y) = -(\delta x, \delta y)$$

Therefore, according to Eq. (5.2), the coordinate lines of this transformation (i.e. the level lines of its components, $g_1(x,y) = -\delta x$ and $g_2(x,y) = -\delta y$) consist of the lines $-\delta x = p$

and $-\delta y = q$ (namely, $x = -p/\delta$ and $y = -q/\delta$) for all integer values of p and q. These two sets of coordinate lines form an orthogonal grid whose period is $1/\delta = 1/\varepsilon + 1$ times larger than the period of the original, untransformed layer. This is, indeed, the interpretation of the moiré pattern in the periodic superposition shown in Fig. 2.1(f).

The above considerations give us, indeed, the mathematical expressions of the different phenomena that occur in the aperiodic and in the periodic cases under the same layer transformations: The dot trajectories that occur in the aperiodic case are expressed mathematically by the transformation $\overline{\mathbf{h}}(x,y) = (\varepsilon x, \varepsilon y)$ (interpreted as a vector field), the Glass pattern is the zone of high correlation between the two superposed layers around their fixed point at the origin, and the moiré bands that occur in the periodic case are expressed mathematically by the transformation $\mathbf{g}_M(x,y) = -(\delta x, \delta y)$ (interpreted as a domain transformation).

So how do we explain the significant difference between the properties of the resulting moiré effects in the periodic and aperiodic cases, under the same layer transformations? Clearly, the effect of the vector field $\overline{\mathbf{h}}(x,y)$ on the individual screen dots is precisely the same in both periodic and aperiodic superpositions. In both cases each screen dot appears in the layer superposition twice: once in its original location, (x,y), and then in its transformed dot location, $\overline{\mathbf{g}}(x,y)$. These dot pairs are located on straight trajectories emanating from the origin, as we clearly see in the superposition of the aperiodic screens (Fig. 2.1(e)). In the periodic case, however, when the transformed dots move from their original locations to their new locations, they also form due to their periodic nature new moiré bands, as shown in Fig. 2.1(f). These moiré bands are much more prominent than the radial dot trajectories, and they completely outweigh and obscur them. And yet, as already explained in Sec. 4.7, the radial dot trajectories *do* exist even in the periodic case, although they are hardly visible: If we look under a magnifying glass at the dot arrangements in the center of the periodic superposition (Fig. 2.1(f)), we can still identify there the radial dot arrangements, although farther away from the origin they are largely masked out by the effects of periodicity. ∎

Example 5.2: Hyperbolic dot trajectories surrounding a fixed point:

Consider the screen superpositions shown in Figs. 2.2(a) and 2.2(b). In both of these superpositions one layer is obtained from the other by a slight scaling of $s_x = 1 - \varepsilon$ in the x direction, and a slight scaling of $s_y = 1 + \varepsilon$ in the y direction (ε being a small positive fraction). This mapping can be expressed either by the direct transformation $\overline{\mathbf{g}}_1(x,y) = ((1-\varepsilon)x, (1+\varepsilon)y)$ or by the domain transformation $\mathbf{g}_1(x,y) = \overline{\mathbf{g}}_1^{-1}(x,y) = (x/(1-\varepsilon), y/(1+\varepsilon)) = ((1+\delta_1)x, (1-\delta_2)y)$, where $\delta_1 = \frac{\varepsilon}{1-\varepsilon}$ and $\delta_2 = \frac{\varepsilon}{1+\varepsilon}$ are small positive fractions that are close to ε. (Note that here we have used the known fact that the inverse of the transformation $\overline{\mathbf{g}}_1(x,y) = (ax, by)$ is given by $\mathbf{g}_1(x,y) = (x/a, y/b)$; this can be derived explicitly using the same method as in Example 5.1 above).

Let us start, once again, with the dot trajectories that occur in the aperiodic case. In our present example the layer transformations $\overline{\mathbf{g}}_1(x,y)$ and $\overline{\mathbf{g}}_2(x,y)$ are given by:

$$\bar{g}_1(x,y) = ((1-\varepsilon)x, (1+\varepsilon)y)$$

$$\bar{g}_2(x,y) = (x,y)$$

so that the vector field $\bar{h}(x,y)$ is, according to Eq. (5.4):

$$\bar{h}(x,y) = \bar{g}_1(x,y) - (x,y) = (-\varepsilon x, \varepsilon y)$$

This vector field assigns to each point (x,y) a vector equal to $(-\varepsilon x, \varepsilon y)$. This vector differs from that of Example 5.1 in the sign of its x component; this results in a hyperbolic vector field surrounding the fixed point at the origin, as shown in Fig. 4.3(a). And indeed, this vector field clearly reflects the dot trajectories that surround the Glass pattern in the aperiodic superposition in Fig. 2.2(a). (Note that this hyperbolic dot arrangement can be also seen under a magnifying glass in the center of the periodic case (Fig. 2.2(b)), although farther away from the center it is masked out by the effects of periodicity.)

We proceed now to the explanation of the macroscopic Glass pattern. The Glass pattern of Fig. 2.2(a) is the zone of high correlation between the two superposed layers which is generated, according to Eq. (5.3), about the point (x,y) that satisfies $g_1(x,y) = g_2(x,y)$. Note that $g_1(x,y)$ and $g_2(x,y)$ express the same layer mappings as *domain* transformations, and are, therefore, the inverse of the transformations $\bar{g}_1(x,y)$ and $\bar{g}_2(x,y)$:

$$g_1(x,y) = ((1+\delta_1)x, (1-\delta_2)y)$$

$$g_2(x,y) = (x,y)$$

where $\delta_1 = \frac{\varepsilon}{1-\varepsilon}$ and $\delta_2 = \frac{\varepsilon}{1+\varepsilon}$. Therefore, the fixed point $g_1(x,y) = g_2(x,y)$ occurs — just as in the previous example — exactly at the origin, $(x,y) = (0,0)$, the only point that satisfies $(1+\delta_1)x = x$ and $(1-\delta_2)y = y$. And indeed, as shown in Fig. 2.2(a), in our present case, too, the Glass pattern is generated around the fixed point at the origin.

Finally, we now explain the straight moiré bands which are obtained when the same transformations $\bar{g}_1(x,y)$ and $\bar{g}_2(x,y)$ are applied to two identical *periodic* dot screens, as shown in Fig. 2.2(b). The domain transformation undergone by the resulting first-order moiré is given according to Eq. (5.1) by:

$$g_M(x,y) = g_1(x,y) - g_2(x,y) = (\delta_1 x, -\delta_2 y)$$

Therefore, according to Eq. (5.2), the coordinate lines of this transformation (i.e. the level lines of its components, $g_1(x,y) = \delta_1 x$ and $g_2(x,y) = -\delta_2 y$) consist of the lines $\delta_1 x = p$ and $-\delta_2 y = q$ (namely, $x = p/\delta_1$ and $y = -q/\delta_1$) for all integer values of p and q. These two sets of coordinate lines form a grid whose orthogonal periods are $1/\delta_1 = 1/\varepsilon - 1$ and $1/\delta_2 = 1/\varepsilon + 1$ times larger than the period of the original, untransformed layer (for example, if $\varepsilon = 0.1$ then these factors are 9 and 11, respectively). This is, indeed, the interpretation of the moiré pattern in the periodic superposition shown in Fig. 2.2(b). Note that the horizontal and vertical scaling factors are identical (up to their sign) in $\bar{h}(x,y)$, but not in $g_M(x,y)$; this point is discussed in Problem 5-6. ■

Example 5.3: Dot trajectories surrounding each of several fixed points:

Consider the screen superpositions shown in Figs. 3.15(a) and 3.15(b). As we have seen in Sec. 3.5 (Example 3.6), the layer transformations $g_1(x,y)$ and $g_2(x,y)$ that were applied to the original layers are given in this case by Eqs. (3.35):

$$g_1(x,y) = (x, y + y_0 - ax^2)$$

$$g_2(x,y) = (x + x_0 - ay^2, y)$$

Note, however, that these transformations are given in Sec. 3.5 as *domain* transformations. In this case the direct transformations are easy to find, and they are given by:

$$\overline{g}_1(x,y) = g_1^{-1}(x,y) = (x, y - y_0 + ax^2)$$

$$\overline{g}_2(x,y) = g_2^{-1}(x,y) = (x - x_0 + ay^2, y)$$

and therefore we have, according to Eq. (5.5):

$$\overline{h}(x,y) = \overline{g}_1(x,y) - \overline{g}_2(x,y) = (x_0 - ay^2, ax^2 - y_0)$$

The vector field $\overline{h}(x,y)$ assigns to each point (x,y) a vector equal to $(x_0 - ay^2, ax^2 - y_0)$, as shown in Fig. 4.6(d). And indeed, as we can see in Fig. 3.15(a), this vector field gives a close approximation to the dot trajectories in the layer superposition.

The Glass pattern, on its part, consists of the zones of high correlation between the two superposed layers, which are generated about the mutual fixed points of the domain transformations $g_1(x,y)$ and $g_2(x,y)$. These mutual fixed points are given by the system of equations:

$$ay^2 - x_0 = 0$$

$$y_0 - ax^2 = 0$$

whose solutions are, as we already saw in Example 3.6:

$$x = \pm\sqrt{y_0/a}, \quad y = \pm\sqrt{x_0/a}$$

This means that in this case we have four different fixed points (see Eq. (3.38)):

$$(x,y) = (\pm\sqrt{y_0/a}, \pm\sqrt{x_0/a})$$

And indeed, as clearly shown by Fig. 3.15(a), in this case four different Glass patterns are generated in the superposition.

Finally, we now explain the straight moiré bands which are obtained when the same transformations $g_1(x,y)$ and $g_2(x,y)$ are applied to two identical periodic dot screens (see Fig. 3.15(b)). The domain transformation undergone by the resulting first-order moiré is given according to Eq. (5.1) by:

$$\mathbf{g}_M(x,y) = \mathbf{g}_1(x,y) - \mathbf{g}_2(x,y) = (ay^2 - x_0,\ y_0 - ax^2)$$

Therefore, according to Eq. (5.2), the coordinate lines of this transformation (i.e. the level lines of its two components, $g_1(x,y) = ay^2 - x_0$ and $g_2(x,y) = y_0 - ax^2$) consist of the lines $x_0 - ay^2 = p$ and $y_0 - ax^2 = q$ (namely: $y = \pm\sqrt{(x_0+p)/a}$ and $x = \pm\sqrt{(y_0-q)/a}$ for all integer values of p and q that give real values in the square roots). These coordinate lines form an orthogonal grid with decreasing line distances to both positive and negative directions, as shown, indeed, by the moiré pattern in the periodic superposition (Fig. 3.15(b)).   ∎

Example 5.4: Dot trajectories with no fixed points at all:

Consider the screen superpositions shown in Figs. 3.8(a) and 3.8(b). The layer transformations that have been applied to the original layers in both of these superpositions are given, as we have seen in Example 3.11, by the *domain* transformations:

$$\mathbf{g}_1(x,y) = (x - ay^2,\ y)$$

$$\mathbf{g}_2(x,y) = (x - x_0,\ y - y_0)$$

In this case, too, the direct transformations are easy to find:

$$\overline{\mathbf{g}}_1(x,y) = \mathbf{g}_1^{-1}(x,y) = (x + ay^2,\ y)$$

$$\overline{\mathbf{g}}_2(x,y) = \mathbf{g}_2^{-1}(x,y) = (x + x_0,\ y + y_0)$$

and therefore we obtain, according to Eq. (5.5):

$$\overline{\mathbf{h}}(x,y) = \overline{\mathbf{g}}_1(x,y) - \overline{\mathbf{g}}_2(x,y) = (ay^2 - x_0,\ -y_0)$$

The vector field $\overline{\mathbf{h}}(x,y)$ assigns to each point (x,y) a vector equal to $(ay^2 - x_0, -y_0)$, as shown in Fig. 4.5(c). And indeed, this vector field consists of inversed S-like trajectories that clearly correspond to the dot trajectories seen in the aperiodic superposition (Fig. 3.8(a)). We have already obtained the explicit equations of these curves in Example 4.10.

However, unlike in the previous examples, in our present case the superposition of the two transformed aperiodic layers has no macroscopic Glass patterns. This can be easily verified by observing Figs. 3.6 and 3.8 from a distance of 3–4 meters, where the individual dots of the layers are no longer discerned by the eye. Under these viewing conditions the dot trajectories (the microstructure dot alignments) are no longer visible, and only the macroscopic structures, i.e. the Glass patterns, may remain visible as brighter areas in the superposition where the gray level average is lower. But indeed, a comparison between the superpositions at the left column of Figs. 3.6 and 3.8 under these viewing conditions clearly shows that in Fig. 3.8(a) there exist no macroscopic Glass patterns. How do we explain this mathematically? As we already know, Glass patterns are generated in the superposition about the mutual fixed points of the domain transformations $\mathbf{g}_1(x,y)$ and $\mathbf{g}_2(x,y)$. In our case the mutual fixed points are given, therefore, by the points (x,y) that solve the system of equations:

$$x_0 - ay^2 = 0$$

$$y_0 = 0$$

However, since the second equation in this system is never satisfied (because we are given here a non-zero vertical shift $y_0 \neq 0$), no points (x,y) solve this set of equations, and the transformations $\mathbf{g}_1(x,y)$ and $\mathbf{g}_2(x,y)$ have no mutual fixed points at all. This explains, indeed, why no Glass patterns are generated in this case in the superposition. Note that in the cases shown in Fig. 3.6 the vertical shift y_0 is zero, and therefore the above system of equations is solved by all the points (x,y) that satisfy $ay^2 = x_0$, i.e. $y = \pm\sqrt{x_0/a}$. This gives, indeed, a pair of horizontal fixed lines that are located at equal distances above and below the x axis, which clearly explain the two linear Glass patterns obtained in Fig. 3.6.

Finally, let us explain the straight moiré bands which are obtained when the same transformations $\mathbf{g}_1(x,y)$ and $\mathbf{g}_2(x,y)$ are applied to two identical periodic dot screens, as shown in Fig. 3.8(b). The domain transformation undergone by the resulting first-order moiré effect is given according to Eq. (5.1) by:

$$\mathbf{g}_M(x,y) = \mathbf{g}_1(x,y) - \mathbf{g}_2(x,y) = (x_0 - ay^2,\, y_0)$$

The coordinate lines of this transformation only consist of the level lines of its first component, $g_1(x,y) = x_0 - ay^2$, since its second component is constant and has no level lines: $g_2(x,y) = y_0$. The coordinate lines in this case are, therefore, $x_0 - ay^2 = p$, namely: $y = \pm\sqrt{(x_0-p)/a}$ for all integer values of p that give real values in the square root. This is a series of horizontal lines whose distances decrease to both positive and negative y directions, as shown, indeed, by the moiré pattern in the periodic superposition (Fig. 3.8(b)). ■

Example 5.5: The high sensitivity of dot trajectories to slight layer shifts:

Figs. 5.1(a),(b) show again the screen superpositions of Figs. 4.13(a),(b) that were synthesized in Example 4.11. Suppose now that in each of these figures one of the superposed layers, say, the first one, is slightly shifted horizontally or vertically, so that the layer transformations undergone by the two layers become:

$$\overline{\mathbf{g}}_1(x,y) = (x,y) + \tfrac{1}{2}\varepsilon(2xy,\, y^2 - x^2) + (x_0,y_0)$$

$$\overline{\mathbf{g}}_2(x,y) = (x,y) - \tfrac{1}{2}\varepsilon(2xy,\, y^2 - x^2)$$

where ε is a small positive fraction that guarantees that both transformations are rather weak (at least within the area covered by the figure). The only difference between the four rows of Fig. 5.1 is in the small shift (x_0,y_0) undergone by the first layer, which equals $(0,0)$ in the first row, $(0,T)$ in the second row, $(0,-T)$ in the third row and $(-T,0)$ in the last row, T being a very small distance (one period of the original, undistorted periodic screen).

How do we explain the dramatic changes that these small layer shifts cause in the dot trajectories of the aperiodic cases, while in their repetitive counterparts the moiré patterns

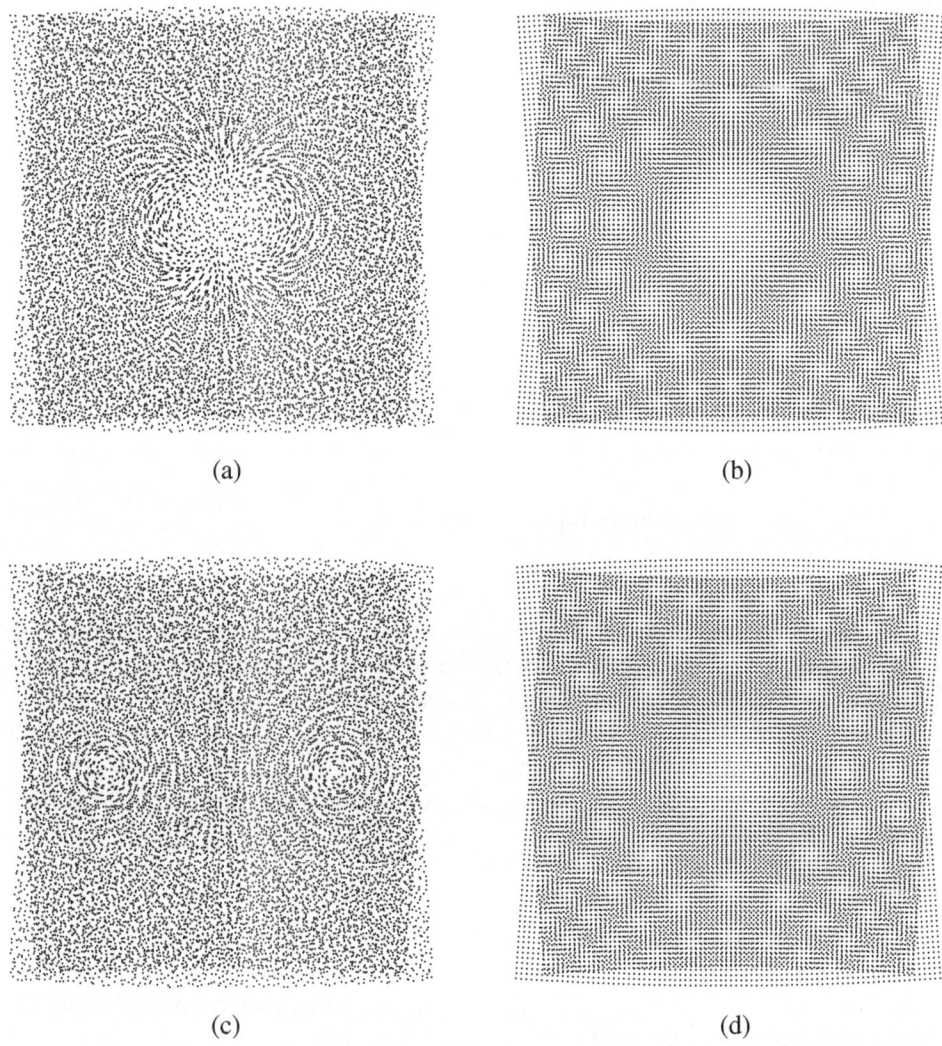

Figure 5.1: (a) The superposition of two identical aperiodic screens that have
undergone the direct transformations $\bar{\mathbf{g}}_1(x,y) = (x,y) + \frac{1}{2}\varepsilon(2xy, y^2 - x^2)$
and $\bar{\mathbf{g}}_2(x,y) = (x,y) - \frac{1}{2}\varepsilon(2xy, y^2 - x^2)$, respectively. (b) The counterpart
of (a) based on periodic screens. (c),(d) Same as (a),(b), but here the
first layer has been slightly shifted upward.

remain basically unchanged? We start, as we did in the previous examples, with the
explanation of the dot trajectories. In the present example we have, according to Eq. (5.5):

$$\bar{\mathbf{h}}(x,y) = \bar{\mathbf{g}}_1(x,y) - \bar{\mathbf{g}}_2(x,y) = \varepsilon(2xy, y^2 - x^2) + (x_0,y_0)$$

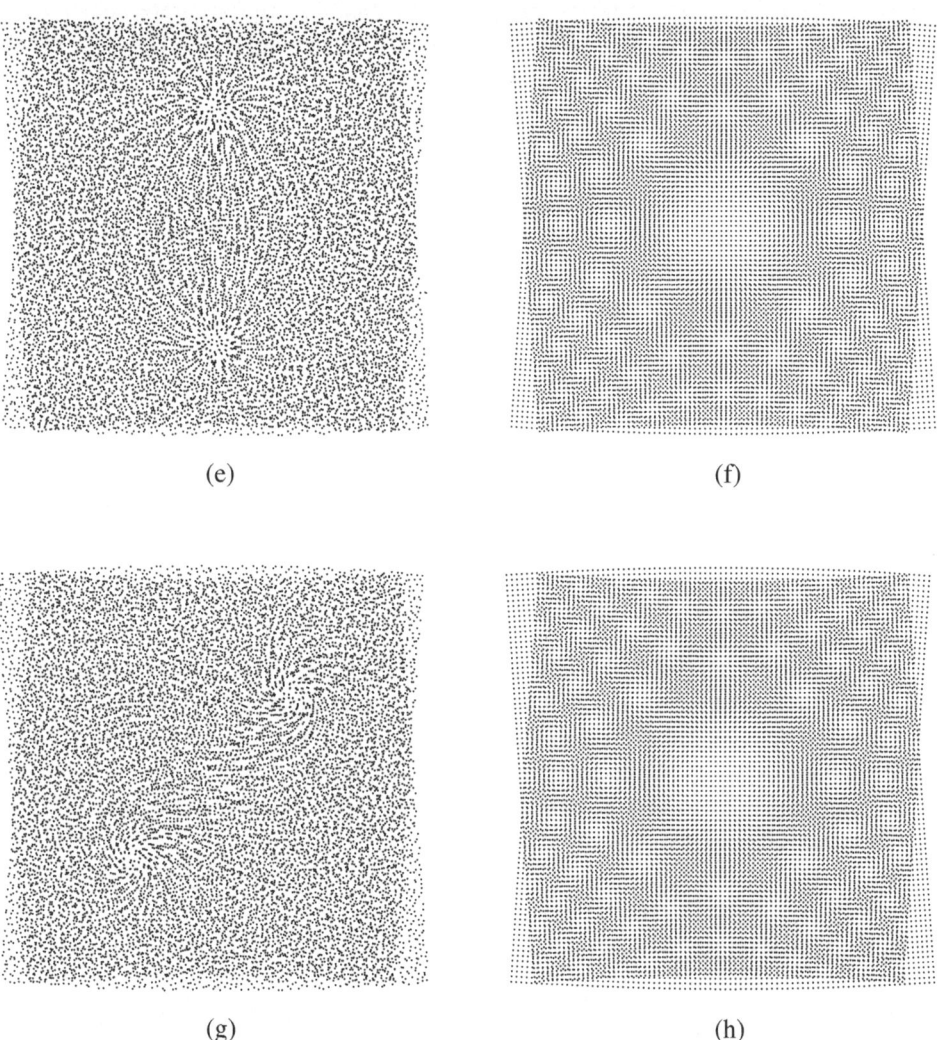

Figure 5.1: (*continued.*) (e),(f) Same as (a),(b), but here the first layer has been slightly shifted downward. (g),(h) Same as (a),(b), but here the first layer has been slightly shifted to the left. Note the dramatic changes in the dot trajectories due to the small layer shifts. In stark contrast, the moiré effects between periodic layers that undergo the same transformations and the same shifts as in the aperiodic cases remain basically unchanged. See Example 5.5.

And indeed, vector field $\overline{\mathbf{h}}(x,y)$ closely approximates the shapes of the dot trajectories, as shown in Fig. 5.2 for the same shifts (x_0,y_0) as in the four rows of Fig. 5.1. Remark the agreement between these vector fields and the dot trajectories in the respective aperiodic layer superpositions. It should be noted that in this case the accurate vector field representations of the dot trajectories, Eqs. (4.8) or (4.9), cannot be used because we do not have the explicit form of the inverse transformations $\mathbf{g}_1(x,y) = \bar{\mathbf{g}}_1^{-1}(x,y)$ and $\mathbf{g}_2(x,y) = \bar{\mathbf{g}}_2^{-1}(x,y)$.

We proceed now to the Glass patterns. Because in the present case the explicit expressions of the inverse transformations $\mathbf{g}_1(x,y)$ and $\mathbf{g}_2(x,y)$ are not available, we can only find the mutual fixed points of the two layers using the approximation provided by Proposition D.10 in Appendix D: $\mathbf{g}_1(x,y) - \mathbf{g}_2(x,y) \approx \bar{\mathbf{g}}_2(x,y) - \bar{\mathbf{g}}_1(x,y)$. The mutual fixed points are given, therefore, by the solutions (x,y) of the system of equations:

$$2\varepsilon xy + x_0 = 0$$

$$\varepsilon(y^2 - x^2) + y_0 = 0$$

A short verification shows that except for the case of $(x_0,y_0) = (0,0)$, in which there exists a single fixed point at the origin, for any other values of the shift (x_0,y_0) there always exists a pair of fixed points that are symmetrically located to both sides of the origin (see Problem 5-2). This explains, indeed, the Glass patterns that we observe in the aperiodic layer superpositions (see the left hand side images in Fig. 5.1).

Now, how do we explain the hyperbolic moiré bands which are generated when the same transformations $\bar{\mathbf{g}}_1(x,y)$ and $\bar{\mathbf{g}}_2(x,y)$ are applied to two identical periodic dot screens, as shown in the right hand side images of Fig. 5.1? Since in this case the explicit expressions of the domain transformations $\mathbf{g}_1 = \bar{\mathbf{g}}_1^{-1}$ and $\mathbf{g}_2 = \bar{\mathbf{g}}_2^{-1}$ are not available we use again the same approximation as above. Hence, the *domain* transformation undergone by the resulting first-order moiré is given by:

$$\mathbf{g}_M(x,y) \approx \bar{\mathbf{g}}_2(x,y) - \bar{\mathbf{g}}_1(x,y) = -\varepsilon(2xy, y^2 - x^2) - (x_0,y_0)$$

And indeed, this domain transformation clearly reflects the hyperbolic moiré shown in the right hand side images of Fig. 5.1. To see this, note that the coordinate lines of the transformation $\mathbf{g}_M(x,y)$ (i.e. the level lines of its two components, $g_1(x,y) = -2\varepsilon xy - x_0$ and $g_2(x,y) = \varepsilon(x^2 - y^2) - y_0$) consist of the curves $-2\varepsilon xy - x_0 = p$ and $\varepsilon(x^2 - y^2) - y_0 = q$ (namely, $y = -\frac{p+x_0}{2x}$ and $y = \pm\sqrt{x^2 - (q+y_0)/\varepsilon}$) for all integer values of p and q. These coordinate lines correspond, indeed, to the two orthogonal sets of hyperbolas that we see in the right hand side images of Fig. 5.1 (see Example D.5 and Fig. D.15(d) in Appendix D, as well as Fig. B.7(d) of Appendix B, which shows the hyperbolic dot screen obtained by applying the domain transformation $\mathbf{g}_M(x,y) = -(2xy, y^2 - x^2)$ to the originally periodic dot screen of Fig. B.7(a)).

Note, however, that unlike the dramatic changes that occur in the *aperiodic* layer superpositions due to the small layer shifts, the moiré effects between the originally *periodic* layers remain basically the same. This is explained by the fact that the values of the shifts x_0 and y_0 do not modify the shapes of the level lines: Clearly, for any integer x_0 and y_0 the two families of hyperbolas obtained for $p,q \in \mathbb{Z}$ remain unchanged. And when

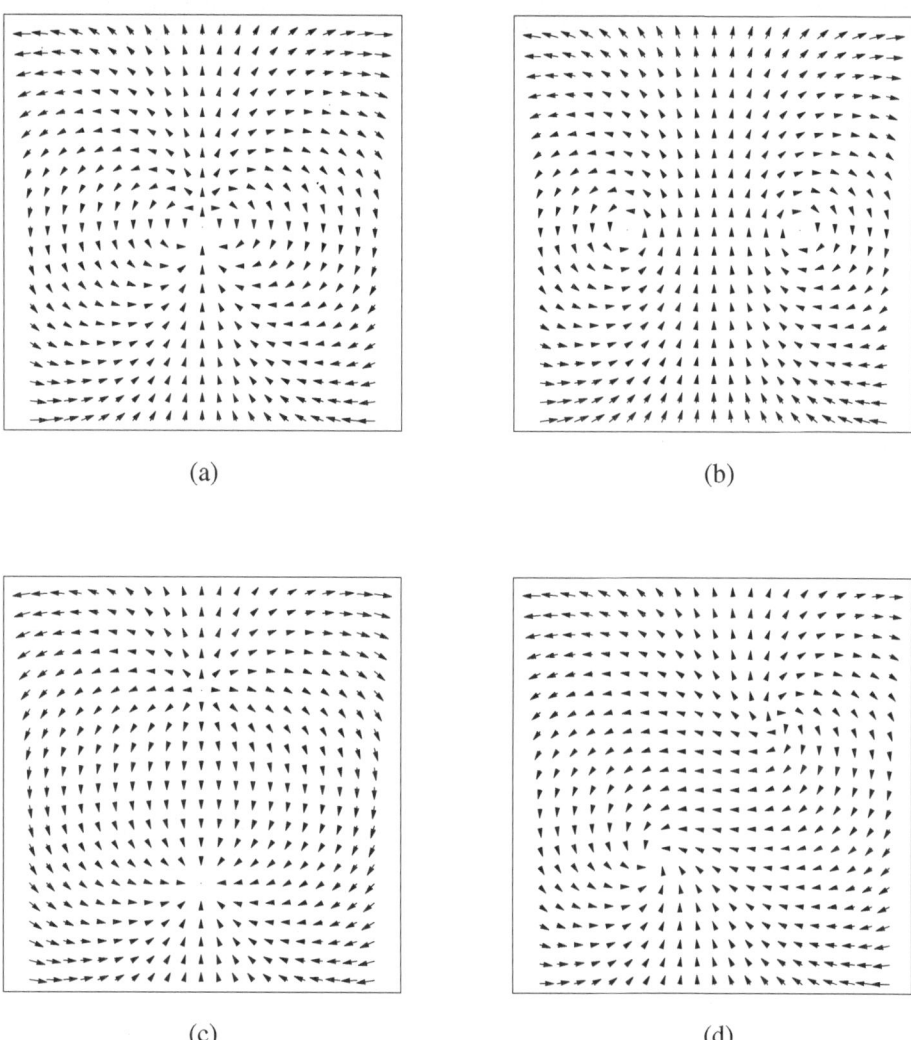

(a)   (b)

(c)   (d)

Figure 5.2: Vector field representations of the relative transformation $\overline{\mathbf{h}}(x,y) = \overline{\mathbf{g}}_1(x,y) - \overline{\mathbf{g}}_2(x,y) = \varepsilon(2xy, y^2 - x^2) + (x_0,y_0)$ of Fig. 5.1, where the layer shifts (x_0,y_0) undergone by the first layer are as follows: (a) No shifts at all. (b) A small vertical shift of y_0 upward. (c) A small vertical shift of y_0 downward. (d) A small horizontal shift of x_0 to the left. Note the agreement with the dot trajectories shown in Figs. 5.1(a),(c),(e) and (g), respectively.

x_0 and y_0 are not integers, they only influence the phase of the curves in each family (meaning that the hyperbolas in each new family will be located between the hyperbolas of the original family), but the hyperbolic shapes of the curves remain unchanged. ■

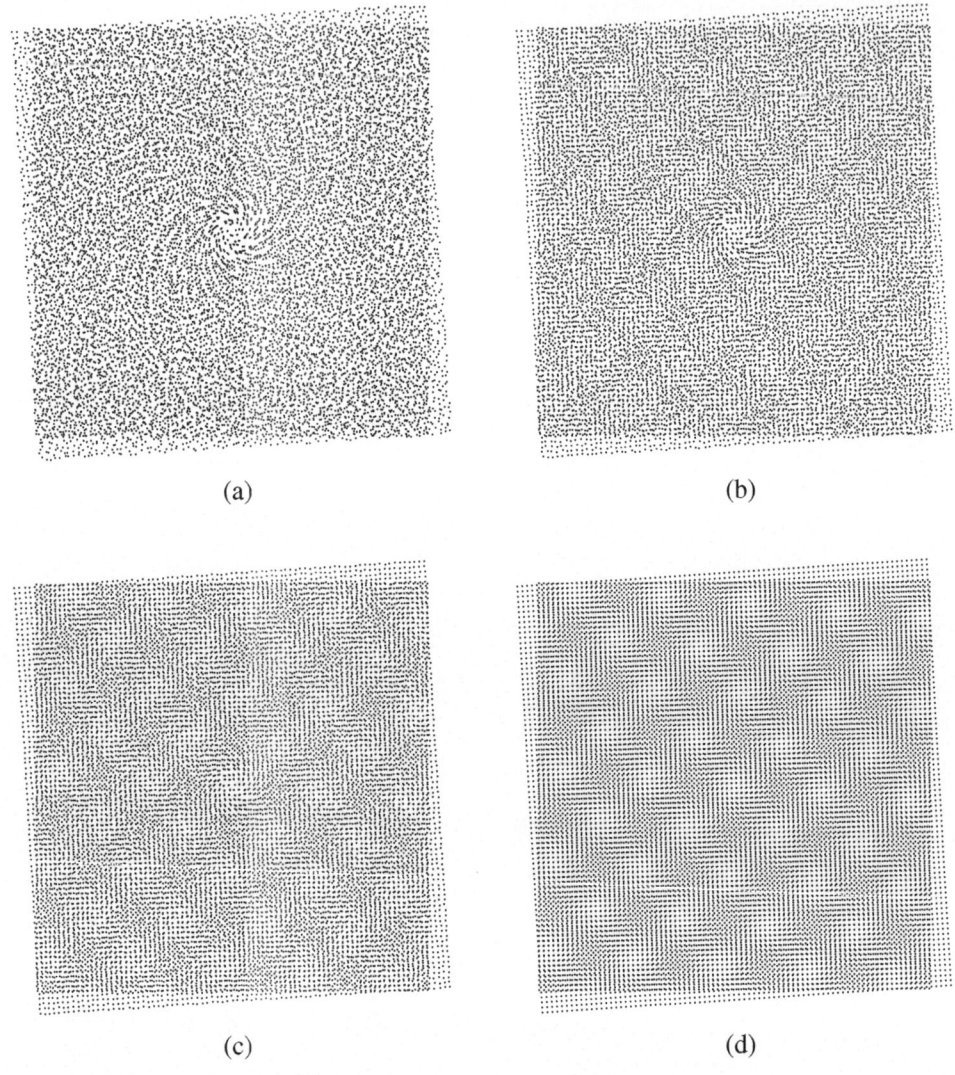

(a)   (b)

(c)   (d)

Figure 5.3: (a) A series of intermediate, partly periodic cases between Figs. 2.1(g) and 2.1(h), which are obtained by gradually varying the degree of randomness of the element locations.

This last example illustrates, indeed, an interesting difference between moiré effects in the superposition of periodic or aperiodic layers: It turns out that the latter are often much more sensitive in their shapes than the former to small layer shifts. This is also illustrated by the superpositions shown in Figs. 3.5–3.9, which, again, differ from each other in the layer shifts alone.

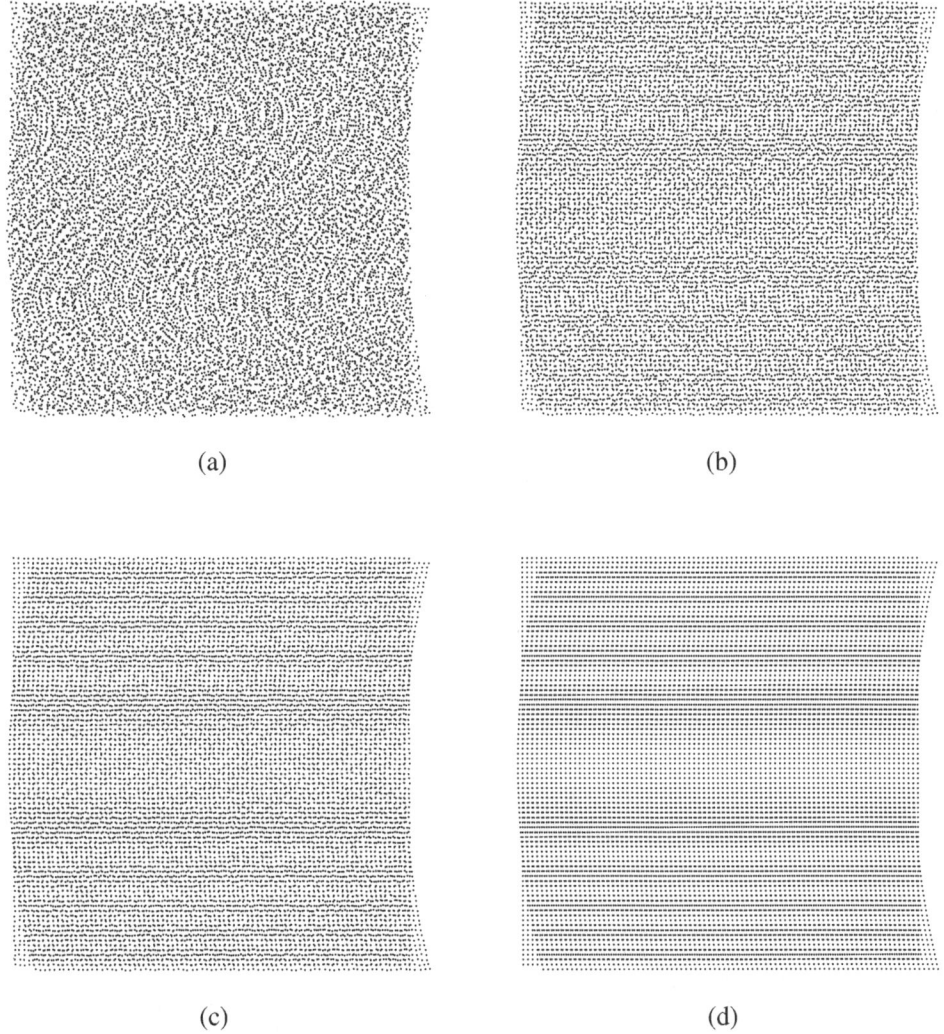

(a)   (b)

(c)   (d)

Figure 5.4: (a) A series of intermediate, partly periodic cases between Figs.
3.8(a) and 3.8(b), which are obtained by gradually varying the
degree of randomness of the element locations.

A further insight into the behaviour of the moiré effects in periodic and aperiodic cases
can be gained by studying intermediate, partly periodic cases. Partly periodic layers can be
obtained, for example, by adding a varying degree of randomness to the element locations
in an initially fully periodic layer. Fig. 5.3 shows what happens to the moiré effect of Figs.
2.1(g),(h) when we gradually vary the degree of randomness in the element locations. A
similar series of intermediate cases is shown in Fig. 5.4, where the degree of randomness
is varied between Figs. 3.8(a) and (b). This subject will be discussed further in Sec. 7.6.

In conclusion, we can summarize our findings as follows:

Proposition 5.1: When we apply weak layer transformations $\mathbf{g}_1(x,y)$ and $\mathbf{g}_2(x,y)$ to two identical *periodic* dot screens that are superposed on top of each other in full coincidence we obtain in the superposition a moiré effect whose geometric transformation is defined by $\mathbf{g}_M(x,y) = \mathbf{g}_1(x,y) - \mathbf{g}_2(x,y)$, as explained in Propositions 10.2 and 10.5 in Chapter 10 of *Vol. I*. (Note that the moiré bands thus obtained can be also regarded as the level lines of $g_{1,1}(x,y) - g_{2,1}(x,y)$ and $g_{2,1}(x,y) - g_{2,2}(x,y)$, the two components of the transformation $\mathbf{g}_M(x,y) = \mathbf{g}_1(x,y) - \mathbf{g}_2(x,y)$.) On the other hand, when we apply the same layer transformations $\mathbf{g}_1(x,y)$ and $\mathbf{g}_2(x,y)$ to two identical *aperiodic* dot screens that are superposed on top of each other, we obtain in the superposition (a) Glass patterns that are located about the fixed points $\mathbf{g}_M(x,y) = (0,0)$ (or the almost fixed points where $\mathbf{g}_M(x,y)$ is minimized), as explained in Secs. 3.3 and 3.7; and/or (b) dot trajectories whose geometric shape is closely approximated by the vector field $\overline{\mathbf{h}}(x,y) = \overline{\mathbf{g}}_1(x,y) - \overline{\mathbf{g}}_2(x,y)$, as explained in Proposition 4.3. Note that $\mathbf{g}_1$ and $\mathbf{g}_2$ are the inverse of the direct transformations $\overline{\mathbf{g}}_1$ and $\overline{\mathbf{g}}_2$. Finally, in intermediate, partly periodic cases the moiré effect shows a weighted blending of both phenomena. ∎

It can be said, therefore, that the effects obtained in the periodic case and in the aperiodic case simply reveal different facets of the same layer transformations that have been applied to the original layers in both cases.

This result can be further simplified thanks to our assumption that the transformations undergone by the original layers are sufficiently weak, because in this case we have by virtue of Proposition D.10 in Appendix D the approximation $\mathbf{g}_M(x,y) \approx -\overline{\mathbf{h}}(x,y)$. (Note that in some particular cases this approximation turns out to be an equality; see Remark D.22 in Appendix D.) Using this approximation we can reformulate Proposition 5.1 in terms of either direct or inverse transformations, as explained below.

Suppose we are given two identical layers $r_1(x,y)$ and $r_2(x,y)$ which are superposed on top of each other in full coincidence, dot on dot, and that we apply to them the geometric (domain) transformations $\mathbf{g}_1(x,y)$ and $\mathbf{g}_2(x,y)$, respectively, to obtain the transformed layers $r_1(\mathbf{g}_1(x,y))$ and $r_2(\mathbf{g}_2(x,y))$. This can be also viewed as the application of the direct transformations $(x,y) \mapsto \overline{\mathbf{g}}_1(x,y)$ and $(x,y) \mapsto \overline{\mathbf{g}}_2(x,y)$, respectively, to the same original layers. As we have seen in Proposition 5.1, the geometric layout of the moiré effect in the superposition of two periodic layers that undergo the domain transformations $\mathbf{g}_1(x,y)$ and $\mathbf{g}_2(x,y)$ (or the direct transformations $\overline{\mathbf{g}}_1(x,y) = \mathbf{g}_1^{-1}(x,y)$ and $\overline{\mathbf{g}}_2(x,y)) = \mathbf{g}_2^{-1}(x,y)$) is expressed by:

$$\mathbf{g}_M(x,y) = \mathbf{g}_1(x,y) - \mathbf{g}_2(x,y) \tag{5.6}$$

But, still according to Proposition 5.1, when we apply the same layer transformations to two identical *aperiodic* dot screens, we obtain in the superposition (a) Glass patterns that are located about the fixed points $\mathbf{g}_M(x,y) = \mathbf{g}_1(x,y) - \mathbf{g}_2(x,y) = (0,0)$; and/or (b) dot trajectories whose geometric shape is closely approximated by the vector field:

$$\bar{\mathbf{h}}(x,y) = \bar{\mathbf{g}}_1(x,y) - \bar{\mathbf{g}}_2(x,y) \tag{5.7}$$

Now, if the transformations being applied to the original layers are sufficiently weak, then by virtue of the approximation $\mathbf{g}_M(x,y) \approx -\bar{\mathbf{h}}(x,y)$ (Proposition D.10 of Appendix D) we obtain the following additional relations:

$$\mathbf{g}_M(x,y) \approx \bar{\mathbf{g}}_2(x,y) - \bar{\mathbf{g}}_1(x,y) \tag{5.8}$$

$$\bar{\mathbf{h}}(x,y) \approx \mathbf{g}_2(x,y) - \mathbf{g}_1(x,y) \tag{5.9}$$

Hence, if our layer transformations are given as domain transformations $\mathbf{g}_1$ and $\mathbf{g}_2$ but the explicit forms of $\bar{\mathbf{g}}_1 = \mathbf{g}_1^{-1}$ and $\bar{\mathbf{g}}_2 = \mathbf{g}_2^{-1}$ are not available, then the only expressions we can use are (5.6) and (5.9). On the other hand, if the original layer transformations are given as *direct* transformations $\bar{\mathbf{g}}_1$ and $\bar{\mathbf{g}}_2$ but the explicit forms of $\mathbf{g}_1 = \bar{\mathbf{g}}_1^{-1}$ and $\mathbf{g}_2 = \bar{\mathbf{g}}_2^{-1}$ are not available, like in Example 5.5, then we can only use expressions (5.7) and (5.8).

5.4 Invariance properties of moiré patterns, Glass patterns and dot trajectories

Because first-order moiré effects between two periodic or repetitive layers are given, according to the classical moiré theory, by Eq. (5.1), it clearly follows that they are invariant under the addition of transformations. That is, if we apply to two originally periodic layers the domain transformations $\mathbf{g}_1(x,y) + \mathbf{f}(x,y)$ and $\mathbf{g}_2(x,y) + \mathbf{f}(x,y)$ rather than $\mathbf{g}_1(x,y)$ and $\mathbf{g}_2(x,y)$, where $\mathbf{f}(x,y)$ represents any arbitrary transformation, the moiré effect generated between the two layers remains unchanged:

$$\mathbf{g}_M(x,y) = (\mathbf{g}_1(x,y) + \mathbf{f}(x,y)) - (\mathbf{g}_2(x,y) + \mathbf{f}(x,y)) = \mathbf{g}_1(x,y) - \mathbf{g}_2(x,y)$$

This property of the first-order moiré effect is illustrated in Figs. 5.5(b),(d) and 5.7(b),(d); but as we can see in parts (a),(c) of these figures, this property does not hold for the dot trajectories between aperiodic layers that undergo the same domain transformations as in the periodic case. Furthermore, as we have seen in Sec. 4.5, dot trajectories are not necessarily invariant even under additions of the same $\bar{\mathbf{f}}(x,y)$ to the *direct* transformations $\bar{\mathbf{g}}_1(x,y)$ and $\bar{\mathbf{g}}_2(x,y)$ that govern the behaviour of the individual dots (see Figs. 4.7(a),(c)). This is due to the fact that although the vector field $\bar{\mathbf{h}}(x,y) = \bar{\mathbf{g}}_1(x,y) - \bar{\mathbf{g}}_2(x,y)$ itself is, of course, invariant under the addition of $\bar{\mathbf{f}}(x,y)$ to $\bar{\mathbf{g}}_1(x,y)$ and $\bar{\mathbf{g}}_2(x,y)$, this vector field does not always give a precise mathematical representation of the dot trajectories, but only provides an approximation (see Proposition 4.3).

On the other hand, as we have seen in Proposition 4.2, the dot trajectories do remain invariant under the *composition* of transformations. More precisely, the dot trajectories that are generated by two originally identical aperiodic layers that undergo the direct transformations $\bar{\mathbf{g}}_1(x,y)$ and $\bar{\mathbf{g}}_2(x,y)$ remain unchanged when the two layers undergo, instead, the transformations $\bar{\mathbf{g}}_1(\bar{\mathbf{f}}(x,y))$ and $\bar{\mathbf{g}}_2(\bar{\mathbf{f}}(x,y))$, where $\bar{\mathbf{f}}(x,y)$ is an arbitrary transformation (see Figs. 4.8(a),(c)). In other words, the dot trajectories obtained by the application of $\bar{\mathbf{g}}_1(x,y)$ and $\bar{\mathbf{g}}_2(x,y)$ to the original aperiodic layers remain unchanged even if

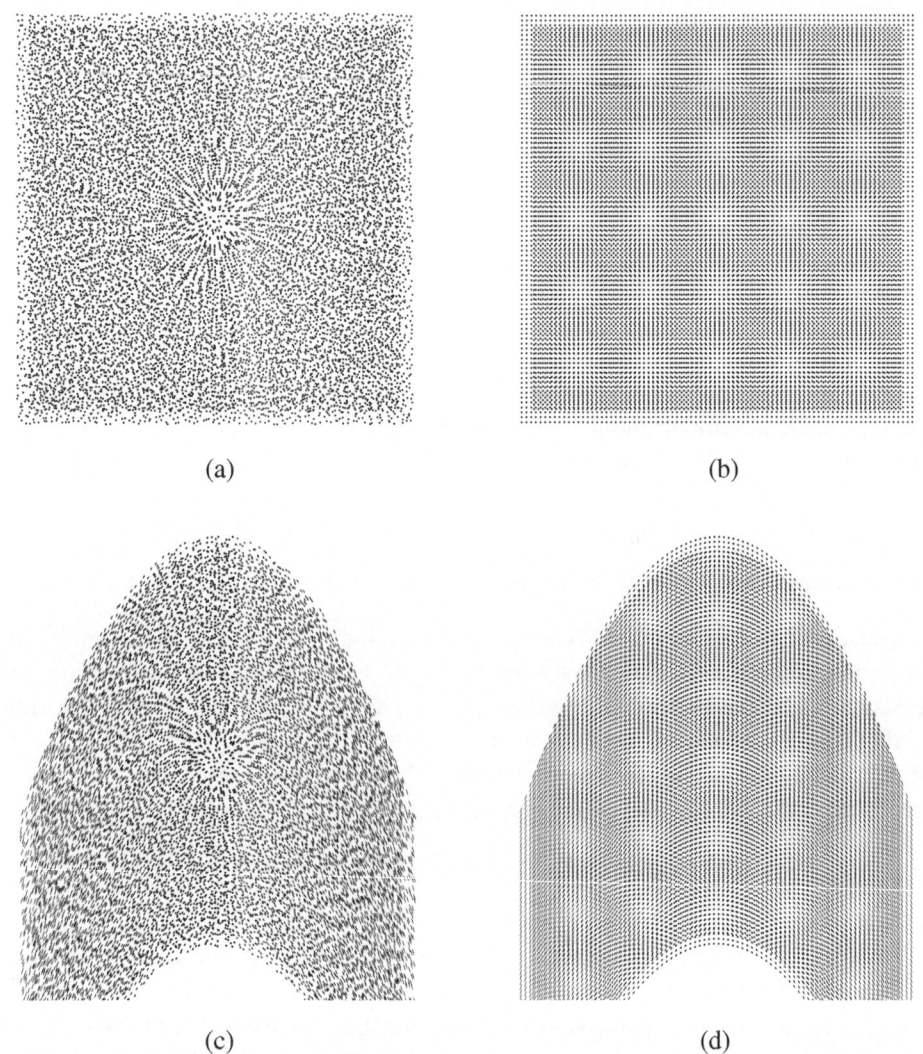

(a)   (b)

(c)   (d)

Figure 5.5: Same as Fig. 4.7, but here the layer transformations have been applied as
domain rather than *direct* transformations (note the inversion of the parabolic
effect in the individual layers with respect to Fig. 4.7). This figure clearly
shows that first-order moiré effects are invariant under the addition of a
transformation $\mathbf{f}(x,y)$ to the two domain transformations $\mathbf{g}_1(x,y)$ and $\mathbf{g}_2(x,y)$
undergone by the original layers, but the dot trajectories are not. (a) The Glass
pattern obtained by applying the domain transformations $\mathbf{g}_1(x,y) = (sx,sy)$ and
$\mathbf{g}_2(x,y) = (x,y)$ to two identical random screens. (b) The first-order moiré
obtained by applying the same domain transformations to two identical periodic
screens. (c) The results obtained by applying to the original random screens the
domain transformations $\mathbf{g}_1(x,y) + \mathbf{f}(x,y)$ and $\mathbf{g}_2(x,y) + \mathbf{f}(x,y)$, where $\mathbf{f}(x,y) =$
$(0,ax^2)$. Note that the Glass pattern remains unchanged, while the dot
trajectories have been bent. (d) The periodic counterpart of (c); note that the
macroscopic effect of the moiré remains unchanged, although the dot alignments in the microstructure have been distorted along with the original layers.

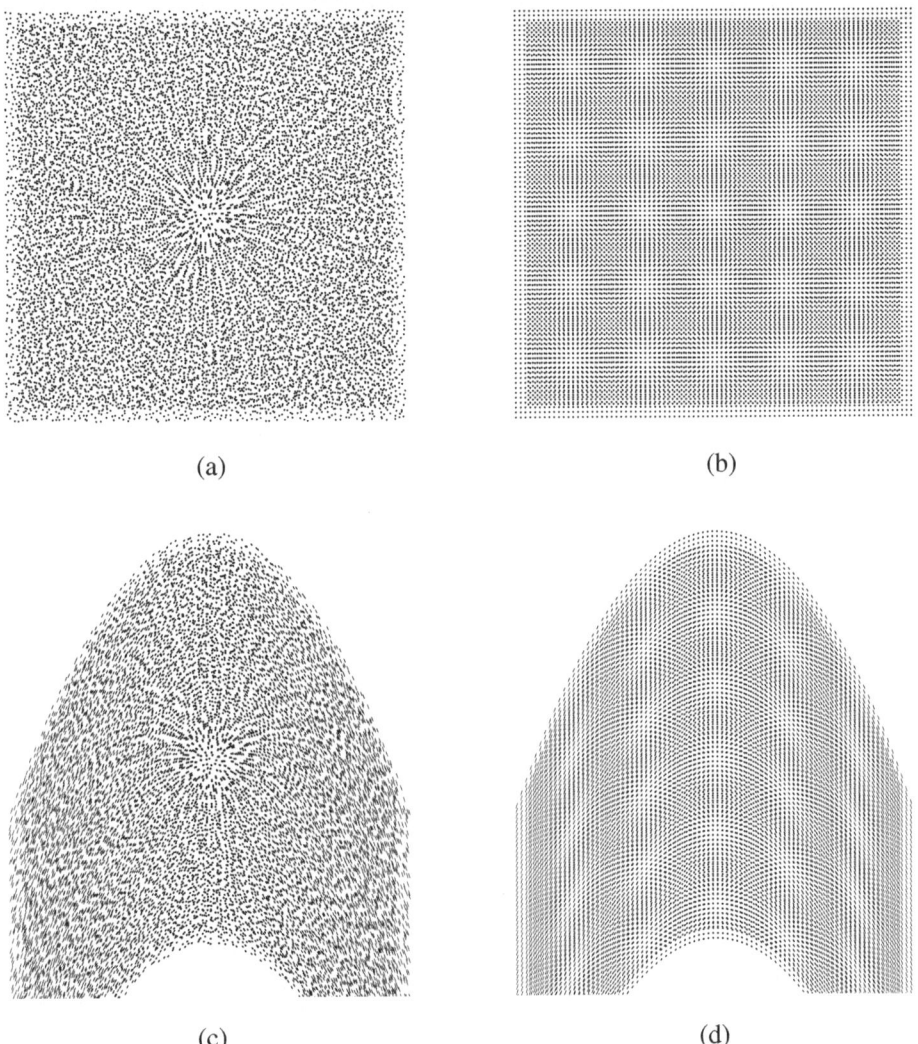

Figure 5.6: Same as Fig. 4.8, but here the layer transformations have been applied as *domain* rather than *direct* transformations. See Problem 5-8.

we first distort the two original layers by the same transformation $\bar{\mathbf{f}}(x,y)$, and only then apply $\bar{\mathbf{g}}_1(x,y)$ and $\bar{\mathbf{g}}_2(x,y)$ to the resulting distorted layers. This is due to the fact that after applying $\bar{\mathbf{f}}(x,y)$ to the two identical aperiodic layers they still remain, of course, two identical aperiodic layers on their own right. However, this property is certainly not shared with the moiré patterns, as shown, indeed, in Figs. 4.8(b),(d).

We see, therefore, that moiré effects and dot trajectories have completely different invariance properties: While moiré effects are invariant under the addition of (domain)

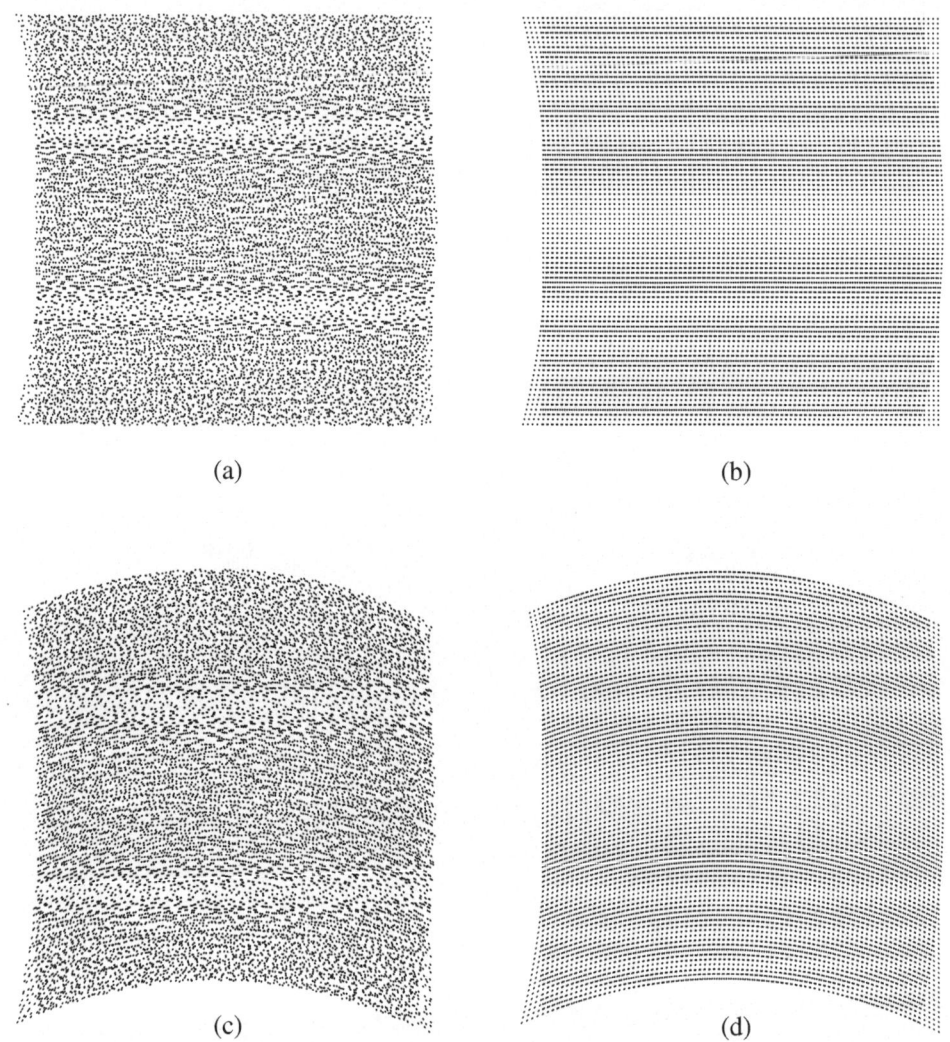

(a)   (b)

(c)   (d)

Figure 5.7: Same as Fig. 4.9, but here the layer transformations have been applied as *domain* rather than *direct* transformations (note the inversion of the parabolic effect with respect to Fig. 4.9). This figure clearly shows that first-order moiré effects and Glass patterns are invariant under the addition of a transformation $\mathbf{f}(x,y)$ to the two domain transformations $\mathbf{g}_1(x,y)$ and $\mathbf{g}_2(x,y)$ undergone by the original layers, but the dot trajectories are not. (a) The Glass pattern obtained by applying the domain transformations $\mathbf{g}_1(x,y) = (x + ay^2, y)$ and $\mathbf{g}_2(x,y) = (x + x_0, y)$ to two identical random screens. (b) The first-order moiré obtained by applying the same domain transformations to two identical periodic screens. (c) The results obtained by applying to the original random screens the domain transformations $\mathbf{g}_1(x,y) + \mathbf{f}(x,y)$ and $\mathbf{g}_2(x,y) + \mathbf{f}(x,y)$, where $\mathbf{f}(x,y) = (0, bx^2)$, $b > a$. Note that the Glass patterns remain unchanged, while the dot trajectories have been bent (see along the right and left borders of the superposition). (d) The periodic counterpart of (c); note that the macroscopic effect of the moiré remains unchanged, although the dot alignments in the microstructure follow the same distortions as the original layers.

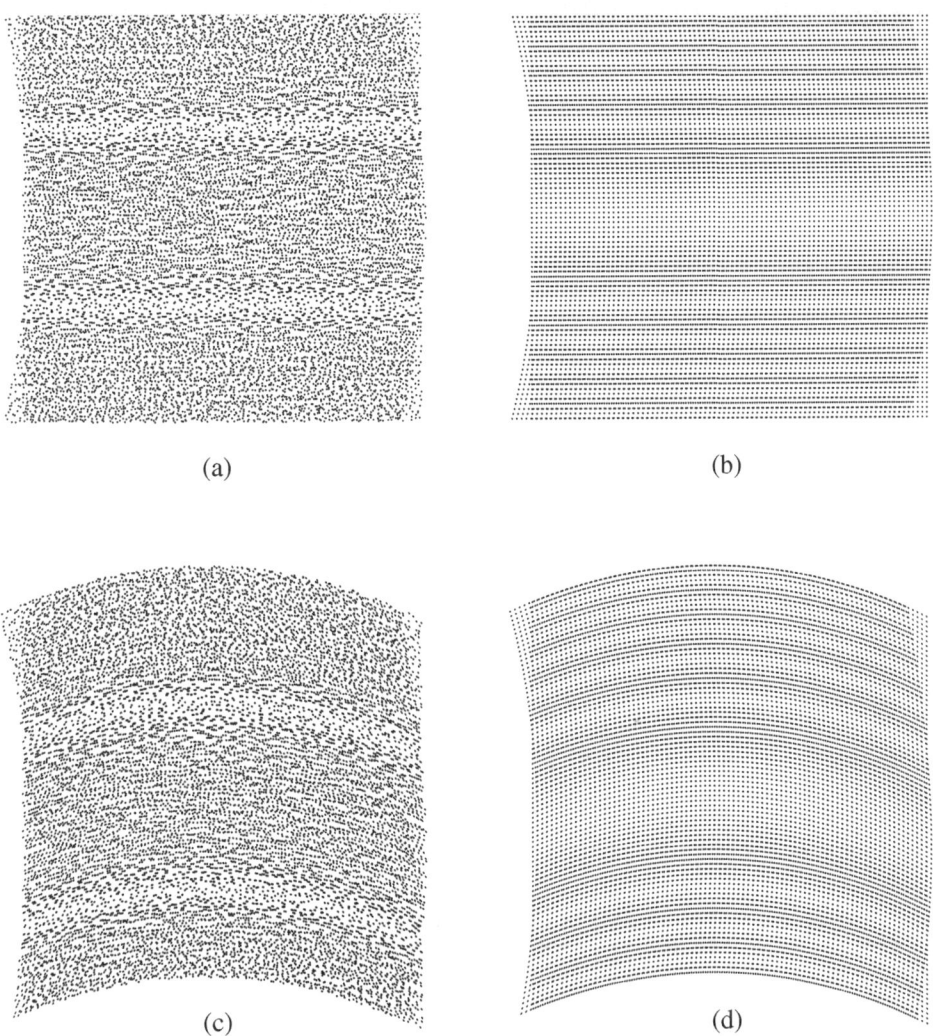

Figure 5.8: Same as Fig. 4.10, but here the layer transformations have been applied as *domain* rather than *direct* transformations. See Problem 5-8.

transformations but not under compositions, dot trajectories are invariant under the composition of transformations but not necessarily under additions. It would be interesting to ask, therefore, how Glass patterns behave in terms of these invariances. For although Glass patterns are macroscopic effects that are closely related to the macroscopic moiré effects between originally periodic layers (remember that Eq. (5.3) corresponds to a subset of Eq. (5.2)), they still occur between aperiodic layers, and they are closely related to the dot trajectories that surround them.

	Aperiodic dot screens	Periodic dot screens
Macro-structure	**Glass patterns:** Their geometry is given by the locus of the mutual fixed points of the two domain transformations $\mathbf{g}_1(x,y)$ and $\mathbf{g}_2(x,y)$, namely, the points (x,y) that satisfy: $$\mathbf{g}_1(x,y) - \mathbf{g}_2(x,y) = (0,0)$$ Invariant under the addition of the same $\mathbf{f}(x,y)$ to both domain transformations, and also under composition.	**First-order moiré patterns:** Determined by the difference between the domain transformations undergone by the two originally identical, periodic layers: $$\mathbf{g}_M(x,y) = \mathbf{g}_1(x,y) - \mathbf{g}_2(x,y)$$ Their geometry is given by the locus of the points of coincidence between the two distorted layers: $$\mathbf{g}_1(x,y) - \mathbf{g}_2(x,y) = (p,q) \quad p,q \in \mathbb{Z}$$ Invariant under the addition of the same $\mathbf{f}(x,y)$ to both domain transformations.
Micro-structure	**Dot trajectories:** Their geometry is determined by the vector field: $$\overline{\mathbf{h}}_1(x,y) = \overline{\mathbf{g}}_1(\mathbf{g}_2(x,y)) - (x,y)$$ and approximated (if the layer transformations are weak) by: $$\overline{\mathbf{h}}(x,y) = \overline{\mathbf{g}}_1(x,y) - \overline{\mathbf{g}}_2(x,y)$$ Invariant under composition (application of $\overline{\mathbf{f}}(x,y)$ to both original layers, and only then applying $\overline{\mathbf{g}}_1(x,y)$ and $\overline{\mathbf{g}}_2(x,y)$ to the two identical distorted layers).	**Dot alignments:** Their geometry follows the distortions of the original periodic layers. Not invariant under additions or compositions.

Table 5.1: Summary of the main properties of the macro- and microstructures that are generated in the superposition of periodic or aperiodic dot screens. Note that the intensity profiles of the periodic and aperiodic macrostructures are not included in this table. They will be studied separately in Chapter 7, using the Fourier-based approach.

And indeed, it turns out that Glass patterns make the best of the two worlds: they remain invariant under both addition (in terms of domain transformations) and composition. These invariance properties of Glass patterns are illustrated in Figs. 5.7 and 4.9, respectively. How do we explain this hybrid behaviour of the Glass patterns mathematically?

The invariance of Glass patterns under *addition* (of domain transformations) is obtained directly from the definition of mutual fixed points between the two domain transformations $\mathbf{g}_1(x,y)$ and $\mathbf{g}_2(x,y)$, since if we have at the point (x_F,y_F):

$$\mathbf{g}_1(x_F, y_F) = \mathbf{g}_2(x_F, y_F)$$

then it follows immediately that we also have at the same point for any arbitrary $\mathbf{f}(x,y)$:

$$\mathbf{g}_1(x_F, y_F) + \mathbf{f}(x_F, y_F) = \mathbf{g}_2(x_F, y_F) + \mathbf{f}(x_F, y_F)$$

Note that such an invariance does not exist under *direct* transformations, as clearly shown in Figs. 4.9(a) and (c).

On the other hand, the invariance of Glass patterns under *composition* follows from the same reasoning that applied to the dot trajectories. If the transformations $\overline{\mathbf{g}}_1(x,y)$ and $\overline{\mathbf{g}}_2(x,y)$ have a mutual fixed point at (x_F, y_F), then a Glass pattern will be generated around this point in the superposition of *any* two identical layers that undergo the transformations $\overline{\mathbf{g}}_1(x,y)$ and $\overline{\mathbf{g}}_2(x,y)$; this includes, of course, the two original layers $r_1(x,y)$ and $r_2(x,y)$ as well as their distorted counterparts $r_1(\mathbf{f}(x,y))$ and $r_2(\mathbf{f}(x,y))$ for any transformation $\mathbf{f}(x,y)$. Note that this reasoning is also true for moiré patterns between periodic layers, but in this case, because of the repetitive nature of the layers, there also appear in the superposition other points of coincidence between the two layers, which are not located at the fixed points of the transformations. As we have seen in Chapter 3, these points of coincidence give the infinitely many repetitions of the moiré that appear in the periodic (or repetitive) case, in addition to the real Glass pattern. But unlike the real Glass pattern, these additional repetitions do not remain invariant under composition; this is clearly shown in Figs. 4.10(b),(d), where only the real Glass patterns (i.e. the moiré bands that are generated about the fixed lines) remain invariant, while all the other moiré bands have been distorted.

In conclusion, we see that because of their particular status, Glass patterns inherit the properties of both first-order moirés and dot trajectories, and they remain invariant under additions of domain transformations as well as under compositions. In the case of a moiré pattern between originally periodic layers, the invariance under composition only holds for those particular moiré bands (or cells) that are generated around real fixed points, but it does not necessarily hold for the other moiré replicas. However, the invariance under addition of domain transformations remains true in all the moiré replicas, too, as shown in Figs. 5.7(b),(d).

Table 5.1 gives a systematic summary of the different macroscopic and microscopic phenomena that may occur in superpositions of two originally identical periodic or aperiodic dot screens, and their behaviour under layer transformations. This comparison will be extended to the case of periodic or aperiodic line gratings in the next chapter (see Table 6.1).

PROBLEMS

5-1. Examples 5.1–5.5 give a full mathematical explanation of the various phenomena which occur in five different periodic and aperiodic layer superpositions. They explain the dot trajectories and the macroscopic Glass patterns in the aperiodic case, and the moiré bands in the periodic case. Following these examples, give the full mathematical explanation of the phenomena which occur in the periodic and aperiodic layer superpositions shown in Fig. 2.3, Fig. 3.16, and any other figures of your choice.

5-2. Calculate the mutual fixed points between the transformations undergone by the aperiodic layers in Figs. 5.1(a),(c),(e) and (g) (see Example 5.5). Because the domain transformations g_1 and g_2 are not available in this case, use the approximation $g_1(x,y) - g_2(x,y) \approx \overline{g}_2(x,y) - \overline{g}_1(x,y)$ provided by Proposition D.10 of Appendix D and find the mutual fixed points between the direct transformations $\overline{g}_1$ and $\overline{g}_2$. Compare your results with the fixed points shown in these superpositions and in the corresponding vector fields of Fig. 5.2. *Hint*: Solving the equations $2xy + x_0 = 0$, $y^2 - x^2 + y_0 = 0$ is straightforward in the case of a horizontal shift ($y_0 = 0$). In the case of a vertical shift ($x_0 = 0$) you must distinguish between two separate cases: if $y_0 > 0$ substitute into the second equation $y = -x_0/(2x)$, while if $y_0 < 0$ you should rather substitute into the second equation $x = -x_0/(2y)$.

5-3. What would you expect to see in the layer superpositions explained in Examples 4.11 and 5.5 (see Figs. 4.13(a),(b) and 5.1) if we replaced there the transformation $\overline{h}(x,y) = (2xy, y^2 - x^2)$ by $\overline{h}(x,y) = (2xy, x^2 - y^2)$? The answer is shown in Fig. 5.9. Redo Example 5.5 for this case, and explain the differences between the aperiodic superpositions of Figs. 5.1 and 5.9. Why do the periodic superpositions remain unchanged?

5-4. Example 5.1 provides the mathematical explanation of the different phenomena that occur in the superposition of two aperiodic or periodic dot screens under the same two layer transformations $\overline{g}_1(x,y) = ((1+\varepsilon)x, (1+\varepsilon)y)$ and $\overline{g}_2(x,y) = (x,y)$. The dot trajectories that occur in the aperiodic case are expressed mathematically by the transformation $\overline{h}(x,y) = (\varepsilon x, \varepsilon y)$ (interpreted as a vector field), the Glass pattern is the zone of high correlation between the two superposed layers around their fixed point at the origin, and the moiré bands that occur in the periodic case are expressed mathematically by the transformation $g_M(x,y) = -(\delta x, \delta y)$ (interpreted as a domain transformation), where $\delta = \frac{\varepsilon}{1+\varepsilon}$. However, if the transformations undergone by the original layers are weak (as we must assume in order not to lose the dot trajectories) then we have by virtue of Proposition D.10 in Appendix D the approximation $g_M(x,y) \approx -\overline{h}(x,y)$. Reformulate the results obtained in Example 5.1 using this approximation, and show that the difference is negligible. *Hint*: Using this approximation we can say that the dot trajectories obtained in the aperiodic case show the trajectories of the radial vector field $(\varepsilon x, \varepsilon y)$, while the moiré effect obtained in the periodic case shows the coordinates of the grid $-(\varepsilon x, \varepsilon y)$, whose period is $1/\varepsilon$ times larger than the period T of the original, periodic layer. And indeed, if the transformation $\overline{g}_1(x,y) = ((1+\varepsilon)x, (1+\varepsilon)y)$ undergone by the first layer is weak, meaning that ε is a small fraction, then the error introduced by this approximation is negligible; for example, the approximated moiré period, $1/\varepsilon$ times T, and the real moiré period, $1/\delta = 1/\varepsilon + 1$ times T, are almost identical (since ε is small).

5-5. Example 5.2 provides the mathematical explanation of the different phenomena that occur in the superposition of two aperiodic or periodic dot screens under the two layer transformations $\overline{g}_1(x,y) = ((1-\varepsilon)x, (1+\varepsilon)y)$ and $\overline{g}_2(x,y) = (x,y)$. The dot trajectories that occur in the aperiodic case are expressed mathematically by the transformation $\overline{h}(x,y) = (-\varepsilon x, \varepsilon y)$ (interpreted as a vector field), the Glass pattern is the zone of high correlation between the two superposed layers around their fixed point at the origin, and the moiré bands that occur in the periodic case are expressed mathematically by the transformation $g_M(x,y) = (\delta_1 x, -\delta_2 y)$ (interpreted as a domain transformation), where $\delta_1 = \frac{\varepsilon}{1-\varepsilon}$ and $\delta_2 = \frac{\varepsilon}{1+\varepsilon}$. However, if the transformations undergone by the original layers are weak (as we must assume in order not to lose the dot trajectories) then we have by virtue of Proposition D.10 in Appendix D the approximation $g_M(x,y) \approx -\overline{h}(x,y)$.

Reformulate the results obtained in Example 5.2 using this approximation, and show that the difference is negligible.

5-6. As we have seen in Example 5.2, when the superposed layers undergo the direct transformations $\overline{\mathbf{g}}_1(x,y) = ((1-\varepsilon)x, (1+\varepsilon)y)$ and $\overline{\mathbf{g}}_2(x,y) = (x,y)$, the two orthogonal periods of the resulting periodic moiré (see Fig. 2.2(b)) are given by $1/\varepsilon - 1$ times T and $1/\varepsilon + 1$ times T, where T is the period of the original periodic layer. If ε is small these orthogonal periods are almost identical, and the tiny difference between them is hardly visible (see Fig. 2.2(b)). And yet, it seems quite strange that applying the layer mappings $\overline{\mathbf{g}}_1(x,y)$ and $\overline{\mathbf{g}}_2(x,y)$ to *aperiodic* layers gives in Fig. 2.2(a) hyperbolic dot trajectories having a full 90° symmetry, while the application of the same layer mappings to *periodic* layers results in Fig. 2.2(b) in moiré bands that do not have such a 90° symmetry (since their two orthogonal periods are not fully identical). How do you explain this fact? *Hint*: Note that the tiny difference between $\mathbf{g}_M(x,y)$ and $-\overline{\mathbf{h}}(x,y)$ is unavoidable; thus, it is not possible to have a full 90° symmetry both in the vector field $\overline{\mathbf{h}}(x,y)$ and in the moiré bands $\mathbf{g}_M(x,y)$: If we apply to the original periodic and aperiodic layers the *domain* transformations $\mathbf{g}_1(x,y) = ((1-\delta)x, (1+\delta)y)$ and $\mathbf{g}_2(x,y) = (x,y)$ we obtain a full 90° symmetry in the moiré bands $\mathbf{g}_M(x,y)$ but not in the vector field $\overline{\mathbf{h}}(x,y)$; and if we apply to the same original layers the *direct* transformations $\overline{\mathbf{g}}_1(x,y) = ((1-\varepsilon)x, (1+\varepsilon)y)$ and $\overline{\mathbf{g}}_2(x,y) = (x,y)$ we obtain a full 90° symmetry in the vector field $\overline{\mathbf{h}}(x,y)$ but not in the moiré bands $\mathbf{g}_M(x,y)$. In other words, if we are given $|\delta_1| = |\delta_2|$ in $\mathbf{g}_M(x,y)$ then we necessarily obtain slightly different $|\varepsilon_1|$ and $|\varepsilon_2|$ in $\overline{\mathbf{h}}(x,y)$, and if we are given $|\varepsilon_1| = |\varepsilon_2|$ in $\overline{\mathbf{h}}(x,y)$ (as in Example 5.2) then we necessarily obtain slightly different $|\delta_1|$ and $|\delta_2|$ in $\mathbf{g}_M(x,y)$.

5-7. A skeptic student claims that it is not really necessary to use both direct and domain transformations to explain the various phenomena which occur in layer superpositions. For instance, all the phenomena described in Example 5.1 (the dot trajectories and the macroscopic Glass patterns in the aperiodic case as well as the moiré bands in the periodic case) could be explained using the direct transformations $\overline{\mathbf{g}}_1(x,y)$ and $\overline{\mathbf{g}}_2(x,y)$ alone, or the domain transformations $\mathbf{g}_1(x,y)$ and $\mathbf{g}_2(x,y)$ alone. What is your opinion? *Hint*: Indeed, in simple cases like Example 5.1 the use of $\overline{\mathbf{g}}_1, \overline{\mathbf{g}}_2$ alone (or $\mathbf{g}_1, \mathbf{g}_2$ alone) in all of the formulas (Eqs. (5.1)–(5.5)) would not cause perceptible errors, since both the direct expression of the scale-up transformation, $\overline{\mathbf{g}}_1(x,y) = ((1+\varepsilon)x, (1+\varepsilon)y)$, and the domain (inverse) expression of the scale-up transformation, $\mathbf{g}_1(x,y) = (x/(1+\varepsilon), y/(1+\varepsilon))$, give similar results. However, a better test case is provided by Example 5.5: As shown in Figs. B.2 and B.3 in Appendix B, or in Figs. D.15(b),(d) in Appendix D, the direct and the inverse interpretations of the transformation $(u,v) = (2xy, y^2 - x^2)$ have completely different geometric shapes, and the shape of the moiré bands in Fig. 5.1(b) clearly indicates that the moiré transformation $\mathbf{g}_M(x,y)$ must be interpreted as an inverse transformation. Furthermore, the dot trajectories in Fig. 5.1(a) clearly reflect the vector field of the *direct* transformation $\overline{\mathbf{h}}(x,y) = (2xy, y^2 - x^2)$; in order to eliminate any possible doubts, try to draw the vector field of the inverse transformation $\overline{\mathbf{h}}^{-1}(x,y)$, whose explicit expression is given in Appendix D in Example D.5, and convince yourself that the vector field which corresponds to our dot trajectories is indeed $\overline{\mathbf{h}}(x,y)$. It should be noted, however, that thanks to approximation (4.10) (see Remark 4.3 and Proposition D.10 in Appendix D) we can, indeed, express our vector field (and the corresponding dot trajectories in the layer superposition) in terms of the *domain* layer transformations: $\overline{\mathbf{h}}(x,y) \approx -\mathbf{g}_M(x,y) = \mathbf{g}_2(x,y) - \mathbf{g}_1(x,y)$. But although we are using here the *domain* layer transformations $\mathbf{g}_1(x,y)$ and $\mathbf{g}_2(x,y)$, $\overline{\mathbf{h}}(x,y)$ itself remains a direct transformation and must be interpreted as such. Similarly, it is also possible to approximate all the superposition phenomena using the *direct* layer transformations, as

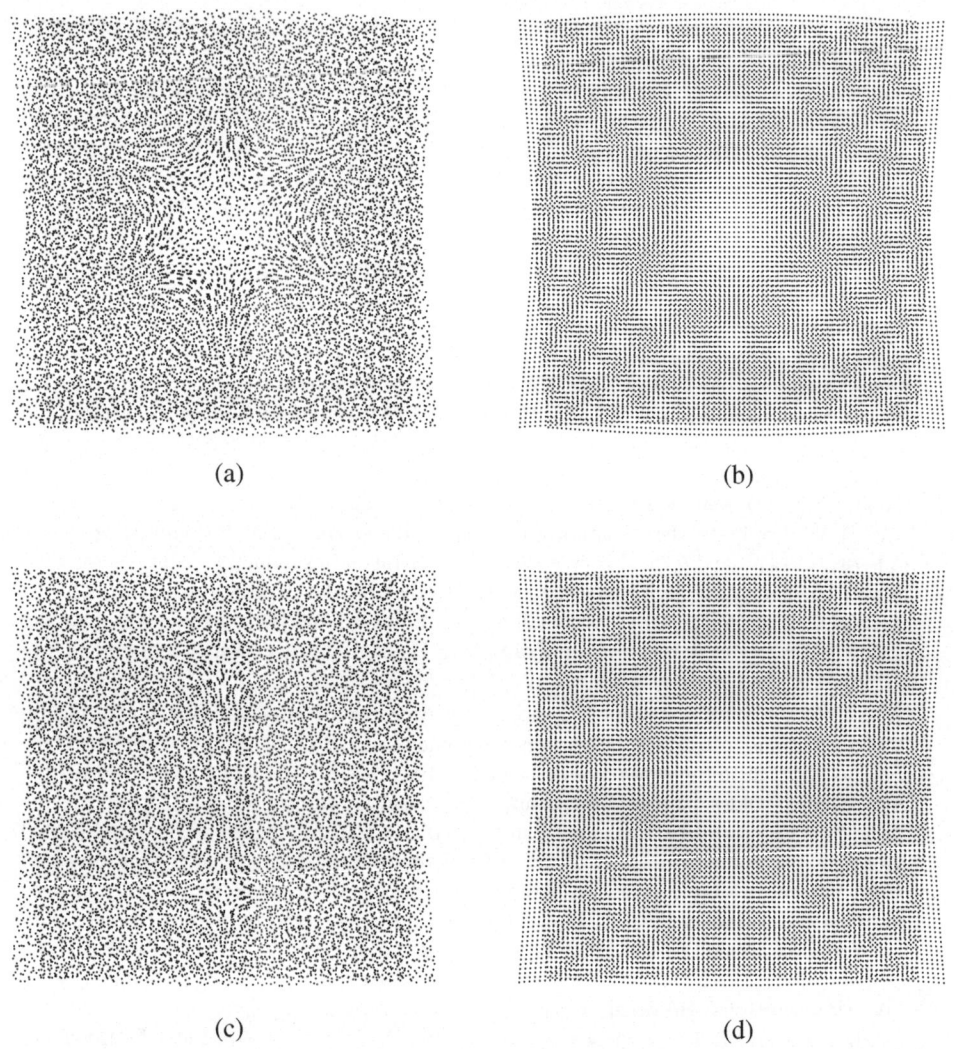

Figure 5.9: (a) The superposition of two identical aperiodic screens that have
undergone the direct transformations $\overline{\mathbf{g}}_1(x,y) = (x,y) + \frac{1}{2}\varepsilon(2xy, x^2 - y^2)$
and $\overline{\mathbf{g}}_2(x,y) = (x,y) - \frac{1}{2}\varepsilon(2xy, x^2 - y^2)$, respectively. (b) The counterpart
of (a) based on periodic screens. (c),(d) Same as (a),(b), but here the
first layer has been slightly shifted upward. Compare with Figs.
5.1(a)–(d).

we have done, indeed, in Example 5.5: $\overline{\mathbf{h}}(x,y) = \overline{\mathbf{g}}_1(x,y) - \overline{\mathbf{g}}_2(x,y)$, $\mathbf{g}_M(x,y) \approx -\overline{\mathbf{h}}(x,y) =$
$\overline{\mathbf{g}}_2(x,y) - \overline{\mathbf{g}}_1(x,y)$. But again, although we are using here the *direct* layer transformations
$\overline{\mathbf{g}}_1(x,y)$ and $\overline{\mathbf{g}}_2(x,y)$, $\mathbf{g}_M(x,y)$ itself remains a domain transformation and must be
interpreted as such, as indicated, indeed, by the hyperbolic shape of the moiré bands in
Fig. 5.1(b) (compare with Figs. B.2 and B.3 in Appendix B).

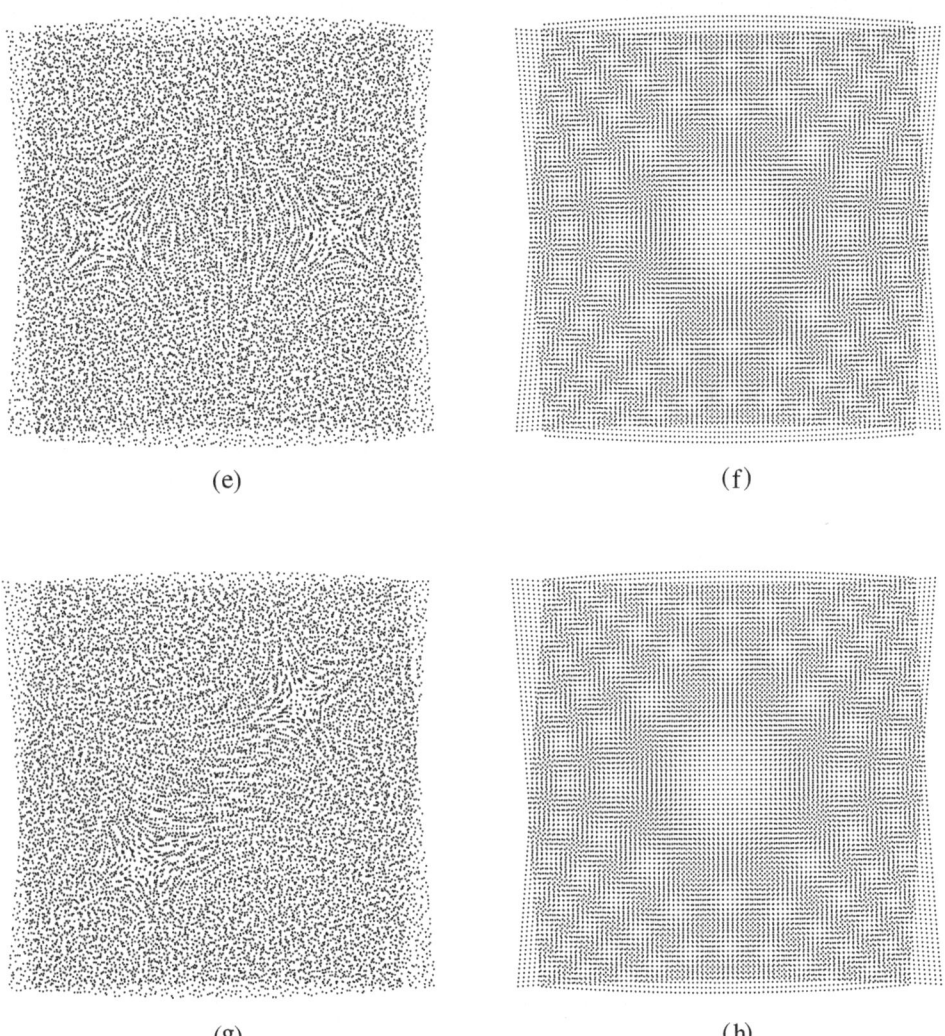

(e)

(f)

(g)

(h)

Figure 5.9: (*continued.*) (e),(f) Same as (a),(b), but here the first layer has been slightly shifted downward. (g),(h) Same as (a),(b), but here the first layer has been slightly shifted to the left. Note the dramatic changes in the dot trajectories due to the small layer shifts. In stark contrast, the moiré effects between periodic layers that undergo the same transformations and the same shifts as in the aperiodic cases remain basically unchanged. Compare with Figs. 5.1(e)–(h).

5-8. Figs. 5.6 and 5.8 show what happens to the superpositions of Figs. 4.8 and 4.10 when the layer transformations are applied as *domain* rather than *direct* transformations (see also Figs. 5.7 and 4.9). How do you explain the behaviour of the moiré patterns, the Glass patterns and the dot trajectories in parts (c) and (d) of these figures? *Hint*: Fig. 4.10 was obtained by applying to the two original layers the *direct* transformations

$\overline{\mathbf{g}}_1(\overline{\mathbf{f}}(x,y))$ and $\overline{\mathbf{g}}_2(\overline{\mathbf{f}}(x,y))$, respectively, by means of a forward mapping algorithm (see Sec. D.9.1 in Appendix D). On the other hand, Fig. 5.8 was obtained by applying to the two original layers a backward mapping algorithm that was given the same two transformations $\overline{\mathbf{g}}_1(\overline{\mathbf{f}}(x,y))$ and $\overline{\mathbf{g}}_2(\overline{\mathbf{f}}(x,y))$. This means, by virtue of the well-known identity $[\mathbf{f}_1 \circ \mathbf{f}_2]^{-1} = \mathbf{f}_2^{-1} \circ \mathbf{f}_1^{-1}$ (see, for example, [Bernstein05 p. 6]), that in Fig. 5.8 the transformations undergone by the layers are $[\overline{\mathbf{g}}_1(\overline{\mathbf{f}}(x,y))]^{-1} = \overline{\mathbf{f}}^{-1}(\overline{\mathbf{g}}_1^{-1}(x,y)) = \mathbf{f}(\mathbf{g}_1(x,y))$ and $[\overline{\mathbf{g}}_2(\overline{\mathbf{f}}(x,y))]^{-1} = \overline{\mathbf{f}}^{-1}(\overline{\mathbf{g}}_2^{-1}(x,y)) = \mathbf{f}(\mathbf{g}_2(x,y))$, and not $\mathbf{g}_1(\mathbf{f}(x,y))$ and $\mathbf{g}_2(\mathbf{f}(x,y))$.

5-9. Table 5.1 summarizes the macroscopic and microscopic properties of the phenomena which may occur in the superposition of two aperiodic dot screens or in the superposition of two periodic dot screens. What would you expect to see in the superposition of two hybrid dot screens that are periodic in one direction and aperiodic in the other direction?

5-10. How do you explain the fact that two of the fixed points in Figs. 3.15(a) and 4.6(d) are surrounded by circular trajectories, while the other two fixed points are surrounded by hyperbolic trajectories? *Hint*: See Example H.4 in Appendix H. A similar phenomenon occurs also in other cases (see, for example, Fig. 3.10).

Chapter 6

Glass patterns in the superposition of aperiodic line gratings

6.1 Introduction

In the previous chapters we saw that when two identical aperiodic dot screens are superposed on top of each other with a small angle or scaling difference, or with any other slight transformation, a typical Glass pattern appears in the superposition (see, for example, Figs. 2.1(c), (e) and (g)). Unlike moiré effects between periodic layers, which extend throughout the entire superposition (see, for example, Fig. 2.1(d), (f) and (h)), Glass patterns are concentrated about the fixed points of the transformation, and farther away they gradually fade out and disappear.

A similar phenomenon also occurs in the 1D case, namely, where the original layers consist of parallel line gratings.[1] The moiré patterns which occur in the aperiodic 1D case differ from their 2D counterparts in several aspects; but although they were already described in the 1960s [Vargady64], their properties have only been investigated in the beginning of the 21st century [Amidror03b]. In fact, the 1D case is conceptually simpler than the 2D case that we have discussed so far, due to its inherently limited structural complexity. As we will see in the examples below, it turns out that even the dot trajectories, which are among the most striking features of Glass patterns in the 2D case, are practically absent in the 1D case. And yet, the investigation of the 1D case is no less interesting than that of the 2D case, and it even yields some quite surprising results.

Our aim in the present chapter is, therefore, to study Glass patterns which occur in the superposition of aperiodic line gratings (for example, between random or pseudo-random gratings), either straight or curved.[2] We will compare these Glass patterns with their 2D counterparts, and explain the similarities and the differences between them. On the other hand, we will compare their behaviour with that of the moiré patterns which are obtained in superpositions of periodic or repetitive line gratings. We will also provide the mathematical derivations of the curve shapes of the Glass patterns in the 1D case, and show how they are related to the 2D case and to the periodic case.

Our discussion in this chapter is organized as follows. In Sec. 6.2 we introduce the Glass patterns which are generated in the superposition of straight line gratings and

---

[1] Note that the use of the term "1D case" is, strictly, an abuse of language, since in fact a line grating is still a 2D layer. The more appropriate term that should be understood here is "a 1D structure (such as a 1D square wave) which has been constantly extended perpendicularly to its own direction within the 2D plane". More generally, we will also consider here non-linear transformations of such straight line gratings, for example, parabolic or circular line gratings.

[2] Although these patterns were not described by Glass, we prefer, for reasons of consistency, to call them *1D Glass patterns,* as a natural extension of their 2D counterparts.

describe their main properties. Then, in Sec. 6.3 we derive mathematically the curve shapes and the locations of these Glass patterns based on a generalization into the aperiodic case of the indicial equations method, that was originally introduced for the study of periodic cases (see Sec. 11.2 in *Vol. I*). Equipped with this mathematical technique, we proceed in Sec. 6.4 to Glass patterns that are generated in the superposition of curved line gratings (i.e. line gratings that underwent non-linear layer transformations). In Sec. 6.5 we see how the various patterns that are generated in the layer superposition are affected when forcing on the original layers constraints such as constancy along one dimension or periodicity. This point of view provides a valuable unifying insight to our understanding of the different moiré phenomena that occur in periodic, aperiodic, 1D or 2D cases. Finally, in Sec. 6.6 we provide another useful approach for illustrating Glass and moiré patterns, based on the intersection of surfaces, and we show how it can be used to simulate the gray levels of the Glass (or moiré) patterns in the superposition.

Note that just as we did in the previous chapters, we assume here, too, that the superposed layers, periodic or not, were identical before the application of the layer mappings. The motivation for this assumption was discussed in Remark 3.1.

6.2 Glass patterns in the superposition of straight line gratings

The simplest aperiodic gratings consist of irregularly spaced parallel straight lines. We will therefore start our study with the superposition of such line gratings, leaving the more interesting cases involving curved gratings to a later section.

Consider the superposition of two identical aperiodic straight gratings with a small angle difference α. Unlike in the superposition of two identical aperiodic dot screens with a small angle difference, where a circular Glass pattern is generated about the origin (Fig. 2.1(c)), in the present case a *linear* Glass pattern is generated in the superposition along a given straight line passing through the origin (Fig. 6.1(a)). This line is, in fact, the locus of the points of coincidence between the two superposed gratings. Along this line the correlation between the superposed gratings is maximal, since the black and white microstructure elements of both gratings fall exactly on top of one another. But as we go farther away from this line the correlation between the layers gradually decreases: the elements of both gratings start falling arbitrarily between each other, and the Glass pattern fades away. Note that just as in the circular Glass pattern between two dot screens, the center of our linear Glass pattern is brighter than its surrounding area. This is, indeed, the macroscopic consequence of the fact that in the center of the linear Glass pattern microstructure elements from both layers fall exactly on top of each other, while farther away they start falling between each other, so that less white area is left in the superposition. Far away from the linear Glass pattern there is no longer any correlation between the two layers, and we obtain a mean gray level which remains constant throughout. This macroscopic view is best appreciated by observing the layer superpositions from a sufficient distance (say, 3–4 meters), where the individual elements (lines) of the original layers are no longer discerned by the eye and what we see is only a gray level average of the microstructure in each area of the superposition.

Let us see now what happens in the superposition when we slightly vary the angle difference α between the two identical line gratings. As shown in Fig. 6.1(c), when α increases, the linear Glass pattern becomes thinner, and its angle is slightly increased (in fact, as we will see in the mathematical derivation below, its orientation is perpendicular to the bisector between the main directions of the two superposed line gratings[3]). When α tends to zero, the linear Glass pattern becomes wider and wider and approaches the angle of zero. And when $\alpha = 0$, i.e. when the two superposed gratings perfectly coincide, the width of the linear Glass pattern becomes infinite, and it completely disappears.

Next, let us see what happens in the superposition when we also scale one of the superposed line gratings. As we can see by comparing Figs. 6.1(a) and 6.2(a), when the scaling ratio s departs from 1 the linear Glass pattern becomes thinner and its orientation is modified.

Finally, let us see what happens when we slightly shift one of the superposed line gratings on top of the other along its main direction. As we can see by comparing Figs. 6.1(a) and 6.2(c), the linear Glass pattern in the superposition undergoes a much larger shift, while its angle remains unchanged. A layer shift along the secondary direction, namely, along the lines of the grating, will obviously have no effect on the superposition.

6.2.1 Superposition of 1D vs. 2D aperiodic layers

As we can see, there exists a fundamental difference between the Glass patterns that are generated between slightly rotated or scaled aperiodic *line gratings* and the Glass patterns which are generated between aperiodic *dot screens* under the same layer transformations: While the Glass patterns between aperiodic dot screens are typically generated about a point of coincidence between the superposed layers, the linear Glass patterns between aperiodic line gratings are generated along a full 1D line of coincidence between the superposed layers (compare, for example, Fig. 2.1(c) with Fig. 6.1(a)).[4] The reason for this difference is explained as follows.

In a superposition of two identical dot screens one of which has been slightly rotated or scaled, the Glass pattern is generated about the *fixed point* of the transformation $\mathbf{g}(x,y)$ in question, namely, the point (x_F,y_F) for which $\mathbf{g}(x_F,y_F) = (x_F,y_F)$. Any rotation or scaling transformation $\mathbf{g}(x,y)$ has exactly one such fixed point.

Suppose now that we superpose, instead of aperiodic dot screens, two identical aperiodic line gratings, and that we apply to one of them exactly the same transformation $\mathbf{g}(x,y)$ as in

---

[3] The main direction of a line grating is defined as the direction perpendicular to the lines.

[4] Although in some particular cases the superposition of aperiodic dot screens may generate Glass patterns along full lines (see, for example, Figs. 2.3 or 3.6), superpositions of aperiodic line gratings *always* give Glass patterns along full lines.

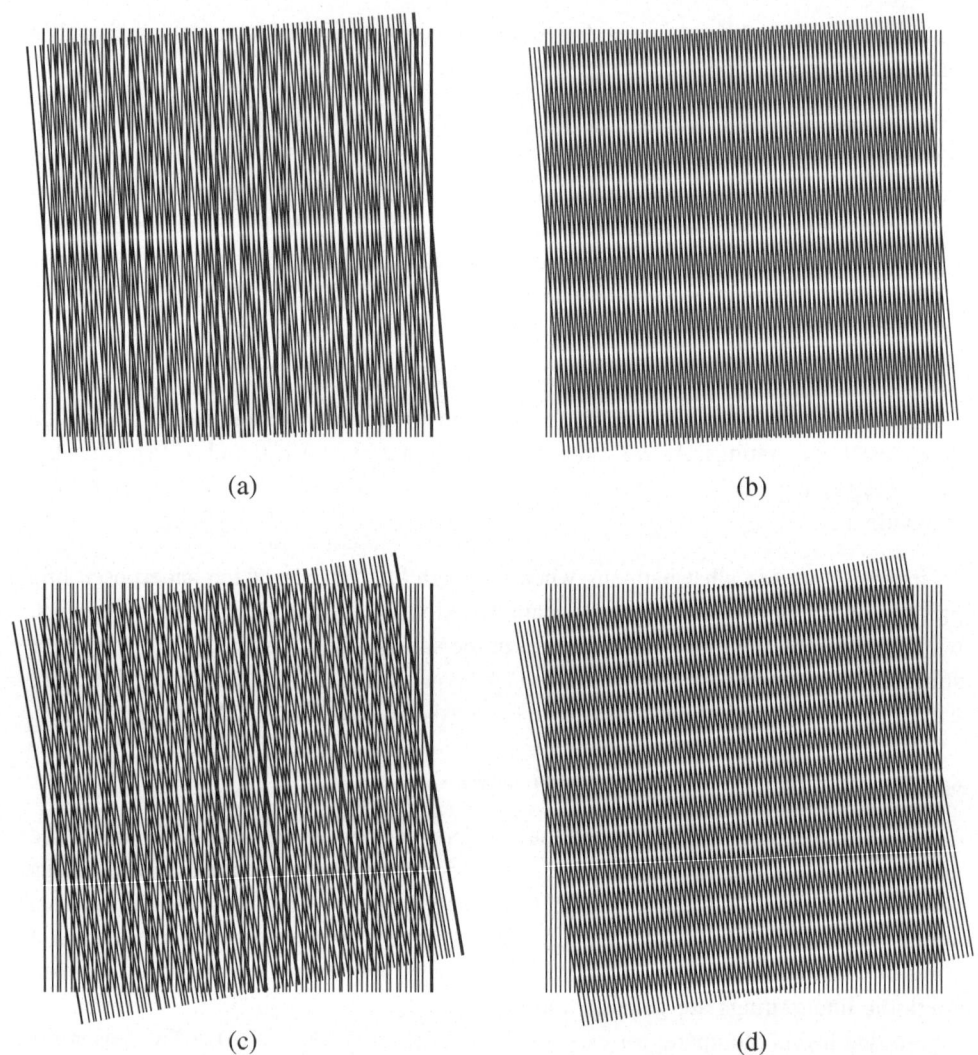

Figure 6.1: Glass patterns in the superposition of straight line gratings. (a) The
superposition of two identical aperiodic line gratings with a small
angle difference of $\alpha = 5°$. (b) The superposition of two identical
periodic line gratings with the same angle difference. In both (a) and
(b) a linear Glass pattern is generated precisely at the same location.
(c),(d) Same as in (a),(b), but with $\alpha = 10°$.

the case of the dot screens, namely, a slight rotation or scaling. Obviously, in both cases
the underlying transformation $\mathbf{g}(x,y)$ has exactly one and the same fixed point. But when
the two superposed layers are line gratings, there also exist between the two superposed
layers infinitely many points of coincidence, which together form a full straight line
passing through the real fixed point. The two superposed gratings coincide along this line

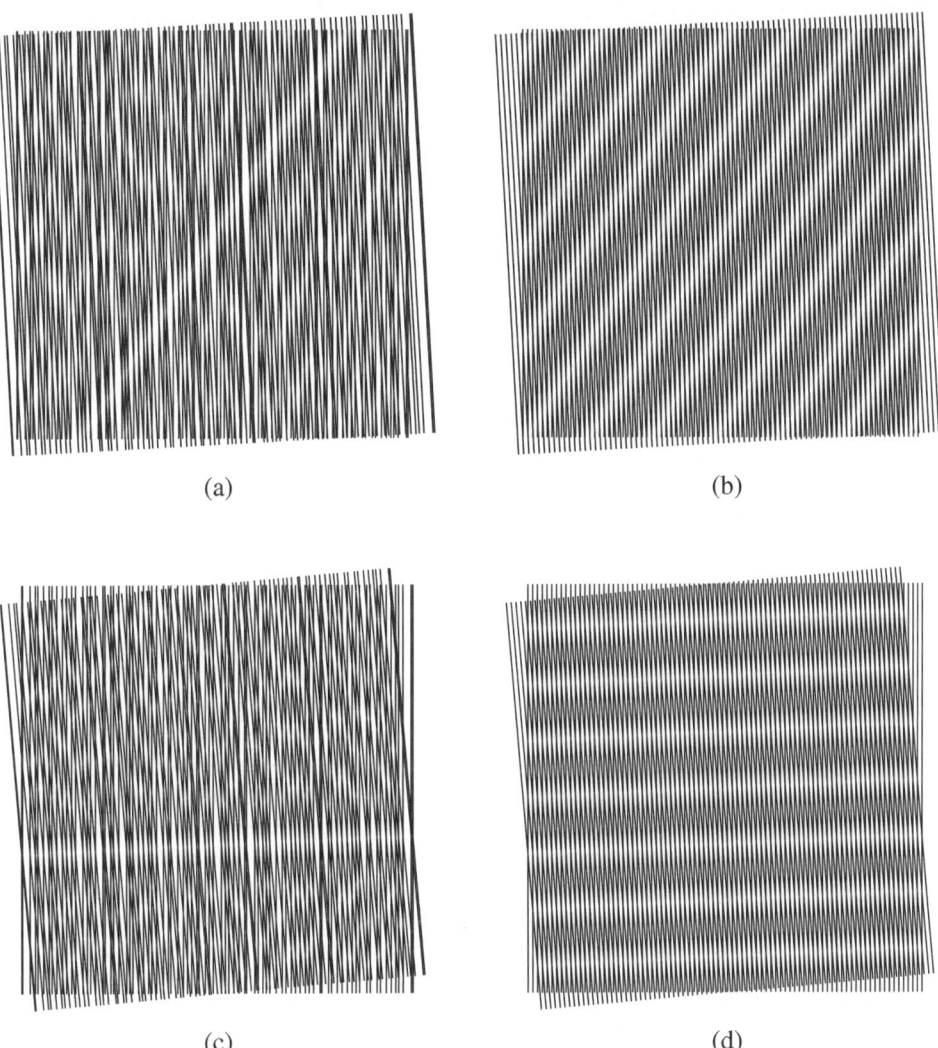

(a)

(b)

(c)

(d)

Figure 6.2: (a),(b) Same as in Figs. 6.1(a),(b), but here the rotated grating has been also slightly scaled up. (c),(d) Same as in Fig. 6.1(a),(b), but here the unrotated grating has been slightly shifted horizontally by $x_0 = T$ to the right, where T equals one period of the periodic grating.

because of the 1D nature of their internal structure. Indeed, an aperiodic line grating is more constrained (i.e., it has fewer degrees of freedom) than an aperiodic dot screen, since it only has full freedom in one dimension, while its other dimension is fully determined (constant). Therefore, due to the additional restriction on the structure of the layers (namely, the fact that their second dimension is fully determined by the first one), the layer superposition has a full line of points of coincidence, which obviously includes the fixed point of the underlying mapping.

We can therefore formulate our result as follows: Glass patterns in the superposition of aperiodic gratings are not only generated about the fixed points of the underlying layer transformations, as in the 2D case, but they extend into full lines. In fact, as we will see later in this chapter, these lines may even be curved when the superposed layers consist of curved line gratings. We will return to these results in more detail in Sec. 6.5.

6.2.2 Superposition of periodic vs. aperiodic line gratings

Figs. 6.1–6.2 show side by side the behaviour of the linear Glass pattern in the superposition of two aperiodic line gratings (in the left hand side of each figure), and the behaviour of the moiré patterns in the superposition of a corresponding pair of *periodic* line gratings having undergone exactly the same rotations, scalings and layer shifts (in the right hand side of each figure). As we can clearly see, in both aperiodic and periodic cases a linear Glass pattern is generated exactly in the same location, along the line of coincidence between the superposed gratings. Note that this line passes through the fixed point of the underlying layer transformation $g(x,y)$. However, while the aperiodic moiré consists of only one linear Glass pattern, the moiré effect in the corresponding superposition of periodic gratings consists of infinitely many repetitions of this Glass pattern. In other words, the Glass pattern which is generated along the line of coincidence is periodically repeated throughout the superposition, forming the bright bands of the periodic moiré pattern. From this point of view, the periods of the periodic moiré pattern are simply duplicates of the main linear Glass pattern which is generated through the fixed point, and the period length of the moiré corresponds to the distance between these duplicates.[5] This does not mean, of course, that our rotation or scaling transformation $g(x,y)$ has more fixed points when the two superposed layers are periodic than when the layers are aperiodic: obviously, in both cases $g(x,y)$ has exactly one fixed point. But unlike in the case of 2D aperiodic layers, such as random screens, in the superposition of line gratings this fixed point is no longer the only point of coincidence between the superposed layers. As we saw in Sec. 6.2.1 above, the superposition of identical gratings with a small angle difference already gives a full line of coincidence between the two superposed layers, which includes the fixed point of $g(x,y)$. But when the two superposed layers are also periodic, we have, in addition, infinitely many new lines of coincidence between the two superposed layers, where the two layers happen to coincide because of the periodicity in their internal structure. These new lines of coincidence are not fixed points of the underlying mapping $g(x,y)$, nor coincidence lines in the case of aperiodic gratings. We can say, therefore, that the fixed point of $g(x,y)$ determines the *main* periodic

---

[5] It is important to note, however, that these duplicates are *not* necessarily identical in their microstructure. The periodicity of the moiré concerns only its macrostructure, namely, the moiré intensity profile (the variation in the mean gray level that is observed from such a distance that the microstructure detail of the original layers is no longer discerned by the eye). In other words, although the microstructure in the superposition of two periodic layers is not always periodic, the intensity profile of the isolated moiré is, indeed, periodic (see Sec. 6.3 in *Vol. I*).

band (or Glass pattern) of the moiré, while all the other moiré bands that are generated in the periodic case are only duplicates which exist due to the periodicity of the superposed layers. This is, indeed, the 1D counterpart of Remark 3.2 from Sec. 3.3.1.

Hence, as we have already seen for 1D vs. 2D cases in Sec. 6.2.1, the fact of having additional constraints on the internal structure of the superposed layers (here, due to their periodicity) causes the addition of new loci of coincidence in the superposition, which are not loci of coincidence in the corresponding superposition of unconstrained layers. Having constrained (fixed) the second dimension of the superposed layers results in the extension of their locus of coincidence from the fixed point of the underlying transformation $\mathbf{g}(x,y)$ into a full line; and adding the constraint of periodicity on the superposed layers causes the extension of this line into a family of infinitely many periodic lines of coincidence. We will return to these interesting results in more detail in Sec. 6.5.

6.3 Mathematical derivations: generalization of the indicial equations method

Let us now proceed to the mathematical derivation of the lines of coincidence in the superposition of two straight aperiodic line gratings. For this end, we generalize the indicial equations method, that was originally devised for the superposition of periodic or repetitive line gratings (see Sec. 11.2 in *Vol. I*), in order to adapt it to the superposition of aperiodic layers.

As a simple example, consider the case of two aperiodic straight line gratings with a small angle difference α. In this case our first line grating is a family of vertical lines that is defined by:

$$x = c_i \qquad\qquad i \in \mathbb{Z}, \quad c_i \in \mathbb{R} \qquad\qquad (6.1)$$

and our second line grating is a copy of the first line family which has been rotated about the origin by angle α:

$$x\cos\alpha + y\sin\alpha = c_i \qquad\qquad i \in \mathbb{Z}, \quad c_i \in \mathbb{R} \qquad\qquad (6.2)$$

This gives us a system of two equations in x and y, whose solutions (x,y) define the locus of the points of coincidence between the two layers. Note that c_i are arbitrary numbers *that are identical in both line families*; this is, indeed, the key point which guarantees the correlation between the two layers. Since c_i are identical in both line families, we can eliminate them from our two equations. We obtain, therefore:

$$x\cos\alpha + y\sin\alpha = x$$

which gives:

$$y = x\frac{1 - \cos\alpha}{\sin\alpha}$$

or, using the identity $\frac{1 - \cos\alpha}{\sin\alpha} = \tan\frac{\alpha}{2}$ [Spiegel68 p. 16]:

$$y = x \tan\frac{\alpha}{2} \tag{6.3}$$

The solution of our system of equations is, therefore, a straight line passing through the origin at the angle of $\frac{\alpha}{2}$. This is, indeed, the locus of the points of coincidence between the two layers, i.e. the center of the linear Glass pattern (see Fig. 6.1(a)). Note that in the periodic case (Fig. 6.1(b)) the arbitrary numbers c_i are simply replaced in Eqs. (6.1) and (6.2) by integer multiples of the respective grating periods, i.e. by mT_1 and nT_2, where $m,n \in \mathbb{Z}$. As explained in Example 11.1 in *Vol. I*, this gives us the family of moiré bands which are generated in the superposition of the two periodic gratings. Note that the Glass pattern we have derived above is simply the 0-th moiré band of this family, which is obtained when $m = n$ (namely, the moiré band whose index is $p = m - n = 0$).

Similar derivations can be also done for superpositions involving any other linear or non-linear transformations in one or both of the aperiodic gratings. Several illustrative examples involving various non-linear layer transformations are given in the next section.

It is interesting to note that the same technique can be also extended in a straightforward way to the 2D case, i.e. to the superposition of aperiodic dot screens. As a simple example, let us reconsider here the superposition we have already discussed in Sec. 3.3.3, in which one of the two aperiodic dot screens has been rotated by angle α, while the other has been shifted in the x and y directions by (x_0, y_0). As we remember, this superposition gives a single circular Glass pattern about a fixed point that is displaced from the origin (see Fig. 3.2). Let us see how this result can be obtained using our new technique.

Clearly, before applying the layer transformations, both of the original dot screens are identical, and they consist of a family of random dots given by:

$$\binom{x}{y} = \binom{c_i}{k_i} \qquad\qquad i \in \mathbb{Z}, \quad (c_i, k_i) \in \mathbb{R}^2 \tag{6.4}$$

After having applied the layer transformations, our first dot screen is rotated by angle α:[6]

$$\binom{x\cos\alpha + y\sin\alpha}{-x\sin\alpha + y\cos\alpha} = \binom{c_i}{k_i} \qquad\qquad i \in \mathbb{Z}, \quad (c_i, k_i) \in \mathbb{R}^2 \tag{6.5}$$

whereas the second dot screen is simply shifted by (x_0, y_0):

$$\binom{x - x_0}{y - y_0} = \binom{c_i}{k_i} \qquad\qquad i \in \mathbb{Z}, \quad (c_i, k_i) \in \mathbb{R}^2 \tag{6.6}$$

But since (c_i, k_i) are arbitrary pairs of constant numbers that are identical in both dot screens, we can eliminate them from Eqs. (6.5) and (6.6). We obtain, therefore:

$$\binom{x\cos\alpha + y\sin\alpha}{-x\sin\alpha + y\cos\alpha} = \binom{x - x_0}{y - y_0}$$

---

[6] Note that we are using here — just as in Chapter 3 — *domain* (i.e. inverse) transformations. This point has been explained in detail in Remark 4.1 and in Sec. 5.2.

which is a system of two equations in x and y. This system is identical to the one we have obtained when we originally discussed this superposition in Sec. 3.3.3 (see Eq. (3.3)), and indeed, as we have already seen there, its solution gives the mutual fixed point of our two transformations, which is the center of the circular Glass pattern in the layer superposition.

In the most general case, if the two originally identical dot screens undergo the transformations $\mathbf{g}_1(x,y) = (g_{1,1}(x,y), g_{1,2}(x,y))$ and $\mathbf{g}_2(x,y) = (g_{2,1}(x,y), g_{2,2}(x,y))$, respectively, then the elimination of (c_i,k_i), as we did above, gives us a system of two equations:

$$\begin{pmatrix} g_{1,1}(x,y) \\ g_{1,2}(x,y) \end{pmatrix} = \begin{pmatrix} g_{2,1}(x,y) \\ g_{2,2}(x,y) \end{pmatrix} \tag{6.7}$$

or in vector notation:

$$\mathbf{g}_1(x,y) - \mathbf{g}_2(x,y) = (0,0) \tag{6.8}$$

And indeed, the solutions (x,y) of this system of equations in x and y give the mutual fixed points of our two transformed dot screens, i.e. the center points of the Glass patterns in the superposition. Note that we have already obtained this result in Eq. (5.3).

A similar generalization can be also done in the case of aperiodic line gratings. If the two originally identical line gratings, given by Eq. (6.1), undergo the layer transformations $\mathbf{g}_1(x,y) = (g_{1,1}(x,y), g_{1,2}(x,y))$ and $\mathbf{g}_2(x,y) = (g_{2,1}(x,y), g_{2,2}(x,y))$, respectively, then the transformed gratings become:

$$g_{1,1}(x,y) = c_i \qquad\qquad i \in \mathbb{Z}, \quad c_i \in \mathbb{R}$$

and: $\qquad\qquad g_{2,1}(x,y) = c_i \qquad\qquad i \in \mathbb{Z}, \quad c_i \in \mathbb{R}$

By eliminating c_i from both equations we obtain, therefore:

$$g_{1,1}(x,y) = g_{2,1}(x,y) \tag{6.9}$$

or, in other words:

$$g_{1,1}(x,y) - g_{2,1}(x,y) = 0 \tag{6.10}$$

Remark 6.1: Note that Eqs. (6.9) and (6.10) simply consist of the first line (the first coordinate) of Eqs. (6.7) and (6.8), respectively. Hence, while the superposition of aperiodic *dot screens* yields a system of two equations in the two variables x and y, the superposition of aperiodic *line gratings* yields a single equation in the same two variables x and y. This explains, indeed, why the solutions (points of coincidence) form in the case of line gratings a full line (or curve) in the x,y plane, while in the case of dot screens the solutions usually consist of isolated points (fixed points). Note, however, that in some cases the superposition of aperiodic dot screens may also give full lines of coincidence: As we have seen, for instance, in Examples 3.1, 3.2 and 3.3 (Figs. 2.3(a), 2.3(c) and 3.5–3.6), when the transformation $\mathbf{g}_1(x,y) - \mathbf{g}_2(x,y)$ is singular or non-linear, the solutions of Eqs. (6.7) or (6.8) may consist of full lines or curves rather than isolated points. ∎

Remark 6.2: As we have seen, the fact that the arbitrary constants c_i (or (c_i,k_i)) are identical in both layers is crucial for the successful generalization of the indicial equations method into the aperiodic case. This means that superpositions in which we add some degree of random noise to the element locations in the individual layers cannot be treated using this technique. Such cases will have to wait until we introduce our Fourier based approach in Chapter 7. ■

6.4 Examples of Glass patterns in the superposition of curved line gratings

As we have already seen at the end of the last section, the use of the indicial equations method extends in a natural way to Glass patterns that occur between curved line gratings. A curved line grating is obtained by applying a non-linear geometric transformation $\mathbf{g}(x,y)$ to an initially straight line grating. If the originally straight grating is periodic, we obtain a curved repetitive grating; and if the original grating is aperiodic, we obtain a curved aperiodic grating.

Moiré effects that occur between curved *repetitive* gratings have been intensively investigated in the past, and their mathematical theory is already fully understood (see, for example, Chapters 10 and 11 in *Vol. I*). Our aim in this section is, therefore, to study the behaviour of Glass patterns between curved *aperiodic* line gratings, namely, straight aperiodic line gratings which have undergone non-linear mappings. To illustrate such cases we will use in the following examples *parabolic gratings*, which are obtained from the original straight gratings using the same non-linear mapping that we have introduced in Sec. 3.4, namely: $\mathbf{g}(x,y) = (x - ay^2,y)$. Fig. 6.3(a) shows such an aperiodic parabolic grating, and Fig. 6.3(b) shows its periodic (or rather repetitive) counterpart. These parabolic gratings offer us here the same advantages as the parabolic screens in Sec. 3.4: They clearly illustrate the main configurations which may occur in superpositions of curved gratings; their mathematical handling is still tractable; and finally, the results for the corresponding repetitive cases, where the original, untransformed gratings were periodic, are already known (see Sections 10.7.3–10.7.4 in *Vol. I*), so that we can easily compare the results obtained for repetitive and aperiodic gratings.

Example 6.1: The superposition of a parabolic grating and a straight grating (compare with Example 3.3 that illustrates the 2D case):

We start by superposing two identical copies of a straight aperiodic grating on top of one another, and we apply to one of them our parabolic mapping. In this case the locus of the points of coincidence consists of a straight line along the x axis; and indeed, as we can see in Fig. 6.3(c), a linear Glass pattern is generated in the superposition along this line.

This result can be derived mathematically, using our generalized indicial equations method, as follows. Our first line grating (the curved one) is a family of parabolas of the form $x = ay^2 + c$, which is defined by:

$$x - ay^2 = c_i \qquad\qquad i \in \mathbb{Z}, \quad c_i \in \mathbb{R} \qquad\qquad (6.11)$$

Our second line grating is a family of straight vertical lines which is given by:

$$x = c_i \qquad\qquad i \in \mathbb{Z}, \quad c_i \in \mathbb{R} \qquad\qquad (6.12)$$

We obtain, therefore, a system of two equations in x and y, whose solutions (x,y) define the locus of the points of coincidence between the two gratings. And indeed, by eliminating c_i from the two equations we obtain the solution:

$$y = 0$$

for all x, which is, as we have expected, a straight horizontal line which coincides with the x axis. This is, indeed, the center of the resulting Glass pattern, as shown in Fig. 6.3(c).

Suppose now that we shift the untransformed grating horizontally, as we did in Example 3.3 in the 2D case of dot screens. Our two line gratings become, therefore:

$$x - ay^2 = c_i \qquad\qquad i \in \mathbb{Z}, \quad c_i \in \mathbb{R} \qquad\qquad (6.13)$$

and: $\qquad\quad x - x_0 = c_i \qquad\qquad i \in \mathbb{Z}, \quad c_i \in \mathbb{R} \qquad\qquad (6.14)$

By eliminating c_i from these two equations we obtain their solution:

$$y = \pm\sqrt{x_0/a} \qquad\qquad (6.15)$$

for all x. However, just as in Example 3.3 (see Eq. (3.21)), we must distinguish here between two possible cases: If the untransformed grating is shifted *to the right*, meaning that $x_0 > 0$, or more precisely: if x_0 and a have the same sign, then the solution of our system of equations is a pair of horizontal lines whose vertical distances from the x axis are given by $\pm\sqrt{x_0/a}$. This is, indeed, the locus of the points of coincidence between the two gratings. Consequently, we obtain in the superposition two linear Glass patterns that are centered about these lines of coincidence. And indeed, as shown in Figs. 6.4(a),(c), when we gradually shift the untransformed grating to the right, the two parallel lines of coincidence simultaneously move away from the x axis; and inversely, as x_0 tends to zero, the two parallel lines of coincidence move closer and closer to the x axis, until they coincide with it when $x_0 = 0$ (Fig. 6.3(c)). But if the untransformed grating is shifted *to the left*, meaning that $x_0 < 0$, or more precisely: if x_0 and a have opposite signs, then no real points satisfy Eq. (6.15), meaning that no points of coincidence may exist. And indeed, as we can see in Fig. 6.5(a), the Glass pattern in this case disappears; this is best appreciated by observing the figure from a sufficient distance, say, 3–4 meters, where microstructure elements are no longer discerned by the eye. Note, however, that in the corresponding repetitive case (Fig. 6.5(b)) the moiré effect is still clearly visible. This happens, again, as we saw in Example 3.3, due to the additional ordering which exists in the internal structure of the repetitive layers: thanks to their repetitivity, the superposed layers have infinitely many lines of coincidence which are not lines of coincidence in the aperiodic case.

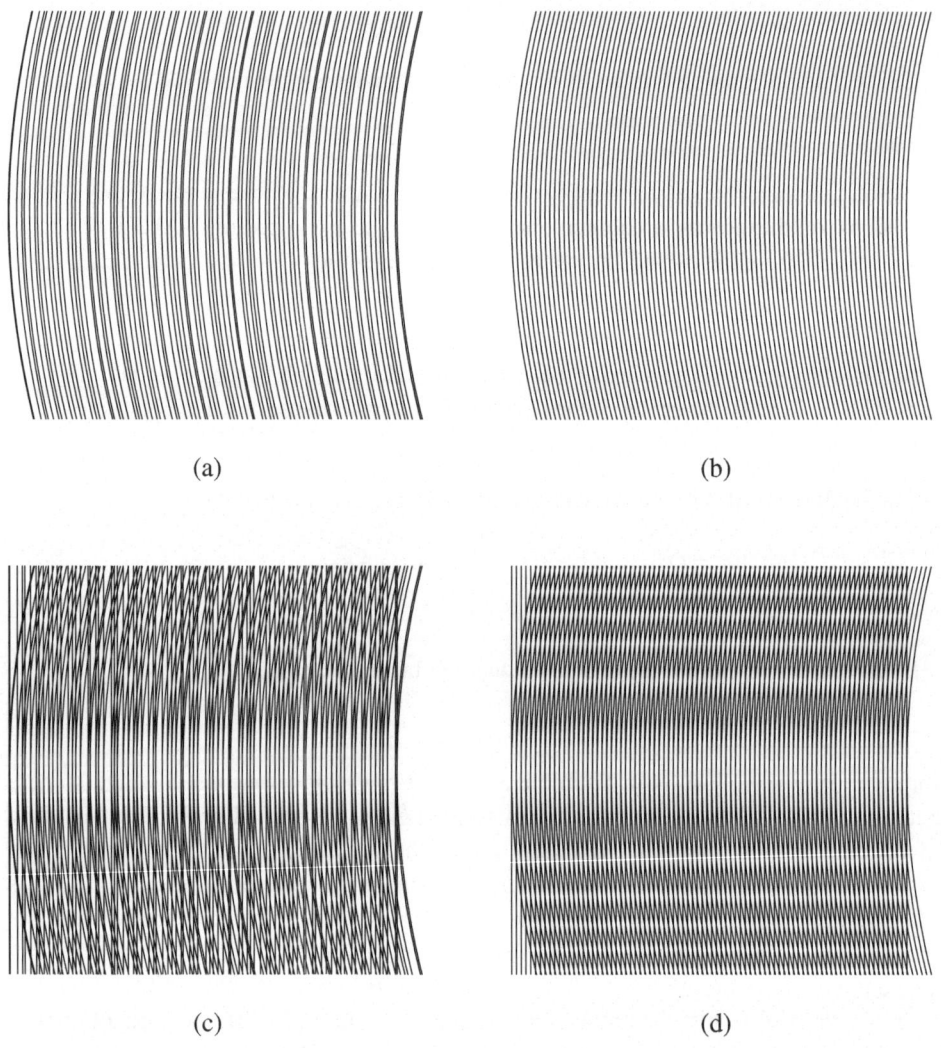

(a)   (b)

(c)   (d)

Figure 6.3: (a) A parabolic aperiodic grating which is obtained by applying
the parabolic transformation $\mathbf{g}(x,y) = (x - ay^2, y)$ to the straight
aperiodic grating of Fig. 6.1(a). (b) A periodic grating which has
undergone the same parabolic transformation. (c) The superposition of two identical aperiodic gratings, one of which has
undergone the parabolic transformation $\mathbf{g}(x,y)$. Since this
transformation does not involve layer shifts, the layer origins are
superposed exactly on top of each other. (d) The superposition
of two identical periodic gratings, one of which has undergone
the same parabolic transformation $\mathbf{g}(x,y)$. In both (c) and (d) a
linear Glass pattern is generated precisely in the same location.
Compare also with Fig. 3.5.

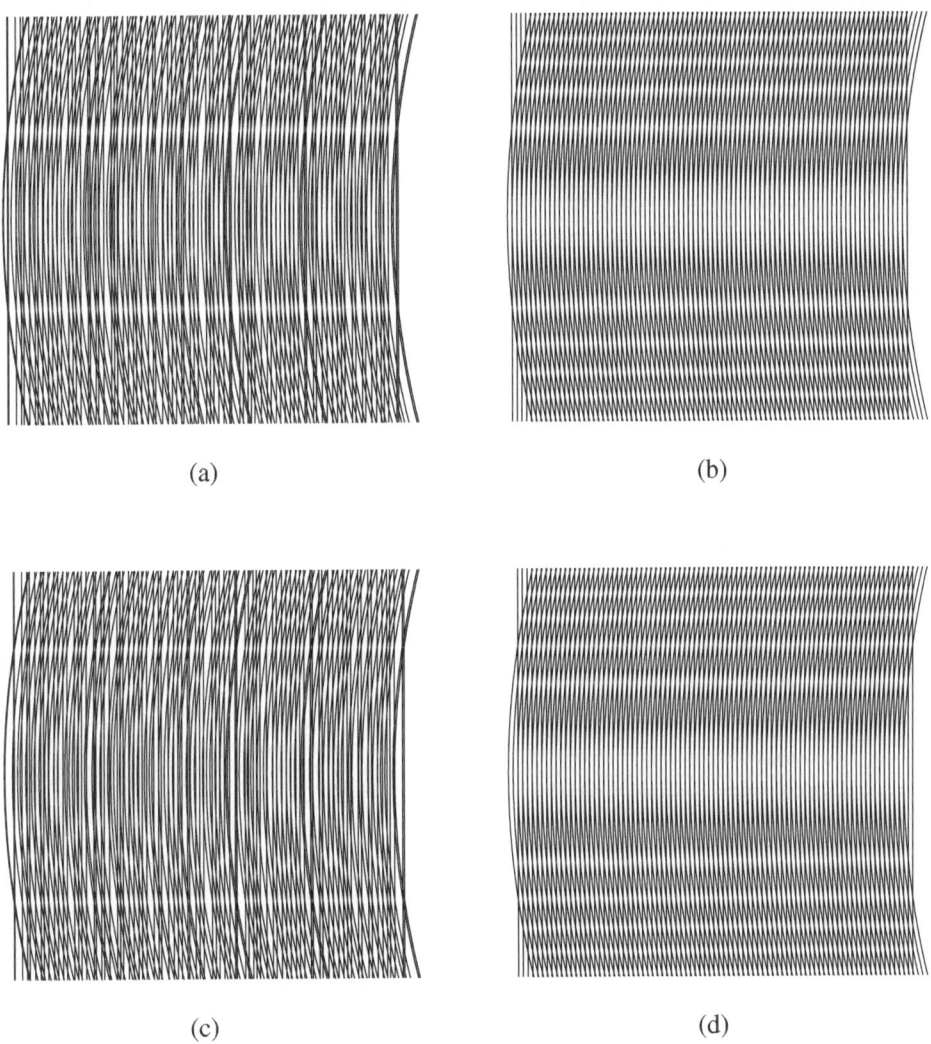

(a)   (b)

(c)   (d)

Figure 6.4: (a),(b) Same as in Figs. 6.3(c),(d) but here the untransformed grating has been slightly shifted by $x_0 = T$ to the right, where T equals one period of the periodic grating of (b). In both (a) and (b) a pair of linear Glass patterns is generated precisely at the same locations. (c),(d) Same as in (a),(b) but with a slight horizontal layer shift of $x_0 = 2T$. Compare with Fig. 3.6.

What happens now if we shift one of the layers vertically? Unlike in the 2D case of Example 3.3 (Figs. 3.8–3.9), in our present case vertical shifts do not have any effect on the straight vertical grating, and therefore the superposition simply remains unchanged. ∎

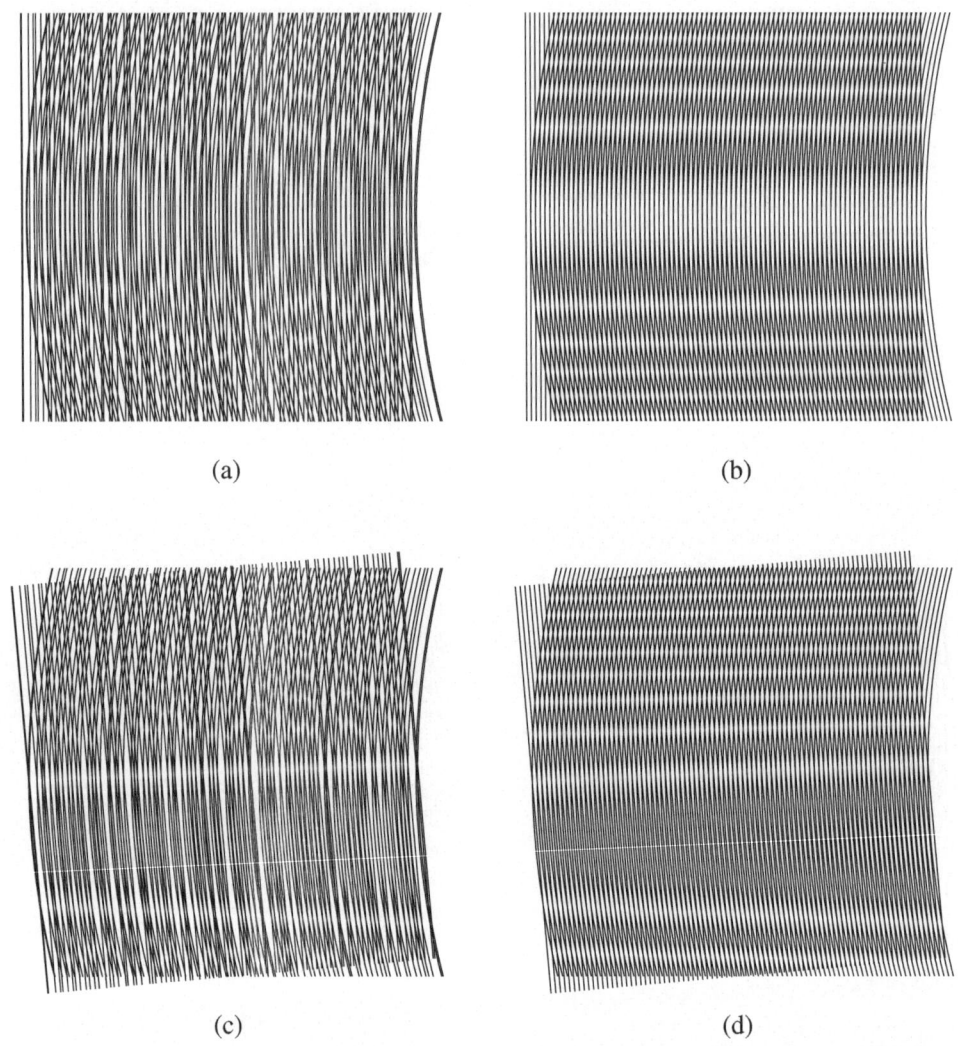

(a)   (b)

(c)   (d)

Figure 6.5: (a),(b) Same as in Figs. 6.4(a),(b) but with a horizontal shift of $x_0 = -T$. Compare with Figs. 3.7(a),(b). (c),(d) Same as in Figs. 6.3(c),(d) but here the untransformed layer has been slightly rotated by angle $\alpha = 5°$. Compare with Figs. 3.10(c),(d).

Example 6.2: The superposition of a parabolic grating and a straight grating with rotation (compare with Example 3.4 that illustrates the 2D case):

Let us now proceed to the more interesting case where one of the superposed layers (say: the untransformed layer) is rotated by a small angle α. As we can see in Fig. 6.5(c), the rotation by α generates in the superposition a parabolic Glass pattern passing through the origin.

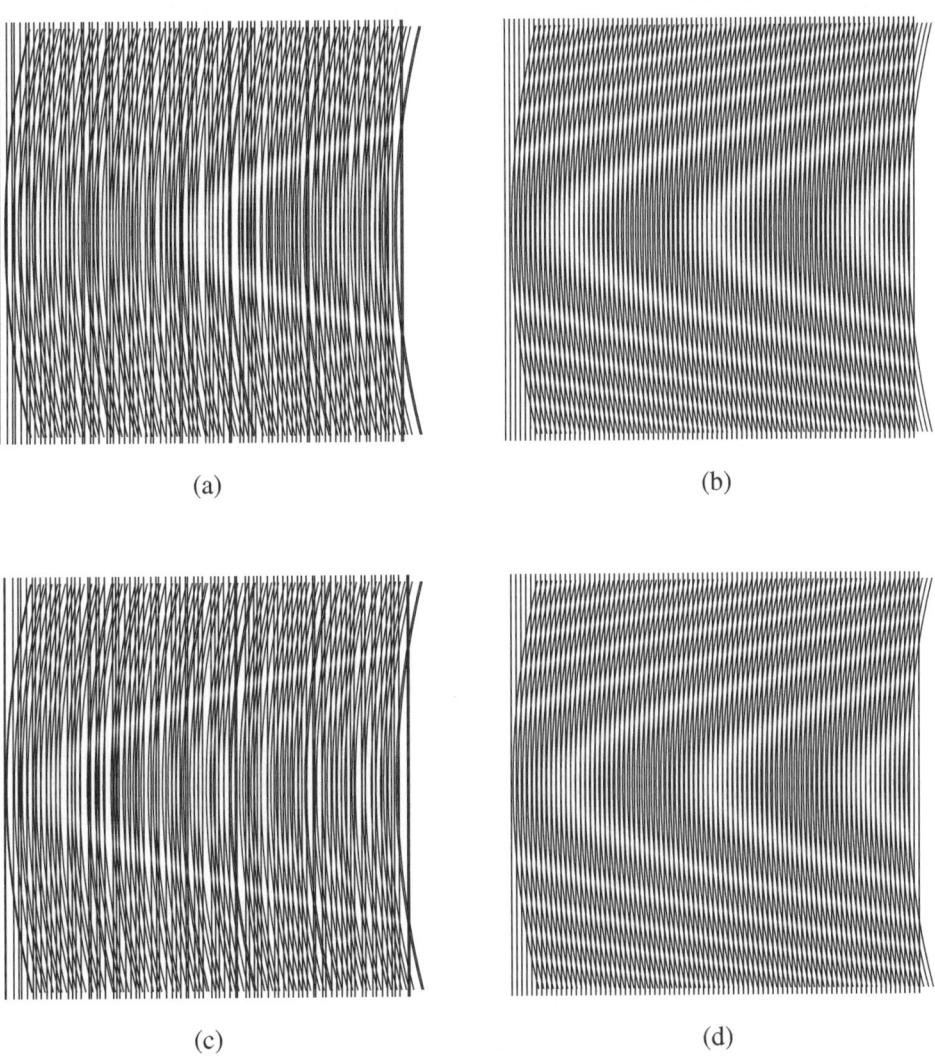

(a)

(b)

(c)

(d)

Figure 6.6: (a),(b) Same as in Figs. 6.3(c),(d) but here the untransformed layer
has been slightly scaled up. (c),(d) Same as in (a),(b) but here the
scaled layer has also undergone a slight horizontal shift of $x_0 = T$ to
the right. Compare with Fig. 3.13.

The mathematical analysis of this case is obtained as follows. Our two gratings are
defined, respectively, by:

$$x - ay^2 = c_i \qquad\qquad i \in \mathbb{Z}, \quad c_i \in \mathbb{R} \qquad\qquad (6.16)$$

and: $$x\cos\alpha + y\sin\alpha = c_i \qquad\qquad i \in \mathbb{Z}, \quad c_i \in \mathbb{R} \qquad\qquad (6.17)$$

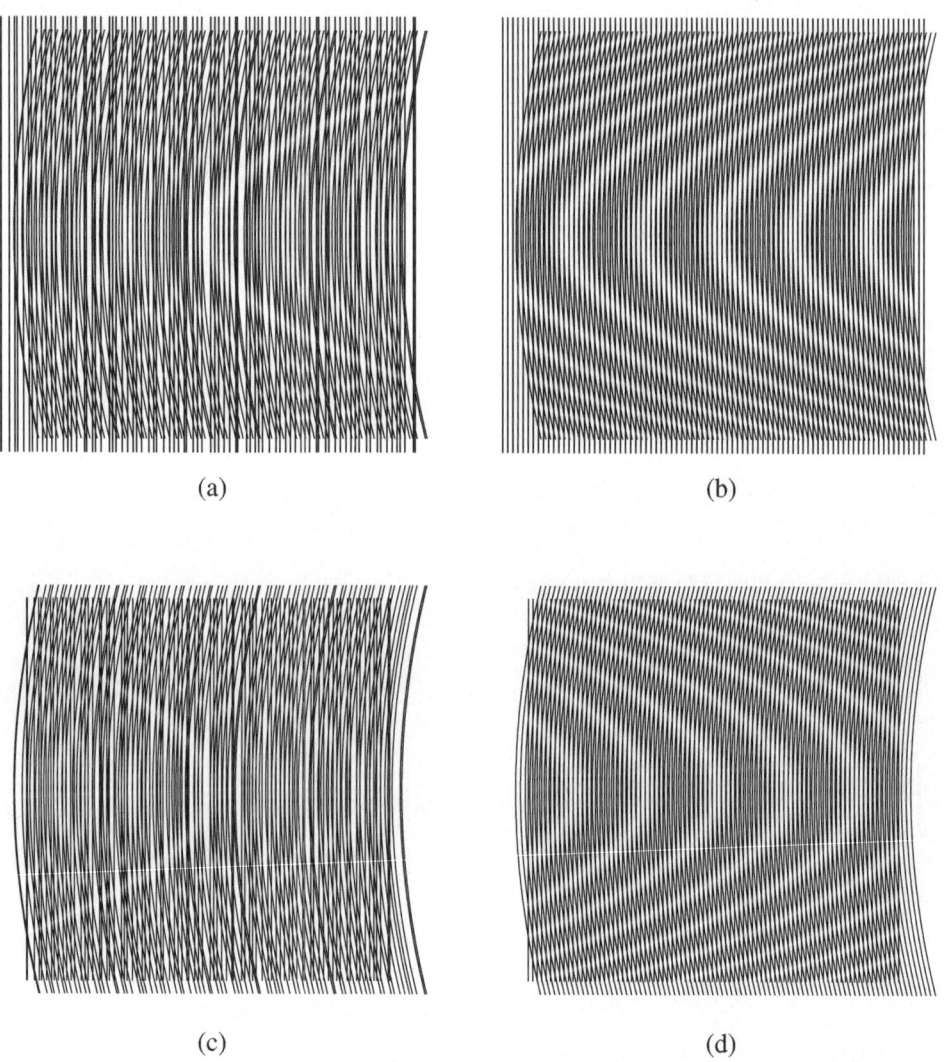

        (a)   (b)

        (c)   (d)

Figure 6.7: (a),(b) Same as in Figs. 6.6(a),(b) but with a larger scaling. (c),(d) Same as in (a),(b) but with a scale-down transformation.

We obtain, therefore, a system of two equations in x and y, whose solutions (x,y) define the locus of the points of coincidence between the two gratings. And indeed, by eliminating c_i from the two equations we obtain:

$$x - ay^2 = x\cos\alpha + y\sin\alpha \tag{6.18}$$

which gives, using the identity $\dfrac{\sin\alpha}{1-\cos\alpha} = \cot\dfrac{\alpha}{2}$:

$$x = \frac{a}{1-\cos\alpha}\, y^2 + \cot\frac{\alpha}{2}\, y \tag{6.19}$$

The solution of our system of equations is, therefore, a horizontally oriented parabola passing through the origin, just as we have expected. Note that this line of coincidence includes, indeed, the two fixed points of our layer transformation that were already found in Example 3.4 (see Eq. (3.28)). This is clearly seen when comparing Figs. 3.10(c) and 6.5(c), and it can be also confirmed by a short trigonometric calculation.[7] ■

Example 6.3: The superposition of a parabolic grating and a straight grating with scaling (compare with Example 3.5 that illustrates the 2D case):

Let us see what happens when one of the superposed layers (say, the untransformed layer) is slightly scaled by a scaling factor of $s \approx 1$. As we can see in Fig. 6.6, the resulting Glass pattern in this case is again parabolic, and the application of a small horizontal layer shift causes this parabolic Glass pattern to move horizontally. Moreover, when $s > 1$ the parabolic Glass pattern is opened to the right, and when $s < 1$ the parabolic Glass pattern is opened to the left (see Fig. 6.7).

In order to analyze this case mathematically we note that our two families of gratings are defined here by:

$$x - ay^2 = c_i \qquad\qquad i \in \mathbb{Z}, \quad c_i \in \mathbb{R} \qquad\qquad (6.20)$$

and: $\qquad sx = c_i \qquad\qquad\qquad i \in \mathbb{Z}, \quad c_i \in \mathbb{R} \qquad\qquad (6.21)$

We obtain, therefore, a system of two equations in x and y, whose solutions (x,y) define the locus of the points of coincidence between the two gratings. And indeed, by eliminating c_i from the two equations we obtain:

$$x - ay^2 = sx$$

which gives:

$$x = \frac{a}{1 - s} y^2 \qquad\qquad (6.22)$$

The solution of our system of equations is, indeed, a horizontally oriented parabola whose extremum is located at the origin, and whose bending rate[8] is determined by $\frac{a}{1-s}$ (a being the bending rate of the original parabolic grating). And indeed, as expected, when $s < 1$ (i.e. when the grating (6.21) is *scaled up*[9]), the Glass pattern is a horizontal parabola opened to the same direction as the original parabolic grating (see Fig. 6.7(a)), and when $s > 1$ (i.e. when the grating (6.21) is *scaled down*), the Glass pattern is a horizontal parabola opened to the opposite direction (see Fig. 6.7(c)). Note that according to Eq. (6.22) the parabolas obtained for $s = 1 + \varepsilon$ and for $s = 1 - \varepsilon$ (where ε is a small positive fraction) are identical, and they only differ in their opening direction. ■

---

[7] In fact, this can be seen even without calculations: Since Eq. (6.18) is simply the first equation of the system (3.26), the curve it defines obviously includes the solutions of system (3.26), which are the intersection points of the two curves defined by the equations of this system.

[8] As explained in Sec. 3.4.1 (see also Example 10.1 in *Vol. I*), the term *bending rate* refers to the slope (or the opening) of the parabola.

[9] Remember that we are using here *domain* (i.e. inverse) transformations, as explained in Remark 4.1.

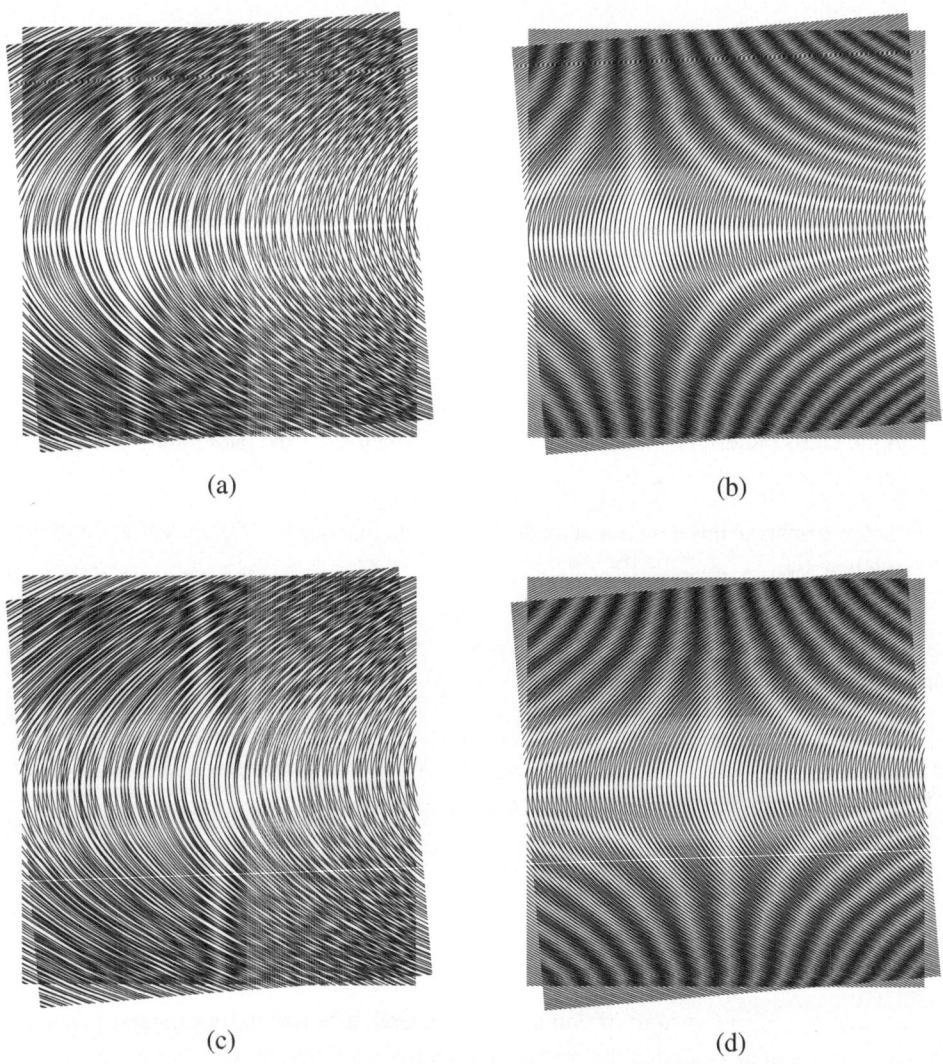

(a)   (b)

(c)   (d)

Figure 6.8: Glass patterns in the superposition of two parabolic
line gratings. (a) The superposition of two identical
aperiodic parabolic line gratings with a small angle
difference of 5°. (b) The superposition of two
identical periodic parabolic line gratings with the same
angle difference. In both (a) and (b) two perpendicular
linear Glass patterns are generated precisely at the
same location. (c),(d) Same as in (a),(b), but here the
unrotated grating has been slightly shifted up by
$y_0 = 1.5T$, where T equals one period of the original
(untransformed) line grating in (b). In both (a) and (b)
two perpendicular linear Glass patterns are generated
precisely at the same location.

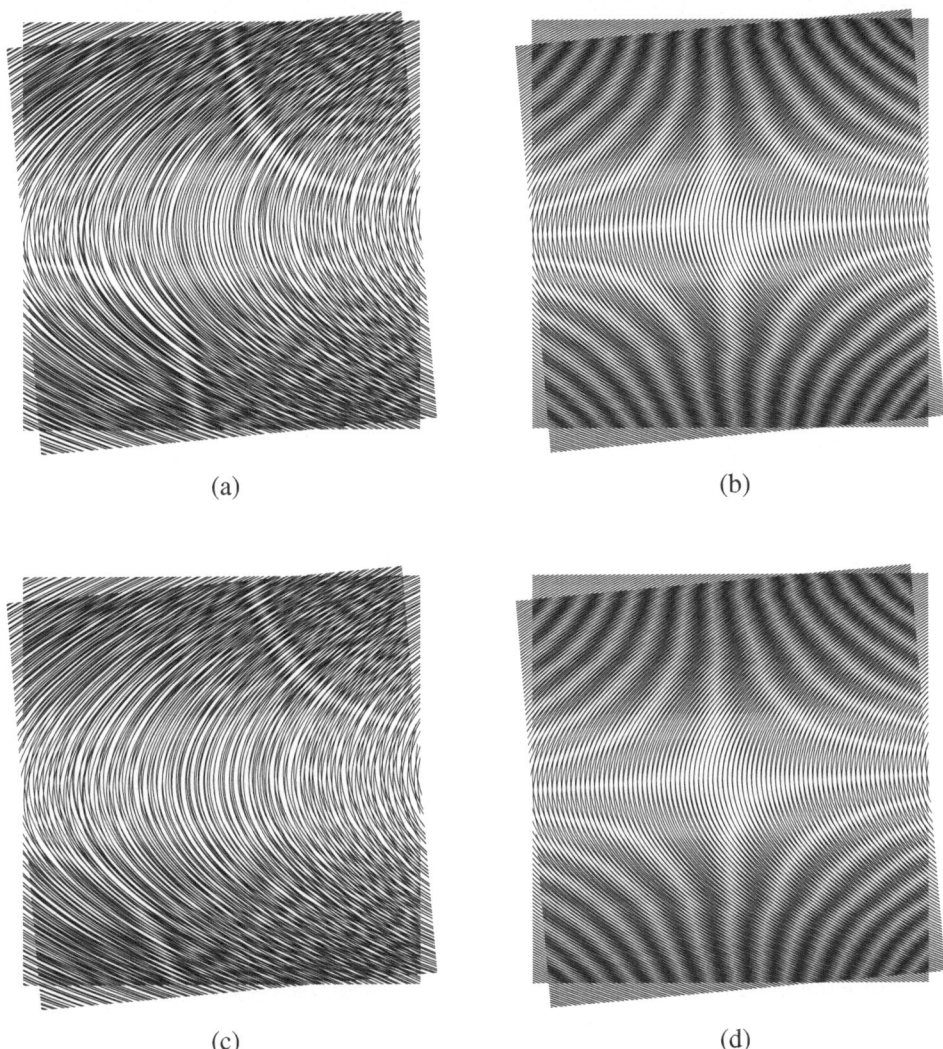

(a)

(b)

(c)

(d)

Figure 6.9: (a),(b) Same as in Figs. 6.8(a),(b), but here the unrotated layer has also undergone a slight layer shift of $(x_0,y_0) = (-T,1.5T)$. In both (a) and (b) a hyperbolic Glass pattern is generated precisely at the same location. (c),(d) Same as in Figs. 6.8(a),(b), but here the untotated layer has also undergone a slight layer shift of $(x_0,y_0) = (-2T,1.5T)$.

Example 6.4: The superposition of two parabolic gratings with rotation:

Having studied the superpositions of a parabolic grating and a straight grating, we proceed now to the superposition of two identical copies of our parabolic grating, one of which has been slightly rotated by angle α. As we can see in Fig. 6.8(a), the locus of the points of coincidence in this case consists of two perpendicular straight lines, and indeed, a Glass pattern is generated in the superposition along these lines.

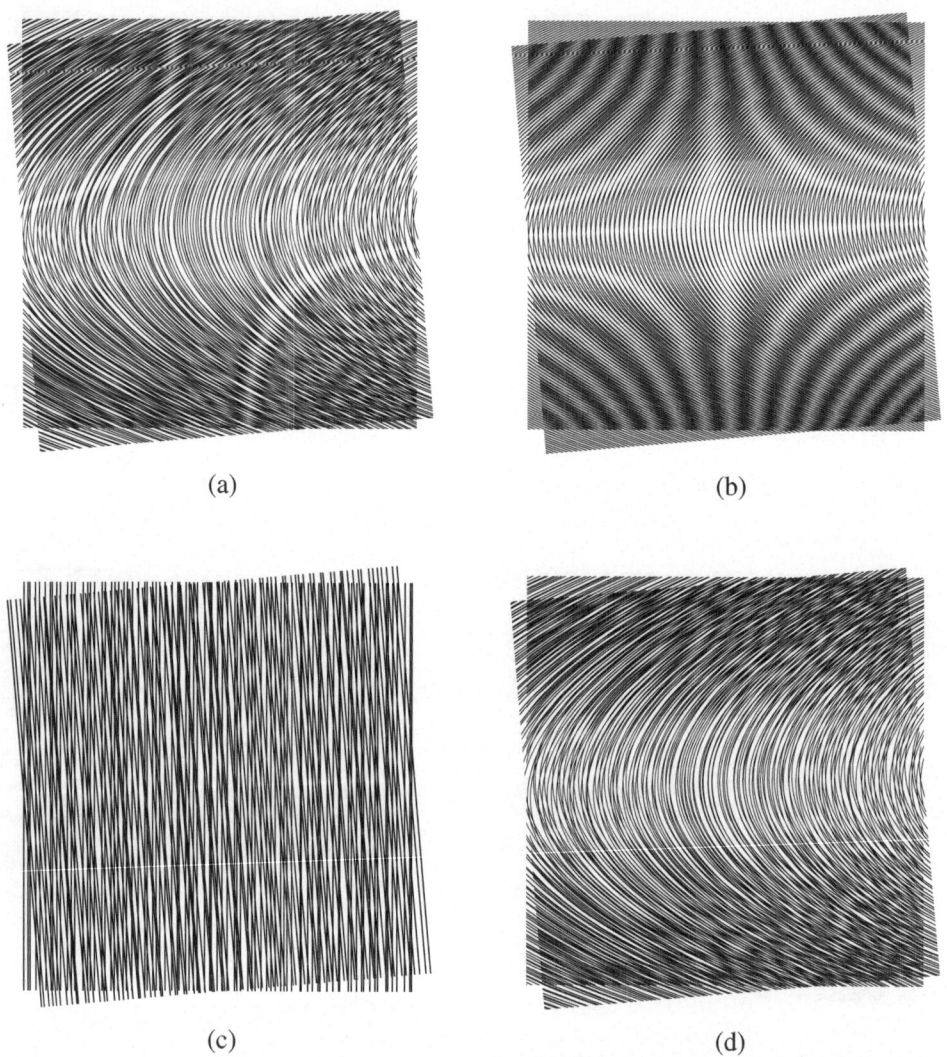

(a)   (b)

(c)   (d)

Figure 6.10: (a),(b) Same as in Figs. 6.8(a),(b), but with a slight layer shift of (x_0, y_0)
= $(T, 1.5T)$. (c),(d) When each of the two superposed layers is
generated with a different seed in the random number generator, there
is no correlation between the two layers and no Glass patterns appear
in their superposition: (c) Same as Fig. 6.1(a), but with a different seed
for each layer. (d) Same as Fig. 6.9(a), but with a different seed for
each layer.

However, as we might expect based on our experience in the previous examples, the
superpotition in the *repetitive* case contains, in addition to this Glass pattern, infinitely
many other curved lines of coincidence between the two layers, which are not lines of
coincidence in the aperiodic case (compare Figs. 6.8(a) and (b)). This happens due to the

additional ordering which exists in the internal structure of repetitive layers: Thanks to the new constraints imposed by their repetitivity, the superposed layers also have infinitely many new lines of coincidence which are not lines of coincidence in the corresponding aperiodic case.

Let us now derive mathematically the lines of coincidence in the aperiodic case. Our first grating is a family of parabolic lines that is defined by:

$$x - ay^2 = c_i \qquad\qquad i \in \mathbb{Z}, \ \ c_i \in \mathbb{R} \qquad\qquad (6.23)$$

Our second grating is a copy of the first line family which has been rotated by angle α:

$$x\cos\alpha + y\sin\alpha - (y\cos\alpha - x\sin\alpha)^2 = c_i \qquad i \in \mathbb{Z}, \ \ c_i \in \mathbb{R} \qquad (6.24)$$

This gives us a system of two equations in x and y, whose solutions (x,y) define the locus of the points of coincidence between the two layers. Because the arbitrary numbers c_i are identical in both line families we can eliminate them from our two equations. We obtain, therefore:

$$x\cos\alpha + y\sin\alpha - (y\cos\alpha - x\sin\alpha)^2 = x - ay^2$$

which gives after some simplifications:

$$a\sin^2\alpha \, y^2 + (\sin\alpha + 2ax\sin\alpha \cos\alpha)y - ax^2\sin^2\alpha + x(\cos\alpha - 1) = 0$$

This second-order equation in y has two solutions:

$$y = x\tan\frac{\alpha}{2} \qquad\qquad (6.25)$$

and:

$$y = -x\cot\frac{\alpha}{2} - \frac{1}{a\sin\alpha}$$

$$= x\tan(90° + \frac{\alpha}{2}) - \frac{1}{a\sin\alpha} \qquad\qquad (6.26)$$

The solution of our system of equations consists, therefore, of two orthogonal straight lines: one with the angle of $\frac{\alpha}{2}$, passing through the origin, and the other with the angle of $90° + \frac{\alpha}{2}$, which crosses the y axis far below the origin, at $y = -\frac{1}{a\sin\alpha}$. These two orthogonal lines are, therefore, the locus of the points of coincidence between the two layers. And indeed, as we can see in Fig. 6.8(a),(b), a Glass pattern is generated in the superposition along these two orthogonal lines. Note that in the repetitive case (Fig. 6.8(b)) the arbitrary numbers c_i are simply replaced in Eqs. (6.23) and (6.24) by integer multiples of the respective grating periods, mT and nT, where $m,n \in \mathbb{Z}$. As explained in Chapter 11 of *Vol. I*, this gives us the family of hyperbolic moiré bands which are generated in the superposition of the two repetitive gratings. Note that the Glass pattern we have derived above is simply the 0-th moiré band of this family, which is obtained when $m = n$ (namely, the moiré band whose index is $p = m - n = 0$).

What happens now in the superposition when we slightly shift one of the parabolic gratings, say, the unrotated one? A vertical shift of y_0 will only cause the moiré effects to move horizontally to the right or to the left, depending on the sign of y_0 (see Fig. 6.8). However, a horizontal shift of x_0 has a more interesting effect: as we can see in Figs. 6.9–6.10(a),(b), it causes the locus of coincidence between the two layers to become a hyperbolic curve; and indeed, a Glass pattern is generated in the layer superposition along the two branches of this curve. Moreover, the value of x_0 determines the parameters of this hyperbolic curve (its orientation and its distance from the center). But once again, due to the additional ordering which exists in the internal structure of repetitive layers, the superposition in the repetitive case contains, in addition to this Glass pattern, other lines of coincidence between the two layers which are not lines of coincidence in the aperiodic case. This explains the repetitive nature of the moiré effect in Figs. 6.9(b),(d) and 6.10(b). ∎

Finally, it should be emphasized that Glass patterns in aperiodic cases are only generated thanks to the correlation which exists between the superposed layers. As shown in Figs. 6.10(c),(d), if we regenerate the same layer superpositions using in each layer a different seed for the random number generator, there is no correlation between the layers and no Glass patterns appear in the superposition.

6.5 The effect of adding constraints to the original layers

As we have already seen, in spite of their quite different appearance, moiré effects that occur between periodic, repetitive or aperiodic layers are, in fact, particular cases of the same basic phenomenon, and all of them satisfy the same fundamental rules. In particular, when the original layers undergo weak transformations, one or more Glass patterns may appear in their superposition. So far, we have only studied the influence of the *layer transformations* on the resulting Glass patterns. In the present section, however, we study the influence of the *internal structure* of the individual layers, while the layer transformations they undergo remain unchanged (see, for example, Fig. 6.11).

As we already know, in the superposition of *aperiodic dot screens* Glass patterns are formed about each of the mutual fixed points of the layer transformations (see Figs. 6.11(a), 6.12(a)); however, in the case of *aperiodic line gratings*, under the same layer transformations, each Glass pattern is extended into a full line, the line of coincidence between the superposed gratings (see Figs. 6.11(c), 6.12(c)). As we have seen in Sec. 6.2.1, this happens because an aperiodic line grating is more constrained (i.e., it has fewer degrees of freedom) than an aperiodic dot screen: it only has full freedom in one dimension, while its other dimension is completely determined (constant). Therefore, due to the additional restriction on the structure of the layers (namely, the fact that their second dimension is completely determined by the first one), the layer superposition has a full line of points of coincidence, which obviously includes the fixed point of the underlying layer transformation. Mathematically, this is, indeed, a consequence of Remark 6.1.

Now, a superposition of *periodic* line gratings in which the superposed layers undergo the same transformation as in the aperiodic case is, in fact, a particular case in which the line distances within each of the gratings are forced to be equal. In this case, as we have seen in Sec. 6.2.2, in addition to the linear Glass pattern described above, an infinite number of new lines of coincidence is generated in the superposition, wherever the two layers happen to coincide because of the repetitivity in their internal structure (see Figs. 6.11(d), 6.12(d)). Thus, due to the additional constraint on the internal structure of the layers (namely, the fact that they are periodic), the layer superposition has a full family of lines of coincidence, which obviously includes the fundamental line of coincidence of the corresponding aperiodic case (which, in turn, includes the mutual fixed points of the underlying layer transformations).

As we can see, in both cases the fact of having additional constraints on the internal structure of the superposed layers causes an addition of new loci of coincidence in the superposition, which are not loci of coincidence in the corresponding superposition of unconstrained layers:

(i) Having constrained (fixed) the second dimension of the original layers results in the extension of their locus of coincidence from isolated fixed points of the underlying layer transformations into full lines;

(ii) Adding the constraint of periodicity to the original layers causes the extension of each such line of coincidence into a family of infinitely many lines of coincidence.

Note, however, that some degenerate or non-linear layer transformations may already have a full *fixed line* (or several fixed lines) rather than *fixed points*. As shown in Fig. 6.13(a), in such cases the superposition of aperiodic dot screens gives a linear Glass pattern along each of the fixed lines of the transformation. The transition from aperiodic dot screens to aperiodic line gratings in such cases does not further increase the dimension of the linear Glass patterns in the superposition, and the Glass patterns remain basically unchanged (see Figs. 6.13(a) and (c)).[10]

The effects of applying successively constraints (i) and (ii) on an aperiodic screen can be schematically illustrated by following the arrows in the left and in the bottom of Fig. 6.14. However, these constraints can be also applied the other way around, namely, constraint (ii) (periodicity) first, and constraint (i) (constancy along one direction) second. This can be illustrated by following the arrows in the top and in the right hand side of Fig. 6.14, as we explain below.

Suppose we are given two aperiodic dot screens that are slightly transformed versions of each other. As predicted by Proposition 3.2, a Glass pattern is generated in the

---

[10] In some trivial degenerate cases, such as the transformation consisting of vertically scaling one of the superposed layers (see Figs. 2.3(a), (b)), the transition from aperiodic dot screens to vertical aperiodic gratings *does* cause the fixed line of the transformation to be extended into a full plane of coincidence between the superposed gratings (which obviously includes the fixed line of the transformation). This causes the linear Glass pattern to extend throughout the entire plane, and hence to disappear.

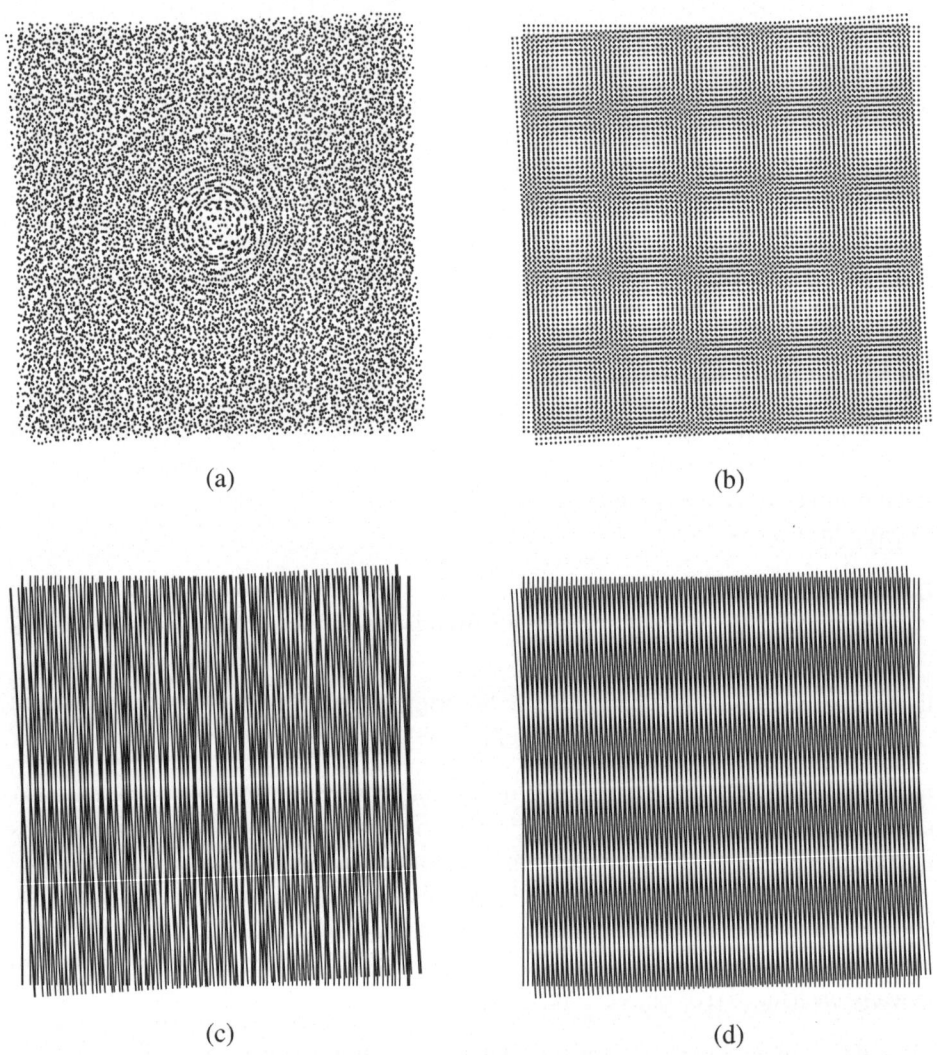

(a)   (b)

(c)   (d)

Figure 6.11: The Glass (or moiré) patterns generated by applying the same slight
layer rotation in the superposition of: (a) two identical aperiodic dot
screens; (b) two identical periodic dot screens; (c) two identical
aperiodic line gratings; and (d) two identical periodic line gratings.

superposition about each of the mutual fixed points (or fixed lines) of the mappings that
were applied to the superposed layers. This is true for aperiodic, periodic and repetitive
cases. However, if we impose constraint (ii) (periodicity) on the original layers, the moiré
effect in the resulting periodic or repetitive cases extends periodically or repetitively far
beyond the range of the Glass pattern in the corresponding aperiodic case (compare, for
example, Figs. 6.11(a) and (b), or Figs. 6.13(a) and (b)). In terms of macrostructures, the
periods of a periodic moiré pattern are simply duplicates of the main Glass pattern which

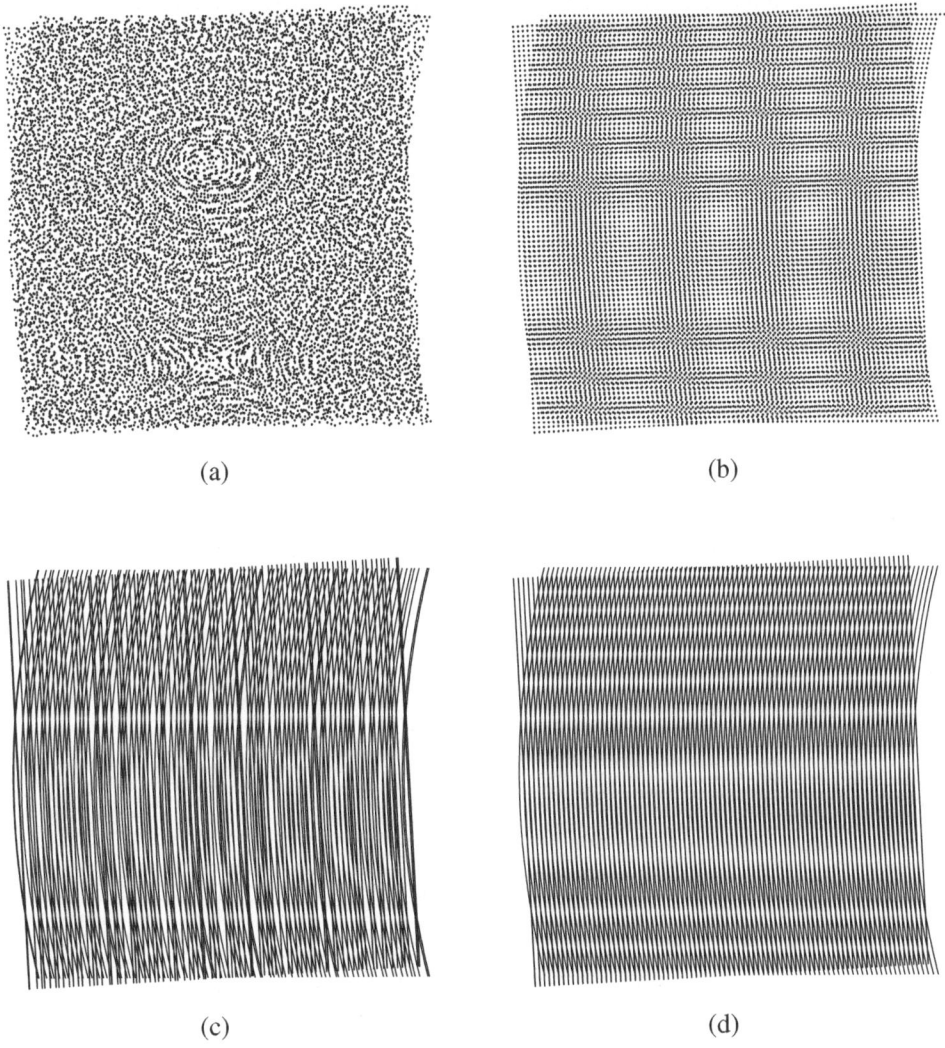

Figure 6.12: The Glass (or moiré) patterns generated by applying the non-linear layer transformation of Figs. 3.13(a),(b) in the superposition of: (a) two identical aperiodic dot screens; (b) two identical periodic dot screens; (c) two identical aperiodic line gratings; and (d) two identical periodic line gratings.

is generated about the fixed point, and the period length of the moiré corresponds to the distance between these duplicates (note, however, that this duplication concerns only the macrostructures, since the microstructure within these duplicates may vary). Similarly, in repetitive cases the repetitive moiré consists of transformed repetitions of the Glass patterns of the corresponding aperiodic cases (see, for example, Figs. 6.12, 3.15). This

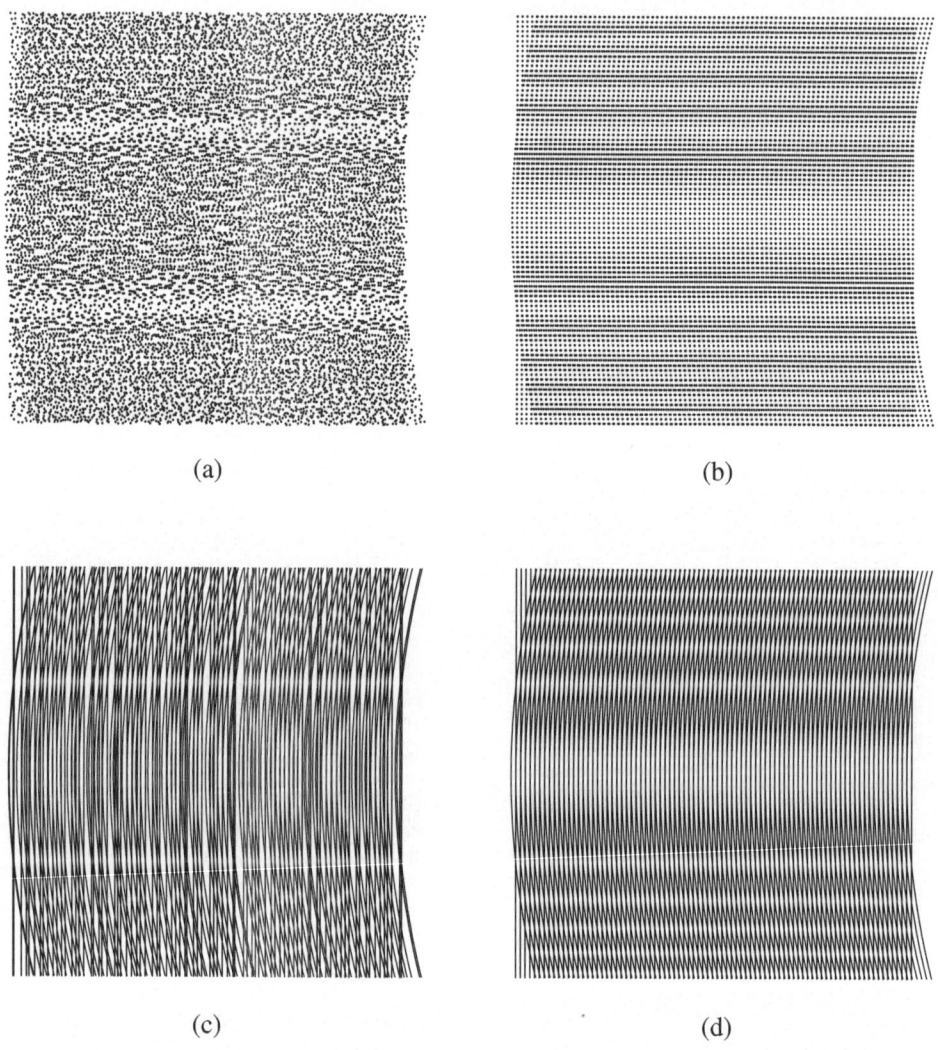

(a)   (b)

(c)   (d)

Figure 6.13: The Glass (or moiré) patterns generated by applying the non-linear layer transformation of Figs. 3.6(a),(b) in the superposition of: (a) two identical aperiodic dot screens; (b) two identical periodic dot screens; (c) two identical aperiodic line gratings; and (d) two identical periodic line gratings.

duplication occurs due to the additional ordering which exists in the internal structure of periodic or repetitive layers: thanks to their periodicity (or repetitivity), the superposed layers also have infinitely many points of coincidence which are not fixed points of the underlying mapping $g(x,y)$. It is interesting to note that for the same reason a moiré pattern may exist in a repetitive superposition (but not in a periodic one!) even if in the

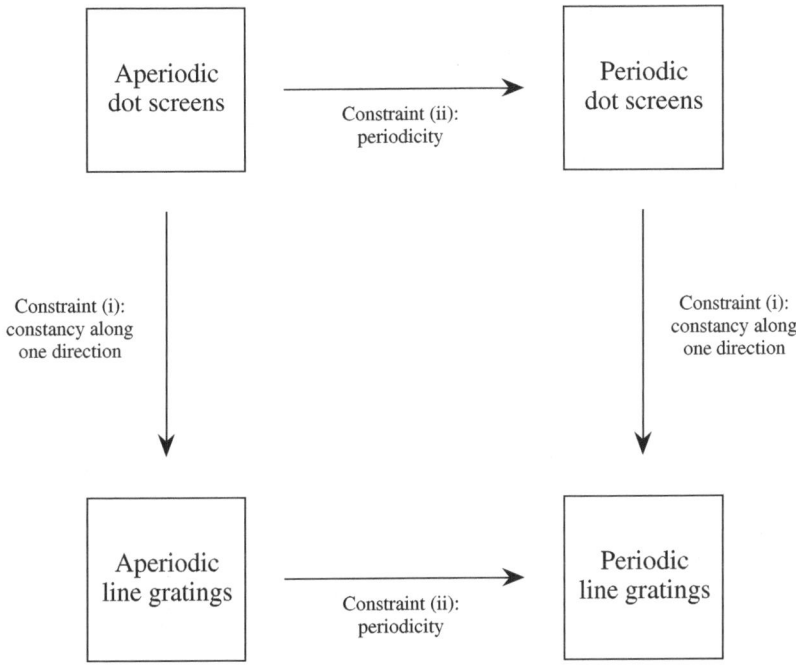

Figure 6.14: A schematic diagram showing the effects of a successive application of constraints (i) and (ii) (or vice versa) on the Glass patterns generated in the superposition. Note that the layer transformation remains identical in all cases, and only the internal structure of the superposed layers is being modified. Specific examples are shown in Figs. 6.11, 6.12 and 6.13.

corresponding aperiodic case there exist no Glass patterns at all. This is illustrated, for example, in Figs. 3.7(a) and (b).

Now, if we also impose constraint (i) on our layers, the original dot screens turn into line gratings, and the periodic family of Glass patterns which is generated in the original superposition turns into a periodic family of lines of coincidence (see Figs. 6.11(b) and (d), or Figs. 6.12(b) and (d)).

Looking at property (ii) the other way around, we can say that by applying the same layer transformations on aperiodic layers instead of on periodic layers, we can select the "real" Glass patterns, i.e. those that are generated about the fixed points of the transformation, from among their infinitely many duplicates which together form the periodic (or repetitive) macroscopic moiré effect. This interesting result can be clearly illustrated by comparing cases (a) and (b) or cases (d) and (c) in Figs. 6.11–6.13. Other examples have been shown in Figs. 6.8(c),(d)–6.10(a),(b).

It should be noted, finally, that thanks to their additional internal structure, periodic cases also satisfy several additional rules, that are expressed in terms of periods or frequencies, as described by the classical periodic moiré theory. But these rules are specific to periodic cases, and they are obviously no longer valid in general aperiodic cases.

6.6 A first step towards the intensity profiles of Glass and moiré patterns

As we already know, Glass patterns are generated in the superposition of aperiodic dot screens around the points (x,y) in which:

$$\mathbf{g}_M(x,y) = \mathbf{g}_1(x,y) - \mathbf{g}_2(x,y) = (0,0) \tag{6.27}$$

(see Eq. (6.8)). So far we have interpreted these points as the mutual fixed points of the layer transformations $\mathbf{g}_1(x,y)$ and $\mathbf{g}_2(x,y)$. But there exists also an alternative point of view, that is based on the interpretation of a transformation $(u,v) = \mathbf{g}(x,y)$ as a pair of surfaces $u = g_1(x,y)$ and $v = g_2(x,y)$ over the x,y plane (see Sec. B.2 in Appendix B). Under this interpretation the points (x,y) that satisfy Eq. (6.27) are simply the points in which the zero level curves of the two surfaces of the difference transformation $\mathbf{g}_1(x,y) - \mathbf{g}_2(x,y)$ intersect, i.e. the points where both of the surfaces $u = g_{1,1}(x,y) - g_{2,1}(x,y)$ and $v = g_{1,2}(x,y) - g_{2,2}(x,y)$ cross the x,y plane simultaneously:

$$u = g_{1,1}(x,y) - g_{2,1}(x,y) = 0$$
$$v = g_{1,2}(x,y) - g_{2,2}(x,y) = 0 \tag{6.28}$$

Based on this interpretation, we present here an alternative way for illustrating moiré and Glass patterns in the superposition of periodic or aperiodic layers. This approach will allow us to make a first step towards the modelization of the intensity profiles (the macroscopic gray levels) of Glass and moiré patterns.

For the sake of simplicity, we start with the case of line gratings. Let $p(x',y')$ be a straight line grating consisting of vertical lines. Because this grating is constant along the y' direction, its definition involves in fact only the x' coordinate; for example, if our grating is cosinusoidal its definition is $p(x') = \cos(2\pi f x')$ or rather $p(x') = \frac{1}{2}\cos(2\pi f x') + \frac{1}{2}$, and if it consists of an aperiodic sequence of white bands on a black background its definition is given by:

$$p(x') = \sum_i \text{rect}\left(\frac{x' - x'_i}{\tau_i}\right)$$

where:

$$\text{rect}(x) = \begin{cases} 1 & |x| < \frac{1}{2} \\ 0 & |x| > \frac{1}{2} \end{cases}$$

and where x'_i and τ_i determine the location and the width of the i-th white band. Each such grating can be considered, of course, as a surface over the plane, whose altitude at each point indicates its gray level there.

Suppose now that we apply to the entire x',y' plane, including our vertical line grating $p(x')$, a 2D transformation $(x',y') = \mathbf{g}(x,y) = (g_1(x,y),g_2(x,y))$. Since our line grating only depends on the first coordinate, x', the second component of $(x',y') = \mathbf{g}(x,y)$, $g_2(x,y)$, has no effect on our grating. The resulting curvilinear grating after the application of the transformation $\mathbf{g}(x,y)$ to the original grating $p(x')$ is, therefore, $r(x,y) = p(g_1(x,y))$. We call the function $x' = g(x,y)$ that bends the straight grating $p(x')$ into the curvilinear grating $r(x,y)$ the *bending function* or *bending transformation* of $r(x,y)$ (see Sec. 10.2 in *Vol. I*). For example, if our original grating is $p(x') = \cos(2\pi f x')$, then after the application of the rotation $(x',y') = \mathbf{g}(x,y) = (x\cos\alpha + y\sin\alpha, -x\sin\alpha + y\cos\alpha)$ we obtain the rotated grating $r(x,y) = \cos(2\pi f[x\cos\alpha + y\sin\alpha])$, whose bending function is $x' = x\cos\alpha + y\sin\alpha$ (see Fig. 10.1(b) in *Vol. I*). Similarly, after applying to our original grating $p(x')$ the transformation $(x',y') = \mathbf{g}(x,y) = (y - ax^2, y)$ we obtain the parabolic grating $r(x,y) = \cos(2\pi f[y - ax^2])$, whose bending function is $x' = y - ax^2$ (see Fig. 10.1(c) and Example 10.1 in *Vol. I*). Clearly, in each of these cases the resulting curvilinear grating can be considered as a surface over the plane, whose altitude at each point indicates its gray level there. This surface is simply the geometrically transformed version of the original surface of the untransformed line grating. However, for the sake of our considerations below, we can also attribute to each such curvilinear grating a second surface, which belongs to its bending function $x' = g(x,y)$. For example, although the parabolic grating $r(x,y) = \cos(2\pi f[y - ax^2])$ itself can be considered as a surface over the x,y plane, we can also consider its bending function, $x' = y - ax^2$, as a surface over the x,y plane. In order to avoid confusion in the discussion which follows we will call the surface $r(x,y)$ the *gray-level surface* of the grating, while the surface $g(x,y)$ will be called the *transformation surface* of the grating.

What happens now in the superposition of two line gratings? As usual, we suppose that before the application of the layer transformations $\mathbf{g}_1(x,y)$ and $\mathbf{g}_2(x,y)$, the two original gratings were identical vertical gratings $p_1(x')$ and $p_2(x')$ (that could only differ in their intensity profiles), and that these original gratings were superposed on top of each other in full coincidence. After applying the transformations $(x',y') = \mathbf{g}_1(x,y) = (g_{1,1}(x,y),g_{1,2}(x,y))$ and $(x',y') = \mathbf{g}_2(x,y) = (g_{2,1}(x,y),g_{2,2}(x,y))$ to the two original gratings they become, respectively, $r_1(x,y) = p_1(g_{1,1}(x,y))$ and $r_2(x,y) = p_2(g_{2,1}(x,y))$.

Consider now the superposition of the two transformed gratings. As we already know, the resulting Glass (or moiré) pattern in the superposition of the grating is determined by the difference transformation (6.27), or more precisely, by its first component:

$$x' = g_{1,1}(x,y) - g_{2,1}(x,y) \tag{6.29}$$

(because our two line gratings are independent of the second components $g_{1,2}(x,y)$ and $g_{2,2}(x,y)$). Now, since $g_{1,1}(x,y)$ and $g_{2,1}(x,y)$ can be considered as surfaces over the x,y plane — the transformation surfaces of the gratings $r_1(x,y)$ and $r_2(x,y)$ — their difference can be also considered as such a surface. This difference surface has the value zero (i.e. it crosses the x,y plane) wherever $g_{1,1}(x,y) = g_{2,1}(x,y)$; these are precisely the points which define the geometric location of the resulting Glass pattern in the grating superposition.

Example 6.5: Consider the grating superposition shown in Fig. 6.3(c). In this case the two original aperiodic vertical gratings $p_1(x')$ and $p_2(x')$ were transformed, respectively, by the 2D transformations $(x',y') = \mathbf{g}_1(x,y) = (x - ay^2, y)$ and $(x',y') = \mathbf{g}_2(x,y) = (x,y)$, where a is a small positive number. The two transformed gratings are given, therefore, by $r_1(x,y) = p_1(x - ay^2)$ and $r_2(x,y) = p_2(x)$, and their transformation surfaces are $g_{1,1}(x,y) = x - ay^2$ and $g_{2,1}(x,y) = x$. The resulting difference surface is, therefore, $g_{1,1}(x,y) - g_{2,1}(x,y) = -ay^2$, which is a bottom-opened parabolic surface that is constant along the x direction. This surface touches the x,y plane along the x axis; and indeed, this is precisely the geometric location of the Glass pattern in the grating superposition, as shown in Fig. 6.3(c).

Considering now the grating superposition of Fig. 6.4(a), where the untransformed grating has been slightly shifted by $x_0 > 0$, we see that in this case the second surface is $g_{1,1}(x,y) = x - x_0$, and therefore the difference surface becomes: $g_{1,1}(x,y) - g_{2,1}(x,y) = -ay^2 + x_0$. This means that the parabolic surface has been elevated by a constant value of x_0, and therefore it now crosses the x,y plane along the two horizontal lines $y = \pm\sqrt{x_0/a}$ to both sides of the x axis. This corresponds, indeed, to the Glass pattern that we observe in Fig. 6.4(a).

Finally, what happens now in Fig. 6.5(a)? In this case the untransformed grating has been shifted by $-x_0$, where $x_0 > 0$, and therefore the difference surface becomes $g_{1,1}(x,y) - g_{2,1}(x,y) = -ay^2 - x_0$. In this case the parabolic surface is entirely below the x,y plane and it has no zeros at all; and indeed, as we can see in Fig. 6.5(a), no Glass patterns are generated in this layer superposition. ∎

At first sight, this new point of view provides exactly the same information as the generalized indicial equations method that we have presented in Secs. 6.3–6.4, namely, the geometric location of the "skeleton" (the center line) of the Glass pattern in the grating superposition. However, its significant advantage is that it represents the phenomenon in question in the form of a continuous surface. This may give sense to points of the surface that are not precisely located at zero altitude, i.e. on the x,y plane, but only close to it (including *almost fixed points*; see Sec. 3.7). Note that the indicial equations method cannot take into consideration such points. Furthermore, if we consider this surface as the *transformation surface* of the resulting Glass pattern, it may also help us get an idea on the gray levels of the Glass pattern. For this end, we use an intensity profile such as:

$$p(x) = \tfrac{1}{2}e^{-2\pi fx^2} + \tfrac{1}{2} \tag{6.30}$$

that has the value 1 wherever $x = 0$ and gradually decays to $\tfrac{1}{2}$ (and hence gets darker) as x goes farther away from zero (see Fig. 6.15(a)); the value of the constant f determines the decaying rate of this profile. The actual gray-level approximation of the curvilinear Glass pattern in the grating superposition can be obtained, then, by replacing x in Eq. (6.30) with the actual difference function $g_{1,1}(x,y) - g_{2,1}(x,y)$, which is, in fact, the bending function of the Glass pattern:

$$r(x,y) = \tfrac{1}{2}e^{-2\pi f[g_{1,1}(x,y) - g_{2,1}(x,y)]^2} + \tfrac{1}{2} \tag{6.31}$$

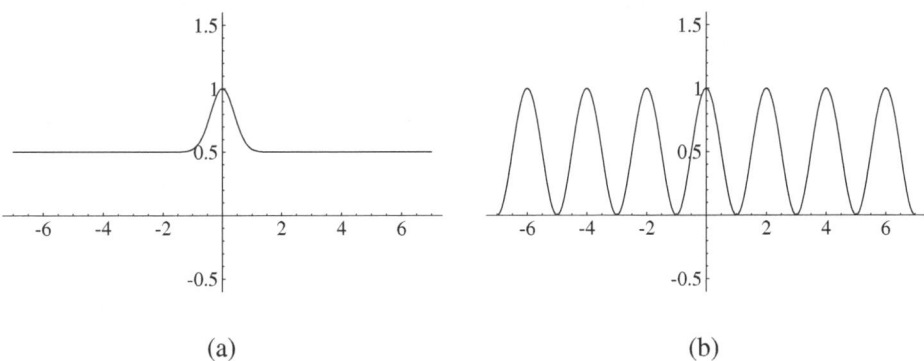

(a)   (b)

Figure 6.15: Typical intensity profiles (a) $p(x) = \frac{1}{2}e^{-2\pi fx^2} + \frac{1}{2}$ and (b) $p(x) = \frac{1}{2}\cos(2\pi fx) + \frac{1}{2}$ that can be used, respectively, for simulating the gray levels of a Glass pattern or a repetitive moiré pattern in the superposition of two gratings. The gray level surface $r(x,y)$ simulating the desired Glass or moiré pattern is obtained by replacing x by the actual difference function $g_{1,1}(x,y) - g_{2,1}(x,y)$.

And indeed, the surface defined by Eq. (6.31) can be used to simulate the gray level surface of the Glass pattern (compare, for example, Fig. 6.16(c) with Fig. 6.11(c)).

It should be noted, however, that the intensity profile thus obtained is only an artificial approximation, since the choice of the profile (6.30) is only based on our visual impression, but not on any solid mathematical considerations (see Problem 6-15). The exact intensity profile of the Glass pattern will be derived in Chapter 7, based on the Fourier theory.

Note that in a similar way we can also simulate the resulting moiré pattern in cases where the original line gratings are periodic: All that we have to do is to replace the aperiodic intensity profile (6.30) by a periodic one, such as:

$$p(x) = \tfrac{1}{2}\cos(2\pi fx) + \tfrac{1}{2} \qquad\qquad (6.32)$$

(Fig. 6.15(b)), where f is the frequency of the two original gratings. When we replace here x by the actual difference function $g_{1,1}(x,y) - g_{2,1}(x,y)$ we obtain the surface:

$$r(x,y) = \tfrac{1}{2}\cos(2\pi f[g_{1,1}(x,y) - g_{2,1}(x,y)]) + \tfrac{1}{2} \qquad\qquad (6.33)$$

which simulates the gray level values of the periodic (or repetitive) moiré pattern (compare, for example, Fig. 6.16(d) with Fig. 6.11(d)). Note that the skeleton of the Glass pattern remains the same as in Eq. (6.31); but the cosinusoidal range-transformation that is applied here to the surface $g_{1,1}(x,y) - g_{2,1}(x,y)$ gives, in addition, all of its repetitive replicas. For example, consider Figs. 6.3(d), 6.4(b) and 6.5(b), which show the periodic counterparts of Figs. 6.3(c), 6.4(a) and 6.5(a) that were discussed in Example 6.5 above.

Remark 6.3: Note that a curvilinear cosinusoidal grating $r(x,y) = \cos(2\pi fg(x,y))$, such as the parabolic cosinusoidal grating $r(x,y) = \cos(2\pi f[y - ax^2])$ (see Fig. 10.1(c) in *Vol. I*), can be interpreted in two different ways (see also Remark D.15 in Appendix D):

(a) Either as a straight cosinusoidal grating $\cos(2\pi fx')$ that has been distorted under the *domain transformation* $x' = g(x,y)$ into a curvilinear grating (in our example, a parabolic cosinusoidal grating);

(b) or as a surface $u = g(x,y)$ over the x,y plane that has undergone the *range transformation* $z = \cos(2\pi fu)$, turning it into an undulating surface whose values only vary within the vertical range of $-1...1$.

These two interpretations also apply to line gratings of any other intensity profiles, such as $r(x,y) = \frac{1}{2}\cos(2\pi fg(x,y)) + \frac{1}{2}$, black and white gratings, etc. In all cases both of the interpretations (a) and (b) are equally correct, and we can always choose the one that best suits our needs. For example, in our discussion above we have used point of view (a) for the interpretation of the two individual gratings $r_1(x,y)$ and $r_2(x,y)$ as geometrically transformed cosinusoidal surfaces, but we have used point of view (b) for the reconstruction (or simulation) of the gray level surfaces (6.31) and (6.33) of the Glass (or moiré) pattern based on its transformation surface $g_{1,1}(x,y) - g_{2,1}(x,y)$. ■

Having understood the new interpretation of Glass patterns between line gratings, we proceed now to the interpretation of Glass patterns between real 2D structures $p(x',y')$ such as dot screens. Unlike line gratings, such structures depend on both coordinates, x' and y', and when we apply to them a 2D transformation $(x',y') = \mathbf{g}(x,y) = (g_1(x,y),g_2(x,y))$ the resulting transformed structure is given by $r(x,y) = p(g_1(x,y),g_2(x,y))$. This means that the first coordinate of the structure has been affected by the bending function $x' = g_1(x,y)$ and its second coordinate has been affected by the bending function $y' = g_2(x,y)$. Because each of these two bending functions can be considered as a surface over the x,y plane, the 2D layer transformation $\mathbf{g}(x,y)$ is represented by *two* transformation surfaces.

What happens now in the superposition of two such transformed layers? As usual, we suppose that the two original layers, before having undergone the transformations, were superposed on top of each other in full coincidence. After the application of the transformations $(x',y') = \mathbf{g}_1(x,y) = (g_{1,1}(x,y),g_{1,2}(x,y))$ and $(x',y') = \mathbf{g}_2(x,y) = (g_{2,1}(x,y),g_{2,2}(x,y))$ the two original layers $p_1(x',y')$ and $p_2(x',y')$ become, respectively, $r_1(x,y) = p_1(g_{1,1}(x,y),g_{1,2}(x,y))$ and $r_2(x,y) = p_2(g_{2,1}(x,y),g_{2,2}(x,y))$.

Consider now the superposition of the two transformed layers. As we already know, the resulting Glass (or moiré) pattern in the superposition is determined by the difference transformation (6.27). This transformation can be interpreted as a pair of surfaces, $g_{1,1}(x,y) - g_{2,1}(x,y)$ and $g_{1,2}(x,y) - g_{2,2}(x,y)$, and Eq. (6.28), which gives the centers of the generated Glass patterns, can be unserstood, therefore, as the set of points in which both of these surfaces are simultaneously zero and cross the x,y plane.

For example, consider the dot screen superposition shown in Fig. 6.11(a) and the line grating superposition shown in Fig. 6.11(c). In both of these figures one of the two

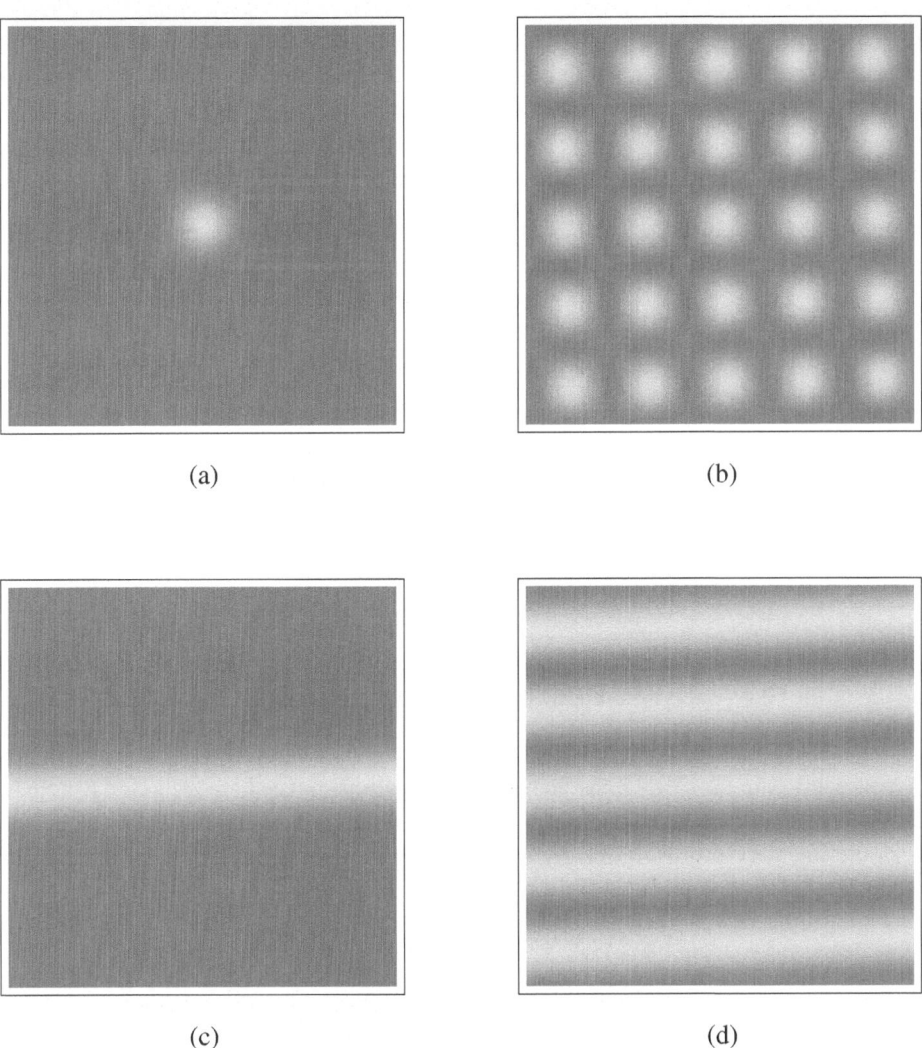

(a)   (b)

(c)   (d)

Figure 6.16: Simulation of the gray levels of the Glass and moiré patterns in the layer superpositions of Fig. 6.11, using the gray level surfaces defined by Eqs. (6.35) and (6.37) in cases (a) and (b), respectively, and the gray level surfaces defined by Eqs. (6.31) and (6.33) in cases (c) and (d), respectively.

originally identical layers has undergone the transformation $\mathbf{g}_1(x,y) = (x\cos\alpha + y\sin\alpha, -x\sin\alpha + y\cos\alpha)$ while the other has undergone the identity transformation $\mathbf{g}_2(x,y) = (x,y)$, meaning that it was left unchanged. In both figures the difference transformation is given by $\mathbf{g}_1(x,y) - \mathbf{g}_2(x,y) = (g_{1,1}(x,y) - g_{2,1}(x,y), g_{1,2}(x,y) - g_{2,2}(x,y))$. But while in the case of line gratings only the first component of this difference transformation influences the resulting

Glass (or moiré) pattern, in the 2D case both of the components influence the resulting effect. And indeed, as we can see in Eq. (6.28), a Glass pattern is generated here about the points where *both* of the difference surfaces $g_{1,1}(x,y) - g_{2,1}(x,y)$ and $g_{1,2}(x,y) - g_{2,2}(x,y)$ cross the x,y plane simultaneously; this happens at the points where $x\cos\alpha + y\sin\alpha = x$ and $-x\sin\alpha + y\cos\alpha = y$, which means, for $\alpha \neq 0$, $(x,y) = (0,0)$.

Now, in order to obtain an idea about the gray levels of the Glass pattern, we use a 2D generalization of the intensity profile (6.30), namely:

$$p(x,y) = \tfrac{1}{2}e^{-2\pi(f_1 x^2 + f_2 y^2)} + \tfrac{1}{2} \tag{6.34}$$

that has the value 1 wherever both $x = 0$ and $y = 0$, and gradually decays to $\tfrac{1}{2}$ (and hence gets darker) as x and y go farther from zero. The actual gray-level approximation of the Glass pattern in the superposition of the original layers after they have undergone the transformations $\mathbf{g}_1(x,y)$ and $\mathbf{g}_2(x,y)$, respectively, can be obtained, then, by replacing x and y in Eq. (6.34) with the actual difference functions $g_{1,1}(x,y) - g_{2,1}(x,y)$ and $g_{1,2}(x,y) - g_{2,2}(x,y)$, respectively:

$$r(x,y) = \tfrac{1}{2}e^{-2\pi(f_1[g_{1,1}(x,y) - g_{2,1}(x,y)]^2 + f_2[g_{1,2}(x,y) - g_{2,2}(x,y)]^2)} + \tfrac{1}{2} \tag{6.35}$$

And indeed, the surface defined by Eq. (6.35) can be used to simulate the gray level surface of the Glass pattern in the 2D case (compare, for example, Fig. 6.16(a) with Fig. 6.11(a)).

Finally, just as in the case of line gratings, we can also simulate here the moiré pattern in cases where the original layers are periodic: All that we have to do is to replace the aperiodic intensity profile (6.34) by a periodic one, such as the 2D generalization of (6.32):

$$p(x,y) = \left(\tfrac{1}{2}\cos(2\pi f_1 x) + \tfrac{1}{2}\right)\left(\tfrac{1}{2}\cos(2\pi f_2 y) + \tfrac{1}{2}\right) \tag{6.36}$$

where f_1 and f_2 are the frequencies in the x and y directions of both of the original layers. Replacing here x and y by the actual difference functions $g_{1,1}(x,y) - g_{2,1}(x,y)$ and $g_{1,2}(x,y) - g_{2,2}(x,y)$ we obtain the surface:

$$r(x,y) = \left(\tfrac{1}{2}\cos(2\pi f_1[g_{1,1}(x,y) - g_{2,1}(x,y)]) + \tfrac{1}{2}\right)\left(\tfrac{1}{2}\cos(2\pi f_2[g_{1,2}(x,y) - g_{2,2}(x,y)]) + \tfrac{1}{2}\right) \tag{6.37}$$

which simulates the gray level values of the periodic (or repetitive) moiré pattern (compare, for example, Fig. 6.16(b) with Fig. 6.11(b)).

It should be remembered, however, that simulations based on Eqs. (6.31), (6.33), (6.35) and (6.37) can only provide approximate results; the exact intensity profile of Glass patterns between line gratings or between dot screens will be derived in Chapter 7 using the Fourier theory.[11]

---

[11] An interesting variant based on line gratings where the intensity profile of each of the moiré bands (or of the linear Glass pattern) carries 2D information is provided in Sec. H.3 of Appendix H.

PROBLEMS

6-1. *Superposition of two identical line gratings having a slightly different scaling.* Fig. 6.1 shows the Glass (or moiré) patterns that are obtained when two identical line gratings (periodic or aperiodic) are superposed with a small angle difference. What would you expect to obtain if the same gratings (periodic or aperiodic) were superposed with a small *scaling* difference instead of the small *angle* difference? Compare with Figs. 2.1(c),(d) and 2.1(e),(f) that illustrate the 2D case.

6-2. *Linear Glass patterns as visual position indicators.* In various metrological applications it is customary to use moiré effects between two identical periodic line gratings, one grating serving as a fixed reference, and the other being attached to the object under observation. What should be the advantages and the drawbacks of using, instead, the linear Glass paterns that are generated between two identical random line gratings? *Hint:* Observing and tracking displacements of a moving periodic moiré in a conventional moiré-based system presents certain difficulties, as the individual moiré fringes are all similar, and they have no distinctive properties that allow us to identify and track a specific fringe within the moving periodic moiré pattern. The use of random line gratings instead of periodic ones solves this problem, since it generates a single fringe (linear Glass pattern), that is easier to track in the superposition [Vargady64]. However, this risks to be a double-edged argument, since once the Glass pattern has moved outside the superposed area, it is not always obvious to find where it has gone and to bring it back; and in the mean time we do not have any other moiré fringes to observe and track. How can we combine the advantages of systems having a unique fringe position indicator with the advantages of systems having periodic fringes?

6-3. What do you expect to obtain in Fig. 6.1 if you replace in each of the superpositions one of the two line gratings by its own negative? What could be the advantages of such cases for practical applications that are based on visual observation of the moiré? (See [Vargady64]).

6-4. *1D equivalent of dot trajectories?* How do you explain the blurred, nebulous structures that occur along the x axis in the superpositions shown in Figs. 6.4(a),(c), 6.5(a) and 6.6(a),(c)? Compare these superpositions with their 2D counterparts that are shown in Figs. 3.6(a),(c), 3.7(a) and 3.13(a),(c), respectively. Can we say that these nebulous structures are the 1D counterparts of the dot trajectories that appear in 2D cases?

6-5. Establish the equations whose solutions give the geometric locations of the Glass patterns in the dot-screen superposition of Fig. 6.11(a) and in the corresponding line-grating superposition of Fig. 6.11(c). Show how Remark 6.1 explains mathematically the relationship between the two cases. *Hint:* Eq. (6.10), whose solutions give the location of the Glass pattern in the line-grating superposition, is precisely the first equation of the system (6.8), whose solution gives the location of the Glass pattern in the dot-screen superposition. This implies that the fixed points (the locations of the Glass patterns in the dot-screen superposition) are a subset of the Glass patterns in the line-grating superposition: Remember that the solution of a system of equations is simply the intersection of the solutions of the individual equations.

6-6. Establish the equations whose solutions give the geometric locations of the Glass patterns for the case of Figs. 6.12(a) and 6.12(c) (see also Example 6.2), and for the case of Figs. 6.13(a) and 6.13(c) (see also Example 6.1). How do you explain the fact

	Aperiodic line gratings	Periodic line gratings
Macro-structure	**Glass patterns:** Their geometry is given by the locus of the *points of coincidence* between the two superposed gratings, namely, the points (x,y) that satisfy: $$g_{1,1}(x,y) - g_{2,1}(x,y) = 0$$ Invariant under the addition of the same $f(x,y)$ to both bending functions $g_{1,1}(x,y)$ and $g_{2,1}(x,y)$.	**First-order moiré patterns:** Determined by the difference between the bending functions undergone by the two originally identical, periodic line gratings: $$g_M(x,y) = g_{1,1}(x,y) - g_{2,1}(x,y)$$ Their geometry is given by the locus of the points of coincidence between the two distorted gratings: $$g_{1,1}(x,y) - g_{2,1}(x,y) = p \quad p \in \mathbb{Z}$$ Invariant under the addition of the same $f(x,y)$ to both bending functions $g_{1,1}(x,y)$ and $g_{2,1}(x,y)$.

Table 6.1: Summary of the main properties of the macrostructures that are generated in the superposition of periodic or aperiodic line gratings (compare with Table 5.1 that is dedicated to the case of dot screens). Note that the intensity profiles of the periodic and aperiodic macrostructures are not included in this table. They will be studied separately in Chapter 7, using the Fourier based approach.

that in the last case the Glass patterns have the same geometric shape in the dot screen superposition (a) and in the line grating superposition (c)?

6-7. Fig. 3.17 shows a Glass pattern consisting of a family of concentric circles, that is generated in the superposition of two aperiodic dot screens. How can you generate a similar Glass pattern in the superposition of two aperiodic line gratings? *Hint*: You may transform both aperiodic line gratings into concentric circular gratings and consider their difference transformation in terms of polar coordinates. Can you also obtain a similar Glass pattern between two aperiodic line gratings that consist of essentially vertical lines?

6-8. *Synthesis of a given Glass pattern between two aperiodic line gratings.* Design a method for synthesizing two aperiodic line gratings that give in their superposition a Glass pattern with a predefined shape. Is the solution unique? How many different pairs of transformed line gratings may give the same desired Glass pattern in their superposition?

6-9. *Synthesis of a given Glass pattern between two aperiodic line gratings (continued).*
(a) Design layer transformations $\mathbf{g}_1(x,y)$ and $\mathbf{g}_2(x,y)$ that produce in the superposition of two initially identical aperiodic line gratings a Glass pattern consisting of horizontal periodic bands, as shown in Fig. 6.17(a). *Hint*: You may try the transformations $\mathbf{g}_1(x,y) = (x, y + a\cos(2\pi f x))$ and $\mathbf{g}_2(x,y) = (x,y)$; note that $\mathbf{g}_1(x,y)$ is a non-linear variant of horizontal shearing.

(b) Do you see any contradiction in the fact that two *aperiodic* layers generate in their superposition a *periodic* Glass pattern? Explain.

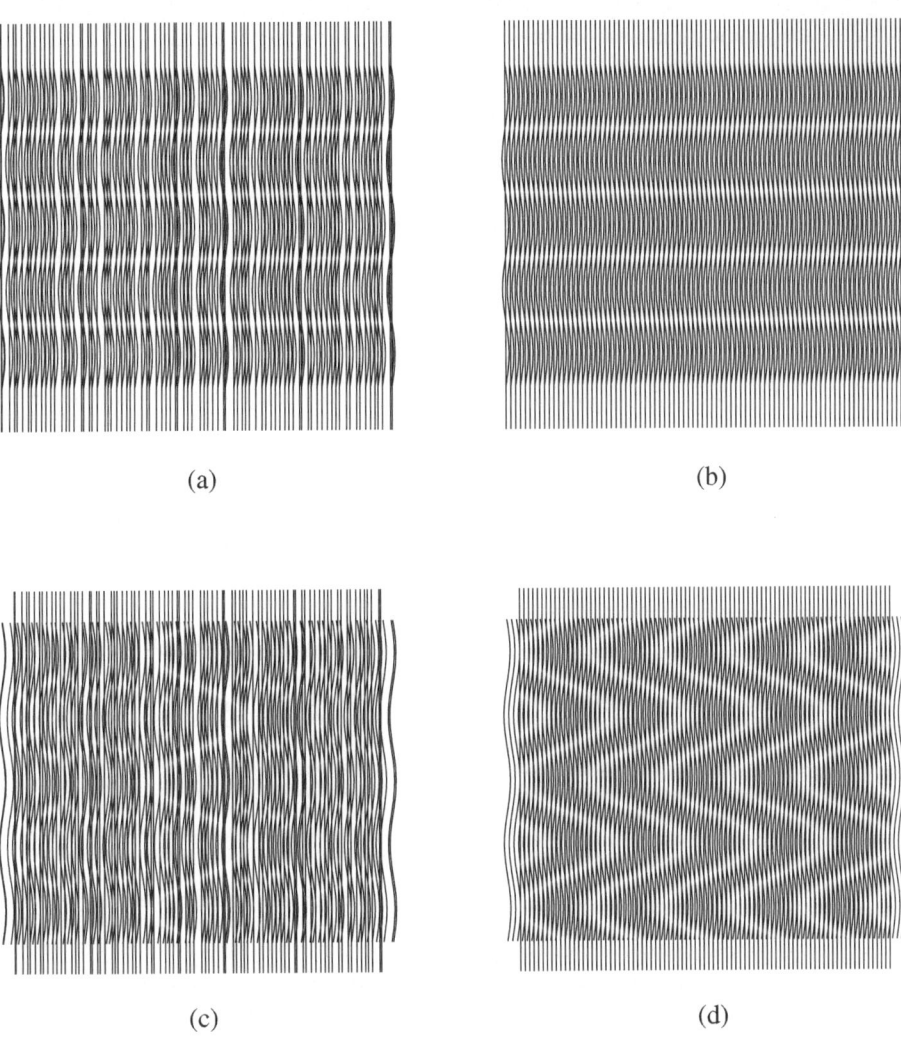

(a)   (b)

(c)   (d)

Figure 6.17: (a) The superposition of two identical aperiodic line gratings one of
which has undergone a slight cosinusoidal shear transformation.
(b) The periodic counterpart of (a). (c),(d) Same as (a),(b) but here
the untransformed layer has undergone a slight scale-down transformation.

(c) Can you design layer transformations $g_1(x,y)$ and $g_2(x,y)$ that give in the superposition of two vertical aperiodic line gratings a Glass pattern consisting of *vertical*
rather than horizontal periodic bands?

6-10. Figs. 6.17(c),(d) show what happens in the superpositions shown in Figs. 6.17(a),(b)
when the untransformed line grating is slightly scaled down. How do you explain the
phenomena that you observe in the periodic and in the aperiodic cases? Why does the
Glass pattern in Fig. 6.17(a) consist of infinitely many bands, while in Fig. 6.17(c),

after the application of a slight scaling to one of the superposed layers, the Glass pattern only consists of a single curve? What would you expect to see in Figs. 6.17(c),(d) if the untransformed layer were slightly scaled up rather than slightly scaled down? See also Fig. 11.3 and Problem 11-6 in *Vol. I*.

6-11. *Latent images in the superposition of aperiodic line gratings.* Fig. 7.8 and Problem 7-8 in *Vol. I* describe a method for generating latent images in the superposition of two identical *periodic* line gratings. The latent image becomes visible in the superposition thanks to the introduction of a small shift (of up to half a period) in one of the gratings within the borders of the desired image. Can you design a similar method for generating latent images in the superposition of *aperiodic* line gratings? How do the properties of such a latent image differ from those of a Glass pattern between two line gratings? *Hint*: What happens to the latent image and to the Glass pattern while you slowly shift one of the superposed gratings along its main direction? And while you slowly rotate it?

6-12. Table 6.1 summarizes the different macroscopic phenomena that may occur in superpositions of two originally identical periodic or aperiodic line gratings, and their behaviour under layer transformations. This can be considered as an extension to Table 5.1 that was dedicated in Chapter 5 to 2D cases alone, such as the superposition of two dot screens. Compare Table 6.1 with Table 5.1 and explain the main differences between them. In particular:

(a) Why are Glass patterns in the 2D case invariant under *composition*, while their 1D counterparts are not?

(b) What can you say about the *microstructures* in the superposition of line gratings?

6-13. Parts (a) and (b) of Figs. 6.11, 6.12 and 6.13 show the effects of adding the constraint of 2D periodicity to originally aperiodic dot screens. What would you expect to see in each of these figures in the intermediate case in which the superposed aperiodic dot screens undergo the weaker constraint of 1D periodicity, i.e. periodicity along only one direction?

6-14. What would you expect to see in Figs. 6.11, 6.12 and 6.13 if the constraints they illustrate (constancy along one direction or periodicity) were applied to only one of the two superposed layers?

6-15. As we have seen in Sec. 6.6, the simulation of the gray levels of a Glass pattern using the intensity profile (6.30) is only an artificial, arbitrary approximation. Although in many cases this simulation gives quite good results (compare, for example, Figs. 6.16 and 6.11), in other cases it does not correspond well to the real gray levels of the Glass pattern. Can you think of such cases? *Hint*: Consider cases in which the profiles of the two original layers are very different. This is explained and illustrated in Chapter 7, where the *exact* intensity profile of the Glass pattern is derived, using the Fourier theory, and is shown to be a normalized cross correlation of the profiles of the two original layers.

6-16. Explain the various Glass and moiré patterns that are generated in Fig. 6.12 using the method described in Sec. 6.6, following Example 6.5. Plot the gray level surfaces that simulate these Glass and moiré patterns, and compare them to the gray levels that you observe in the real layer superpositions.

6-17. Redo the last problem, considering this time Fig. 6.13. How do you explain the fact that in this case, unlike in Fig. 6.12, the phenomena observed in (a) and (b) are not different from those observed in (c) and (d), respectively?

Chapter 7

Quantitative analysis and synthesis of Glass patterns

7.1 Introduction

In all of the Glass patterns that we have studied so far (with only a few exceptions in Sec. 4.8) the two original aperiodic layers, before the application of the transformations, were assumed to be identical (see Remark 3.1); in particular, they were assumed to have the same element shapes, usually black elements (dots or lines) on a white background. As we have seen, this gives Glass patterns that are brighter in their center, because in that area the black elements from both layers fall almost exactly on top of each other. However, it turns out that by properly choosing the dot shapes of the two superposed aperiodic screens one may obtain in the superposition a Glass pattern of any desired intensity profile, as illustrated in Fig. 7.1(a). This may remind us of a similar result that is already known for periodic dot screens (see Fig. 7.1(b)). But while in the superposition of periodic screens the resulting moiré profile is periodically repeated throughout the super-position, in the present case the Glass pattern consists of only one moiré profile (compare Figs. 7.1(a) and (b)). This surprising phenomenon was already known to Joe Huck in his artistic work since the late 1970s [Walker80a, Huck03], but it had to wait until its rediscovery in [Amidror03] before it was investigated and fully understood [Amidror03c].

Our aim in the present chapter is to provide a full qualitative and quantitative theoretical explanation of Glass patterns in general, and of this surprising phenomenon in particular. However, such questions cannot be treated using the mathematical tools that we have presented so far in the previous chapters, since these tools are not adapted for a quantitative treatment of Glass patterns in terms of their gray level variations (intensity profiles). All that we know for the time being is that in the superposition of originally identical aperiodic layers the resulting Glass pattern is brighter in its center (see, for example, Figs. 2.1(c),(e) and (g)); but this result is only qualitative, and we cannot yet confirm it quantitatively. As we already know from the moiré theory between *periodic* layers, quantitative information can be best obtained using the Fourier-based approach (see, for example, Chapters 2 and 4 in *Vol. I*). Therefore, based on this previous knowledge, we try in the present chapter to generalize the Fourier-based approach into the case of aperiodic layers. And indeed, it turns out that this approach will allow us to predict quantitatively the intensity profile of the resulting Glass patterns, and furthermore, it will also show us how to synthesize Glass patterns having any desired intensity profiles.

We start our work in Sec. 7.2 with a short review of the Fourier-based approach in the periodic case. This section gives the basic notions that are required for understanding the rest of the chapter, but it can be skipped by readers who are already familiar with this material. In order to see how the Fourier-based approach can be extended into the aperiodic case, we first consider in Sec. 7.3 the superposition of aperiodic line gratings, which is much simpler to understand, and only then, armed with these results, we proceed in Sec. 7.4 to the case of aperiodic screens. In both cases we analyze the resulting Glass patterns and their intensity profiles both in the image domain and in the Fourier domain, as a generalization of the moiré theory between periodic layers. These considerations also lead us to the synthesis of Glass patterns having intensity profiles of any desired shapes. Then, in Sec. 7.5 we explain why, in stark contrast to the periodic case, higher-order moirés cannot exist in the superposition of aperiodic layers. In Sec. 7.6 we deepen our understanding of the connection between periodic and aperiodic cases by explaining what happens in hybrid cases, where the participating layers are intermediate between fully periodic and fully random. In Sec. 7.7 we see what happens when the superposed layers are only partly correlated, and in Sec. 7.8 we explain the connection between the Glass patterns in the superposition and the cross correlation between the two original layers.

It should be noted, however, that in spite of its extreme usefulness, the Fourier approach is only adapted to the investigation of global, macroscopic phenomena, and it cannot be used in the study of microstructure properties such as dot trajectories. The reason is that Fourier considerations only treat the global aspects of the structures, but they do not go down to the level of individual screen dots and their local behaviour. Therefore the approaches described here and in Chapter 4 remain complementary, and they can be used either independently of each other or combined. We will return to this point in Sec. 7.4.6.

7.2 Brief review of the Fourier approach in the periodic case

As we already know from our experience in the periodic case (see *Vol. I*), spectral domain considerations prove to be extremely helpful in the understanding of the moiré effects between periodic or repetitive layers. They open the way to many new important insights into the phenomena in question, and they provide powerful tools for analyzing them both qualitatively and quantitatively. It is but natural, therefore, to ask if spectral domain considerations can also explain moiré phenomena in our more general case, i.e. in the superposition of arbitrary, not necessarily periodic layers. As we will see below, this generalization into the aperiodic case is not really trivial, because the Fourier approach that was developed for the periodic case highly depends on properties of periodic layers, such as their periods and frequencies, properties that do not exist in the case of aperiodic layers.

One of the most fundamental results of the Fourier theory is the convolution theorem (see Eqs. (2.1) and (2.2) in Chapter 2). Being independent of the individual layer structures, this theorem remains true in the aperiodic case, too. This means that even in the aperiodic case the spectrum of the superposition is a convolution of the spectra of the individual layers. However, unlike in the case of periodic layers, the spectra in aperiodic cases are no longer impulsive but rather continuous (or even diffuse, in stochastic layers such as random dot screens). Such spectra are clearly much more difficult to manage and

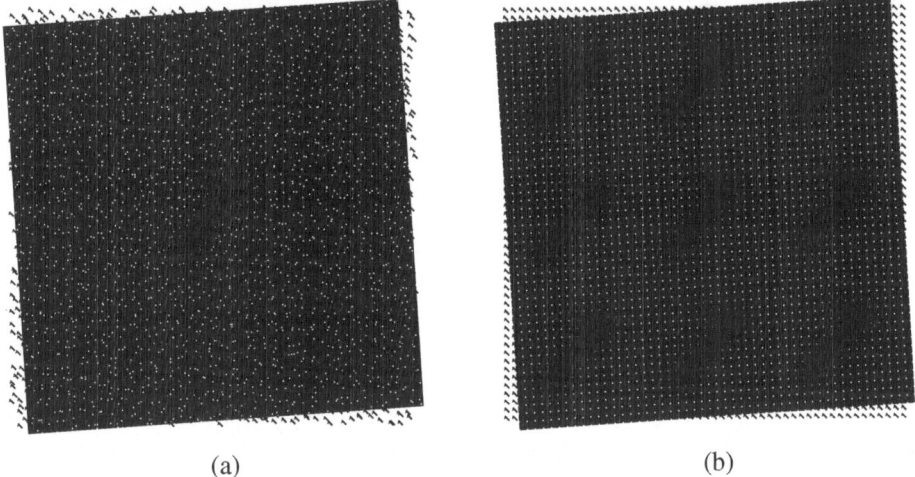

(a)   (b)

Figure 7.1: (a) The superposition with a small angle difference of two random dot screens having the same dot locations, one consisting of "1"-shaped dots and the other consisting of small white dots (pinholes) on a black background, gives a single "1"-shaped moiré intensity profile (Glass pattern). (b) The periodic counterpart of (a): The superposition with a small angle difference of a periodic dot screen consisting of "1"-shaped dots and a periodic dot screen consisting of small white dots (pinholes) on a black background gives a periodic "1"-shaped moiré intensity profile.

to interpret. For example, it is no longer obvious to identify in the spectrum of the super-position the contribution of each individual layer. Moreover, although the spectral contri-bution of the moiré (or Glass) pattern is certainly present in the spectrum of the superpo-sition, it can no longer be isolated and extracted as easily as in the periodic case. For these reasons, it was long believed that the use of the spectral approach in aperiodic cases was rather hopeless, or at least not as practical as it proved to be in the periodic case.

However, more recent investigations [Amidror03c] showed that in spite of these difficulties Glass patterns between aperiodic layers can be analyzed using a judicious extension of the same Fourier-based theory that governs the classical moiré patterns between periodic layers. Surprisingly, it turns out that even spectral domain considerations can be extended in a natural way to the aperiodic case, with just a few straightforward adaptations. The main underlying idea here is to consider the spectra in question as continuous extensions of their periodic counterparts by "filling the gaps" between the discrete impulses of the periodic spectra. And indeed, as shown below, this approach proves to be very fruitful: Not only it allows us to predict quantitatively the intensity profile of the resulting Glass patterns, but moreover, it also opens the way to the synthesis of Glass patterns having any desired shapes and intensity profiles.

But before we continue, we first review in the rest of this section the basic notions and terminology that are required for the understanding of our new results. For more details and illustrated examples the reader is referred to Chapters 2 and 4 of *Vol. I*. Readers who are already familiar with this material may skip directly to Sec. 7.3.

7.2.1 Spectra of periodic and aperiodic layers

Although the structures that we study in the present chapter are not periodic, it would be helpful to start with a short reminder from the periodic case, whose mathematical behaviour is already fully understood (see *Vol. I*). Suppose we are given a *periodic* image defined on the continuous x,y plane, such as a line grating or a dot screen. The spectrum of a periodic image in the u,v frequency plane is not continuous, and it rather consists of impulses corresponding to the frequencies which appear in the Fourier series decomposition of the image. In the case of a 1-fold periodic image, such as a line grating, the spectrum consists of a 1D "comb" of impulses centered about the origin; in the case of a 2-fold periodic image the spectrum is a 2D "nailbed" of impulses centered about the origin. Note that we sometimes use the more general term "cluster" for a comb or a nailbed.

Each impulse in the 2D spectrum is characterized by its *geometric location* and its *amplitude*. The geometric location of an impulse is represented by a *frequency vector* $\mathbf{f}$ in the spectrum plane, which connects the spectrum origin to the geometric location of the impulse. This vector can be expressed by its polar coordinates (f,θ), where θ is the direction of the impulse and f is its distance from the origin (i.e. its frequency in that direction). In terms of the original image, the geometric location of an impulse in the spectrum determines the frequency f and the direction θ of the corresponding periodic component in the image, and the amplitude of the impulse represents the intensity of that periodic component in the image. (Note that if the original image is not symmetric about the origin, the amplitude of each impulse in the spectrum may also have a non-zero imaginary component.)

However, the question of whether an impulse in the spectrum represents a *visible* periodic component in the image depends strongly on properties of the human visual system. The fact that the eye cannot distinguish fine details above a certain frequency (i.e. below a certain period) suggests that the human visual system model includes a low-pass filtering stage. For the sake of simplicity, this low-pass filter can be approximated by the *visibility circle,* a circular step function around the spectrum origin whose radius represents the *cutoff frequency* (i.e. the threshold frequency beyond which fine detail is no longer detected by the eye). Obviously, the radius of the visibility circle depends on several factors such as the contrast of the observed details, the viewing distance, light conditions, etc. If the frequencies of the original image elements are beyond the border of the visibility circle in the spectrum, the eye can no longer see them; but if a strong enough impulse in the spectrum of the image superposition falls inside the visibility circle, then a moiré effect becomes visible in the superposed image.

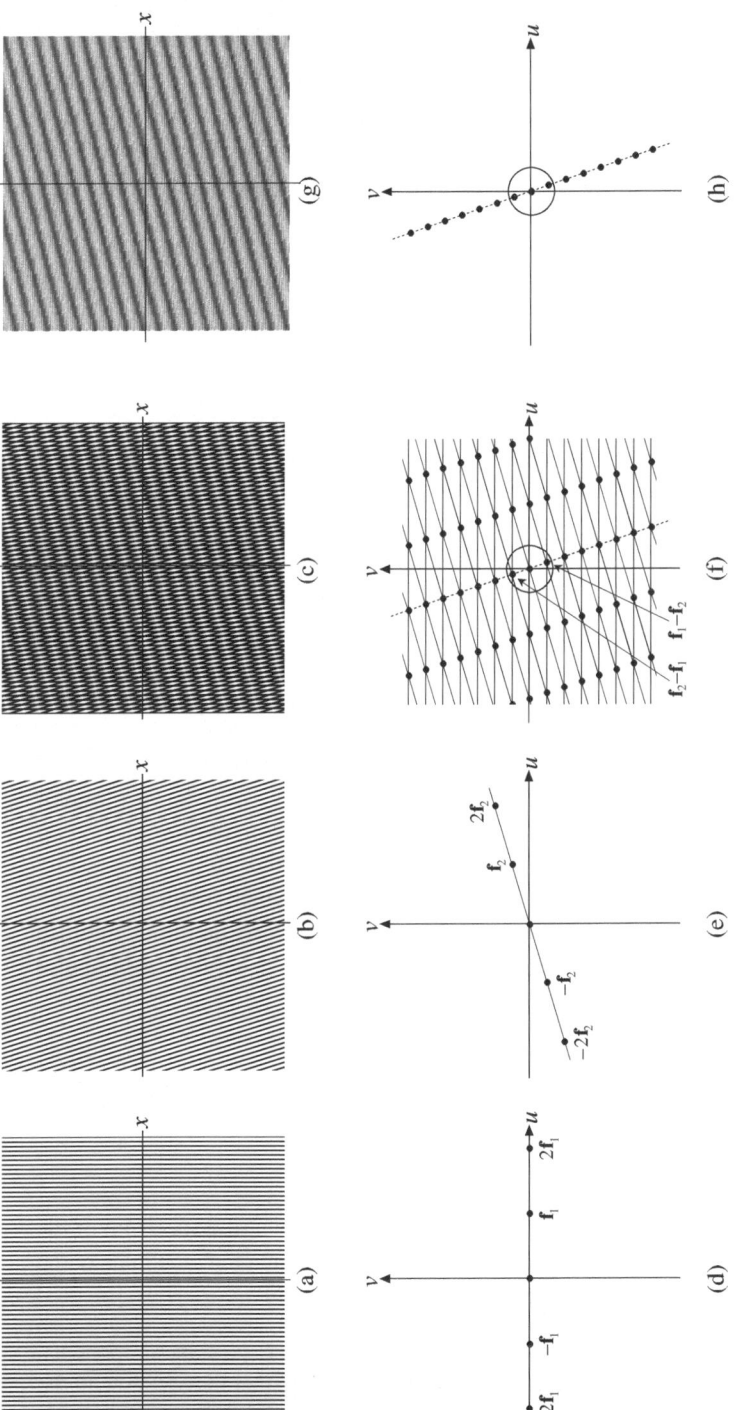

Figure 7.2: Periodic line gratings (a) and (b) and their superposition (c) in the image domain; their respective spectra are the infinite impulse combs shown in (d) and (e) and their convolution (f). The circle in the center of the spectrum (f) represents the visibility circle. It contains the impulse pair whose frequency vectors are $\mathbf{f}_1 - \mathbf{f}_2$ and $\mathbf{f}_2 - \mathbf{f}_1$ and whose indices are $(1,-1)$ and $(-1,1)$; this is the fundamental impulse pair of the $(1,-1)$-moiré seen in (c). The dotted line in (f) indicates the infinite impulse comb that represents this moiré. This $(1,-1)$-moiré comb is shown isolated in (h), after its extraction from the spectrum convolution (f). The impulse amplitudes of this comb are the term-by-term products of the respective impulse amplitudes from the combs (d) and (e) taken head to tail. In (g) is shown the image-domain function that corresponds to the spectrum (h); this is the intensity profile of the $(1,-1)$-moiré shown in (c). Black dots in the spectra indicate the impulse locations; the straight lines connecting them have been added only to clarify the geometric relations. Impulse amplitudes are not shown.

Finally, a short reminder about the Fourier spectra of aperiodic layers. Unlike the spectra of periodic layers, which are purely impulsive, the Fourier transform of an aperiodic layer is basically continuous. For example, the Fourier transform of a unit cube is a 2D sinc function; see, for example, [Bracewell95 pp. 150–151]. But when the layer's structure is very complex, as in the case of a random dot screen, its Fourier spectrum becomes very jumpy or noisy and has a typical *diffuse* appearance (see [Bracewell95 pp. 586–590; 600–601]). This is further discussed and illustrated in Problems 2-1 to 2-10.

7.2.2 Moiré effects in the superposition of periodic gratings

The simplest moiré effects occur in the superposition of two straight periodic gratings, as shown in Fig. 7.2. Let $r_1(x,y)$ and $r_2(x,y)$ be such periodic gratings; their frequencies and orientations are given by their frequency vectors $\mathbf{f}_1$ and $\mathbf{f}_2$, respectively. The spectrum $R_i(u,v)$ of each of the original gratings consists of a comb whose impulses are located at integer multiples of the fundamental frequency, $n\mathbf{f}_i$:

$$R_1(u,v) = \sum_{n=-\infty}^{\infty} a^{(1)}_n \, \delta_{n\mathbf{f}_1}(u,v) \qquad (7.1)$$

$$R_2(u,v) = \sum_{n=-\infty}^{\infty} a^{(2)}_n \, \delta_{n\mathbf{f}_2}(u,v) \qquad (7.2)$$

Here, $\delta_{n\mathbf{f}_i}(u,v)$ denotes an impulse located in the spectrum at the frequency vector $n\mathbf{f}_i$, and $a^{(i)}_n$ is its amplitude.

When we superpose (i.e., multiply) the line gratings $r_1(x,y)$ and $r_2(x,y)$, the spectrum of the superposition is, according to the convolution theorem, the convolution of the two original combs, $R_1(u,v)**R_2(u,v)$, which gives an oblique nailbed of impulses (see Fig. 7.2(f)). This convolution of combs can be seen as an operation in which frequency vectors from the individual spectra are added vectorially, while the corresponding impulse amplitudes are multiplied. Therefore the geometric location of the general (k_1,k_2)-impulse in the spectrum convolution is expressed by the vectorial sum:

$$\mathbf{f}_{k_1,k_2} = k_1\mathbf{f}_1 + k_2\mathbf{f}_2 \qquad (7.3)$$

and its amplitude is expressed by:

$$a_{k_1,k_2} = a^{(1)}_{k_1} a^{(2)}_{k_2} \qquad (7.4)$$

where $\mathbf{f}_i$ denotes the frequency vector of the fundamental impulse in the spectrum of the i-th grating, and $k_i\mathbf{f}_i$ and $a^{(i)}_{k_i}$ are, respectively, the frequency vector and the amplitude of the k_i-th harmonic impulse in the spectrum of the i-th grating.

The vectorial sum of Eq. (7.3) can also be written in terms of its Cartesian components. If f_i are the frequencies of the original gratings and θ_i are the angles that they form with the positive horizontal axis, then the coordinates (f_u, f_v) of the general (k_1,k_2)-impulse in the spectrum convolution are given by:

$$f_{u\,k_1,k_2} = k_1 f_1 \cos\theta_1 + k_2 f_2 \cos\theta_2$$
$$f_{v\,k_1,k_2} = k_1 f_1 \sin\theta_1 + k_2 f_2 \sin\theta_2$$
(7.5)

Therefore the frequency, the period and the angle of the general impulse are given by the length and the direction of the vector $\mathbf{f}_{k_1,k_2}$ as follows:

$$f = \sqrt{f_u^{\,2} + f_v^{\,2}} \qquad T = 1/f \qquad \varphi = \arctan(f_v/f_u)$$
(7.6)

Now, if one of the new impulses in the spectrum convolution, say, the (k_1,k_2)-impulse, falls (together with its symmetric twin) close to the origin, inside the visibility circle, this implies the existence in the superposed image of a moiré effect with a visible period. This moiré is represented in the spectrum convolution by a full comb of impulses, centered on the origin, which contains the (k_1,k_2)-impulse as well as all its harmonics. We call this moiré a (k_1,k_2)-moiré since the fundamental impulse of its comb is the (k_1,k_2)-impulse of the spectrum convolution. For example, in the case of Fig. 7.2(f), the $(1,-1)$-impulse (as well as its symmetric $(-1,1)$-impulse) falls inside the visibility circle; this indicates that the moiré effect that is clearly visible in the grating superposition (Fig. 7.2(c)) is the first-order $(1,-1)$-moiré. The frequency and the period of this moiré can be found from Eqs. (7.5) and (7.6) using $k_1 = 1$ and $k_2 = -1$; equivalently, they can be found by a simple geometric consideration, as shown in Fig. 7.3 for the case with $f_1 = f_2$:

$$\sin(\alpha/2) = \frac{f_M/2}{f}$$

whence: $f_M = 2f \sin(\alpha/2)$
(7.7)

and thus: $T_M = \dfrac{1}{2f \sin(\alpha/2)} = \dfrac{T}{2\sin(\alpha/2)}$
(7.8)

The orientation of the moiré in this case is perpendicular to the bisector of the original gratings.

These equations can be generalized to the case in which the original gratings have different frequencies $f_1 \neq f_2$; in this case the respective formulas are (see Sec. C.1 of Appendix C in *Vol. I*):

$$f_M = \sqrt{f_1^2 - 2f_1 f_2 \cos\alpha + f_2^2}$$
(7.9)

and thus: $T_M = \dfrac{T_1 T_2}{\sqrt{T_1^2 + T_2^2 - 2T_1 T_2 \cos\alpha}}$
(7.10)

and the moiré orientation is:

$$\varphi_M = \arctan\left(\frac{T_2 \sin\theta_1 - T_1 \sin\theta_2}{T_2 \cos\theta_1 - T_1 \cos\theta_2}\right)$$
(7.11)

Returning to the general (k_1,k_2)-moiré in the superposition of two gratings, the location $\mathbf{f}_{k_1,k_2}$ of its fundamental impulse in the spectrum is given by Eq. (7.3). The n-th impulse of

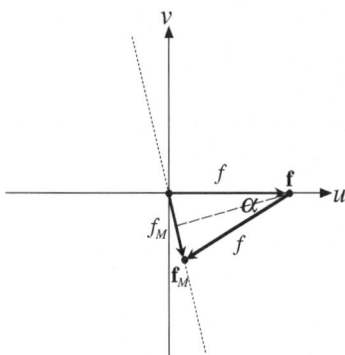

Figure 7.3: Geometric consideration in the spectral domain for finding the frequency f_M and the period T_M of the (1,-1)-moiré effect between two gratings with identical frequencies f and an angle difference of α. The dotted line indicates the infinite impulse comb that represents the (1,-1)-moiré (see Fig. 7.2(f)); the fundamental impulse of this moiré has the frequency f_M.

the comb of this moiré is the (nk_1,nk_2)-impulse in the spectrum convolution; its location is $n\mathbf{f}_{k_1,k_2}$, and its amplitude d_n is given by $d_n = a_{nk_1, nk_2}$, which means, according to Eq. (7.4):

$$d_n = a^{(1)}{}_{nk_1}\, a^{(2)}{}_{nk_2} \tag{7.12}$$

where $a^{(1)}{}_i$ and $a^{(2)}{}_i$ are the respective impulse amplitudes from the combs of the first and of the second line gratings. For example, in the case of the simplest first-order moiré between two gratings, the (1,-1)-moiré (Fig. 7.2(f)), the amplitudes of the moiré-comb impulses are:

$$d_n = a^{(1)}{}_n\, a^{(2)}{}_{-n}. \tag{7.13}$$

In other words, we can say:

Proposition 7.1: (Reformulation of Proposition 4.1 in *Vol. I*): The impulse amplitudes of the comb of the (1,-1)-moiré in the spectrum convolution are obtained by a term-by-term multiplication of the combs of the original superposed gratings, one of which is being inverted (rotated by 180°) before the multiplication. ∎

The moiré comb can be considered, in fact, as a product of the two original combs, after they have been normalized (rotated and stretched) to fit the impulse locations of the resulting moiré comb. This moiré comb can be easily extracted from the spectrum convolution, as shown in Fig. 7.2(h). Thus, by taking its inverse Fourier transform, we can reconstruct, back in the image domain, the isolated contribution of the moiré in question to the image superposition; this is the intensity profile of the moiré (see Fig. 7.2(g)). Note that this moiré is visible both in the layer superposition (Fig. 7.2(c)) and in the extracted moiré intensity profile (Fig. 7.2(g)); and yet, the latter does not contain the fine structure of the original layers $r_1(x,y)$ and $r_2(x,y)$ but only the contribution of the moiré itself.

However, this term-by-term multiplication of the original combs, as defined by Eq. (7.13), can be also interpreted, back in the image domain, using the T-convolution theorem, which is the periodic counterpart of the convolution theorem. The full details can be found in Sec. 4.2 of *Vol. I*; here we will only give the following result for the (1,-1)-moiré (as an illustration, refer to Fig. 7.2):

Proposition 7.2: (Reformulation of Proposition 4.2 in *Vol. I*): The intensity profile of the (1,-1)-moiré that is generated in the superposition of two periodic line gratings with periods T_1 and T_2 and an angle difference α can be seen from the image-domain point of view as the result of a 3-stage process:

(1) Normalization of the original gratings (by linear stretching and rotation transformations) in order to bring both of them to a common period and orientation (that is, to make their periods, or their impulse combs in the spectrum, coincide).

(2) T-convolution of the two normalized line gratings.

(3) Stretching and rotating the resulting normalized moiré intensity profile into its actual scale and orientation, as determined by Eqs. (7.10) and (7.11). ∎

Thus, while the period and the orientation of the (1,-1)-moiré bands are determined by Eqs. (7.10) and (7.11), their intensity profile is governed by Proposition 7.2. Note that in the particular case where $T_1 = T_2$ and $\theta_1 \approx \theta_2$ the (1,-1)-moiré bands are approximately perpendicular to the original gratings, and their period is given by Eq. (7.8).

In conclusion, the T-convolution theorem allows us to express the moiré profile that we have extracted in the spectral domain in terms of the image domain, too. The importance of the image-domain interpretation of the moiré profile as a T-convolution will become clearer in the following; in essence, it allows us to express the 1D profile shape of the resulting moiré bands in terms of the 1D profile shapes of the individual lines of the two original gratings. An interesting variant in which the intensity profile of the (1,-1)-moiré bands may carry 2D rather than 1D information is provided in Sec. H.3 of Appendix H.

7.2.3 Moiré effects in the superposition of periodic dot screens

The generalization of the above results into the 2D case of periodic dot screens is of particular importance for our needs. We therefore summerize below the main results that will be needed later. The full developments can be found in Chapter 4 of *Vol. I*.

Let $r_1(x,y)$ be a periodic dot screen whose frequencies and orientations are given by the two perpendicular frequency vectors $\mathbf{f}_1$, $\mathbf{f}_2$; and let $r_2(x,y)$ be a second periodic dot screen whose frequencies and orientations are given by the two perpendicular frequency vectors $\mathbf{f}_3$, $\mathbf{f}_4$. The spectrum $R_1(u,v)$ of the screen $r_1(x,y)$ consists of a 2D impulse nailbed, whose impulses are located at integer linear combinations of the two fundamental frequency vectors of $r_1(x,y)$, $m\mathbf{f}_1 + n\mathbf{f}_2$. Similarly, The spectrum $R_2(u,v)$ consists of a 2D nailbed whose (m,n)-th impulse is located at the integer linear combinations $m\mathbf{f}_3 + n\mathbf{f}_4$. The amplitude of the (m,n)-th impulse of the i-th nailbed is denoted by $a^{(i)}_{m,n}$.

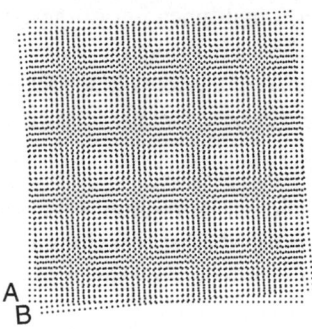

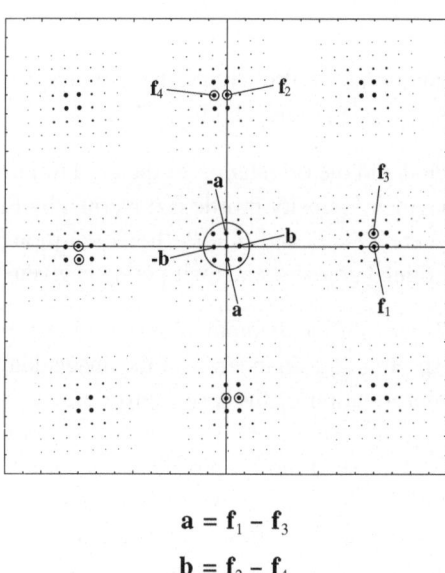

$$a = f_1 - f_3$$
$$b = f_2 - f_4$$

Figure 7.4: The superposition of two dot screens with identical frequencies and with an angle difference of $\alpha = 5°$ (top), and the corresponding spectrum (bottom). Only impulse locations are shown in the spectrum, but not their amplitudes. Encircled dots denote the locations of the fundamental impulses of the two original dot screens. Large dots represent convolution impulses of the first order (i.e., (k_1,k_2,k_3,k_4)-impulses with $k_i = 1$, 0, or −1); smaller dots represent convolution impulses of higher orders. (Note that only impulses of the first few orders are shown; in reality, each impulse cluster extends in all directions *ad infinitum*.) The circle around the spectrum origin represents the visibility circle. Note that the spectrum origin is closely surrounded by the impulse cluster of the (1,0,-1,0)-moiré.

Assume now that we superpose (i.e., multiply) $r_1(x,y)$ and $r_2(x,y)$. According to the convolution theorem the spectrum of the superposition is the convolution of the nailbeds $R_1(u,v)$ and $R_2(u,v)$; this means that a centered copy of one of the nailbeds is placed on top of each impulse of the other nailbed (the amplitude of each copied nailbed being scaled down by the amplitude of the impulse on top of which it has been copied). This convolution gives a "forest" of impulses scattered throughout the spectrum, as shown in Fig. 7.4. As we can see in the figure, the spectrum origin is closely surrounded by a full cluster of impulses. The cluster impulses closest to the origin, inside the visibility circle, include the (k_1,k_2,k_3,k_4)-impulse, the fundamental impulse of the moiré in question,[1] and its perpendicular counterpart, the $(-k_2,k_1,-k_4,k_3)$-impulse, which is the fundamental impulse of the same moiré in the perpendicular direction. Naturally, each of these two impulses is also accompanied by its respective symmetrical twin to the opposite side of the origin. The locations (frequency vectors) of these four impulses are marked in Fig. 7.4 by **a**, **−a**, **b** and **−b**.

If we look attentively at the impulse cluster surrounding the origin, we can see that this cluster is in fact a nailbed whose support is the regular lattice which is spanned by **a** and **b**, the geometric locations of the fundamental moiré impulses (k_1,k_2,k_3,k_4) and $(-k_2,k_1,-k_4,k_3)$. This infinite impulse nailbed represents in the spectrum the 2D (k_1,k_2,k_3,k_4)-moiré, and its basis vectors **a** and **b** determine the period and the two perpendicular directions of the moiré. This impulse nailbed is the 2D generalization of the moiré comb that we had in the case of line grating superpositions. Note that the impulse nailbed shown in Fig. 7.4 belongs to the simplest first-order moiré between two dot-screens, the $(1,0,-1,0)$-moiré, which is the 2D generalization of the $(1,-1)$-moiré between two gratings (Fig. 7.2).

The full expressions for the location and the amplitude of each of the impulses of the (k_1,k_2,k_3,k_4)-moiré nailbed can be found in *Vol. I*, Sec. 4.3. Here we will only give them for the $(1,0,-1,0)$-moiré. The location of the (m,n)-th impulse in the spectrum is given in this case by:

$$m\mathbf{a} + n\mathbf{b} = m\mathbf{f}_1 + n\mathbf{f}_2 - m\mathbf{f}_3 - n\mathbf{f}_4 \qquad (7.14)$$

For instance, the $(1,0)$-th impulse of the moiré nailbed is the $(1,0,-1,0)$-th impulse of the convolution, and it is located in the spectrum at the point $\mathbf{a} = \mathbf{f}_1 - \mathbf{f}_3$. Similarly, the $(0,1)$-th impulse of this moiré-nailbed is the $(0,1,0,-1)$-th impulse of the convolution, and it is located in the spectrum at the point $\mathbf{b} = \mathbf{f}_2 - \mathbf{f}_4$.

The amplitude $d_{m,n}$ of the (m,n)-th impulse in the $(1,0,-1,0)$-moiré cluster is given by:

$$d_{m,n} = a^{(1)}_{m,n} a^{(2)}_{-m,-n} \qquad (7.15)$$

This means that the $(1,0,-1,0)$-moiré-nailbed is simply a term-by-term product of the nailbeds $R_1(u,v)$ and $R_2(-u,-v)$ of the original screens, namely, where $R_2(u,v)$ is inverted (rotated by 180°) before the multiplication.

---

[1] Note that this impulse is generated in the convolution by the (k_1,k_2)-impulse in the spectrum $R_1(u,v)$ of the first image and the (k_3,k_4)-impulse in the spectrum $R_2(u,v)$ of the second image.

Since we already know the exact locations of the impulses of the moiré-nailbed, this nailbed can be considered, in fact, as a product of the two original nailbeds, after they have been normalized (rotated and stretched) to fit the impulse locations of the resulting moiré nailbed. Now, this moiré nailbed can be extracted from the spectrum convolution. Thus, by taking its inverse Fourier transform, we can reconstruct, back in the image domain, the isolated contribution of the moiré in question to the image superposition; this is the intensity profile of the moiré, a function in the image domain whose value at each point (x,y) indicates quantitatively the intensity level (or more precisely, the reflectance or the transmittance) of the moiré in question. Note that although this moiré is visible both in the layer superposition $r_1(x,y)r_2(x,y)$ and in the extracted moiré intensity profile, the latter does not contain the fine structure of the original layers $r_1(x,y)$ and $r_2(x,y)$ but only the pure contribution of the extracted moiré itself.

Now, just as in the case of grating superposition (Sec. 7.2.2), the spectral-domain term-by-term multiplication of the moiré nailbeds as defined by Eq. (7.15) can be also interpreted, back in the image domain, using the 2D version of the T-convolution theorem (Sec. 4.3 in *Vol. I*). This gives the following result, which is the 2D generalization of Proposition 7.2:

Proposition 7.3: (Particular case of Proposition 4.5 in *Vol. I*): The intensity profile of the $(1,0,-1,0)$-moiré in the superposition of two periodic dot screens with frequencies $\mathbf{f}_1$, $\mathbf{f}_2$ and $\mathbf{f}_3$, $\mathbf{f}_4$ and an angle difference of α can be seen from the image-domain point of view as the result of a 3-stage process:

(1) Normalization of the original screens (by linear stretching and rotation transformations) in order to bring both of them to a common period and orientation (that is, to make their periods, or their impulse nailbeds in the spectrum, coincide).

(2) T-convolution of the two normalized screens.

(3) Stretching and rotating the resulting normalized moiré intensity profile into its actual scale and orientation, as determined by the vectors $\mathbf{a} = \mathbf{f}_1 - \mathbf{f}_3$ and $\mathbf{b} = \mathbf{f}_2 - \mathbf{f}_4$, or more explicitly, by Eqs. (7.5) and (7.6). ■

Let us see now how this T-convolution sheds a new light on the profile shape of the $(1,0,-1,0)$-moiré and explains the striking visual effects observed in superpositions of dot screens such as those in Fig. 7.1(b).

7.2.4 Shape of the intensity profile of the moiré pattern

Case 1: Suppose, first, that one of the superposed screens consists of dots of a given shape (such as the digit "1"), and that the other screen consists of tiny white (or transparent) pinholes on a black background. Such pinholes play in the T-convolution the role of very narrow pulses with amplitude 1. As shown in Fig. 7.5(a), the T-convolution of such narrow pulses (from one of the screens) and dots of any shape (from the other screen) gives dots of the latter shape, in which the zero values remain at zero, the 1 values are scaled down to the value A (the volume of the narrow white pulse divided by the total

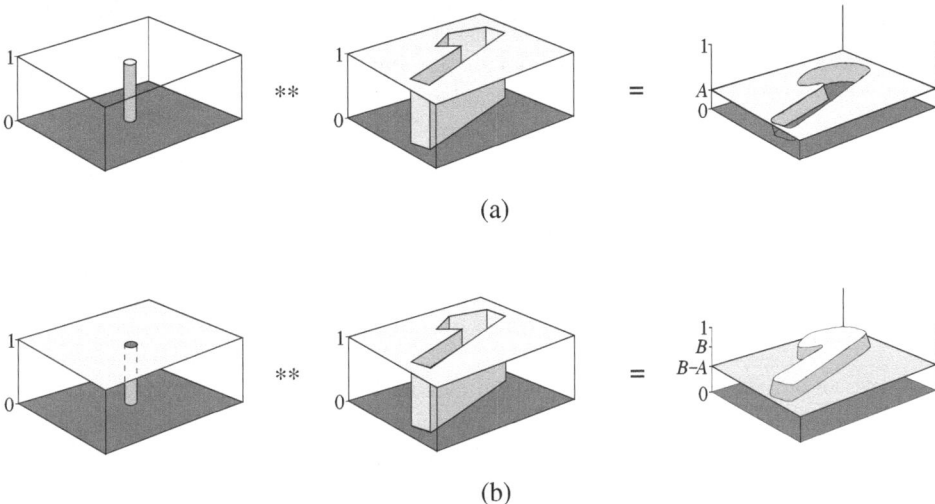

(a)

(b)

Figure 7.5: (a) The convolution of tiny white dots (from the first screen) with dots of
any given shape (from the other screen) gives dots of essentially the
same given shape. (b) The convolution of tiny black dots (from the first
screen) with dots of any given shape (from the other screen) gives dots
of essentially the same shape, but in inverse video.

period area), and the sharp step transitions are replaced by slightly softer ramps. This
means that the dot shape received in the normalized moiré period is practically identical to
the dot shape of the other screen (in our example: "1"), except that its white areas turn
darker. This normalized moiré period is stretched back into the real size of the moiré
period, T_M, as it is determined by Eqs. (7.5) and (7.6). Hence, the moiré form in this case
is essentially a magnified version of the screen element (the digit "1"), where the
magnification rate is controlled by the angle α between the two screens.

Case 2: A similar effect, albeit somewhat less impressive, occurs in the superposition
where one of the two screens contains tiny *black* dots (see Fig. 4.4(c) and (d) in *Vol. I*).
Tiny black dots on a white background can be interpreted as "inverse" pulses of
0-amplitude on a constant background of amplitude 1. As we can see in Fig. 7.5(b), the
T-convolution of such inverse pulses (from one of the screens) and dots of any shape
(from the other screen) gives dots of the latter shape, where the zero values are replaced by
the value B (the volume under a one-period cell of the second screen divided by the period
area) and the 1 values are replaced by the value $B-A$ (where A is the volume of the "hole"
of the narrow black pulse divided by the period area). This means that the dot shape of the
normalized moiré period is similar to the dot shape of the second screen, except that it
appears in inverse video and with slightly softer ramps.

Case 3: When none of the two superposed screens contains tiny dots, either white
or black, the profile form of the resulting moiré is still a magnified version of the

T-convolution of the two original screens. As before, this *T*-convolution gives some kind of blending between the two original dot shapes, but this time the resulting shape has a rather blurred or smoothed-out appearance and the moiré looks less attractive to the eye.

7.2.5 Orientation and size of the moiré cells

As we can see in Fig. 7.1(b), although the (1,0,-1,0)-moiré cells inherit the forms of the original screen cells, they do not inherit their orientations. Rather than having the same direction as the cells of the original screens (or an intermediate orientation), the moiré cells appear in a perpendicular direction. This fact is explained as follows.

As we already know, the orientation and the size of the moiré are determined by the location of the fundamental impulses of the moiré nailbed in the spectrum, i.e., by the location of the basis vectors **a** and **b**. We have seen following Eq. (7.14) that in the case of the (1,0,-1,0)-moiré these vectors are given by:

$$\mathbf{a} = \mathbf{f}_1 - \mathbf{f}_3$$
$$\mathbf{b} = \mathbf{f}_2 - \mathbf{f}_4 \tag{7.16}$$

And indeed, as we can see in Figs. 7.4 and 7.6, when the two original screens have the same frequency, these basis vectors are perpendicular to the bisectors of the angles formed between the frequency vectors $\mathbf{f}_1, \mathbf{f}_3$ and $\mathbf{f}_2, \mathbf{f}_4$. This means that in this case the (1,0,-1,0)-moiré nailbed (and the corresponding moiré profile in the image domain) are closely perpendicular to the original screens $r_1(x,y)$ and $r_2(x,y)$. Note that this is, in fact, a generalization of the results obtained in the 1D case (see Sec. 7.2.2 and Fig. 7.2). The period of our moiré can be found by Eq. (7.8) which was derived for the (1,-1)-moiré between two line-gratings with identical periods *T* and angle difference of α.

7.3 Intensity profile of Glass patterns in the superposition of aperiodic gratings

Having reviewed the basic concepts of the peridic cases, we are ready now to investigate their aperiodic counterparts. We start our study with the superposition of aperiodic gratings consisting of parallel straight lines (see Fig. 6.1). This case is simpler and easier to understand than the superposition of aperiodic dot screens, and it will serve us as a useful introduction to the case of aperiodic dot screens that we will study in Sec. 7.4.

But before we start, let us introduce a few terms that will be needed in the following.

Definition 7.1: Two layers (line gratings, dot screens, etc.) will be called *isometric* if the individual elements in the two layers (lines, dots, etc.) have the same locations. ■

For example, if dot screen $r_1(x,y)$ consists of arbitrarily located circular dots, and dot screen $r_2(x,y)$ is obtained by replacing each of the circular dots of $r_1(x,y)$ by a triangular dot that is centered at the same location, then $r_1(x,y)$ and $r_2(x,y)$ are isometric. Note that the

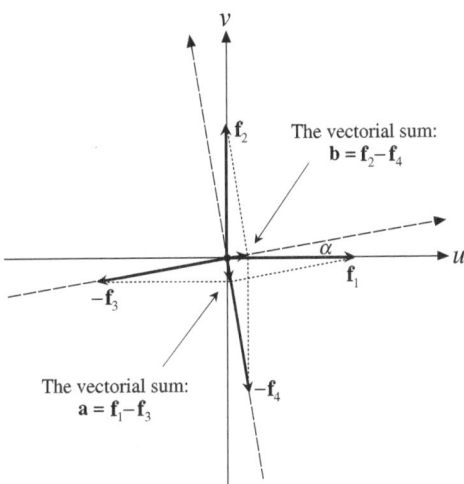

Figure 7.6: A detail from Fig. 7.4 showing the spectral interpretation (vector diagram) of the $(1,0,-1,0)$-moiré between two dot screens with identical frequencies and a small angle difference α. The low-frequency vectorial sums **a** and **b** (which are the geometric locations of the two fundamental impulses of the $(1,0,-1,0)$-moiré cluster) are closely perpendicular to the directions of the two original screens: **a** is perpendicular to the bisecting direction between $\mathbf{f}_1$ and $\mathbf{f}_3$, and **b** is perpendicular to the bisecting direction between $\mathbf{f}_2$ and $\mathbf{f}_4$.

size of the triangular dots need not be identical to the size of the original circular dots; all that we require is that they be centered at the same locations.

Definition 7.2: Two layers (line gratings, dot screens, etc.) will be called *congruent* if they can be made isometric (namely, brought into coincidence, element on element) by rotations, translations, or by any combination thereof. ■

For example, any two periodic layers having the same periodicity (or the same frequencies) are congruent.

Definition 7.3: Two layers (line gratings, dot screens, etc.) will be called *similar* if they can be made congruent by a linear spatial scaling. (Note that by *spatial* scaling we mean a spatial expansion or contraction, and not a scaling in the function's amplitude.) ■

It follows, therefore, that similar layers can be brought to coincidence, element on element, by rotations, translations, spatial scalings, or by any combinations thereof. Congruent layers, however, have the same spatial scaling, and they can only have different rotations or translations.

For example, any two periodic line gratings with periods $T_1 = T_2$ are congruent; but if their periods T_1 and T_2 are different, they are no longer congruent but only similar.

Note that we will generally be interested in congruent or similar layers that differ from each other only by slight rotation or spatial-scaling transformations, so that a moiré effect (or Glass pattern) becomes visible in their superposition.

Finally, some cases of particular interest are introduced by the following definition:

Definition 7.4: An aperiodic layer with *fixed element shapes* is a layer that is composed of identical elements whose shapes and profiles are fixed but whose locations are arbitrary (random, pseudo-random or deterministic). In particular:

An aperiodic dot screen is said to have a *fixed dot shape* if it is composed of dots whose shapes and profiles are identical while their locations are arbitrary (random, pseudo-random or deterministic). Such screens can be obtained, for example, by randomizing the dot locations of an initially periodic dot screen.

An aperiodic line grating is said to have a *fixed line shape* if it consists of parallel lines having an identical intensity profile but varying distances; the line distances may be random, pseudo-random or deterministic. ■

7.3.1 Superposition of correlated gratings

Suppose we are given two aperiodic gratings $r_1(x,y)$ and $r_2(x,y)$ as shown in Figs. 7.7(a),(b). We assume, at first, that the two gratings $r_1(x,y)$ and $r_2(x,y)$ are congruent, and only their orientations are slightly different. As shown in Figs. 6.1(a) and 7.7(c), such aperiodic gratings give a clearly visible linear Glass pattern in their superposition. Because each of the original gratings is constant along its own lines, i.e. perpendicularly to its main direction, its spectrum $R_i(u,v)$ consists of a line impulse (a "blade") passing through the origin that is oriented along the grating's main direction (see Figs. 7.7(d),(e)). A line impulse is a generalized function that runs along a 1D line through the plane, and is null everywhere else. A line impulse can be graphically illustrated as a blade whose behaviour is continuous (or diffuse; see end of Sec. 7.2.1) along its 1D line support, but impulsive in the perpendicular direction.

As our aperiodic gratings in the image domain are real valued but they are not necessarily symmetric about the origin, it follows that their spectra are Hermitian [Bracewell86 p. 15]. This means that the amplitude of each of the blades in Figs. 7.7(d),(e) is complex valued, where the real part is symmetric with respect to the origin and the imaginary part is antisymmetric (see Fig. 7.8).

Consider now the superposition of our aperiodic gratings $r_1(x,y)$ and $r_2(x,y)$ (Fig. 7.7(c)). Since this superposition is the product of the two original gratings, it follows according to the convolution theorem that the spectrum of the superposition is the convolution of the original spectra. The convolution of two 2D functions can be illustrated graphically by the "move and multiply" method (see, for example, [Rosenfeld82 pp. 13–14] or [Gaskill78 pp. 291–292]: We first rotate one of the original functions by 180°, and then we determine the value of the convolution at any point (u,v) in the plane as the volume

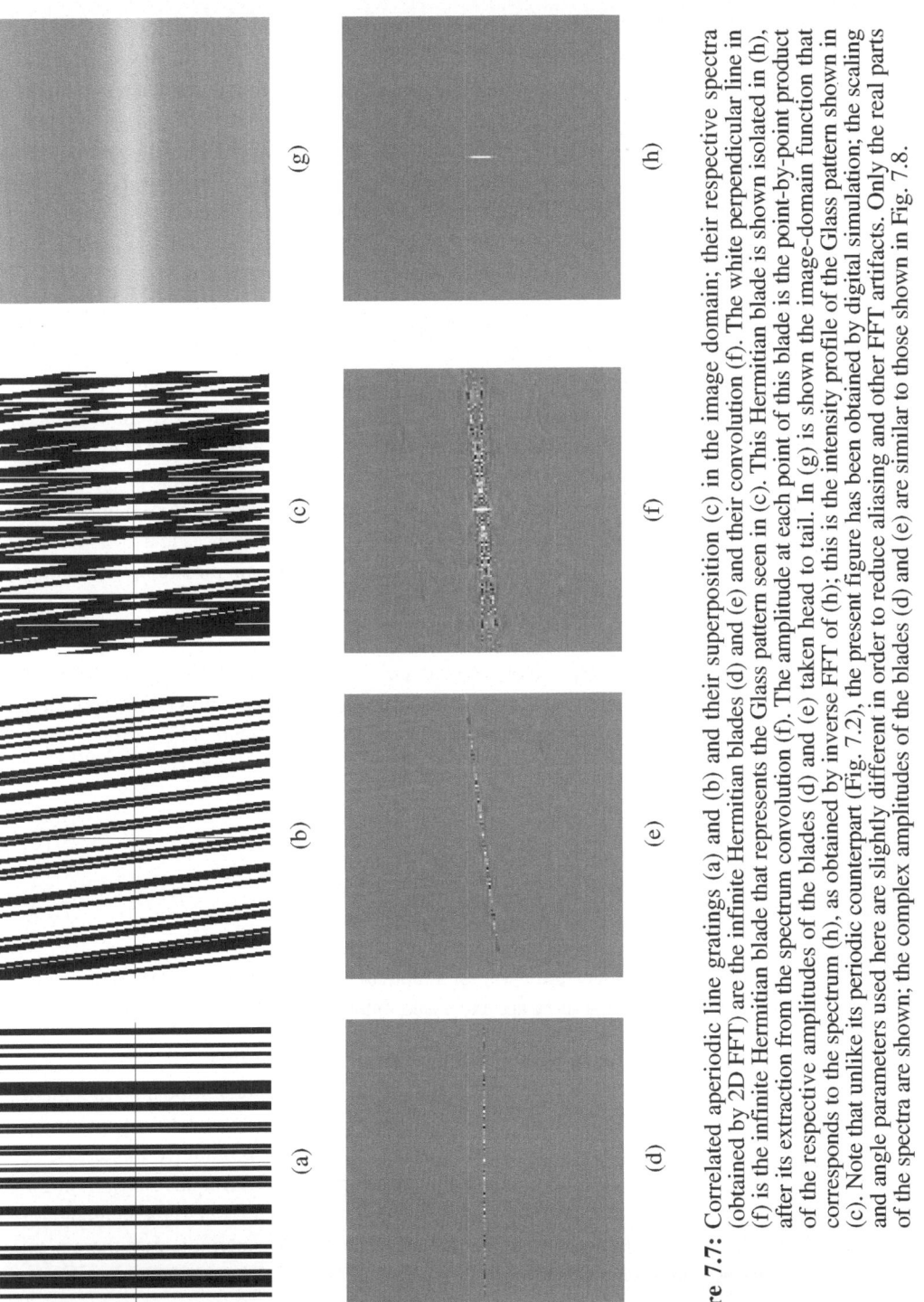

Figure 7.7: Correlated aperiodic line gratings (a) and (b) and their superposition (c) in the image domain; their respective spectra (obtained by 2D FFT) are the infinite Hermitian blades (d) and (e) and their convolution (f). The white perpendicular line in (f) is the infinite Hermitian blade that represents the Glass pattern seen in (c). This Hermitian blade is shown isolated in (h), after its extraction from the spectrum convolution (f). The amplitude at each point of this blade is the point-by-point product of the respective amplitudes of the blades (d) and (e) taken head to tail. In (g) is shown the image-domain function that corresponds to the spectrum (h), as obtained by inverse FFT of (h); this is the intensity profile of the Glass pattern shown in (c). Note that unlike its periodic counterpart (Fig. 7.2), the present figure has been obtained by digital simulation; the scaling and angle parameters used here are slightly different in order to reduce aliasing and other FFT artifacts. Only the real parts of the spectra are shown; the complex amplitudes of the blades (d) and (e) are similar to those shown in Fig. 7.8.

under the product of the two functions when the origin of the moving function is located at the point (u,v). In our case, the convolution is performed between two blades $R_1(u,v)$ and $R_2(u,v)$ that have different orientations, so that the value of the convolution at any point in the (u,v) plane is simply the product of the two blades $R_1(u,v)$ and $R_2(-u,-v)$ at their intersection point. It follows, therefore, that unlike in the periodic case, the spectrum of the grating superposition (i.e. the convolution of the two line impulses; see Fig. 7.7(f)) is no longer impulsive, but rather a 2D continuous (or diffuse) function. We will call this function, for the sake of our discussion, a "hump". This hump is Hermitian, since it is the spectrum of the grating product $r_1(x,y)r_2(x,y)$ which is obviously real valued.

Consider now the cross section (infinitely thin slice) of this hump that passes through the spectrum origin perpendicularly to the line bisecting the original line spectra of Figs. 7.7(d),(e). This section, which appears in Fig. 7.7(f) as a white dotted line, can be extracted by setting all the rest of the spectrum convolution to zero. Clearly, this isolated section (see Fig. 7.7(h)) is a line impulse; moreover, the amplitude of this line impulse is, by construction, a spatially scaled version of the product of the two original line impulses, one of which has been inverted (rotated by 180°) before the multiplication. Thus, if we consider in the spectral domain each of the two original line impulses P_1 and P_2 as well as our new line impulse P as a 1D function running along its own main direction, we obtain:

$$P(f_M) = P_1(f)P_2(-f) \qquad (7.17)$$

where $f_M = cf$, c being a scaling factor which depends on the angle difference α between the two original line impulses (i.e. the angle between the original gratings). The value of f_M for any given point f along the original blades can be found as shown in Fig. 7.9:

$$\sin(\alpha/2) = \frac{f_M/2}{f}$$

hence:

$$f_M = 2f\sin(\alpha/2) \qquad (7.18)$$

and:

$$c = \frac{f_M}{f} = 2\sin(\alpha/2) \qquad (7.19)$$

Note that the functions $P_1(f)$ and $P_2(f)$ are identical to the spectra of the original layers, $R_1(u,v)$ and $R_2(u,v)$, except that they are expressed in terms of different basis vectors.

We immediately recognize that Eq. (7.17) is, in fact, a generalization of Eq. (7.13) of the (1,-1)-moiré between two periodic gratings: It generalizes the purely impulsive spectra of periodic gratings into the continuous (or diffuse) line-impulse spectra of aperiodic gratings. Furthermore, Eq. (7.18) is identical to Eq. (7.7) that was obtained in the case of the (1,-1)-moiré between two periodic gratings having the same frequency ($f_1 = f_2$); the only difference is that in our case we cannot proceed from Eq. (7.18) to the language of periods, as we did in the periodic case (Eq. (7.8)), because in an aperiodic case there is no equivalent to the relation $T = 1/f$.

(a)   (b)

Figure 7.8: The amplitude of the Hermitian blade of Fig. 7.7(d). (a) The real part of
the amplitude; (b) the imaginary part of the amplitude.

It should be noted that Eqs. (7.17) and (7.18) above were derived for the case in which
the original gratings have undergone only rotations, but not scalings. (Recall our
assumption in the beginning of Sec. 7.3.1, that our two aperiodic gratings are congruent,
and only their orientations are slightly different.) If, in addition to the rotation, each of the
gratings undergoes a scaling transformation, Eqs. (7.17) and (7.18) become:

$$P(f_M) = P_1(f_1)P_2(-f_2) \tag{7.20}$$

and $$f_M = \sqrt{f_1^2 - 2f_1 f_2 \cos\alpha + f_2^2} \tag{7.21}$$

These are generalizations into the aperiodic case of Eqs. (7.13) and (7.9) of the $(1,-1)$-
moiré between two periodic gratings with different frequencies $f_1 \neq f_2$. Note that in this
case the cross section (7.20) through the spectrum convolution is no longer oriented
perpendicularly to the bisector of the original line impulses; it can be shown that its
orientation is given, just like in the periodic case (see Eq. (7.11)), by:

$$\varphi_M = \arctan\left(\frac{T_2\sin\theta_1 - T_1\sin\theta_2}{T_2\cos\theta_1 - T_1\cos\theta_2}\right) \tag{7.22}$$

We see, therefore, that Eqs. (7.17) and (7.20) are the aperiodic counterparts of Eq.
(7.13) that we have obtained in Sec. 7.2.2 in the case of the $(1,-1)$-moiré between periodic
gratings. The $(1,-1)$-moiré is, indeed, the simplest and most common type of a (k_1,k_2)-
moiré between two periodic gratings; the reason that this particular moiré is the basis for
the generalization into the aperiodic case will be explained later, in Sec. 7.5.

Hence, interestingly, the extension of the periodic case into the aperiodic case is most
naturally done in the *spectral* domain, where instead of considering three impulse combs
with discrete frequencies $n\mathbf{f}_1$, $n\mathbf{f}_2$, and $n\mathbf{f}_M = n(\mathbf{f}_1 - \mathbf{f}_2)$, $n \in \mathbb{Z}$, as we did in the periodic case,
we consider three line impulses with continuous frequencies $n\mathbf{f}_1$, $n\mathbf{f}_2$, and $n\mathbf{f}_M = n(\mathbf{f}_1 - \mathbf{f}_2)$,
where $n \in \mathbb{R}$. Each of these line impulses is therefore a continuous extension of the
corresponding impulse comb, where the gaps between the discrete impulse locations have
been filled in. In this continuous case, the basis vectors $\mathbf{f}_1$, $\mathbf{f}_2$ and $\mathbf{f}_M$ along the respective

line impulses simply indicate the corresponding unit frequency, rather than a discrete fundamental frequency that determines a periodicity in the image domain.

The relationship between the discrete spectra of the periodic case and the continuous spectra of the aperiodic case can be shown even better by rewriting Eq. (7.20) in an equivalent form, using the basis vectors $\mathbf{f}_M$, $\mathbf{f}_1$ and $\mathbf{f}_2$:

$$P(n\mathbf{f}_M) = P_1(n\mathbf{f}_1)P_2(-n\mathbf{f}_2) \qquad n \in \mathbb{R} \tag{7.23}$$

where f_1, f_2 and f_M of Eq. (7.20) are the lengths of the vectors $n\mathbf{f}_1$, $n\mathbf{f}_2$ and $n\mathbf{f}_M$. This can be written in a more concise form, where the coordinate n of each of the line impulses is expressed in terms of its own basis vector, $\mathbf{f}_M$, $\mathbf{f}_1$ or $\mathbf{f}_2$:

$$P(n) = P_1(n)P_2(-n) \qquad n \in \mathbb{R} \tag{7.24}$$

This is clearly the continuous counterpart of Eq. (7.13). Note that in Eq. (7.13), too, the index n in each of the three components refers to the n-th impulse of a different comb, and the corresponding impulse location in the u,v plane is determined in terms of the fundamental frequency vector of its own comb: $\mathbf{f}_M$, $\mathbf{f}_1$ or $\mathbf{f}_2$.

This relationship between the periodic and the aperiodic cases can be best appreciated by comparing the spectra of Fig. 7.2 (the periodic case) and Fig. 7.7 (its aperiodic counterpart).

Extending this relationship between Fig. 7.2 and Fig. 7.7 one step further, we may guess that the line-impulse of Fig. 7.7(h), which we have extracted from the spectrum convolution of Fig. 7.7(f), is the spectrum of the moiré effect that is generated in the superposition of the two gratings, namely, in our case, the linear Glass pattern that is clearly visible in Fig. 7.7(c). And indeed, if we apply an inverse Fourier transform to the extracted line impulse of Fig. 7.7(h), we obtain back in the image domain (see Fig. 7.7(g)) the intensity profile of the isolated Glass pattern of Fig. 7.7(c). Note that just like in the periodic case (Fig. 7.2), this extracted Glass pattern no longer contains the fine structure of the original layers, but only the pure contribution of the Glass pattern itself.

However, since the extracted line impulse of Fig. 7.7(h) is a normalized *product* of the line impulses of the original spectra, as stated by Eq. (7.20), it follows from the convolution theorem that the extracted intensity profile of the Glass pattern, shown in Fig. 7.7(g), is simply a normalized *convolution* of the intensity profiles of the two original gratings, one of which has been rotated by 180° due to the minus sign in Eq. (7.20).

This result can be formulated, therefore, as an extension of Proposition 7.2 of Sec. 7.2.2 into the aperiodic case:

Proposition 7.4: The intensity profile of the linear Glass pattern that is generated in the superposition of two similar aperiodic line gratings (namely, isometric gratings that have undergone linear rotation and spatial-scaling transformations) can be seen from the image-domain point of view as the result of a 3-stage process:

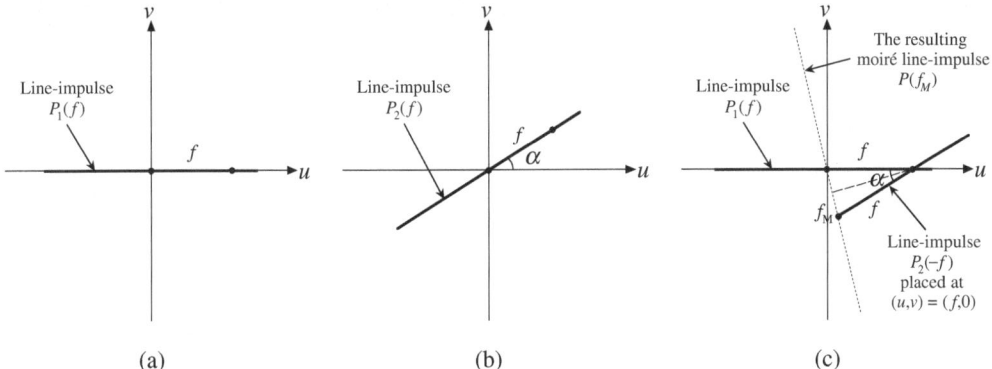

(a)   (b)   (c)

Figure 7.9: Geometric consideration in the frequency domain illustrating the scaling ratio between the original line spectra shown in (a) and (b) and the isolated line-spectrum shown as a dotted line in (c).

(1) Normalization of each of the original gratings by applying to it the inverse of its rotation and spatial-scaling transformation (plus a 180° rotation in the second layer).

(2) Convolution of the two normalized gratings, giving the normalized linear Glass pattern.

(3) Rotating and spatially scaling the normalized Glass pattern into its actual scale and orientation, as determined by Eqs. (7.21) and (7.22). ∎

This can be stated more formally as the aperiodic counterpart of Proposition 10.2 of *Vol. I*:

Proposition 7.5: Let $r_1(x,y)$ and $r_2(x,y)$ be two aperiodic line gratings that are obtained by applying linear transformations (bending functions) $g_1(x,y)$ and $g_2(x,y)$, respectively, to two isometric aperiodic line gratings having the intensity profiles $p_1(x')$ and $p_2(x')$:

$$r_1(x,y) = p_1(g_1(x,y)), \qquad r_2(x,y) = p_2(g_2(x,y))$$

Then, the Glass pattern $m(x,y)$ in the superposition of $r_1(x,y)$ and $r_2(x,y)$ is given by:

$$m(x,y) = p(g(x,y))$$

where:

(1) $p(x')$, the normalized intensity profile of the Glass pattern, is the convolution of the normalized intensity profiles of the original gratings:

$$p(x') = p_1(x') * p_2(-x') \tag{7.25}$$

(2) $g(x,y)$, the linear transformation (bending function) which brings $p(x')$ back into the actual scale and orientation of the Glass pattern $m(x,y)$, is given by:

$$g(x,y) = g_1(x,y) - g_2(x,y) \quad ∎ \tag{7.26}$$

Note that if the explicit expressions of the linear transformations $g_1(x,y)$ and $g_2(x,y)$ are given by:

$$g_1(x,y) = u_1x + v_1y = \mathbf{f}_1 \cdot \mathbf{x} \quad \text{with} \quad \mathbf{f}_1 = (u_1,v_1), \quad \mathbf{x} = (x,y)$$

$$g_2(x,y) = u_2x + v_2y = \mathbf{f}_2 \cdot \mathbf{x} \quad \text{with} \quad \mathbf{f}_2 = (u_2,v_2), \quad \mathbf{x} = (x,y)$$

then we have:

$$g(x,y) = (u_1 - u_2)x + (v_1 - v_2)y = \mathbf{f}_M \cdot \mathbf{x} \quad \text{with} \quad \mathbf{f}_M = \mathbf{f}_1 - \mathbf{f}_2, \quad \mathbf{x} = (x,y)$$

This general formulation englobes formulas (7.21) and (7.22) and their particular case, Eq. (7.18), which simply give explicit expressions for the length and the orientation of the vector $\mathbf{f}_M = \mathbf{f}_1 - \mathbf{f}_2$. But while in the periodic case $\mathbf{f}_1$, $\mathbf{f}_2$ and $\mathbf{f}_M = \mathbf{f}_1 - \mathbf{f}_2$ were the frequency vectors of the original periodic gratings and of the resulting moiré, here, they are simply the basis vectors of the respective spectra, and they convey only the scaling and the orientation of the aperiodic gratings — but not any notion of periodicity.[2]

Thus, by extending our moiré theory from the periodic case to the aperiodic case, we have succeeded in extracting the isolated Glass pattern from the grating superposition both in the spectral domain and in the image domain. The full significance of these results will be better appreciated in Sec. 7.4, when we will discuss their 2D counterparts.

Remark 7.1: Note that since the intensity profiles $p_i(\ )$ are not periodic, we are dealing here with convolution, and not as in Sec. 7.2, with T-convolution (the periodic counterpart of convolution). While this convolution gives a single Glass pattern, T-convolution gives in the periodic case infinitely many moiré replicas. The reason is that in the periodic case, each time the moving layer in the "move and multiply" convolution process advances by a full period, the same values are recorded in the result on a periodic basis. ■

7.3.2 Superposition of uncorrelated gratings

What would have happened now if the original gratings $r_1(x,y)$ and $r_2(x,y)$ were not isometric (or at least correlated) before undergoing the linear rotation and scaling transformations $g_1(x,y)$ and $g_2(x,y)$? As shown in Fig. 7.10, two uncorrelated gratings (a) and (b) do not generate a Glass pattern in their superposition (c). Let us see how we can explain this fact using the theory we have developed above.

First of all, we note that the convolution in Eq. (7.25) is, in fact, the cross correlation of $p_1(x')$ and $p_2(x')$ (see [Gaskill78 p. 172]):

$$p_1(x') \star p_2(x') = p_1(x') * p_2(-x') \tag{7.27}$$

---

[2] It is interesting to note that just as its periodic counterpart (see Sec. 10.9 in *Vol. I*), this proposition remains true for non-linear bending functions $g_i(x,y)$, too, namely, when the original aperiodic layers undergo any given geometric transformations. In such cases, part (2) of the proposition simply gives the bending function which defines the shape of the resulting Glass pattern, as explained in Sec. 6.6. The most general form of this proposition is given at the end of Sec. 7.8.

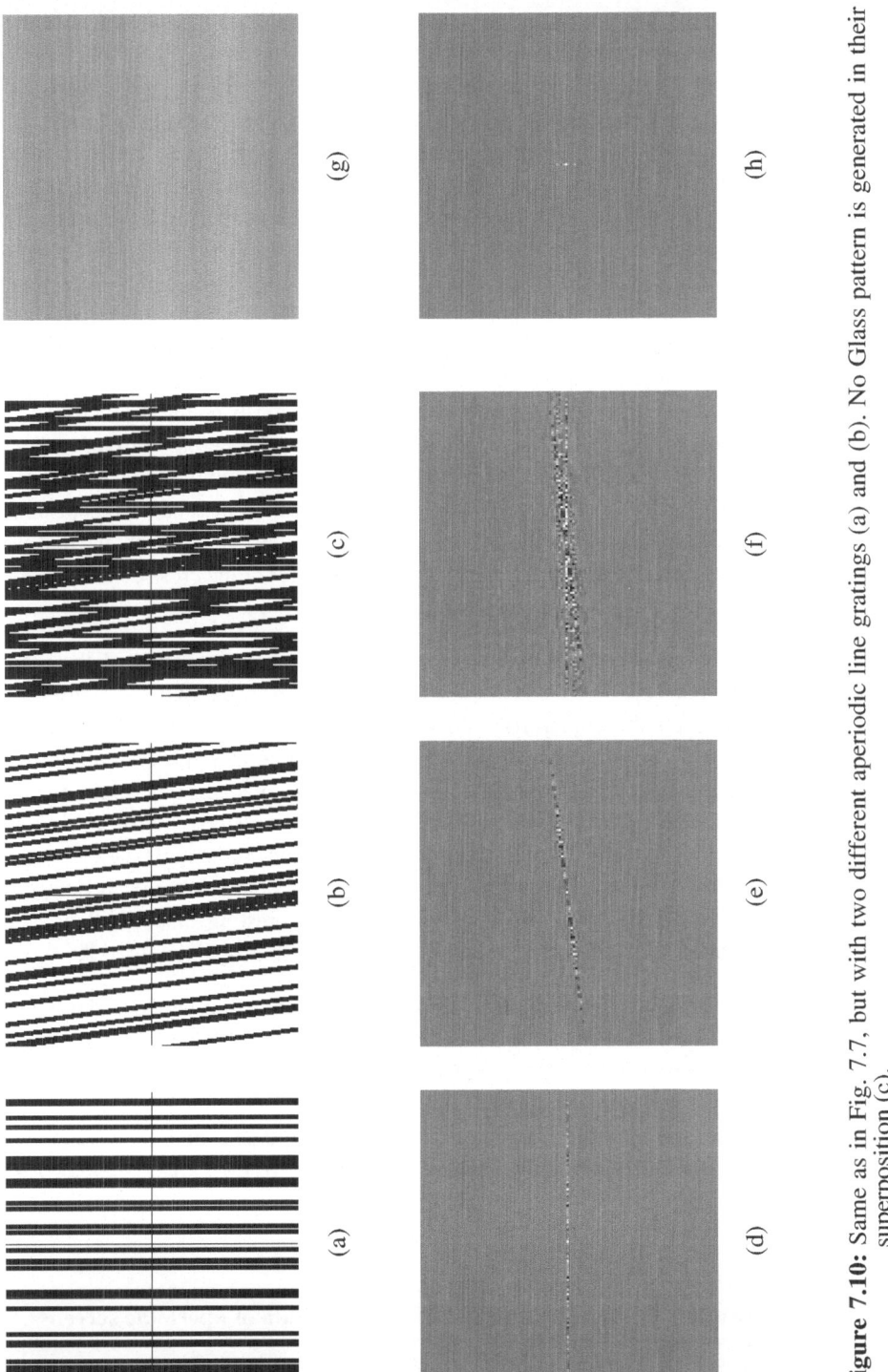

Figure 7.10: Same as in Fig. 7.7, but with two different aperiodic line gratings (a) and (b). No Glass pattern is generated in their superposition (c).

Intuitively, the cross correlation of two functions is obtained by a "move and multiply" process similar to that of convolution, except that none of the two original functions is inverted (or rotated by 180°) before the operation. Suppose, first, that our two original layers are isometric. Therefore when $p_1(x')$ and $p_2(x')$ are located, during the "move and multiply" process, in full or almost full coincidence, the resulting cross-correlation value (the volume under the product function) is high; but when the two layers are out of coincidence, the resulting values fluctuate arbitrarily around some mean value. This results in much higher values about the center of the cross correlation (see Fig. 7.11(a)). And indeed, this area of higher values in the center of the cross correlation simply represents the brighter zone in the center of the intensity profile of the Glass pattern (see Fig. 7.7(g)). A similar result will be obtained whenever the two original layers are well correlated, giving a "privileged" area of brighter values in the center, which corresponds to the Glass pattern. But when the original layers are not correlated, there exists no mutual locus where all of their elements fall on top of one another, and therefore their cross correlation does not contain such a "privileged" area, and it simply fluctuates around the same mean level throughout (see Fig. 7.11(b)). Hence, no visible macrostructure (Glass pattern) appears in the superposition. Note, however, that the terms "high values" and "brighter" that we have used above to describe the "privileged" area in the center of the cross correlation are not always appropriate. This is explained in Remark 7.2.

Remark 7.2: It should be noted that the "privileged" area in the center of the cross correlation of two correlated layers is not necessarily brighter. If the two layers involved are correlated and each of them has *fixed element shapes*, it follows that when the layers are in full or almost full coincidence during the "move and multiply" process, all of their individual elements coincide simultaneously in the same manner. Therefore in this "privileged" area, the values of the full-layer convolution (or cross correlation) are determined basically by the convolution (or cross correlation) of a single element from each layer. But although in many cases this indeed gives high values in the "privileged" area, in other circumstances the result may have a more interesting shape. In Sec. 7.4.2 this will be illustrated in the 2D case through Fig. 7.5. See also Problem 7-24. ■

In conclusion, we have shown that the fundamental relationship between the image domain and the spectral domain holds also in the generalized case where the superposed gratings are no longer periodic. This is clearly illustrated by Figs. 7.2 and 7.7. The main differences between periodic and aperiodic cases are that in aperiodic cases our spectra consist of continuous structures instead of discrete impulses, and that frequency considerations in the spectral domain, such as those leading to Eq. (7.18) (see Fig. 7.9), can no longer be interpreted in the image domain in terms of periods.

7.4 Intensity profile of Glass patterns in the superposition of aperiodic screens

Having understood the simpler case of aperiodic line gratings, we are ready now to proceed to the superposition of aperiodic dot screens.

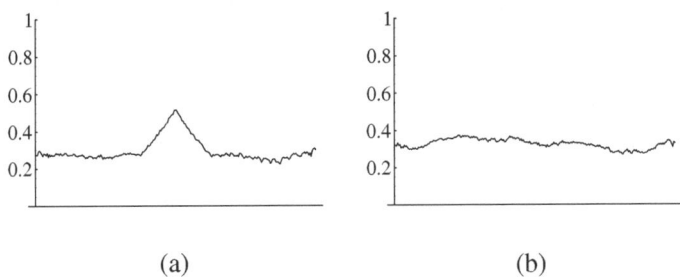

(a)   (b)

Figure 7.11: (a) A cross section through the extracted Glass pattern of Fig. 7.7(g),
showing its intensity profile. (b) A cross section through Fig. 7.10(g).
Clearly, when the original gratings are not correlated, no Glass pattern is
visible in their superposition and in the extracted intensity profile.

7.4.1 Superposition of correlated screens

Suppose we are given two aperiodic dot screens $r_1(x,y)$ and $r_2(x,y)$. We assume, at first,
that the two screens are either congruent or similar, and only their orientations (and
possibly their spatial scalings) are slightly different. As shown in Fig. 2.1(c), such
aperiodic dot screens give a clearly visible Glass pattern in their superposition.

Unlike in the case of aperiodic gratings, the spectrum of an aperiodic screen is no longer
a line impulse, but a 2D continuous (or diffuse) hump, which is, of course, Hermitian,
since our aperiodic screen is real valued. Such a spectrum is shown in Fig. 2.10(e).

Consider now the superposition of our aperiodic screens $r_1(x,y)$ and $r_2(x,y)$. Since this
superposition is the product of the two original gratings, it follows according to the
convolution theorem that the spectrum of the superposition is the convolution of the
original spectra. Hence, unlike in the periodic case, this spectrum convolution is not
impulsive but rather a 2D continuous (or diffuse) hump; see, for example, Fig. 2.12.

Following our experience from the previous cases, and as an extension of Fig. 7.2(h)
and Fig. 7.7(h), we would like now to extract from this convolution the 2D spectrum that
belongs to the isolated Glass pattern. Let us try, therefore, to identify the spectrum that we
would like to extract. As we did in the case of aperiodic gratings (Sec. 7.3), we consider
the simplest moiré between two periodic screens, namely, the (1,0,-1,0)-moiré. The reason
that this particular moiré is the only one which can be generalized into the aperiodic case
will be explained later, in Sec. 7.5. As we already know, the impulse amplitudes of the
(1,0,-1,0)-moiré between periodic screens are given by Eq. (7.15):

$$d_{m,n} = a^{(1)}_{m,n} \, a^{(2)}_{-m,-n} \qquad \text{with} \qquad m,n \in \mathbb{Z}$$

Note that in this equation the indices m,n in each of the three components refer to the
m,n-th impulse of a different nailbed. The location of this impulse in the u,v plane is

determined in terms of the fundamental frequency vectors of its own nailbed, which are, respectively, $\mathbf{a} = \mathbf{f}_1 - \mathbf{f}_3$ and $\mathbf{b} = \mathbf{f}_2 - \mathbf{f}_4$, $\mathbf{f}_1$ and $\mathbf{f}_2$, and $\mathbf{f}_3$ and $\mathbf{f}_4$ (see Fig. 7.4).

Now, being inspired by what we did in the case of aperiodic gratings (Eq. (7.24)), we consider the continuous extension of the spectrum given by Eq. (7.15), namely:

$$P(m,n) = P_1(m,n)\, P_2(-m,-n) \qquad \text{with} \qquad m,n \in \mathbb{R} \tag{7.28}$$

where the coordinates m,n of each of the three functions is expressed in terms of its own basis vectors. Note that $P_1(m,n)$ and $P_2(m,n)$ are identical to the spectra of the original screens, $R_1(u,v)$ and $R_2(u,v)$, except that they are expressed in terms of different basis vectors, as explained below.

As we remember, we have assumed that our superposed screens are congruent (or similar). This means that the original screens were initially isometric, but before being superposed, each of them was linearly transformed (scaled and rotated). Thus, the coordinates m,n in $P_1(m,n)$ are expressed in terms of the basis vectors $\mathbf{f}_1$, $\mathbf{f}_2$ of the first spectrum after it has been scaled and rotated; and similarly, the coordinates m,n in $P_2(m,n)$ are expressed in terms of the basis vectors $\mathbf{f}_3$, $\mathbf{f}_4$ of the second spectrum after it has been scaled and rotated. The resulting spectrum, $P(m,n)$, is defined as the product of $P_1(m,n)$ and $P_2(-m,-n)$, but its coordinates m,n are expressed in terms of the basis vectors $\mathbf{a} = \mathbf{f}_1 - \mathbf{f}_3$ and $\mathbf{b} = \mathbf{f}_2 - \mathbf{f}_4$.

Therefore, Eq. (7.28) can be written, back in our usual u,v coordinate system, as follows:

$$P(m\mathbf{a} + n\mathbf{b}) = P_1(m\mathbf{f}_1 + n\mathbf{f}_2)\, P_2(-m\mathbf{f}_3 - n\mathbf{f}_4) \tag{7.29}$$

Now, as we did in the previous cases, we wish to extract $P(m,n)$ from the spectrum convolution. However, in our present case this seems to be rather hopeless, because unlike in the cases we have studied previously, both the full convolution and the spectrum we wish to extract from it are 2D humps that overlap each other and are not separable. In other words, $P(m,n)$ cannot be separated or isolated by setting the rest of the spectrum convolution to zero, as was the case in the previous sections (see Figs. 7.2, 7.4 and 7.7). Nevertheless, we can still synthesize this isolated spectrum using Eq. (7.28). And indeed, if we apply an inverse Fourier transform to the extracted hump $P(m,n)$, we obtain, back in the image domain, the intensity profile of the isolated Glass pattern. Note that just like in the periodic case, this extracted Glass pattern does not contain the fine structure of the original layers, but only the pure contribution of the Glass pattern itself.

But since the extracted function $P(m,n)$ is a normalized product of the spectra of the original screens, as stated by Eq. (7.28), it follows from the convolution theorem that the extracted intensity profile of the Glass pattern is simply a normalized convolution of the intensity profiles of the two original screens. This result can therefore be formulated as an extension of Proposition 7.3 (Sec. 7.2.3) into the aperiodic case:

Proposition 7.6: The intensity profile of the Glass pattern that is generated in the superposition of two similar aperiodic dot screens (i.e., isometric screens which have undergone linear rotation and spatial-scaling transformations) can be seen from the image-domain point of view as the result of a 3-stage process:

(1) Normalization of each of the original screens by applying to it the inverse of its rotation and spatial-scaling transformation (plus a 180° rotation in the second layer).

(2) Convolution of the two normalized screens, giving the normalized Glass pattern.

(3) Rotating and spatially scaling the normalized Glass pattern into its actual scale and orientation, as determined by the vectors $\mathbf{a} = \mathbf{f}_1 - \mathbf{f}_3$ and $\mathbf{b} = \mathbf{f}_2 - \mathbf{f}_4$, or more explicitly, by Eqs. (7.21) and (7.22). ∎

Note that the aim of the normalization is to bring all of the three functions in Eq. (7.28) to the same common basis (or coordinate system) before we can perform the multiplication. The need for this normalization was more obvious in the periodic case (Propositions 7.2 and 7.3), where the entities to be multiplied in the spectral domain were impulse combs or impulse nailbeds, whose multiplication was done term-by-term; but it remains essential in the aperiodic case, too, although the entities to be multiplied are continuous.

This result can be stated more formally as the aperiodic counterpart of Proposition 10.5 of *Vol. I*; note that functions denoted by boldface letters indicate mappings of $\mathbb{R}^2$ onto itself:

Proposition 7.7: Let $r_1(\mathbf{x})$ and $r_2(\mathbf{x})$ be two aperiodic screens that are obtained by applying linear mappings (scalings, rotations, etc.) $\mathbf{g}_1(\mathbf{x})$ and $\mathbf{g}_2(\mathbf{x})$, respectively, to two isometric aperiodic screens having the intensity profiles $p_1(\mathbf{x}')$ and $p_2(\mathbf{x}')$:

$$r_1(\mathbf{x}) = p_1(\mathbf{g}_1(\mathbf{x})), \qquad r_2(\mathbf{x}) = p_2(\mathbf{g}_2(\mathbf{x}))$$

Then, the Glass pattern $m(\mathbf{x})$ in the superposition of $r_1(\mathbf{x})$ and $r_2(\mathbf{x})$ is given by:

$$m(\mathbf{x}) = p(\mathbf{g}(\mathbf{x}))$$

where:

(1) $p(\mathbf{x}')$, the normalized intensity profile of the Glass pattern, is the convolution of the normalized intensity profiles of the original screens:

$$p(\mathbf{x}') = p_1(\mathbf{x}') ** p_2(-\mathbf{x}') \tag{7.30}$$

(2) $\mathbf{g}(\mathbf{x})$, the linear transformation which brings $p(\mathbf{x}')$ back into the actual scale and orientation of the Glass pattern $m(\mathbf{x})$, is given by:

$$\mathbf{g}(\mathbf{x}) = \mathbf{g}_1(\mathbf{x}) - \mathbf{g}_2(\mathbf{x}) \quad\blacksquare \tag{7.31}$$

Note that if the explicit expressions of the linear mappings $\mathbf{g}_1(\mathbf{x})$ and $\mathbf{g}_2(\mathbf{x})$ are given by:

$$\mathbf{g}_1(\mathbf{x}) = \begin{pmatrix} u_1 x + v_1 y \\ u_2 x + v_2 y \end{pmatrix} = F_1 \cdot \mathbf{x} \quad \text{with} \quad F_1 = \begin{pmatrix} u_1 & v_1 \\ u_2 & v_2 \end{pmatrix} = \begin{pmatrix} \mathbf{f}_1 \\ \mathbf{f}_2 \end{pmatrix}, \quad \mathbf{x} = (x,y)$$

$$\mathbf{g}_2(\mathbf{x}) = \begin{pmatrix} u_3 x + v_3 y \\ u_4 x + v_4 y \end{pmatrix} = F_2 \cdot \mathbf{x} \quad \text{with} \quad F_2 = \begin{pmatrix} u_3 & v_3 \\ u_4 & v_4 \end{pmatrix} = \begin{pmatrix} \mathbf{f}_3 \\ \mathbf{f}_4 \end{pmatrix}, \quad \mathbf{x} = (x,y)$$

then we have:

$$\mathbf{g}(\mathbf{x}) = \begin{pmatrix} (u_1-u_3)x + (v_1-v_3)y \\ (u_2-u_4)x + (v_2-v_4)y \end{pmatrix} = F_M \cdot \mathbf{x} \qquad \text{with} \qquad F_M = \begin{pmatrix} u_1-u_3 & v_1-v_3 \\ u_2-u_4 & v_2-v_4 \end{pmatrix} = \begin{pmatrix} \mathbf{a} \\ \mathbf{b} \end{pmatrix}$$

where $\quad F_M = F_1 - F_2, \quad$ namely: $\quad \begin{pmatrix} \mathbf{a} \\ \mathbf{b} \end{pmatrix} = \begin{pmatrix} \mathbf{f}_1 - \mathbf{f}_3 \\ \mathbf{f}_2 - \mathbf{f}_4 \end{pmatrix}.$

This indeed gives us the connection between Eq. (7.31) and the basis vectors of the individual screens and their resulting moiré, which are, respectively: $\mathbf{f}_1$ and $\mathbf{f}_2$, $\mathbf{f}_3$ and $\mathbf{f}_4$, and $\mathbf{a}$ and $\mathbf{b}$, where $\mathbf{a} = \mathbf{f}_1 - \mathbf{f}_3$ and $\mathbf{b} = \mathbf{f}_2 - \mathbf{f}_4$, exactly as in the periodic case (see Eq. (7.16)).[3]

Thus, by extending our moiré theory from the 2D periodic case to the 2D aperiodic case, we have succeeded in extracting the isolated Glass pattern from the screen superposition both in the spectral domain and in the image domain. Let us now see the full significance of this result.

7.4.2 Shape of the intensity profile of the Glass pattern

Based on our previous experience with periodic dot screens (see Sec. 7.2.4), we divide our discussion into three different cases.

Case 1: Suppose, first, that one of the superposed layers is an aperiodic screen consisting of arbitrarily positioned dots having *a fixed shape* (such as the digit "1"), and that the second layer is an aperiodic screen consisting of tiny pinholes on a black background, where the dot locations in both screens are almost identical (or slightly transformed). Such pinholes play in the convolution the role of very narrow pulses with amplitude 1. According to Remark 7.2 in Sec. 7.3.2, the shape of the Glass pattern that is generated in the layer convolution is determined basically by the convolution of one element from each layer. As shown in Fig. 7.5(a), this means that in this case, just as it happens in the superposition of periodic layers (Sec. 7.2.4), the moiré (or Glass) pattern that appears in the superposition is essentially a magnified and rotated version of an individual dot of the first screen. But as already explained in Remark 7.1 at the end of Sec. 7.3.1, the Glass pattern generated in the aperiodic case is not periodically repeated throughout the superposition, as in the periodic case, and it consists of only one copy of the magnified dot shape (compare Figs. 7.1(a) and (b)).

This surprising result seems at first to contradict the basic properties of Glass patterns as generally known until now. A "classical" Glass pattern is expected to be brighter in its center than in areas farther away, due to the partial overlapping of the dots of the two

---

[3] Note that just as its periodic counterpart (see Sec. 10.9 of *Vol. I*), this proposition remains true for non-linear transformations $\mathbf{g}_i(x,y)$, too, namely: when the original aperiodic layers undergo any given geometric transformations. In such cases, part (2) of the proposition simply gives the geometric transformation which defines the resulting Glass pattern, as explained in Secs. 3.5 and 6.6. Note that in the case of non-linear transformations several Glass patterns may be generated in the superposition simultaneously; this situation is discussed further in Sec. 7.4.4. The most general form of this proposition is given at the end of Sec. 7.8 as *the fundamental Glass pattern theorem*, the aperiodic counterpart of the *fundamental moiré theorem* of Sec. 10.9 in *Vol. I*. Note, however, that the case of non-linear transformations is based on *generalized* Fourier considerations (see Appendix G).

layers in this area (see Fig. 2.1(c)). But the Glass pattern of Fig. 7.1(a) seems to completely contradict this property.

In reality, however, there is no contradiction at all. The key point is that in "classical" Glass patterns the two superposed layers are identical, both consisting of black dots on a white background. But if, as shown in Fig. 7.1(a), one of the screens consists of tiny pinholes on a black background and the other screen has a fixed dot shape, then the convolution of the dot shape of one screen with the dot shape of the other screen gives, indeed, a Glass pattern that has the intensity profile of the dot shape of the other screen. Note also the absence of dot trajectories in Fig. 7.1(a), as already discussed in Sec. 4.8.

Case 2: Suppose now that we replace our pinhole screen by an inverse-video copy of itself, consisting of tiny black dots on a white (or, rather, transparent) background. This time, the convolution of the individual dot shapes of the two layers basically gives an inverse-video version of the result obtained in the first case, as shown in Fig. 7.5(b). Hence, if one of the screens contains tiny black dots, and the other screen has a fixed dot shape, then the moiré intensity profile that we obtain is a magnified version of the individual dot shape of the other screen, but this time in inverse video. In our example, we will obtain a single "1"-shaped Glass pattern that is brighter inside the digit shape and darker outside. Note, however, that this moiré intensity profile is weaker and less impressive than that of the previous case.

Case 3: Finally, when none of the superposed layers consists of tiny dots (either white or black), or when the superposed layers do not have fixed element shapes, the intensity profile form of the resulting moiré (or Glass pattern) is still a magnified version of the convolution of the dot shapes of the two layers. This convolution gives some kind of blending between the original dot shapes, but the resulting shape has a blurred or smoothed-out appearance resembling a 2D Gaussian. As we can now understand, this is exactly what happens in "classical" Glass patterns, where the two superposed layers are identical copies of each other (as in Fig. 2.1(c)). This is the reason for which Glass patterns in the previous chapters did not have the shape of a magnified element of one of the superposed layers. See also Problem 7-26.

7.4.3 Orientation and size of the Glass pattern

Looking at Figs. 7.1(a) and (b), we already know by now how to explain their common "1"-shaped moiré (or Glass) pattern: In both of the periodic and the aperiodic cases, this is simply a normalized convolution of a single pinhole from one layer and a single "1"-element from the other layer. But how can we explain the fact that in both cases the resulting moiré (or Glass) patterns have the same orientation and the same size?

As we have seen, both the $(1,0,-1,0)$-moiré in the periodic case and the Glass pattern in the aperiodic case are governed by the same basis vectors in the spectrum (Eq. (7.16)):

$$\mathbf{a} = \mathbf{f}_1 - \mathbf{f}_3$$

$$\mathbf{b} = \mathbf{f}_2 - \mathbf{f}_4$$

Although in the aperiodic case these vectors cannot be interpreted in terms of periodicities as in the periodic case, they still determine the orientation and the size of the resulting moiré effect. For example, when the two superposed screens are congruent (meaning that they have the same scaling, or, in the periodic case, the same frequencies or periodicities), the basis vectors **a** and **b** are perpendicular to the bisectors of the angles formed between the frequency vectors $\mathbf{f}_1$, $\mathbf{f}_3$ and $\mathbf{f}_2$, $\mathbf{f}_4$. This means that the (1,0,-1,0)-moiré (or the Glass pattern) are closely perpendicular to the original screens $r_1(x,y)$ and $r_2(x,y)$ (see Figs. 7.1(a) and (b)). As for the size of the moiré (or Glass) patterns, they are obtained by multiplying the original screen element size by the spatial scaling (magnification) of the moiré. For example, if the length of each of the basis vectors $\mathbf{f}_1,...,\mathbf{f}_4$ of the original screens is f and the length of each of the two basis vectors **a** and **b** of the moiré (or Glass) pattern is $f_M = 0.1f$, then the size of the "1"-shaped moiré in both Figs. 7.1(a) and (b) will be 10 times the size of the element "1" in the original screen. But while in the periodic case the magnification rate can be also interpreted in terms of periods: $T_M = 10T$, in the aperiodic case the only possible interpretation is in terms of the *scaling ratio*, which is clearly the aperiodic counterpart of the periodic notion of *frequency ratio* (compare Eqs. (7.18) and (7.19) of the aperiodic case with Eqs. (7.7) and (7.8) in the periodic case).

We therefore obtain the following remarkable result:

Proposition 7.8: If the respective dot screens in the periodic case and in the aperiodic case consist of the same microstructure elements and they have undergone the same rotation and scaling transformations (or more generally, the same linear or affine transformations), then the resulting (1,0,-1,0)-moiré (or Glass) patterns in both cases have the same shape, the same size and the same orientation. The only difference is that in the aperiodic case only one such pattern is generated, while in the periodic case there appear infinitely many replicas, as was already explained in Remark 7.1. ■

This result is clearly illustrated in Figs. 7.1(a) and (b).

7.4.4 Cases with several fixed points or with continuous fixed lines

So far we have considered the shape and the orientation of the intensity profile in cases where only one Glass pattern is generated in the screen superposition. But as we already know from chapter 3, when the superposed screens undergo non-linear transformations that happen to have more than a single mutual fixed point, several Glass patterns may be generated simultaneously, one about each of these fixed points. Such a case with 4 fixed points has been presented in Chapter 3 in Example 3.6 and in Fig. 3.15(a).

However, in Chapter 3 we only considered superpositions of two originally identical layers, and therefore all the Glass patterns we obtained there consisted of a brighter area about the fixed point, and they did not have any interesting intensity profile shapes. What would have happenned now if we repeated the same example as in Fig. 3.15(a), but this time using original layers having different dot shapes, for example "2"-shaped dots in one layer and tiny pinholes on a black background in the other layer?

This case is illustrated in Fig. 7.12(a). As we can clearly see, all of the 4 Glass patterns that are generated in this case have a "2"-shaped intensity profile, but each of them is oriented differently; furthermore, two of them are also mirror imaged. How can we explain this fact? As we have seen in Example 3.6 and in Example 5.3, the domain transformations undergone in this case by the two original layers are given by:

$$\mathbf{g}_1(x,y) = (x, y + y_0 - ax^2)$$

$$\mathbf{g}_2(x,y) = (x + x_0 - ay^2, y)$$

and therefore the difference transformation governing the Glass pattern is:

$$\mathbf{g}_M(x,y) = \mathbf{g}_1(x,y) - \mathbf{g}_2(x,y) = (ay^2 - x_0, y_0 - ax^2) \tag{7.32}$$

As we already know, this transformation has 4 equidistant fixed points, one in each quadrant (see Eq. (3.39)):

$$(x,y) = (\pm\sqrt{y_0/a}, \pm\sqrt{x_0/a})$$

Now, remember that the local reflection properties of a transformation at any point (x,y) are determined by the sign of its Jacobian at that point (see Sec. C.2 of Appendix C). In our case, the Jacobian of the transformation $\mathbf{g}_M(x,y)$ is:

$$J\mathbf{g}_M(x,y) = \begin{vmatrix} 0 & 2ay \\ -2ax & 0 \end{vmatrix} = 4a^2xy$$

Therefore, as explained in Sec. C.2 of Appendix C, the transformation $\mathbf{g}_M(x,y)$ is non-reflecting at the points (x,y) where $J\mathbf{g}_M(x,y) > 0$, i.e. at the first and third quadrants of the plane, and it is reflecting at the points (x,y) where $J\mathbf{g}_M(x,y) < 0$, i.e. at the second and fourth quadrants of the plane. This explains, indeed, why the "2"-shaped Glass patterns in the second and fourth quadrans are mirror imaged, while the other two are not. Note that the two Glass patterns that are mirror imaged in Fig. 7.12(a) are precisely the ones that are surrounded by hyperbolic dot trajectories in Fig. 3.15(a); see also Problem 7-7 and Example H.4 in Appendix H.

Fig. 7.12(b), on its part, shows the periodic counterpart of Fig. 7.12(a), using the same dot shapes and layer transformations $\mathbf{g}_1(x,y)$ and $\mathbf{g}_2(x,y)$ as in Fig. 7.12(a). Note the non-linear magnification rate of the moiré, and its varying orientations; both can be determined by the local properties of the non-linear domain transformation (7.32), as explained in detail in Appendix C.[4]

---

[4] As explained and illustrated at the end of Sec. 10.9.2 of *Vol. I*, when $\mathbf{g}_M(x,y) = \mathbf{g}_1(x,y) - \mathbf{g}_2(x,y)$ happens to be linear or affine, then even if $\mathbf{g}_1$ and $\mathbf{g}_2$ themselves are non-linear, the resulting moiré effect is purely periodic, with a constant magnification rate and a constant orientation throughout the entire plane. But this condition is clearly not satisfied in our present example.

Figure 7.12: (a) Same as in Fig. 3.15(a), but here the first random dot screen consists
of "2"-shaped dots and the second random dot screen consists of tiny
pinholes on a black background. After the application of the non-linear
layer transformations we obtain in the superposition 4 Glass patterns
having the intensity profile of "2", but two of them are mirror-imaged.

Remark 7.3: It is interesting to note that the difference transformation (7.32) that was
obtained according to our fixed-point considerations in Chapter 3 (see Sec. 3.5) is
precisely the transformation that, according to part (2) of Proposition 7.7, brings back the

Figure 7.12: (*continued.*) (b) The periodic counterpart of (a), using the same dot shapes and layer transformations as in (a). The local orientation of the moiré at any point (x,y) can be determined by applying Eqs. (C.10) of Appendix C to the moiré transformation (7.32).

normalized intensity profile of the Glass pattern into its actual, transformed shape in the superposition. We see, therefore, that Proposition 7.7 englobes the fixed-point considerations of Chapter 3, but it also goes much beyond them; indeed, using the Fourier-based approach we not only obtain the geometric locations (center points) of the Glass patterns, as in Chapter 3, but also their precise intensity profiles. ■

Finally, note that a Glass pattern can only have a meaningful intensity profile shape (basically, the magnified and rotated dot shape of one of the original screens) when it is generated about a non-singular mutual fixed point of the transformations $g_1(x,y)$ and $g_2(x,y)$ (see also Sec. A.4 in Appendix A). In singular cases, where a Glass pattern is generated along a fixed line or a fixed curve (see, for example, Figs. 2.3 or 3.6), one of the two dimensions is lost, and therefore even if the original layers consist respectively of "2"-shaped dots and of tiny pinholes, the resulting intensity profile will be infinitely stretched along one dimension, and it will loose its "2"-like shape.

7.4.5 Superposition of uncorrelated screens

What would have happened now if the original screens $r_1(x,y)$ and $r_2(x,y)$ had not been isometric (or at least correlated) before undergoing the linear rotation and scaling transformations $g_1(x,y)$ and $g_2(x,y)$? It is well known that two uncorrelated screens do not generate a Glass pattern in their superposition. This can be explained, just as in the case of aperiodic gratings (Sec. 7.3.2), using the fact that the convolution in Eq. (7.30) is the cross correlation of $p_1(\mathbf{x}')$ and $p_2(\mathbf{x}')$:

$$p_1(\mathbf{x}') \star\star p_2(\mathbf{x}') = p_1(\mathbf{x}') ** p_2(-\mathbf{x}') \tag{7.33}$$

Therefore, as we have already seen in Sec. 7.3.2, if the original screens are correlated, their cross correlation contains in its center a "privileged" area that corresponds to the Glass pattern. But when the original screens are not correlated, their cross correlation does not contain such a "privileged" area, and it simply fluctuates around the same mean level throughout. Hence, no visible macrostructure (Glass pattern) appears in the screen superposition.

7.4.6 Discussion

As we can see, the fundamental moiré relationship between the image domain and the spectral domain holds also in the generalized case where the superposed screens are no longer periodic. The difficulty in the case of aperiodic screens is that its spectral considerations are much less intuitive, and indeed we have obtained the generalization of our results to this case thanks to the insights that we have gained by analyzing the simpler periodic cases in Sec. 7.2.

Thus, in spite of their different appearance, moiré effects that occur between periodic or aperiodic structures are particular cases of the same basic phenomenon, and all of them satisfy the same fundamental rules. Superpositions of periodic structures are simply a particular case in which the elements within each of the layers are arranged periodically, so that the resulting moiré effect is periodic, too.

But the importance of our generalization of the periodic moiré theory into the case of aperiodic layers is not only theoretical: Our results also allow to predict quantitatively the intensity profile of the resulting Glass patterns, and they open the way to the synthesis of Glass patterns having intensity profiles of any desired shapes.

These results may find application in several scientific and technological disciplines. For example, they can be used in vision research, in physiological experiments such as form perception, and in the study of the human visual system, where "classical" Glass patterns have already been used since long [Glass73, Dakin97]. Possible technological fields of application may include precision optical alignment, image registration [Dey91], measurement of microscopic displacements, analogic magnifiers, and even document security and authentication [Amidror04a].

It should be noted that the synthesis of Glass patterns having a predetermined intensity profile (gray level distribution) is completely independent of the synthesis of dot trajectories in the microstructure, which was described in Chapter 4. In the present chapter we explain how to design two random dot screens having a similar dot distribution but different dot shapes (for example: "1"-shaped dots in one layer and tiny pinholes on a black background in the other layer), that generate in their superposition a Glass pattern having a given macroscopic intensity profile (in our example: a Glass pattern having the shape of a largely magnified "1"). In Chapter 4, on the contrary, we were not interested in the shape of the individual dots in the superposed layers nor in the macroscopic intensity profile of the resulting Glass pattern in the superposition. Rather, we were interested in the geometric transformations that were applied to each of the superposed layers, and in their influence on the shapes of the resulting microstructures (dot trajectories) that are generated around the Glass pattern. Therefore these two approaches are complementary, and they can be used independently of each other. See also Problem 7-2.

Finally, note that although we have limited our discussion here to monochrome black and white layers, the extension of our results to the polychromatic case is straightforward, and it can be done exactly as in the periodic case (see Chapter 9 of *Vol. I*).

7.5 Higher order moirés

It is interesting to note that the moiré theory we have developed here for aperiodic line gratings and for aperiodic dot screens is in fact based on a generalization of the first-order $(1,-1)$-moiré between periodic line gratings, or its 2D counterpart, the $(1,0,-1,0)$-moiré between periodic dot screens. The reason is that in aperiodic structures higher-order moirés simply do not exist: While in periodic cases new higher-order moirés may occur when a scaling ratio of $s = 2, 3$, etc. is applied to one of the superposed layers (due to interferences between higher harmonics of the original frequencies), in aperiodic cases no higher-order moirés can exist, since at such scaling ratios no correlation exists between the superposed layers (for instance, a random screen $r(x,y)$ is not correlated with $r(2x,2y)$).

Furthermore, it turns out that even the other first-order moirés from the periodic case do not have equivalents in the aperiodic case: As we have seen throughout this chapter, our generalization from the periodic to the aperiodic case concerns only the "subtractive" $(1,-1)$-moiré (or its 2D counterpart, the $(1,0,-1,0)$-moiré, in the case of dot screens), but not the "additive" $(1,1)$-moiré (or its 2D counterpart, the $(1,0,1,0)$-moiré). The aperiodic equivalent of the $(1,1)$-moiré would be obtained theoretically as a continuous extension of $d_n = a^{(1)}_n a^{(2)}_n$, namely: $P(n) = P_1(n)P_2(n)$ (compare with Eqs. (7.13) and (7.24)). But due to the lack of inversion in $P_2(n)$, the result in the image domain is the cross correlation of $p_1(x')$ and $p_2(-x')$, which are not correlated (the correlated functions are $p_1(x')$ and $p_2(x')$). Therefore in such cases no Glass pattern is generated in the superposition.

The subtractive $(1,-1)$-moiré (and its 2D counterpart, the $(1,0,-1,0)$-moiré) are indeed the strongest, the simplest and the most frequent of all the (k_1,k_2)- or (k_1,k_2,k_3,k_4)-moirés between periodic gratings (or dot screens). They give the most spectacular moiré effects between periodic layers, like those of Fig. 7.1(b), and their explanation, both in the image domain and in the spectral domain, is the most straightforward (see Chapter 4 in *Vol. I*). It is therefore interesting to see that when we randomize the element locations in two originally periodic layers, the only moiré which survives in the resulting layer superposition is precisely the $(1,-1)$-moiré (in the case of gratings) or the $(1,0,-1,0)$-moiré (in the case of screens). Note that even in this surviving moiré, only one of the infinitely many moiré replicas from the periodic case still survives in the aperiodic case; as we have seen in Chapter 3, this is precisely the one which is generated about the mutual fixed point of the two superposed screens.

7.6 Intermediate, partly aperiodic cases

Having presented our generalized Fourier-based approach, which explains quantitatively the moiré (or Glass) patterns between both periodic and aperiodic layers, it would be interesting to see now whether this general approach could also predict the behaviour of moiré patterns in intermediate cases, that is, when the element locations in the two superposed layers are neither fully periodic nor fully aperiodic. Such partly aperiodic layers can be obtained, for example, by adding a varying degree of randomness to the element locations in an initially fully periodic layer (of course, each dot must be perturbed in the same way in both original layers in order to preserve the correlation). As we can see in Fig. 7.13, the resulting Glass patterns in such cases indeed have an intermediate look: Depending on the degree of randomness being added, they may have around the center one or more oscillations between darker and brighter areas. But because the correlation between the layers decreases with the distance, these oscillations are not repeated *ad infinitum* as in the periodic case, and they gradually fade out as we go farther from the center of the Glass pattern. Clearly, the higher the degree of randomness, the lower the number of oscillations around the center. These oscillations are, in fact, replicas of the central Glass pattern, that gradually become more and more blurred and noisy as we go away from the center, until they finally disappear within the background noise.

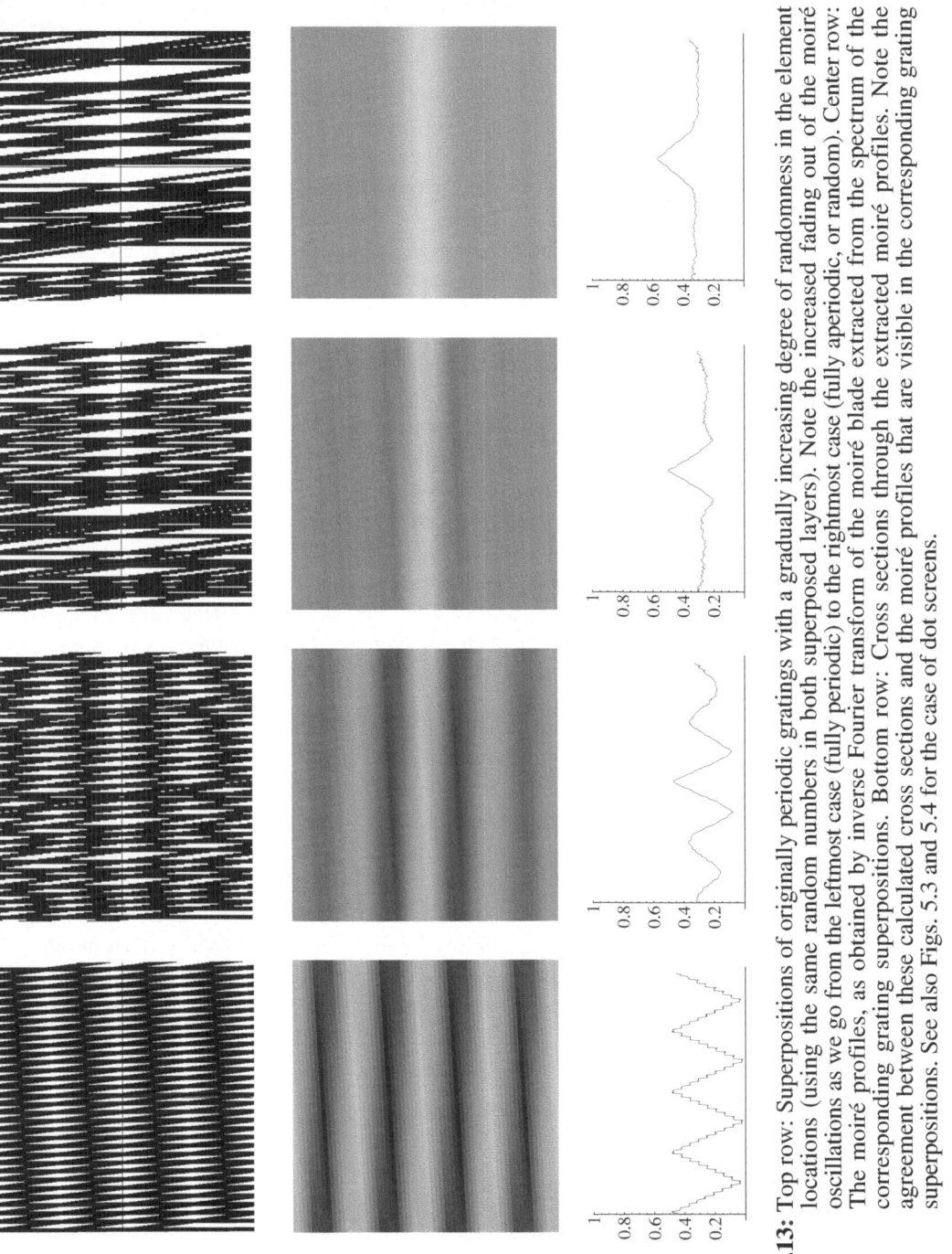

Figure 7.13: Top row: Superpositions of originally periodic gratings with a gradually increasing degree of randomness in the element locations (using the same random numbers in both superposed layers). Note the increased fading out of the moiré oscillations as we go from the leftmost case (fully periodic) to the rightmost case (fully aperiodic, or random). Center row: The moiré profiles, as obtained by inverse Fourier transform of the moiré blade extracted from the spectrum of the corresponding grating superpositions. Bottom row: Cross sections through the extracted moiré profiles. Note the agreement between these calculated cross sections and the moiré profiles that are visible in the corresponding grating superpositions. See also Figs. 5.3 and 5.4 for the case of dot screens.

From the image-domain point of view, the center of the Glass patterns in Fig. 7.13 consists of a gray level that is brighter than areas farther away, due to the partial overlapping of the microstructure elements of both layers in this area. Because of the residual periodicity in the superposed layers, black elements from the two layers that are half a period away from the center are more likely to fall side by side, thus increasing there the covering rate and darkening the macroscopic gray level. Depending on the degree of periodicity in the superposed layers, several such oscillations between brighter and darker areas may occur in the superposition. But as we go farther away from the center the correlation between the two layers becomes lower, and the elements from both layers start falling in an arbitrary, non-correlated manner; in this area the Glass pattern fades out, and we obtain a mean gray level that remains constant throughout.

And indeed, these image-domain considerations are fully confirmed in the spectral domain, too. As one would expect, the Fourier spectrum of such partly periodic layers is intermediate between the spectra of periodic and aperiodic cases: Its blade consists of partly blurred impulses corresponding to the frequencies in question and their harmonics, plus a diffuse background that is typical to random cases. Moreover, as shown in Fig. 7.13, the moiré intensity profile obtained by an inverse Fourier transform of the extracted moiré blade shows clearly the oscillations between darker and brighter areas, which fade out and disappear as we go farther away from the center; and the number of oscillations clearly depends on the degree of randomness. This confirms that our generalized Fourier-based approach indeed predicts the quantitative behaviour of moiré patterns in all cases — periodic, random and intermediate.

It should be noted that although the case of partly random layers has been illustrated here by line gratings for the sake of simplicity, the same results remain true for dot screens, too, and in particular for all the special cases enumerated in Sec. 7.4.2. For example, partly randomized versions of Fig. 7.1(b) indeed look intermediate between Figs. 7.1(a) and (b), with a sharp Glass pattern in the center that is surrounded by several blurred replicas depending on the degree of periodicity in the superposed layers. Note that the influence of the degree of randomness on the microstructures (dot trajectories) in the superposition has been shown earlier in Figs. 5.3–5.4. These figures also illustrate the behaviour of the *macrostructures* in the superposition, but they do so only qualitatively; our Fourier-based approach allows us now to understand this phenomenon quantitatively, in terms of the gray level variations (intensity profiles), as illustrated in Fig. 7.13 for the case of line gratings.

7.7 Intermediate, partly correlated cases

Let us see now how our generalized Fourier-based approach explains quantitatively what happens in the superposition of intermediate, partly correlated cases.

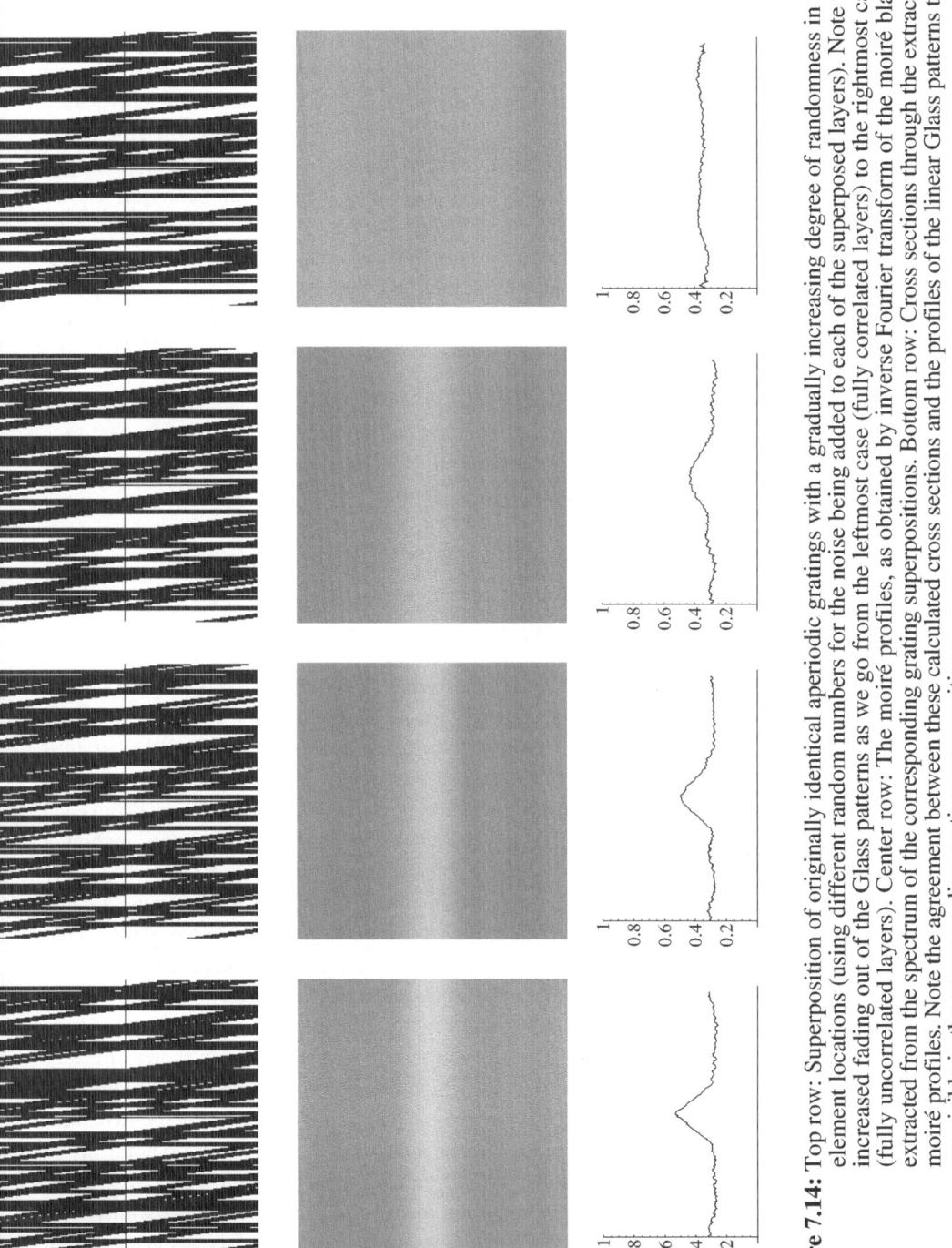

Figure 7.14: Top row: Superposition of originally identical aperiodic gratings with a gradually increasing degree of randomness in the element locations (using different random numbers for the noise being added to each of the superposed layers). Note the increased fading out of the Glass patterns as we go from the leftmost case (fully correlated layers) to the rightmost case (fully uncorrelated layers). Center row: The moiré profiles, as obtained by inverse Fourier transform of the moiré blade extracted from the spectrum of the corresponding grating superpositions. Bottom row: Cross sections through the extracted moiré profiles. Note the agreement between these calculated cross sections and the profiles of the linear Glass patterns that are visible in the corresponding grating superpositions.

Suppose that we superpose two originally identical aperiodic layers, and that we gradually decrease the correlation between them (for example, by adding gradually increasing randomness to the element locations in each of the layers, using different random numbers for each of the layers). Based on our experience so far, we would expect that the Glass pattern in the layer superposition gradually fade out as the correlation between the superposed layers is reduced, until it completely disappears when the layers are no longer correlated. And indeed, the Fourier-based approach that we have developed in the present chapter allows us now to understand this phenomenon quantitatively, as illustrated in Fig. 7.14. Note, once again, that although Fig. 7.14 has been drawn, for the sake of simplicity, using line gratings, the same results remain true for dot screens, too, and in particular for all the cases enumerated in Sec. 7.4.2. For example, partly randomized versions of Fig. 7.1(a) indeed look as blurred, noisy versions of the figure, where the blur rate depends on the degree of randomness that was added to the superposed layers.

Similarly, if we decrease the correlation between two aperiodic dot screens that give visible dot trajectories in the microstructure of their superposition, the dot trajectories gradually fade out until they completely disappear when the layers are no longer correlated.

7.8 Glass patterns and cross correlation

Another interesting question that could be asked at this point is: what kind of aperiodic layers give a Glass pattern in their superposition? As we already know, a necessary condition on the superposed layers in order to generate a Glass pattern is that they must be correlated, or at least partly correlated. At first sight, however, it seems that this condition is not sufficient: For example, the superposition of two images consisting each of a single black dot is unlikely to give any visible Glass pattern even if the dots are highly correlated, while the superposition of two correlated screens made of such dots can clearly generate a Glass pattern. Intuitively, this means that the original images must also be "structured" (i.e. consist of a distinct microstructure) in order to generate a visible Glass pattern. Now that we already understand the mathematical explanation of Glass patterns, let us see what is the correct interpretation of this observation. Consider, for example, the superposition of one black straight line with a slightly rotated copy of itself (Fig. 7.15). At first sight, no Glass pattern can be expected in this superposition, even though the two layers are correlated. But a second thought, which is indeed confirmed by mathematical evidence, shows that this reasoning is not correct. Clearly, in the area where the two black lines cross each other, there is less black than in areas where the two lines do not overlap; and indeed, if we observe the superposition at a grazing angle along the x axis, in order to obtain visually a better spatial integration of the surface, we will see a slightly whiter zone around the overlapping area. This is indeed confirmed by mathematical evidence (see Fig. 7.15(g)): the inverse Fourier transform of the extracted moiré blade gives, as expected, an

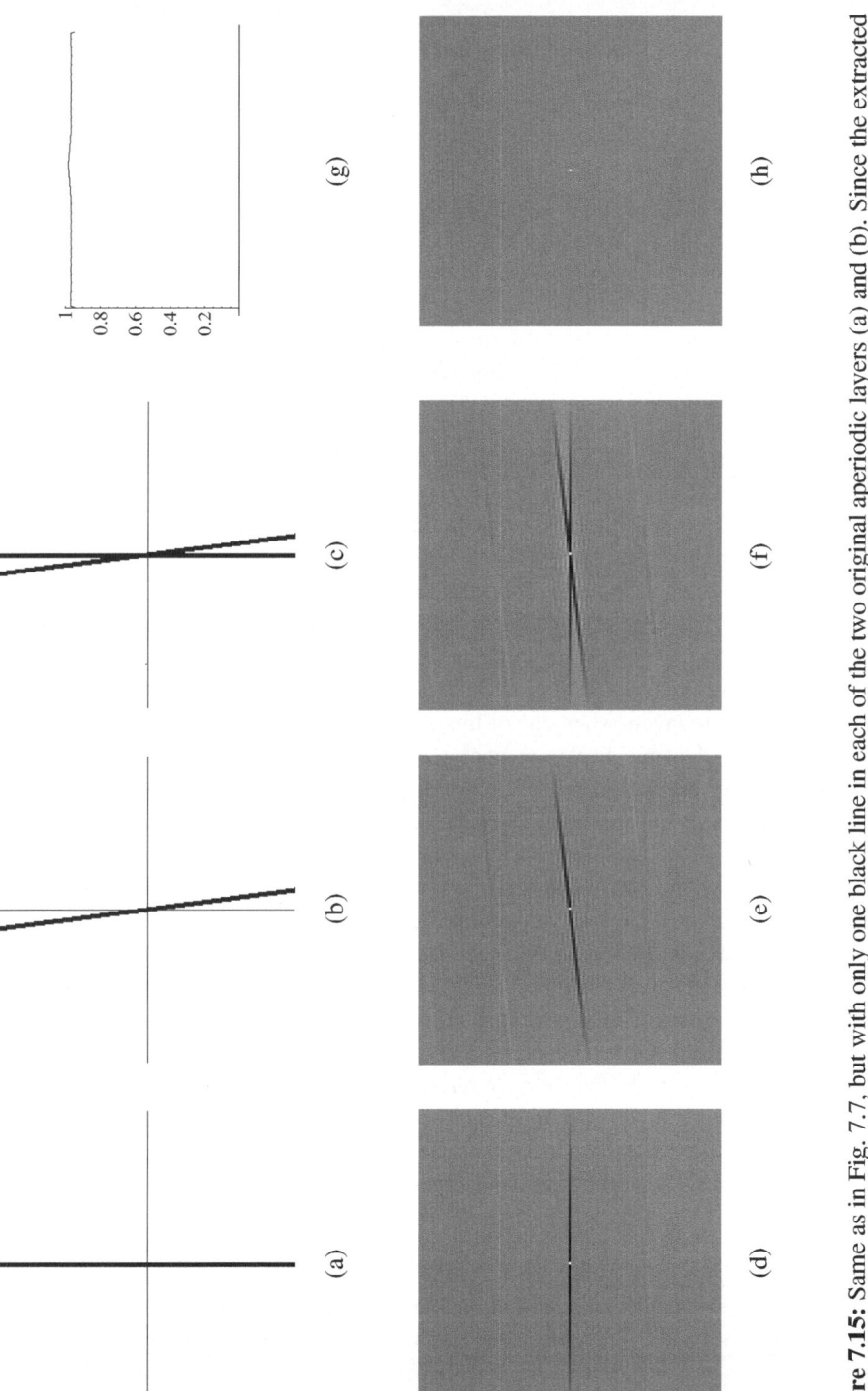

Figure 7.15: Same as in Fig. 7.7, but with only one black line in each of the two original aperiodic layers (a) and (b). Since the extracted Glass pattern obtained by inverse FFT of (h) is almost fully white, and hence hardly visible, we show in (g) instead a vertical cross section through this image, showing more clearly its intensity profile; note the slightly whiter zone in its center. Compare with the intensity profile of Fig. 7.7(g) which is shown in Fig. 7.11(a).

almost uniformly white image, but the central zone of this image is slightly whiter. This whiter zone is, indeed, a Glass pattern, although it can hardly be perceived visually.

Having understood this point, we arrive to the following important conclusion: Correlation (or at least partial correlation) between the original layers is a *necessary and sufficient* condition for the generation of a Glass pattern in the superposition. It is clear, however, that when the original layers are also highly structured, the generated Glass pattern becomes much more visible (compare Fig. 7.7 and Fig. 7.15). For example, when the superposition consists of many crossing lines, as in Fig. 7.7(c), the accumulated effect of the whiter zone due to the crossing lines is significantly stronger, and the Glass pattern becomes clearly visible. This is, indeed, the correct mathematical interpretation of our initial intuition regarding "structured" layers (see also Problem 7-12).

The role of correlation between the original layers as a necessary and sufficient condition for the generation of a Glass pattern can be better understood by reformulating Propositions 7.5 and 7.7 in terms of cross correlation instead of convolution, using Eqs. (7.27) and (7.33). These reformulated propositions express the isolated intensity profile of the Glass pattern (as shown, for example, in Fig. 7.7(g), i.e. without considering the microstructure phenomena) in terms of normalized cross correlation. It is important to note, however, that the intensity profile of the Glass pattern is *not* equal to the cross correlation between the two superposed layers. To see this, note that the value of the cross correlation between two given layers indicates in each point (x,y) the *global* agreement between the two entire layers when one of them is shifted by x,y; thus, cross correlation inherently involves a dynamic displacement of one layer on top of the other (remember the "move and multiply" graphical presentation of cross correlation). The Glass pattern, on the other hand, indicates the zones of *local* agreement between the two layers in the given *static* superposition, without applying to the layers any further displacements (note that a Glass pattern may significantly change or even disappear when one of the layers is displaced). This gap between the Glass patterns (i.e. the zones of high local correlation) and the cross correlation is precisely bridged by our two propositions: As indicated in their first part, these propositions are not formulated in terms of the superposed layers themselves but in terms of their untransformed counterparts. The cross correlation between the two *untransformed* layers (which are highly similar, due to the required condition of isometry) gives the *normalized* intensity profile of the resulting Glass pattern. Part (2) of each of these propositions serves to bring this normalized intensity profile into its actual scaling and orientation (or more generally, into its actual geometric transformation), which corresponds to the actual Glass pattern as it appears in our given layer superposition (see an example for Figs. 7.12 and 7.16 in Problem 7-13).

In order to better understand this point, let us consider its simpler, 1D counterpart. Suppose we are given an aperiodic time signal $p_1(t)$ and its copy $p_2(t)$ (whose amplitude may be slightly distorted). The cross correlation between these two signals is a new function of t, $p(t) = p_1(t) \star p_2(t)$, that has a strong peak at $t = 0$ (because the signals $p_1(t)$ and $p_2(t)$ are highly correlated along the entire t axis when their mutual displacement is

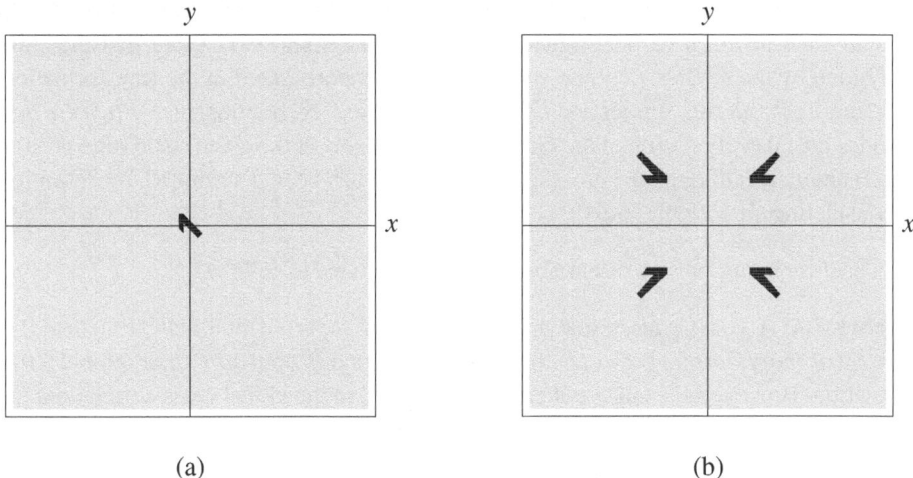

Figure 7.16: The effect of applying the transformation $\mathbf{g}(x,y) = (ay^2 - x_0, y_0 - ax^2)$ (where $a, x_0, y_0 > 0$) as a domain transformation. (a) The 2D function $f(x,y) = \text{one}(x,y)$, that takes the value 0 (black) inside the borders of the "1"-shaped element and 0.9 (light gray) everywhere else. (b) The transformed function $f(\mathbf{g}(x,y))$. Note that the domain transformation $\mathbf{g}(x,y)$ maps the origin onto the four points $(x,y) = (\pm\sqrt{y_0/a}, \pm\sqrt{x_0/a})$, which are the solutions of $\mathbf{g}(x,y) = (0,0)$; see Problems 7-15 and 7-16.

zero). Now, suppose that we apply to our signals two different transformations g_1 and g_2 along the t axis (for example, linear time scaling). The resulting signals are $r_1(t) = p_1(g_1(t))$ and $r_2(t) = p_2(g_2(t))$. Clearly, these new signals are no longer *globally* correlated, since no displacement between them will result in a global match along the entire t axis. When the displacement between these transformed signals is zero, there exists a certain narrow zone of local correlation between them (in our example of linear scaling this zone is located around the point $t = 0$), but farther away to both directions of the t axis the local correlation is obviously destroyed. When we displace one of these signals, the zone of local correlation between them will simply slide along the t axis, but it will still be located in a certain narrow zone. Clearly, this zone of local correlation between the two signals is the 1D analog of our Glass pattern. The global cross correlation between the two transformed signals $r_1(t)$ and $r_2(t)$ will not have a significant peak at any value of t, since no displacement t can bring them into full match, as was the case in the original *untransformed* signals $p_1(t)$ and $p_2(t)$. However, if we normalize our time signals before we perform the cross correlation, i.e. if we first undo the effects of the transformations g_1 and g_2 (by applying to our signals the inverse transformations g_1^{-1} and g_2^{-1}), then the cross correlation will give, again, the same peak around $t = 0$ as before we applied g_1 and g_2. At this stage the analogy with our 2D case becomes clear: What we have done here is simply the 1D analog of part (1) of propositions 7.5 and 7.7. Now, according to part (2) of these propositions, if we apply the difference transformation $g(t) = g_1(t) - g_2(t)$ to this normalized cross correlation, we will obtain the intensity profile of the "Glass pattern" in the real superposition of our transformed signals $r_1(t)$ and $r_2(t)$. Note that the shifts undergone by the signals $r_1(t)$ and $r_2(t)$ are already incorporated in the transformations $g_1(t)$ and $g_2(t)$, so the application of $g(t)$ to the cross correlation $p(t) = p_1(t) \star p_2(t)$ provides the intensity profile of the Glass pattern precisely at the given static superposition of $r_1(t)$ and $r_2(t)$ (although the cross correlation itself has been determined, by definition, by considering all possible shifts).

In conclusion, our present discussion can be summarized as follows:

Remark 7.4: A Glass pattern that appears in a layer superposition indicates zone(s) of high *local correlation* between the superposed layers, but *not* their global cross correlation. However, the Glass pattern is also related to the global cross correlation; this relationship is expressed by Propositions 7.5 and 7.7 (or rather by their versions formulated in terms of cross correlation instead of convolution). ■

Note that we have expressed Propositions 7.5 and 7.7 in terms of convolution rather than in terms of cross correlation in order to stress the similarity with their periodic counterparts. In the periodic case both subtractive and additive moirés may exist, and hence the minus sign does not always appear in the periodic counterpart of Eq. (7.17). It is therefore natural to formulate the periodic case in terms of convolution (see also Footnote 4 in Sec. 4.2 of *Vol. I*). But since in the aperiodic case we are only concerned with subtractive moirés (see Sec. 7.5), it may be more convenient to formulate this case directly in terms of cross correlation. This is done in the following reformulation of Propositions 7.5 and 7.7, which also incorporates their generalization to cases with non-linear layer transformations:

The fundamental Glass-pattern theorem (for aperiodic line gratings): Let $r_1(x,y)$ and $r_2(x,y)$ be two aperiodic line gratings that are obtained by applying the bending functions (linear or not) $g_1(x,y)$ and $g_2(x,y)$, respectively, to two isometric aperiodic line gratings having the intensity profiles $p_1(x')$ and $p_2(x')$:

$$r_1(x,y) = p_1(g_1(x,y)), \qquad\qquad r_2(x,y) = p_2(g_2(x,y))$$

Then, the Glass pattern $m(x,y)$ in the superposition of $r_1(x,y)$ and $r_2(x,y)$ is given by:

$$m(x,y) = p(g(x,y))$$

where:

(1) $p(x')$, the normalized intensity profile of the Glass pattern, is the cross correlation of the normalized intensity profiles of the original gratings:

$$p(x') = p_1(x') \star p_2(x') \tag{7.34}$$

(2) $g(x,y)$, the bending function which brings $p(x')$ back into the actual Glass pattern $m(x,y)$ as it appears in the superposition of the two *transformed* layers, is given by:

$$g(x,y) = g_1(x,y) - g_2(x,y) \quad \blacksquare \tag{7.35}$$

Thus, the resulting linear Glass pattern is obtained along the zero-level curve of the surface $g(x,y)$ (the solution of the equation $g(x,y) = 0$). If $g(x,y)$ has several zero-level curves, then a Glass pattern is generated along each of these level curves, and if $g(x,y)$ has no zero-level curves then no Glass patterns are generated in the superposition.

The fundamental Glass-pattern theorem (for aperiodic dot screens): Let $r_1(\mathbf{x})$ and $r_2(\mathbf{x})$ be two aperiodic screens that are obtained by applying the mappings (linear or not) $\mathbf{g}_1(\mathbf{x})$ and $\mathbf{g}_2(\mathbf{x})$, respectively, to two isometric aperiodic screens having the intensity profiles $p_1(\mathbf{x}')$ and $p_2(\mathbf{x}')$:

$$r_1(\mathbf{x}) = p_1(\mathbf{g}_1(\mathbf{x})), \qquad\qquad r_2(\mathbf{x}) = p_2(\mathbf{g}_2(\mathbf{x}))$$

Then, the Glass pattern $m(\mathbf{x})$ in the superposition of $r_1(\mathbf{x})$ and $r_2(\mathbf{x})$ is given by:

$$m(\mathbf{x}) = p(\mathbf{g}(\mathbf{x}))$$

where:

(1) $p(\mathbf{x}')$, the normalized intensity profile of the Glass pattern, is the cross correlation of the normalized intensity profiles of the original, untransformed screens:

$$p(\mathbf{x}') = p_1(\mathbf{x}') \star\star p_2(\mathbf{x}') \tag{7.36}$$

(2) $\mathbf{g}(\mathbf{x})$, the transformation which brings $p(\mathbf{x}')$ back into the actual Glass pattern $m(\mathbf{x})$ as it appears in the superposition of the two *transformed* layers, is given by:

$$\mathbf{g}(\mathbf{x}) = \mathbf{g}_1(\mathbf{x}) - \mathbf{g}_2(\mathbf{x}) \quad\blacksquare \tag{7.37}$$

Thus, the resulting Glass pattern is obtained about the zero of $\mathbf{g}(x,y)$ (the solution of the system of two equations $\mathbf{g}(x,y) = (0,0)$). If $\mathbf{g}(x,y)$ has several zeros, then a Glass pattern is generated about each of them, and if $\mathbf{g}(x,y)$ has no zeros then no Glass patterns are generated in the superposition (see, however, Sec. 3.7 on almost fixed points).

As explained in Appendix G, these two theorems are based on *generalized* Fourier considerations. In fact, they are equivalent to particular cases of their respective periodic counterparts, which are simply limited to the subtractive first-order moiré (taking $(k_1,k_2) = (1,-1)$ or $(k_1,k_2,k_3,k_4) = (1,0,-1,0)$, respectively), and where only the fundamental periods of the moiré (the ones that are located about the mutual fixed points) are taken into account, without the duplicates due to periodicity. See also Remark 3.2 and Secs. 7.5 and 6.5.

As we can see, the Fourier (or generalized Fourier) approach allows us to obtain full, quantitative information on the Glass patterns and their macroscopic gray levels. This clearly englobes the geometric information obtained by fixed-point considerations (see Sec. 3.5) or by indicial equations (see Sec.6.3), that is limited to the locations of the center points (or center lines) of the Glass patterns. Furthermore, unlike the approach presented in Sec. 6.6 that only provides an *approximation* to the gray levels of the Glass patterns, the Fourier-based approach provides the precise mathematical values. It should be noted, however, that just as in the periodic case, Fourier-based considerations cannot treat microstructure phenomena in the layer superposition (such as dot trajectories); such phenomena should be investigated using other mathematical tools (see Chapter 4).

PROBLEMS

7-1. *Synthesis of a linear Glass pattern with a given intensity profile in the superposition of two line gratings*. Design two aperiodic line gratings that would give in their superposition a linear Glass pattern whose profile shape is: (a) trapezoidal; (b) triangular; (c) cosinusoidal.

7-2. Is it possible to combine the approaches described in Chapters 4 and 7 in order to synthesize a Glass pattern with a predefined intensity profile (say, a "1"-shaped profile, like in Fig. 7.1(a)), having specified dot trajectories in its microstructure? For example, can you regenerate the "1"-shaped Glass pattern of Fig. 7.1(a) with circular, radial, spiral or hyperbolic dot trajectories surrounding it, as in Figs. 2.1(c),(e),(g) and 2.2(a)? *Hint*: Consider Fig. 4.15, and remember that the dot trajectories in Figs. 2.1 and 2.2 were obtained in the superposition of two originally identical dot screens consisting of black dots on white background, while for the synthesis of a Glass pattern with a given intensity profile like in Fig. 7.1(a) one of the superposed layers should consist of tiny pinholes on a black background. Can you obviate this problem by generating a Glass pattern using Case 2 of Sec. 7.4.2 rather than Case 1?

7-3. *Document security*. Can you think of an application of Glass patterns having a predetermined intensity profile (gray level distribution) that appear between aperiodic dot screens, as shown, for example, in Fig. 7.1(a), for document authentication and anti-counterfeiting? *Hint*: The document can be protected by means of a halftoned image, logo, etc. that is printed at high resolution using a specially designed pseudo-random halftone screen. This special screen consists of tiny halftone dots having a predefined shape that remains unchanged throughout a wide range of gray levels; for example, the halftone dots may have the shape of the letter pair "US" in varying sizes and linewidths, to allow for the various gray levels of the image. The locations of the screen dots are determined by a sequence of numbers that is generated by a pseudo-random number generator. When the aperiodic halftoned image thus obtained is superposed by an appropriate aperiodic pinhole screen whose pinhole locations are determined by the *same* pseudo-random number sequence, a highly visible Glass pattern of the same shape (in our example, a Glass pattern having the shape "US") will be visible in the image within the superposed area. Since the detail of the tiny halftone dots will not resist photocopying, scanning, or any other digital or analog copying method, a counterfeited document will be immediately recognized by the absence or by the corrupted shape of the Glass pattern when the pinhole screen is superposed on the document. Compare this method with its periodic counterpart (see Problem 4-13 in *Vol. I*). What are the main differences between the two methods? A significant advantage of the aperiodic method is in its intrinsically incorporated encryption system due to the arbitrary choice of the random number sequence for the generation of the two pseudo-random screens being used. Clearly, a Glass pattern is only visible in the superposition if the revealing pinhole screen is generated with the same pseudo-random number sequence as the aperiodic halftone screen that is printed on the document. Thus, keeping this sequence secret prevents fraudulent production of documents carrying the encrypted screen, or unauthorized production of the required revealing screen. Furthermore, if the pseudo-random number sequence depends on the serial number of the document, it becomes impossible for a potential counterfeiter to generate an appropriate pinhole screen that will be able to reveal the Glass pattern. For more details see [Amidror04a].

7-4. It is shown in Sec. 7.5 that in stark contrast to periodic-layer superpositions, in a superposition of *aperiodic* layers no higher-order moirés can be generated. To better understand this, suppose you are given a superposition of two periodic line gratings that generates a clearly visible second-order moiré (see, for example, Fig. 2.6 in *Vol. I*). What do you think would happen to this second-order moiré when you gradually randomize the line distances of the original gratings, using the same randomization in both original layers?

7-5. What would you expect to obtain in the superposition in cases such as Fig. 2.3 or Fig. 3.6 if the two original layers consisted of "1"-shaped dots and white pinholes on a black background, respectively?

7-6. Find the local orientations and the local stretching factors of the moiré pattern of Fig. 7.12(b) at different points (x,y) throughout the plane. Verify the agreement of your results with the figure. *Hint*: Apply Eqs. (C.10) in Sec. C.4 of Appendix C to the moiré transformation (7.32). Remember that $\mathbf{g}_M(x,y)$ is a domain transformation.

7-7. How do you explain the fact that the two Glass patterns that are mirror imaged in Fig. 7.12(a) are precisely the ones that are surrounded in Fig. 3.15(a) by hyperbolic dot trajectories? *Hint*: The moiré transformation $\mathbf{g}_M(x,y) = \mathbf{g}_1(x,y) - \mathbf{g}_2(x,y)$ is given here by Eq. (7.32), while the dot trajectories are determined by the transformation $\overline{\mathbf{h}}(x,y) = \overline{\mathbf{g}}_1(x,y) - \overline{\mathbf{g}}_2(x,y) = (x_0 - ay^2, ax^2 - y_0)$ (remember that if $\mathbf{g}_1(x,y) = (x, y + y_0 - ax^2)$ and $\mathbf{g}_2(x,y) = (x + x_0 - ay^2, y)$ then $\overline{\mathbf{g}}_1(x,y) = (x, y - y_0 + ax^2)$ and $\overline{\mathbf{g}}_2(x,y) = (x - x_0 + ay^2, y)$). Show that the Jacobians of $\mathbf{g}_M(x,y)$ and $\overline{\mathbf{h}}(x,y)$ are both equal to $4a^2xy$. What does this say on the reflecting properties of the four Glass patterns, and on the shapes of the dot trajectories that surround each of them? See also Example H.4 in Appendix H.

7-8. What would you expect to obtain in the four superpositions shown in Fig. 3.10 if in each of them the dots of one of the two layers were replaced by "2"-shaped dots, and the dots of the other layer were replaced by tiny pinholes on a black background? Given that in this case the layer transformations are $\mathbf{g}_1(x,y) = (x - ay^2, y)$ and $\mathbf{g}_2(x,y) = (x\cos\alpha + y\sin\alpha, -x\sin\alpha + y\cos\alpha)$ (see Example 3.4), find the moiré transformation $\mathbf{g}_M(x,y) = \mathbf{g}_1(x,y) - \mathbf{g}_2(x,y)$ and its Jacobian, and show that this transformation is locally reflecting at one of the two mutual fixed points of $\mathbf{g}_1(x,y)$ and $\mathbf{g}_2(x,y)$, and locally non-reflecting at the second fixed point. What does this say about the shape of the Glass patterns in this superposition? *Hint*: The locations of the two mutual fixed points of $\mathbf{g}_1(x,y)$ and $\mathbf{g}_2(x,y)$ have been found in Example 3.4 (see Eq. 3.27)).

7-9. When the moiré transformation $\mathbf{g}_M(x,y) = \mathbf{g}_1(x,y) - \mathbf{g}_2(x,y)$ is linear or affine, the local orientation and the local stretching factor of the moiré pattern between two periodic screens remains constant throughout the entire plane. Use Eqs. (C.10) in Sec. C.4 of Appendix C to find these constant values for a general linear or affine moiré transformation. Compare with the results of Sec. 7.4.3 and with the particular case shown in Fig. 7.1.

7-10. Although the operations of convolution and cross correlation have similar definitions, their properties are quite different. For example, while convolution is commutative, $f(x,y) ** g(x,y) = g(x,y) ** f(x,y)$, cross correlation is not: if $c_{f,g}(x,y) = f(x,y) \star\star g(x,y)$ and $c_{g,f}(x,y) = g(x,y) \star\star f(x,y)$ then $c_{g,f}(x,y) = c_{f,g}(-x,-y)$ [Gaskill78 pp. 172–173]. Now, our explanation of the Glass patterns in the superposition of aperiodic layers (see, for example, Sec. 7.3.2 and Sec. 7.4.5) is based on the cross correlation between the original layers. And yet, although the cross correlation between our two layers is not commutative, the resulting Glass pattern does not depend on the order in which the two layers are superposed. How do you explain this contradiction? *Hint*: Note that the cross correlation determines the profile shape of the Glass pattern, but its orientation is determined by the vectors $\mathbf{a} = \mathbf{f}_1 - \mathbf{f}_3$ and $\mathbf{b} = \mathbf{f}_2 - \mathbf{f}_4$, as explained in Sec. 7.4.3. When exchanging the order of the superposed layers the cross correlation is indeed rotated by 180°, but the vectors $\mathbf{a}$ and $\mathbf{b}$ also change their signs, thus cancelling out the 180° rotation of the intensity profile.

7-11. What happens in Propositions 7.5 and 7.7 when both of the original layers undergo the same transformation (and in particular, when they both undergo the identity transformation, meaning that the layers remain untransformed)? What is the meaning of the difference transformations $g(x,y) = g_1(x,y) - g_2(x,y) = 0$ or $\mathbf{g}(x,y) = \mathbf{g}_1(x,y) - \mathbf{g}_2(x,y) = (0,0)$ that are obtained in such cases? *Hint*: In such situations, where the two superposed layers perfectly coincide on top of each other, the Glass pattern is infinitely big and hence invisible. As we already know from Sec. 2.3.2, this is precisely the aperiodic counterpart of a *singular moiré* in the periodic case (see Sec. 2.9 in *Vol. I*). In order to understand what happens in such situations, start with the familiar non-singular case in which one of the two original layers has undergone a slight scaling transformation, and then try to see what happens while the scaling rate tends to 1 and the two layers gradually approach perfect coincidence.

7-12. To better understand why the superposition of highly structured aperiodic layers gives more visible Glass patterns (see Sec. 7.8), sketch using the "move and multiply" method the cross correlation between two copies of an image consisting of one vertical black line on a white background, and then the cross correlation between two copies of an image consisting of several black vertical lines with different line distances (since we assume that our images are aperiodic). Because all the images are constant along the y direction, you can simply draw their 1D cross correlations along the x axis. What is the difference between the two cross correlations? How does this explain the fact that Glass patterns are more visible in the superposition of highly structured aperiodic layers?

7-13. Looking at Fig. 7.12 a student makes the following reasoning: Within the four dark "2"-shaped areas of this figure the two superposed layers are well correlated (the pinholes of the second layer coincide with the tiny "2"-shaped dots of the first layer), while in the remaining areas of the figure the correlation between the two layers is low. Therefore, the isolated intensity profile of the Glass pattern (i.e. the four macroscopic "2"-shaped forms, without taking into account the microstructures) shows the cross correlation between the two superposed layers. Do you agree with this reasoning? *Hint*: In fact, this student is confusing the notions of *cross correlation* between two functions (or images) and *local correlation* between two functions. The cross correlation between two 2D functions is a third 2D function that indicates, for each relative x,y shift, the global volume under the product of the two functions (remember the "move and multiply" graphical presentation of cross correlation). Thus, points (x,y) in which the value of the cross correlation function is high indicate shifts of x,y between the two layers in which the *global* volume under the product of the two entire layers is high. Obviously, cross correlation can detect displacements x,y in which the two *entire* layers are *globally* well correlated, but it cannot detect isolated zones of *local* correlation between the two layers, since the contribution of such local zones is relatively small, and it may be buried and lost within the global volume under the product of the entire layers. Hence, the values of the cross correlation function do *not* indicate the location of the Glass patterns, which are precisely the zones of high *local* correlation within the given static superposition. In order to better illustrate this, note that the original displacements of the two layers cannot change the shape of the resulting cross correlation (this follows immediately from the definition of cross correlation). However, the shapes observed in Fig. 7.12 clearly do depend on the displacements of the two layers (for example, think what happens in the superposition when the displacements x_0 and y_0 are positive, and what happens when they are negative). The actual contribution of cross correlation in Fig. 7.12 is explained by the fundamental Glass pattern theorem for aperiodic screens: If we first undo the transformations $g_1(x)$ and $g_2(x)$ undergone by the two superposed layers, we obtain the two original *untransformed* layers (which are highly similar, due to the required condition of isometry). The cross correlation between them gives the *normalized* intensity profile (7.36) of the Glass pattern that is generated in the superposition. But in order to obtain the actual intensity profile of the Glass pattern as it appears in the given superposition of the two *transformed* layers, we have to apply to the normalized intensity profile (7.36) the difference transformation $g(x) = g_1(x) - g_2(x)$. In the case of Fig. 7.12 this difference transformation is given by $g(x,y) = (ay^2 - x_0, y_0 - ax^2)$ (see Eq. (7.32)); the effect of this domain transformation on an object that is located about the origin (as is the normalized intensity profile (7.36)) is illustrated in Fig. 7.16. Note that although the cross correlation has been performed, by definition, by considering all possible shifts, the specific displacement between the two layers in the given static superposition are taken care of by the transformations $g_1(x)$ and $g_2(x)$, and the corresponding displacement of the Glass pattern in the superposition is provided by the transformation $g(x)$.

7-14. Can the local correlation between two images $r_1(x,y)$ and $r_2(x,y)$ within a certain zone be quantified by the volume under the product $r_1(x,y) r_2(x,y)$ within this zone (possibly normalized by the area of the zone)? *Hint*: Note that in Fig. 7.12 the zones of high local correlation are *darker* than zones of low local correlation, while in Fig. 3.15 the zones of high local correlation are *lighter* than zones of low local correlations.

7-15. Explain the behaviour of the domain transformation $g(x,y) = (ay^2 - x_0, y_0 - ax^2)$ shown in Fig. 7.16. For this end, consider Fig. 7.17, and compare it with Example D.3 and Fig. D.12 in Appendix D (where $x_0 = y_0 = 0$ and the roles of x and y are not inversed). What is the effect of the layer shifts x_0 and y_0? What do you expect to see in Fig. 7.16 when the layer shifts are zero? And when they are negative? How does this explain the Glass patterns that would appear in Fig. 7.12 with such layer shifts? Where are the zeros of the transformation $g(x,y)$ (i.e. the points (x,y) for which $(ay^2 - x_0, y_0 - ax^2) = (0,0)$) in each of these cases, and how are they related to the location of the Glass patterns? (See Sec. 3.5).

7-16. A student argues that the transformation $g(x,y) = (ay^2 - x_0, y_0 - ax^2)$ maps the origin $(0,0)$ onto the single point $(-x_0, y_0)$, and not onto four different points as claimed in the legend of Fig. 7.16. What is the source of his error? *Hint*: When speaking about a transformation $g(x,y)$ one should carefully distinguish between its role as a *direct* transformation and its role as a *domain* (and hence, inverse) transformation. When considered as a direct transformation, $g(x,y)$ indeed maps the origin onto the single point $(-x_0, y_0)$, as shown in Figs. 7.17(a),(b). But in Fig. 7.16 $g(x,y)$ is applied to the function $f(x,y)$ as a *domain* (and hence, inverse) transformation, as shown in Figs. 7.17(c),(d). Find the explicit form of this inverse transformation, and show that it indeed maps the origin onto four different points. See Appendix D for a more detailed discussion.

7-17. *The fundamental Glass-pattern theorems and their periodic (or rather repetitive) counterparts*. Compare the fundamental Glass-pattern theorems given at the end of Sec. 7.8 with the fundamental moiré theorems that are given in Secs. 10.9.1 and 10.9.2 of *Vol. I*. What are the main differences between them? Can you think of a general formulation that englobes both the periodic (or repetitive) case and the aperiodic case?

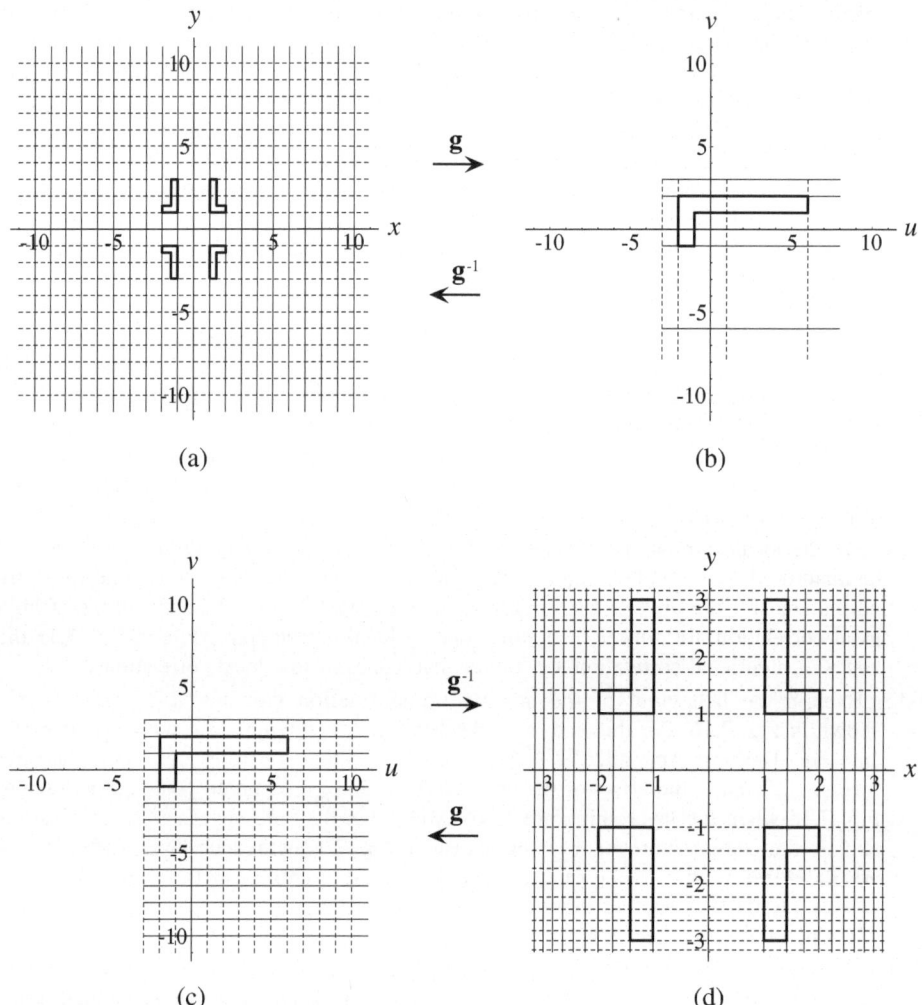

Figure 7.17: (a),(b) The effect of the transformation $(u,v) = (y^2 - 3, 3 - x^2)$ on the unit grid consists of a non-linear expansion, a 90° rotation due to the $-x,y$ inversion, and a shift of $(-3,3)$. Each of the four objects in (a) is mapped into the same image in (b). (c),(d) The effect of the inverse transformation on the unit grid consists of a non-linear contraction, a $-90°$ rotation and a shift. In both cases the plain grid lines are mapped into the respective plain lines, and the dashed grid lines are mapped into the respective dashed lines. Note that part (d) has been magnified to better show details. Compare with Fig. D.12 and Example D.3 in Appendix D.

7-18. *Advantages of the Fourier-based approach.* In Chapter 3 we have seen that Glass patterns are generated about the mutual fixed points of the layer transformations $\mathbf{g}_1(x,y)$ and $\mathbf{g}_2(x,y)$, namely, about the zeros of the difference transformations, the points

(x,y) for which $\mathbf{g}(x,y) = \mathbf{g}_1(x,y) - \mathbf{g}_2(x,y) = (0,0)$ (see Proposition 3.2). How does the Fundamental Glass-pattern theorem for aperiodic dot screens encompass this result and generalize it using the Fourier-based theory? Enumerate the main advantages of the Fourier-based approach, but also its main limitations (such as its inability to deal with microstructure phenomena like dot trajectories).

7-19. As explained in Sec. 2.4, most of the aperiodic figures throughout this book were generated with a pseudo-random element distribution rather than with a purely random element distribution, in order to avoid the inherent blotchiness due to the clustering and avoidance effects of purely random layers. This is illustrated in Fig. 7.18, that shows side by side the superpositions of pseudo-random layers and their purely random counterparts. How do you explain the darker bands that surround the central bright area of the Glass pattern in the pseudo-random cases? *Hint*: As explained in Sec. 2.4, pseudo-random algorithms are based on perturbing the element locations in an originally periodic structure by some degree of random noise. This means that the pseudo-random distribution introduces into our aperiodic layers some residual periodicity. The effect of such a residual periodicity on the intensity profile of the resulting Glass pattern is shown in Fig. 7.13 (compare the images in the last two columns).

7-20. Can you suggest a method for visually verifying if an aperiodic image (or signal) contains some periodic noise or some periodic residues, without using Fourier transformations? Can you also determine the degree of periodicity using this method? *Hint*: Consider the superposition of the given image with a slightly rotated or scaled copy of itself (for example, a photocopy on transparency).

7-21. What would you expect to see in Fig. 7.1(a) if the "1"-shaped elements of the first layer were oriented in random directions? And if they had random sizes?

7-22. *Artificial Glass patterns.* Since their discovery, Glass patterns (and in particular their dot trajectories) have been widely used in the research of the human visual system (see, for example, [Glass69], [Dakin97], [Wilson98], [Cardinal03], [Barlow04]). However, many authors in this field prefer to use "artificial" Glass patterns that are not obtained in a real superposition of random dot screens, but rather generated artificially on computer. To construct such a pattern, pairs of dots *having a constant dot distance* are placed at random such that the orientation of the pair falls along the desired curves. For example, in order to create a circular or a hyperbolic Glass pattern the dot pairs are randomly placed along the contours defined by the equation $x^2 + y^2 = k$ or $xy = k$, respectively, where the constant k is determined by the x,y coordinates of the first dot in each pair (see, for example, [Wilson98] where a detailed and illustrated explanation is provided). One of the reasons for using such artificial Glass patterns resides in the ease of their generation: Indeed, this method allows one to generate patterns having any desired dot trajectories directly from the curve equation, without having to manipulate two superposed layers and without mastering the theory presented in Chapter 4 for the generation of two layers that would give in their superposition the desired dot trajectories. Can you think of other reasons for using such artificial Glass patterns? How do they differ from their true counterparts (such as those shown in Fig. 2.1)? *Hint*: One of the main differences between such artificial Glass patterns and their real counterparts is that in the former the distance between the two dots within each pair is kept constant throughout the image, while in the latter the distance between the dots of each pair is zero at the center of the Glass pattern and it gradually increases with the distance from the center. This also implies an important difference in terms of the macrostructure: In artificial Glass patterns the global gray level remains constant throughout the image, and it does not become brighter in the center (as it does in real Glass patterns, due to the partial overlapping of dots from both layers in that area).

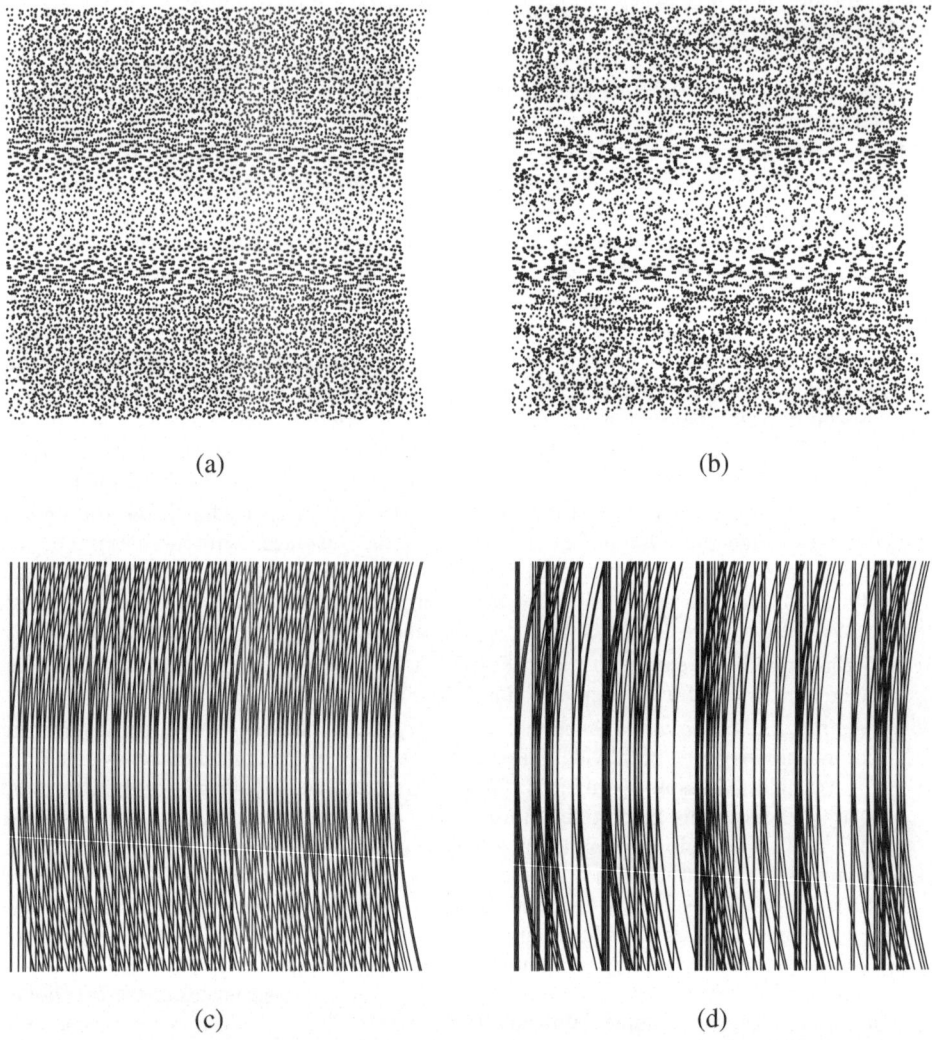

(a)   (b)

(c)   (d)

Figure 7.18: A comparison between the Glass patterns that are generated in the
superposition of two pseudo-random layers (left) and in the
superposition of their fully random counterparts (right). (a) Same as
Fig. 3.5(c). (b) The fully random counterpart of (a). (c) Same as Fig.
6.3(c). (d) The fully random counterpart of (c). Notice in (a) and in (c)
the darker bands that appear above and below the central bright area of
the Glass pattern; these darker bands may be seen better when
observing the figure from a distance of 2–3 meters so that the
microstructure details are no longer discerned by the eye.

And indeed, a further reason for using such artificial Glass patterns in the research of
the human visual system is that they provide the desired microstructure effects while
eliminating the macrostructure effects; this allows one to concentrate on the role of the

dot trajectories in human perception without being distracted by irrelevant global intensity variations within the pattern.

7-23. *Artificial Glass patterns (continued)*. Suppose that while generating an artificial Glass pattern as explained in the previous problem you place the first dot of each of the dot pairs on one layer, and the second dot of each pair on a second layer. Suppose, further, that you replace each of the tiny dots in one of the layers by a tiny "1"-shaped dot, and that you take the inverse video of the other layer so that it becomes a pinhole screen consisting of tiny pinholes on a black background. Would you expect to obtain in the superposition of such layers macroscopic effects like in Fig. 7.1(a)?

7-24. Explain the triangular shape of Fig. 7.11(a), namely, of the cross section through the intensity profile of the linear Glass pattern in the superposition of two slightly rotated copies of an aperiodic line grating. *Hint*: According to Propositions 7.4 and 7.5 the normalized intensity profile of this Glass pattern is given by the convolution (7.25) of the two original, normalized layers, *before* they undergo the layer transformations, i.e. while they are fully superposed on top of each other, line on line. The final shape of the Glass pattern *after* the application of the layer transformations (in our case, the layer rotation) is then obtained by applying to this normalized profile the geometric transformation (7.26); this brings the intensity profile of the Glass pattern to its actual scaling and orientation in the layer superposition. Now, according to Remark 7.2, convolution (7.25) is determined by the convolution of one element from each layer. In our case, the individual element in each of the layers is a straight vertical black line on a white background. And indeed, if the widths of the two lines are identical, their convolution gives a vertical line having a triangular intensity profile, as shown in Fig. 7.11(a); if the two line widths are not identical, their convolution has, instead, a trapezoidal profile. This triangular or trapezoidal convolution is, indeed, the *normalized* Glass pattern as explained above. The actual Glass pattern that we see in the layer superposition is obtained by applying to this normalized Glass pattern the geometric transformation defined by part (2) of Proposition 7.5 (see Eq. (7.26)).

7-25. Explain the triangular shapes of the cross sections through the intensity profiles shown in the bottom row of Fig. 7.13.

7-26. *The intensity profile of the Glass pattern in the superposition of two similar dot screens having the same fixed dot shape*. As an example for Case 3 in Sec. 7.4.2, suppose we are given two identical aperiodic dot screens both of which consist of the same dot elements. What is the shape of the macroscopic Glass pattern (the intensity profile) that becomes visible when we superpose the two layers with a small angle difference like in Fig. 2.1(c)? And when we superpose them with a small scaling difference like in Fig. 2.1(e)? *Hint*: According to Proposition 7.7, the shape of the resulting intensity profile in both cases is identical, and it is simply determined by the convolution (7.30) or, equivalently, by the cross correlation (7.33). But because in our case the dot shapes of both layers are identical, this reduces, in fact, to the autocorrelation of a single screen dot. For example, if both screens consist of identical white square dots on a black background, the autocorrelation has the shape of a "lazy pyramid", and if both screens consist of identical white circular dots on a black background, the autocorrelation has the shape of a "Chinese hat" (see [Bracewell95 pp. 187–192]). What happens when both screens consist of black dots on a white background, as is the case in Figs. 2.1(c),(e)? And what happens when one screen consists of identical black square dots and the other consists of identical black circular dots?

7-27. *The radius of the macroscopic Glass pattern*. Fig. 2.1(c) shows a Glass pattern that is generated when two identical aperiodic dot screens are superposed with a small angle difference α. As we have seen in Secs. 2.3 and 3.3.1, this Glass pattern is brightest at its center (the fixed point of the rotation), and it gradually fades out and disappears as we go away from this point. But looking at Fig. 2.1, it is difficult to judge visually if the intensity profile of the Glass pattern has a finite radius, beyond which it has no more influence, or if its influence continuously decays until it fully disappears at infinity. What is your opinion? *Hint*: As we have seen in the previous problem, the shape of the macroscopic Glass pattern (i.e. the shape of the intensity profile) is determined by the convolution (or, rather, by the cross correlation) of a single dot from each layer. For example, if both layers consist of identical circular dots, as is the case in Fig. 2.1(c), the intensity profile of the resulting Glass pattern has the shape of a "Chinese hat". Now, because the individual screen dots are finite, it is clear that their convolution (or cross correlation) is also finite, and therefore the resulting Glass pattern is finite. Note, however, that this convolution (or cross correlation) determines the shape of the *normalized* Glass pattern, but the actual Glass pattern that we see in the layer superposition is obtained by applying to this normalized Glass pattern the transformation defined by part (2) of Proposition 7.7 (see Eq. (7.31)). The resulting Glass pattern is, therefore, finite (except in particular, singular cases, where the transformation (7.31) gives an infinite magnification).

In the simple case of Fig. 2.1(c), where the layer transformations merely consist of a rotation of one layer by angle α, the radius of the actual Glass pattern in the layer superposition can be found by a simple geometric consideration: In this case the macroscopic Glass pattern occurs because close to the center of rotation each rotated black dot almost fully overlaps its unrotated counterpart from the other layer, giving in this zone less black than farther away, where the overlapping of the corresponding rotated and unrotated dots gradually decreases. Now, it is clear that starting from a certain distance d from the center of rotation, which depends on the radius of the black dots and on the rotation angle α, a rotated dot no longer touches its unrotated counterpart from the other layer. It follows, therefore, that the influence of the macroscopic Glass pattern stops at the distance d from the center of rotation. Show that if the dot radius is r and the rotation angle is α, then the distance d from the center of rotation starting from which a rotated dot no longer touches its unrotated counterpart is determined by $|\sin(\alpha/2)| = r/d$. This means that the radius of the macroscopic Glass pattern is:

$$d = \frac{r}{|\sin(\alpha/2)|}$$

Note, however, that when $\alpha = 0°$ we obtain a singular superposition, in which the Glass pattern is infinitely big (and therefore invisible).

Can you find a similar geometric consideration for the case of Fig. 2.1(e), in which one of the two aperiodic dot screens is scaled by a factor $s \approx 1$?

7-28. *The radius of the microstructure in a Glass pattern.* As we have seen in the previous problem, the radius of the intensity profile (the macroscopic structure) of the Glass pattern in Fig. 2.1(c) is finite, except in the singular case which occurs when $\alpha = 0°$. However, as we have already noticed earlier, the influence of the microstructure extends beyond that of the macrostructures (and the microstructure may even persist after the macrostructure has fully vanished, as we have seen in Figs. 3.19–3.20). And indeed, in Fig. 2.1(c) the circular dot trajectories are still visible even beyond the radius of the macroscopic Glass pattern. However, looking at Fig. 2.1(c) it is difficult to judge visually if the microstructure of this Glass pattern has a finite radius, too, or if its radius is infinite and the visibility of the dot trajectories gradually decays untill they fully disappear at intinity. What is your opinion?

7-29. Compare the results obtained in Problem 7-27 with those obtained in Problem 3-14 and Fig. 3.23. Is there any connection between the radius d of the macroscopic Glass pattern in the aperiodic layer superposition and the period T_M of the moiré in the corresponding periodic layer superposition (see Figs. 2.1(c),(d) or Figs. 4.16(a),(b))? *Hint*: There is no direct connection between the two because there is no direct connection between the radius of the individual screen dot and the period of the periodic screen (see Fig. 2.1(b)). All that we know is that one period of the periodic screen contains at least two dot radiuses, so that $T > 2r$. This gives us, therefore:

$$T_M = \frac{T}{2\left|\sin(\alpha/2)\right|} > \frac{2r}{2\left|\sin(\alpha/2)\right|} = d$$

which means that the period T_M of the moiré in the periodic case is larger than the radius d of the macroscopic Glass pattern. This is confirmed by Figs. 2.1(c),(d) and Figs. 4.16(a),(b), which also show that the radius of the microstructures (the dot trajectories) is clearly larger than T_M, and hence much larger than d, as we have already seen in Problem 7-28.

7-30. *Extension of periodic-case notions to the general case.* We have seen throughout the present and the previous chapters how the moiré theory, which was originally developed for studying the superposition of periodic layers, can be extended to cover aperiodic cases, too. This extension involves the generalization of several concepts from their original, narrow meaning in the periodic case to a new, wider sense in the extended theory that covers both periodic and aperiodic cases. For example, the original image-domain notion of a "period" is generalized into the notion of a "distance" (see also the last paragraph in Sec. 3.3.3). Thus, many formulas in the periodic case that contain the period T are, in fact, particular cases of their more general counterparts that deal with distances (see, for example, Sec. 3.3.3). In the spectral domain, the discrete spectra of periodic structures (*impulse combs* and *impulse nailbeds*) are generalized, by filling the gaps between the impulses, into continuous *line-impulses* and *humps*. Consequently, the original periodic-case notion of "fundamental frequency vectors" in the spectral domain is generalized into "basis frequency vectors" (see the discussions following Eq. (7.22) and Proposition 7.5 in Sec. 7.3.1), which convey only the scaling and the orientation of the layers in question, but no longer the notion of periodicity. (Note that the notion of *frequency* exists, of course, in any spectrum, both in periodic and in aperiodic cases. What we are generalizing here to the aperiodic case is the notion of *fundamental frequency*, i.e. the frequency of the first harmonic impulse in the spectrum of a periodic structure, which is also the reciprocal value of the period T.) Similarly, the original periodic-case notion of "frequency ratio" is generalized into the wider-sense notion of "scaling ratio" (see the last paragraph before Proposition 7.8 in Sec. 7.4.3), which obviously reduces in the periodic case to its original meaning. And finally, T-convolution or T-cross correlation are generalized into convolution or cross correlation (see Remark 7.1). Note, however, that this list of examples is not exhaustive, and there exist other periodic-case notions that get a wider sense in the generalization process. Can you find some further examples and discuss their significance?

On the other hand, there exist also some notions that are specific to the periodic case, and that cannot be generalized to the aperiodic case. An obvious such example is the (k_1,k_2)-moiré for $|k_i| > 1$ (see Sec. 7.5). Can you think of other such examples?

7-31. *Probabilistic and geometric approaches for the quantitative derivation of the intensity profile of Glass patterns.* A student suggests a probabilistic approach as an alternative to our Fourier-based quantitative derivation of the intensity profile of Glass patterns: Given two black and white structures with known stochastic distributions, one can calculate the probability of seeing white (i.e. the value 1) at any point (x,y) of the superposition. This probability is precisely the intensity profile of the Glass pattern in the layer superposition. What do you think of this proposal? *Hint*: The probability of seeing white at any point (x,y) is simply the mean relative area that is occupied by white in the zone surrounding the point (x,y). This relative area can be calculated using geometric considerations in many simple cases, for example in a single dot screen or line grating, or in a superposition of two periodic gratings (see Fig. 2.9 in *Vol. I*). But this relative area becomes much more difficult to calculate in a superposition of two aperiodic layers such as random dot screens. Note also that the probability of seeing white at any point (x,y) of this superposition cannot be easily determined from the respective probabilities for the two individual layers. The reason is that our two superposed layers are *not* statistically independent of each other; on the contrary, they are very dependent (remember the high correlation that is required between the two layers in order that a Glass pattern be generated). And indeed, it is precisely where the probabilistic or geometric calculations become intractable that the Fourier-based approach comes to the rescue. This was already true in the periodic case (see Chapter 4 in *Vol. I*), and it is allthemore so in the aperiodic case, as described in the present chapter.

7-32. A well-known approach for extending the Fourier theory from the periodic case (discrete Fourier series) into the aperiodic case (continuous Fourier transforms) consists of investigating what happens when the period tends to infinity ($T \to \infty$), or equivalently, when the frequency tends to zero ($f \to 0$); see, for example, [Cartwright90 pp. 101–103], [Bracewell86 pp. 208–209] or [Gaskill78 pp. 111–112]. Based on this idea, one may try to extend the already known results of the moiré theory between periodic layers to the aperiodic case by a similar limiting process, letting the layer periods tend to infinity (or equivalently, letting the frequencies tend to zero). What do you think of this approach?

7-33. *The influence of the superposition rule.* We have seen in Secs. 4.8 and 4.9 and in the figures therein that the superposition rule between the layers has a major influence on the microstructures (dot trajectories). What is the influence of the superposition rule on the resulting macroscopic effects (see, for example, Fig. 4.15)? How can it be explained using the Fourier approach?

7-34. *Synthesis of the various moiré-theory phenomena and the interconnections between them.* After having understood the mathematical basis of each of the various phenomena that occur in the moiré theory, we always try to *synthesize* it, i.e. to design layers that will generate this phenomenon in their superposition. For example, it follows from Proposition 5.1 that in order to synthesize a repetitive moiré whose geometric layout is given by the level lines of the functions $k_1(x,y)$ and $k_2(x,y)$ we have to apply to the originally identical periodic layers domain transformations $g_1(x,y)$ and $g_2(x,y)$ such that $g_1(x,y) - g_2(x,y) = k(x,y) = (k_1(x,y), k_2(x,y))$. Similarly, we have seen in Sec. 3.6 that in order to synthesize a Glass pattern having a desired fixed locus, we have to find a mapping $k(x,y)$ whose zeros give precisely the desired fixed locus, and to apply to the two originally identical aperiodic dot screens domain transformations $g_1(x,y)$ and $g_2(x,y)$ such that $g_1(x,y) - g_2(x,y) = k(x,y)$. In Sec. 4.6 we saw that in order to synthesize a Glass pattern whose dot trajectories correspond to the field lines of the vector field $\bar{h}(x,y)$ we have to apply to the two originally identical aperiodic dot screens direct transformations $\bar{g}_1(x,y)$ and $\bar{g}_2(x,y)$ such that $\bar{g}_1(x,y) - \bar{g}_2(x,y) = \bar{h}(x,y)$. Finally, in the present chapter we saw how to synthesize Glass (or moiré) patterns having a specified intensity profile. Can you comment on the relationship between these different synthesized phenomena? (see also Problem 7-2).

Appendix A

Fixed point theorems for first- and second-order polynomial mappings

A.1 Introduction

In this appendix we provide a deeper insight into the fixed point theorem that was presented and used in Chapter 3. In particular, we concentrate on two of the simplest but most important special cases of this theorem, one for linear or affine mappings (in Sec. A.2), and the other for second-order polynomial mappings (in Sec. A.3). The importance of these particular cases is twofold: On the one hand, these cases cover many of the layer transformations being used in the figures and in the examples throughout this book; but on the other hand, these cases may also serve as simple illustrations to the general case involving more complex mappings. A further generalization to the case of mutual fixed points between two mappings is discussed in Sec. A.4.

A.2 The fixed point theorem for linear or affine mappings

As we have seen in Sec. 3.2, the affine fixed point theorem states that all non-degenerate affine mappings $\mathbf{g}(x,y)$ from $\mathbb{R}^2$ onto itself have a single fixed point.

In order to more deeply understand this theorem, we start by analyzing the different possible types of affine mappings $\mathbf{g}(x,y)$. The most general form of an affine mapping is:

$$x' = a_1 x + b_1 y + x_0$$
$$y' = a_2 x + b_2 y + y_0 \tag{A.1}$$

Let us first consider the homogeneous mapping that is associated with $\mathbf{g}(x,y)$, i.e. the corresponding linear transformation where the shift (x_0, y_0) is zero:

$$x' = a_1 x + b_1 y$$
$$y' = a_2 x + b_2 y \tag{A.2}$$

Such a linear transformation may either:

(a) map $\mathbb{R}^2$ onto the whole of $\mathbb{R}^2$ (this occurs, for example, in rotations, scalings, flipping over an axis, etc.);

(b) map $\mathbb{R}^2$ onto $\mathbb{R}$ (this occurs, for example, in a projection onto the x axis); or

(c) map $\mathbb{R}^2$ onto the origin $(0,0)$ (this occurs in the zero transformation that maps all the points (x,y) onto $(0,0)$).

Cases (b) and (c) occur when the linear transformation (A.2) is singular, i.e. when its determinant equals zero:

$$\begin{vmatrix} a_1 & b_1 \\ a_2 & b_2 \end{vmatrix} = a_1 b_2 - a_2 b_1 = 0 \qquad (A.3)$$

The same three cases occur also in the affine mapping (A.1), except that here, in cases (b) and (c) $\mathbb{R}^2$ is mapped, respectively, onto a shifted line $(x_0,y_0) + \mathbb{R}$ or onto a shifted point (x_0,y_0).

The degenerate linear or affine transformations of types (b) and (c) do not interest us, of course, in our study on superpositions of transformed layers, since their application would completely destroy our 2D layers. Consequently, we are only interested in linear or affine transformations belonging to type (a), namely, cases in which determinant (A.3) is non-zero. These cases are called *regular* or *non-singular*.

Let us now proceed to the fixed points of such non-singular transformations. A non-singular linear transformation (A.2) may either:

(1) have a single fixed point, located at the origin (this occurs, for example, in rotations, scalings, etc.);

(2) have a full line of fixed points that passes through the origin (this occurs, for example, in transformations such as flipping over the x axis, or scaling in the y direction alone, in both of which all points of the x axis are fixed points); or

(3) have all the points of the entire x,y plane as fixed points (this occurs in the identity transformation).

When does each of these cases occur? As we know, the fixed points of the linear transformation (A.2) are those points of the plane which satisfy $(x',y') = (x,y)$, namely:[1]

$$x = a_1 x + b_1 y$$
$$y = a_2 x + b_2 y \qquad (A.4)$$

This gives us the following linear set of equations for x and y:

$$(1 - a_1)x \ - \ b_1 y = 0$$
$$-a_2 x + (1 - b_2)y = 0 \qquad (A.5)$$

Clearly, cases (2) and (3) above occur when this linear set of equations is singular, i.e. when:

$$\begin{vmatrix} 1 - a_1 & -b_1 \\ -a_2 & 1 - b_2 \end{vmatrix} = 1 - a_1 - b_2 + a_1 b_2 - a_2 b_1 = 0 \qquad (A.6)$$

---

[1] Note that here $\mathbf{g}(x,y) = (x,y)$ (or equivalently, in terms of matrices, $A\mathbf{x} = \mathbf{x}$) does not mean that $\lambda = 1$ is an eigenvalue of $\mathbf{g}(x,y)$ [Kreyszig93 p. 157], since in our case it may certainly happen that $\mathbf{x} = (0,0)$ is the only solution of $\mathbf{g}(\mathbf{x}) = \mathbf{x}$ (as in the case of a scaling or rotation transformation $\mathbf{g}$).

In such cases the associated affine mapping (A.1), which is obtained by adding to (A.2) a shift of (x_0,y_0), may have either infinitely many fixed points, or no fixed points at all. Only in case (1), i.e. when determinant (A.6) is non-zero, the associated affine mapping (A.1) has precisely one single fixed point; its location is given then by Eq. (3.13).

Let us first consider transformations that satisfy both conditions (a) and (1), i.e. where both of the determinants (A.3) and (A.6) are non-zero. We call such affine mappings *non-degenerate affine mappings*.

It is clear, therefore, that all non-degenerate affine mappings $\mathbf{g}(x,y)$ from $\mathbb{R}^2$ onto itself have a single fixed point; this is, indeed, precisely what is claimed by our affine fixed point theorem in Sec. 3.2. But this theorem does not say anything about degenerate transformations that do not satisfy condition (a) or condition (1).

As an illustration, let us mention that mappings such as rotations, scalings, etc. as well as their combinations have, indeed, a single fixed point. This is also true for all of their combinations with translation, but not for pure translations. Note that pure translations are excluded, since their determinant (A.6) is zero; this can be understood more intuitively as follows: The homogeneous transformation (A.2) associated with a pure translation is the identity transformation, that belongs to class (3) above and has all the points of the x,y plane as fixed points. But the addition of a translation to the identity transformation destroys all of its fixed points, so that a pure translation has no fixed points at all.

As a second example, let us consider the linear transformation which consists of vertical scaling. This transformation belongs to class (2) above, and has all the points of the x axis as fixed points. What happens now when we add to this linear transformation a translation? In this case, the answer depends on the direction of the translation: If the translation is horizontal, it is clear that all the fixed points on the x axis are destroyed, and the resulting affine mapping $\mathbf{g}(x,y)$ has no fixed points. But if the translation is vertical, the resulting affine mapping $\mathbf{g}(x,y)$ will still have a full line of fixed points, which is parallel to the x axis. Note, however, that such cases are not treated by our affine fixed point theorem, since their determinant (A.6) is zero: As we have seen, this theorem only considers non-degenerate affine mappings, but it does not say anything about degenerate affine mappings. Indeed, some degenerate affine mappings have a full line of fixed points, while others have no fixed points at all.

Thus, in order to cover all of the possible cases we need to introduce a more general version of our theorem, that treats all affine mappings from $\mathbb{R}^2$ onto itself, including degenerate cases such as vertical scalings and translations:

Generalized affine fixed point theorem: An affine mapping $\mathbf{g}(x,y)$ from $\mathbb{R}^2$ onto itself has a fixed point (either one or infinitely many) *iff* rankA = rankB, where A is the 2×2 coefficient matrix of the homogeneous system of Eqs. (A.5) and B is the 2×3 extended matrix that includes x_0 and y_0 in its third column:

$$A = \begin{pmatrix} 1 - a_1 & -b_1 \\ -a_2 & 1 - b_2 \end{pmatrix} \qquad B = \begin{pmatrix} 1 - a_1 & -b_1 & x_0 \\ -a_2 & 1 - b_2 & y_0 \end{pmatrix}$$

Moreover, if the rank of both A and B is 2, the fixed point is unique; if their rank is 1, there exists a full line of fixed points; and if their rank is 0, all the points of the x,y plane are fixed points of $g(x,y)$ (this occurs if $g(x,y)$ is the identity mapping without translation). But if rankA $\neq$ rankB, the affine mapping $g(x,y)$ has no fixed points at all. ∎

This generalized theorem is, in fact, an application to the particular case of Eqs. (A.5) of the algebraic theorem on the dimension of the solution space of a system of linear equations [Bronshtein97 p. 143].

A.3 The fixed point theorem for second-order polynomial mappings

As we have seen in Secs. 3.2 and A.1, the affine fixed point theorem states that all non-degenerate affine mappings $g(x,y)$ from $\mathbb{R}^2$ onto itself have a single fixed point.

This theorem can be generalized to the case of second-order polynomial mappings. By a second-order polynomial mapping (or simply, a mapping of order 2) we mean a mapping $g(x,y)$ that is defined by a pair of algebraic equations of order 2, namely:

$$x' = a_1x^2 + b_1xy + c_1y^2 + d_1x + e_1y + x_0$$

$$y' = a_2x^2 + b_2xy + c_2y^2 + d_2x + e_2y + y_0$$

(A.7)

Just as in the affine case, such transformations do not necessarily map $\mathbb{R}^2$ onto the entire $\mathbb{R}^2$: in some degenerate cases the transformation (A.7) maps $\mathbb{R}^2$ onto a straight or a curved line, or even into a single point. For example, the second order transformation $g(x,y) = (x,x^2)$ maps $\mathbb{R}^2$ onto the parabola $y = x^2$. Such situations occur when the mapping $g(x,y)$ defined by the set of equations (A.7) is *singular*, namely, when the two equations forming the set are not independent. Two equations $g_1(x,y) = 0$ and $g_2(x,y) = 0$ are said to be *independent* (or *functionally independent*) if there exists no function $f(u,v)$ other than $f(u,v) \equiv 0$ such that $f(g_1(x,y),g_2(x,y)) = 0$ is satisfied for all (x,y). Equivalently, this means that the Jacobian:

$$J(x,y) = \begin{vmatrix} \dfrac{\partial g_1}{\partial x} & \dfrac{\partial g_1}{\partial y} \\ \dfrac{\partial g_2}{\partial x} & \dfrac{\partial g_2}{\partial y} \end{vmatrix} = \dfrac{\partial g_1}{\partial x}\dfrac{\partial g_2}{\partial y} - \dfrac{\partial g_2}{\partial x}\dfrac{\partial g_1}{\partial y}$$

(A.8)

is not identically zero (see [Bronstein90 pp. 226, 430–431] or [Courant88 pp. 154–155]). If $g_1(x,y)$ and $g_2(x,y)$ are *dependent*, for instance if $g_2(x,y) = g_1(x,y)^2$, they are a consequence of each other, and hence the 2D transformation $g(x,y) = (g_1(x,y),g_2(x,y))$ they define is singular, and it maps $\mathbb{R}^2$ onto a 1D curve or even a single point in $\mathbb{R}^2$.

In the particular case of polynomial mappings of order 2, the condition for Eq. (A.7) to be singular is, therefore, that the Jacobian (A.8) be identically zero, namely:

$$(2a_1x + b_1y + d_1)(b_2x + 2c_2y + e_2) - (2a_2x + b_2y + d_2)(b_1x + 2c_1y + e_1) \equiv 0 \qquad (A.9)$$

which gives:

$$2(a_1b_2 - a_2b_1)x^2 + 4(a_1c_2 - a_2c_1)xy + 2(b_1c_2 - b_2c_1)y^2 + (b_2d_1 - b_1d_2 + 2a_1e_2 - 2a_2e_1)x$$
$$+ (2c_2d_1 - 2c_1d_2 + b_1e_2 - b_2e_1)y + (d_1e_2 - d_2e_1) \equiv 0$$

But since this expression must be identically zero for any values of x and y, this means that Eq. (A.7) is singular when all of the following conditions are simultaneously satisfied:

$$a_1b_2 = a_2b_1$$

$$a_1c_2 = a_2c_1$$

$$b_1c_2 = b_2c_1$$

$$b_2d_1 + 2a_1e_2 = b_1d_2 + 2a_2e_1 \qquad (A.10)$$

$$2c_2d_1 + b_1e_2 = 2c_1d_2 + b_2e_1$$

$$d_1e_2 = d_2e_1$$

Such degenerate cases do not interest us, of course, in our study on superpositions of transformed layers, since the mappings $g(x,y)$ they represent do not map $\mathbb{R}^2$ onto itself; we will only be interested in *non-singular* mappings of order 2, namely, cases in which Eq. (A.7) is non-singular.

We now proceed to discuss the fixed points of non-singular mappings of order 2. The fixed points of transformation (A.7) are those points of the plane for which (x',y') equals (x,y), namely:

$$x = a_1x^2 + b_1xy + c_1y^2 + d_1x + e_1y + x_0$$
$$y = a_2x^2 + b_2xy + c_2y^2 + d_2x + e_2y + y_0 \qquad (A.11)$$

This gives us the following system of equations for x and y:

$$a_1x^2 + b_1xy + c_1y^2 + (d_1 - 1)x + e_1y + x_0 = 0$$
$$a_2x^2 + b_2xy + c_2y^2 + d_2x + (e_2 - 1)y + y_0 = 0 \qquad (A.12)$$

A general rule in algebra states that a system of two equations $p_1(x,y) = 0$, $p_2(x,y) = 0$, where $p_1(x,y)$ is an m-order polynomial in x and y and $p_2(x,y)$ is an n-order polynomial in x and y, has mn solutions (x,y), real or complex [Bronstein90 pp. 226–227]. This means that in our case the system of equations (A.12) has up to 4 *real* solutions (x,y). This agrees, indeed, with our geometric intuition, since each equation of order 2 represents in fact a conic curve in the plane, and the intersection of two such conics clearly gives up to 4 real solutions. However, depending on the locations and orientations of the two conic curves, they may have only 3, 2, 1 or even 0 intersection points. Moreover, just like in the affine case, there may exist here, too, degenerate systems (A.12) with infinitely many solutions, meaning that the corresponding mapping $\mathbf{g}(x,y)$ defined by (A.7) has infinitely many fixed points. In general, the system of equations (A.12) may either:

(1) have between 0 and 4 solutions (which are fixed points of the mapping (A.7));

(2) have a full straight or curved line of solutions (fixed points of (A.7)); or

(3) have the full x,y plane as solutions (fixed points of (A.7)).

Cases (2) and (3) occur when the set of equations (A.12) is singular, or in other words, if the two equations forming the set are not independent. This happens when their Jacobian is identically zero:[2]

$$J(x,y) = \begin{vmatrix} \dfrac{\partial g_1}{\partial x} & \dfrac{\partial g_1}{\partial y} \\[2mm] \dfrac{\partial g_2}{\partial x} & \dfrac{\partial g_2}{\partial y} \end{vmatrix} = \frac{\partial g_1}{\partial x}\frac{\partial g_2}{\partial y} - \frac{\partial g_2}{\partial x}\frac{\partial g_1}{\partial y} =$$

$$= (2a_1x + b_1y + d_1 - 1)(b_2x + 2c_2y + e_2 - 1)$$

$$- (2a_2x + b_2y + d_2)(b_1x + 2c_1y + e_1) \equiv 0 \tag{A.13}$$

which gives:

$$2(a_1b_2 - a_2b_1)x^2 + 4(a_1c_2 - a_2c_1)xy + 2(b_1c_2 - b_2c_1)y^2$$

$$+ [b_2(d_1 - 1) - b_1d_2 + 2a_1(e_2 - 1) - 2a_2e_1]x$$

$$+ [2c_2(d_1 - 1) - 2c_1d_2 + b_1(e_2 - 1) - b_2e_1]y$$

$$+ [(d_1 - 1)(e_2 - 1) - d_2e_1] \equiv 0$$

But since this expression must be identically zero for any values of x and y, it follows that Eq. (A.12) is singular when all of the following conditions are simultaneously satisfied:

$$a_1b_2 = a_2b_1$$

$$a_1c_2 = a_2c_1$$

$$b_1c_2 = b_2c_1$$

$$b_2(d_1 - 1) + 2a_1(e_2 - 1) = b_1d_2 + 2a_2e_1 \tag{A.14}$$

$$2c_2(d_1 - 1) + b_1(e_2 - 1) = 2c_1d_2 + b_2e_1$$

$$(d_1 - 1)(e_2 - 1) = d_2e_1$$

---

[2] Note that $g_1(x,y)$ and $g_2(x,y)$ in this Jacobian are the functions in the left hand side of Eqs. (A.12), while $g_1(x,y)$ and $g_2(x,y)$ in the Jacobian (A.8) are the functions in the right hand side of Eqs. (A.7).

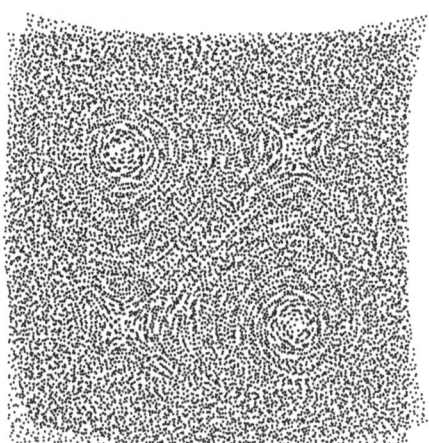

Figure A.1: An example with 4 fixed points: The superposition of two originally identical aperiodic dot screens, one of which has undergone the second-order transformation $\mathbf{g}(x,y) = (x - ay^2 + x_0, y - ax^2 + y_0)$. Fig. 3.15 shows a slightly different variant in which both of the layers have undergone second-order transformations.

We call second-order polynomial mappings $\mathbf{g}(x,y)$ for which both the Jacobians (A.9) and (A.13) are not identically zero *non-degenerate* second order mappings. We obtain, therefore, the following result:

The fixed point theorem for second-order polynomial mappings: A non-degenerate second-order polynomial mapping $\mathbf{g}(x,y)$ from $\mathbb{R}^2$ onto itself may have up to 4 fixed points. ∎

An example of a layer superposition in which one of the two layers has undergone a second-order mapping $\mathbf{g}(x,y)$ having 4 fixed points is shown in Fig. A.1. In this case we have $a_1 = 0$, $b_1 = 0$, $c_1 = -a$, $d_1 = 1$, $e_1 = 0$, $a_2 = -a$, $b_2 = 0$, $c_2 = 0$, $d_2 = 0$ and $e_2 = 1$, so that $a_1c_2 \neq a_2c_1$ in conditions (A.10) and (A.14), meaning that both of the Jacobians (A.9) and (A.13) are not identically zero. Note, however, that if $\mathbf{g}(x,y)$ is only *non-singular* (meaning that only the Jacobian (A.9) is not identically zero), it will have infinitely many fixed points, for example one or two full lines of fixed points. Such cases are illustrated in Figs. 3.5(c),(d) and 3.6; the mappings in these cases are clearly non-singular (they map $\mathbb{R}^2$ onto the whole of $\mathbb{R}^2$), and yet they have infinitely many fixed points. (Explanation: in these mappings, given by Eq. (3.19), we have $a_1 = 0$, $b_1 = 0$, $c_1 = -a$, $d_1 = 1$, $e_1 = 0$, $a_2 = 0$, $b_2 = 0$, $c_2 = 0$, $d_2 = 0$ and $e_2 = 1$. Therefore we have in conditions (A.10) $d_1e_2 \neq d_2e_1$, while in conditions (A.14) we have instead $(d_1 - 1)(e_2 - 1) = d_2e_1$ and all the equalities are satisfied.)

As we have seen above, a detailed analysis of all the different possible cases for mappings of order 2 amounts to an analysis of the intersection points between two conics in the plane. A full discussion on the intersection of conics can be found, for example, in [Barrett97].

A.4 Mutual fixed points between two mappings; application to the moiré theory

As we have seen in Chapter 3, when we superpose two originally identical aperiodic layers (such as random or pseudo-random screens) that have undergone transformations $\mathbf{g}_1(x,y)$ and $\mathbf{g}_2(x,y)$, respectively, we may obtain in the superposition a visible Glass pattern about each mutual fixed point of the transformations $\mathbf{g}_1(x,y)$ and $\mathbf{g}_2(x,y)$ (see Sec. 3.5). The fixed point theorems described in Secs. A.2 and A.3 above correspond, in fact, to the case where one of the two layer transformations, say, $\mathbf{g}_2(x,y)$, is the identity transformation, meaning that only one of the two layers has been transformed. In such cases the mutual fixed points of the two layers are obtained at the points (x,y) where $\mathbf{g}_1(x,y) = (x,y)$, which is precisely the situation described in Eqs. (A.4) and (A.11). In the more general case where both of the superposed layers have been transformed, the mutual fixed points of $\mathbf{g}_1(x,y)$ and $\mathbf{g}_2(x,y)$ are the points that satisfy $\mathbf{g}_1(x,y) = \mathbf{g}_2(x,y)$, namely:

$$\mathbf{g}_M(x,y) = \mathbf{g}_1(x,y) - \mathbf{g}_2(x,y) = (0,0) \tag{A.15}$$

In componentwise notation, these points are the solutions of the system of equations:

$$g_{M_1}(x,y) = g_{1,1}(x,y) - g_{2,1}(x,y) = 0$$
$$g_{M_2}(x,y) = g_{1,2}(x,y) - g_{2,2}(x,y) = 0 \tag{A.16}$$

where $\mathbf{g}_1(x,y) = (g_{1,1}(x,y),g_{1,2}(x,y))$, $\mathbf{g}_2(x,y) = (g_{2,1}(x,y),g_{2,2}(x,y))$ and $\mathbf{g}_M(x,y) = (g_{M_1}(x,y), g_{M_2}(x,y))$.

Therefore, in cases where both of the layers have been transformed, conditions (1)–(3) in Secs. A.2 and A.3 apply, in fact, to the solutions of equations (A.15) or (A.16). For example, the layer superposition may have a *linear* Glass pattern when $\mathbf{g}_1(x,y)$ and $\mathbf{g}_2(x,y)$ have a full line of mutual fixed points, i.e. when $\mathbf{g}_M(x,y) = (0,0)$ has a full continuous line (or curve) of solutions within the x,y plane. Note that these points are *not* fixed points of the transformation $\mathbf{g}_M(x,y)$ itself (the fixed points of $\mathbf{g}_M(x,y)$ are given by the solutions of $\mathbf{g}_M(x,y) - (x,y) = (0,0)$, not by the solutions of $\mathbf{g}_M(x,y) = (0,0)$).

This generalization to the case of two transformed layers is valid for *any* transformations $\mathbf{g}_1(x,y)$ and $\mathbf{g}_2(x,y)$, and not only for first- or second-order polynomial mappings.

Appendix B

The various interpretations of a 2D transformation

B.1 Introduction

Consider a system of two equations in two independent variables x and y:

$$u = g_1(x,y)$$
$$v = g_2(x,y)$$

(B.1)

or in vector notation:

$$\mathbf{u} = \mathbf{g}(\mathbf{x})$$

(B.2)

where $\mathbf{x} = (x,y)$, $\mathbf{u} = (u,v)$, and $\mathbf{g}(x,y) = (g_1(x,y),g_2(x,y))$. Clearly, both $g_1(x,y)$ and $g_2(x,y)$ are *scalar functions*, i.e. functions that return for each point $(x,y) \in \mathbb{R}^2$ a single real value:

$$g_1: \mathbb{R}^2 \to \mathbb{R}, \qquad g_2: \mathbb{R}^2 \to \mathbb{R}$$

whereas $\mathbf{g}(x,y)$ is a *mapping* (or, equivalently, a *transformation*), i.e. a *vector function* that returns for each point $(x,y) \in \mathbb{R}^2$ a new point $(u,v) \in \mathbb{R}^2$:

$$\mathbf{g}: \mathbb{R}^2 \to \mathbb{R}^2$$

We denote this function by a boldface letter $\mathbf{g}$ since the value it returns, $\mathbf{g}(x,y)$, is a vector.

The mathematical relationship defined by (B.1) (or alternatively by (B.2)) can be interpreted in several different yet completely equivalent ways, as explained in the following sections. Because all of these interpretations are mathematically equivalent, we are free in each application to choose any of them according to our convenience. It is important, however, to be aware of the different interpretations, and to know which of them is being used in each case, in order to avoid any possible confusions.

B.2 Interpretation as two surfaces over the plane or as two sets of level lines

Each of the two real valued functions of the system (B.1) defines a surface (manifold) $z = g(x,y)$ over the x,y plane, where z represents the altitude of the surface at the point (x,y) in terms of the vertical axis, perpendicularly to the x,y plane. Therefore, the 2D mapping $\mathbf{g}(x,y)$ can be interpreted geometrically as a pair of surfaces (see Fig. B.1).

The level lines (or level curves) of each of these surfaces are given by:

$$g_i(x,y) = const.$$

In particular, the set of points (x,y) satisfying $g_i(x,y) = 0$ (i.e. the solution of the equation $g_i(x,y) = 0$) can be interpreted as the zero level line of the surface $g_i(x,y)$, namely, the intersection of the surface with the x,y plane. Therefore, the set of points (x,y) satisfying the two equations:

$$g_1(x,y) = 0$$
$$g_2(x,y) = 0$$
(B.3)

(i.e. the simultaneous solution of both equations) consists of the points of intersection between the zero level curves of $g_1(x,y)$ and the zero level curves of $g_2(x,y)$. Depending on the case, there may exist zero, one, several, or even infinitely many such intersection points. For example, if the zero level curves of the two functions are two intersecting straight lines, then the system (B.3) has a single solution (intersection point). As a further example, if the zero level lines of the two functions are second-order curves such as parabolas or ellipses, they may have up to 4 intersection points (see Sec. A.3 in Appendix A). Finally, in the case where the zero level curves of $g_1(x,y)$ and $g_2(x,y)$ coincide there exist infinitely many intersection points; and in the case where the two zero level curves are parallel to each other (or when at least one of the two surfaces does not intersect the x,y plane at all) the system (B.3) has no solutions.

In the example shown in Fig. B.1, where $\mathbf{g}(x,y) = (2xy, y^2 - x^2)$, the zero level lines of the surface $g_1(x,y)$ are two perpendicular lines that coincide with the x and y axes, and the zero level lines of the surface $g_2(x,y)$ are two perpendicular lines that coincide with the main diagonals. Their intersection consists, therefore, of a single point at the origin. But if we consider, instead, the two functions defined by $\mathbf{g}(x,y) = (2xy - 1, y^2 - x^2 - 1)$, the zero level lines of the two surfaces become hyperbolic (see in Fig. B.1 the level lines corresponding to the altitude $z = 1$), and their intersection consists of two points.

It should be noted that the system of equations (B.3) is not equivalent to (B.1) because it only takes into consideration the subsets of the surfaces $u = g_1(x,y)$ and $v = g_2(x,y)$ where the surface altitude is zero. The generalization to any other altitudes c,k is straightforward (by considering the equations $g_1(x,y) = c$ and $g_2(x,y) = k$); but none of these equation pairs is equivalent to (B.1), either.

B.3 Interpretation as a mapping from the plane into itself

So far, we considered (B.1) as a system consisting of two scalar functions, $g_1(x,y)$ and $g_2(x,y)$. However, using an alternative interpretation, we may consider the vector function $\mathbf{g}(x,y)$ of (B.2) as a mapping from the x,y plane onto the u,v plane (or a subset thereof). Thus, the transformation $\mathbf{g}$ maps each point (x,y) of the domain of $\mathbf{g}$ in the x,y plane into its image point $(u,v) = \mathbf{g}(x,y)$ in the u,v plane. This is concisely expressed by the notation: $(x,y) \mapsto \mathbf{g}(x,y)$. The interpretation of $\mathbf{g}$ as a mapping is illustrated in Fig. B.2 for the same transformation as in Fig. B.1.

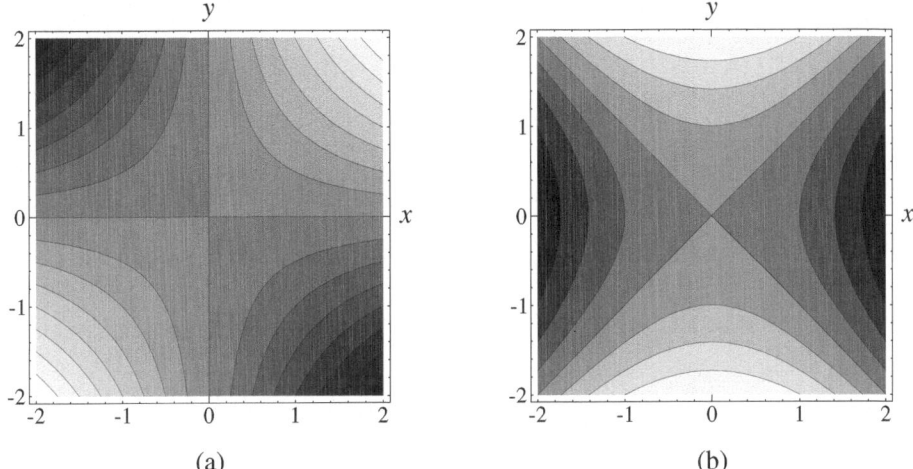

Figure B.1: Illustration of the transformation $\mathbf{g}(x,y) = (2xy, y^2 - x^2)$ as a pair of surfaces. (a) The surface $g_1(x,y) = 2xy$. (b) The surface $g_2(x,y) = y^2 - x^2$. Gray levels indicate the surface altitude: brighter shades represent higher values and darker shades represent lower values. The curves plotted on each of the surfaces are its level lines.

Note, however, that a mapping $\mathbf{g}(x,y)$ does not necessarily map $\mathbb{R}^2$ onto the entire $\mathbb{R}^2$, or even onto a 2D subregion thereof: In some degenerate cases $\mathbf{g}(x,y)$ may map $\mathbb{R}^2$ into a straight or a curved line, into a single point, or even into an empty set. For example, the second order transformation $\mathbf{g}(x,y) = (x,x^2)$ (namely, $u = x$, $v = x^2$) maps $\mathbb{R}^2$ onto the parabola $v = u^2$ which is only a 1D curve within the u,v plane. As another example, the transformation $u = \sqrt{x}$, $v = \sqrt{-x}$ maps $\mathbb{R}^2$ onto the single point $(0,0)$, while the transformation $u = \sqrt{x}$, $v = 1/\sqrt{-x}$ maps $\mathbb{R}^2$ onto an empty set. Such situations occur when the mapping $\mathbf{g}(x,y)$ (or equivalently, the system of equations (B.1)) is *singular*, namely, when the two equations forming the system are not *independent*.

Definition B.1: Two functions (or equations) $u = g_1(x,y)$ and $v = g_2(x,y)$ are said to be *independent* (or *functionally independent*) if there exists no function $f(u,v)$ other than $f(u,v) \equiv 0$ such that $f(g_1(x,y),g_2(x,y)) = 0$ is satisfied for all (x,y). Equivalently, this means that the Jacobian:

$$J(x,y) = \begin{vmatrix} \dfrac{\partial g_1}{\partial x} & \dfrac{\partial g_1}{\partial y} \\ \dfrac{\partial g_2}{\partial x} & \dfrac{\partial g_2}{\partial y} \end{vmatrix} = \frac{\partial g_1}{\partial x}\frac{\partial g_2}{\partial y} - \frac{\partial g_2}{\partial x}\frac{\partial g_1}{\partial y} \tag{B.4}$$

is not identically zero (see, for example, [Bronstein90 pp. 226, 430–431] or [Courant88 pp. 154–155]). ∎

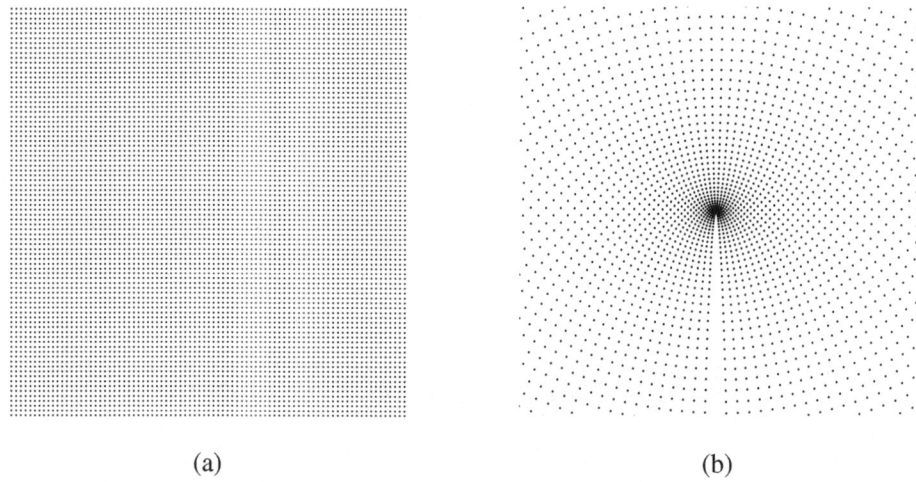

(a)   (b)

Figure B.2: Illustration of the transformation $g(x,y) = (2xy, y^2 - x^2)$ as a direct mapping $(x,y) \mapsto g(x,y)$. (a) A periodic dot screen in the original x,y plane. (b) The transformed dot screen after each of its dots (x,y) has been moved by the mapping g to its new location $g(x,y)$. The "seam" along the negative part of the y axis in (b) was left intentionally, to clearly illustrate how the upper half plane of (a) is deformed around the origin in (b), and covers the entire destination plane. The lower half plane of (a) is deformed upwards in a similar way, and it covers once again the entire destination plane. See also Fig. B.4(a),(b).

If $g_1(x,y)$ and $g_2(x,y)$ are *dependent*, for instance if $g_2(x,y) = g_1(x,y)^2$, they are a consequence of each other, and hence the 2D transformation $g(x,y) = (g_1(x,y),g_2(x,y))$ they define is singular, and it maps $\mathbb{R}^2$ onto a 1D curve or even into a single point or an empty set in $\mathbb{R}^2$. This can be easily seen from Definition B.1: If there exists a function $f(u,v)$ other than $f(u,v) \equiv 0$ such that $f(g_1(x,y),g_2(x,y)) = 0$ for all x,y, then the image of our transformation $(u,v) = g(x,y)$ satisfies $f(u,v) = 0$, which means, indeed, that the image of g is a 1D curve (or a point, or even an empty set) within the 2D u,v space.

Thus, in order to avoid such degenerate cases, we must request that the functions $g_1(x,y)$ and $g_2(x,y)$ be independent.

Example B.1: Consider the system consisting of the two functions:

$$u = f(x) \cos y$$

$$v = f(x) \sin y$$

This transformation is, in fact, a generalization of the polar to Cartesian coordinate transformation $u = r\cos\theta$, $v = r\sin\theta$ where the radius length r is replaced by its modulated version $f(r)$. It is easy to see that the Jacobian of this transformation is given by:

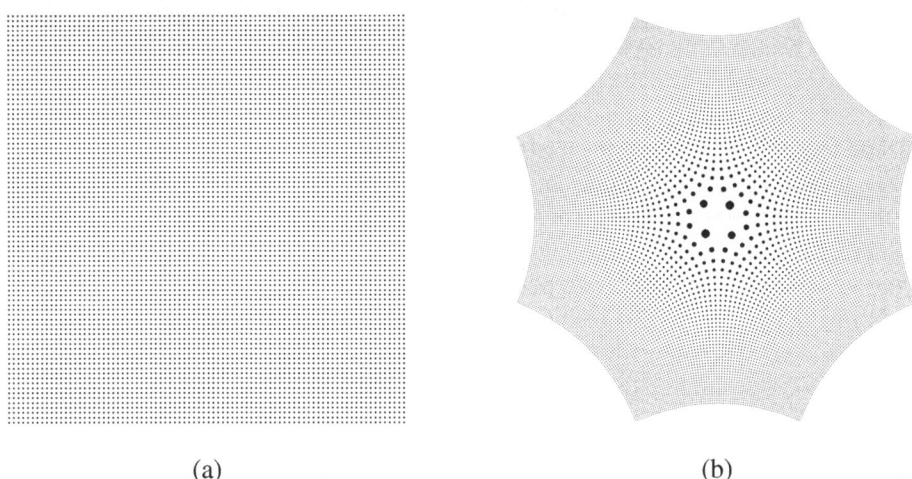

(a)   (b)

Figure B.3: Illustration of $\mathbf{g}(x,y) = (2xy, y^2 - x^2)$ as a domain transformation. It operates on the original image $r(x,y)$ shown in (a), and gives the transformed image $r(\mathbf{g}(x,y))$ shown in (b). The original image $r(x,y)$ in (a) is the same periodic dot screen as in Fig. B.2(a). Note that the left half plane of (a) is mapped twice into the two left quadrants of (b); similarly, the right half plane of (a) is also mapped twice into the two right quadrants of (b). See also Fig. B.4(c),(d).

$$J(x,y) = f(x) \frac{d}{dx} f(x) \cos^2 y + f(x) \frac{d}{dx} f(x) \sin^2 y = f(x) \frac{d}{dx} f(x)$$

If we take, for example, $f(x) = \sin x$ we obtain the system:

$$u = \sin x \cos y$$

$$v = \sin x \sin y \tag{B.5}$$

These two functions are clearly independent, since their Jacobian is not identically zero; and indeed, they map $\mathbb{R}^2$ onto a 2D subregion of $\mathbb{R}^2$, the entire unit disk. On the other hand, if we take $f(x) = 1$ then the Jacobian becomes identically zero, meaning that the two functions $u = \cos x$, $v = \sin x$ are dependent; and indeed, this system maps $\mathbb{R}^2$ onto a 1D curve within $\mathbb{R}^2$, the *perimeter* of the unit circle.

It may be also instructive to see what happens to the independent system $u = r\cos\theta$, $v = r\sin\theta$ (that maps $\mathbb{R}^2$ onto $\mathbb{R}^2$) as it gradually approaches its dependent counterpart $u = \cos\theta$, $v = \sin\theta$ (that maps $\mathbb{R}^2$ onto a 1D curve). This can be done, for example, by observing the system $u = (1+\varepsilon r)\cos\theta$, $v = (1+\varepsilon r)\sin\theta$ while ε gradually tends to zero. ∎

Remark B.1: When $(u,v) = \mathbf{g}(x,y)$ is a *linear* transformation, the two surfaces defined over the x,y plane by the functions $u = g_1(x,y)$, $v = g_2(x,y)$ (see Sec. B.2) are planes that pass through the origin. Hence, the equations $g_1(x,y) = 0$, $g_2(x,y) = 0$ may have either a single common solution at the origin (if the zero level lines of the two planes have a single intersection point), a full straight line of solutions passing through the origin (if the two planes have a common zero level line), or a full plane of solutions (in the degenerate case where both planes fully coincide with the x,y plane).[1] Now, because $\mathbf{g}$ is a linear transformation, it must satisfy the relationship $\dim \operatorname{Ker} \mathbf{g} + \dim \operatorname{Im} \mathbf{g} = 2$ (see Sec. 5.4.1 in *Vol. I*). It follows, therefore, that $g_1(x,y)$ and $g_2(x,y)$ are independent (i.e. $\dim \operatorname{Im} \mathbf{g} = 2$) *iff* they have a single solution point (i.e. $\dim \operatorname{Ker} \mathbf{g} = 0$). Thus, if $g_1(x,y)$ and $g_2(x,y)$ have a full continuous line of solutions in common, they must be dependent and the transformation $\mathbf{g}(x,y)$ necessarily maps $\mathbb{R}^2$ onto a 1D line. For example, in the case of the linear transformation $\mathbf{g}(x,y) = (x,2x)$ the planes $u = g_1(x,y) = x$ and $u = g_2(x,y) = 2x$ have a full continuous line of solutions in common along the y axis; and indeed, $\mathbf{g}(x,y)$ maps the entire plane $\mathbb{R}^2$ into the line $y = 2x$.[2]

It is interesting to note, however, that the situation in non-linear transformations is more flexible than in the linear case: The two functions $u = g_1(x,y)$, $v = g_2(x,y)$ may have a full continuous curve of solutions in common (or even several or infinitely many such curves) without necessarily being dependent (and hence, without implying that the system (B.1) is singular and maps its entire 2D domain into a 1D curve). For example, both of the functions $u = x$, $v = xe^y$ have a full line of solutions along the y axis; but they are still independent (their Jacobian is not identically zero), and $\mathbf{g}(x,y) = (x,xe^y)$ still maps $\mathbb{R}^2$ onto a 2D subrange of $\mathbb{R}^2$ (the first and third quadrants of the plane). As a second example, consider the two functions of (B.5). These functions are clearly independent, and indeed, they map $\mathbb{R}^2$ onto a 2D subrange of $\mathbb{R}^2$ (the unit disk). And yet, they have infinitely many continuous lines of solutions in common (all the vertical lines $x = n\pi$, $n \in \mathbb{Z}$). Incidentally, in this case each of the two functions has also infinitely many additional zeros that are not shared with the other: The first function has the horizontal lines $y = (m + \frac{1}{2})\pi$, $m \in \mathbb{Z}$ as zeros, while the second function has the horizontal lines $y = m\pi$, $m \in \mathbb{Z}$ as zeros.

It turns out that if the Jacobian is non-zero at a solution point (x_0, y_0) of the system $(u,v) = \mathbf{g}(x,y)$ then that solution point is isolated [Howse95 p. 14]. Note, however, that although the converse is true for linear transformations, it is not necessarily true in the general case. For example, the point $(0,0)$ is clearly an isolated solution of the system $(u,v) = (2xy, y^2 - x^2)$, and yet the Jacobian $J(x,y) = 4x^2 + 4y^2$ vanishes at this point. ∎

B.4 Interpretation as a domain transformation $r(\mathbf{g}(x,y))$

Suppose we are given a scalar function (i.e. a surface) $r(u,v)$, and that we apply to it the transformation $(u,v) = \mathbf{g}(x,y)$. The resulting distorted function (or surface) is given,

---

[1] Note that the case with no solutions at all (where the two planes, and hence their zero level lines, are parallel to each other) is excluded when $\mathbf{g}$ is linear, since both planes of a linear transformation must pass through the origin. This case may occur, however, if $\mathbf{g}$ is an affine transformation.

[2] More generally, this situation occurs in all linear transformations having the form $u = ax + by$, $v = s(ax + by)$. The common zeros of these two planes (i.e. $\operatorname{Ker} \mathbf{g}$) consist of the entire line $ax + by = 0$, and the image of the transformation ($\operatorname{Im} \mathbf{g}$) consists of the line $v = su$.

therefore, by $r(\mathbf{g}(x,y))$. This is illustrated in Fig. B.3, where the original function $r(u,v)$ is shown in (a), and the resulting distorted function $r(\mathbf{g}(x,y))$ is shown in (b). This figure uses, again, the same transformation $\mathbf{g}(x,y) = (2xy, y^2 - x^2)$ as in Fig. B.2, and yet, the effect of the transformation seems to be completely different: While in Fig. B.2 the distortion generated by $\mathbf{g}(x,y)$ seems to be parabolic, in Fig. B.3 the distortion seems to be hyperbolic. How can we explain this fact?

As pointed out in Sec. D.6 of Appendix D, each transformation $\mathbf{g}(x,y)$ can be used in two different ways: either as a direct transformation, or as a domain, inverse transformation. Consider, for example, the transformation $\mathbf{g}(x,y) = (2x,2y)$. Clearly, this transformation maps each point (x,y) to the new location $(2x,2y)$, and thus it expands the original layer by two. This is, indeed, the interpretation of $\mathbf{g}(x,y)$ as a *direct* transformation. However, when the same transformation $\mathbf{g}(x,y)$ is used as a *domain* transformation, for example, when it acts on the original layer $r(u,v)$ to give $r(2x,2y)$, its effect is inversed: $r(2x,2y)$ is a two-fold shrinked version of $r(u,v)$, while the two-fold expansion of $r(u,v)$ is expressed by $r(x/2,y/2)$, i.e. by using the *inverse* transformation $\mathbf{g}^{-1}(x,y) = (x/2,y/2)$. This inversion effect of domain transformations is explained in detail in Sections D.6 and D.10 of Appendix D; see also Remark 4.1 in Sec. 4.4.

Returning now to our case, we see that the transformation $\mathbf{g}(x,y) = (2xy, y^2 - x^2)$ is used in Fig. B.2 as a direct transformation $(x,y) \mapsto (2xy, y^2 - x^2)$, while in Fig. B.3 it is used as a domain (and hence, inverse) transformation that distorts $r(u,v)$ into $r(2xy, y^2 - x^2)$. This explains, indeed, the different geometric shapes that are obtained by the same transformation in Figs. B.2 and B.3. The effects of the same transformation $\mathbf{g}(x,y) = (2xy, y^2 - x^2)$ as a direct transformation and as an inverse transformation are also illustrated in Fig. B.4.

The lesson is, therefore, that in situations where $\mathbf{g}(x,y)$ can be used in both ways, it is important to clearly indicate which of the interpretations of $\mathbf{g}(x,y)$ is intended, in order to avoid any possible confusion.

B.5 Interpretation as a coordinate change

If we consider the level lines of the function $u = g_1(x,y)$ and the level lines of the function $v = g_2(x,y)$ as two sets of curvilinear coordinates, we may interpret system (B.1) or its vector representation (B.2) as a transformation from Cartesian to curvilinear coordinates in the plane. In other words, we can regard (B.1) or (B.2) as a mapping from $\mathbb{R}^2$ onto itself that defines a new curvilinear coordinate system u,v in the plane, instead of the original Cartesian coordinate system x,y (see Fig. B.1 and Fig. B.4(d)).

According to this interpretation of $\mathbf{g}(x,y)$, the zero level curves $g_1(x,y) = 0$ and $g_2(x,y) = 0$ are simply the curvilinear axes $u = 0$ and $v = 0$ of the new curvilinear coordinate system defined by the transformation $\mathbf{g}(x,y)$. The other coordinate curves $u = m$ and $v = n$ for all

$m,n \in \mathbb{Z}$ are given by the two curve families defined by $g_1(x,y) = m$ and $g_2(x,y) = n$, which are, respectively, integer level lines of $g_1(x,y)$ and integer level lines of $g_2(x,y)$.

It is interesting to note, however, that the curvilinear coordinate system obtained from the level lines of $u = g_1(x,y)$ and $v = g_2(x,y)$ corresponds to the effect of $\mathbf{g}(x,y)$ as an *inverse* transformation (for example, in the case shown in Fig. B.1 we obtain a hyperbolic coordinate net rather than a parabolic one). This phenomenon is explained in detail in Sec. D.4 of Appendix D.

Obviously, in order to provide a useful coordinate system the transformation $(u,v) = \mathbf{g}(x,y)$ must satisfy the condition $J(x,y) \not\equiv 0$ set by definition B.1. We have seen in Sec. B.3 that this condition eliminates the risk that $\mathbf{g}(x,y)$ maps $\mathbb{R}^2$ into a degenerate subset of $\mathbb{R}^2$ such as a 1D curve, a single point, or an empty set. In terms of coordinate lines, the condition $J(x,y) \not\equiv 0$ guarantees that the two components of $(u,v) = \mathbf{g}(x,y)$, namely, $u = g_1(x,y)$ and $v = g_2(x,y)$, do not have exactly the same level curves [Kaplan03 p. 162]. But while this condition is obviously *necessary*, it is not yet *sufficient* to guarantee that the resulting coordinate system is useful, and it does not exclude other pathologic situations that can make our new coordinate system useless. For instance, each of the surfaces $u = g_1(x,y)$, $v = g_2(x,y)$ still may have several distjoint zero level curves, meaning that each of the resulting curvilinear coordinate axes $g_1(x,y) = 0$ and $g_2(x,y) = 0$ consists of several disjoint branches. This occurs, for example, in the case of $\mathbf{g}(x,y) = (2xy, y^2 - x^2)$, where each of the surfaces has two perpendicular zero level lines (see Fig. B.1), or in its variant $\mathbf{g}(x,y) = (2xy - 1, y^2 - x^2 - 1)$, where each of the surfaces has a hyperbolic zero level line composed of two disjoint branches. Furthermore, the curvilinear axes $g_1(x,y) = 0$ and $g_2(x,y) = 0$ may have several intersection points even if the Jacobian is not identically zero; this may happen, for example, if the axes are second order curves such as parabolas, hyperbolas or ellipses, since such curves may have up to 4 intersection points. Even worse, as we have already seen in Remark B.1, the two functions $u = g_1(x,y)$, $v = g_2(x,y)$ may have coinciding zero level curves (and hence generate a useless, degenerate coordinate system) even if their Jacobian is not identically zero: Indeed, the condition $J(x,y) \not\equiv 0$ excludes the possibility that *all* the level curves of $g_1(x,y)$ and $g_2(x,y)$ be identical, but $g_1(x,y)$ and $g_2(x,y)$ still may have *some* (and even infinitely many) coinciding level lines. For example, as we have seen in Remark B.1, they may have coinciding zero level lines, and thus give coinciding coordinate axes. It is clear, therefore, that in order to obtain a useful coordinate system we need a stronger condition than simply having a non-identically zero Jacobian $J(x,y)$. For example, we may require that $g_1(x,y)$ and $g_2(x,y)$ satisfy also the following identities, which are known as the Cauchy-Riemann conditions:

$$(a) \quad \frac{\partial g_1}{\partial x} = \frac{\partial g_2}{\partial y}, \quad \frac{\partial g_1}{\partial y} = -\frac{\partial g_2}{\partial x} \qquad \text{or} \quad (b) \quad \frac{\partial g_1}{\partial x} = -\frac{\partial g_2}{\partial y}, \quad \frac{\partial g_1}{\partial y} = \frac{\partial g_2}{\partial x} \qquad (\text{B.6})$$

In this case, the transformation $\mathbf{g}(x,y)$ is called *conformal* [Courant88 pp. 166–167], and it maps the straight lines $x = const.$, $y = const.$ into curve families $u = const.$ and $v = const.$ which intersect *at right angles*. This orthogonality is clearly stronger than the mere independence of $g_1(x,y)$ and $g_2(x,y)$; and indeed, condition (a) implies $J(x,y) > 0$, and

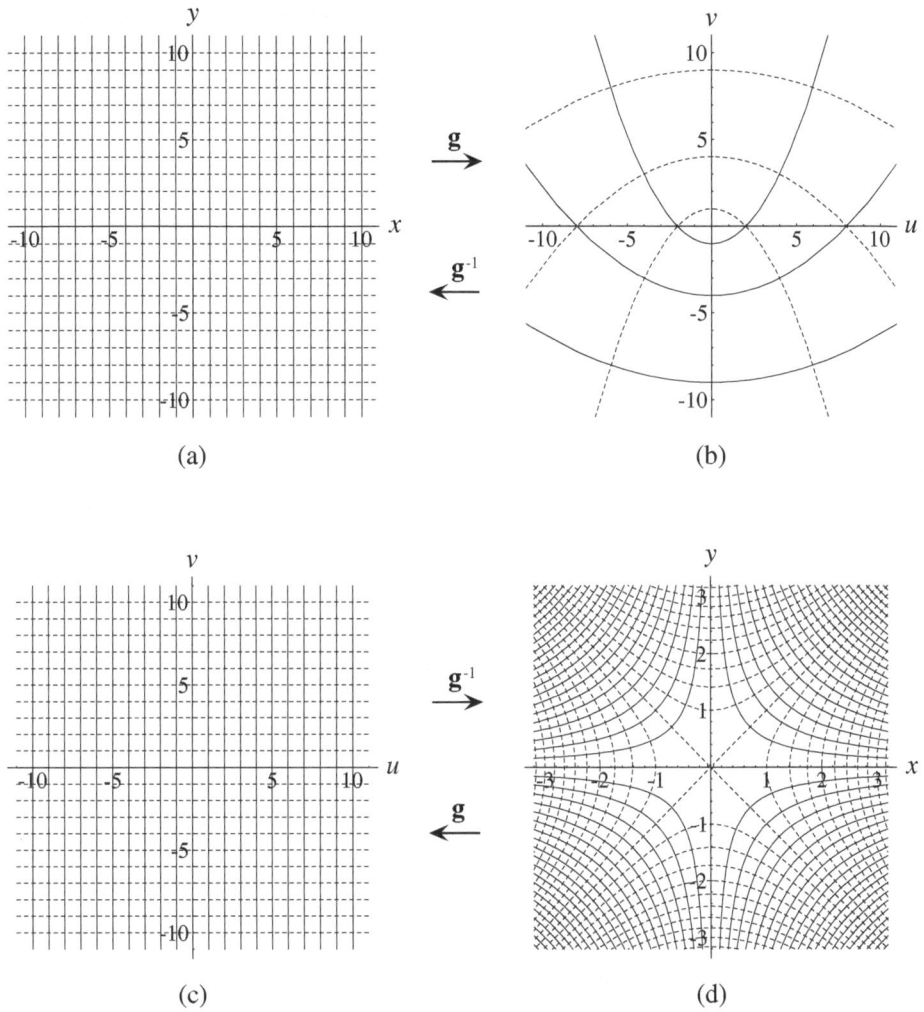

Figure B.4: (a),(b) Representation of the same transformation $\mathbf{g}(x,y) = (2xy, y^2 - x^2)$ as a coordinate change in $\mathbb{R}^2$. (c),(d) Representation of the inverse transformation $\mathbf{g}^{-1}$ as a coordinate change in $\mathbb{R}^2$ (see Example D.5 in Appendix D).

condition (b) implies $J(x,y) < 0$. Such an orthogonality is not *required* for having a useful coordinate system (see for instance Fig. 10.2(b) in *Vol. I*), but it is certainly advantageous. But on the other hand, this condition is not yet sufficient for excluding cases with multiple-branch coordinate axes or with coordinate axes having several intersection points. For example, although the transformation $\mathbf{g}(x,y) = (2xy, y^2 - x^2)$ shown in Fig. B.1 is clearly conformal, each of its new coordinate axes $g_1(x,y) = 0$ and $g_2(x,y) = 0$ consists of two perpendicular lines. Usually such pathologies can be resolved, however, by considering our transformation on a suitable subrange of $\mathbb{R}^2$.

B.6 Interpretation as a 2D vector field

Another useful interpretation is obtained by considering the vector function $\mathbf{g}(x,y)$ of Eq. (B.2) as a *vector field*. A vector field in $\mathbb{R}^2$ is a function $\mathbf{g}(x,y)$ that assigns to each point (x,y) in the x,y plane a vector $(u,v) = \mathbf{g}(x,y)$. Well known examples in physics include electric or magnetic fields as well as the gravitation field of the earth, all of which are vector fields that are defined in the 3D space $\mathbb{R}^3$.

A vector field in $\mathbb{R}^2$ can be illustrated graphically by drawing an arrow emanating from each point (x,y) of the x,y plane (or, more practically, from some representative points on a given grid within the x,y plane), where the length and the orientation of each arrow indicate the length and the orientation of the vector $\mathbf{g}(x,y)$ that has been assigned by $\mathbf{g}$ to the point (x,y) (see Fig. B.5(a)).[3] It is important to note, however, that each such arrow does not connect the point (x,y) to its destination $\mathbf{g}(x,y)$ under the transformation $\mathbf{g}$, but rather to the point $(x,y) + \mathbf{g}(x,y)$. Note also that the null vector $(0,0)$ is assigned to a point (x,y) *iff* (x,y) is a solution of the system (B.3). As we have seen, depending on the case there may exist one such point, several such points, infinitely many, or even none at all. It is important to stress, however, that these points are *not* fixed points of $\mathbf{g}(x,y)$, since they do not satisfy $\mathbf{g}(x,y) = (x,y)$ (they *are*, however, fixed points of the transformation $\mathbf{g}(x,y) + (x,y)$, since they do satisfy, of course, $\mathbf{g}(x,y) + (x,y) = (x,y)$).

The vector field interpretation of $\mathbf{g}(x,y)$ is closely related to the interpretation of $\mathbf{g}(x,y)$ as a direct mapping. For example, by comparing Fig. B.5 with Fig. B.4 it can be seen that the vector $\mathbf{g}(x,y)$ attached to the point $(x,y) = (1,0)$ of the vector field is precisely the destination of the point $(1,0)$ under the *direct* transformation, namely $(0,-1)$ (Fig. B.4(b)), and not the destination of $(1,0)$ under the *inverse* transformation, which is $(\frac{\sqrt{2}}{2},\frac{\sqrt{2}}{2})$ (Fig. B.4(d)). And yet, as we can clearly see in the figures, the geometric shapes of the direct transformation and of its corresponding vector field may look significantly different. In fact, all the various representations of the same transformation $\mathbf{g}(x,y)$ — as a vector field, as a direct mapping and as an inverse mapping — can be completely different from each other. We will return to this point in more detail in Sec. B.7.

As an alternative way of visualizing a vector field one may also draw its *trajectories* (also known as *field lines*). Loosely speaking, these are the curves obtained by following the arrows in Fig. B.5(a) and joining them into continuous curves in the x,y plane (see Fig. B.5(b)). More precisely, trajectories (or field lines) are curves for which the tangent vector to the curve at each point (x,y) is exactly $\mathbf{g}(x,y)$. Thus, at every point (x,y), the direction in

---

[3] For practical reasons it is customary to scale the arrow length in the drawing by a constant factor, in order to avoid drawings with too short, hard-to-see arrows, or drawings with too long, overlapping arrows.

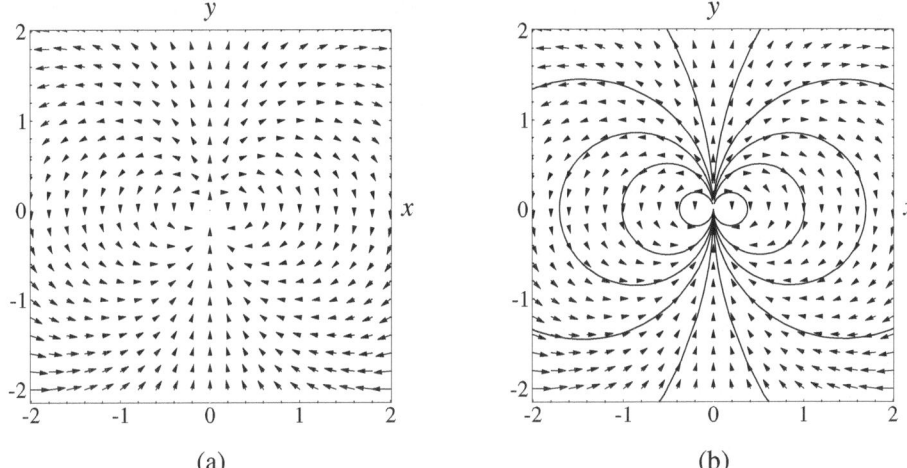

(a)   (b)

Figure B.5: (a) Illustration of the same transformation $\mathbf{g}(x,y) = (2xy, y^2 - x^2)$ as a vector field in $\mathbb{R}^2$. (b) Some trajectories of this vector field. Note that these trajectories are given by the family of the vertically tangent circles $(x - c)^2 + y^2 = c^2$ (i.e. $y = \sqrt{2cx - x^2}$) for all possible values of the constant $c \in \mathbb{R}$; this can be easily shown by calculating dx/dy and verifying that it satisfies Eq. (B.9), namely: $dx/dy = 2xy/(y^2 - x^2)$.

which the trajectory runs is determined by the vector $\mathbf{g}(x,y)$. Note that except for points where $\mathbf{g}(x,y)$ is not defined or where $\mathbf{g}(x,y) = \mathbf{0}$, every point in the plane belongs to one and only one trajectory. This also means, up to the same exceptions, that trajectories do not intersect. The advantage of drawing the trajectories of a vector field is that they are easier to picture visually. However, although the trajectories clearly show the directions of a vector field, they do not convey the information on the length of its vectors, and hence they are not a full representation of $\mathbf{g}(x,y)$. We will return to this point in Remark B.2 below.

The trajectories of a vector field $\mathbf{g}(x,y)$ are given in the parametric form by a family of curves $(x(t),y(t))$ whose members differ from each other by a constant c; these curves are the solutions of the system of differential equations (see [Bronshtein97 p. 526]):

$$\frac{d}{dt}x(t) = g_1(x(t),y(t))$$
$$\frac{d}{dt}y(t) = g_2(x(t),y(t))$$

(B.7)

where t is the parameter of each of the curves.[4] Using the vector notation $\mathbf{x}(t) = (x(t),y(t))$ this can be written more compactly as:

---

[4] Remember that a curve in the x,y plane can be generally defined in Cartesian coordinates in three equivalent forms: *explicitly* by $y = h(x)$, *implicitly* by $f(x,y) = 0$, or *parametrically* by $x = f_1(t)$, $y = f_2(t)$, where the parameter t varies continuously throughout an interval such as $-\infty < t < \infty$ [Bronshtein97 pp. 75–76]. Conversions between these forms can be done as explained in [Bronshtein97 p. 551], but they are not always possible [Harris98 p. 121].

$$\mathbf{x}'(t) = \mathbf{g}(\mathbf{x}(t)) \tag{B.8}$$

Note the close similarity between the equations of the vector field $\mathbf{g}(x,y)$ (Eqs. (B.1) or (B.2)) and the parametric equations defining its trajectories (Eqs. (B.7) or (B.8), respectively). But if we prefer the explicit form $y = f(x)$ of the trajectories instead of their parametric form given by $(x(t),y(t))$, then the corresponding differential equation that defines them is:[5]

$$\frac{dy}{dx} = \frac{g_2(x,y)}{g_1(x,y)} \tag{B.9}$$

Ways of solving systems of differential equations such as (B.7) can be found, along with many illustrative examples and figures showing their trajectories, in Chapter 4 of [Kreyszig93]. Equivalent ways for the solution of the corresponding differential equation (B.9) can be found, for example, in [Bronshtein97 pp. 395–398]. A complete classification of the different trajectory shapes for *linear* differential equation systems, including nodes, saddle points, center points, spirals, etc. can be found in [Kreyszig93 pp. 176–178], [Gray97 pp. 551–567] or in Sec. 1.4 of [Tabor89]. A similar classification for *non-linear* differential equation systems can be found in [Gray97 pp. 586–602].

Remark B.2: It should be remembered that because the trajectories do not contain all the information of the vector field (they only convey the vector directions but not their lengths), a vector field $\mathbf{g}(x,y)$ cannot be uniquely determined by its trajectories $y = f(x)$. For example, the two vector fields $\mathbf{g}(x,y) = (2y,-2x)$ and $\mathbf{g}(x,y) = (-y/(x^2+y^2), x/(x^2+y^2))$ have the same trajectories (a family of circles $x^2 + y^2 = c^2$ for any constant c), the difference being only in the vector lengths along these circles. More generally, if $\mathbf{g}(x,y)$ has circular or radial trajectories, it is clear that all the vector fields $h(r)\mathbf{g}(x,y)$ with $r = \sqrt{x^2 + y^2}$ have the same trajectories $y = f(x)$ as $\mathbf{g}(x,y)$. But if we use the parametric form of the trajectories, $(x(t),y(t))$, then the trajectories may be expressed differently for each of these vector fields, the difference being not in the shape of the curves but in their tracing speed in terms of the parameter t. For example, $(\cos t, \sin t)$ and $(\cos 2t, \sin 2t)$ represent the same circle, the difference being only in the parametrization (the speed of drawing the curves as t advances). Thus, if we consider the tracing speed of the trajectories as an indication to the vector lengths in $\mathbf{g}(x,y)$, the parametric form of the trajectories may convey both the vector directions and the vector lengths of the vector field (up to a constant).

An alternative way for sorting out this problem consists of drawing the trajectories so that the strength of the vector field (i.e. the vector lengths) is represented by the density of the trajectories [Needham97 pp. 453 and 494]: the closer together the trajectories, the

---

[5] Note that the explicit form of the trajectories, which is obtained by solving the differential equation (B.8), may have singular points wherever $g_1(x,y) = 0$ (i.e. at vertical tangencies of the solution). An advantage of the parametric form of the trajectories, which is obtained by solving the system of differential equation (B.7), is that such points are no longer singular points [Birkhoff89 p. 134]. Another advantage of the parametric form is that it explicitly indicates the *direction* of each trajectory: The positive sense of a trajectory is defined as the sense in which the curve is traced out for increasing values of t [Kreyszig93 p. 457].

stronger the vector field (just as the density of the level lines in a topographic map indicates the steepness of the ground). ∎

Remark B.3: Note that in physics the trajectories of a vector field are often interpreted as curves which trace the motion of particles under the influence of the field. If the parameter t is understood as time, the trajectory $(x(t),y(t))$ gives the path of the particle, namely, the location of the particle at each moment t. The derivative of this curve, $(\frac{d}{dt}x(t), \frac{d}{dt}y(t))$, given by Eqs. (B.7), defines the *velocity* of the particle (which is a vectorial entity, too) at each moment t. For other possible physical interpretations see [Needham97 pp. 451–454]. ∎

Remark B.4: Although in many cases it is easy to guess intuitively the trajectories (solution curves) of a differential equation from its vector field, it is a well known fact that in some cases this task is not as easy as it sounds. More details on this subject, as well as several illustrated examples, can be found in [Schwalbe97, pp. 39–42 and 69]. ∎

Remark B.5: There exists another remarkable visual difference between $\mathbf{g}(x,y)$ as a transformation and $\mathbf{g}(x,y)$ as a vector field, which concerns their behaviour under various symmetry operations: When considered as transformations, $\mathbf{g}(-x,-y)$ is a global 180° rotation of $\mathbf{g}(x,y)$ and $\mathbf{g}(x,-y)$ is a global vertical reflection of $\mathbf{g}(x,y)$ (see Sec. C.2 in Appendix C and the figures therein). However, when considered as vector fields, $\mathbf{g}(-x,-y)$ only differs from $\mathbf{g}(x,y)$ in the sense of its vectors, while $\mathbf{g}(x,-y)$ looks completely different. For example, the vector field $\mathbf{g}(x,y) = (x,y)$ consists of radial trajectories emanating from the origin, and the vector field $\mathbf{g}(x,y) = (-x,-y)$ consists of radial trajectories pointing to the origin; but the vector field $\mathbf{g}(x,y) = (x,-y)$ has a completely different shape, and it consists of hyperbolic trajectories. One should be aware of such differences in order to avoid mistakes when trying to interpret intuitively the meaning of $\mathbf{g}(-x,-y)$, $\mathbf{g}(x,-y)$, etc. ∎

B.7 Relationship between the different representations of $\mathbf{g}(x,y)$

As we have seen, any transformation $\mathbf{g}(x,y)$ can be interpreted in several different ways, whose graphical representations can be very different. For example, Figs. B.4(b), B.4(d) and B.5 show the graphical representations of the same transformation $\mathbf{g}(x,y) = (2xy, y^2 - x^2)$ when it is interpreted, respectively, as a direct transformation, as a domain transformation, and as a vector field. What are the mathematical relationships between the curve families that represent the same transformation $\mathbf{g}(x,y) = (g_1(x,y),g_2(x,y))$ in its various interpretations? The answer is as follows: The level lines of $\mathbf{g}(x,y)$ when it is viewed as a *domain* transformation (in our example, the two families of curves shown in Figs. B.1 or B.4(d)) are given by the curve families $g_1(x,y) = const.$ and $g_2(x,y) = const.$ The level lines of $\mathbf{g}(x,y)$ when it is viewed as a *direct* transformation (in our example, the two families of curves shown in Fig. B.4(b)) are given by the curve families $g_1^{-1}(x,y) = const.$ and $g_2^{-1}(x,y) = const.$, where $g_1^{-1}(x,y)$ and $g_2^{-1}(x,y)$ are the two components of the inverse transformation $\mathbf{g}^{-1}(x,y)$, namely: $\mathbf{g}^{-1}(x,y) = (g_1^{-1}(x,y),g_2^{-1}(x,y))$. And finally, the trajectories (field lines) of $g(x,y)$ when it is viewed as a vector field (in our example, the curve family shown in Fig. B.5(b)) are given by the family of parametric curves $(x(t),y(t))$ that are the solutions of the system of differential equations (B.7).

It is interesting to note that given a real valued function (i.e. a surface) $z = g(x,y)$ over the x,y plane, the gradient of $g(x,y)$, denoted by $\nabla g(x,y)$, gives at each point (x,y) a vector defining the maximal slope of $g(x,y)$ at this point. This is, indeed, a vector field whose definition is:

$$\nabla g(x,y) = (\frac{\partial}{\partial x}g(x,y), \frac{\partial}{\partial y}g(x,y)) \tag{B.10}$$

For example, for the paraboloid $g(x,y) = x^2 + y^2$ we have: $\mathbf{g}(x,y) = \nabla g(x,y) = (2x,2y)$. The trajectories of this vector field are the lines of maximal slope of $g(x,y)$; they are called *gradient lines* or *gradient curves* of $g(x,y)$. Note that the gradient lines of $g(x,y)$ are orthogonal to the level lines of $g(x,y)$. In our example of the paraboloid the trajectories of the vector field (B.10) are given by the following system of linear differential equations:

$$\frac{d}{dt}x(t) = 2x(t)$$
$$\frac{d}{dt}y(t) = 2y(t) \tag{B.11}$$

whose solution curves consist of the family of straight lines that is given in parametric form by $x(t) = c_1e^t$, $y(t) = c_2e^t$ for any constants c_1, c_2, or in explicit form by $y = cx$ for any constant c [Kreyszig93 p. 168]). These lines are, indeed, the gradient lines of our paraboloid.

Note, however, that while for every reasonably well behaved surface $g(x,y)$ there exists a vector field $\mathbf{g}(x,y)$ such that $\mathbf{g}(x,y) = \nabla g(x,y)$, the converse is not necessarily true: Not every transformation $\mathbf{g}(x,y) = (g_1(x,y), g_2(x,y))$ can be represented as a gradient field of some surface $g(x,y)$. For example, the transformation $\mathbf{g}(x,y) = (2y,-2x)$ has no surface $g(x,y)$ such that $\frac{\partial}{\partial x}g(x,y) = 2y$ and $\frac{\partial}{\partial y}g(x,y) = -2x$, since this would imply by integration that $g(x,y) = 2xy + c_1(y)$ and $g(x,y) = -2xy + c_2(x)$, but these two conditions on $g(x,y)$ are contradictory.[6] On the other hand, the transformation $\mathbf{g}(x,y) = (2x,2y)$ does have a surface

---

[6] This can be also explained geometrically: The trajectories of $\mathbf{g}(x,y) = (2y,-2x)$ are concentric circles about the origin; their parametric and implicit expressions are given below after Eq. (B.14). Note, however, that it is only the additional information conveyed by the parametric expression of these circles (their relative "tracing speed" as a function of the parameter t) or, equivalently, the information provided by the vector lengths within the vector field $\mathbf{g}(x,y)$, that prevents these circles from being gradient lines of any surface $g(x,y)$. For if we only consider the geometric shape of these circles, as it is conveyed by their implicit expression ($x^2 + y^2 = c^2$ for any constant c), there do exist surfaces $g(x,y)$ having these circles as gradient lines. For example, these circles are the gradient lines of the helicoid $g(x,y) = a \arctan(y/x)$ [Weisstein99 p. 810]: indeed, the gradient field of this surface is $\nabla g(x,y) = (-y/(x^2 + y^2), x/(x^2 + y^2))$, whose trajectories have the same implicit expression $x^2 + y^2 = c^2$ as the trajectories of our vector field $\mathbf{g}(x,y) = (2y,-2x)$. The difference lies in the evolution of the tracing speed (or of the vector lengths) between the inner and the outer circles, information which is only conveyed by the parametric expression of these circles. To see this, note that in the helicoid the steepness of the gradient curves along the surface increases as their radius gets smaller (meaning that the inner vectors of the vector field are longer), while in the surface that would have as its gradient lines the circular trajectories of $\mathbf{g}(x,y) = (2y,-2x)$, the steepness of the gradient curves along the surface would remain identical for all radiuses (since the vector lengths within the vector field increase linearly with the radius); but this is geometrically impossible.

$g(x,y)$ such that $\frac{\partial}{\partial x}g(x,y) = 2x$ and $\frac{\partial}{\partial y}g(x,y) = 2y$, since by integration we have $g(x,y) = x^2 + c_1(y)$ and $g(x,y) = y^2 + c_2(x)$, and indeed, taking $c_1(y) = y^2$ and $c_2(x) = x^2$ we obtain $g(x,y) = x^2 + y^2$.

A vector field $\mathbf{g}(x,y)$ for which there exists a surface $g(x,y)$ such that $\mathbf{g}(x,y) = \nabla g(x,y)$ is said to be a *conservative* vector field [Kreyszig93 p. 479; Weisstein99 p. 311]; in this case $g(x,y)$ is said to be a *potential function* of $\mathbf{g}(x,y)$. And it turns out [Kaplan03 pp. 326–327] that if the domain of $\mathbf{g}(x,y)$ is simply connected (a region without holes) then $\mathbf{g}(x,y)$ is conservative *iff* it is irrotational, i.e. $\text{curl}\,\mathbf{g} = 0$. In the 2D case $\text{curl}\,\mathbf{g} = 0$ means:

$$\frac{\partial}{\partial x}g_2(x,y) - \frac{\partial}{\partial y}g_1(x,y) = 0 \tag{B.12}$$

which is precisely the second part of the Cauchy-Riemann condition (b) (see Eq. (B.6)) for the transformation $\mathbf{g}(x,y) = (g_1(x,y), g_2(x,y))$.

Note, however, that even when such a surface $g(x,y)$ does exist it is clearly not unique, since any surface $g(x,y) + c$ has the same gradient field as $g(x,y)$. It follows, therefore, that if $g(x,y)$ exists then it is unique up to an additive constant [Ivanov95 p. 247].

In a similar way, one may also define for the same surface $z = g(x,y)$ a different vector field, that we denote here by $Hg(x,y)$, in which the trajectories coincide with the level lines of the surface, $g(x,y) = const$. In this vector field the trajectories are perpendicular at each point to the gradient of the surface. This vector field is given by (see [Birkhoff89 p. 135]):

$$Hg(x,y) = (\frac{\partial}{\partial y}g(x,y), -\frac{\partial}{\partial x}g(x,y)) \tag{B.13}$$

This follows, indeed, from (B.10) if we remember that the vector $(b,-a)$ is perpendicular to the vector (a,b). Note that we could also take the vector field $-Hg(x,y)$, whose trajectories follow the same level lines but in the opposite direction; which of the two is used is just a matter of convention.

In our present example of the paraboloid $g(x,y) = x^2 + y^2$, the vector field (B.13) is $\mathbf{h}(x,y) = Hg(x,y) = (2y, -2x)$. Its trajectories are given by the following system of linear differential equations:

$$\frac{d}{dt}x(t) = 2y(t)$$
$$\frac{d}{dt}y(t) = -2x(t) \tag{B.14}$$

whose solution curves consist of the family of concentric circles that is given in parametric form by $x(t) = c_1\cos t + c_2\sin t$, $y(t) = -c_1\sin t + c_2\cos t$ for any constants c_1, c_2, or in implicit form by $x^2 + y^2 = c^2$ for any constant c [Kreyszig93 p. 170]. These circles are, indeed, the level lines of our paraboloid.

However, just as in the case of gradient fields, it turns out that while for every reasonably well behaved surface $g(x,y)$ there exists a vector field $\mathbf{g}(x,y)$ such that $\mathbf{g}(x,y) = Hg(x,y)$, the converse is not necessarily true: Not every transformation $\mathbf{g}(x,y) = (g_1(x,y), g_2(x,y))$ can be represented as a vector field $Hg(x,y)$ of some surface $g(x,y)$. For example, the transformation $\mathbf{g}(x,y) = (2x,2y)$ has no surface $g(x,y)$ such that $\frac{\partial}{\partial y}g(x,y) = 2x$ and $-\frac{\partial}{\partial x}g(x,y) = 2y$, since this would imply by integration that $g(x,y) = 2xy + c_1(y)$ and $g(x,y) = -2xy + c_2(x)$, but these two conditions on $g(x,y)$ are contradictory. On the other hand, the transformation $\mathbf{g}(x,y) = (2y,-2x)$ does have a surface $g(x,y)$ such that $\frac{\partial}{\partial y}g(x,y) = 2y$ and $-\frac{\partial}{\partial x}g(x,y) = -2x$, since this implies by integration that $g(x,y) = y^2 + c_1(x)$ and $g(x,y) = x^2 + c_2(y)$, and indeed, by taking $c_1(x) = x^2$ and $c_2(y) = y^2$ we obtain $g(x,y) = x^2 + y^2$.

It can be shown that for a given $\mathbf{g}(x,y)$ there exists a surface $g(x,y)$ such that $\mathbf{g}(x,y) = Hg(x,y)$ *iff* $\mathbf{g}(x,y)$ is a *solenoidal* vector field, which means (see [Kaplan03 p. 184; Weisstein99 p. 1671–1672]) that $\operatorname{div}\mathbf{g} = 0$, or in the 2D case:

$$\frac{\partial}{\partial x}g_1(x,y) + \frac{\partial}{\partial y}g_2(x,y) = 0 \tag{B.15}$$

Such a surface $g(x,y)$ is often called a *Hamiltonian potential function* of $\mathbf{g}(x,y)$ [Howse95]. Note that condition (B.15) is precisely the first part of the Cauchy-Riemann condition (b) (see Eq. (B.6)) for the transformation $\mathbf{g}(x,y) = (g_1(x,y), g_2(x,y))$.

These results allow us to answer the following interesting questions: Given a transformation $\mathbf{g}(x,y) = (g_1(x,y),g_2(x,y))$ with two known families of level lines, can we find vector fields $\mathbf{v}_1(x,y)$ and $\mathbf{v}_2(x,y)$ having the same two curve families as trajectories (field lines)? And conversely, given a vector field $\mathbf{v}(x,y)$ with known trajectories, can we find a transformation $\mathbf{g}(x,y) = (g_1(x,y),g_2(x,y))$ whose level lines (the level lines of $g_1(x,y)$ or of $g_2(x,y)$) are identical to these trajectories? This would give us an interesting connection between the *level lines* of transformations (that are viewed as a pair of surfaces over the x,y plane) and the *trajectories* of other transformations (that are considered as vector fields).

The answer to the first question is, indeed, affirmative: As we have seen above, the two level line families of the *domain* transformation $\mathbf{g}(x,y) = (g_1(x,y),g_2(x,y))$ can be also regarded as the trajectories of the vector fields $Hg_1(x,y)$ and $Hg_2(x,y)$. And similarly, the two level lines of the *direct* transformation $\mathbf{g}(x,y)$ can be also regarded as the trajectories of the vector fields $Hg_1^{-1}(x,y)$ and $Hg_2^{-1}(x,y)$, where $g_1^{-1}(x,y)$ and $g_2^{-1}(x,y)$ are the two components of the inverse transformation $\mathbf{g}^{-1}(x,y)$, namely: $\mathbf{g}^{-1}(x,y) = (g_1^{-1}(x,y),g_2^{-1}(x,y))$. For example, in the case of the domain transformation $\mathbf{g}(x,y) = (2xy, y^2 - x^2)$, whose level lines are given by the two families of hyperbolas $2xy = m$ and $y^2 - x^2 = n$ for any m and n (see Figs. B.1 and B.4(d)), these two hyperbolic curve families are also the trajectories of the vector fields $Hg_1(x,y) = (2x,-2y)$ and $Hg_2(x,y) = (2y,2x)$ (see Fig. B.6). However, going the other way around is not always possible: Given a vector field $\mathbf{v}(x,y)$ it is not always possible to find a transformation $\mathbf{g}(x,y) = (g_1(x,y),g_2(x,y))$ whose level lines (i.e. the level lines of one of the surfaces $g_1(x,y)$ or $g_2(x,y)$) correspond to the trajectories of the vector field $\mathbf{v}(x,y)$; as we have just seen, this is only possible if $\mathbf{v}(x,y)$ is solenoidal.

Any vector field $\mathbf{v}(x,y)$ can be classified as conservative, solenoidal, both conservative and solenoidal, or neither conservative nor solenoidal. For example, the vector field

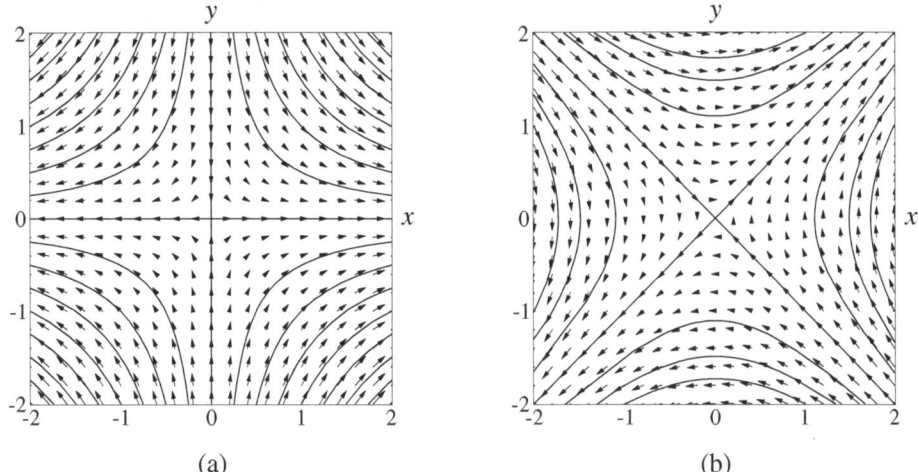

Figure B.6: The level lines of the domain transformation $g(x,y) = (g_1(x,y),g_2(x,y)) = (2xy, x^2 - y^2)$ (see Figs. B.1 or B.4(d)) are identical to the trajectories of the vector fields (a) $Hg_1(x,y) = (2x,-2y)$ and (b) $Hg_2(x,y) = (2y,2x)$.

$g(x,y) = (2xy, x^2 - y^2)$ is both conservative and solenoidal, while the vector field $g(x,y) = (x, y + x)$, which corresponds to a linear shear transformation, is neither conservative nor solenoidal. A vector field that is both conservative and solenoidal is called a *harmonic* vector field [Ivanov95 p. 242]. If $v(x,y)$ is a harmonic vector field, it is possible to find a transformation $g(x,y) = (g_1(x,y),g_2(x,y))$ where $g_1(x,y)$ is a surface whose *level lines* correspond to the trajectories of the given vector field, and $g_2(x,y)$ is a surface whose *gradient lines* correspond to the trajectories of the given vector field. In this case, the level lines of the surfaces $g_1(x,y)$ and $g_2(x,y)$ of $g(x,y)$ give two orthogonal sets of curves that correspond respectively to the trajectories and to the *equipotential lines*[7] of the harmonic vector field $v(x,y)$ [Kreyszig93 pp. 886–889]. Thus, a harmonic vector field $v(x,y)$ can be also depicted by the net consisting of its trajectories and its equipotential lines [Ivanov95 p. 248]; but this graphic representation of the vector field $v(x,y)$ should not be confused with the curvilinear net that represents $v(x,y)$ as a transformation (compare Figs. B.5(b) and B.4(b); note that the equipotential lines are not shown in Fig. B.5(b), but they are a 90° rotated copy of the trajectory lines).

Clearly, if $v(x,y) = (v_1(x,y),v_2(x,y))$ is harmonic it satisfies conditions (B.12) and (B.15), i.e. the two Cauchy-Riemann conditions (b) (see Eq. (F.6)), and hence it is a *conformal* transformation (see Sec. B.5). However, the converse is not necessarily true: Although the transformation $u(x,y) = (v_1(x,y),-v_2(x,y))$ is conformal, since it satisfies the two Cauchy-Riemann conditions (a), it is not a harmonic vector field. For example, consider

---

[7] The equipotential lines are to the trajectories (field lines) of a harmonic vactor field $g(x,y)$ what the level lines are to the gradient lines of the surface $g(x,y)$.

$\mathbf{v}(x,y) = (2xy, x^2-y^2)$ and $\mathbf{u}(x,y) = (2xy, y^2-x^2)$: Although both are conformal, $\mathbf{v}(x,y)$ is harmonic but $\mathbf{u}(x,y)$ is not (to convince oneself, it is easy to verify that there exists no surface $g(x,y)$ such that $\mathbf{u}(x,y) = \nabla g(x,y)$, and no surface $g(x,y)$ such that $\mathbf{u}(x,y) = Hg(x,y)$). On the other hand, the transformation $\mathbf{w}(x,y) = (-v_2(x,y),v_1(x,y))$ is harmonic, since it satisfies the Cauchy-Riemann conditions (b). This transformation is said to be *harmonically conjugate* to $\mathbf{v}(x,y)$ [Ivanov95 p. 242]; its equipotential lines are the trajectories of $\mathbf{v}(x,y)$, and vice versa [Ivanov95 p. 248; Needham97 p. 509].

Interestingly, these results also suggest that there may exist two different ways for representing a general 2D transformation $\mathbf{g}(x,y)$ as a pair of surfaces: Either, as explained in Sec. B.2, as the pair of surfaces $g_1(x,y)$ and $g_2(x,y)$ which are the two Cartesian components of $\mathbf{g}(x,y)$:

$$\mathbf{g}(x,y) = (g_1(x,y), 0) + (0, g_2(x,y)) = (g_1(x,y), g_2(x,y)) \tag{B.16}$$

or as a pair of surfaces $g(x,y)$ and $h(x,y)$ such that:

$$\mathbf{g}(x,y) = \nabla g(x,y) + Hh(x,y) \tag{B.17}$$

which means, in componentwise notation:

$$= (\frac{\partial}{\partial x}g(x,y) + \frac{\partial}{\partial y}h(x,y), \frac{\partial}{\partial y}g(x,y) - \frac{\partial}{\partial x}h(x,y))$$

This last decomposition of $\mathbf{g}(x,y)$ is known as a *gradient-Hamiltonian decomposition*, and the functions $g(x,y)$ and $h(x,y)$ are called, respectively, a *gradient potential function* of $\mathbf{g}(x,y)$ and a *Hamiltonian potential function* of $\mathbf{g}(x,y)$. However, this representation of $\mathbf{g}(x,y)$ is not necessarily unique. More details on the gradient-Hamiltonian decomposition and on its limitations can be found in [Howse95 pp. 60, 67–68]. Note that this decomposition sheds a new light on the fact already mentioned above that if $\mathbf{g}(x,y)$ is not a conservative vector field, there is no surface $g(x,y)$ such that $\mathbf{g}(x,y) = \nabla g(x,y)$: As we can now understand, this simply means that the gradient-Hamiltonian decomposition of such transformations $\mathbf{g}(x,y)$ must have a non-vanishing Hamiltonian component. The Hamiltonian component vanishes *iff* $\mathbf{g}(x,y)$ is conservative and hence can be represented as $\mathbf{g}(x,y) = \nabla g(x,y)$, and the gradient component vanishes *iff* $\mathbf{g}(x,y)$ is solenoidal and hence can be represented as $\mathbf{g}(x,y) = Hh(x,y)$. If $\mathbf{g}(x,y)$ is harmonic, i.e. both conservative and solenoidal, its gradient-Hamiltonian decomposition cannot be unique, since in this case we have both $\mathbf{g}(x,y) = \nabla g(x,y)$ with $h(x,y) \equiv 0$, and $\mathbf{g}(x,y) = Hh(x,y)$ with $g(x,y) \equiv 0$.

B.8 Remark on the local reflection of a 2D transformation

A transformation $\mathbf{g}(x,y)$ is said to be locally reflecting or locally non-reflecting around the point (x,y) according to whether the Jacobian at that point is positive or negative (see Appendix C). For example, the transformation $\mathbf{g}(x,y) = (2xy, y^2 - x^2)$ is non-reflecting throughout the plane (since $J(x,y) = 4(x^2 + y^2) > 0$), whereas the transformation $\mathbf{g}(x,y) = (2xy, x^2 - y^2)$ is reflecting throughout the plane (since $J(x,y) = -4(x^2 + y^2) < 0$). An

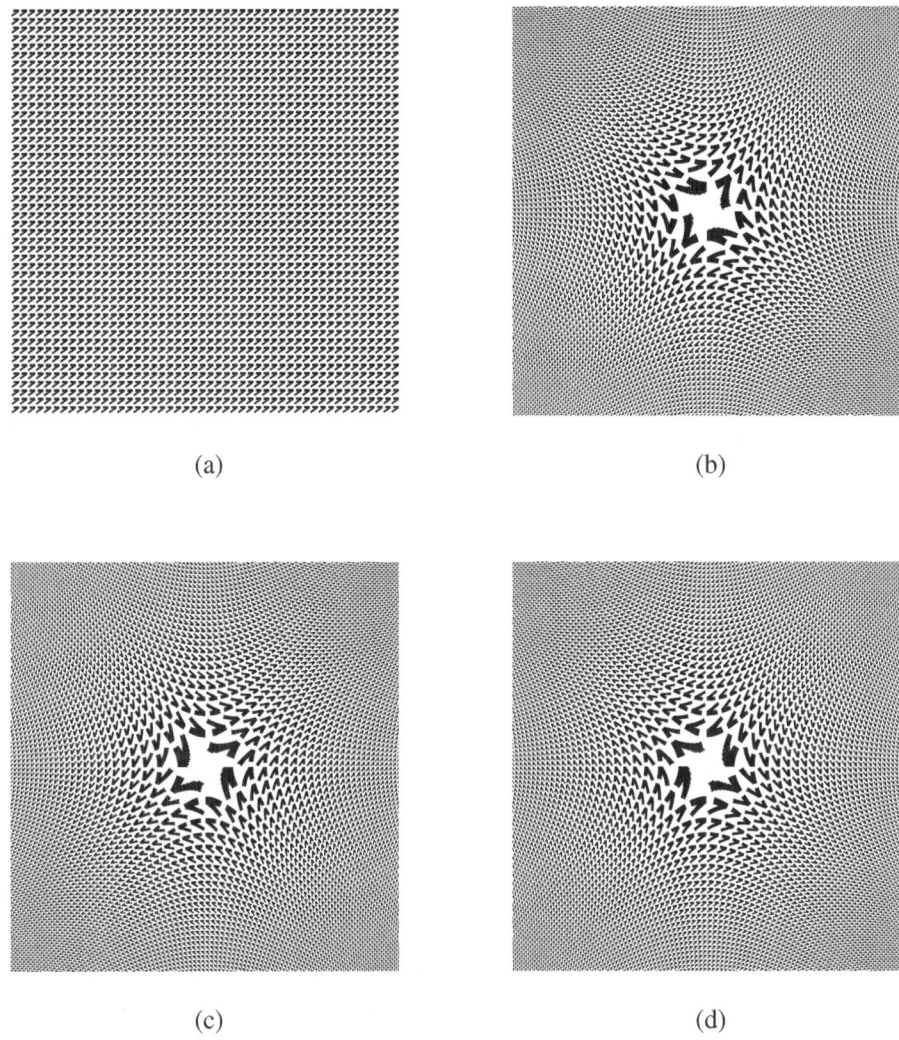

(a)   (b)

(c)   (d)

Figure B.7: Same as Fig. B.3, but this time using in the original image (a) a dot screen with the asymmetric element "1" rather than the symmetric element "•". This allows to clearly show the local orientation of the transformed plane at each point, and thus to distinguish between transformations such as: (b) $g(x,y) = (2xy, y^2 - x^2)$, (c) $g(x,y) = (2xy, x^2 - y^2)$, and (d) $g(x,y) = -(2xy, y^2 - x^2)$. Note that these transformations only differ in their local orientation at each point (x,y), but not in their global geometry.

example of a transformation that is reflecting in some parts of the plane and non-reflecting in other parts of the plane is given by Eq. (7.32), which corresponds to the moiré effect shown in Fig. 7.12(b).

The information about the local orientation of a transformation $\mathbf{g}(x,y)$ at each point (x,y) is obviously lost when we represent $\mathbf{g}(x,y)$ graphically as a coordinate change, like in Fig. B.4. But when we illustrate the effect of $\mathbf{g}(x,y)$ as we do in Figs. B.2 and B.3, there exists a simple "trick" that allows us to clearly show the local orientation of $\mathbf{g}(x,y)$ at any point of the plane. All that we need to do is to apply $\mathbf{g}(x,y)$ to a structure made of *asymmetric* rather than symmetric elements. When we do so, the local orientation of the transformation at each point (x,y) is indicated by the local orientation of the corresponding asymmetric element in the transformed plane. This is clearly illustrated in Fig. B.7, where three different transformations are applied to the same periodic dot screen (a) that consists of asymmetric "1"-shaped dots. The transformations being applied are: (b) $\mathbf{g}(x,y) = (2xy, y^2 - x^2)$; (c) $\mathbf{g}(x,y) = (2xy, x^2 - y^2)$; and (d) $\mathbf{g}(x,y) = -(2xy, y^2 - x^2)$. Note that in (c) all the "1"-shaped elements are mirror-imaged, while in (b) and (d) they are just rotated, but not mirror-imaged. This illustrates the fact that transformation (c) is reflecting throughout the plane, while the transformations (b) and (d) are non-reflecting.

More details on the connection between the Jacobian of a transformation $\mathbf{g}(x,y)$ and the properties of the transformation can be found in Appendix C.

Appendix C

The Jacobian of a 2D transformation and its significance

C.1 Introduction

Let $\mathbf{g}(x,y)$ be the 2D transformation whose two components are:

$$u = g_1(x,y)$$

$$v = g_2(x,y)$$

(C.1)

The *Jacobian matrix* of this transformation is the matrix:

$$\begin{pmatrix} \dfrac{\partial g_1}{\partial x} & \dfrac{\partial g_1}{\partial y} \\ \dfrac{\partial g_2}{\partial x} & \dfrac{\partial g_2}{\partial y} \end{pmatrix}$$

(C.2)

Note that the rows of this matrix correspond to the gradients of $g_1(x,y)$ and $g_2(x,y)$ (see Eq. (B.10) in Appendix B), while the columns of this matrix correspond to the directional derivatives of $\mathbf{g}(x,y)$ in the x and y directions, respectively. The *Jacobian determinant* (or simply the *Jacobian*) of the transformation $\mathbf{g}(x,y)$ is the scalar function $J: \mathbb{R}^2 \to \mathbb{R}$ that is defined as the determinant of this matrix:[1]

$$J(x,y) = \begin{vmatrix} \dfrac{\partial g_1}{\partial x} & \dfrac{\partial g_1}{\partial y} \\ \dfrac{\partial g_2}{\partial x} & \dfrac{\partial g_2}{\partial y} \end{vmatrix} = \dfrac{\partial g_1}{\partial x}\dfrac{\partial g_2}{\partial y} - \dfrac{\partial g_2}{\partial x}\dfrac{\partial g_1}{\partial y}$$

(C.3)

We have already seen in Appendix B that the Jacobian is tightly related to the mathematical properties of the transformation $\mathbf{g}(x,y)$. We will now explain the geometric interpretation of the Jacobian, and see in more detail its special role in connection with the transformation $\mathbf{g}(x,y)$. Other properties of $\mathbf{g}(x,y)$ that can be deduced from its Jacobian matrix (C.2) are discussed later in Secs. C.4–C.5.

C.2 Geometric interpretation of the Jacobian

Consider a 2D transformation $\mathbf{g}: \mathbb{R}^2 \to \mathbb{R}^2$. Clearly, this transformation maps any square element of the original x,y plane into its distorted image in the destination u,v plane. The area of the new distorted element can be smaller, equal or larger than the area of the original, undistorted element, depending on the local properties of the transformation $\mathbf{g}$ at

---

[1] Confusingly, some references use the term "Jacobian" for the Jacobian matrix, while other references use it, as we do, for the Jacobian determinant.

that point. In order to investigate how **g** influences the area at each point of the plane, we consider an infinitesimal square area-element $dxdy$ within the original x,y plane, and its distorted image $dudv$ in the transformed u,v plane. It turns out that at any point (x,y) we have (see [Weisstein99 p. 950] or [Colley98 p. 347]):

$$dudv = J(x,y)\, dxdy \tag{C.4}$$

More generally, if $A[S]$ and $A[\mathbf{g}(S)]$ denote, respectively, the area of a closed region S of the x,y plane and the area of its image $\mathbf{g}(S)$ in the u,v plane, then we have at the limit when $A[S]$ approaches zero [Spiegel63 p. 108]:

$$\lim \frac{A[\mathbf{g}(S)]}{A[S]} = J(x,y) \tag{C.5}$$

This means that the Jacobian is, in fact, an infinitesimal scale factor that indicates the local area scaling caused by the transformation $\mathbf{g}(x,y)$ at each point (x,y). Negative scaling values indicate that local scaling at (x,y) is also accompanied by local reflection (mirror imaging). If the Jacobian equals zero at a certain point (x,y), it means that the transformation **g** maps elements with non-zero area around that point to a zero-area image; the point (x,y) is called, then, a singular point of the transformation **g**.

Note that the Jacobian matrix of a transformation $(u,v) = \mathbf{g}(x,y)$ and the Jacobian matrix of the inverse transformation $(x,y) = \mathbf{g}^{-1}(u,v)$ are inverse matrices of each other [Kaplan03 p. 120; Courant89 p. 252]. This implies that the Jacobian $J_{\mathbf{g}}(x,y)$ of **g** and the Jacobian $J_{\mathbf{g}^{-1}}(x,y)$ of $\mathbf{g}^{-1}$ are reciprocals of each other: $J_{\mathbf{g}^{-1}}(x,y) = 1/J_{\mathbf{g}}(x,y)$. Hence, if one Jacobian is non-singular (different from zero) at the point (x,y), so is the other. Furthermore, a transformation $\mathbf{g}(x,y)$ is invertible over $\mathbb{R}^2$ or a subregion thereof (meaning that there exists a transformation $\mathbf{g}^{-1}(u,v)$ such that $\mathbf{g}^{-1}(\mathbf{g}(x,y)) = (x,y)$ within that region) *iff* the Jacobian of **g** is not identically zero within that region. Other important theorems on transformations and their Jacobians can be found, for example, in [Spiegel63 p. 108].

Remark C.1: When **g** is a linear transformation the two functions $g_1(x,y)$ and $g_2(x,y)$ of Eq. (C.1) are linear:

$$u = a_1 x + b_1 y$$
$$v = a_2 x + b_2 y \tag{C.6}$$

and their slopes $\frac{\partial g_1}{\partial x}, \frac{\partial g_1}{\partial y}, \frac{\partial g_2}{\partial x}, \frac{\partial g_2}{\partial y}$ reduce into the constant values a_1, b_1, a_2, b_2. This means that when **g** is linear, its Jacobian matrix is simply reduced to the matrix of the linear transformation, and the Jacobian is reduced to the determinant of this matrix:

$$J(x,y) = \begin{vmatrix} a_1 & b_1 \\ a_2 & b_2 \end{vmatrix} \tag{C.7}$$

so that the Jacobian $J(x,y)$ becomes a constant number, $J(x,y) = a_1 b_2 - a_2 b_1$.

Viewed the other way around, non-linear transformations can be seen as a generalization of linear transformations into the case where the slopes of $g_1(x,y)$ and $g_2(x,y)$ are no longer constant but rather vary with x and y. And indeed, if the transformation is not linear, then the Jacobian matrix will vary from point to point. Nevertheless, it may still be regarded as a local linear approximation to the true mapping (plus a relocation of the origin), and the Jacobian $J(x,y)$ can still be interpreted as the local area scale factor, which may now change from point to point. In other words, the Jacobian matrix of a non-linear transformation $g(x,y)$ can be viewed at any given point (x,y) as the matrix of a linear transformation $g'(x,y)$ that approximates the non-linear transformation $g(x,y)$ at the point (x,y), and whose constant coefficients a_1, b_1, a_2, b_2, are, respectively, the values of $\frac{\partial g_1}{\partial x}, \frac{\partial g_1}{\partial y}, \frac{\partial g_2}{\partial x}, \frac{\partial g_2}{\partial y}$ at that particular point.

It is interesting to note in this context that in linear algebra determinants play the role of area (or volume) functions [Lay03 pp. 204–209]. For example, the determinant (C.7) of the matrix of the linear transformation (C.6) gives the area of the parallelogram that is determined by the columns of the matrix, i.e. by the vectors (a_1,a_2) and (b_1,b_2) [Lay03 p. 205]. But these vectors are precisely the images under our linear transformation of the standard basis vectors $(1,0)$ and $(0,1)$ [Lang87 pp. 394–395]. Thus, if we denote by S the parallelogram determined by the basis vectors $(1,0)$ and $(0,1)$, and by $A[R]$ the area of the region R, then we have for any linear transformation g:

$$A[g(S)] = \begin{vmatrix} a_1 & b_1 \\ a_2 & b_2 \end{vmatrix} A[S] \tag{C.8}$$

Furthermore, as long as g is a linear transformation this property remains valid for any parallelogram S in the plane [Lang87 p. 457], and even for any arbitrary closed region S of the plane [Lay03 pp. 207–209]. Eq. (C.8) is, indeed, the linear equivalent of Eqs. (C.4) and (C.5).

In non-linear transformations the area scaling effect of g may be different at each point (x,y) of the plane, and therefore Eq. (C.8) is no longer globally valid for all arbitrary closed regions S; but it remains valid locally, for any *infinitesimal* region, as expressed, indeed, by Eqs. (C.4) and (C.5). ∎

C.3 Properties of the transformation g(x,y) that can be deduced from its Jacobian

As we have seen in the previous section, a 2D transformation $g(x,y)$ is closely related to its Jacobian. Therefore, it is not surprising that the Jacobian can provide precious information on the nature of the transformation in question. In the present section we provide a summary of various properties of the transformation $g(x,y)$ that can be deduced directly from its Jacobian. We start with a few results concerning the global nature of the Jacobian and its effect on the global properties of the transformation $g(x,y)$, and then we proceed to some of their local counterparts. Note that $g(x,y)$ is considered here as a direct transformation; but if $g(x,y)$ is used as a domain transformation (for example, if it is applied to an image $r(x,y)$ to give the distorted image $r(\mathbf{g}(x,y))$, as in Figs. C.1, C.2 or B.6), then we are concerned in fact with the inverse transformation $\mathbf{g}^{-1}$, whose properties can be deduced from the nature of the reciprocal Jacobian $1/J(x,y)$.

(1) If $|J(x,y)| < 1$ everywhere, then $\mathbf{g}(x,y)$ is *area-contracting*.

(2) If $|J(x,y)| > 1$ everywhere, then $\mathbf{g}(x,y)$ is *area-expanding*.

(3) If $J(x,y) > 0$ everywhere, then $\mathbf{g}(x,y)$ is *non-reflecting*. Such a mapping can locally represent rotations, scalings and shearing deformations and it can globally represent "rubber-sheet" distortions, but it will nowhere cause reflection. A string of text subjected to such a mapping would remain legible (although possibly highly distorted), and it would not be converted into a mirror image of itself. For example, when such a transformation $\mathbf{g}$ is applied to a periodic dot screen consisting of asymmetric "1"-shaped elements (Fig. C.1(a)), it maps each of these elements into a distorted and possibly rotated "1"-shaped element, but none of the resulting distorted "1"s is mirror imaged (see Figs. C.1(b),(c),(f),(g),(h) and Figs. B.6(a),(d)). Note that such transformations are also called in literature *orientation-preserving* [Courant89 p. 260], but this term may be somewhat misleading since the image of our "1"-shaped elements under such transformations may still be rotated. A better term would be *sense-preserving*.

(4) If $J(x,y) < 0$ everywhere, then $\mathbf{g}(x,y)$ is *reflecting*. Such a mapping will locally include reflection (possibly also combined with rotations, scalings and shearing deformations) and it can globally represent a "rubber-sheet" distortion combined with reflection. A string of text subjected to such a mapping would be converted into a mirror image of itself (in addition to any other distortions). For example, when such a transformation $\mathbf{g}$ is applied to a periodic dot screen consisting of asymmetric "1"-shaped elements (Fig. C.1(a)), it maps each of these elements into a mirror imaged and possibly otherwise distorted and rotated "1"-shaped element (see Figs C.1(d),(e) and Fig. B.6(c)). Note that such transformations are also called in literature *orientation-reversing* [Courant89 p. 260], but this term may be confusing since pure rotations that are not accompanied by reflection (including a rotation by 180°) are not orientation-reversing transformations. A better term would be *sense-reversing*.

(5) If $J(x,y) = const.$ everywhere, then $\mathbf{g}(x,y)$ has a constant scaling factor throughout. This obviously occurs if $\mathbf{g}(x,y)$ is a linear or affine transformation, but it may also occur in non-linear transformations (such as (C.9) below; see Fig. C.1(b)).

(6) If $J(x,y) = 1$ everywhere, then $\mathbf{g}(x,y)$ is *area-preserving*. This occurs, for example, if $\mathbf{g}(x,y)$ is a rotation or a shift transformation, but it may occur also in non-linear transformations (such as (C.9) below; see Fig. C.1(b)).

(7) If $J(x,y) = -1$ everywhere, then $\mathbf{g}(x,y)$ is *area-preserving* and *reflecting*. This occurs, for example, if $\mathbf{g}(x,y)$ is a reflection or a rotoinversion (namely, rotation combined with reflection [Cantwell02 p. 9]), but it may occur also in non-linear transformations.

(8) If $J(x,y) = 0$ everywhere, then $\mathbf{g}(x,y)$ is *degenerate*, meaning that it maps $\mathbb{R}^2$ or any 2D subregion thereof onto a 1D curve, a single point, or an empty set (see Sec. B.3 in Appendix B). This occurs *iff* the two components of $\mathbf{g}(x,y)$, i.e. $g_1(x,y)$ and $g_2(x,y)$, are dependent (see Definition B.1 in Appendix B). For example, the transformation $(u,v) = \mathbf{g}(x,y) = (x,x^2)$ maps the entire x,y plane onto the 1D parabola $v = u^2$.

(9) If $J(x,y) \neq 0$ and $J(x,y) \neq \pm\infty$ everywhere, then $\mathbf{g}(x,y)$ has no singularities, and it is one-to-one and invertible.

The remaining points of the list concern the *local* properties of the Jacobian around a given point (x_0,y_0) and their local influence on the transformation $\mathbf{g}(x,y)$:

(10) If $|J(x_0,y_0)| < 1$ at the point (x_0,y_0), then $\mathbf{g}(x,y)$ is *area-contracting* near that point.

(11) If $|J(x_0,y_0)| > 1$ at the point (x_0,y_0), then $\mathbf{g}(x,y)$ is *area-expanding* near that point.

(12) If $J(x_0,y_0) > 0$ at the point (x_0,y_0), then $\mathbf{g}(x,y)$ is *non-reflecting* (orientation-preserving) near that point [Courant89 p. 260]. For example, an asymmetric "1"-shaped element near that point will be mapped by $\mathbf{g}$ into a distorted but not mirror-imaged "1".

(13) If $J(x_0,y_0) < 0$ at the point (x_0,y_0), then $\mathbf{g}(x,y)$ is *reflecting* (orientation-reversing) near that point [Courant89 p. 260]. For example, an asymmetric "1"-shaped element near that point will be mapped by $\mathbf{g}$ into a distorted, mirror-imaged "1".

(14) If $J(x_0,y_0) = 0$ at the point (x_0,y_0), then $\mathbf{g}(x,y)$ has a local area scaling of zero at that point. This means that $\mathbf{g}$ maps elements with non-zero area around the point (x_0,y_0) to a zero area (or an almost-zero area) image. This occurs, for example, at the point $(0,0)$ in Fig. B.2(b).

(15) If $J(x_0,y_0) = \infty$ at the point (x_0,y_0), then $\mathbf{g}(x,y)$ has an infinitely big local area scaling at that point. This occurs, for example, at the point $(0,0)$ in the transformation shown in Fig. B.3(b), whose expression as a *direct* transformation is given by $\mathbf{g}^{-1}(x,y) = (\sqrt{(\sqrt{u^2 + v^2} - v)}/2, \sqrt{(\sqrt{u^2 + v^2} + v)}/2$); see Example D.5 in Appendix D.

(16) If $J(x_0,y_0) = -\infty$ at the point (x_0,y_0), then $\mathbf{g}(x,y)$ has an infinitely big local area scaling at that point, and, in addition, $\mathbf{g}(x,y)$ is also reflecting at that point.

In all of the cases (14)–(16) the transformation $\mathbf{g}(x,y)$ is said to be singular at the point (x_0,y_0), or, equivalently, to have a singular point at (x_0,y_0).

Figs. C.1 and C.2 illustrate some of these cases by applying various domain transformations to a periodic dot screen $r(x,y)$ consisting of asymmetric "1"-shaped elements (Fig. C.1(a)). The transformations being used are all variants of the parabolic transformation $\mathbf{g}(x,y)$ that is given by:

$$u = x$$

$$v = y - a(x - c)^2$$

(C.9)

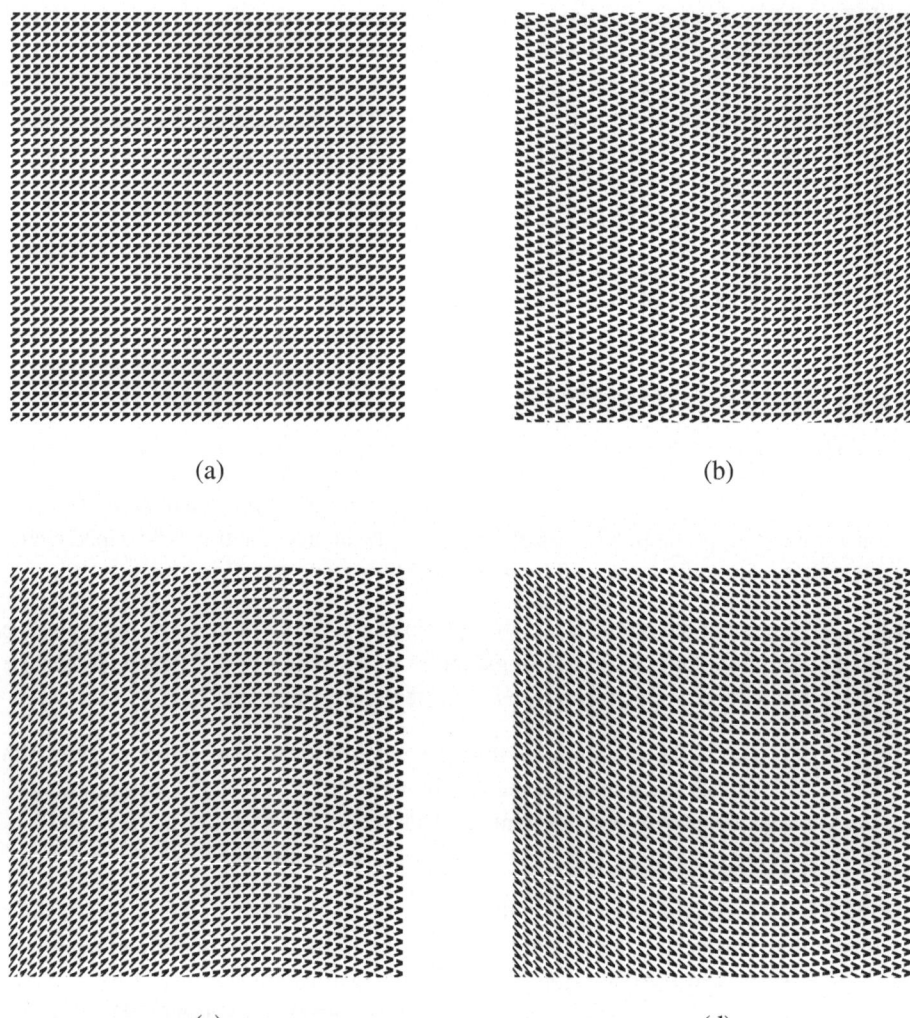

(a)   (b)

(c)   (d)

Figure C.1: A dot screen $r(x,y)$ composed of asymmetric "1"-shaped dots (a), and its deformations $r(\mathbf{g}(x,y))$ under different variants of the domain transformation (C.9). Note the influence of each of the transformations on the orientation of the parabolas, on the local orientation of the "1"-shaped elements (cells), and on the global orientation of the distorted image. The transformations being used (and their effects) are:

(b) $u = x, \quad v = y - a(x-c)^2$ (top-opened parabolas, upright "1"s);

(c) $u = x, \quad v = y + a(x-c)^2$ (bottom-opened parabolas, upright "1"s);

(d) $u = x, \quad v = -y + a(x-c)^2$ (top-opened parabolas, reflected "1"s);

(e) $u = x, \quad v = -y - a(x-c)^2$ (global vertical reflection of (b));

(f) $u = y - a(x-c)^2, \quad v = -x$ (top-opened parabolas, rotated "1"s);

(g) $u = -x, \quad v = -y + a(x-c)^2$ (top-opened parabolas, rotated "1"s);

(h) $u = -y + a(x-c)^2, \quad v = x$ (top-opened parabolas, rotated "1"s).

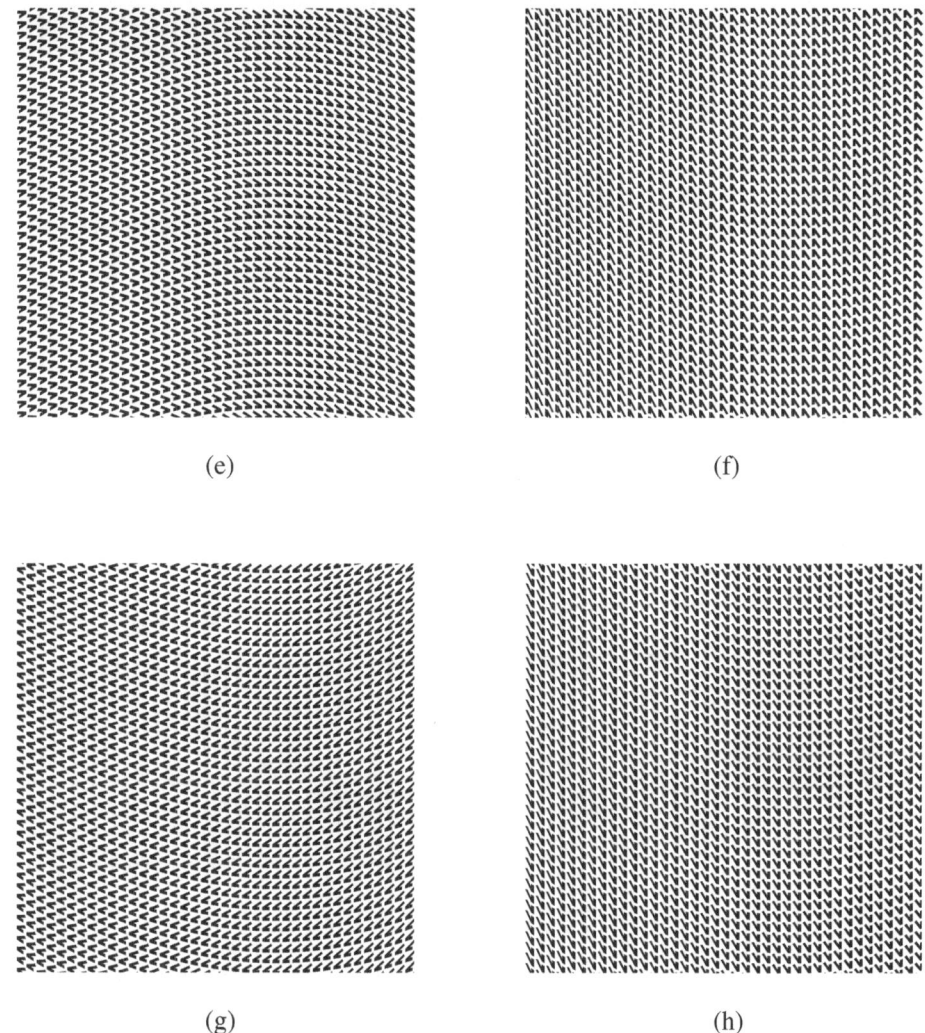

Figure C.1: (*continued.*) Note that for practical reasons the upright orientation of the "1"s in the original dot screen (a) is not vertical, but rather rotated by −45° (in order to avoid vertical collisions between the "1" elements in successive rows). This implies that each of the transformed "1"s, too, is in fact a distorted version of the rotated "1".

When (C.9) is applied as a domain transformation, it bends the horizontal coordinate lines of the *x,y* pane into equispaced top-opened parabolas that are shifted by a constant *c* to the right (see Fig. C.1(b)). We have chosen this transformation because of its global asymmetry with respect to the origin and with respect to the main axes. This asymmetry allows us to investigate different variants of this transformation, and to clearly visualize

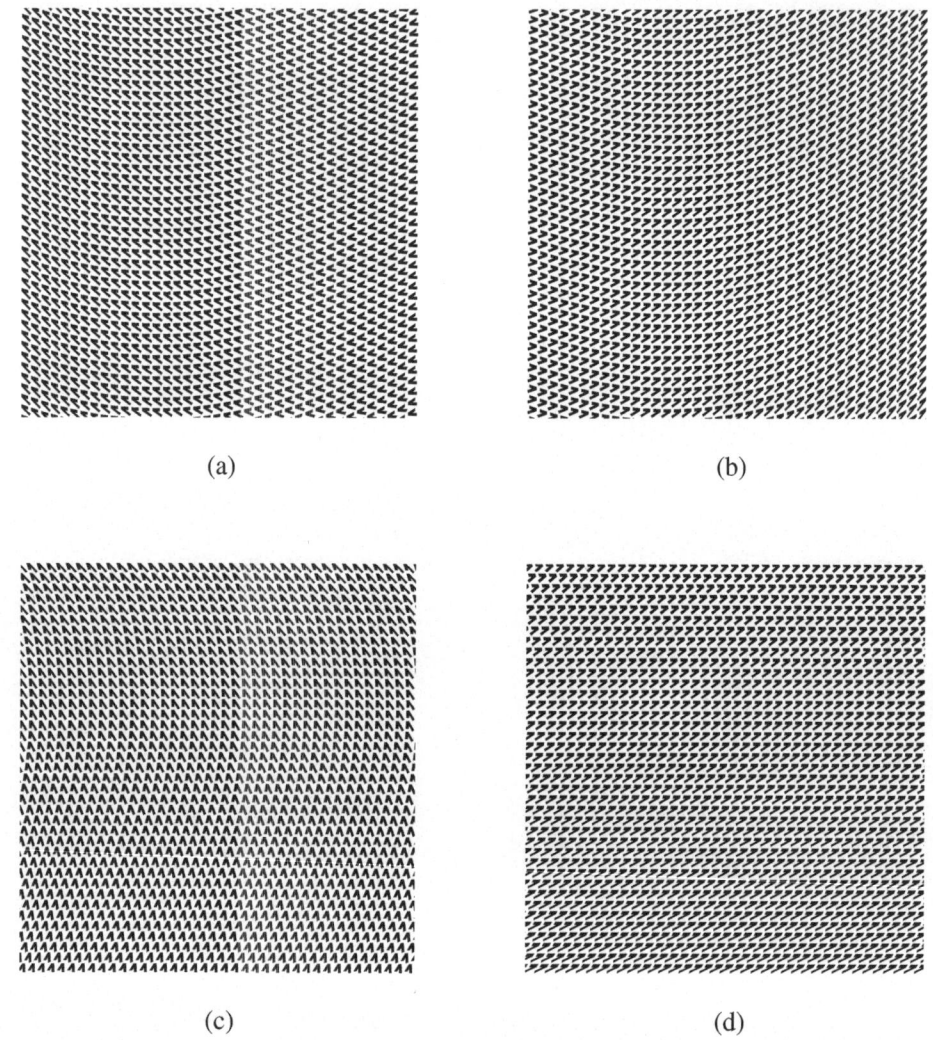

(a)   (b)

(c)   (d)

Figure C.2: Same as in Fig. C.1, with other variants of the domain transformation (C.9):
     (a) $u = -x, \quad v = y - a(x + c)^2$ (global horizontal reflection of Fig. C.1(b));
     (b) $u = x, \quad v = y - a(x + c)^2$ (horizontally reflected parabolas, upright "1"s);
     (c) $u = y, \quad v = -x - a(y - c)^2$ (global 90° rotation of Fig. C.1(b));
     (d) $u = x + a(y - c)^2, \quad v = y$ (90° rotated parabolas, upright "1"s);

their effect on the resulting transformed plane. In particular, it allows us to easily distinguish between: (1) Global reflections and rotations of the entire plane (see Figs. C.1(b) and C.1(e)); (2) reflections or rotations of the parabolic geometry alone, with no influence on the local orientation of the "1"-shaped elements (see Figs. C.1(b) and

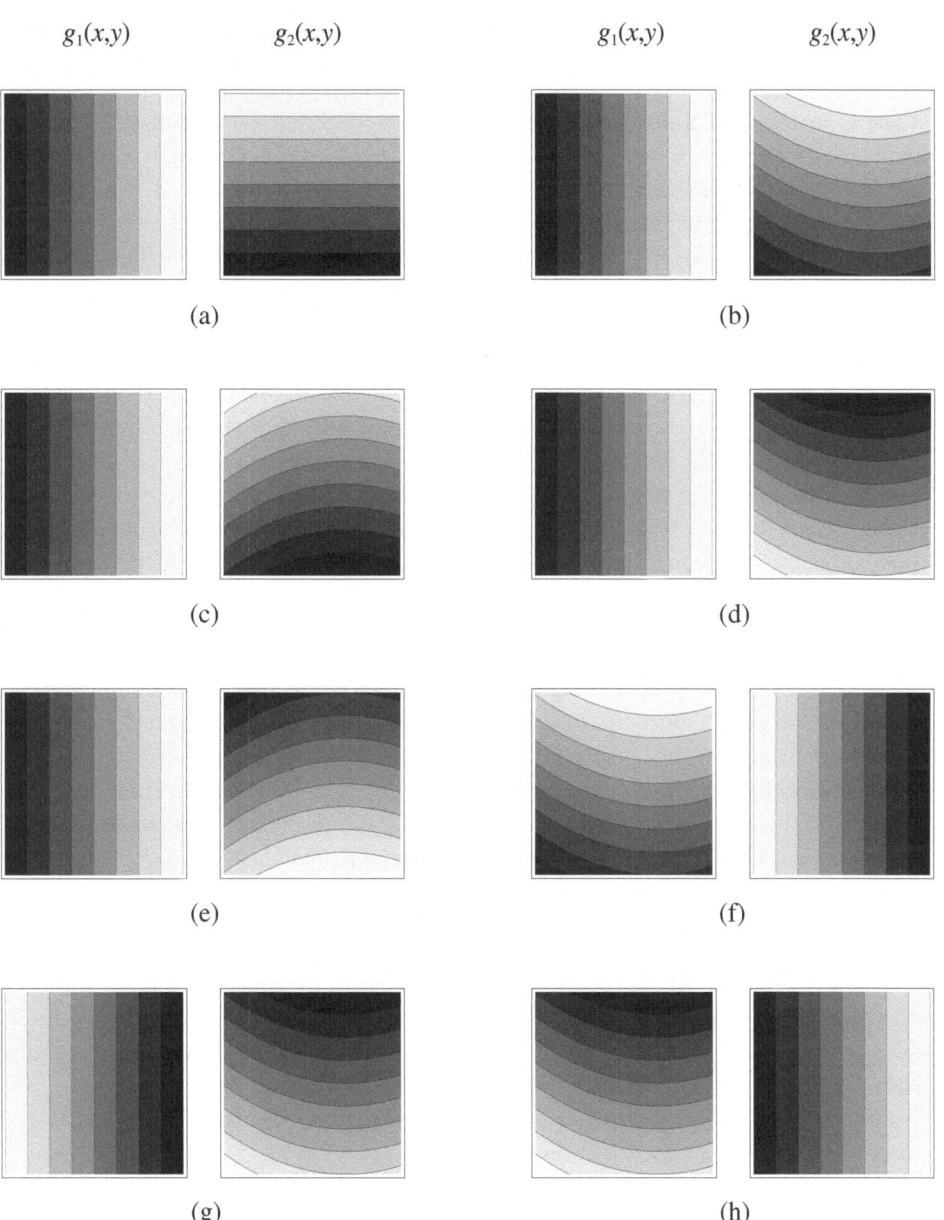

Figure C.3: The two components (surfaces) $g_1(x,y)$ and $g_2(x,y)$ of each of the transformations shown in Figs. C.1(a)–(h). The curves plotted on each of the surfaces are level lines, and the gray levels show the surface altitude: brighter shades indicate higher values and darker shades indicate lower values.

C.1(c)); and (3) local reflections or rotations of the "1"-shaped elements alone, with no global reflections or rotations (see Figs. C.1(b) and C.1(d),(f)–(h)).

The global and local orientation properties of the different variants of (C.9) are best explained by Fig. C.3, which shows for each of the cases of Fig. C.1 the two components (surfaces) $g_1(x,y)$ and $g_2(x,y)$ of the transformation being used (see Sec. B.2 in Appendix B). The cells shown in Fig. C.1 (each of which contains a "1"-shaped element) are, in fact, the curved quadrilaterals that are formed by the curvilinear coordinate lines, i.e. by the level curves of the surfaces $g_1(x,y)$ and $g_2(x,y)$. The cell orientations in the transformed plane simply reflect the orientations of these two families of level curves, where the orientation of each curve family corresponds to the direction of increasing curve altitudes. The ascending order of the level curves in each family is clearly indicated in Fig. C.3 by the gray levels that show the corresponding surface altitude, going from dark (lower levels) to bright (higher levels). Note that the difference between cases (b) and (c) is only in the *orientation* of the parabolic level curves, but not in their order; while the difference between cases (b) and (d) is only in the *order* of the parabolic level curves, but not in their orientation. The difference between cases (b) and (e) is both in the orientation and in the order of the parabolic level curves. Note also the difference between cases (d) and (h), which simply consists of interchanging $g_1(x,y)$ and $g_2(x,y)$.

Thus, as we can see by comparing Figs. C.1(a)–(h) with their corresponding families of level curves in Figs. C.3(a)–(h), the global geometry of the cells in the transformed plane is due to the *shape* of the corresponding level curves, while the local orientation of the cells is due to the *ordering* of the level curves. For example, when the level curves are identical in their shape but their order is inversed, the result is a local sense inversion (compare Figs. C.1(b) and C.1(d)). As another example, interchanging $g_1(x,y)$ and $g_2(x,y)$ does not modify the shape of the level curves, and it only causes a reflection of the cells (compare Figs. C.1(d) and C.1(h)).

Table C.1 gives the full list of the different variants of transformation (C.9) involving top-opened and bottom-opened parabolas. A similar table can be also constructed for the cases involving a global horizontal reflection (whose top-opened parabolas are shifted from the center to the left rather than to the right) and for the cases involving global rotations by 90° or 270° (which give left-opened and right-opened parabolas, respectively). Some of these cases are shown in Fig. C.2.

In the general case of a transformation $u = g_1(x,y)$, $v = g_2(x,y)$ the total number of different possible symmetric variants is 512: There exist 16 possible sign combinations for x and y (such as $u = g_1(-x,y)$, $v = g_2(x,-y)$, etc.); for each of these we have 4 possible sign variations of g_1 and g_2 themselves (such as $u = -g_1(x,y)$, $v = g_2(x,y)$); then we have 4 possible permutations of x and y within g_1 and g_2 (for example, $u = g_1(y,x)$, $v = g_2(x,y)$); and for each of the 256 combinations we have so far there still exist two possible permutations of g_1 and g_2 themselves (such as $u = g_2(x,y)$, $v = g_1(x,y)$). The total number of variants may significantly reduce in simple cases such as $g_1(x,y) = x$, but on the other hand it may further increase if we also consider the signs of the individual terms within $g_1(x,y)$

and $g_2(x,y)$, as we did in the case of $g_2(x,y) = y - a(x - c)^2$. Note, however, that if the transformation $\mathbf{g}(x,y)$ is symmetric, as was the case in Fig. B.7, then many of its different variants may give the same result. In Fig. C.1 we have intentionally chosen a highly asymmetric transformation $\mathbf{g}(x,y)$, which allows us to clearly distinguish in the figure between its different variants.

C.4 The local orientation properties of a transformation g(x,y)

As we have seen in Sec. C.3, the Jacobian of a transformation $\mathbf{g}(x,y)$ contains in a nutshell all the information about the local *magnification* and *reflection* properties of the transformation. It does not provide, however, any information related to the local *orienration* (or rotation) properties of the transformation; for example, it does not account for the difference between the transformations that generate Figs. C.1(b) and C.1(f)–(h). What other mathematical construct or criterion can be used to provide this missing information?

In order to answer this question, suppose that each of the cells generated by the coordinate grid of the original, untransformed image (the "1"-shaped elements in Fig. C.1(a)) is drawn by a laser beam or by a plotter, that scans the entire cell line-by-line. We assume that the beam scans the cell in horizontal lines that follow the positive x direction, and that successive horizontal lines are ordered from the bottom upward, following the positive y direction. When the transformation $\mathbf{g}(x,y)$ is applied, each of the original square elements is distorted into a curvilinear quadrilateral. Consider the distorted element located at the point (x,y). When the laser beam draws this distorted element, each of the originally straight scanlines is distorted into a curvilinear line, and the vertical step of the beam, as it advances between the successive curvilinear scanlines, is also distorted (see Fig. C.4).

When we proceed to the limit, each of the curvilinear quadrilaterals becomes infinitesimally small, and the direction of the scanlines and the direction of the steps between successive scanlines reduce into the local tangent slopes of the curvilinear coordinates at the point (x,y). It turns out that these directions are given by the two following vectors:

Scanline direction: $\mathbf{v}_1(x,y) = \dfrac{1}{J(x,y)} (\frac{\partial}{\partial y} g_2(x,y), -\frac{\partial}{\partial x} g_2(x,y))$

$$(C.10)$$

Interline direction: $\mathbf{v}_2(x,y) = \dfrac{1}{J(x,y)} (-\frac{\partial}{\partial y} g_1(x,y), \frac{\partial}{\partial x} g_1(x,y))$

The orientations of these two vectors give us the local scanline direction and the local interline direction at the point (x,y), and their lengths indicate the local stretching factor of the distorted cell in these two directions. Note that the scanline direction $\mathbf{v}_1(x,y)$ and the interline direction $\mathbf{v}_2(x,y)$ are perpendicular to the gradients $\nabla g_1(x,y)$ and $\nabla g_2(x,y)$ (see Fig. C.4(b)); but they are not necessarily perpendicular to each other.

	Transformation	$J(x,y)$	$\mathbf{v}_1(x,y)$ $\mathbf{v}_2(x,y)$	Parabola orientation	Element orientation	Fig.	Remarks
1	$u = x$ $v = y - a(x-c)^2$	1	$(1, 2a(x-c))$ $(0, 1)$	↑	↗	C.1(b)	(1)
2	$u = y - a(x-c)^2$ $v = -x$	1	$(0, 1)$ $(-1, -2a(x-c))$	↑	↖	C.1(f)	(2)
3	$u = -x$ $v = -y + a(x-c)^2$	1	$(-1, -2a(x-c))$ $(0, -1)$	↑	↙	C.1(g)	(2)
4	$u = -y + a(x-c)^2$ $v = x$	1	$(0, -1)$ $(1, 2a(x-c))$	↑	↘	C.1(h)	(2)
5	$u = y - a(x-c)^2$ $v = x$	-1	$(0, 1)$ $(1, 2a(x-c))$	↑	↗R		
6	$u = -x$ $v = y - a(x-c)^2$	-1	$(-1, -2a(x-c))$ $(0, 1)$	↑	↖R		
7	$u = -y + a(x-c)^2$ $v = -x$	-1	$(0, -1)$ $(-1, -2a(x-c))$	↑	↙R		
8	$u = x$ $v = -y + a(x-c)^2$	-1	$(1, 2a(x-c))$ $(0, -1)$	↑	↘R	C.1(d)	(3)
9	$u = x$ $v = y + a(x-c)^2$	1	$(1, -2a(x-c))$ $(0, 1)$	↓	↗	C.1(c)	(4)
10	$u = y + a(x-c)^2$ $v = -x$	1	$(0, 1)$ $(-1, 2a(x-c))$	↓	↖		
11	$u = -x$ $v = -y - a(x-c)^2$	1	$(-1, 2a(x-c))$ $(0, -1)$	↓	↙		
12	$u = -y - a(x-c)^2$ $v = x$	1	$(0, -1)$ $(1, -2a(x-c))$	↓	↘		
13	$u = y + a(x-c)^2$ $v = x$	-1	$(0, 1)$ $(1, -2a(x-c))$	↓	↗R		
14	$u = -x$ $v = y + a(x-c)^2$	-1	$(-1, 2a(x-c))$ $(0, 1)$	↓	↖R		
15	$u = -y - a(x-c)^2$ $v = -x$	-1	$(0, -1)$ $(-1, 2a(x-c))$	↓	↙R		
16	$u = x$ $v = -y - a(x-c)^2$	-1	$(1, -2a(x-c))$ $(0, -1)$	↓	↘R	C.1(e)	(5)

Table C.1: (*continued on the opposite page*)

Legend:

Parabolas orientation: ↑ = top-opened parabolas; ↓ = bottom-opened parabolas. Cells orientation: ↗ = upright orientation; ↖ = rotated by 90°; ↙ = rotated by 180°; ↘ = rotated by 270°; ↗ = reflected (i.e. mirror-imaged); ↖ = reflected and rotated by 90°; ↙ = reflected and rotated by 180°; ↘ = reflected and rotated by 270°. All cases involving reflection are indicated by R.

Remarks:

(1) All the "1"-shaped cells preserve their original orientation, and the parabolas have an upright orientation. This case serves us as a reference when comparing between cases.

(2) The "1"-shaped cells are rotated, but the parabolas preserve their original upright orientation.

(3) The "1"-shaped cells are reflected in the vertical sense, but the parabolas preserve their upright orientation.

(4) The parabolas are reflected vertically, but the "1"-shaped cells preserve their original orientation.

(5) A global reflection about the horizontal axis: both the parabolas and the "1"-shaped cells are vertically reflected.

Note that the upright orientation of the "1"-shaped elements is ↗, and all rotations and reflections are considered with respect to this original orientation. $J(x,y)$ is the Jacobian of the transformation in question, and $\mathbf{v}_1(x,y)$ and $\mathbf{v}_2(x,y)$ are the scanline and interline orientation vectors (see Sec. C.4).

Table C.1: (*continued.*) Different variants of the parabolic transformation (C.9) and some of their local and global properties. The legend and the remarks for the table are given above. Cases 1–4: Local rotations of the cells. Cases 5–8: Local rotoinversions of the cells. Cases 9–12: Same as cases 1–4 but with vertically reflected parabolas. Cases 13–16: Same as cases 5–8 but with vertically reflected parabolas. The different transformations $g(x,y)$ are applied to a periodic dot screen $r(x,y)$ (see Fig. C.1(a)) as *domain* (inverse) transformations, and the distorted result is $r(g(x,y))$ as shown in Figs. C.1(b)–(h).

To illustrate this result, let us return to Figs. C.1(b) and C.1(f)–(h). We see that in all of these figures the level curves are identical in their geometric shape, and what distinguishes between the 4 cases is only the order of the level curves, which determines the local orientation vectors $\mathbf{v}_1(x,y)$ and $\mathbf{v}_2(x,y)$ and hence the scanning order of the "1"-shaped elements. The local orientation vectors $\mathbf{v}_1(x,y)$ and $\mathbf{v}_2(x,y)$ are given in Table C.1 for each of the transformations in question. Another case illustrating the use of this result for the explanation of the orientation of moiré patterns is shown in Fig. 7.12.

As we can see, the vectors $\mathbf{v}_1(x,y)$ and $\mathbf{v}_2(x,y)$ are, in fact, the two columns of the Jacobian matrix of the *inverse* transformation $\mathbf{g}^{-1}(x,y)$:

$$
\begin{pmatrix} \dfrac{\partial g_1}{\partial x} & \dfrac{\partial g_1}{\partial y} \\[2mm] \dfrac{\partial g_2}{\partial x} & \dfrac{\partial g_2}{\partial y} \end{pmatrix}^{-1} = \frac{1}{J(x,y)} \begin{pmatrix} \dfrac{\partial g_2}{\partial y} & -\dfrac{\partial g_1}{\partial y} \\[2mm] -\dfrac{\partial g_2}{\partial x} & \dfrac{\partial g_1}{\partial x} \end{pmatrix}
\tag{C.11}
$$

It should be noted, however, that Eqs. (C.10) are based on the assumption that $\mathbf{g}(x,y)$ is used as a *domain* (inverse) transformation, i.e. that it is applied to an original image $r(x,y)$ to give the distorted image $r(\mathbf{g}(x,y))$, as in Figs. C.1 and C.2. But if $\mathbf{g}(x,y)$ is used as a *direct* transformation, then the vectors $\mathbf{v}_1(x,y)$ and $\mathbf{v}_2(x,y)$ are given by the columns of the *inverse* of matrix (C.11), namely, by the columns of the Jacobian matrix (C.2) of $\mathbf{g}(x,y)$ itself:

Scanline direction: $\mathbf{v}_1(x,y) = (\dfrac{\partial}{\partial x}g_1(x,y), \dfrac{\partial}{\partial x}g_2(x,y))$

Interline direction: $\mathbf{v}_2(x,y) = (\dfrac{\partial}{\partial y}g_1(x,y), \dfrac{\partial}{\partial y}g_2(x,y))$

$$\tag{C.12}$$

This is illustrated in Fig. C.5 for the simple case of a linear transformation. The top row of the figure shows the effect of the direct transformation $(u,v) = (x+y, 2y)$, while the bottom row shows the effect of its inverse, $(x,y) = (u - \frac{1}{2}v, \frac{1}{2}v)$. As we can clearly see, the Jacobian matrix of the direct transformation $\mathbf{g}$ is $\left(\begin{smallmatrix} 1 & 1 \\ 0 & 2 \end{smallmatrix}\right)$ and the Jacobian matrix of the inverse transformation $\mathbf{g}^{-1}$ is $\frac{1}{2}\left(\begin{smallmatrix} 2 & -1 \\ 0 & 1 \end{smallmatrix}\right)$. And indeed, the scanline and interline directions of the direct transformation are given, in accordance with Eqs. (C.12), by $\mathbf{v}_1 = (1,0)$, $\mathbf{v}_2 = (1,2)$ (see Fig. C.5(b)), while those of the inverse transformation are given, in accordance with Eqs. (C.10), by $\mathbf{v}_1 = (1,0)$, $\mathbf{v}_2 = (-\frac{1}{2}, \frac{1}{2})$ (see Fig. C.5(d)).

C.5 Other properties of $\mathbf{g}(x,y)$ that can be deduced from its Jacobian matrix

The Jacobian matrix of a transformation $\mathbf{g}(x,y)$ can provide further information on the nature of $\mathbf{g}(x,y)$, in addition to the local scaling, reflection and orientation properties already mentioned so far. Some of these additional properties are briefly summarized below.

(1) If all the entries of the Jacobian matrix are constant numbers (rather than functions of x and y), the first-order derivatives of $\mathbf{g}(x,y)$ are constant and they do not vary from point to point. This means that $\mathbf{g}(x,y)$ is either *affine* or *linear* (depending on whether or not it also shifts the origin). Such transformations always map straight lines into straight lines.

(2) If there is at most one element in each row and column of the Jacobian matrix which is not identically zero, then the transformation $(u,v) = \mathbf{g}(x,y)$ is *independent*. This means that a change in each of its input variables (x or y) causes a corresponding change in only a single distinct output variable (u or v). Such a mapping will preserve the independence of the coordinate axes. A simple example would be the interchange of the two axes, $\mathbf{g}(x,y) = (y,x)$.

(3) If the Jacobian matrix is diagonal (i.e. all its elements which are not located on its main diagonal are identically zero), then the transformation $(u,v) = \mathbf{g}(x,y)$ is *diagonal*. This means that each output variable of the transformation (u or v) depends only on the corresponding input variable (respectively, x or y), so that the coordinate axes are preserved. A simple example would be the non-linear scaling transformation $\mathbf{g}(x,y) =$ ($\log x$, $\log y$). Note that a diagonal mapping is more strongly constrained than an independent mapping in which the coordinate axes may be interchanged. A diagonal mapping is necessarily independent.

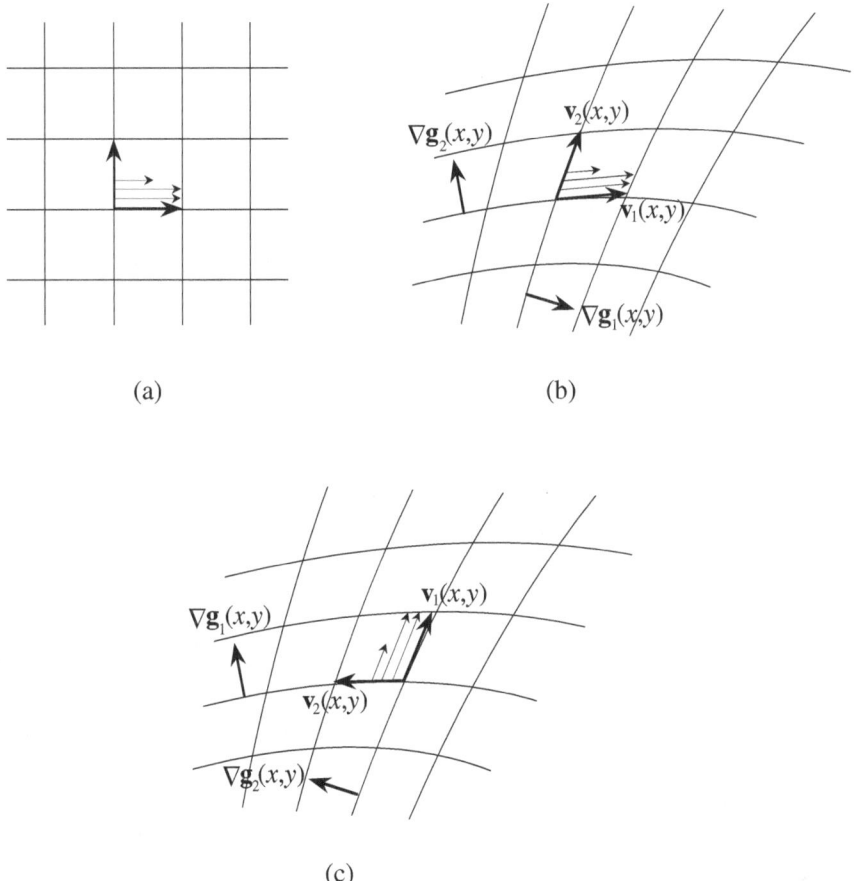

(a)

(b)

(c)

Figure C.4: Explanation of the local orientation properties of a transformation $\mathbf{g}(x,y)$. (a) The original coordinate grid before the application of $\mathbf{g}(x,y)$. (b) The distorted grid after the application of $\mathbf{g}(x,y)$. Each grid cell in (a) is drawn by a succession of scanlines, which is distorted in (b) into a succession of curvilinear scanlines. The scanline direction and the interline direction in (b) determine the local cell orientations in the distorted grid. (c) Another variant of the transformation $\mathbf{g}(x,y)$, in which the global geometry remains the same as in (b) but the cells are rotated by 90°.

(4) The Cauchy-Riemann conditions that determine whether a given transformation $\mathbf{g}(x,y)$ is conformal (see at the end of Sec. B.5 in Appendix B) can be also expressed in terms of the Jacobian matrix of $\mathbf{g}(x,y)$. In these terms, a transformation $\mathbf{g}(x,y)$ is conformal *iff* its Jacobian matrix has the form:

$$\text{(a)} \quad \begin{pmatrix} a & -b \\ b & a \end{pmatrix} \qquad \text{or:} \qquad \text{(b)} \quad \begin{pmatrix} a & b \\ b & -a \end{pmatrix} \qquad\qquad \text{(C.13)}$$

where a and b are functions $f_1(x,y)$ and $f_2(x,y)$ (note that the signs of a and b can be also negative). When the elements a and b are constant these matrices correspond to a linear (or affine) *similarity* transformation, namely, a transformation that is composed of a rotation and a uniform scaling [Casselman04 p. 144; Lay03 p. 339], and possibly also a reflection,[2] and (in the affine case) a shift. And indeed, in sufficiently small regions a conformal mapping looks like a linear (or affine) similarity transformation: it locally preserves shapes and angles (possibly up to a reflection) — even though it may distort large shapes wildly. We can say, therefore, that a conformal transformation is a transformation that behaves *locally* as a linear (or affine) similarity transformation, although at each point (x,y) the similarity transformation in question may be different. Note that the Jacobian determinant of a conformal transformation $\mathbf{g}(x,y)$ can be zero at some *isolated* points (x,y); these are the singular points of the transformation.

A conformal transformation $\mathbf{g}(x,y)$ can be also characterised as *isotropic*. This means that $\mathbf{g}(x,y)$ may apply a local scale factor to the distances between neighbouring points, but this factor does not depend on the orientation of the line between the two points, although it may vary from point to point. An isotropic mapping $\mathbf{g}(x,y)$ will convert a circle at any point in the plane into another circle (but possibly of a different size and in a different place), whereas a non-isotropic mapping would produce an ellipse. If the mapping is also linear then circles of any size will behave in this way, whereas with a non-linear mapping this may only be true for circles of infinitely small size.

Note that an alternative condition on the Jacobian matrix J in order that the transformation $\mathbf{g}(x,y)$ be conformal is that J satisfies $J^TJ = f(x,y)I$, where J^T denotes the transpose of the matrix J, $f(x,y)$ is a function and I is the identity matrix. The equivalence of this condition with the Cauchy-Riemann conditions (a) or (b) of Eq. (C.13) can be easily obtained by performing explicitly the matrix multiplication J^TJ:

$$\begin{pmatrix} a & c \\ b & d \end{pmatrix}\begin{pmatrix} a & b \\ c & d \end{pmatrix} = \begin{pmatrix} a^2 + c^2 & ab + cd \\ ab + cd & b^2 + d^2 \end{pmatrix}$$

where a, b, c, d stand for functions $f_1(x,y),...,f_4(x,y)$. By equating the elements of the product matrix with the elements of a diagonal matrix having equal elements on the diagonal we obtain the two identities:

---

[2] Note that case (a) in Eq. (C.13) corresponds to a *direct similarity*, that preserves shapes and angles, while case (b) corresponds to an *opposite similarity*, that preserves shapes and angles up to a reflection (meaning that angles are reversed).

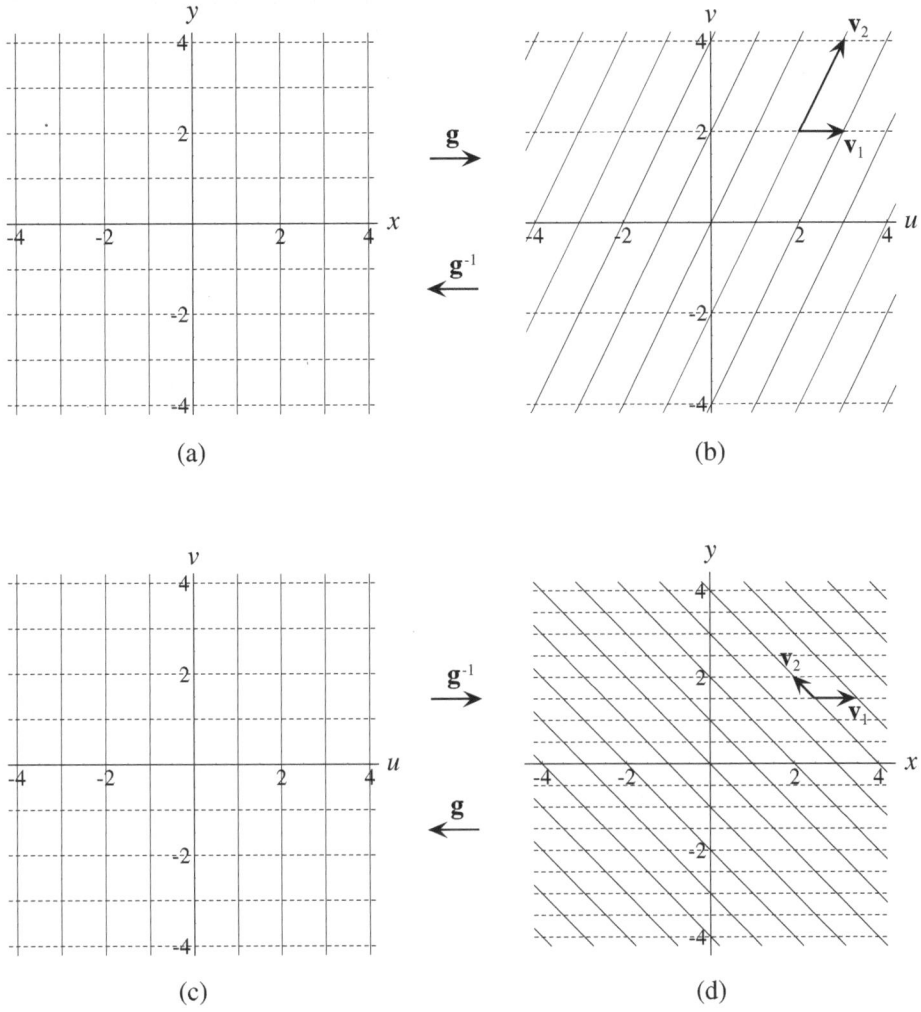

Figure C.5: Top row: the scanline and interline directions of the direct transformation
g given by $(u,v) = (x+y, 2y)$ are $\mathbf{v}_1 = (1,0)$, $\mathbf{v}_2 = (1,2)$. Bottom row: the
scanline and interline directions of the inverse transformation $\mathbf{g}^{-1}$,
$(x,y) = (u - \tfrac{1}{2}v, \tfrac{1}{2}v)$, are $\mathbf{v}_1 = (1,0)$, $\mathbf{v}_2 = (-\tfrac{1}{2}, \tfrac{1}{2})$.

$$ab + cd = 0$$

$$a^2 + c^2 = b^2 + d^2$$

Substituting $d = -ab/c$ from the first identity into the second identity we get:

$$c^2(a^2 + c^2) = b^2(a^2 + c^2)$$

which means, in conjunction with the first identity, that either $c = -b$, $a = d$ or $c = b$, $a = -d$; these are, indeed, the Cauchy-Riemann conditions of Eq. (C.13).

Appendix D

Direct and inverse spatial transformations

D.1 Introduction

Because spatial transformations (i.e. 2D functions of the form $(u,v) = \mathbf{g}(x,y)$) are widely used in science and technology, one cannot overestimate the importance of their full understanding. And yet, there exist several different potential sources of confusion in the handling of such transformations. The risk of confusion is increased even further due to the existence of different notation standards, as well as different paradigms for the software algorithms which implement these transformations (or rather their discrete forms) in computer applications. It is therefore our aim in this appendix to develop an intuitive understanding of such transformations, to shed some additional light on their behaviour, and to explain the main sources of confusion and how to avoid them.

We start our discussion in Sec. D.2 with a general reminder whose aim is to put our spatial transformations $(u,v) = \mathbf{g}(x,y)$ in their right mathematical context and to help us understand their various graphical representations. In Sec. D.3 we deepen our understanding of the interconnections between the domain and range planes of a transformation $\mathbf{g}$. Then, in Sec. D.4 we introduce $\mathbf{g}^{-1}$, the inverse transformation of $\mathbf{g}$, and discuss the relationship between these two transformations and their respective coordinate systems. In Sec. D.5 we explain the active and passive interpretations of a transformation $\mathbf{g}$, and then, in Sec. D.6 we discuss the very important notions of domain and range transformations. In Sec. D.7 we present the relative point of view, which explains the relationship between object deformations and coordinate deformations. In Sec. D.8 we provide several examples that show the various graphical representations of some typical linear and non-linear transformations, to illustrate the main subjects that were discussed so far. In Sec. D.9 we proceed to the explanation of some other possible sources of confusion, including forward and backward transformations in computer applications, and the use of pre-multiplication or post-multiplication formalisms. Then we discuss the implications of all these results to the moiré theory, and in particular to the preparation of our moiré figures (in Sec. D.10), and to the fixed points between the superposed layers (in Sec. D.11). Finally, in Sec. D.12 we derive some useful approximations that allow us to formulate our main results in terms of either direct or inverse transformations.

D.2 Background and basic notions

A transformation (or mapping) from $D \subset \mathbb{R}^m$ to $R \subset \mathbb{R}^n$ is a function $\mathbf{g}: D \rightarrow R$ that returns for each point $(x_1,...,x_m) \in D$ a new point $(y_1,...,y_n) \in R$. The set D is called the *domain* of the transformation $\mathbf{g}$, the set R is called the *range* of $\mathbf{g}$ (or the *image* of the

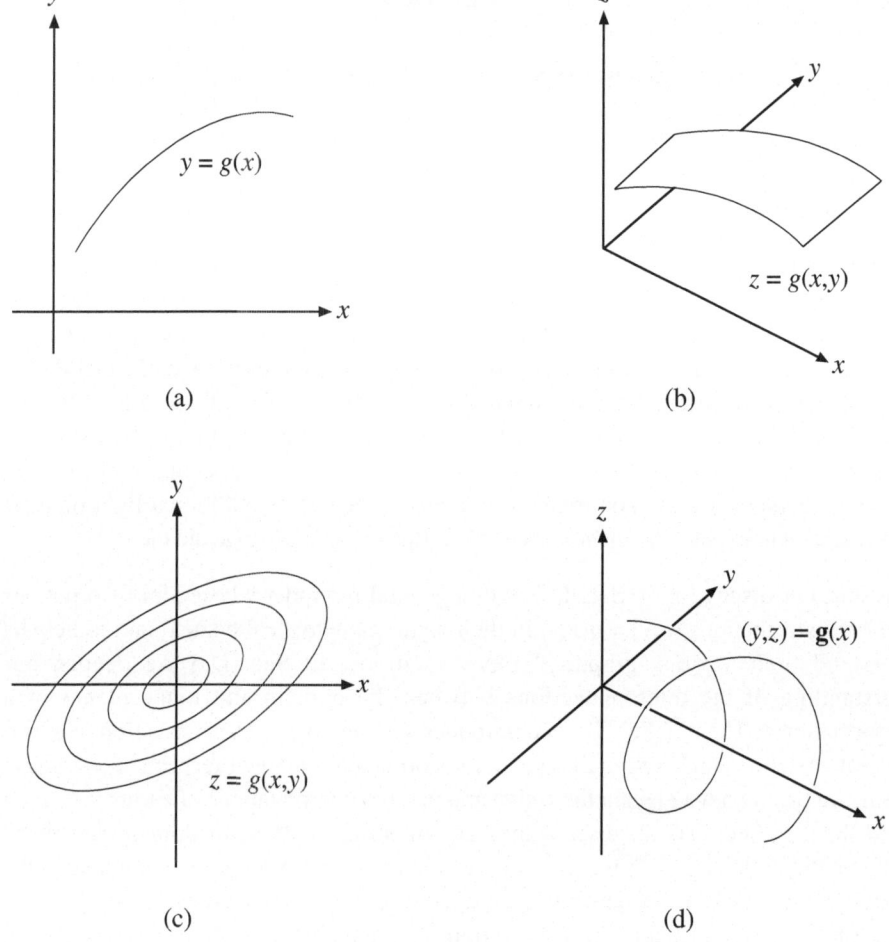

Figure D.1: (a) Schematic graphic representation of a function $y = g(x)$ as a curve
in the 2D plane. (b) Schematic graphic representation of a function
$z = g(x,y)$ as a surface in the 3D space. (c) Schematic graphic representation of a function $z = g(x,y)$ as a topographic map. (d) Schematic
graphic representation of a function $(y,z) = \mathbf{g}(x)$ as a curve in the 3D
space.

domain D under $\mathbf{g}$), and the new point $(y_1,...,y_n)$ is said to be the *image* of the original point
$(x_1,...,x_m)$ under the transformation $\mathbf{g}$; symbolically, this is denoted by:

$$(y_1,...,y_n) = \mathbf{g}(x_1,...,x_m)$$

In the moiré theory we usually have $m = n = 2$, and therefore we will be mainly
interested in transformations of the form $(u,v) = \mathbf{g}(x,y)$. However, for didactic reasons, let
us first briefly review here the simpler cases in which $m < 2$ or $n < 2$.

The simplest possible case is that of the functions of the form $y = g(x)$, which have a 1D domain and a 1D range. Such functions can be illustrated pictorially as a graph in the x,y plane that shows, for each original point x along the horizontal axis, the image y to which it is mapped by g (see Fig. D.1(a)). Thus, if g is a continuous function, it can be viewed as a curve in the x,y plane. Simple examples include $y = 2x$ or the non-linear function $y = x^2$.

The next simple case is that of the functions of the form $z = g(x,y)$, which have a 2D domain and a 1D range. Such functions can be illustrated pictorially as a 3D graph (or rather as a 2D perspective view of such a 3D graph) that shows for each original point (x,y) in the x,y plane the value z to which it is mapped by g (see Fig. D.1(b)). Thus, if g is a continuous function, it can be interpreted as a surface in the 3D x,y,z space. A function $z = g(x,y)$ can be also represented graphically as a topographic map in the x,y plane; in this case the relief of the surface is represented by level lines (see Fig. D.1(c)).

Yet another simple case, although less frequently encountered, is that of the functions $(y,z) = \mathbf{g}(x)$, which have a 1D domain and a 2D range. We denote such a function by a boldface letter $\mathbf{g}$ since the value $\mathbf{g}(x)$ it returns for each original scalar x is a vector. Such a function can be illustrated, once again, as a 3D graph (or rather as its 2D perspective view). But this time the graph shows, for each given point along the x axis, the value (y,z) to which it is mapped by $\mathbf{g}$ (see Fig. D.1(d)). Thus, if $\mathbf{g}$ is a continuous function, it can be interpreted as a curve in the 3D x,y,z space.

Having reviewed the simpler cases with $m < 2$ or $n < 2$, we arrive now to our main case of interest, that of the transformations $(u,v) = \mathbf{g}(x,y)$, which have a 2D domain and a 2D range. Clearly, a full graphic representation of such a transformation requires a 4D drawing in the four coordinates x,y,u,v, which shows for each original point (x,y) in the 2D domain its corresponding image (u,v) in the 2D range. But because such a 4D drawing is not realizable, several different more-or-less tricky methods exist to allow us represent the transformations $(u,v) = \mathbf{g}(x,y)$ graphically, within the limits of the possible. These different representations of the transformation $\mathbf{g}$ and the interconnections between them have been reviewed in detail in Appendix B; here, for the sake of our introductory survey, we only give a short reminder of the main representations, and we refer the reader to the appropriate sections in Appendix B for further details.

(1) The most straightforward method for representing the transformation $(u,v) = \mathbf{g}(x,y)$ graphically consists of drawing its 2D domain and its 2D range separately, as two different planes (see, for example, Fig. D.4 in the next section). The 2D domain is drawn with its standard Cartesian x,y coordinate system, and the 2D range is drawn with its own standard Cartesian u,v coordinate system. This representation illustrates the action of the transformation $(u,v) = \mathbf{g}(x,y)$ by showing within its range (the u,v plane) how the original x,y coordinate grid of the domain plane has been affected. This gives in the u,v plane a distorted, curvilinear grid (the image of the original x,y grid under the transformation $\mathbf{g}$), in addition to the standard Cartesian u,v grid of the u,v plane itself. More details on this method can be found in Sec. B.3 of Appendix B.

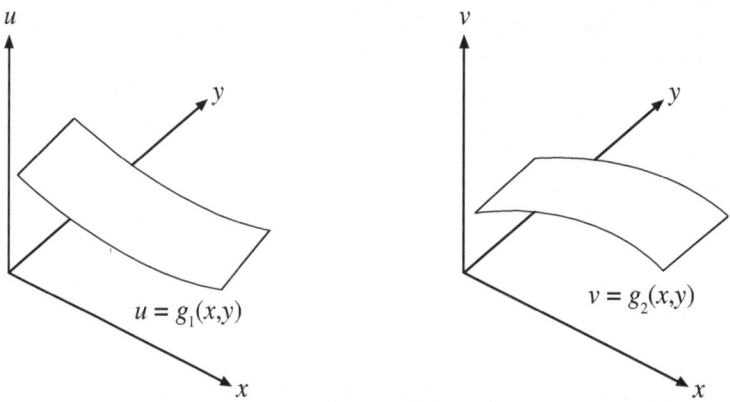

Figure D.2: Schematic graphic representation of a transformation $(u,v) = \mathbf{g}(x,y)$ as a pair of surfaces $u = g_1(x,y)$, $v = g_2(x,y)$ in the 3D space.

(2) Another variant of method (1) consists of drawing the two planes, the 2D domain and the 2D range, superposed on top of each other within a single plot (see, for example, Fig. D.3 in the next section). Although this graphic representation of $(u,v) = \mathbf{g}(x,y)$ may become overcrowded with details from both planes, it still allows to compare easily the original x,y grid with its distorted image, and it can be used as a "dictionary" (correspondance map) between the x,y and the u,v coordinate systems. For more details on this variant see Sec. B.5 in Appendix B.

(3) A third representation of the transformation $(u,v) = \mathbf{g}(x,y)$ is based on its component-wise notation:

$$u = g_1(x,y)$$

$$v = g_2(x,y)$$

This notation allows us to interpret the transformation $(u,v) = \mathbf{g}(x,y)$ as a pair of functions of the form $z = g(x,y)$, each of which can be illustrated, as we have seen above, as a surface in the 3D space (see Fig. D.2). More details on this method can be found in Sec. B.2 of Appendix B.

(4) In another variant of method (3), each of the two surfaces is drawn as a separate topographic map with its own level lines (see, for example, Fig. B.1 in Appendix B). The axes in both of the maps are x and y.

(5) A further method for representing the transformation $(u,v) = \mathbf{g}(x,y)$ graphically consists of showing its effect within a single planar plot as a *vector field*. This is done by drawing an arrow emanating from each point (x,y) of the x,y plane (or more practically, from some representative points on a given grid within the x,y plane), where the length and the orientation of each arrow indicate the length and the orientation of the vector $(u,v) = \mathbf{g}(x,y)$ that is assigned by $\mathbf{g}$ to the point (x,y); see, for example, Fig. D.9(f)

below.[1] It is important to stress, however, that each such arrow does not connect the point (x,y) to its image $(u,v) = \mathbf{g}(x,y)$, but rather to the point $(x,y) + \mathbf{g}(x,y)$. The axes of the vector field remain, therefore, x and y. For more detail on this method see Sec. B.6 in Appendix B.

(6) In another variant of this method, successive vectors of the vector field are connected into continuous curves, known as the *trajectories* (or *field lines*) of the vector field (see Fig. B.5(b) in Appendix B). This allows to illustrate the effect of the vector field $(u,v) = \mathbf{g}(x,y)$ along its trajectories, much like the graphical description of an electric or magnetic field in physics. More details can be found in Sec. B.6 of Appendix B.

There also exist other, more exotic methods for representing the transformation $(u,v) = \mathbf{g}(x,y)$. Although they are quite rarely used, it may still be interesting to mention some of them briefly:

(7) In a different variant of the vector field (method (5)), each arrow emanates from the point (x,y) and points to its image $(u,v) = \mathbf{g}(x,y)$. This variant is rarely if ever used in the literature; but in fact it is equivalent to the vector field of the relative transformation $\mathbf{k}(x,y) = \mathbf{g}(x,y) - (x,y)$, where each arrow (before its length is possibly being scaled) connects the point (x,y) to the point $(x,y) + \mathbf{k}(x,y) = \mathbf{g}(x,y)$.

(8) Another unusual representation of the transformation $(u,v) = \mathbf{g}(x,y)$ is a variant of method (1) in which the action of the transformation is not demonstrated by the way it affects the standard Cartesian x,y grid, but rather by the way it affects some other curves in the x,y plane. This can be advantageous when the transformation $\mathbf{g}$ maps the straight lines of the x,y grid into too complicated curves, or in cases in which it is easier or more interesting to see how $\mathbf{g}$ acts on some other curve families. Perhaps the most classical example of this type is the Cartesian to polar coordinate transformation $(r,\theta) = (\sqrt{x^2 + y^2}, \arctan(y/x))$, which is represented in the literature by its effect on the concentric circles and the radius lines that surround the origin of the x,y plane; we will return to this case in more detail in Sec. D.8, Example D.4. Several other examples can be found in [Needham97]; see, for example, the figures in pp. 58, 100 and 163 there. Yet another example appears in [Kreyszig93 p. 745, Fig. 304].

All these different representations of the transformation $(u,v) = \mathbf{g}(x,y)$ are in fact equivalent, although each of them focuses on some different facets of the transformation. It is therefore up to us to choose in each case the most suitable representation of $\mathbf{g}$, depending on the circumstances. For example, if we are mainly interested in the effect of the transformation $\mathbf{g}$ on the original Cartesian coordinate system, the most natural representation to consider is (1); but if we want to visually detect the critical points of $\mathbf{g}$ (the points where $\mathbf{g}(x,y) = (0,0)$), then the most suitable representation would certainly be (5) or (6). In the remaining sections of this appendix we will mainly use representations

---

[1] For practical reasons it is customary to scale the arrow lengths in the drawing by a constant factor, in order to avoid drawings with too short, hard-to-see arrows, or drawings with too long, overlapping arrows.

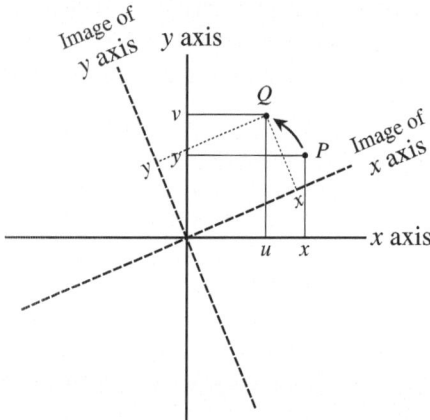

Figure D.3: The effect of a transformation $(u,v) = \mathbf{g}(x,y)$ (in the present example: a rotation by angle α), shown within the original x,y coordinate system. Domain coordinates x,y and range coordinates u,v are marked along the same original axes. Note that the dashed lines *do not* represent the u and v axes (see Remark D.3).

(1) and (2) and their counterparts for the inverse transformation $(x,y) = \mathbf{g}^{-1}(u,v)$. Note, however, that some of the other methods are also frequently used in this book; it is therefore important to recognize the different methods correctly and to understand which of them is being used in each case, in order to avoid confusion. In particular, we should be attentive to the fact that one and the same transformation may look completely different in each of its various graphic representations; several illustrative examples are provided in Sec. D.8 below.

D.3 A deeper look into the domain and range planes of the mapping $(u,v) = \mathbf{g}(x,y)$

Suppose we are given a transformation $(u,v) = \mathbf{g}(x,y)$ that operates on the x,y plane: $\mathbf{g}: \mathbb{R}^2 \to \mathbb{R}^2$. As a simple example we may consider the transformation which rotates the plane by angle α about the origin. When the transformation $\mathbf{g}$ is applied, it moves each point $P = (x,y)$ of the x,y plane to a new point Q whose location in the same x,y plane is given by $\mathbf{g}(x,y)$. This mapping effect of $\mathbf{g}$ is often denoted in literature by $(x,y) \mapsto \mathbf{g}(x,y)$ or $(x,y) \mapsto (u,v)$ (see, for example, [Lang87 p. 386]). Fig. D.3 shows the image Q of a point P under the transformation of rotation by angle α, as well as the images of the original x and y axes under the same transformation (the dashed lines).

Note that in Fig. D.3 we have drawn the effect of the transformation $(u,v) = \mathbf{g}(x,y)$ within the original x,y coordinate system (also called the x,y *space*). This means that Fig. D.3 shows simultaneously the situations before and after the application of the transformation $\mathbf{g}$ (see method (2) above). But for the sake of clarity, in order not to overload the drawing

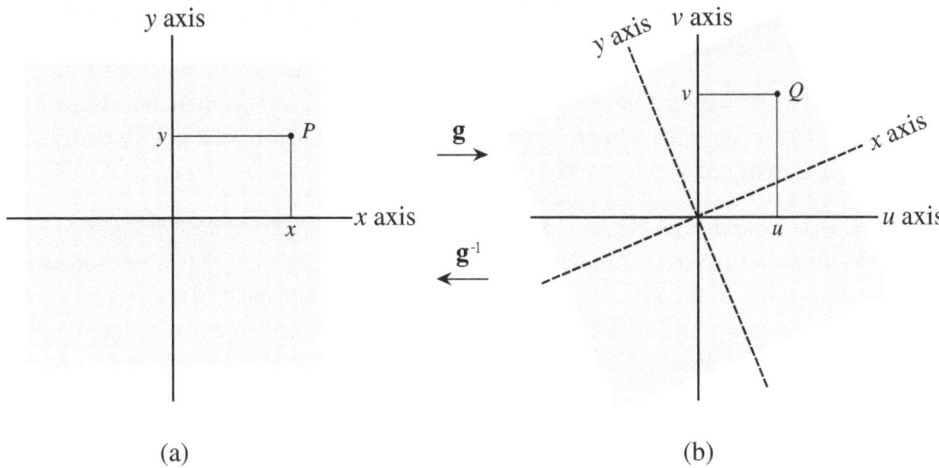

(a)   (b)

Figure D.4: The effect of a transformation $(u,v) = \mathbf{g}(x,y)$ (the same transformation as in Fig. D.3), shown in two separate plots: (a) The domain of $\mathbf{g}$ (the x,y plane before the application of the transformation). (b) The range of $\mathbf{g}$ (the u,v plane after the application of the transformation). A more detailed version of this figure is provided in Fig. D.5.

with details, it is often preferable to use method (1), and to illustrate the original and the transformed planes (i.e. the domain and the range of $\mathbf{g}$) in separate plots, as shown in Fig. D.4. In this case the standard axes of the domain of $\mathbf{g}$ are denoted x and y, as shown in Fig. D.4(a); but after the transformation has been applied, i.e. in the range of $\mathbf{g}$, the x and y axes are already transformed (distorted or simply moved to new locations), and the role of the standard axes is taken over by u and v (see Fig. D.4(b)). Note that the new standard axes u and v are identical to the old standard x and y axes as they existed before the application of $\mathbf{g}$. A more detailed version of Fig. D.4 is shown in Fig. D.5; this figure clearly shows the effect of the rotation transformation $\mathbf{g}$, which simply rotates the entire plane shown in (a) into the plane shown in (b). The images of the original x and y axes under the transformation $\mathbf{g}$ are called in Fig. D.5(b) the x and y axes, since they correspond, respectively, to the curves $y = 0$ and $x = 0$ in the u,v coordinate system of Fig. D.5(b). These axes form, indeed, the transformed coordinate system of the u,v plane, and in the general case they may be curvilinear (see, for example, Figs. D.9(a),(b) below).

It may be sometimes helpful to describe the effect of the transformation $(u,v) = \mathbf{g}(x,y)$ using the following physical interpretation (see Figs. D.9(a),(b)): Imagine that the original x,y coordinate system is printed on a flat sheet of flexible rubber, and that this sheet undergoes a planar transformation while remaining flat. Depending on the forces that are applied to that rubber sheet, the result may appear rotated, scaled, or otherwise distorted; but it always remains flat. Now, we copy the resulting distorted x,y plane on a new sheet of paper; this sheet corresponds to the range of the transformation $\mathbf{g}(x,y)$ (see Fig. D.9(b)). We draw on this sheet a new Cartesian coordinate system, *identical* to the original untransformed x,y coordinate system of Fig. D.9(a); these new undistorted axes are the u and v coordinates of the range of our transformation (see Fig. D.9(b)).

Remark D.1: The fact that the new u,v axes are identical to the original, undistorted x,y axes will allow us later to compare any object in the plane *before* and *after* it undergoes the transformation $(u,v) = \mathbf{g}(x,y)$. ∎

Remark D.2: Note the double role of the x,y coordinate system: on the one hand it refers to the original, *undistorted* coordinate system *before* the application of the transformation $(u,v) = \mathbf{g}(x,y)$, i.e. in the *domain* of $\mathbf{g}(x,y)$; but on the other hand it also refers to the *distorted* coordinate system *after* the application of the transformation, i.e. in the *range* of $\mathbf{g}(x,y)$, whose new undistorted coordinates are u,v. ∎

Having understood the coordinate systems involved in the domain and in the range of the transformation $\mathbf{g}$, we now present the terminology that is used to refer to them.[2] The u,v coordinate system of the range of $\mathbf{g}$ (see Fig. D.5(b)) is called the u,v *space*, the *target space* or the *destination space* of the transformation $\mathbf{g}$. The x,y coordinate system of Fig. D.5(a), showing the situation before the transformation $\mathbf{g}$ has been applied, is called the *original x,y space*, and the distorted x,y coordinate system of Fig. D.5(b), showing the situation after the transformation $\mathbf{g}$ has been applied, is called the $\mathbf{g}$-*transformed x,y space* or simply the *transformed x,y space*. Note that the values of the x,y coordinates are not affected by the transformation: If point P has coordinates (x,y) in the original x,y space, its image Q has the same (possibly curvilinear) coordinates (x,y) in the transformed x,y space (while its Cartesian coordinates in terms of the u,v space are $(u,v) = \mathbf{g}(x,y)$).[3]

As we can see, Fig. D.5 has an important advantage over Fig. D.4 in that it allows us to easily visualize for any given point in the plane the correspondence between its x,y coordinate values and its u,v coordinate values. We will return to this point in Sec. D.5, where we discuss the active and passive interpretations of a transformation $(u,v) = \mathbf{g}(x,y)$.

Remark D.3: Since the transformation $(u,v) = \mathbf{g}(x,y)$ maps any original point (x,y) into its image (u,v), it also maps in Fig. D.3 the original x and y axes into the two respective dashed lines. Therefore, it may be tempting to call these dashed lines in Fig. D.3 "the u and v axes". However, this reasoning is wrong since the points $(x,0)$, which form the x axis, are not mapped by the transformation $\mathbf{g}$ into the points $(u,0)$, which form the u axis, but rather into the points $(u,v) = \mathbf{g}(x,0)$. And indeed, as clearly shown in Fig. D.5(a), the u and v axes are obtained by applying to the x and y axes the *inverse* transformation (in our case: a rotation in the *negative* direction). To better understand this, consider our example of a rotation by angle α. This transformation is given by:

$$\begin{pmatrix} u \\ v \end{pmatrix} = \begin{pmatrix} \cos\alpha & -\sin\alpha \\ \sin\alpha & \cos\alpha \end{pmatrix} \begin{pmatrix} x \\ y \end{pmatrix} = \begin{pmatrix} x\cos\alpha - y\sin\alpha \\ x\sin\alpha + y\cos\alpha \end{pmatrix} \tag{D.1}$$

---

[2] Unfortunately, as we will see later in Sec. D.6.2, a different convention also coexists, and is being used.

[3] Note that although u and v are obtained by transforming the original x,y coordinates through $(u,v) = \mathbf{g}(x,y)$, they appear in Fig. D.5(b) as *untransformed* (because they are the standard coordinates of the range of $\mathbf{g}$), while the transformed axes in this figure belong to the transformed x,y coordinates.

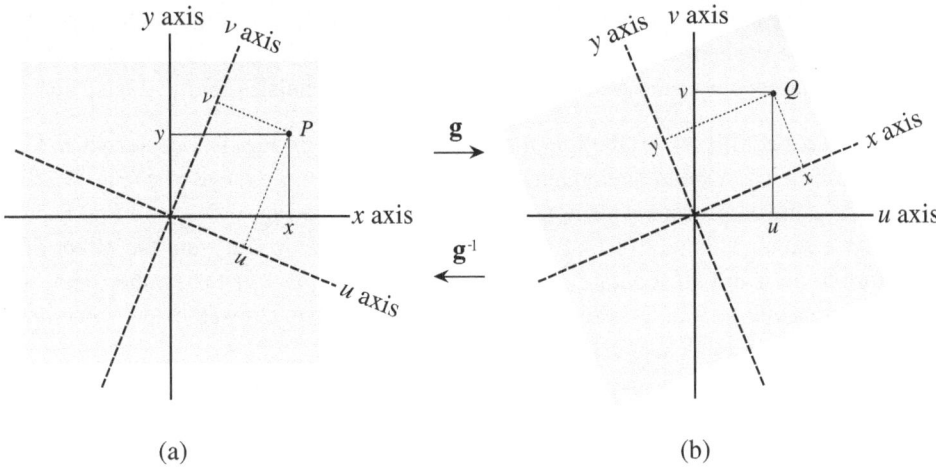

(a)   (b)

Figure D.5: Same as Fig. D.4, but this time showing both the x,y and the u,v coordinates in (a) as well as in (b). This figure clearly illustrates the active interpretation of the transformation $(u,v) = g(x,y)$, which amounts to the rotation of the entire plane of (a) into the plane shown in (b). Each of the views (a) and (b) also illustrates the passive interpretation of the same transformation, which consists of the conversion of x,y coordinates into u,v coordinates. (The active and passive interpretations are discussed in Sec. D.5.)

Its inverse transformation, $(x,y) = g^{-1}(u,v)$, is the rotation by angle $-\alpha$, namely:

$$\begin{pmatrix} x \\ y \end{pmatrix} = \begin{pmatrix} \cos\alpha & \sin\alpha \\ -\sin\alpha & \cos\alpha \end{pmatrix}\begin{pmatrix} u \\ v \end{pmatrix} = \begin{pmatrix} u\cos\alpha + v\sin\alpha \\ -u\sin\alpha + v\cos\alpha \end{pmatrix} \tag{D.2}$$

Now, the u axis is given by definition by the curve $v = 0$. This curve can be also expressed in our example, using the bottom row of Eq. (D.1), as:

$$x\sin\alpha + y\cos\alpha = 0$$

namely: $y = -x\tan\alpha = x\tan(-\alpha)$

This means that the u axis is represented in Fig. D.3 by a line whose angle is $-\alpha$, and not by the dashed line whose angle is α. This fact is clearly shown in Fig. D.5(a).

On the other hand, the dashed lines in Fig. D.5(b) do represent the x and y axes. To see this, remember that the x axis is, by definition, the line $y = 0$, which can be also expressed in our case, using the bottom row of Eq. (D.2), as:

$$-u\sin\alpha + v\cos\alpha = 0$$

namely: $v = u\tan\alpha$

This is clearly a line passing through the origin of the u,v plane at the angle of α. This means that the dashed line located above the u axis in the u,v system of Fig. D.5(b) represents, indeed, the x axis. A similar reasoning can be formulated for the y axis, too.

The possible confusion in labelling the transformed axes in Fig. D.3 occurs since Fig. D.3 attempts to show both the domain of **g** (Fig. D.5(a)) and the range of **g** (Fig. D.5(b)) within the same coordinate system, while the axis names in Fig. D.5(a) and Fig. D.5(b) are not the same. This fact further justifies why we prefer to illustrate the effect of a transformation **g** in two separate plots, as shown in Figs. D.4 or D.5, rather than in a single plot, as in Fig. D.3. Drawing a single plot is, of course, allowable, and sometimes even advantageous, but it should be done with care.[4] ■

D.4 2D transformations and their inverse

In order to further clarify the situation in the 2D case, let us describe in more detail the effect of the 2D transformation $(u,v) = \mathbf{g}(x,y)$ and the effect of its inverse $(x,y) = \mathbf{g}^{-1}(u,v)$ on each point of the plane (see Fig. D.9).[5] Note that in the following we will sometimes need the componentwise notation of the transformation $(u,v) = \mathbf{g}(x,y)$:

$$u = g_1(x,y)$$
$$v = g_2(x,y) \tag{D.3}$$

Similarly, the componentwise notation of the inverse transformation $(x,y) = \mathbf{g}^{-1}(u,v)$ is:

$$x = g_1^{-1}(u,v)$$
$$y = g_2^{-1}(u,v) \tag{D.4}$$

where $g_1^{-1}(u,v)$ and $g_2^{-1}(u,v)$ are the components of the inverse transformation $\mathbf{g}^{-1}(u,v)$. Note that the inverse of the inverse transformation $(x,y) = \mathbf{g}^{-1}(u,v)$ is the original transformation $(u,v) = \mathbf{g}(x,y)$ itself. The original transformation **g** is also called the *direct* transformation (as opposed to the *inverse* transformation $\mathbf{g}^{-1}$).

Remark D.4: Once we have defined the inverse transformation $(x,y) = \mathbf{g}^{-1}(u,v)$, it becomes a transformation in its own right, and we can obviously write it with any variables we may wish. Thus, if we prefer to maintain our convention of using the x,y variables for the transformation's domain and the u,v variables for its range, we may write our inverse

---

[4] For example, if one insists on naming the rotated, dashed axes of Fig. D.3 by u and v, then the transition from x,y to u,v values must be expressed mathematically by the *inverse* transformation, in our case: a rotation by angle $-\alpha$. This practice can be found, for example, in [Knopp74 pp. 401–407] or in [Spiegel68 p. 36]. Note that in references using this convention the transformation $(u,v) = \mathbf{g}(x,y)$ is often called "a transition from u,v to x,y" (see, for example, [Knopp74 p. 406]), in order to avoid the inversion effect between the figure and the formula (see also Remark D.16 in Sec. D.6.2).

[5] For the sake of the present discussion we suppose that the transformation $(u,v) = \mathbf{g}(x,y)$ is sufficiently well behaved, and that it has a unique inverse transformation $(x,y) = \mathbf{g}^{-1}(u,v)$.

transformation as: $(u,v) = \mathbf{g}^{-1}(x,y)$. This also allows us to compare $\mathbf{g}$ and $\mathbf{g}^{-1}$ by plotting them together in the same drawing without having to bother about the axes names. Moreover, this even allows us to define new combined transformations such as $\mathbf{h}(x,y) = \mathbf{g}(x,y) - \mathbf{g}^{-1}(x,y)$. In fact, for any rational polynomial $r(x) = \Sigma a_n x^n$ with positive and negative powers $n \in \mathbb{Z}$ we can define a corresponding functional polynomial in $\mathbf{g}(x,y)$, $\mathbf{r}(x,y) = \Sigma a_n \mathbf{g}^{[n]}(x,y)$, where the "powers" in brackets indicate composition of transformations (or inverse transformations), as follows: $\mathbf{g}^{[1]} = \mathbf{g}$, $\mathbf{g}^{[2]} = \mathbf{g} \circ \mathbf{g}$, $\mathbf{g}^{[3]} = \mathbf{g} \circ \mathbf{g} \circ \mathbf{g}$, etc., $\mathbf{g}^{[0]} = \mathbf{i}$ (the identity transformation $\mathbf{i}(x,y) = (x,y)$), $\mathbf{g}^{[-1]} = \mathbf{g}^{-1}$ (the inverse transformation of $\mathbf{g}$), $\mathbf{g}^{[-2]} = \mathbf{g}^{-1} \circ \mathbf{g}^{-1}$, etc. For example, the rational polynomial $r(x) = 3x^2 + 5x + 2 + \frac{4}{x}$ defines the functional polynomial:

$$\mathbf{r}(x,y) = 3\mathbf{g}^{[2]}(x,y) + 5\mathbf{g}^{[1]}(x,y) + 2\mathbf{g}^{[0]}(x,y) + 4\mathbf{g}^{[-1]}(x,y)$$

$$= 3[\mathbf{g} \circ \mathbf{g}](x,y) + 5\mathbf{g}(x,y) + 2(x,y) + 4\mathbf{g}^{-1}(x,y)$$

Thus, one should not be shocked if we occasionally say that the inverse of the transformation $\mathbf{g}(x,y)$ is $\mathbf{g}^{-1}(x,y)$ (rather than $\mathbf{g}^{-1}(u,v)$). In fact, it would be desirable to stick systematically to either of these two conventions; however, it turns out that each of the two may be more suitable in some different situations. We therefore have to live with both conventions, but whenever this may cause confusion we will add an adequate remark to clarify our intentions. ■

D.4.1 The image of the standard Cartesian grid under the transformations g and g⁻¹

Suppose now that a moving point $P = (x,y)$ describes a curve in the domain of our transformation $\mathbf{g}(x,y)$, i.e. within the x,y plane. As we already know, the image of this point in the range of our transformation will likewise describe a curve in the u,v plane, which is called the *image curve* of the original curve.[6] For example, to the line $x = c$, which is parallel to the y axis, there corresponds in the u,v plane the image curve given in parametric form by the pair of equations [Courant88 pp. 134–135]:

$$u = g_1(c,y)$$

$$v = g_2(c,y)$$

or, more concisely:

$$(u,v) = \mathbf{g}(c,y) \tag{D.5}$$

where y is the parameter of the curve. Similarly, to the line $y = k$ there corresponds in the u,v plane the image curve given in parametric form by the pair of equations:

$$u = g_1(x,k)$$

$$v = g_2(x,k)$$

---

[6] We could also draw the resulting curve within the original x,y plane, like in Fig. D.3; but as already mentioned above, we prefer to draw it in a separate figure in order not to overload the original figure, and in order to avoid confusion in the axis names.

or, more concisely:

$$(u,v) = \mathbf{g}(x,k) \tag{D.6}$$

where x is the parameter of the curve.[7] Note that the image curves (D.5) and (D.6) can be also expressed in the implicit form, in terms of the *inverse* transformation $(x,y) = \mathbf{g}^{-1}(u,v)$. Since they are the image curves of the lines $x = c$ and $y = k$, they can be written, respectively, using Eqs. (D.4), as follows [Courant88 p. 135]:

$$g_1^{-1}(u,v) = c \tag{D.7}$$

and:

$$g_2^{-1}(u,v) = k \tag{D.8}$$

Now, if we assign to c and k sequences of equidistant values c_1, c_2, c_3, ... and k_1, k_2, k_3, ... (for instance, consecutive integer values), then the rectangular coordinate grid consisting of the lines $x = c_i$ and $y = k_i$ in the x,y plane (see Fig. D.9(a)) gives rise to a corresponding curvilinear grid consisting of two families of curves (D.7) and (D.8) in the u,v plane (see Fig. D.9(b)). This curvilinear grid furnishes a useful geometric picture of the mapping $(u,v) = \mathbf{g}(x,y)$ that clearly shows how it distorts the original x,y plane.

Remark D.5: It follows from Eqs. (D.7) and (D.8) that the two families of curvilinear lines in the u,v plane, that are the image under $\mathbf{g}$ of the standard unit grid of the x,y plane, are simply the level lines of the two surfaces defined, respectively, by the two components of the *inverse* transformation $\mathbf{g}^{-1}$: The images of the lines $x = m$, $m \in \mathbb{Z}$ under $(u,v) = \mathbf{g}(x,y)$ are the level lines $g_1^{-1}(u,v) = m$ of the surface $z = g_1^{-1}(u,v)$ (see the plain lines in Fig. D.9(b)), and the images of the lines $y = n$, $n \in \mathbb{Z}$ under $(u,v) = \mathbf{g}(x,y)$ are the level lines $g_2^{-1}(u,v) = n$ of the surface $z = g_2^{-1}(u,v)$ (see the dashed lines in Fig. D.9(b)). ■

Consider now the inverse transformation, $(x,y) = \mathbf{g}^{-1}(u,v)$. Obviously, this transformation maps the curvilinear lines in the u,v plane that are defined by Eq. (D.5) or, equivalently, by Eq. (D.7) (see the plain curves in Fig. D.9(b)) back into the original straight vertical lines $x = c$ in the x,y plane (Fig. D.9(a)). Likewise, it maps the curvilinear lines in the u,v plane that are defined by Eq. (D.6) or, equivalently, by Eq. (D.8) (see the dashed curves in Fig. D.9(b)) back into the original straight horizontal lines $y = k$ in the x,y plane. Thus, the inverse transformation undoes the effects of the original transformation $\mathbf{g}(x,y)$, and brings each distorted entity in the u,v plane back into its undistorted state in the original x,y plane. This can be clearly seen by comparing Figs. D.9(b) and D.9(a).

However, it is also possible to consider the inverse transformation $(x,y) = \mathbf{g}^{-1}(u,v)$ as a transformation in its own right, as shown in Figs. D.9(c),(d).[8] From this point of view, the

---

[7] In a similar way we can also find the image of a *curve*, say, $y = x^2 + k$, by plugging its equation (instead of the equation $y = k$) into $(u,v) = \mathbf{g}(x,y)$. For example, the image of the curve $y = x^2 + k$ in the u,v plane is given in parametric form by $(u,v) = \mathbf{g}(x, x^2 + k)$. This is further explained in Remark D.8 below.

[8] As we have seen in Remark D.4, because $\mathbf{g}^{-1}$ is a transformation in its own right, we could also consider it in the x,y space. But in order to avoid confusion we prefer to use here the variables x,y for the domain of $\mathbf{g}$ (and the range of $\mathbf{g}^{-1}$), and the variables u,v for the range of $\mathbf{g}$ (and the domain of $\mathbf{g}^{-1}$).

transformation $\mathbf{g}^{-1}(u,v)$ maps the standard Cartesian coordinate grid of its own domain, the u,v plane, into a curvilinear grid within its range, the x,y plane. Thus, to each line $u = p$, which is parallel to the v axis in the u,v plane, there corresponds in the x,y plane a curve, which is given in parametric form by the pair of equations:

$$x = g_1^{-1}(p,v)$$

$$y = g_2^{-1}(p,v)$$

or, more concisely:

$$(x,y) = \mathbf{g}^{-1}(p,v) \tag{D.9}$$

where v is the parameter of the curve. Similarly, to the line $v = q$ in the u,v plane there corresponds in the x,y plane a curve, which is given in parametric form by the pair of equations:

$$x = g_1^{-1}(u,q)$$

$$y = g_2^{-1}(u,q)$$

or, more concisely:

$$(x,y) = \mathbf{g}^{-1}(u,q) \tag{D.10}$$

where u is the parameter of the curve. Note that the curves (D.9) and (D.10) can be also expressed in the implicit form, in terms of the *original* transformation $\mathbf{g}(x,y)$. Since they are the image curves of the lines $u = p$ and $v = q$, they can be written, respectively, using Eqs. (D.3):

$$g_1(x,y) = p \tag{D.11}$$

and: $\qquad\qquad\qquad\qquad g_2(x,y) = q \tag{D.12}$

If we assign to p and q sequences of equidistant values $p_1, p_2, p_3, \ldots$ and $q_1, q_2, q_3, \ldots$ (for instance, consecutive integer values), then the rectangular coordinate grid consisting of the lines $u = p_i$ and $v = q_i$ in the u,v plane (Fig. D.9(c)) gives rise to a corresponding curvilinear grid in the x,y plane (Fig. D.9(d)) which consists of the two families of curves (D.9) and (D.10). This curvilinear grid furnishes a useful geometric picture of the inverse mapping $(x,y) = \mathbf{g}^{-1}(u,v)$ that clearly shows how it distorts its original u,v plane.

Remark D.6: It follows from Eqs. (D.11) and (D.12) that the two families of curvilinear lines in the x,y plane, that are the image under $\mathbf{g}^{-1}$ of the standard unit grid of the u,v plane, are simply the level lines of the two surfaces defined, respectively, by the two components of the *direct* transformation $\mathbf{g}$: The images of the lines $u = m$, $m \in \mathbb{Z}$ under $(x,y) = \mathbf{g}^{-1}(u,v)$ are the level lines $g_1(x,y) = m$ of the surface $z = g_1(x,y)$ (see the plain lines in Fig. D.9(d)), and the images of the lines $v = n$, $n \in \mathbb{Z}$ under $(x,y) = \mathbf{g}^{-1}(u,v)$ are the level lines $g_2(x,y) = n$ of the surface $z = g_2(x,y)$ (see the dashed lines in Fig. D.9(d)). ∎

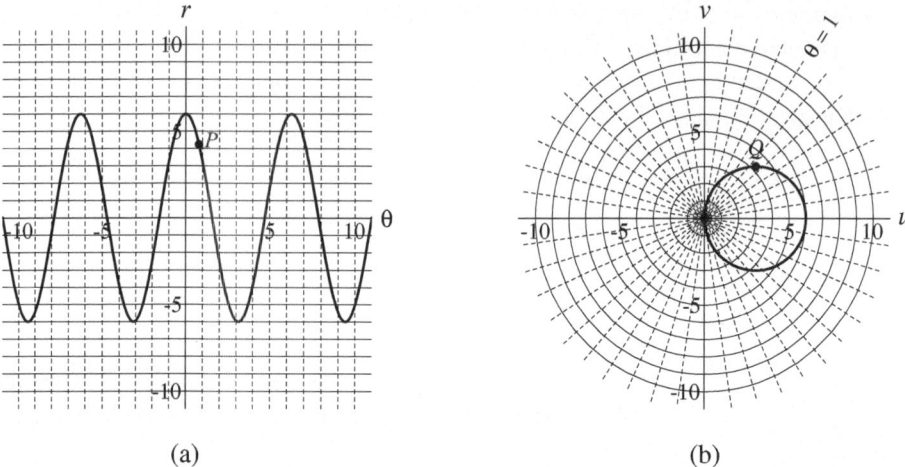

(a)   (b)

Figure D.6: Transformation (D.13) maps the cosinusoidal curve $r = 6\cos\theta$ in the θ,r plane (a) into the circle $(u-3)^2 + v^2 = 9$ in the u,v plane (b). Note that the radial and circular grid lines in (b) are the images of the vertical and horizontal θ,r grid lines in (a) under the transformation (D.13); they represent, therefore, the transformed θ,r plane. The values along the u and v axes in (b) correspond to the Cartesian coordinates of the destination u,v plane, i.e. to the vertical and horizontal grid lines $u = c$, $v = k$; but these lines are not shown in the figure in order not to overload it with details. Note that all angles are measured in radians.

Remark D.7: Note that one can express the curvilinear grid lines of both $\mathbf{g}(x,y)$ and its inverse $\mathbf{g}^{-1}(u,v)$ without knowing the *explicit* expression of the inverse transformation: The curvilinear grid lines of $\mathbf{g}(x,y)$ can be expressed in *parametric* form by Eqs. (D.5) and (D.6), and the curvilinear grid lines of $\mathbf{g}^{-1}(u,v)$ can be expressed in *implicit* form by Eqs. (D.11) and (D.12). All of these equations only require the knowledge of the direct transformation $\mathbf{g}(x,y)$. ■

This last fact allows us in Fig. D.9 to plot the effects of both $\mathbf{g}$ and $\mathbf{g}^{-1}$ even if the explicit expression of $\mathbf{g}^{-1}$ is unavailable. This technique has been used, indeed, to generate the figures that accompany the examples in Sec. D.8.

D.4.2 The image of a general curve under the transformations g and g⁻¹

We have seen above in detail how the transformation $(u,v) = \mathbf{g}(x,y)$ acts on the straight grid lines $x = c$ and $y = k$ of the x,y plane, and how the images under $\mathbf{g}$ of these grid lines are expressed as curves in the u,v plane. Similarly, we also seen how the inverse transformation $(x,y) = \mathbf{g}^{-1}(u,v)$ acts on the straight grid lines $u = c$ and $v = k$ of the u,v plane, and how the images under $\mathbf{g}^{-1}$ of these grid lines are expressed as curves back in the x,y plane. But because the transformation $(u,v) = \mathbf{g}(x,y)$ maps the x,y plane into the u,v plane, it is clear that it also maps any subset of the x,y plane into a subset of the u,v plane.

For example, the rotation transformation (D.1) maps any subset of the x,y plane into a similar subset of the u,v plane which has only been rotated by angle α about the origin. A more interesting example is given by the non-linear transformation:[9]

$$u = r\cos\theta$$
$$\qquad\qquad\qquad\qquad\qquad\qquad\qquad\qquad\text{(D.13)}$$
$$v = r\sin\theta$$

This transformation converts polar coordinates θ,r into Cartesian coordinates u,v. Knowing the inverse of this transformation (see Example D.4 in Sec. D.8 below), $r = \sqrt{u^2 + v^2}$, $\theta = \arctan(v/u)$, we can see from Eqs. (D.7) and (D.8) how transformation (D.13) acts on the straight grid lines of the θ,r plane: It maps each horizontal line $r = c$ of the θ,r plane (where c is an arbitrary constant) into a circle centered about the origin with radius c in the u,v plane, that is expressed by $\sqrt{u^2 + v^2} = c$. Similarly, it maps each vertical line $\theta = k$ of the θ,r plane into a radial line in the u,v plane which emanates from the origin at the angle of k radians, and whose expression is $\arctan(v/u) = k$.

Proceeding with the same example, consider now the planar curve defined by the equation $r = 6\cos\theta$. When plotted in the θ,r plane this equation gives a cosinusoidal curve (see Fig. D.6(a)).[10] However, the image of this curve in the u,v plane under the transformation (D.13) is a circle tangent to the vertical axis, as shown in Fig. D.6(b) [Colley98 pp. 68–69]. In other words, when the point P in Fig. D.6(a) traces out the cosinusoidal curve, its image Q traces out the circle shown in Fig. D.6(b). We see, therefore, that the non-linear transformation (D.13) maps cosinusoidal curves in the polar θ,r plane into circles tangent to the vertical axis in the Cartesian u,v plane. Other interesting examples showing how various curves in the θ,r plane are transformed into the u,v plane under the transformation (D.13) can be found in [Lang87 pp. 254–257].

So how can we express mathematically the new curve which is obtained in the u,v plane as the image under transformation **g** of a given curve $f(x,y) = 0$ in the x,y plane? Just as we did in the case of the straight grid lines, we replace each occurrence of x and y in the curve equation $f(x,y) = 0$ by the respective component from Eq. (D.4), and we obtain: $f(g_1^{-1}(u,v), g_2^{-1}(u,v)) = 0$, or more concisely, using vector notation: $f(\mathbf{g}^{-1}(u,v)) = 0$. This is, indeed, the implicit form of our image curve in the u,v plane.

Returning to our example, the image in the u,v plane of the original cosinusoidal curve $r = 6\cos\theta$ under the transformation (D.13) is obtained by plugging in this curve equation the two components of the inverse transformation, $r = \sqrt{u^2 + v^2}$, $\theta = \arctan(v/u)$. We get, therefore:

---

[9] Note that the names of the variables have no real importance, and they can be chosen as desired. In this case we have preferred to keep using the range coordinates u,v in accordance with our usual convention, but to use the classical polar coordinate names r,θ rather than our usual domain coordinate names x,y. Depending on the context we may prefer in other circumstances to make other choices, such as $(u,v) = (x\cos y, x\sin y)$ or $(x,y) = (r\cos\theta, r\sin\theta)$.

[10] Note that because in this case r is considered as a function of θ, it is more natural to talk about the θ,r plane, in which θ is the horizontal axis and r is the vertical axis. But in other situations it is often more convenient to refer to the polar coordinates plane as the r,θ plane.

$$\sqrt{u^2 + v^2} = 6\cos(\arctan(v/u))$$

and using the identity $\cos x = 1/\sqrt{1 + \tan^2 x}$ [Spiegel68 p. 15] and hence $\cos(\arctan(v/u)) = u/\sqrt{u^2 + v^2}$ we obtain the desired curve equation in the u,v plane:

$$u^2 + v^2 = 6u$$

or equivalently, $(u-3)^2 + v^2 = 9$

This is, indeed, a circle tangent to the vertical v axis (see Fig. D.6(b)).

Similar results can be also obtained for the image of a curve $f(u,v) = 0$ under the inverse transformation $(x,y) = \mathbf{g}^{-1}(u,v)$. We have, therefore, the following general result:

Proposition D.1: Let $f(x,y) = 0$ be a curve in the x,y plane. The image in the u,v plane of this curve after the application of the direct transformation $(u,v) = \mathbf{g}(x,y)$ is $f(g_1^{-1}(u,v), g_2^{-1}(u,v)) = 0$, or in vector notation, $f(\mathbf{g}^{-1}(u,v)) = 0$. Conversely, if $f(u,v) = 0$ is a curve in the u,v plane, then the image in the x,y plane of this curve under the inverse transformation $(x,y) = \mathbf{g}^{-1}(u,v)$ is $f(g_1(x,y), g_2(x,y)) = 0$, or in vector notation, $f(\mathbf{g}(x,y)) = 0$. ∎

Remark D.8: It is interesting to note that if the curve $f(x,y) = 0$ is also known in parametric form, $x = f_1(t)$, $y = f_2(t)$, then the image in the u,v plane of this curve under the direct transformation $(u,v) = \mathbf{g}(x,y)$ can be also expressed, in parametric form, by $(u,v) = \mathbf{g}(f_1(t), f_2(t))$.[11] Similarly, if the curve $f(x,y) = 0$ is known in the explicit form $y = h(x)$, then the image in the u,v plane of this curve under the same direct transformation $(u,v) = \mathbf{g}(x,y)$ can be also expressed in the parametric form $(u,v) = \mathbf{g}(x, h(x))$.[12]

Conversely, if the curve $f(u,v) = 0$ is also known in parametric form, $u = f_1(s)$, $v = f_2(s)$, then the image in the x,y plane of this curve under the inverse transformation $(x,y) = \mathbf{g}^{-1}(u,v)$ can be also expressed, in parametric form, by $(x,y) = \mathbf{g}^{-1}(f_1(s), f_2(s))$. Similarly, if the curve $f(u,v) = 0$ is known in the explicit form $v = h(u)$, then the image in the x,y plane of this curve under the same inverse transformation $(x,y) = \mathbf{g}^{-1}(u,v)$ can be also expressed in the parametric form $(x,y) = \mathbf{g}^{-1}(u, h(u))$.

We have already met a few such parametric examples earlier in Sec. D.4 (see, for instance, Eqs. (D.5), (D.6) and the footnote thereafter, and Eqs. (D.9) and (D.10)). Note that in Proposition D.1 we have to plug the two components of the (inverse) transformation into the curve's definition, while here we have to plug the two components of the curve into the spatial transformation's definition. ∎

---

[11] A curve in the plane can be generally defined in three equivalent forms: *explicitly* by $y = h(x)$, *implicitly* by $f(x,y) = 0$, or *parametrically* by $x = f_1(t)$, $y = f_2(t)$, where the parameter t varies continuously throughout an interval such as $-\infty < t < \infty$ [Bronshtein97 pp. 75–76]. Conversions between these forms can be done as explained in [Bronshtein97 p. 551]. Depending on the case, one or the other of these equivalent forms may have a simpler expression or be more convenient to use; note, however, that not every curve can be expressed in the explicit form.

[12] Note that this is equivalent to the parametric form $(u,v) = \mathbf{g}(f_1(t), f_2(t))$ where $f_1(t) = t$ and $f_2(t) = h(t)$.

D.5 The active and passive interpretations of a transformation[13]

A transformation $(u,v) = \mathbf{g}(x,y)$ can be interpreted in two different ways (see [Courant88 pp. 133–140]): either as a mapping (the active interpretation), or as a coordinate change (the passive interpretation).

According to the *active* interpretation, the one we have tacitly adopted thus far, the transformation $\mathbf{g}$ moves (or maps) each point $P = (x,y)$ of the original x,y plane into its image point $Q = (u,v)$. This destination point can be represented either in the original x,y coordinate system, where the values (u,v) returned by the transformation are understood as new values along the original x and y axes (as in Fig. D.3), or in the target u,v coordinate plane, as in Fig. D.5(b). Note that $\mathbf{g}$ moves any point (x,y) of the original x,y plane (Fig. D.5(a)) to the point having the same coordinates (x,y) in the distorted x,y plane (Fig. D.5(b)). The active point of view is also illustrated in Figs. D.9(a),(b), where the non-linear transformation $\mathbf{g}$ maps the original axes of the x,y plane (a) into the distorted x,y axes in the destination u,v plane (b).

On the other hand, in the *passive* interpretation of the transformation $\mathbf{g}$ we only concentrate on the plane after its deformation by $\mathbf{g}$ has been completed (Fig. D.5(b)), but we consider this plane through two different coordinate nets: the distorted x,y coordinate net and the new undistorted u,v coordinate net.[14] This allows us to interpret the transformation $(u,v) = \mathbf{g}(x,y)$ as a coordinate change in the plane, or, in other words, as a "dictionary" that translates the position of any given point in the plane from the x,y language to the u,v language, without actually moving the given point from one location in the plane to another; see Fig. D.5(b).[15] If $(u,v) = \mathbf{g}(x,y)$ is a one-to-one transformation we can in general assign to each point (x,y) the corresponding values (u,v) as new coordinates, because each pair of values (x,y) uniquely determines the pair (u,v), and vice versa. Thus, both (x,y) and (u,v) uniquely determine the position of any given point in the plane. The direct transformation $(u,v) = \mathbf{g}(x,y)$ translates from the x,y language to the u,v language, and the inverse transformation $(x,y) = \mathbf{g}^{-1}(u,v)$ translates from the u,v language to the x,y language.[16]

---

[13] The material in this section is only used within the present appendix, but it is not required elsewhere in the book. It is given here for the sake of completeness only, and may be skipped if desired.

[14] It may be helpful to imagine that these coordinate nets are printed on two different transparencies, that can be superposed on top of the same distorted plane. This is clearly illustrated in Fig. 9: Fig. D.9(b) shows the distorted plane superposed by the distorted x,y coordinate net, and Fig. D.9(c) shows the same plane superposed by the undistorted u,v coordinate net.

[15] Note that the passive interpretation of the transformation can be also considered in Fig. D.5(a) which shows the domain of $\mathbf{g}$, i.e the situation before the transformation has been applied. However, this may be somewhat less natural since in order to draw the u and v axes in Fig. D.5(a) we need to apply to the x and y axes the *inverse* transformation $\mathbf{g}^{-1}$.

[16] Note the inherent inversion that exists in the passive interpretation: The coordinates (x,y) of any given point in terms of the x,y coordinate system are translated by the transformation $(u,v) = \mathbf{g}(x,y)$ into the coordinates (u,v) of the same point in terms of the u,v coordinate system; and yet, the u,v coordinate system is obtained from the x,y system by applying the *inverse* transformation, $\mathbf{g}^{-1}$ (see Remark D.3). For example, when $\mathbf{g}$ represents a rotation by angle α (Fig. D.5), the u,v axes are obtained from the x,y axes by a rotation by $-\alpha$.

The active and passive interpretations of a transformation $(u,v) = \mathbf{g}(x,y)$ are concisely summarized as follows:

Proposition D.2: A transformation $(u,v) = \mathbf{g}(x,y)$ can be interpreted in two different ways: either as a mapping which actually moves each point (x,y) into a new location (u,v) within one and the same coordinate system (the *active* interpretation), or as a coordinate change which converts each point from the x,y coordinate system to the u,v coordinate system (the *passive* interpretation).[17] ∎

The active and passive interpretations of a transformation $(u,v) = \mathbf{g}(x,y)$ are very closely interrelated. The curves in the u,v plane that are, according to the active interpretation, the images under $\mathbf{g}$ of straight lines parallel to the axes in the x,y plane (see Eqs. (D.7) and (D.8)), can be also regarded according to the passive interpretation as the coordinate curves for the curvilinear coordinates $x = g_1^{-1}(u,v)$, $y = g_2^{-1}(u,v)$ in the u,v plane (see Fig. D.9(b)). Similarly, the curves in the x,y plane that are, according to the active interpretation, the images under $\mathbf{g}^{-1}$ of straight lines parallel to the axes in the u,v plane (see Eqs. (D.11) and (D.12)), can be also regarded as the coordinate curves for the curvilinear coordinates $u = g_1(x,y)$, $v = g_2(x,y)$ in the x,y plane (see Fig. D.9(d)).

Thus, the difference between the two interpretations is mainly in the point of view. If we are mainly interested in the x,y plane, we regard u and v simply as a new means of locating points in the x,y plane, and the u,v plane becomes then merely subsidiary (as in Fig. D.3). But if we are equally interested in the two planes, the x,y plane and the u,v plane, it is preferable to regard the transformation $(u,v) = \mathbf{g}(x,y)$ as specifying a correspondence between the two planes, i.e., as a mapping of one on the other (as in Fig. D.4).[18] It is, however, often desirable to keep the two interpretations in mind at the same time.

Let us consider as a final example the polar to Cartesian transformation (D.13) which is illustrated in Fig. D.6. In this case we say, according to the *active* interpretation, that the transformation maps any point $P = (\theta,r)$ into its image $Q = (u,v) = (r\cos\theta, r\sin\theta)$; for example, the point $P = (\pi/4, 3\sqrt{2})$ is mapped into the point $Q = (3,3)$. Furthermore, by considering successive points P along the curve $r = 6\cos\theta$ shown in Fig. D.6(a), we can see that our transformation maps (or distorts) this curve from a cosinusoidal line in the original θ,r plane into a circle $(u-3)^2 + v^2 = 9$ in the transformed θ,r plane (whose new standard coordinate axes are u and v; see Fig. D.6(b)). What is, then, the *passive* interpretation of the same transformation (D.13)? Consider again point Q in Fig. D.6(b): If we regard (D.13) as a dictionary, we see that without actually moving the point Q, our transformation $(u,v) = (r\cos\theta, r\sin\theta)$ converts the point coordinates from the θ,r language, $(\pi/4, 3\sqrt{2})$, to the u,v language, $(3,3)$. By considering now the entire circle of Fig. D.6(b),

---

[17] These two interpretations can be also described as "different objects viewed in the same coordinates" and "the same object viewed in different coordinates", respectively.

[18] This difference of point of view is often reflected by the notation being used to express the given transformation. For example, the notation $\mathbf{g}(x,y) = (2xy, y^2 - x^2)$, which does not explicitly mention the u,v coordinates, suggests that one is mainly interested in the x,y plane, while the equivalent notation $(u,v) = (2xy, y^2 - x^2)$ suggests that one is interested in both the x,y and u,v planes.

we see that our transformation (D.13) converts the mathematical expression of our circle (without distorting the circle itself!) from the θ,r language, $r = 6\cos\theta$, into the u,v language, $u^2 + v^2 = 6u$, or equivalently $(u-3)^2 + v^2 = 9$ (see the mathematical derivation in Sec. D.4.2). Similar results can be obtained for any other curve or object in the plane.

Remark D.9: It should be noted that there does exist one real difference between the two points of view: $(u,v) = g(x,y)$ always defines a mapping, no matter how many points (x,y) it maps to one point (u,v); but it cannot define a meaningful coordinate change if the correspondence is not one-to-one (such a case is illustrated in Example D.7 below). ∎

Remark D.10: To see how all this is related to the well-known theorems on coordinate (or basis) changes in linear algebra, consider the particular case in which $(u,v) = g(x,y)$ is a linear transformation. Let $\mathbf{e}_1$, $\mathbf{e}_2$ be the standard basis of the original x,y space: $\mathbf{e}_1 = (1,0)$, $\mathbf{e}_2 = (0,1)$, and let $\mathbf{f}_1$, $\mathbf{f}_2$ be the standard basis of the g-transformed x,y space (see Fig. D.5(b)), expressed in terms of the basis vectors $\mathbf{e}_1$ and $\mathbf{e}_2$: $\mathbf{f}_1 = (f_{1,1}, f_{1,2})$, $\mathbf{f}_2 = (f_{2,1}, f_{2,2})$, namely:

$$\mathbf{f}_1 = f_{1,1}\mathbf{e}_1 + f_{1,2}\mathbf{e}_2$$
$$\mathbf{f}_2 = f_{2,1}\mathbf{e}_1 + f_{2,2}\mathbf{e}_2 \tag{D.14}$$

Then, according to well known results in linear algebra (see, for example, [Lang87 p. 394–395] or [Lay03 p. 249]), the matrix representation of the transformation $\mathbf{g}$ is given by:

$$\begin{pmatrix} u \\ v \end{pmatrix} = \begin{pmatrix} f_{1,1} & f_{2,1} \\ f_{1,2} & f_{2,2} \end{pmatrix} \begin{pmatrix} x \\ y \end{pmatrix} \tag{D.15}$$

where the components of $\mathbf{f}_1$ and $\mathbf{f}_2$ form the *columns* of the matrix. For instance, if $(u,v) = g(x,y)$ represents rotation by angle α we have: $\mathbf{f}_1 = (\cos\alpha, \sin\alpha)$, $\mathbf{f}_2 = (-\sin\alpha, \cos\alpha)$, and hence the transformation is given by:

$$\begin{pmatrix} u \\ v \end{pmatrix} = \begin{pmatrix} \cos\alpha & -\sin\alpha \\ \sin\alpha & \cos\alpha \end{pmatrix} \begin{pmatrix} x \\ y \end{pmatrix} \tag{D.16}$$

which agrees, indeed, with Eq. (D.1).

In linear algebra books transformation (D.15) is considered as a change of coordinates from the basis $\mathbf{f}_1$, $\mathbf{f}_2$ to the standard basis $\mathbf{e}_1$, $\mathbf{e}_2$ (see, for example, [Lay03 p. 249]): Given a point (x,y) in terms of the basis $\mathbf{f}_1$, $\mathbf{f}_2$ of the transformed x,y space, $\mathbf{g}$ returns the coordinates (u,v) of the same point in terms of the standard basis $\mathbf{e}_1$, $\mathbf{e}_2$. Restated in our terms, we can say that transformation (D.15) is considered in linear algebra as a "dictionary" that translates coordinates of any given point in terms of the transformed x,y space into the coordinates of the same point in terms of the u,v space (see Fig. D.5(b)). For example, in the case of rotation, the point $(1,0)$ in terms of the rotated x,y coordinates is converted by the transformation (D.16) into $(\cos\alpha, \sin\alpha)$, which specifies the coordinates of the same point in terms of the u,v space. This corresponds, indeed, to the *passive* interpretation of the linear transformation $(u,v) = g(x,y)$.

On the other hand, it is also possible to consider the *active* interpretation of the same linear transformation: According to this point of view, the transformation moves (or maps) any point $x\mathbf{e}_1 + y\mathbf{e}_2$ in the original plane to its image point $x\mathbf{f}_1 + y\mathbf{f}_2$ in the same plane (see Fig. D.3) or, equivalently, to the point (u,v) in terms of the target plane (see Fig. D.5). Note that just as in the general case, a linear transformation $\mathbf{g}$ moves any point (x,y) given in the original x,y coordinate system to the point having the same coordinates (x,y) in the transformed x,y coordinate system. More details on active and passive linear transformations in a given vector space can be also found in [Wolf79 Sec. 1.3].[19] ■

Remark D.11: Note that the matrix in Eq. (D.15) is called in some references "the transition matrix from the old basis $\mathbf{e}_1, \mathbf{e}_2$ to the new basis $\mathbf{f}_1, \mathbf{f}_2$" (see, for example, [Lipschutz68 p. 153] and the remark following Theorem 7.4 there), while in other references it is called "the change-of-coordinates matrix from the basis $\mathbf{f}_1, \mathbf{f}_2$ to the basis $\mathbf{e}_1, \mathbf{e}_2$" (see, for example, [Lay03 pp. 249, 273]). The reason for this terminological inconsistency is that the matrix in question contains the coefficients $f_{i,j}$ that are used to convert the old basis vectors $\mathbf{e}_1, \mathbf{e}_2$ into the new basis vectors $\mathbf{f}_1, \mathbf{f}_2$ (see Eq. (D.14)); and yet, a multiplication by this matrix, as shown in Eq. (D.15), converts the coordinates (x,y) given in terms of the new basis $\mathbf{f}_1, \mathbf{f}_2$ back into coordinates (u,v) given in terms of the old basis $\mathbf{e}_1, \mathbf{e}_2$. ■

Remark D.12: Another possible source of confusion exists due to terminological inconsistencies in the literature regarding the active and passive interpretations of a transformation. In some references such as [Harris98 pp. 351–353] the active transformation has the same meaning as in our definition, i.e. moving a point from its original location P to its image location Q within the same coordinate system (like in Fig. D.3); but the passive transformation means moving the *coordinate axes* from their original location to their image location under $\mathbf{g}$ while the point P itself does not move. In this case, the coordinates of the point P in terms of the new coordinate axes are given by the inverse transformation $\mathbf{g}^{-1}$. For example, if the active transformation $\mathbf{g}$ consists of displacing any point (x,y) to the point $(x+a, y+b)$:

$$u = x + a$$
$$v = y + b \tag{D.17}$$

then the passive transformation, according to this interpretation, consists of displacing the coordinate system by (a,b), while the point (x,y) itself remains in its original location. The new coordinates of the point (x,y) are expressed, therefore, by the *inverse* transformation:

---

[19] It is interesting to note that the active interpretation of linear transformations is rarely mentioned in linear algebra books (with [Mansfield76 pp. 202–206] being a remarkable exception), while in calculus books the active interpretation is used frequently (see, for example, [Colley98 pp. 326–327, 334–335]). The reason is that calculus begins with a single coordinate system, while linear algebra is at the other extreme: its main concern is in considering all possible bases (or linear coordinate systems) for a given vector space, and then choosing the one which is most convenient depending on the nature of the problem at hand [Mansfield76 p. 60], for example the one which gives the simplest matrix representation for a given linear transformation.

$$u = x - a$$

$$v = y - b \tag{D.18}$$

[Harris98 p. 351]. This does not agree with our terminology, based on [Courant88 pp. 133–140], according to which the active and the passive points of view are merely two different interpretations of the *same* transformation $(u,v) = \mathbf{g}(x,y)$. For example, according to *our* passive interpretation the transformation (D.17) simply translates the transformed x,y coordinates of any given point back to the u,v coordinates of the same point: $u = x + a$, $v = y + b$, and no use is made of the inverse transformation (D.18).

The difference between these two views of the passive interpretation of $\mathbf{g}$ can be illustrated by splitting Fig. D.5(b) into two different figures, one showing the point Q superposed by the standard u,v coordinates and the other showing the same point Q superposed by the transformed x,y coordinates. According to *our* passive point of view, both figures show the range of the transformation $\mathbf{g}$ after its action has already been completed. But according to the other point of view, the first figure shows the initial situation, and the other figure shows how the transformation $\mathbf{g}$ actively distorts the original coordinate system, while the point Q remains in its original location. The coordinates of Q with respect to the new distorted coordinate system are obtained, therefore, by applying the inverse transformation $\mathbf{g}^{-1}$ to the coordinates of Q in the undistorted coordinate system. ■

Remark D.13: Although the terms "mapping" and "transformation" are synonyms, they are often used in slightly different connotations. The term "mapping" usually implies the active interpretation, where points P of one space are mapped to corresponding points Q in another space. On the other hand, the term "transformation" is often utilized when the target space is the same as the source space, and it refers to a deformation or a rearrangement of that space. ■

D.6 Domain and range transformations of a function

So far we have considered the influence of transformations $(u,v) = \mathbf{g}(x,y)$ on the plane or on its 2D or 1D subsets. We now proceed to the influence of such transformations on *functions* that are defined on the plane. This will bring us to the important notions of domain and range transformations, which have a central role in our work on the moiré theory.

Let $z = f(u,v)$ be a function $f: \mathbb{R}^2 \to \mathbb{R}$. As we already know from Sec. D.2, such a function can be interpreted geometrically as a surface over the u,v plane. Since a function is always defined between two spaces, its *domain* and its *range*,[20] each function $z = f(u,v)$

---

[20] The domain of the function $z = f(u,v)$ is the 2D set consisting of the points (u,v) on which the function is defined, and the range of the function is the 1D set consisting of the values z it returns. The reason we use here the domain variables u,v rather than x,y is explained in Footnote 24 below.

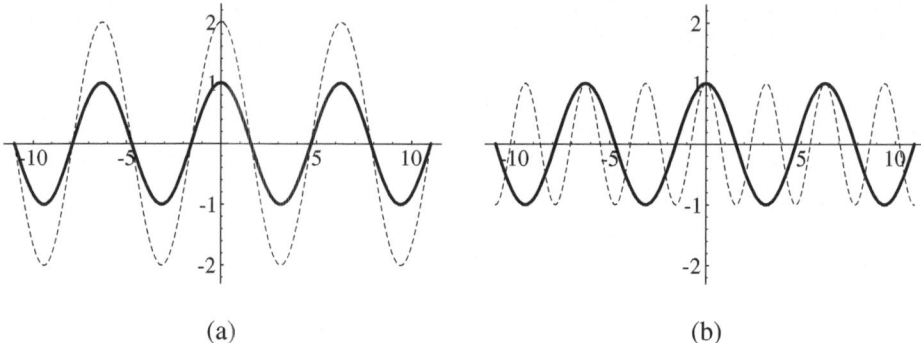

                    (a)   (b)

Figure D.7: (a) The influence of the range transformation $s = 2z$ on the function $z = \cos u$. The dashed curve represents the resulting function, $s = 2\cos u$. (b) The influence of the domain transformation $u = 2x$ on the same function $z = \cos u$. The dashed curve represents the resulting function, $z = \cos(2x)$. In both (a) and (b) the transformed and untransformed functions are plotted on the same axes, in order to allow the comparison between them.

can be distorted in two different ways: either by transforming its domain, or by transforming its range. We can distinguish, therefore, between domain and range transformations as follows:[21]

(1) A *domain transformation* of the function $z = f(u,v)$ is a transformation $(u,v) = \mathbf{g}(x,y)$, $\mathbf{g}: \mathbb{R}^2 \to \mathbb{R}^2$, which is applied to the domain of $f(u,v)$. Following this operation the original function $f(u,v)$ is transformed into the new function $f(\mathbf{g}(x,y))$.

(2) A *range transformation* of the same function $z = f(u,v)$ is a transformation $s = t(z)$, $t: \mathbb{R} \to \mathbb{R}$, which is applied to the range of $f(u,v)$ (i.e. to the values z it returns). Following this operation the original function $f(u,v)$ is transformed into the new function $t(f(u,v))$.

Geometrically speaking, a domain transformation distorts the surface $f(u,v)$ spatially (for example: by rotation, translation, scaling along the u,v directions, etc.). In other words, it moves each point (u,v) to a new location $(x,y) = \mathbf{g}^{-1}(u,v)$, without affecting the z coordinate assigned to the point. On the other hand, a range transformation distorts the surface $f(u,v)$ vertically (for example: by scaling it by 2 in the z direction, by taking the cosine of the altitude z, etc.). In other words, it changes the z coordinate assigned to each point into $s = t(z)$, without affecting the point's u,v coordinates (see Fig. D.7(a)).

It is important to note that domain and range transformations are not different types of transformations, but rather different uses or applications of a transformation. Thus, in

---

[21] Although these definitions are given here for the case of a function $z = f(u,v)$, they are, in fact, completely general, and apply to any other types of functions, including $z = f(u)$, $(x,y) = \mathbf{g}(u)$, etc. (see Sec. D.2), with the appropriate adaptations to the dimensionalities of their domain and range.

cases where the domain and the range of f have the same dimensionality (like in the 1D case $z = f(u)$ that we discuss in Sec. D.6.1 below) the very same transformation can be applied either to the domain of f or to the range of f. In the first case it will play the role of a domain transformation of the function f, while in the second case it will play the role of a range transformation of f.

Remark D.14: As we have already seen, transformations can be applied not only to functions defined over the plane, but also directly to the plane itself or to objects that are subsets of the plane. For example, if $z = f(u,v)$ is a function (surface) over the u,v plane, $f(u,v) = 0$ defines a curve in the u,v plane (which corresponds to the zero level line of the surface $z = f(u,v)$). But unlike the surface $z = f(u,v)$, which has both a domain and a range, the planar curve $f(u,v) = 0$ only has a domain (since its range is reduced to the degenerate space $\{0\}$), and therefore it can only undergo *domain* transformations. ■

While the effect of a range transformation on the original function f is rather straightforward, the effect of a domain transformation on the function f may be less obvious and sometimes even quite confusing. It is therefore our aim here to help in developing an intuitive understanding of range and domain transformations and to point out the main pitfalls in their use.

D.6.1 The 1D case

In order to better understand the situation, let us start with the simpler, 1D case. Suppose that we are given a function $z = f(u)$, for example $z = \cos u$. Note that in this case both the range and the domain of the function f are one-dimensional, so that any function $g: \mathbb{R} \to \mathbb{R}$ may be applied to f either as a range transformation, giving $g(f(u))$, or as a domain transformation, giving $f(g(x))$.

We start with the case of a range transformation. Suppose that the transformation $s = t(z)$ is applied to our function $z = f(u)$ as a range transformation. The effect of this transformation on $f(u)$ is straightforward: for example, if we apply to $f(u)$ the range transformation $s = 2z$ we obtain a vertically stretched version of $f(u)$, namely, $2f(u)$. This transformation maps each value z on the vertical axis to $2z$, without affecting the u coordinate; this can be clearly seen by plotting the two functions on top of each other (see Fig. D.7(a)).[22]

Now, suppose that instead of the range transformation $s = t(z)$ we apply to our function $z = f(u)$ a domain transformation $u = g(x)$; for example, we may choose once again the same two-fold magnification transformation, $u = 2x$. Clearly, this transformation maps each value x on the horizontal axis to the value $2x$, without affecting the z coordinate. Based on our experience with range transformations it could be natural, therefore, to expect that the application of this transformation would stretch our function $z = f(u)$ laterally by a factor of 2. However, in reality $f(2x)$ is not a stretched version of $f(u)$, but

---

[22] Note that in order to compare the two functions we must plot both of them on the same axes. Therefore, in Fig. D.7(a) the vertical axis represents both of the variables z and s (see Remark D.1).

rather a condensed version of $f(u)$ which has been squeezed laterally by a factor of 2. For example, $z = \cos(2x)$ is a laterally condensed version of $z = \cos u$; this can be clearly seen by plotting the two functions on top of each other (see Fig. D.7(b)).[23]

This difference between the influence of range and domain transformations can be explained as follows. The application of a domain transformation $u = g(x)$ to the given function $z = f(u)$ gives $z = f(g(x))$, i.e. it moves each point u in the domain of the original function $z = f(u)$ into the new location x which is determined by the *inverse* of the transformation g, namely, $x = g^{-1}(u)$. But when we apply to the given function $z = f(u)$ a range transformation $s = t(z)$, it simply modifies the z coordinate assigned to each point into the new value $s = t(z)$, and no inversion is involved. This is schematically illustrated by the commutative diagram shown in Fig. D.8(a). The original function $z = f(u)$ is represented in this figure by the top horizontal arrow. After applying to this function both a domain transformation $u = g(x)$ and a range transformation $s = t(z)$ we obtain the resulting function $s = f_r(x)$, whose variables are x and s rather than u and z; this function is represented in our figure by the bottom horizontal arrow. The new function f_r can be expressed in terms of the known functions g, f and t by:

$$f_r(x) = t(f(g(x))) \tag{D.19}$$

as shown by the circular arrow in the figure. (Note the order inversion in (D.19): although t appears first in the equation, in reality it operates last, after g and f.) In particular, if g is the identity transformation, f_r is simply a range transformation of f; while if t is the identity transformation, f_r is a domain transformation of f: $f_r(x) = f(g(x))$.

Fig. D.8(a) illustrates the fundamental difference which exists between the two transformations that we have applied to our original function $z = f(u)$ to obtain the new function $s = f_r(x)$: While the range transformation $s = t(z)$ maps the variable z of our original function into the transformed variable s of the new function $s = f_r(x)$, the domain transformation $u = g(x)$ maps the variable x (the variable of the new, distorted function $s = f_r(x)$) back into the variable u (the variable of our original, undistorted function $z = f(u)$). The transformation from the original variable u into the new, distorted variable x is given, therefore, by the *inverse* transformation, $x = g^{-1}(u)$, and not by the transformation $u = g(x)$ that we actually plug into the given function $f(u)$. And indeed, by drawing the functions $f(g(x))$ and $f(u)$ on the same coordinate system we can see that the influence of the domain transformation $u = g(x)$ on $f(u)$ is inverse. For example, $f(2x)$ is a *squeezed* version of $f(u)$; $f(x+1)$ is a unit translation of $f(u)$ to the *negative* direction; etc. If we wish to stretch our function $f(u)$ laterally by a factor of 2 within the original coordinate system, we need to take $f(x/2)$ and not $f(2x)$.

---

[23] Once again, in order to compare the two functions we must plot both of them on the same axes. Therefore, in Fig. D.7(b) the horizontal axis represents both of the variables u and x (see Remark D.1). In fact, we could say that "the function $z = f(2u)$ is a laterally condensed version of $z = f(u)$"; but in order to avoid any confusion due to the use of the original variable u in the transformed function, too, it would be better to use instead a more "neutral" variable name, such as w, and say that "the function $z = f(2w)$ is a laterally condensed version of $z = f(w)$". In this case the horizontal axis would be simply named w.

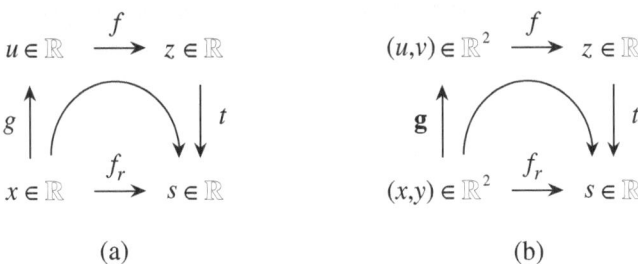

(a)	(b)

Figure D.8: (a) Commutative diagram illustrating the influence of a domain transformation $u = g(x)$ and a range transformation $s = t(z)$ on a given 1D function $z = f(u)$. (b) Commutative diagram illustrating the influence of a domain transformation $(u,v) = \mathbf{g}(x,y)$ and a range transformation $s = t(z)$ on a given 2D function $z = f(u,v)$. Note that while the range transformation t maps the original variable z of our function f to the new, distorted variable s, the domain transformation maps the new, distorted variables x,y back to the original variables u,v of our function f.

Remark D.15: Note that any expression of the form $f(g(x))$ can be interpreted in two different ways: either as the application of g (as a domain transformation) to the given function f, or as the application of f (as a range transformation) to the given function g. Although both interpretations give, of course, the same result, in each situation one or the other may be easier to understand. The same is also true for expressions of the form $f(g(x,y))$ (where $f(u)$ is still a 1D function): For example, the expression $\cos(2\pi\sqrt{x^2 + y^2})$ can be understood either as the result of bending the original straight cosinusoidal surface $\cos(2\pi u)$ into a circular cosinusoidal surface, or as the result of applying the function $\cos(2\pi z)$ to the z values (altitude) of the original conic surface $z = \sqrt{x^2 + y^2}$. ∎

D.6.2 The 2D case

The same considerations hold also in the 2D case. Suppose that we are given a function $z = f(u,v)$. Just as in the 1D case, the influence of a range transformation $s = t(z)$ on this function is straightforward: it simply maps each z value (altitude) of $z = f(u,v)$ to the new value $t(z)$, without affecting the u and v coordinates. Thus, if we apply to $f(x,y)$ the range transformation $s = 2z$, we obtain a vertically stretched version of $f(u,v)$, namely, $2f(u,v)$.

Now, let us proceed to the influence of a domain transformation $(u,v) = \mathbf{g}(x,y)$ on our function $z = f(u,v)$. As a simple example we may consider the domain transformation $(u,v) = (2x,2y)$. Clearly, this transformation maps each point (a,b) of the plane to the point $(2a,2b)$. But in spite of this stretching effect, it turns out that plugging this transformation into the function $z = f(u,v)$ gives $z = f(2x,2y)$, which is not a spatially stretched version of $f(u,v)$ but rather a spatially condensed version of $f(u,v)$. The reason for this inversion is that the application of a domain transformation $(u,v) = \mathbf{g}(x,y)$ to the given function $z = f(u,v)$ gives $z = f(\mathbf{g}(x,y))$, i.e. it moves each point (u,v) in the domain of the original function

$z = f(u,v)$ into the new location (x,y) which is determined by the *inverse* of the transformation $\mathbf{g}$, namely, $(x,y) = \mathbf{g}^{-1}(u,v)$. This is illustrated by the commutative diagram shown in Fig. D.8(b). The original function $z = f(u,v)$ is represented here by the top horizontal arrow, and the resulting function f_r obtained by applying to f both a domain transformation $\mathbf{g}$ and a range transformation t is represented by the bottom horizontal arrow. But the effect of f_r can be also expressed, by following the circular arrow, as:

$$f_r(x,y) = t(f(\mathbf{g}(x,y)))\tag{D.20}$$

Clearly, if $\mathbf{g}$ is the identity transformation, f_r is simply a range transformation of the function f; and similarly, if t is the identity transformation, f_r is a domain transformation of the function f: $f_r(x,y) = f(\mathbf{g}(x,y))$.

Fig. D.8(b) shows that just as in the 1D case, the influence of the transformation $(u,v) = \mathbf{g}(x,y)$ when it is applied as a domain transformation to our function $f(u,v)$ is, in fact, inverse: it maps the transformed variables x,y of the new function $s = f_r(x,y)$ back into the untransformed variables u,v of the original function $z = f(u,v)$. Hence, the transition from the original u,v space of $f(u,v)$ into the new, distorted x,y space of $f(\mathbf{g}(x,y))$ is represented by the *inverse* transformation, $\mathbf{g}^{-1}$, although we have plugged into f the transformation $\mathbf{g}$ itself and not its inverse $\mathbf{g}^{-1}$.

To better illustrate this, let us consider again the planar transformation $(u,v) = \mathbf{g}(x,y) = (2x,2y)$, which corresponds to a two-fold expansion of the x,y plane. When we apply $\mathbf{g}$ as a domain transformation to the function $f(u,v) = u^2 + v^2 - 1$ (a top opened paraboloid), we clearly obtain the inverse effect since $f(2x,2y) = (2x)^2 + (2y)^2 - 1$ is a spatially shrinked version of the original function $f(u,v)$. The expansion effect is obtained by applying to $f(u,v)$ the domain transformation $(u,v) = \mathbf{g}^{-1}(x,y) = (x/2,y/2)$ (see Remark D.4 on the variable names).

Note that the same rule applies also to any subsets of the function $z = f(u,v)$, for example to the curve defined by its zero level line $f(u,v) = 0$. Thus, proceeding with the same example, if we apply our two-fold magnification $(u,v) = \mathbf{g}(x,y) = (2x,2y)$ as a domain transformation to the circle $f(u,v) = u^2 + v^2 - 1 = 0$, we obtain:

$$(2x)^2 + (2y)^2 - 1 = 0$$

namely: $x^2 + y^2 = (\tfrac{1}{2})^2$

which is clearly a two-fold reduced circle. Note, however, that when we consider our transformation $(u,v) = \mathbf{g}(x,y) = (2x,2y)$ as a direct transformation that operates on the original, undistorted x,y plane, it indeed magnifies the entire x,y plane, including the circle $x^2 + y^2 = (\tfrac{1}{2})^2$ that is embedded in it, into the target u,v plane and its circle $u^2 + v^2 = 1$. (Remember that according to Proposition D.1 the image of the curve $f(x,y) = 0$ under the direct transformation $(u,v) = \mathbf{g}(x,y)$ is given by the implicit equation $f(g_1^{-1}(u,v),g_2^{-1}(u,v)) = 0$. In our case, plugging $x = u/2$ and $y = v/2$ into the circle's equation $x^2 + y^2 = (\tfrac{1}{2})^2$ gives, indeed, $u^2 + v^2 = 1$.)

Proposition D.3: Suppose we are given a transformation $(u,v) = \mathbf{g}(x,y)$. When this transformation is applied to the original, undistorted x,y plane (or any subset thereof) as a *direct* transformation, we obtain in the target u,v plane the $\mathbf{g}$-transformed copy of the x,y plane (or of its subset). For example, the transformation $(u,v) = (2x,2y)$ gives us in the target u,v plane a two-fold magnified version of the original x,y plane (or any subset thereof). But when the same transformation is applied to the function $z = f(u,v)$ (or any subset thereof, such as the level line $f(u,v) = const.$) as a *domain* transformation, we obtain a $\mathbf{g}^{-1}$-transformed copy of the function f (or of its subset). For example, when our transformation $(u,v) = (2x,2y)$ is applied as a domain transformation to the function $z = f(u,v)$, we obtain a spatially condensed version of this function, $z = f(2x,2y)$. ∎

Note that although the transformation $(u,v) = \mathbf{g}(x,y)$ acts on the original, undistorted x,y space and yields its distorted image in the target u,v space, when $\mathbf{g}$ is applied as a domain transformation to a function $z = f(u,v)$ the roles are inversed: The original undistorted function f is given in the u,v coordinate system, while the domain-transformed function $f(\mathbf{g}(x,y))$ subsists over the x,y space.[24]

As a further example, let $(u,v) = \mathbf{g}(x,y)$ be the planar transformation which corresponds to a rotation by 90° counterclockwise about the origin:

$$\begin{pmatrix} u \\ v \end{pmatrix} = \begin{pmatrix} 0 & -1 \\ 1 & 0 \end{pmatrix} \begin{pmatrix} x \\ y \end{pmatrix} = \begin{pmatrix} -y \\ x \end{pmatrix}$$

Clearly, this transformation rotates each point (x,y) of the plane by 90° counterclockwise (for example, it maps the point $(1,0)$ on the horizontal axis to the point $(0,1)$ on the vertical axis, etc.). Thus, the vector (u,v) is a copy rotated by +90° of the vector (x,y). However, for any given function $f(u,v)$, its transformed counterpart $f(\mathbf{g}(x,y))$, that is, $f(-y,x)$, is a −90° rotated version of $f(u,v)$. This can be easily verified by plotting the original and the transformed functions.

More generally, the transformation $(u,v) = \mathbf{g}(x,y)$ consisting of a rotation by angle α counterclockwise about the origin is defined by Eq. (D.1):

$$\begin{pmatrix} u \\ v \end{pmatrix} = \begin{pmatrix} \cos\alpha & -\sin\alpha \\ \sin\alpha & \cos\alpha \end{pmatrix} \begin{pmatrix} x \\ y \end{pmatrix} = \begin{pmatrix} x\cos\alpha - y\sin\alpha \\ x\sin\alpha + y\cos\alpha \end{pmatrix}$$

but the function $f(x\cos\alpha - y\sin\alpha, x\sin\alpha + y\cos\alpha)$ is a rotated version of $f(u,v)$ by $-\alpha$, not by α. Thus, if we wish to rotate $f(u,v)$ by angle α within the original coordinate system, the transformation we need to apply to its domain is the *inverse* transformation which corresponds to a rotation by $-\alpha$. The rotated version of $f(u,v)$ by angle α is, therefore, $f(x\cos\alpha + y\sin\alpha, -x\sin\alpha + y\cos\alpha)$.

---

[24] This agrees with our general convention in the moiré theory, that the geometrically transformed layers subsist in the x,y space (see, for example, the transformed gratings and screens in Chapters 3, 6 and 7 of this volume or in Chapter 10 of *Vol. I*; for instance, the top-opened parabolic cosinusoidal grating of Fig. 10.1(c) in *Vol. I* is expressed by $r(x,y) = \cos(2\pi f[y - ax^2])$). Note, however, that in our discussions on the moiré theory we usually denote the original, undistorted coordinate space by the variables x',y' rather than u and v, since u and v are reserved in our work to the Fourier, spectral domain.

Similarly, applying the direct transformation (D.1) to any curve $f(x,y) = 0$ in the x,y plane gives in the u,v plane a copy rotated by α of the original curve (see Proposition D.1). But applying (D.1) as a domain transformation to the curve $f(u,v) = 0$ rotates this curve by $-\alpha$. Note that in the first case the untransformed curve is $f(x,y) = 0$, while in the second case the untransformed curve is given by $f(u,v) = 0$.

Obviously, if we wish to leave $f(u,v)$ unchanged and to consider the influence of $\mathbf{g}(x,y)$ on the *coordinate system* and on its axes, the result will be inversed. For example, in the case of rotation we will obtain a rotation of the u,v axes by angle α to the positive direction [Weisstein99 p. 1580]. Similarly, if $\mathbf{g}(x,y)$ is defined by $\mathbf{g}(x,y) = (2x,2y)$ the result can be seen as a twofold expansion of the axes while the surface $f(u,v)$ itself remains unchanged; and if $\mathbf{g}(x,y) = (x+1,y)$ the result can be seen as a unit translation of the horizontal axis to the *positive* direction. This point will be addressed in Sec. D.7 below.

Remark D.16: We may mention at this point yet another possible source of confusion due to terminological inconsistencies in the literature, this time related to the naming conventions in domain transformations. A transformation $(u,v) = \mathbf{g}(x,y)$ is usually called in the literature a transformation from the x,y space to the u,v space, because it translates the coordinates of any given point in the plane from the x,y language to the u,v language. For instance (see Example D.4 below), the transformation $(r,\theta) = (\sqrt{x^2+y^2},\ \arctan(y/x))$ is known as the *Cartesian to polar* coordinate transformation, and its inverse, $(x,y) = (r\cos\theta,\ r\sin\theta)$, is called the *polar to Cartesian* coordinate transformation [Colley98 p. 68]. Note, however, that in some references the naming conventions are inversed, and the transformation $(u,v) = \mathbf{g}(x,y)$ is considered as a mapping from the u,v space to the x,y space (see, for example, [Spiegel63 pp. 108, 124, 182]). This naming convention can be explained by the fact that if we are given a function $z = f(u,v)$ in the u,v space, and we apply to it $(u,v) = \mathbf{g}(x,y)$ as a domain transformation, we obtain the function $z = f(\mathbf{g}(x,y))$ in the x,y space. This situation occurs, for example, when changing variables under a double integral in order to facilitate its calculation (see a few such examples in [Colley98 pp. 336–338]). Thus, the transformation $(u,v) = \mathbf{g}(x,y)$ translates the function f from the u,v language to the x,y language, although it actually converts the x,y space into the u,v space.[25] This is, again, a consequence of the inversion property inherent to domain transformations.[26] ∎

Remark D.17: The extension of Remark D.15 to the case of a 2D function f is rather straightforward. Thus, any expression of the form $f(\mathbf{g}(x,y))$ can be interpreted either as the application of $(u,v) = \mathbf{g}(x,y)$ (as a domain transformation) to the given function $z = f(u,v)$, or as the application of $z = f(u,v)$ (as a range transformation) to $(u,v) = \mathbf{g}(x,y)$. Note, however, that in this 2D case the second interpretation is less useful, since it maps the 2D range of $\mathbf{g}$ into a new 1D range. And yet, both interpretations are completely equivalent and give the same result. ∎

---

[25] In computer graphics (image warping) it is often said that $\mathbf{g}$ transforms the image f from the x,y transformed, destination space back into the u,v original, undistorted space.

[26] A useful trick for avoiding confusion (or lengthy explanations) in the case of polar to Cartesian or Cartesian to polar coordinate transformations is to use the ambiguous term "polar coordinate transformation".

D.6.3 The effect of transformation g on objects and on their characteristic functions

We conclude this section with the following result, which illustrates from a slightly different angle Propositions D.1 and D.3 and the potentially confusing nature of domain transformations.

Suppose we are given a point P in the x,y plane, say, the point $P = (1,0)$ on the x axis. The *characteristic function* of this point (i.e. the function that takes the value 1 at this point and remains zero everywhere else) is defined over the x,y plane by:

$$f(x,y) = \begin{cases} 1 & (x,y) = (1,0) \\ 0 & \text{otherwise} \end{cases} \tag{D.21}$$

Now, suppose that we apply to the x,y plane a 2D transformation, say, a rotation by 90° counterclockwise:

$$\begin{pmatrix} u \\ v \end{pmatrix} = g\begin{pmatrix} x \\ y \end{pmatrix} = \begin{pmatrix} 0 & -1 \\ 1 & 0 \end{pmatrix}\begin{pmatrix} x \\ y \end{pmatrix}$$

As a result of this transformation our point P is rotated to a new location, the point $(0,1)$ on the vertical axis:

$$\begin{pmatrix} 0 & -1 \\ 1 & 0 \end{pmatrix}\begin{pmatrix} 1 \\ 0 \end{pmatrix} = \begin{pmatrix} 0 \\ 1 \end{pmatrix}$$

However, if we wish to express the very same result using the *characteristic function* of the point P, Eq. (D.21), we need to apply to it the *inverse* transformation, that is given by:

$$\begin{pmatrix} x \\ y \end{pmatrix} = g^{-1}\begin{pmatrix} u \\ v \end{pmatrix} = \begin{pmatrix} 0 & 1 \\ -1 & 0 \end{pmatrix}\begin{pmatrix} u \\ v \end{pmatrix} = \begin{pmatrix} v \\ -u \end{pmatrix}$$

By applying g^{-1} to $f(x,y)$ as a domain transformation we obtain:

$$h(u,v) = f(g^{-1}(u,v)) = f(v,-u) =$$

$$= \begin{cases} 1 & (v,-u) = (1,0) \\ 0 & \text{otherwise} \end{cases}$$

$$= \begin{cases} 1 & (u,v) = (0,1) \\ 0 & \text{otherwise} \end{cases}$$

This is, indeed, the characteristic function in the u,v plane of our rotated point, $(0,1)$.

Since this result holds for any given point, it obviously remains true for any object on the plane. For example, suppose we draw the standard unit grid on the x,y plane. This unit grid can be represented by its characteristic function which is defined by:

$$f(x,y) = \begin{cases} 1 & x \in \mathbb{Z} \text{ or } y \in \mathbb{Z} \\ 0 & \text{otherwise} \end{cases} \tag{D.22}$$

If we plot this function, we get a white grid on black background. (We could also interchange the 1 and 0 values in this definition in order to obtain a black grid on white background, in order to be consistent with our figures.) This function is, indeed, the mathematical representation of our unit grid. Now, suppose that we apply to the x,y plane a 2D transformation $\mathbf{g}(x,y)$, say, the transformation defined by $(u,v) = (2x,2y)$. As a result of this transformation the unit grid undergoes a two fold magnification, as shown in Figs. D.10(a),(b). However, if we wish to express the resulting distorted grid using its definition by Eq. (D.22), the domain transformation we have to apply to the characteristic function $f(x,y)$ is the *inverse* transformation $\mathbf{g}^{-1}(u,v)$. In other words:

Proposition D.4: If the characteristic function of an object in the x,y space is given by $f(x,y)$, then the characteristic function of the same object after it has been distorted by the transformation $\mathbf{g}(x,y)$ is given by $f(\mathbf{g}^{-1}(u,v))$. Thus, although the object itself is distorted by the original transformation $\mathbf{g}(x,y)$, its mathematical expression, $f(x,y)$, is affected by $\mathbf{g}^{-1}$, not by $\mathbf{g}$. ■

D.7 The relative point of view: object deformations vs. coordinate deformations[27]

It may be sometimes useful to focus our attention on an object $z = f(u,v)$ which is defined on the u,v plane, or any subset thereof, such as $f(u,v) = 0$. (The 1D equivalent would be the object $z = f(u)$ which is defined on u.) We may then think of the range transformation $s = t(z)$ and of the domain transformation $(u,v) = \mathbf{g}(x,y)$ (or $u = g(x)$, in the 1D case) in three different ways:[28]

(1) We can consider them as mappings which affect the entire space (2D plane or 1D line), including the object and the coordinate system.

(2) Alternatively, we can consider them as mappings which only affect the object itself, without influencing the underlying coordinate system.

(3) Finally, we can also consider them as transformations which only affect the coordinate system, while our object itself (say, a rigid physical object) remains unchanged.

For example, in the 1D case we may consider the curve $z = \cos u$ either (1) as a subset of the z,u plane that undergoes the same domain or range transformation as the entire plane; (2) as a free, flexible curve which undergoes the distortions defined by $u = g(x)$ and $s = t(z)$ within the original, unchanged coordinate system (see Fig. D.7); or (3) as a fixed, rigid curve which remains unchanged while the transformations $u = g(x)$ and $s = t(z)$ are applied to the coordinate system. In the more interesting 2D case, we may consider $z = f(u,v)$ either (1) as an object in space which undergoes the same transformation as the

---

[27] The material in this section is only used within the present appendix, but it is not required elsewhere in the book. It is given here for the sake of completeness only, and may be skipped if desired.

[28] As we saw in Remark D.14, if the object in question is a planar curve or any other subset of the plane, it can only undergo *domain* transformations since its range is reduced to the degenerate space {0}.

entire space; (2) as a free, flexible object which undergoes the distortions defined by $(u,v) = \mathbf{g}(x,y)$ (rotation, spatial scaling, etc.) and by $s = t(z)$ (vertical scaling) within the original, unchanged coordinate system; or (3) as a fixed, rigid object which remains unchanged while the transformations $(u,v) = \mathbf{g}(x,y)$ and $s = t(z)$ are applied to the coordinate system. Note that in all of these cases the transformations in question are viewed as active transformations, since they distort the given object, the coordinate system, or both.

Clearly, point of view (3) implies that the new coordinates of our given object, after the transformation has been applied, are obtained by the *inverse* transformation $\mathbf{g}^{-1}$. For example, if the transformation $\mathbf{g}$ consists of rotation by angle α, the application of $\mathbf{g}$ to the axes implies that our object's coordinates with respect to the new axes are obtained by a rotation by $-\alpha$.

The inversion effect due to the relative point of view (object or coordinate distortion) also occurs in range transformations. For instance, the effect of the range transformation $s = 2z$ is inversed between points of view (2) and (3) (our object becomes larger or smaller with respect to the vertical axis).

It is important to note, however, that the inversion effect due to the relative point of view *does not* account for the fundamental inversion effect which is inherent to domain transformations (Sec. D.6). Indeed, as we have just seen, the inversion effect due to the relative point of view affects both domain and range transformations in the same way.

All of the three points of view (1)–(3) are used in the literature; for example, [Courant88 p. 135] uses the first convention, while [Cantwell02 p. 14] uses the second convention. The first convention is used, for example, when plotting data on a logarithmic paper. The third convention is mainly used in the case of linear or affine transformations, such as rotations, scaling and translations. In the case of non-linear transformations, keeping the object unchanged while the coordinate system is being distorted may seem rather unusual; and yet, this is routinely done, for example, when one wishes to consider a given physical object in terms of polar rather than Cartesian coordinates.

More details on object transformations, coordinate changes and the conversions between them, along with several illustrated examples (for linear and affine cases only), can be found in [Foley90 Sec. 5.8].

D.8 Examples

Let us consider now a few examples to better illustrate our discussion. Most of these examples show simple transformations $(u,v) = \mathbf{g}(x,y)$ that are well known and widely used. And yet, it turns out that some of these transformations are mainly known in the literature through one particular representation (which may differ from case to case), while their other representations often remain unfamiliar and sometimes even quite surprising. But because in the present book we often need to consider a given transformation through several of its different representations, we must get acquainted with all of them, even those that are rarely if ever mentioned in the literature. These hidden facets of our transformations are revealed here through a systematic graphical comparison, which is obtained by showing each of the transformations in its different representations, keeping in all cases the same structural framework as in the model shown in Fig. D.9.

Fig. D.9 illustrates a general, arbitrary transformation $(u,v) = \mathbf{g}(x,y)$. We have intentionally chosen for this purpose a non-linear reminiscent of the rotation by angle α, in order to facilitate the understanding of the general case based on the intuition we have acquired in the case of pure rotation. Fig. D.9(a) shows the original x,y space before the application of the direct transformation $\mathbf{g}$, and Fig. D.9(b) shows the target u,v space and the distorted x,y space after the application of this transformation. Similarly, Figs. D.9(c) and (d) show the effect of the inverse transformation, $(x,y) = \mathbf{g}^{-1}(u,v)$, which corresponds to a "non-linear rotation" by angle $-\alpha$. Fig. D.9 serves us as a model, and all of the figures throughout this section are constructed according to this model.

The following list contains the main questions that can be answered by each of these figures (followed by the parts of the figure that illustrate the answer, with an arrow indicating the direction of the effect, when applicable):[29]

1. What is the effect of $\mathbf{g}$ on the standard Cartesian grid? (a) $\rightarrow$ (b)

2. What is the effect of $\mathbf{g}^{-1}$ on the standard Cartesian grid? (c) $\rightarrow$ (d)

3. To what curves $x = g_1^{-1}(u,v) = m$ and $y = g_2^{-1}(u,v) = n$ in the u,v plane does $\mathbf{g}$ map the straight lines $x = m$ and $y = n$ of the x,y plane? (a) $\rightarrow$ (b)

4. How does $\mathbf{g}^{-1}$ map these curves in the u,v plane back into the original straight lines $x = m$ and $y = n$ of the x,y plane? (a) $\leftarrow$ (b)

5. To what curves $u = g_1(x,y) = m$ and $v = g_2(x,y) = n$ in the x,y plane does $\mathbf{g}^{-1}$ map the straight lines $u = m$ and $v = n$ of the u,v plane? (c) $\rightarrow$ (d)

6. How does $\mathbf{g}$ map these curves in the x,y plane back into the original straight lines $u = m$ and $v = n$ of the u,v plane? (c) $\leftarrow$ (d)

7. How do the curves $x = c$ and $y = k$ look like when plotted on the u,v plane? (b)

8. How do the curves $u = c$ and $v = k$ look like when plotted on the x,y plane? (d)

9. How do the level lines of the surfaces $z = g_1(x,y)$ and $z = g_2(x,y)$ look like? (d)

10. How do the level lines of the surfaces $z = g_1^{-1}(u,v)$ and $z = g_2^{-1}(u,v)$ look like? (b)

11. How does $\mathbf{g}$ distort an object? (a) $\rightarrow$ (b)

12. How does $\mathbf{g}^{-1}$ undo this object deformation? (a) $\leftarrow$ (b)

---

[29] Note that parts (g) and (h) of the figures are only provided in cases where the explicit expression of $\mathbf{g}^{-1}$ is available. Consequently, they are missing in Figs. D.9 and D.18.

13. What happens when the object is rigid and can't be distorted by $\mathbf{g}$, and instead, to obtain an equivalent effect, the *inverse* transformation $\mathbf{g}^{-1}$ is applied to the x,y coordinates? (a) $\rightarrow$ (d)

14. What is the effect of $(u,v) = \mathbf{g}(x,y)$ as a domain transformation, and how does $f(\mathbf{g}(x,y))$ look like? For example, how does $\cos(g_1(x,y))$ look like? (e)

15. What is the effect of $(x,y) = \mathbf{g}^{-1}(u,v)$ as a domain transformation, and how does $f(\mathbf{g}^{-1}(u,v))$ look like? For example, how does $\cos(g_1^{-1}(x,y))$ look like? (g)

16. What is the effect of $\mathbf{g}$ as a vector field? (f)

17. What is the effect of $\mathbf{g}^{-1}$ as a vector field? (h)

Note that in all the following figures, both parts (a) and (d) show the same original x,y plane *before* the application of the transformation $\mathbf{g}$; the only difference is that part (a) shows this plane covered by the x,y coordinate net, while part (d) shows it covered by the u,v coordinate net. In fact, we could have plotted the two coordinate nets in both figures (compare with Fig. D.5(a)), but for the sake of clarity we prefer to show them separately. In a similar way, both parts (b) and (c) show the same target u,v plane *after* the application of the transformation $\mathbf{g}$, but the former shows it with the x,y coordinate net while the latter shows it with the u,v coordinate net (compare with Fig. D.5(b)). Hence, in all the following figures parts (a) and (d) represent the domain of the transformation $\mathbf{g}$ (and the range of its inverse $\mathbf{g}^{-1}$), while parts (b) and (c) represent the range of $\mathbf{g}$ (and the domain of $\mathbf{g}^{-1}$).

Remark D.18: In the figures which accompany the following examples we provide, whenever possible, the graphic representations of both $\mathbf{g}$ and its inverse $\mathbf{g}^{-1}$. However, what we want to stress by doing so is not merely the fact that a transformation and its inverse look different; this fact is, indeed, rather obvious. The important point here is that *the very same mathematical expression* $\mathbf{g}(x,y)$ (for example, $\mathbf{g}(x,y) = (2xy, y^2 - x^2)$, which is shown in Fig. D.15) may have completely different "incarnations" depending on how it is being used: When it is used as a direct transformation it bends the original unit grid into a certain curvilinear shape; when it is used as a domain (and hence, inverse) transformation it bends the unit grid (or any other rectilinear structure) into a different curvilinear shape, which corresponds, in fact, to the inverse transformation; and when it is used as a vector field, it yields yet a different graphical representation. Each of these "incarnations" simply reveals a different facet of the same transformation $\mathbf{g}(x,y)$, and one should always be sure to understand which of them is being used in any particular situation in order to avoid confusion. ∎

Finally, before we proceed to the examples themselves, let us remind here that the effect of the inverse transformations, shown in parts (c),(d) in each of the following figures, has been obtained using the technique described in Remark D.7. This allows us to draw parts (c) and (d) of the figures in *all* cases, even in cases where the explicit expression of the inverse transformation $\mathbf{g}^{-1}$ is not available (like in Figs. D.9 and D.18). However, parts (g)

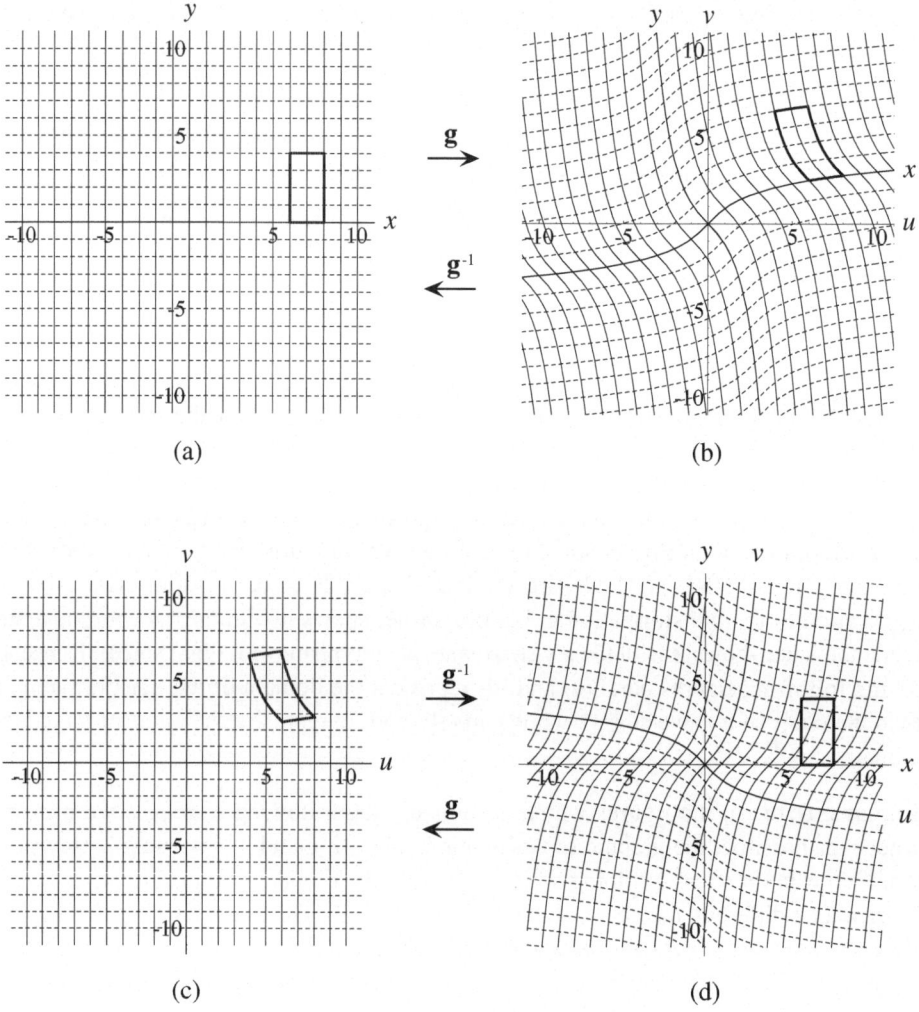

Figure D.9: A generic transformation $(u,v) = \mathbf{g}(x,y)$ (top row) and its inverse $(x,y) = \mathbf{g}^{-1}(u,v)$ (bottom row) for illustrating the general concepts. In both cases the vertical plain grid lines are mapped into the respective plain curves, and the horizontal dashed grid lines are mapped into the respective dashed curves. For the interested readers, the actual transformation used to obtain this "non-linear rotation" effect is $(u,v) = (x - \arg\sinh(y), y + \arg\sinh(x))$. Note that (b) and (c) show the *same* distorted u,v plane (the range of the transformation $\mathbf{g}$) which is only covered by different coordinate nets: the distorted x,y coordinate net in (b), and the standard, undistorted u,v coordinate net in (c). Similarly, (a) and (d) show the *same* undistorted x,y plane (the domain of the transformation $\mathbf{g}$) which is only covered by different coordinate nets: the standard, undistorted x,y coordinate net in (a), and the distorted u,v coordinate net in (d).

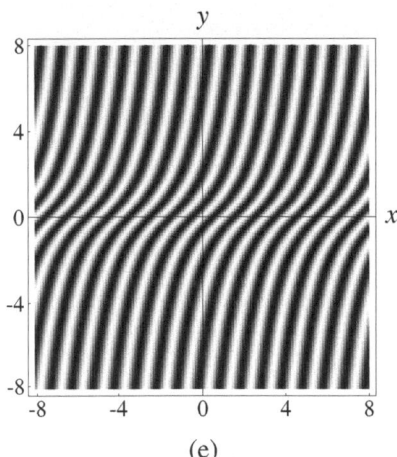
(e)

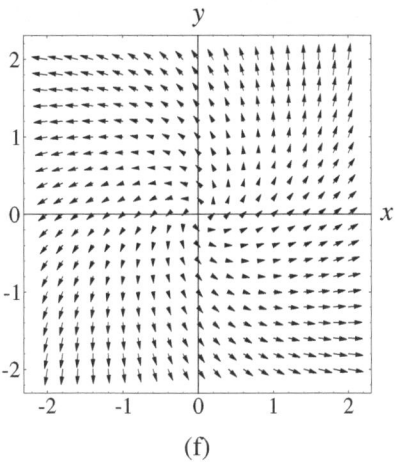
(f)

Figure D.9: (*continued.*) (e) The application of $(u,v) = (x - \arg\sinh(y), y + \arg\sinh(x))$ as a domain transformation to the vertical unit-period cosinusoidal grating $z = \cos(2\pi u)$ gives $z = \cos(4\pi[x - \arg\sinh(y)])$, a "non-linearly rotated" copy of the original function; compare with the effect of the *inverse* transformation on the plain grid lines in (d). (f) The effect of the same transformation as a vector field. Note that in the present example the explicit form of the inverse transformation is not readily available.

and (h) of the figures do not make use of this technique, and hence, as already mentioned, they are only provided when the explicit expression of $\mathbf{g}^{-1}$ is available.

Example D.1: We start with the simple linear transformation $(u,v) = \mathbf{g}(x,y) = (2x, 2y)$, that is illustrated in Figs. D.10(a),(b). Fig. D.10(a) shows the unit grid made of the lines $x = m$ and $y = n$, $m,n \in \mathbb{Z}$ before the application of this transformation, and Fig. D.10(b) shows the image of this grid after the application of the transformation. For example, the image of the vertical line $x = 1$ of Fig. D.10(a) is $u = 2$.[30] As we can clearly see, this transformation uniformly expands the plane by a factor of 2.

The effect of the inverse transformation $\mathbf{g}^{-1}$, which is given by $(x,y) = (u/2, v/2)$, is illustrated in Figs. D.10(c),(d). Fig. D.10(c) shows the unit grid made of the lines $u = m$ and $v = n$, $m,n \in \mathbb{Z}$ before the application of this inverse transformation, and Fig. D.10(d) shows the image of this grid after the application of the inverse transformation. For example, the image of the vertical line $u = 1$ of Fig. D.10(c) is $x = \frac{1}{2}$. As we can see, this transformation uniformly shrinks the u,v plane by a factor of 2.

Note, however, that when we apply the transformation $(u,v) = \mathbf{g}(x,y)$ to a given function $z = f(u,v)$ as a domain transformation, the resulting function $z = f(\mathbf{g}(x,y))$ is distorted in

---

[30] Although the present example is rather trivial, it may still be instructive to see how this result is formally obtained using Proposition D.1. Indeed, the image of the original curve $x = 1$ under $\mathbf{g}$ is simply $u/2 = 1$, which means $u = 2$.

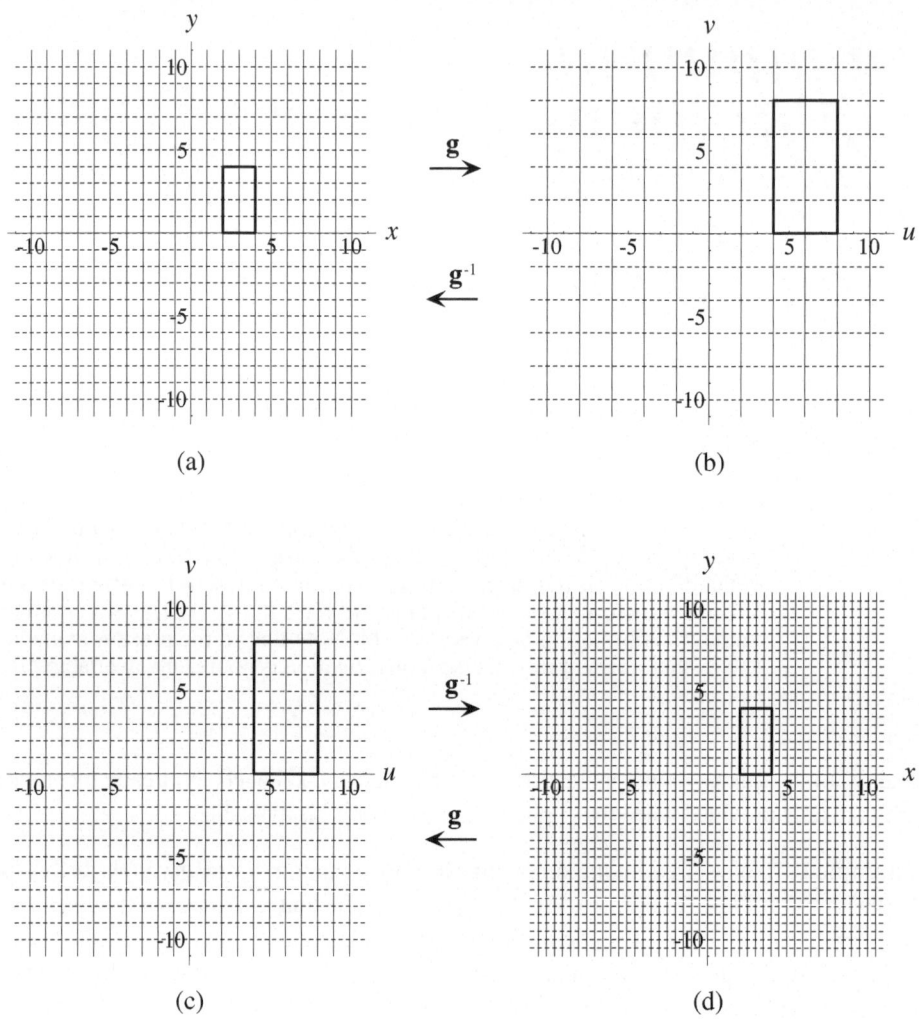

Figure D.10: (a),(b) The effect of the transformation $(u,v) = (2x,2y)$ on the unit grid
consists of uniformly expanding it by a factor of 2. (c),(d) The effect
of the inverse transformation $(x,y) = (u/2,v/2)$ on the unit grid consists
of uniformly shrinking it by a factor of 2. In both cases the vertical
plain grid lines are mapped into the respective plain lines, and the
horizontal dashed grid lines are mapped into the respective dashed
lines.

accordance with the *inverse* transformation, $\mathbf{g}^{-1}$. For example, considering the vertical unit-
period cosinusoidal grating $z = \cos(2\pi u)$, it is clear that $z = \cos(4\pi x)$ is its *shrinked*
counterpart with half the period (see Fig. D.10(e)). If we wish to *double* the period of
$z = \cos(2\pi x)$, we need to apply to it the inverse transformation $\mathbf{g}^{-1}$, which gives
$z = \cos(\pi u)$ (see Fig. D.10(g), and Remark D.4 on the variable names).

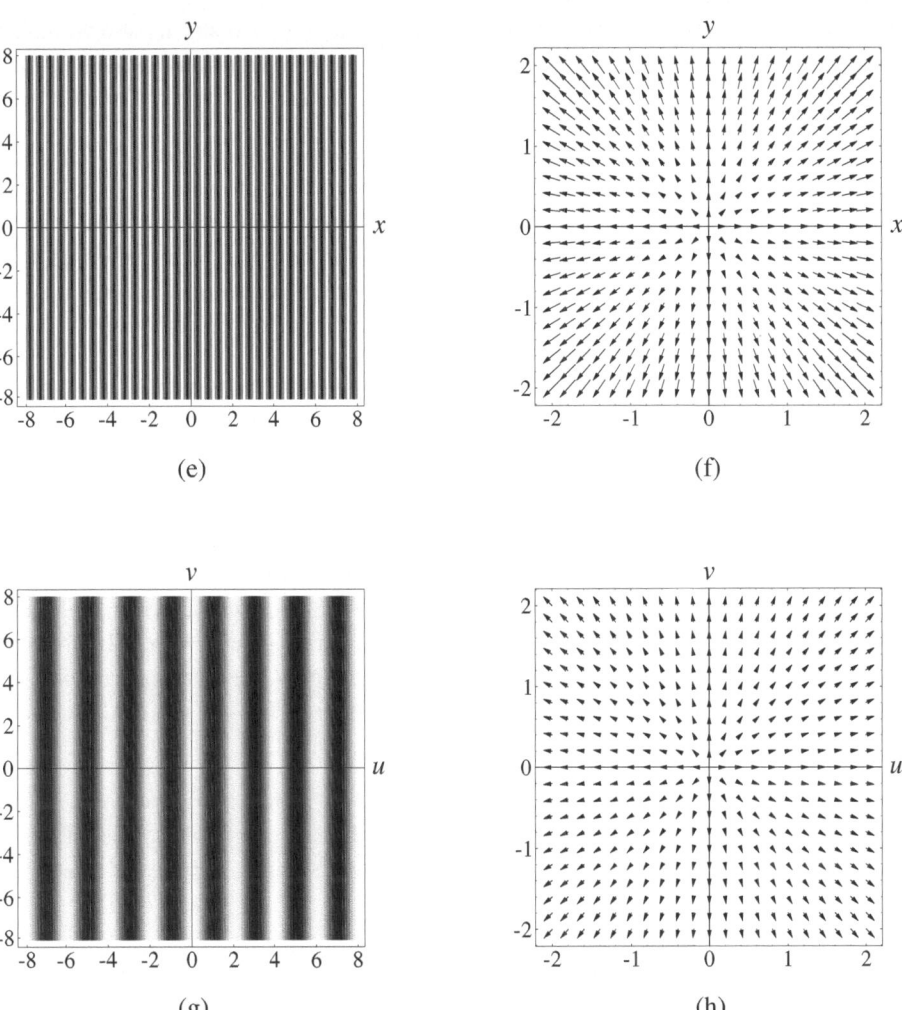

Figure D.10: (*continued.*) (e) The application of $(u,v) = (2x,2y)$ as a domain transformation to the vertical unit-period cosinusoidal grating $z = \cos(2\pi u)$ gives $z = \cos(4\pi x)$, a spatially two-fold shrinked version of the original function; compare with the effect of the *inverse* transformation on the plain grid lines in (d). (f) The effect of the same transformation as a vector field. (g) The application of the inverse transformation $(x,y) = (u/2,v/2)$ as a domain transformation to the vertical unit-period cosinusoidal grating $z = \cos(2\pi x)$ gives $z = \cos(\pi u)$, a spatially two-fold expanded version of the original function; compare with the effect of the *direct* transformation on the plain grid lines in (b). (h) The effect of the same inverse transformation as a vector field. Note that the vectors in (f) and (h) have the same radial orientations, and they only differ in their lengths; but the arrow lengths in both drawings have been scaled down in order to avoid overlappings.

Finally, the effects of the transformations **g** and **g**⁻¹ as vector fields are shown in Figs. D.10(f) and D.10(h), respectively. Note that both of these vector fields consist of radial vectors that emanate from the origin, and the difference is only in the vector lengths. ■

Example D.2: Consider now the linear transformation $(u,v) = \mathbf{g}(x,y)$ given by Eq. (D.1): $(u,v) = (x\cos\alpha - y\sin\alpha, x\sin\alpha + y\cos\alpha)$. The effect of this transformation on the original x,y plane is illustrated by Figs. D.11(a),(b). Fig. D.11(a) shows the unit grid made of the lines $x = m$ and $y = n$, $m,n \in \mathbb{Z}$ before the application of this transformation, and Fig. D.11(b) shows the image of this grid after the application of the transformation. As we can see, this transformation rotates the plane by angle α counterclockwise.

The effect of the inverse transformation, $(x,y) = (u\cos\alpha + v\sin\alpha, -u\sin\alpha + v\cos\alpha)$, is illustrated by Figs. D.11(c),(d). Clearly, this transformation rotates the plane by angle $-\alpha$.

Note that when we apply the transformation $(u,v) = \mathbf{g}(x,y)$ to a given function $z = f(u,v)$ as a domain transformation, the resulting function $z = f(\mathbf{g}(x,y))$ is distorted in accordance with the *inverse* transformation, **g**⁻¹. For example, $z = \cos(2\pi[x\cos\alpha - y\sin\alpha])$ is the counterpart of the cosinusoidal surface $z = \cos(2\pi u)$ which has been rotated by angle $-\alpha$ (see Fig. D.11(e)). In order to rotate the cosinusoidal surface $z = \cos(2\pi x)$ by angle α, we need to apply to it the inverse transformation **g**⁻¹, which gives $z = \cos(2\pi[u\cos\alpha + v\sin\alpha])$ (see Fig. D.11(g)). Similarly, the rotated version of the planar parabolic curve $y = x^2$ by angle α is given, in implicit form, by: $(-u\sin\alpha + v\cos\alpha) = (u\cos\alpha + v\sin\alpha)^2$ (see Proposition D.1).

Finally, the effects of the transformations **g** and **g**⁻¹ as vector fields are shown in Figs. D.11(f) and D.11(h), respectively. Note that both of these vector fields consist of vectors that follow spiral trajectories, and the difference between them is only in the direction of the spirals. ■

Example D.3: Let us consider, as our first non-linear example, the transformation $(u,v) = \mathbf{g}(x,y)$ defined by $(u,v) = (x^2,y^2)$. As we can see in Figs. D.12(a),(b), this transformation maps any x value into $u = x^2$, and any y value into $v = y^2$. The result is a non-linear expansion of the plane, whose effect increases as we move away from the origin. Note that this transformation "folds" the entire plane into the first quadrant; this is explained and illustrated in detail in [Callahan74 p. 232].

The effect of the inverse transformation, $(x,y) = (\sqrt{u}, \sqrt{v})$, is illustrated by Figs. D.12(c),(d). As we can clearly see, this transformation non-linearly shrinks the plane, where the shrinking effect becomes stronger as we move away from the origin. If we take into account both positive and negative values of the roots, the three quadrants that are "lost" when applying $\mathbf{g}(x,y)$ "reappear" under $\mathbf{g}^{-1}(u,v)$.

Hence, the range of the transformation $\mathbf{g}(x,y)$ and the domain of its inverse, $\mathbf{g}^{-1}(u,v)$, only consist here of the first quadrant of the plane. Such "losses" in the range or in the domain occur quite often in non-linear transformations.

This example can be also used to illustrate Remarks D.5 and D.6 in Sec. D.4.1: The level lines of the surfaces $z = x^2$ and $z = y^2$ (the two components of the *original* transformation $\mathbf{g}(x,y)$) are represented, respectively, by the plain and dashed lines of Fig. D.12(d), a figure which illustrates the effect of the *inverse* transformation $\mathbf{g}^{-1}(u,v)$. Note that these level lines become closer to each other as we move away from the origin, reflecting the increasing steepness of the surfaces $z = x^2$ and $z = y^2$ (remember that level lines represent constant differences in the altitude z). Similarly, the level lines of the surfaces $z = \sqrt{u}$ and $z = \sqrt{v}$ (the two components of the *inverse* transformation $\mathbf{g}^{-1}(u,v)$) are represented, respectively, by the plain and dashed lines of Fig. D.12(b), a figure which illustrates the effect of the *direct* transformation $\mathbf{g}(x,y)$.

When we apply the transformation $(u,v) = \mathbf{g}(x,y)$ to a given function $z = f(u,v)$ as a domain transformation, the resulting function $z = f(\mathbf{g}(x,y))$ is distorted in accordance with the *inverse* transformation, $\mathbf{g}^{-1}$. For example, the 2D function $z = \cos(2\pi x^2)$ is a non-linearly *shrinked* version of $z = \cos(2\pi u)$, whose corrugations become narrower as we move away from the origin, just like the vertical lines of Fig. D.12(d) (which are, indeed, as we have just seen above, the level lines of $z = x^2$). This is clearly shown in Fig. D.12(e). If we wish to obtain a non-linearly *expanded* version of $z = \cos(2\pi x)$, whose corrugations behave like the vertical lines of Fig. D.12(b), we need to apply to it the inverse transformation $\mathbf{g}^{-1}$, which gives $z = \cos(2\pi\sqrt{u})$.

Finally, the effects of the transformations $\mathbf{g}$ and $\mathbf{g}^{-1}$ as vector fields are shown in Figs. D.12(f) and D.12(h), respectively. ■

Remark D.19: To visualize the level lines $z = n$ of a surface $z = f(x,y)$ one may draw the surface $\cos(2\pi f(x,y))$; its maxima, which are given by the locus $f(x,y) = n$, $n \in \mathbb{Z}$ in the x,y plane, represent these level lines. ■

Example D.4: Consider the well known Cartesian to polar coordinate transformation $(r,\theta) = (\sqrt{x^2+y^2}, \arctan(y/x))$. (Note that we have replaced here our usual range axis names u,v by the more familiar ones, r,θ.) Although this transformation is widely used in mathematics and in engineering, it still may reserve us some surprises. To start with, let us consider the effect of this transformation on the original x,y plane. This effect is shown in Figs. D.13(a),(b): Part (a) of the figure shows the unit grid made of the lines $x = m$ and $y = n$, $m,n \in \mathbb{Z}$ before the application of this transformation, while part (b) of the figure shows the image of this grid after the application of the transformation.

The effect of the inverse transformation, the polar to Cartesian coordinate transformation $(x,y) = (r\cos\theta, r\sin\theta)$, is illustrated by Figs. D.13(c),(d). As we can see, this transformation maps the vertical lines $r = m$, $m \in \mathbb{Z}$ into the concentric circles $\sqrt{x^2+y^2} = m$, and the horizontal dashed lines $\theta = n$, $n \in \mathbb{Z}$ into the radial dashed lines $\arctan(y/x) = n$.[31]

---

[31] Note that some references use to limit the values of r and θ to $r \geq 0$ and $0 \leq \theta < 2\pi$, so that to each point (x,y) except for the origin there corresponds a unique point (r,θ) and vice versa. However, this restriction is not always advantageous [Colley98 pp. 67–69], and we prefer not to impose it here. Thus, the vertical lines $r = m$ and $r = -m$ are both mapped into the same circle; similarly, the horizontal lines $\theta = n$ and $\theta = 2\pi k n$ for any $k \in \mathbb{Z}$ are mapped into the same radial line.

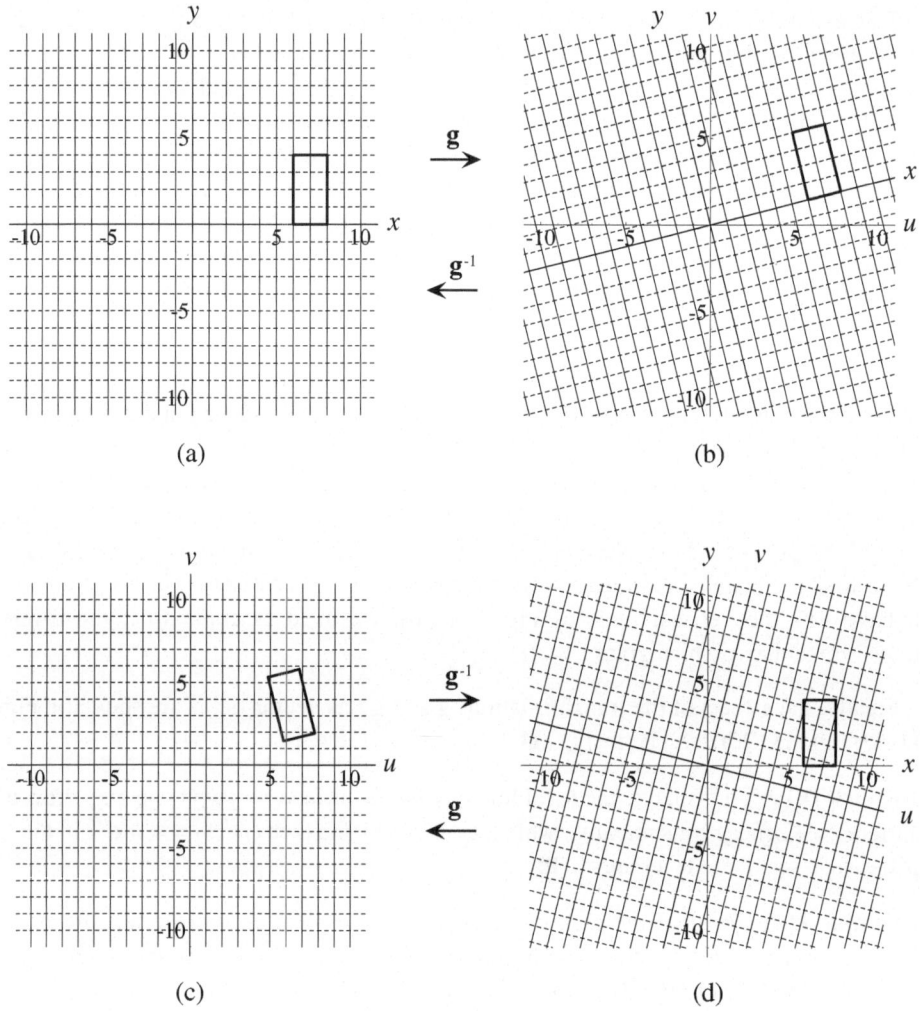

(a)   (b)

(c)   (d)

Figure D.11: (a),(b) The effect of the transformation $(u,v) = (x\cos\alpha - y\sin\alpha,$ $x\sin\alpha + y\cos\alpha)$ on the unit grid consists of rotating it counter-clockwise by angle α. (c),(d) The effect of the inverse transformation $(x,y) = (u\cos\alpha + v\sin\alpha, -u\sin\alpha + v\cos\alpha)$ on the unit grid consists of rotating it by angle $-\alpha$. In both cases the vertical plain grid lines are mapped into the respective plain lines, and the horizontal dashed grid lines are mapped into the respective dashed lines.

Interestingly, in this case it is the effect of the *inverse* transformation $\mathbf{g}^{-1}$ on the unit grid that is widely known (Figs. D.13(c),(d)), while the effect of $\mathbf{g}$ itself on the unit grid looks quite unfamiliar (Figs. D.13(a),(b)), and is rarely if ever mentioned in the literature. It can be easily verified, using Eqs. (D.7) and (D.8), that $\mathbf{g}$ maps the vertical grid lines $x = m$, $m \in \mathbb{Z}$ in Fig. D.13(a) into the family of curves $r\cos\theta = m$, namely, the secant curves

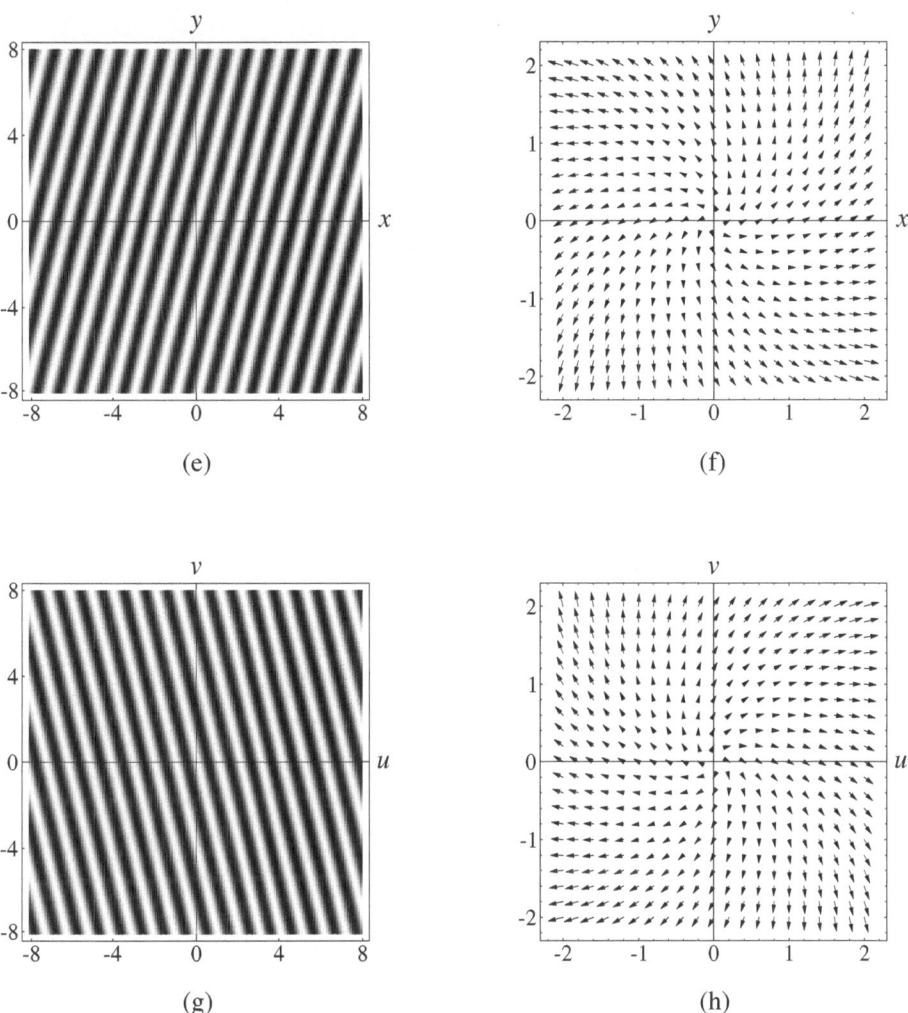

Figure D.11: (*continued.*) (e) The application of $(u,v) = (x\cos\alpha - y\sin\alpha,\ x\sin\alpha +$
$y\cos\alpha)$ as a domain transformation to the vertical unit-period
cosinusoidal grating $z = \cos(2\pi u)$ gives $z = \cos(4\pi[x\cos\alpha - y\sin\alpha])$,
a copy of the original function that is rotated by angle $-\alpha$; compare
with the effect of the *inverse* transformation on the plain grid lines in
(d). (f) The effect of the same transformation as a vector field; note
the left-oriented spiral shape of this vector field. (g) The application of
the inverse transformation $(x,y) = (u\cos\alpha + v\sin\alpha,\ -u\sin\alpha + v\cos\alpha)$
as a domain transformation to the vertical unit-period cosinusoidal
grating $z = \cos(2\pi x)$ gives $z = \cos(2\pi[u\cos\alpha + v\sin\alpha])$, a copy of the
original function that is rotated by angle α; compare with the effect of
the *direct* transformation on the plain grid lines in (b). (h) The effect
of the same inverse transformation as a vector field; note the right-
oriented spiral shape of this vector field.

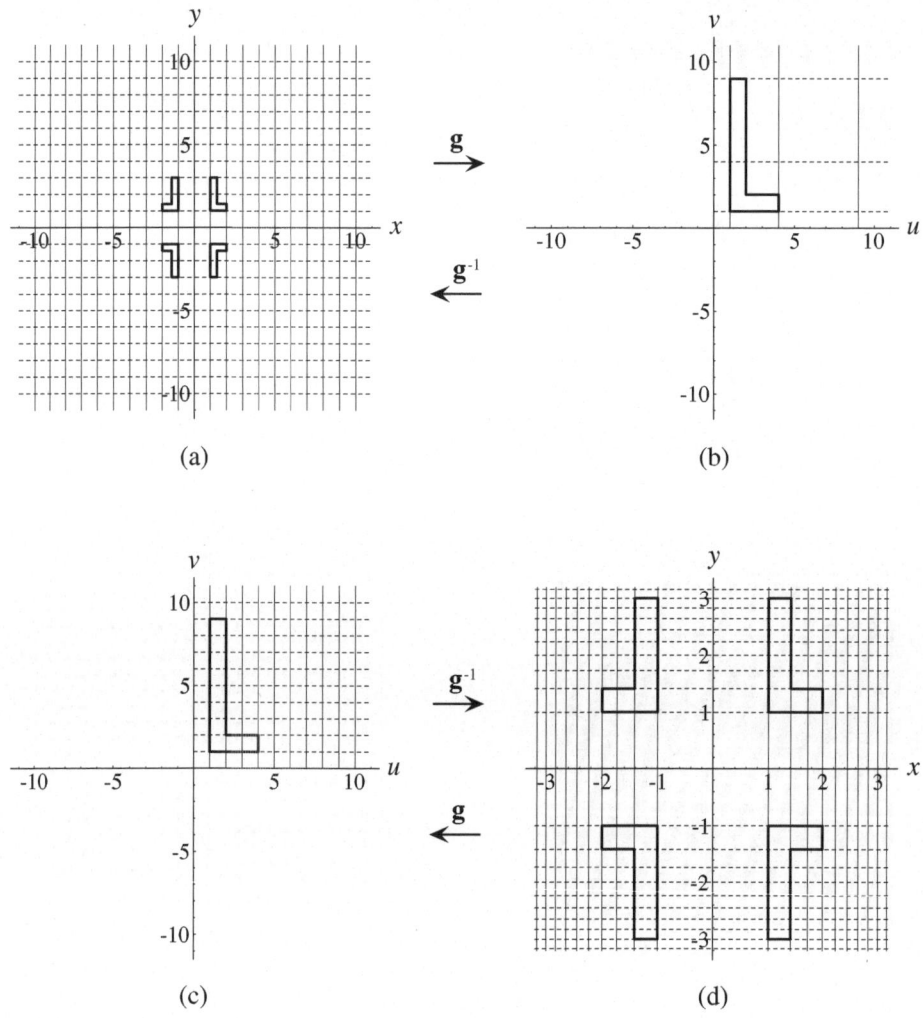

Figure D.12: (a),(b) The effect of the transformation $(u,v) = (x^2, y^2)$ on the unit grid consists of a non-linear expansion. Each of the four objects in (a) is mapped into the same image in (b). (c),(d) The effect of the inverse transformation $(x,y) = (\sqrt{u}, \sqrt{v})$ on the unit grid consists of a non-linear contraction. In both cases the vertical plain grid lines are mapped into the respective plain lines, and the horizontal dashed grid lines are mapped into the respective dashed lines. Note that part (d) has been magnified to better show details.

$r = m \sec\theta$; similarly, **g** maps the dashed horizontal grid lines $y = n$, $n \in \mathbb{Z}$ in Fig. D.13(a) into the family of dashed curves $r\sin\theta = n$, which are the cosecant curves $r = n\csc\theta$. (The shapes of the multi-branched curves $y = \sec x$ and $y = \csc x$ can be found in any mathematical handbook such as [Bronshtein79 p. 69] or [Harris98 p. 301].)

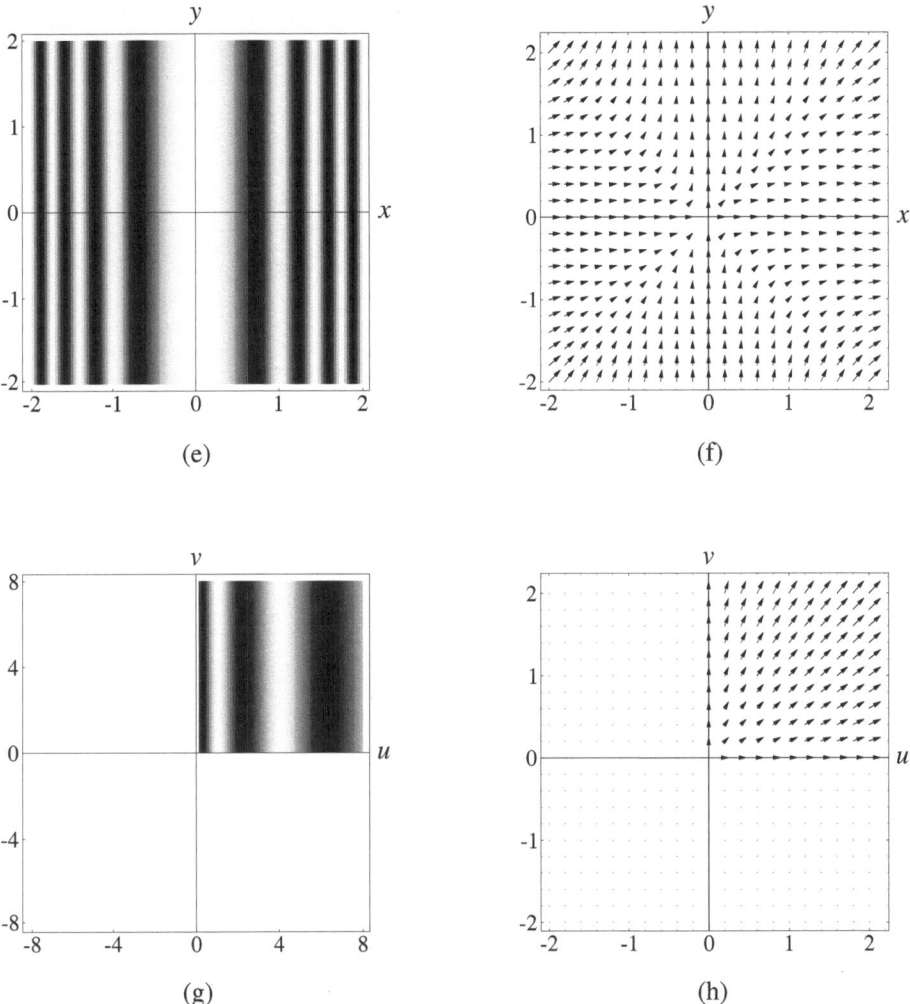

Figure D.12: (*continued.*) (e) The application of $(u,v) = (x^2,y^2)$ as a domain transformation to the vertical unit-period cosinusoidal grating $z = \cos(2\pi u)$ gives $z = \cos(2\pi x^2)$, a non-linearly shrinked version of the original function; compare with the effect of the *inverse* transformation on the plain grid lines in (d). (f) The effect of the same transformation as a vector field. (g) The application of the inverse transformation $(x,y) = (\sqrt{u}, \sqrt{v})$ as a domain transformation to the vertical unit-period cosinusoidal grating $z = \cos(2\pi x)$ gives $z = \cos(2\pi\sqrt{u})$, a non-linearly expanded version of the original function; compare with the effect of the *direct* transformation on the plain grid lines in (b). (h) The effect of the same inverse transformation as a vector field. Note that part (e) has been magnified to better show details.

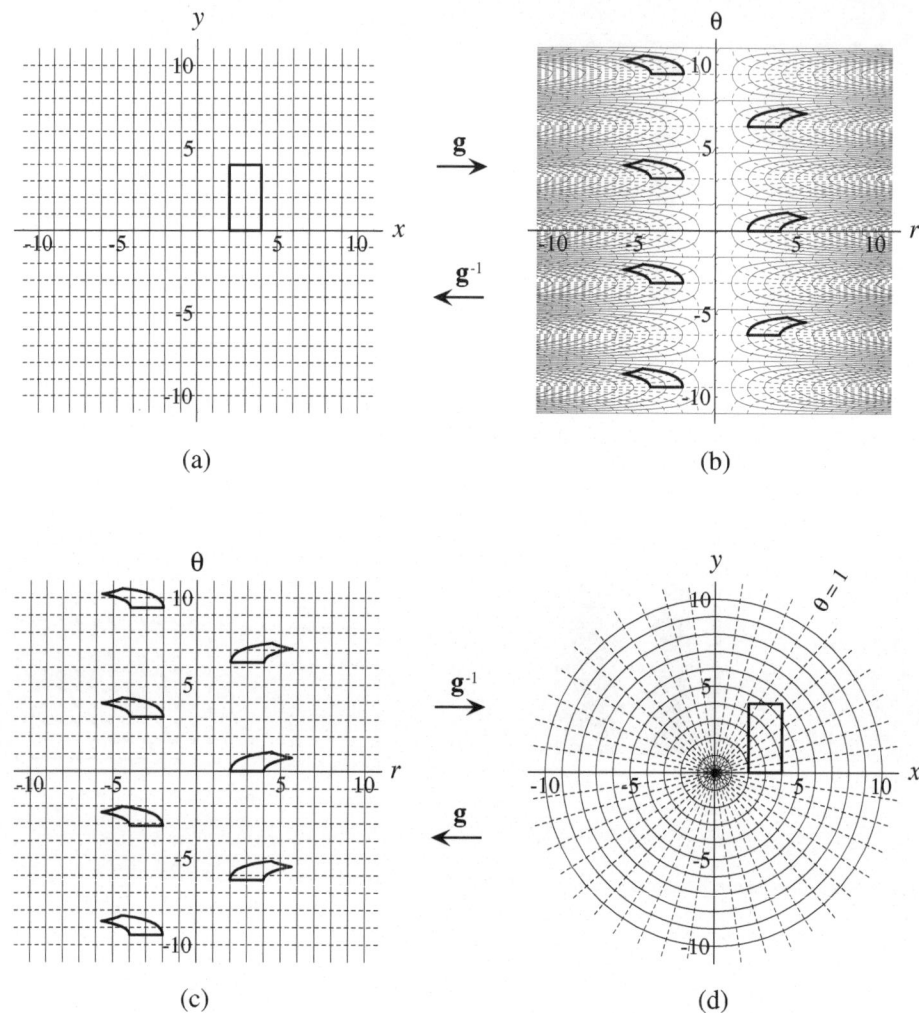

Figure D.13: (a),(b) The effect of the transformation $(r, \theta) = (\sqrt{x^2 + y^2}, \arctan(y/x))$ on the unit grid: Each vertical line $x = m$ is mapped into a secant curve $r = m \sec\theta$, and each horizontal line $y = n$ is mapped into a cosecant curve $r = n \csc\theta$. Note that the rectangular object is mapped into infinitely many images. (c),(d) The effect of the inverse transformation $(x, y) = (r\cos\theta, r\sin\theta)$ on the unit grid: Each vertical line is mapped into a circle, and each horizontal line is mapped into a radial line. Note that the horizontal line $\theta = 1$ is mapped into the radial line whose angle is 1 radian (i.e. $180°/\pi \approx 57.2958°$), the line $\theta = 2$ is mapped into the line at 2 radians, and so forth.[32]

---

[32] Note that the lines $\theta = -10$, $\theta = -9$, ... , $\theta = 10$ of Fig. D.13(c) are mapped into almost equispaced radial lines in Fig. D.13(d). This happens since by pure coincidence 11 radians almost exactly coincide with an integer multiple of 270° (3 quadrants), so that the images of the lines $\theta = k$ and $\theta = k + 22$ fall almost precisely on the same radial line. But in reality, the image of the infinite family of lines $\theta = k$, $k \in \mathbb{Z}$ is everywhere dense in the x,y plane (see the lemma in [Arnold73 pp. 163–164]).

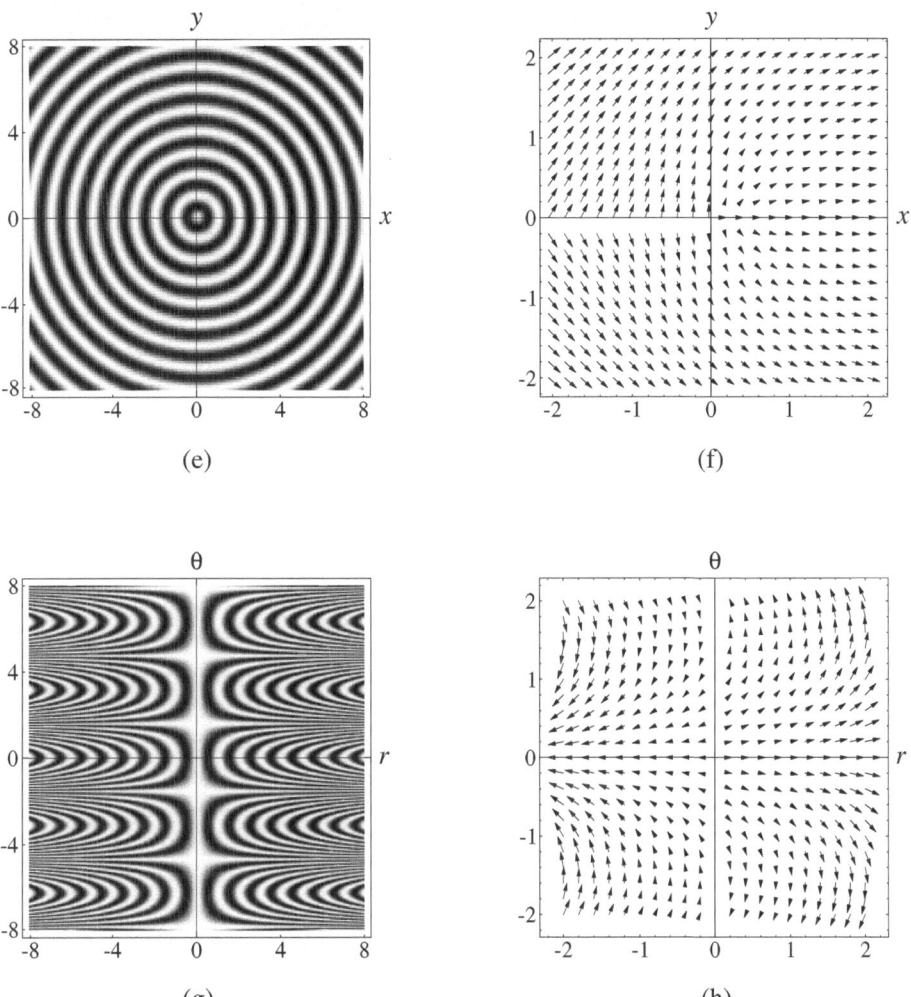

Figure D.13: (*continued.*) (e) The application of $(r,\theta) = (\sqrt{x^2 + y^2}, \arctan(y/x))$ as a domain transformation to the vertical unit-period cosinusoidal grating $z = \cos(2\pi r)$ gives $z = \cos(2\pi\sqrt{x^2 + y^2})$, a circular version of the original function; compare with the effect of the *inverse* transformation on the plain grid lines in (d). (f) The effect of the same transformation as a vector field. (g) The application of the inverse transformation $(x,y) = (r\cos\theta, r\sin\theta)$ as a domain transformation to the vertical unit-period cosinusoidal grating $z = \cos(2\pi x)$ gives $z = \cos(2\pi r\cos\theta)$, a curved version of the original function whose corrugations have the shape of multi-branch secant curves; compare with the effect of the *direct* transformation on the plain grid lines in (b). (h) The effect of the same inverse transformation as a vector field.

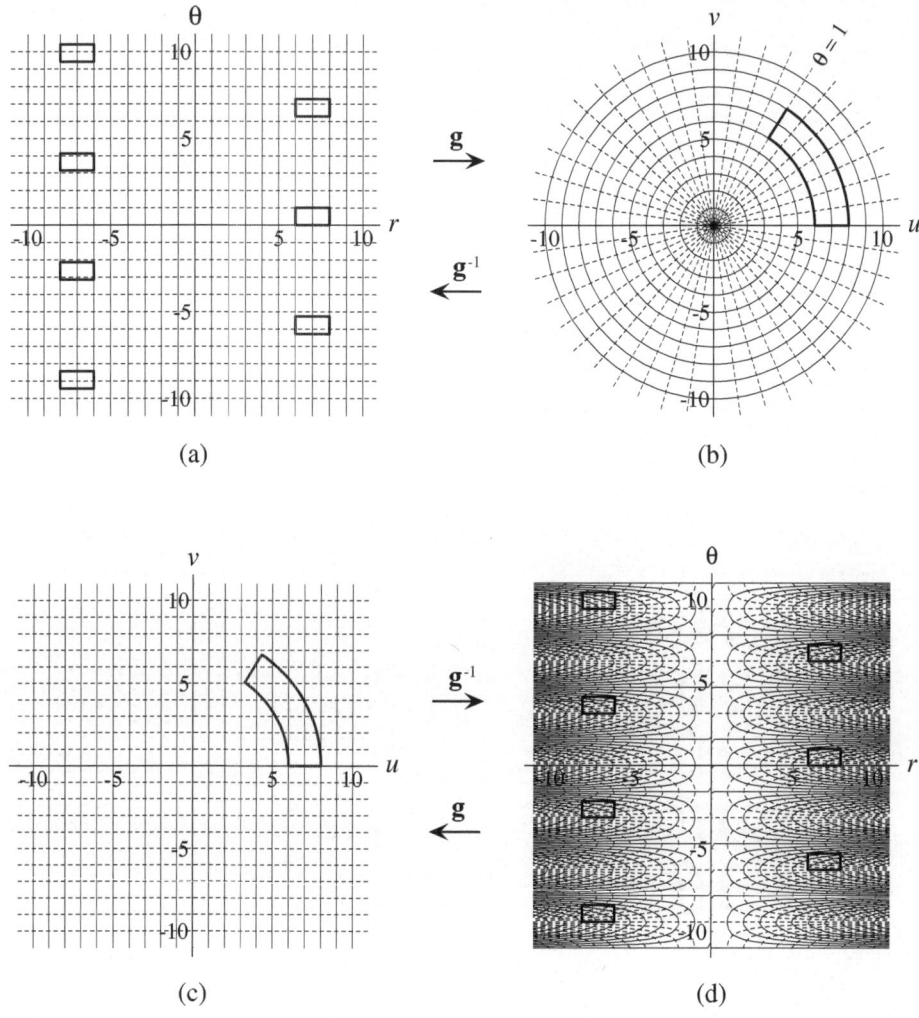

(a)   (b)

(c)   (d)

Figure D.14: (a),(b) The effect of the transformation $(u,v) = (r\cos\theta, r\sin\theta)$ on the
unit grid: Each vertical line is mapped into a circle, and each
horizontal line is mapped into a radial line. Note that infinitely many
objects in (a) are mapped into the same object in (b). (c),(d) The
effect of the inverse transformation $(r,\theta) = (\sqrt{u^2 + v^2}, \arctan(v/u))$ on
the unit grid: Each vertical line is mapped into a multi-branch secant
curve, and each horizontal line is mapped into a multi-branch
cosecant curve. All angles are measured in radians.

So far, in all the previous examples, we described the effect of a transformation $\mathbf{g}(x,y)$ by
means of its influence on the unit grid of the x,y plane. However, we can also describe the
effect of $\mathbf{g}(x,y)$ by studying its influence on other curves in the x,y plane (see Sec. D.4.2).

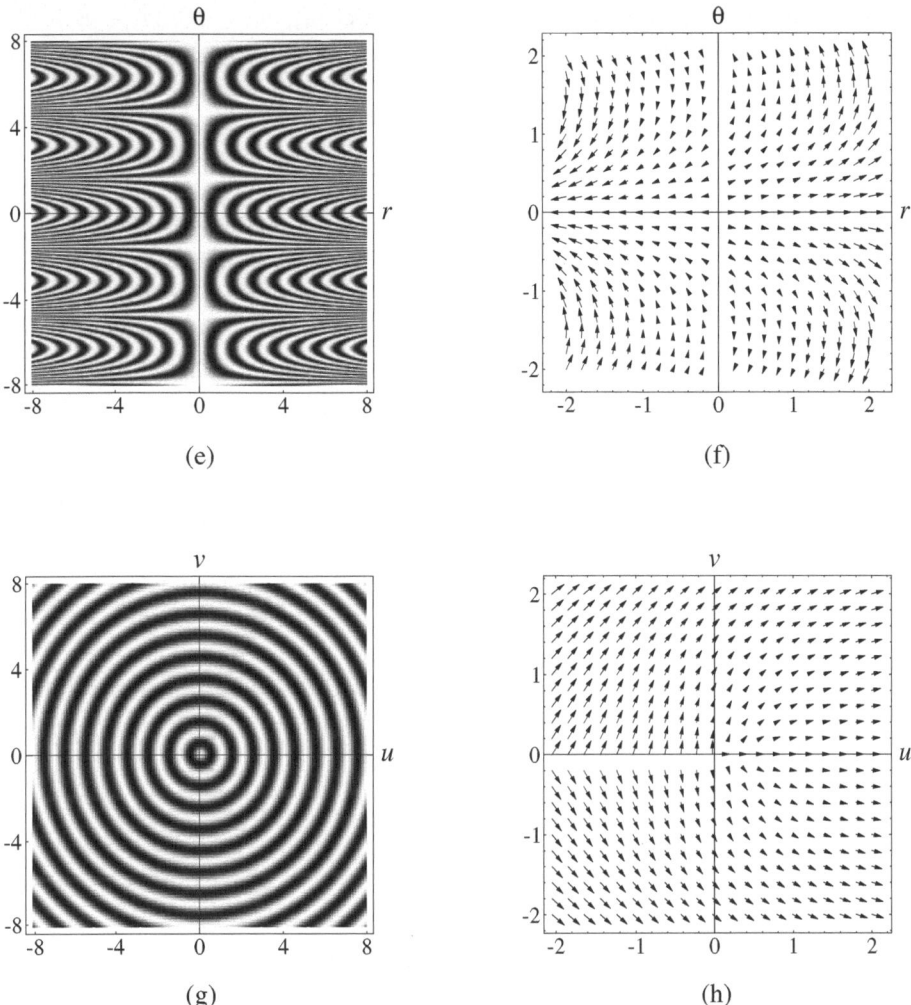

Figure D.14: (*continued.*) (e) The application of $(u,v) = (r\cos\theta, r\sin\theta)$ as a domain transformation to the vertical unit-period cosinusoidal grating $z = \cos(2\pi u)$ gives $z = \cos(2\pi r\cos\theta)$, a curved version of the original function whose corrugations have the shape of multi-branch secant curves; compare with the effect of the *inverse* transformation on the plain grid lines in (d). (f) The effect of the same transformation as a vector field. (g) The application of the inverse transformation $(r,\theta) = (\sqrt{u^2 + v^2},$ $\arctan(v/u))$ as a domain transformation to the vertical unit-period cosinusoidal grating $z = \cos(2\pi r)$ gives $z = \cos(2\pi\sqrt{u^2 + v^2})$, a circular version of the original function; compare with the effect of the *direct* transformation on the plain grid lines in (b). (h) The effect of the same inverse transformation as a vector field.

For example, our transformation $(r,\theta) = (\sqrt{x^2 + y^2}, \arctan(y/x))$, which maps vertical or horizontal straight lines into secant or cosecant curves, also maps the circles $\sqrt{x^2 + y^2} = m$, $m \in \mathbb{Z}$ into the vertical lines $r = m$, and the radial lines $\arctan(y/x) = n$, $n \in \mathbb{Z}$ into the horizontal lines $\theta = n$. This can be seen by drawing circles and radial lines on top of Fig. D.13(a) and tracing out their images in Fig. D.13(b), or, equivalently, by considering the image under $\mathbf{g}$ of the circles and of the radial lines of Fig. D.13(d), as shown in Fig. D.13(c). Note that the mapping from Fig. D.13(d) back into Fig. D.13(c) corresponds to the inverse of $\mathbf{g}^{-1}$, which is $\mathbf{g}$ itself. Interestingly, this effect of the Cartesian to polar coordinate change $(r,\theta) = (\sqrt{x^2 + y^2}, \arctan(y/x))$ is much more familiar than its effect on the Cartesian x,y grid that is shown in Figs. D.13(a),(b). This is, in fact, the way this transformation is described in the literature (see, for example, [Courant88 pp. 137–139]). Indeed, this case is a classical example of a transformation that is not illustrated in the literature by its effect on the standard x,y grid, but rather by its effect on some other curves in the x,y plane (see representation (8) in Sec. D.2).

Note that when we apply the transformation $(u,v) = \mathbf{g}(x,y)$ to a given function $z = f(u,v)$ as a domain transformation, the resulting function $z = f(\mathbf{g}(x,y))$ is distorted in accordance with the *inverse* transformation, $\mathbf{g}^{-1}$. For example, after the application of $\mathbf{g}(x,y)$ the straight vertical corrugations of the cosinusoidal function $z = \cos(2\pi r)$ in the r,θ space become in the x,y space, in the resulting function $z = \cos(2\pi\sqrt{x^2 + y^2})$, a family of concentric circular corrugations that surround the origin (see Fig. D.13(e), and compare with Fig. D.13(d)).[33] On the other hand, applying the inverse transformation $\mathbf{g}^{-1}$ to the function $z = \cos(2\pi x)$ gives the unfamiliar function $z = \cos(2\pi r\cos\theta)$ which is shown in Fig. D.13(g); compare also with Fig. D.13(b).

Finally, the effects of the transformations $\mathbf{g}$ and $\mathbf{g}^{-1}$ as vector fields are shown in Figs. D.13(f) and D.13(h), respectively. As we can see, both of these vector fields look rather unfamiliar, and they are rarely if ever mentioned in the literature.

Because of the particular importance of the transformation discussed in this example, we also show its inverse (the polar to Cartesian coordinate conversion) as a transformation in its own right in a separate figure, Fig. D.14. Note that Fig. D.14 shows how the inverse transformation distorts a given rectangular object, whereas Fig. D.13 shows a much less familiar result, the way in which $\mathbf{g}(x,y)$ itself distorts a rectangular object. ∎

Example D.5: Figs. D.15(a),(b) illustrate the influence on the x,y plane of the transformation $\mathbf{g}(x,y)$ given by $(u,v) = (2xy, y^2 - x^2)$ [Courant88 pp. 136–137]. The vertical lines $x = m$, $m \in \mathbb{Z}$ in Fig. D.15(a) are mapped by this transformation into the family of top-opened parabolas $v = \frac{1}{4m^2}u^2 - m^2$, while the horizontal dashed lines $y = n$, $n \in \mathbb{Z}$ in Fig. D.15(a) are mapped into the family of bottom-opened parabolas $v = n^2 - \frac{1}{4n^2}u^2$ shown in Fig. D.15(b) by dashed lines.[34]

---

[33] This explains, indeed, why in some references the transformation $\mathbf{g}(x,y)$ is called *polar to Cartesian* rather than *Cartesian to polar* coordinate transformation (see Remark D.16).

[34] Note that the vertical lines $x = m$ and $x = -m$ are both mapped into the same top-opened parabola; similarly, the horizontal lines $y = n$ and $y = -n$ are both mapped into the same bottom-opened parabola.

In this case the explicit expression of the inverse transformation is quite complicated: $(x,y) = \left(\sqrt{\left(\sqrt{u^2+v^2}-v\right)/2}, \sqrt{\left(\sqrt{u^2+v^2}+v\right)/2}\right)$. We therefore use this example to illustrate how the influence of the inverse transformation on the u,v plane can be found without even knowing its explicit form (see Remark D.7): As shown in Figs. D.15(c),(d), the vertical lines $u = m$, $m \in \mathbb{Z}$ are mapped by the inverse transformation into the family of hyperbolas which are given according to Eq. (D.11) by $u = 2xy = m$, namely, $y = \frac{m}{2x}$. Similarly, the horizontal dashed lines $v = n$, $n \in \mathbb{Z}$ are mapped by the inverse transformation into the dashed family of hyperbolas which are given according to Eq. (D.12) by $v = y^2 - x^2 = n$, namely, $y = \pm\sqrt{n + x^2}$. Thus, we avoided using here the explicit form of $\mathbf{g}^{-1}$.

Note that just as we did in the previous example, we may also consider the effects of $\mathbf{g}(x,y)$ and of its inverse $\mathbf{g}^{-1}(u,v)$ on *curved* line families. For example, we can see by comparing Fig. D.15(d) with Fig. D.15(c) that $\mathbf{g}(x,y)$ (the inverse of $\mathbf{g}^{-1}(u,v)$) maps the hyperbolic lines $2xy = m$, $m \in \mathbb{Z}$ into the vertical lines $u = m$ and the dashed hyperbolic lines $y^2 - x^2 = n$, $n \in \mathbb{Z}$ into the dashed horizontal lines $v = n$. Similarly, we can see by comparing Fig. D.15(b) with Fig. D.15(a) that the inverse transformation $\mathbf{g}^{-1}(u,v)$ maps the top-opened parabolas of Fig. D.15(b) into the vertical lines of Fig. D.15(a) and the dashed, bottom-opened parabolas into horizontal lines.

This example can be also used to illustrate Remarks D.5 and D.6 in Sec. D.4.1: The level lines of the surfaces $z = 2xy$ and $z = y^2 - x^2$ (the two components of the *original* transformation $\mathbf{g}(x,y)$) are represented, respectively, by the plain and dashed hyperbolas of Fig. D.15(d), a figure which illustrates the effect of the *inverse* transformation $\mathbf{g}^{-1}(u,v)$. Similarly, the level lines of the surfaces $z = \sqrt{\left(\sqrt{u^2+v^2}-v\right)/2}$ and $z = \sqrt{\left(\sqrt{u^2+v^2}+v\right)/2}$ (the two components of the *inverse* transformation $\mathbf{g}^{-1}(u,v)$) are represented, respectively, by the plain and dashed parabolas of Fig. D.15(b), a figure which illustrates the effect of the *direct* transformation $\mathbf{g}(x,y)$.

Once again, note that when we apply the transformation $(u,v) = \mathbf{g}(x,y)$ to a given function $z = f(u,v)$ as a domain transformation, the resulting function $z = f(\mathbf{g}(x,y))$ is distorted in accordance with the *inverse* transformation, $\mathbf{g}^{-1}$. For example, following the application of $\mathbf{g}(x,y)$ the straight vertical corrugations of the function $z = \cos(2\pi u)$ become in the resulting function, $z = \cos(4\pi xy)$, a family of hyperbolic corrugations, just like the plain curves in Fig. D.15(d) (which are, indeed, as we have just seen above, the level lines of $z = 2xy$). This is clearly shown in Fig. D.15(e). On the other hand, applying the inverse transformation $\mathbf{g}^{-1}$ to the function $z = \cos(2\pi x)$ gives a family of parabolic corrugations (see Fig. D.15(g)), just like the plain curves in Fig. D.15(b).

Finally, the effects of the transformations $\mathbf{g}$ and $\mathbf{g}^{-1}$ as vector fields are shown in Figs. D.15(f) and D.15(h), respectively. While the vector field of $\mathbf{g}$ is widely known (see, for example, Fig. 4.14(a) in Chapter 4), the vector field of $\mathbf{g}^{-1}$ looks rather unfamiliar. ∎

Example D.6: Suppose we wish to construct a transformation $\mathbf{g}(x,y)$ that bends vertical lines into right-opened parabolas. This transformation shifts each point (x,y) to the right

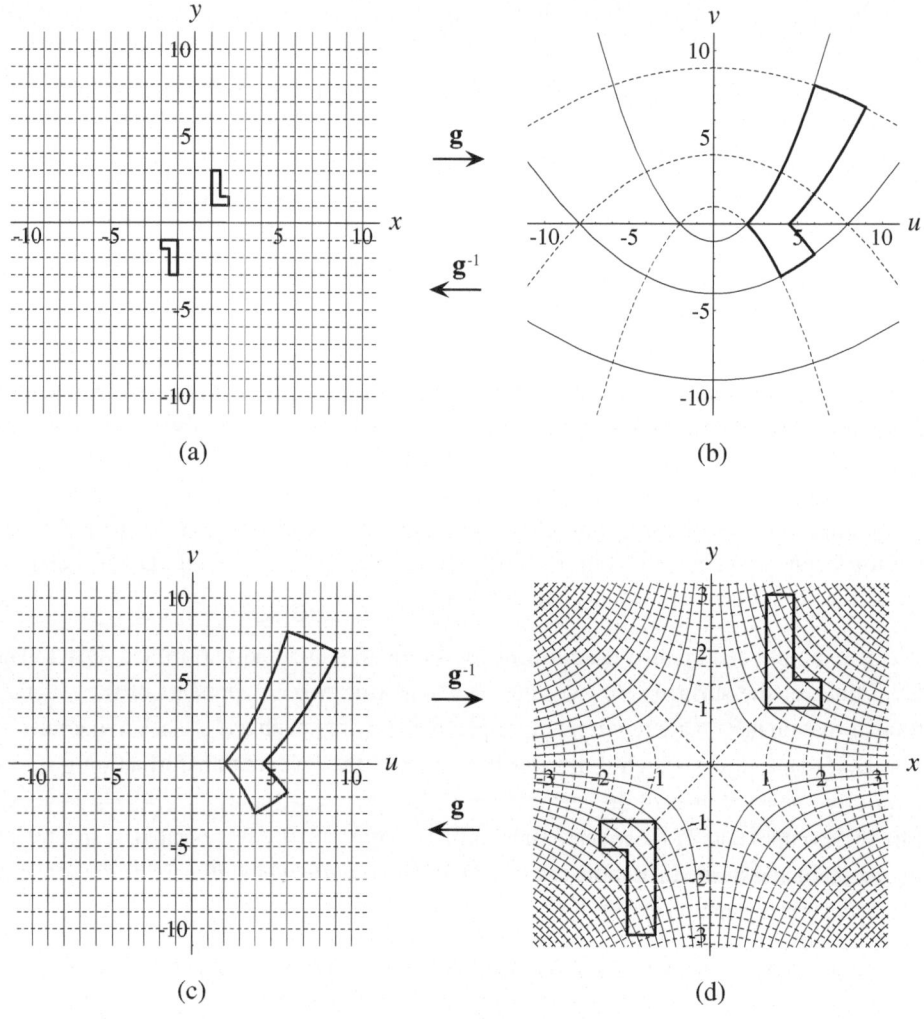

(a)

(b)

(c)

(d)

Figure D.15: (a),(b) The effect of the transformation $(u,v) = (2xy, y^2 - x^2)$ on the unit grid: Each vertical line is mapped into a top-opened parabola, while each horizontal line is mapped into a bottom-opened parabola. Both of the objects in (a) are mapped into the same object in (b). (c),(d) The effect of the inverse transformation on the unit grid: Each vertical line is mapped into a hyperbola which is asymptotic to the axes, while each horizontal line is mapped into a hyperbola which is asymptotic to the main diagonals. Note that part (d) has been magnified to better show details.

by ay^2, a being a positive constant, while the y coordinate remains unchanged (see Fig. 3.4 in Chapter 3). It is clear, therefore, that this transformation is given by $(u,v) = (x + ay^2, y)$. And indeed, as we can clearly see by comparing Fig. D.16(a) with Fig. D.16(b), this

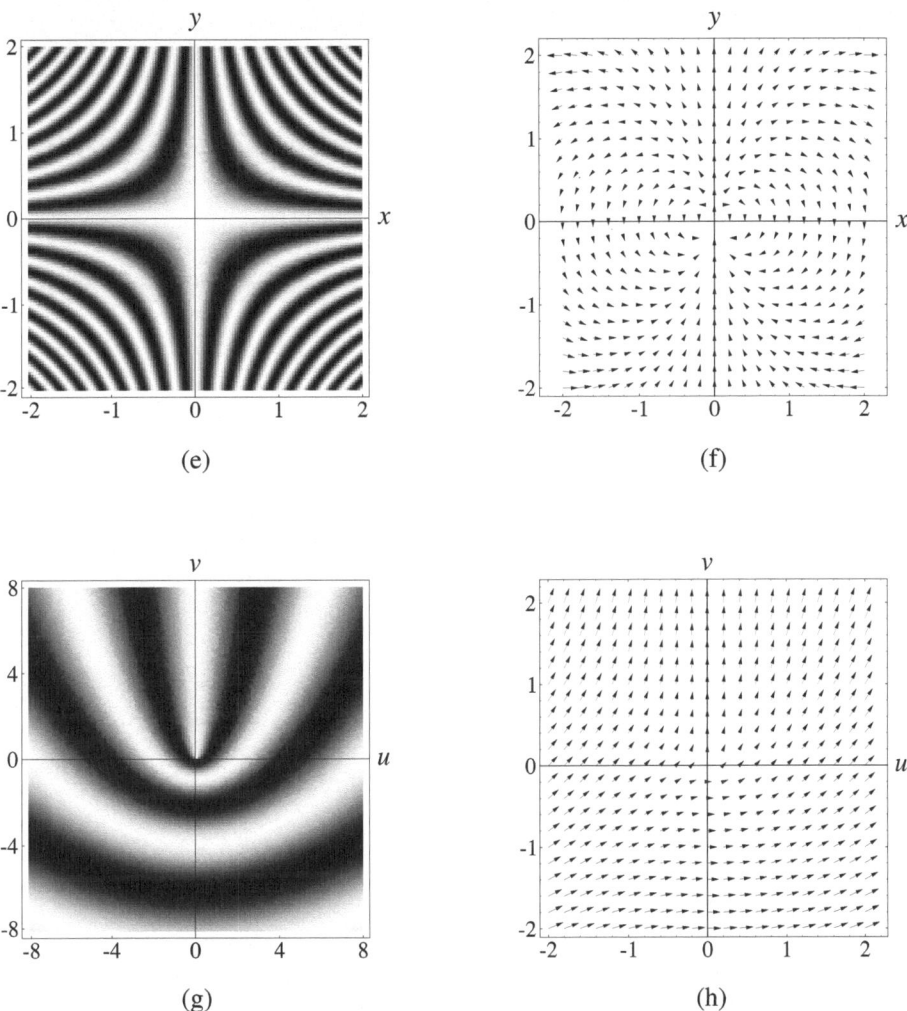

(e)

(f)

(g)

(h)

Figure D.15: (*continued.*) (e) The application of $(u,v) = (2xy, y^2 - x^2)$ as a domain transformation to the vertical unit-period cosinusoidal grating $z = \cos(2\pi u)$ gives $z = \cos(4\pi xy)$, a hyperbolic version of the original function; compare with the effect of the *inverse* transformation on the plain grid lines in (d). (f) The effect of the same transformation as a vector field. (g) The application of the inverse transformation as a domain transformation to the vertical unit-period cosinusoidal grating $z = \cos(2\pi x)$ gives a parabolic version of the original function; compare with the effect of the *direct* transformation on the plain grid lines in (b). (h) The effect of the same inverse transformation as a vector field. Note that part (e) has been magnified to better show details.

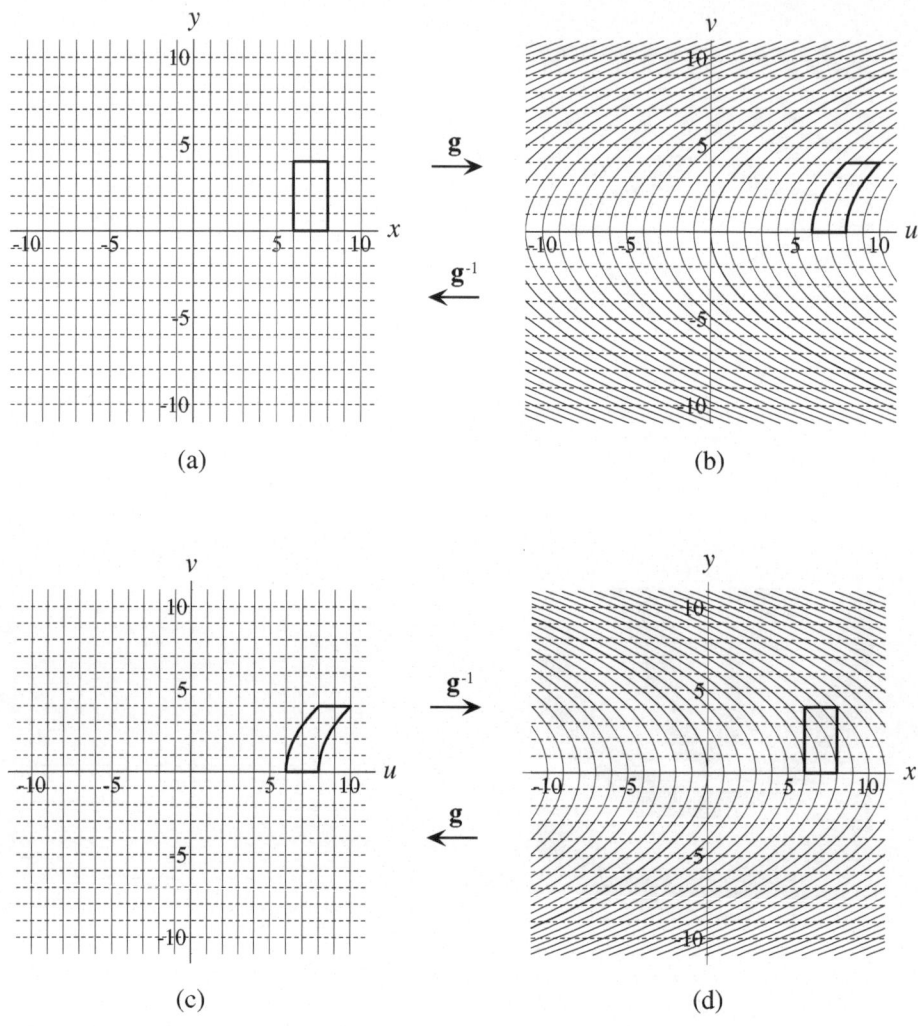

Figure D.16: (a),(b) The effect of the transformation $(u,v) = (x + ay^2, y)$ on the unit grid: Each vertical line is mapped into a right-opened parabola, while each horizontal line is simply shifted to the right. (c),(d) The effect of the inverse transformation $(x,y) = (u - av^2, v)$ on the unit grid: Each vertical line is mapped into a left-opened parabola, while each horizontal line is simply shifted to the left.

transformation bends each vertical line $x = m$, $m \in \mathbb{Z}$ into a right-opened parabola $x = ay^2 + m$, while each horizontal dashed line $y = n$, $n \in \mathbb{Z}$ is simply shifted to the right, i.e. mapped into itself. Consequently, this transformation maps the unit grid of the x,y plane (Fig. D.16(a)) into a right-opened parabolic grid (Fig. D.16(b)); this is, in fact, a non-linear horizontal shearing operation. Similarly, the inverse transformation, which is

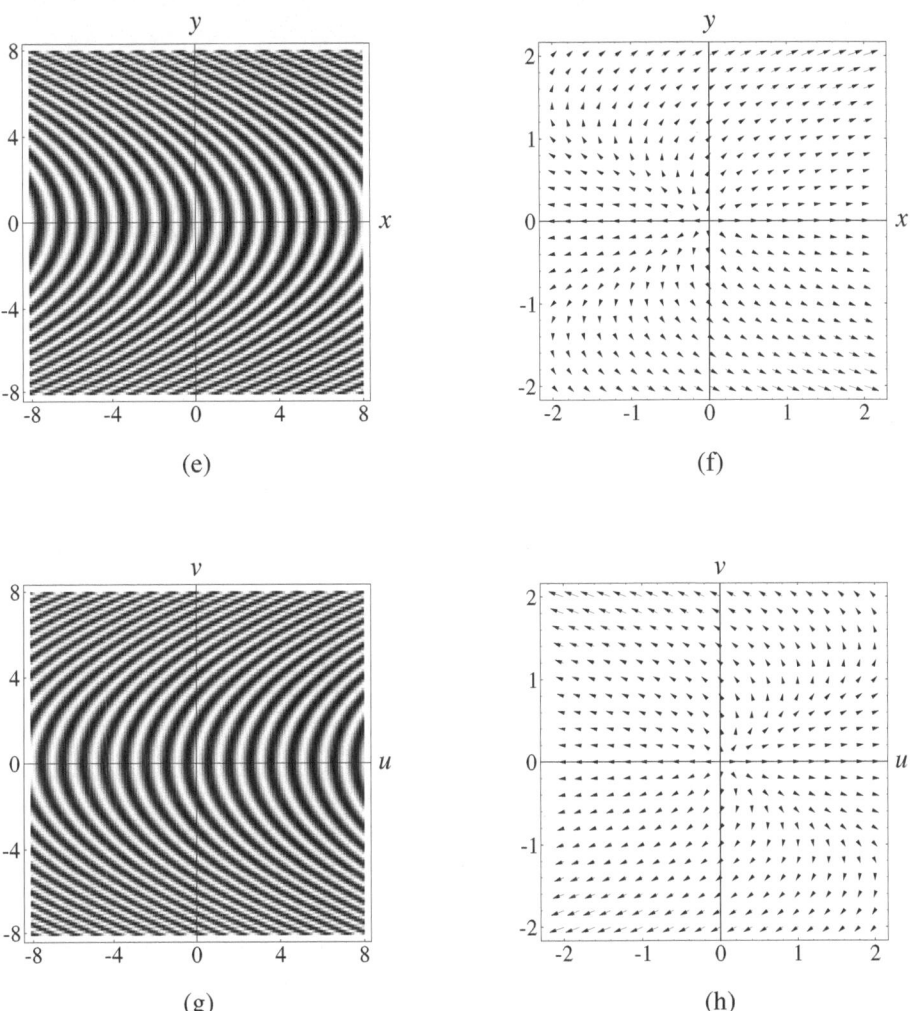

(e)

(f)

(g)

(h)

Figure D.16: (*continued.*) (e) The application of $(u,v) = (x + ay^2, y)$ as a domain transformation to the vertical unit-period cosinusoidal grating $z = \cos(2\pi u)$ gives $z = \cos(x + ay^2)$, a left-opened parabolic version of the original function; compare with the effect of the *inverse* transformation on the plain grid lines in (d). (f) The effect of the same transformation as a vector field. (g) The application of the inverse transformation $(x,y) = (u - av^2, v)$ as a domain transformation to the vertical unit-period cosinusoidal grating $z = \cos(2\pi x)$ gives $z = \cos(2\pi[u - av^2])$, a right-opened parabolic version of the original function; compare with the effect of the *direct* transformation on the plain grid lines in (b). (h) The effect of the same inverse transformation as a vector field.

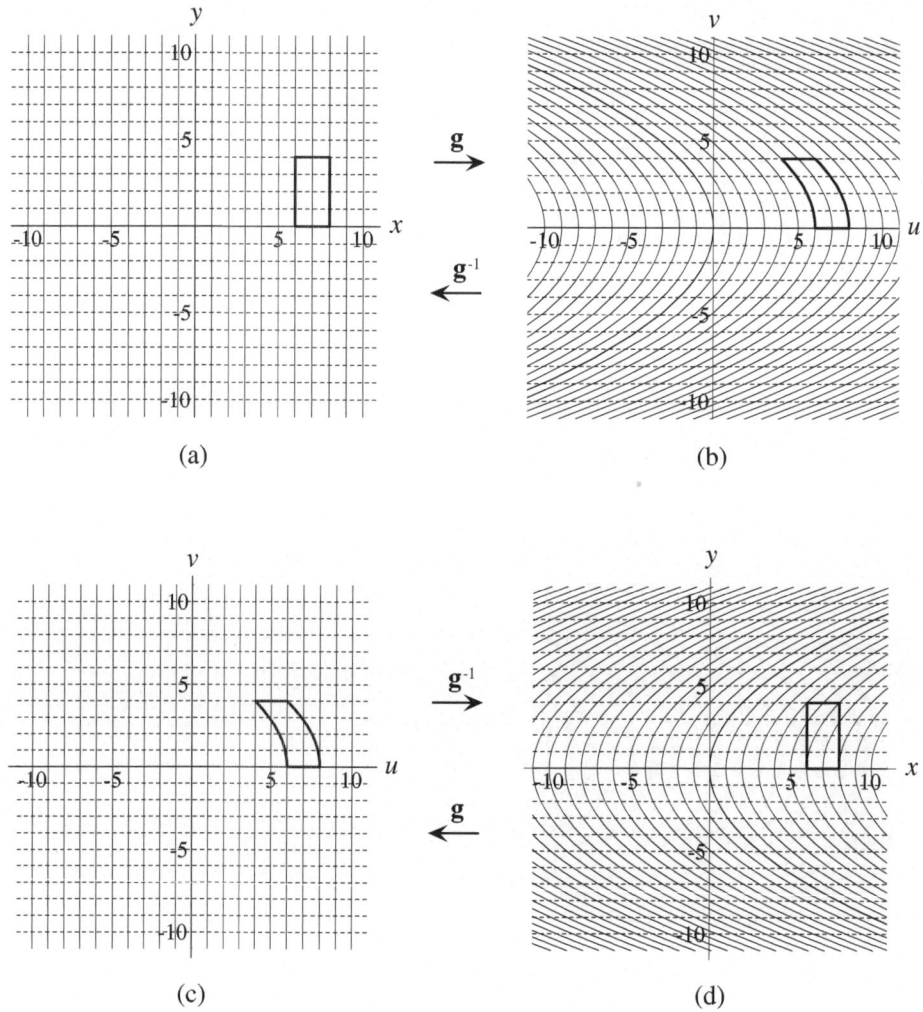

Figure D.17: (a),(b) The effect of the transformation $(u,v) = (x - ay^2, y)$ on the unit
grid: Each vertical line is mapped into a left-opened parabola, while
each horizontal line is simply shifted to the left. (c),(d) The effect of
the inverse transformation $(x,y) = (u + av^2, v)$ on the unit grid: Each
vertical line is mapped into a right-opened parabola, while each
horizontal line is simply shifted to the right.

given by $(x,y) = (u - av^2, v)$, maps the unit grid (Fig. D.16(c)) into a left-opened parabolic
grid (Fig. D.16(d)).

However, if our aim is to construct a transformation that bends the vertical corrugations
of the function $z = \cos(2\pi u)$ into right-opened parabolas, as shown in Fig. D.16(g), then

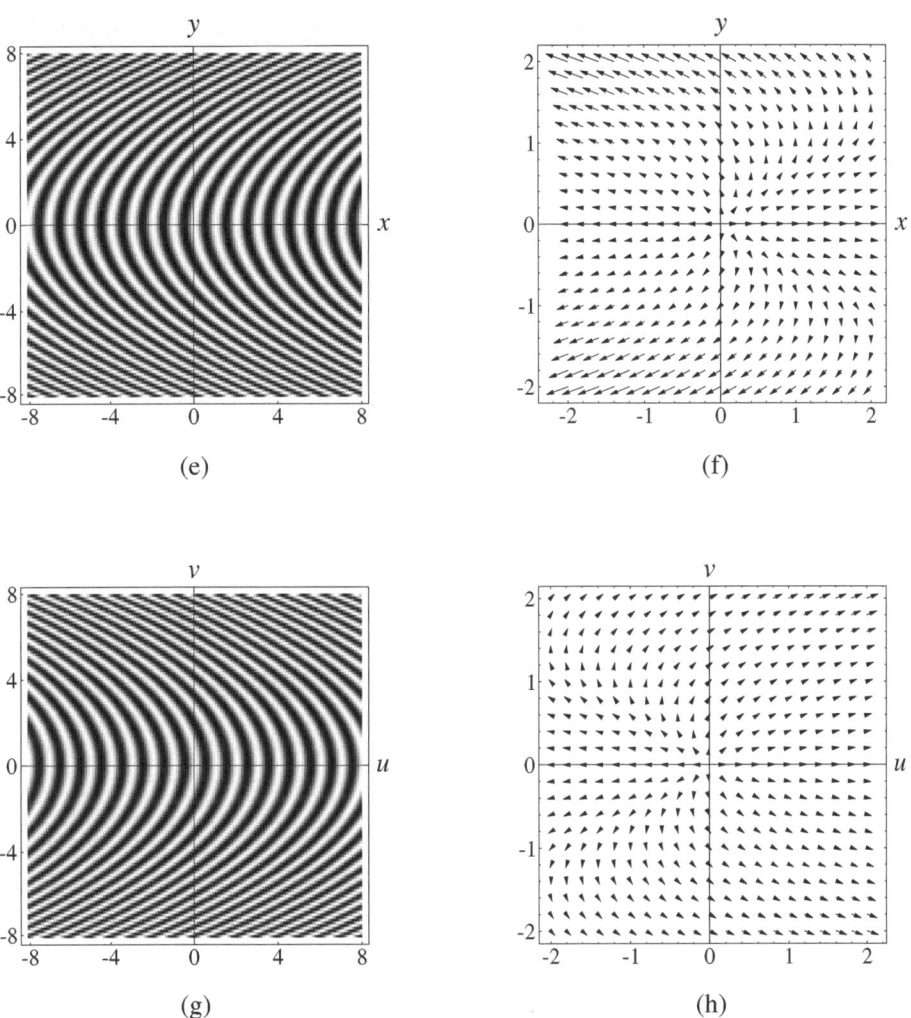

(e)   (f)

(g)   (h)

Figure D.17: (*continued.*) (e) The application of $(u,v) = (x - ay^2, y)$ as a domain transformation to the vertical unit-period cosinusoidal grating $z = \cos(2\pi u)$ gives $z = \cos(2\pi[x - ay^2])$, a right-opened parabolic version of the original function; compare with the effect of the *inverse* transformation on the plain grid lines in (d). (f) The effect of the same transformation as a vector field. (g) The application of the inverse transformation $(x,y) = (u + av^2, v)$ as a domain transformation to the vertical unit-period cosinusoidal grating $z = \cos(2\pi x)$ gives $z = \cos(2\pi[u + av^2])$, a left-opened parabolic version of the original function; compare with the effect of the *direct* transformation on the plain grid lines in (b). (h) The effect of the same inverse transformation as a vector field.

the transformation we need is the *inverse* of $g(x,y)$, which is given, back in the original coordinate system (see Remark D.4 on the variable names), by $(u,v) = (x - ay^2, y)$. This transformation is shown as a transformation in its own right in a separate figure, Fig. D.17. As shown in Figs. D.17(a),(b), this transformation bends each vertical line $x = m$, $m \in \mathbb{Z}$ into a *left*-opened parabola, while each horizontal dashed line $y = n$, $n \in \mathbb{Z}$ is simply shifted to the left, i.e. mapped into itself. In other words, this transformation maps the unit grid of the x,y plane (Fig. D.17(a)) into a *left*-opened parabolic grid (Fig. 17(b)). But because of the inversion effect that is inherent to domain transformations, when this transformation is applied to the function $z = \cos(2\pi u)$, it bends its vertical corrugations into *right*-opened parabolas, in accordance with the effect of the inverse of this transformation, $(x,y) = (u + av^2, v)$, shown in Figs. D.17(c),(d). The resulting function, $z = \cos 2\pi(x - ay^2)$, corresponds, therefore, to our requirements, as shown in Fig. D.17(e).

Finally, the effects of the transformations g and g^{-1} as vector fields are shown in Figs. D.16(f) and D.16(h), respectively. ■

Example D.7: Suppose now we wish to construct a 2-fold counterpart of the previous example, which bends the vertical corrugations of $z = \cos 2\pi x$ into right-opened parabolas, as shown in Fig. D.17(e), and the horizontal corrugations of $z = \cos 2\pi y$ into top-opened parabolas. Using the same logic as at the end of the previous example, taking into account the inversion effect of domain transformations, it is clear that the transformation $g(x,y)$ we seek here is defined by $(u,v) = (x - ay^2, y - ax^2)$. But although this transformation seems to be a straightforward generalization of the transformation $(u,v) = (x - ay^2, y)$ that is shown in Fig. D.17, it is in fact much more interesting. As we can see in Figs. D.18(a),(b), this transformation maps the vertical lines $x = m$, $m \in \mathbb{Z}$ into a family of shifted, left-opened parabolas, and the horizontal dashed lines $y = n$, $n \in \mathbb{Z}$ into a family of shifted, bottom-opened parabolas. However, it turns out that this transformation is quite unusual in several aspects: First, the individual parabolas within each family intersect each other. Moreover, as we can see in Fig. D.18(b), this transformation is neither surjective (its image does not cover the entire u,v plane) nor injective (it maps different points (x,y) to the same image (u,v)). And finally, as a consequence of all these pathologies, the inverse transformation does not have a simple explicit expression. Nevertheless, as we already know from Remark D.7, the effect of the inverse transformation $(x,y) = g^{-1}(u,v)$ on the plane can be determined, without even knowing its explicit expression. And indeed, as clearly shown in Figs. D.18(c),(d), this inverse transformation maps the vertical lines $u = m$, $m \in \mathbb{Z}$ into the family of equispaced right-opened parabolas $x - ay^2 = m$, and the horizontal dashed lines $v = n$, $n \in \mathbb{Z}$ into the family of equispaced top-opened parabolas $y - ax^2 = n$.

Viewed the other way around, this also means, of course, that $g(x,y)$ maps each of the parabolas of Fig. D.18(d) into the corresponding straight line in Fig. D.18(c). Note, however, that as the point $P = (x,y)$ traces out the parabola, its image $Q = (u,v)$ in Fig. D.18(c) moves along the corresponding straight line slowlier and slowlier, until it reaches the border of the range of the transformation, where it stops and starts moving backwards along the same straight line.

In accordance with Remark D.6, the plain and dashed parabolas of Fig. D.18(d) are, respectively, the level lines of the surfaces $z = x - ay^2$ and $z = y - ax^2$ (the two components of the *direct* transformation **g**). Similarly, in accordance with Remark D.5, the plain and dashed parabolas of Fig. D.18(b) are the level lines of the two surfaces defined by the *inverse* transformation $\mathbf{g}^{-1}$. However, since the level lines within each of these two surfaces intersect each other, it follows that neither of these surfaces is single valued: As shown in Fig. D.19, each of these surfaces may have 0, 1, 2, 3 or 4 z-values in different parts of the plane. In fact, each of these surfaces resembles a handkerchief that has been folded twice; see also [Callahan74 p. 232] or [Koenderink90 pp. 332–334].

Note that in this example neither **g** nor $\mathbf{g}^{-1}$ can be used to define a meaningful coordinate system, since these transformations are not one-to-one (see Remark D.9).

Now, as we already know, when we apply the transformation $(u,v) = \mathbf{g}(x,y)$ to a given function $z = f(u,v)$, the resulting function $z = f(\mathbf{g}(x,y))$ is distorted in accordance with the *inverse* transformation, $\mathbf{g}^{-1}$. For example, following the application of $\mathbf{g}(x,y)$ the straight vertical corrugations of $z = \cos(2\pi u)$ become in the resulting function, $z = \cos(2\pi[x - ay^2])$, a family of equispaced right-opened parabolic corrugations (see Fig. D.18(d)). And indeed, it is mainly for this inverse effect of the domain transformation (although we don't even know the explicit expression of $\mathbf{g}^{-1}$) that we are interested in the transformation $(u,v) = (x - ay^2, y - ax^2)$ in our work (see, for example, Figs. A.1 and 3.15).

Finally, the effect of the transformation **g** as a vector field is shown in Fig. D.18(f). Note that this vector field clearly shows the two critical points of the transformation **g** (i.e. the points where $\mathbf{g}(x,y) = (0,0)$; see Sec. H.1 in Appendix H), which are $(0,0)$ and $(1/a,1/a)$. The critical points of a transformation **g** are clearly revealed by its vector field representation, but they are not as easily seen in the other representations of **g**. ■

It is interesting to note that the transformation $(u,v) = (x - ay^2, y - ax^2)$ of the last example belongs, in fact, to a larger family of transformations, whose graphical behaviour can be easily understood using the following result:

Proposition D.5: Suppose we are given a transformation $(u,v) = \mathbf{g}(x,y)$ that is defined by:

$$\begin{pmatrix} u \\ v \end{pmatrix} = \begin{pmatrix} x + f_1(y) \\ y + f_2(x) \end{pmatrix} \tag{D.23}$$

where $z = f_1(s)$ and $z = f_2(s)$ are arbitrary 1D functions. Then, the geometric effects of the transformation **g** and of its inverse $\mathbf{g}^{-1}$ on the coordinate grid of the plane can be pictured as follows:

(a) The inverse transformation $(x,y) = \mathbf{g}^{-1}(u,v)$ maps the vertical lines $u = m$, $m \in \mathbb{Z}$ of the u,v plane into a family of parallel curves $x = -f_1(y) + m$ in the x,y plane, that are unit-spaced copies of the curve $x = -f_1(y)$ shifted along the x axis. Similarly, $\mathbf{g}^{-1}$ distorts the

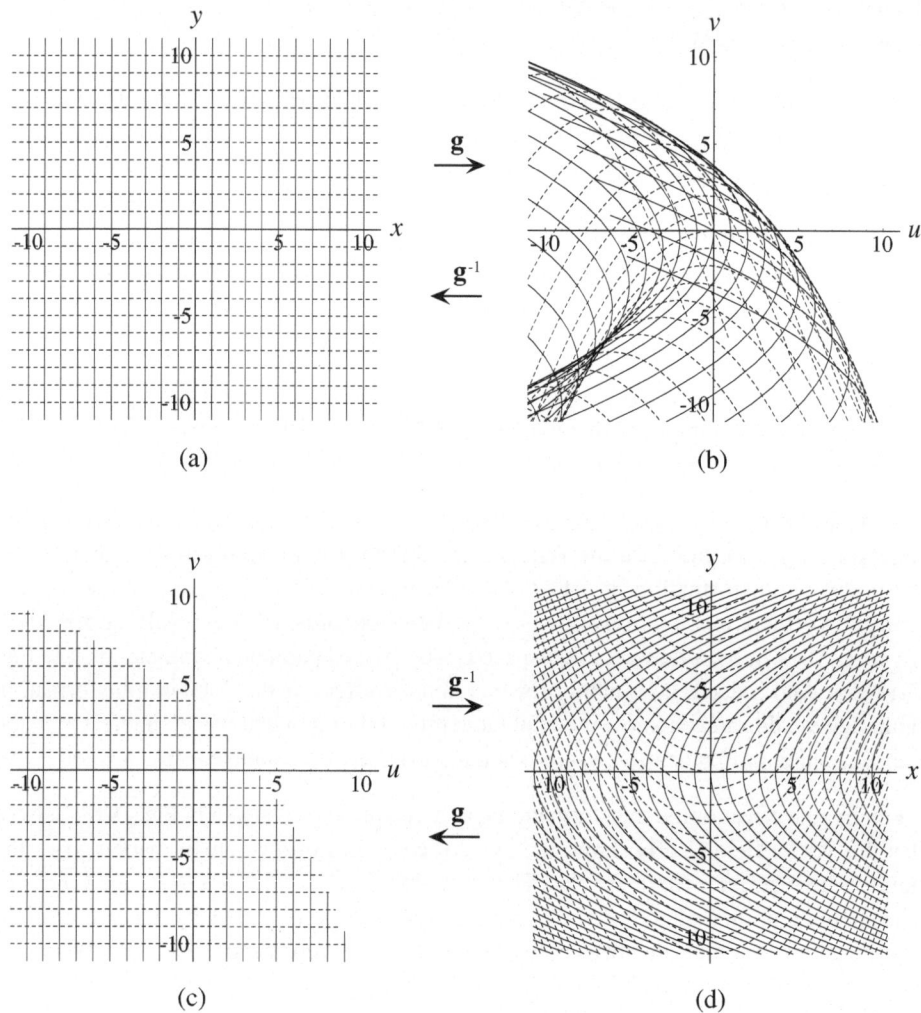

(a)   (b)

(c)   (d)

Figure D.18: (a),(b) The effect of the transformation $(u,v) = (x - ay^2, y - ax^2)$ on the unit grid: The vertical lines of the grid are mapped into a self-intersecting family of shifted, left-opened parabolas, while the horizontal lines of the grid are mapped into a self-intersecting family of shifted, bottom-opened parabolas. (c),(d) The effect of the inverse transformation on the unit grid: The vertical lines of the grid are mapped into a family of right-opened parabolas, while the horizontal lines of the grid are mapped into a family of top-opened parabolas. Note that the unprinted area of the u,v plane is never accessed by these transformations.

horizontal lines $v = n$, $n \in \mathbb{Z}$ of the u,v plane into a family of parallel curves $y = -f_2(x) + n$ in the x,y plane, that are unit-spaced copies of the curve $y = -f_2(x)$ shifted along the y axis.

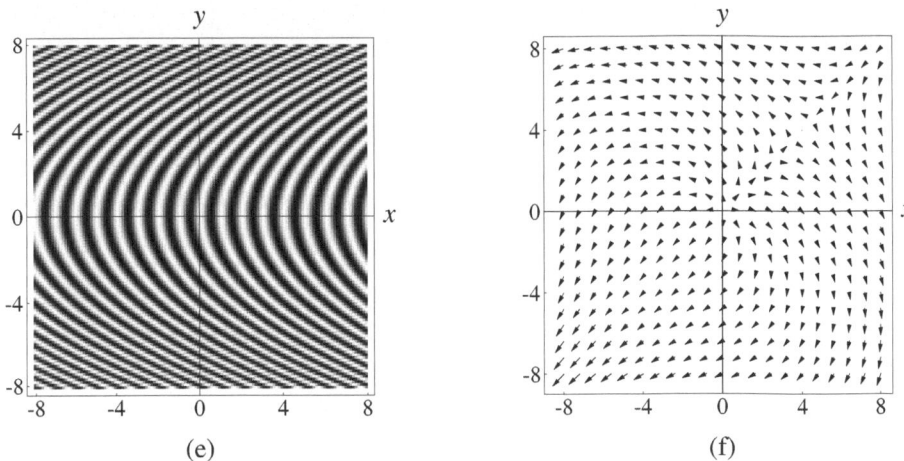

(e) (f)

Figure D.18: (*continued.*) (e) The application of $(u,v) = (x - ay^2, y - ax^2)$ as a domain transformation to the vertical unit-period cosinusoidal grating $z = \cos(2\pi u)$ gives $z = \cos(2\pi[x - ay^2])$, a right-opened parabolic version of the original function; compare with the effect of the *inverse* transformation on the plain grid lines in (d). (f) The effect of the same transformation as a vector field. Note that in the present example the explicit form of the inverse transformation is not readily available.

(b) The direct transformation $(u,v) = \mathbf{g}(x,y)$ maps the vertical lines $x = m$, $m \in \mathbb{Z}$ of the x,y plane into a family of curves $u = f_1(v - f_2(m)) + m$ in the u,v plane, that are parallel copies of the curve $u = f_1(v)$, and whose origins are centered at the points $(u,v) = (f_2(m), m)$ along the curve $v = f_2(u)$. Similarly, $\mathbf{g}$ distorts the horizontal lines $y = n$, $n \in \mathbb{Z}$ of the u,v plane into a family of curves $v = f_2(u - f_1(n)) + n$ in the x,y plane, that are parallel copies of the curve $v = f_2(u)$, and whose origins are centered at the points $(u,v) = (f_1(n), n)$ along the curve $u = f_1(v)$. ■

The mathematical demonstration of this result is quite simple: Part (a) follows from the fact that for any $u = m$ we have from the first line of (D.23) $x = -f_1(y) + m$; and similarly, for any $v = n$ we have from the second line of (D.23) $y = -f_2(x) + n$. Part (b) follows from the fact that for any $x = m$ we have from (D.23) $u = f_1(y) + m$, $v = f_2(m) + y$, which gives $y = v - f_2(m)$ and hence $u = f_1(v - f_2(m)) + m$; similarly, for any $y = n$ we have from (D.23) $u = f_1(n) + x$, $v = f_2(x) + n$, which gives $x = u - f_1(n)$ and hence $v = f_2(u - f_1(n)) + n$.

Note that several of the transformations we have seen in the examples above have the form (D.23). This includes the transformations $(u,v) = (x - \operatorname{argsinh}(y), y + \operatorname{argsinh}(x))$ (see Fig. D.9), $(u,v) = (x + ay^2, y)$ (Fig. D.16), $(u,v) = (x - ay^2, y)$ (Fig. D.17) and $(u,v) = (x - ay^2, y - ax^2)$ (Fig. D.18). And indeed, the geometric behaviour of each of these transformations and of its inverse, as shown in the figures, is clearly explained by Proposition D.5 (even if the explicit form of the inverse transformation is not available).

D.9 Other possible sources of confusion

As we can see, the handling of spatial transformations or coordinate changes can be often quite confusing. In particular, it may be sometimes unclear whether one should use the direct transformation $\mathbf{g}$ or its inverse $\mathbf{g}^{-1}$ to obtain the desired effect. In many cases (like in Figs. D.11 and D.16) the penalty for using the wrong direction is just a sign inversion. Such sign inversions are often mistakenly attributed to a "forgotten sign", and rectified by inverting the sign without even caring to understand why. But in other cases the difference between the results obtained by using $\mathbf{g}$ or $\mathbf{g}^{-1}$ is much more significant, and it cannot be simply dismissed as a "forgotten sign". For example, the transformation $(u,v) = (2xy, y^2 - x^2)$ (see Example D.5 above) distorts the standard Cartesian grid into two families of parabolas as in Fig. D.15(b), while the inverse transformation distorts the standard Cartesian grid into two families of hyperbolas as in Fig. D.15(d). And indeed, it turns out that in some circumstances the effect of the transformation $(u,v) = (2xy, y^2 - x^2)$ is depicted by two families of parabolas (see, for example, [Spiegel68 p. 126]), while in other circumstances it is depicted by two families of hyperbolas (see, for example, [Spiegel63 p. 185] or Example 10.23 and Fig. 10.36 in Chapter 10 of *Vol. I*). In fact, both representations are correct — depending on the point of view and on the application: When $(u,v) = (2xy, y^2 - x^2)$ is used as a direct transformation it indeed distorts the original Cartesian grid into two families of parabolas, but when it is used as a domain (and hence inverse) transformation, it distorts the original Cartesian grid (or any other rectilinear structure) into two families of hyperbolas (see Proposition D.3). In fact, the same transformation may also have a third graphical representation when it is interpreted as a vector field; in this case its graphical representation resembles the physical illustration of a magnetic field (see Fig. D.15(f)). On the other hand, although polar coordinates are introduced in the literature either by $(r,\theta) = (\sqrt{x^2 + y^2}, \arctan(y/x))$ (see, for example, [Courant88 p. 138]), or by its inverse, $(x,y) = (r\cos\theta, r\sin\theta)$ (see, for example, [Kreyszig93 p. 672]), both of these transformations are systematically illustrated in the literature by Figs. D.14(a)(b), and rarely — if ever — by Figs. D.13(a),(b).

All these potential sources of confusion are intrinsic to the use of transformations and coordinate changes, and they originate from pure mathematical considerations. But as if these were not enough, there exist also several extraneous potential sources of confusion. One of these is related to the discrete representation of images on digital computers, and another is related to the existence of two different standards of notation in the literature. These additional sources of confusion are shortly described in the following subsections.

D.9.1 Forward and backward mapping algorithms in digital imaging

Although spatial transformations are basically continuous mathematical entities, in most modern applications they are implemented by digital computers, and they operate on digital, discrete images. This means that the input and the output spaces are represented as 2D (or more generally, N-dimensional) arrays of pixels lying on an integer grid. This discrete nature of digital images and of their transformations introduces several complica-

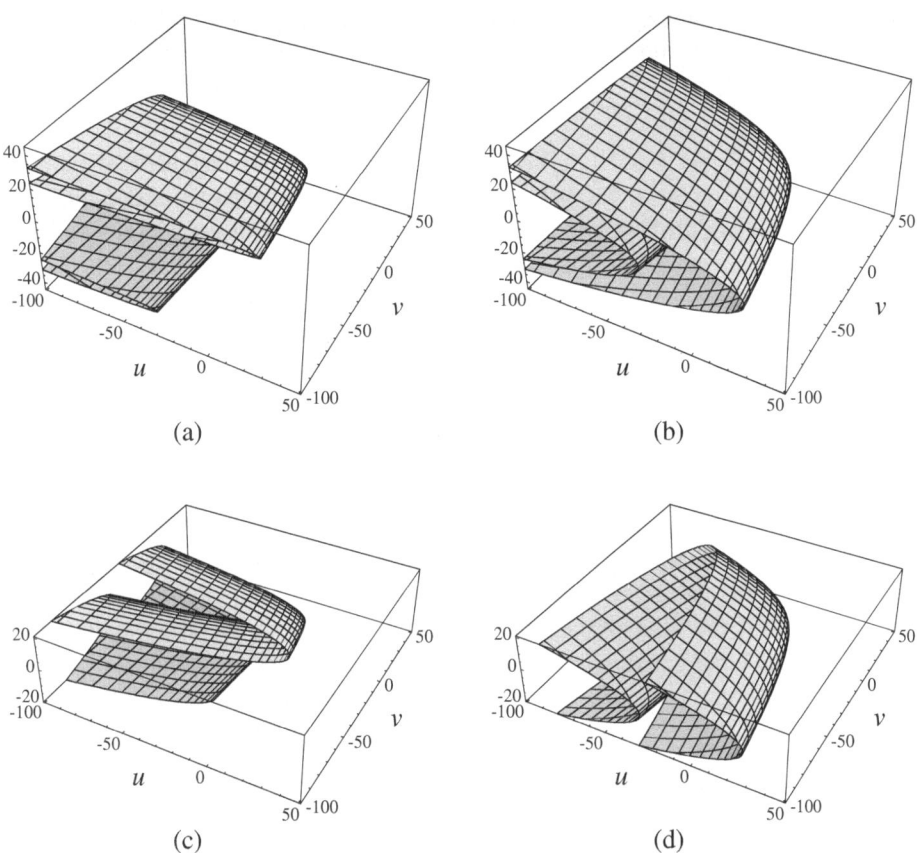

Figure D.19: The multivalued surfaces (a) $z = g_1^{-1}(u,v)$ and (b) $z = g_2^{-1}(u,v)$ of Example D.7. These are the surfaces whose level lines are shown, respectively, by the continuous and dashed parabolas of Fig. D.18(b). Note that for the sake of clarity the surfaces have been drawn for a larger span of u and v values than Fig. D.18(b). Parts (c) and (d) show the same surfaces as in the upper figures, but here they have been vertically truncated at the levels $z = 20$ and $z = -20$ in order to clearly show the parabolic shape of their level lines.

tions, that are addressed in various ways by the algorithms which perform digital transformations. These algorithms can be divided into two main families, known as *forward mapping* and *backward mapping*, each having its own advantages and shortcomings [Wolberg90 pp. 42–45].[35]

A *forward mapping* operates by scanning the original input image pixel by pixel, and copying the value of the image at each input pixel location (x,y) onto the corresponding

---

[35] Note that these terms should not be confused with the terms *direct mapping* and *inverse mapping*, which refer to a transformation $(u,v) = \mathbf{g}(x,y)$ and its inverse $(x,y) = \mathbf{g}^{-1}(u,v)$.

position (u,v) in the output image. The output position (u,v) is determined by passing the x and y coordinates of each input pixel through the transformation:

$$(u,v) = \mathbf{g}(x,y)$$

Note, however, that in general the coordinates u and v thus obtained are real numbers even if x and y are integers. This would not be a problem if the output domain were continuous. But in our discrete case the positions u and v must be discretized (for example by rounding or truncation) in order to fit to the underlying discrete grid of the output image, so that every pixel in the input image can be copied into a pixel in the output image. This seemingly innocent discretization may give rise to two types of problems in the output image: holes and overlaps. Holes occur in the output image if contiguous pixels in the input image are mapped into sparse positions in the output grid, and overlaps occur when different input pixels happen to fall on the same destination pixel in the output image (see [Wolberg90 pp. 42–44] for a more detailed explanation).

Due to these as well as other shortcomings of forward mapping, the use of *backward mapping* is more common in most digital image applications. A backward mapping operates by scanning the target output image pixel by pixel, and copying onto each output pixel location (u,v) the value of the input pixel at the corresponding (x,y) position in the input image. The input position (x,y) that corresponds to a given output pixel location is determined by passing the u and v coordinates of each output pixel through the *inverse* transformation:

$$(x,y) = \mathbf{g}^{-1}(u,v)$$

Just as in the previous case, the continuous x and y locations thus obtained are, in general, real numbers, even though u and v are integers. This means that, here too, the values x and y must be discretized in order to fit to the underlying discrete grid of the input image.[36] Note, however, that unlike in the forward mapping scheme, the backward mapping scheme guarantees that all output pixels are computed. For this reason, backward mapping proves to be a much more convenient approach than forward mapping, although it requires that the explicit expression of the inverse transformation $\mathbf{g}^{-1}$ be available. Another advantage of backward mapping that is often overlooked in the literature is that it behaves exactly as a domain transformation, meaning that if our input image corresponds to the function $z = f(x,y)$, the resulting output image after applying the transformation is exactly $f(\mathbf{g}(x,y))$.

This last point brings us back to the potential confusion which may occur due to the use of forward or backward algorithms in digital imaging. Suppose we are given two different image-processing programs, one based on a forward-mapping algorithm and the other based on a backward-mapping algorithm. If we use in both programs the *same*

---

[36] In fact, a better solution consists of interpolating between the values of the input image at the pixel locations surrounding the non-integer location (x,y), in order to determine the theoretic value of the underlying continuous input image at the point (x,y).

transformation $\mathbf{g}(x,y)$, say, $(u,v) = (2xy, y^2 - x^2)$, we will get two completely different transformed images: one will look like Fig. D.15(b), and the other will look like Fig. D.15(d). The situation is even worse if we do not know which type of algorithm is implemented in the software being used, since then we may obtain the wrong result without even being aware of the risk. In cases such as rotation transformations the difference is merely in the direction of the rotation, so that we may be tempted to "rectify" the error by inverting the sign of a variable within the program. But this "remedy" will not help in other cases; on the contrary, it may even worsen the situation when other transformations will be used.

Finally, as already mentioned, it may happen in some cases that the explicit formula of the inverse transformation $\mathbf{g}^{-1}(u,v)$ is unknown or unavailable. In such cases the use of a backward mapping scheme for applying $\mathbf{g}$ to the input image is not possible. However, in such cases one can still apply $\mathbf{g}$ to the input image by supplying $\mathbf{g}$ itself to a program which is based on a forward mapping algorithm.

Remark D.20: The layer transformations in most of the figures throughout this volume have been obtained by a *forward-mapping algorithm*, a PostScript program that moves each dot of the given dot screen (periodic or not) from its original location (x,y) to its new location $(u,v) = \mathbf{g}(x,y)$. However, some of the figures — notably those that show the transformation of more complicated images — have been obtained by a *backward-mapping algorithm*, a C language program that takes an input image $f(x,y)$ and generates its distorted version $f(\mathbf{g}(x,y))$. In all cases care has been taken to supply to the program in question the right transformation, either $\mathbf{g}$ or $\mathbf{g}^{-1}$, in order to obtain the intended effect in the figure. Note, however, that our figure legends do not usually mention the implementation-dependent technicalities that were involved in the figure-generation process, and they only concentrate on the mathematical and visual effects that are demonstrated by each figure. ■

We will return to some of these issues in more detail in Sec. D.10.

D.9.2 Pre-multiplication and post-multiplication based notations

The last source of confusion we discuss here concerns transformations that can be expressed in matrix form. In addition to linear transformations, this includes also affine, bilinear, and some other types of transformations [Wolberg90 pp. 45–61]. The problem arises here from the two different standard notations that are widely used in the literature for expressing such transformations, using pre-multiplication or post-multiplication form [Lipschutz68 p. 157]. In the first case the transformation matrix precedes the vector to which it is applied:

$$\begin{pmatrix} u \\ v \end{pmatrix} = \begin{pmatrix} a & b \\ c & d \end{pmatrix} \begin{pmatrix} x \\ y \end{pmatrix} \tag{D.24}$$

while in the second case the transformation matrix is written after the vector:

$$(u,v) = (x,y) \begin{pmatrix} e & f \\ g & h \end{pmatrix} \tag{D.25}$$

Both notation standards are fully equivalent, provided that we consistently stick to the same notation. It should be noted, however, that the matrix representation of a given transformation is different in each of these two notation standards, so when we copy a formula from a book, we must carefully check whether the copied matrix agrees with our notation system. For example, in the present work we always use the pre-multiplication notation (Eq. (D.24)); in this notation the matrix form of a rotation by positive angle α is given by:

$$\begin{pmatrix} \cos\alpha & -\sin\alpha \\ \sin\alpha & \cos\alpha \end{pmatrix}$$

and the inverse transformation is given by:

$$\begin{pmatrix} \cos\alpha & \sin\alpha \\ -\sin\alpha & \cos\alpha \end{pmatrix}$$

However, in books using the post-multiplication notation these matrices are interchanged (see, for example, [Wolberg90 pp. 49 and 217]). The rule is that if we wish to use a matrix taken from the other notation standard, we must first take the transpose of the matrix, as illustrated below:

$$\begin{pmatrix} u \\ v \end{pmatrix} = \begin{pmatrix} a & b \\ c & d \end{pmatrix} \begin{pmatrix} x \\ y \end{pmatrix} = \begin{pmatrix} ax + by \\ cx + dy \end{pmatrix}$$

$$(u,v) = (x,y) \begin{pmatrix} a & b \\ c & d \end{pmatrix}^T = (x,y) \begin{pmatrix} a & c \\ b & d \end{pmatrix} = (ax + by, \, cx + dy)$$

D.10 Implications to the moiré theory: issues related to the figures

In the previous sections of this appendix we have presented various results about *direct*, *inverse*, *domain* and *range* transformations, as well as about *forward* and *backward-mapping* algorithms in software applications; but the reasons that we actually need all these results in our work may still be unclear. In the present section and in the following one we therefore try to show some of the actual implications of these results to the moiré theory: In the present section we will see their practical implications to the generation of our moiré figures, and in the following section we will see the implications to the fixed points between the superposed layers.

As we already know (see Remark 4.1 in Chapter 4), the classical moiré theory between periodic or repetitive layers is based, as a rule, on the interpretation of the layer transformations as *domain* (and hence, *inverse*) transformations. This rule holds, for example, in the explanation of the moiré effects that are obtained when we apply the transformations $\mathbf{g}_1(x,y)$ and $\mathbf{g}_2(x,y)$ to the original periodic layers $p_1(x',y')$ and $p_2(x',y')$, giving the distorted layers $p_1(\mathbf{g}_1(x,y))$ and $p_2(\mathbf{g}_2(x,y))$ (see Propositions 10.2 and 10.5 in *Vol. I*). In fact, domain transformations are used throughout Chapters 10 and 11 of *Vol. I*, which present the theory of curvilinear moiré effects between geometrically transformed periodic layers (see Sec. 10.2 and in particular Footnote 1 and Remark 10.2 there, as well as Secs. 11.2.2, 11.3 and 11.4, all in *Vol. I*). Domain transformations are also used in the case of aperiodic layers, as we have seen throughout the present volume, with just one outstanding exception: When it comes to dot trajectories, the transformation $\mathbf{g}(x,y)$ that is applied to a dot screen is understood as an operation that moves each point (x,y) of the original dot screen to its new destination under $\mathbf{g}$, namely: $(x,y) \mapsto \mathbf{g}(x,y)$. This means, as explained in Chapter 4, that in the context of dot trajectories we are interested in the effect of $\mathbf{g}$ as a *direct* transformation. (Note that in *Vol. I*, too, there exist some circumstances where the transformations $\mathbf{g}_i(x,y)$ are not used as domain transformations, and the expressions involved are formulated in terms of direct rather than inverse transformations; see, for example, Eq. (3.1) and its vector form in Sec. 3.4.1 of *Vol. I*.)

However, the fact that we need to use the same transformation alternately as a domain transformation or as a direct transformation, depending on the context, may cause us trouble in the generation and in the interpretation of our figures. The following examples will help us understand this point.

Example D.8: Consider the following cases, which illustrate the use of the inverse mapping as a domain transformation:

(a) Suppose we are given a layer $p(x',y')$. In order to rotate it by positive angle α counterclockwise we have to apply to $p(x',y')$ the following domain transformation:

$$x' = x\cos\alpha + y\sin\alpha$$

$$y' = -x\sin\alpha + y\cos\alpha$$

which corresponds, in fact, to the inverse transformation (rotation by angle $-\alpha$; see Example D.2 above). For instance, in order to rotate the vertical cosinusoidal grating $\cos(2\pi f x')$ by angle α (see Fig. D.11(g)) we have to plot the function:

$$r(x,y) = \cos(2\pi f[x\cos\alpha + y\sin\alpha])$$

where the original variable x' has been replaced by $x\cos\alpha + y\sin\alpha$.

(b) Bending a periodic layer $p(x',y')$ into a right-opened parabolic shape is obtained by applying to $p(x',y')$ the domain transformation:

$$x' = x - ay^2$$

$$y' = y$$

which corresponds, in fact, to the inverse transformation (see Example D.6 above). Taking again the example of the vertical cosinusoidal grating $\cos(2\pi f x')$, we can bend it into a right-opened parabolic cosinusoidal grating (see Fig. D.16(g)) by plotting the function:

$$r(x,y) = \cos(2\pi f[x - ay^2])$$

Another example consists of the distorted periodic and aperiodic layers shown in Figs. 3.5(a),(b), which are expressed mathematically by $r(x,y) = p(x - ay^2, y)$.

(c) Bending a two-fold periodic dot screen $p(x',y')$ into a 2D hyperbolic dot screen is obtained by applying to $p(x',y')$ the domain transformation:

$$x' = 2xy$$

$$y' = y^2 - x^2$$

(see Example D.5 above). The resulting geometrically transformed dot screen is expressed, therefore, by $p(2xy, y^2-x^2)$ (see Figs. B.7(a),(b) in Appendix B). ∎

Naturally, it would be best to generate all the distorted gratings and dot screens in the figures that accompany our discussions on the classical moiré theory by an algorithm based on backward mapping. As we have seen in Sec. D.9.1, backward-mapping algorithms operate on the original image exactly as a domain transformation does, and therefore the explicit expressions that we have to provide to the algorithm are identical to those used in our equations, and the resulting moirés obtained in the figures fully correspond to our mathematical results as we indeed obtain them in terms of domain transformations. However, in reality three different problems may occur:

Problem 1: Because in the context of dot trajectories (in Chapter 4) we need to express the transformed layers in terms of direct transformations while elsewhere (in Chapters 3, 6 and 7) we need to express the same transformed layers in terms of domain (and hence, inverse) transformations, we are facing a dilemma in our figure legends: Given that each figure can be used in all chapters, which of the two expressions should we present in the figure legend? For example, taking the parabolically distorted dot screens shown in Fig. 3.5, should we say in the figure legend that they were obtained by applying the transformation $\mathbf{g}(x,y) = (x - ay^2, y)$, as we would do in Chapters 3, 6 and 7 — or by applying the transformation $\overline{\mathbf{g}}(x,y) = (x + ay^2, y)$, as we would do in Chapter 4? The solution we have adopted is to provide in the figure legends just a verbal description of the transformations, without giving their explicit formulas. But when we still wish to add the formulas, the convention we adopt in the figure legends throughout this volume corresponds to the normal usage of transformations in the classical moiré theory as *domain* transformations. In cases where the transformation should be understood as a *direct* transformation, we either say it explicitly, or use the barred notation $\overline{\mathbf{g}}(x,y)$, as explained in Sec. 4.4, Remark 4.1.

Problem 2: When we generate our figures using a program based on a backward-mapping algorithm the transformations we have to supply to the program correspond to those being used as domain transformations (in Chapters 3, 6 and 7), but not to those being used as direct transformations (in Chapter 4). Therefore, in order to generate using such a backward-mapping algorithm figures that come to illustrate dot trajectories, as in Chapter 4, we must supply to the program the *inverse* of the transformations being used in the text. Note, however, that this is just a technical problem that should be taken into account when generating the figures, but we can forget about it once the figures are done.

Of course, if we generate our figures using a program that is based on a forward-mapping algorithm, the inverse problem will occur. And indeed, as mentioned in Remark D.20, most of the transformed dot screens shown in the figures throughout the present volume were produced, for technical reasons, by an algorithm based on *forward mapping*. This algorithm uses the most straightforward and natural approach for distorting a random or periodic dot screen: it simply maps each dot of the original, undistorted dot screen from its original location (x,y) to the new, transformed location $(u,v) = \mathbf{g}(x,y)$.[37] This forces us, in order to obtain the correct results in the figures, to provide to the forward-mapping program the *direct* transformation of each of the layers, namely, the inverse of the transformation $\mathbf{g}$ that is used in the mathematical expression $p(\mathbf{g}(x,y))$. But once again, this technicality in the figure generation does not affect our mathematical developments, and once the figures are ready we continue to assume that the transformed layers in question have been obtained by applying $\mathbf{g}$ as a domain (and hence inverse) transformation, as usual.

Example D.9: The following cases illustrate the use of direct transformations in the forward-mapping program that generates most of the layer superpositions in this volume:

(a) In order to rotate a dot screen by positive angle α counterclockwise we have to provide to the forward-mapping program the following direct transformation:

$$u = x\cos\alpha - y\sin\alpha$$

$$v = x\sin\alpha + y\cos\alpha$$

This means that a dot that should have been printed in the original, untransformed dot screen at the location (x,y) will be printed in reality at the location (u,v) as defined above (see, for example, Fig. 2.1(d)). And yet, assuming that the original, undistorted dot screen is given by $p(x',y')$, the rotated dot screen continues to be expressed mathematically by $p(x\cos\alpha + y\sin\alpha, -x\sin\alpha + y\cos\alpha)$, using the *inverse* mapping $(x',y') = (x\cos\alpha + y\sin\alpha, -x\sin\alpha + y\cos\alpha)$ as a domain transformation.

(b) In order to bend an original dot screen into a right-opened parabolic dot screen we have to provide to the forward-mapping program the following direct transformation:

$$u = x + ay^2$$

$$v = y$$

---

[37] The reason for using a forward-mapping algorithm to generate the figures in this volume is basically practical. Note that in the case of random screens the use of a backward-mapping algorithm may be more difficult.

(see, for example, Fig. 3.5). And yet, assuming that the original, undistorted dot screen is given by $p(x',y')$, the distorted dot screen of the resulting figure is expressed mathematically, as explained in Sec. 3.4.1, by $p(x-ay^2, y)$, using the *inverse* mapping $(x',y') = (x-ay^2, y)$ as a domain transformation.

(c) Finally, if we bend an original dot screen $p(u,v)$ by using in the forward-mapping program the transformation:

$$u = 2xy$$

$$v = y^2 - x^2$$

the resulting dot screen will be distorted into a *parabolic* shaped screen (see Fig. D.15(b)). If we wish to obtain the *hyperbolic* screen which is expressed by $p(2xy, y^2 - x^2)$, as shown in Fig. D.15(d) or in Fig. B.7 of Appendix B, we have to provide to our forward-mapping program the inverse transformation, i.e. the rather complicated expression with the nested roots that is given in Example D.5 above. ∎

Problem 3: The fact that the transformations used to draw the figures are sometimes inversed with respect to the conventions used in the theoretic results may raise an additional problem. According to Proposition 5.1 in Sec. 5.3 (see also the second part of Proposition 10.5 in Sec. 10.9.2 of *Vol. I*), when we apply the domain transformations $g_1(x,y)$ and $g_2(x,y)$, respectively, to two periodic layers, the resulting $(1,-1)$-moiré in the layer superposition undergoes the domain transformation $g_M(x,y) = g_1(x,y) - g_2(x,y)$. Thus, in order to obtain in the layer superposition a $(1,-1)$-moiré having the domain transformation $k(x,y)$, we have to apply to the two original layers, respectively, such domain transformations $g_1(x,y)$ and $g_2(x,y)$ that give the difference $k(x,y)$; for example $g_1(x,y) = (x,y) + k(x,y)$ and $g_2(x,y) = (x,y)$, or alternatively, $g_1(x,y) = (x,y) + \frac{1}{2}k(x,y)$ and $g_2(x,y) = (x,y) - \frac{1}{2}k(x,y)$ (see also Eqs. (3.41) and (3.42)). This would pose no problems if we were to draw the figure using a program that is based on a backward-mapping algorithm. But as we have seen in Problem 2 above, in order to draw these distorted layers using our forward-mapping algorithm we have to provide to our program the inverse expressions $[(x,y) + k(x,y)]^{-1}$ and $[(x,y)]^{-1} = (x,y)$, or $[(x,y) + \frac{1}{2}k(x,y)]^{-1}$ and $[(x,y) - \frac{1}{2}k(x,y)]^{-1}$; but these expressions may turn out to be too complex or simply unavailable.

What happens, then, if we "forget" the inversions and provide to the forward-mapping program the known, non-inverted expressions (namely, $(x,y) + k(x,y)$ and (x,y), or $(x,y) + \frac{1}{2}k(x,y)$ and $(x,y) - \frac{1}{2}k(x,y)$)? The respective domain transformations become, in this case, $[(x,y) + k(x,y)]^{-1}$ and $[(x,y)]^{-1} = (x,y)$, or $[(x,y) + \frac{1}{2}k(x,y)]^{-1}$ and $[(x,y) - \frac{1}{2}k(x,y)]^{-1}$. Hence, the differences between the layer's domain transformations become, respectively:

$$[(x,y) + k(x,y)]^{-1} - (x,y) \tag{D.26}$$

and:
$$[(x,y) + \tfrac{1}{2}k(x,y)]^{-1} - [(x,y) - \tfrac{1}{2}k(x,y)]^{-1} \tag{D.27}$$

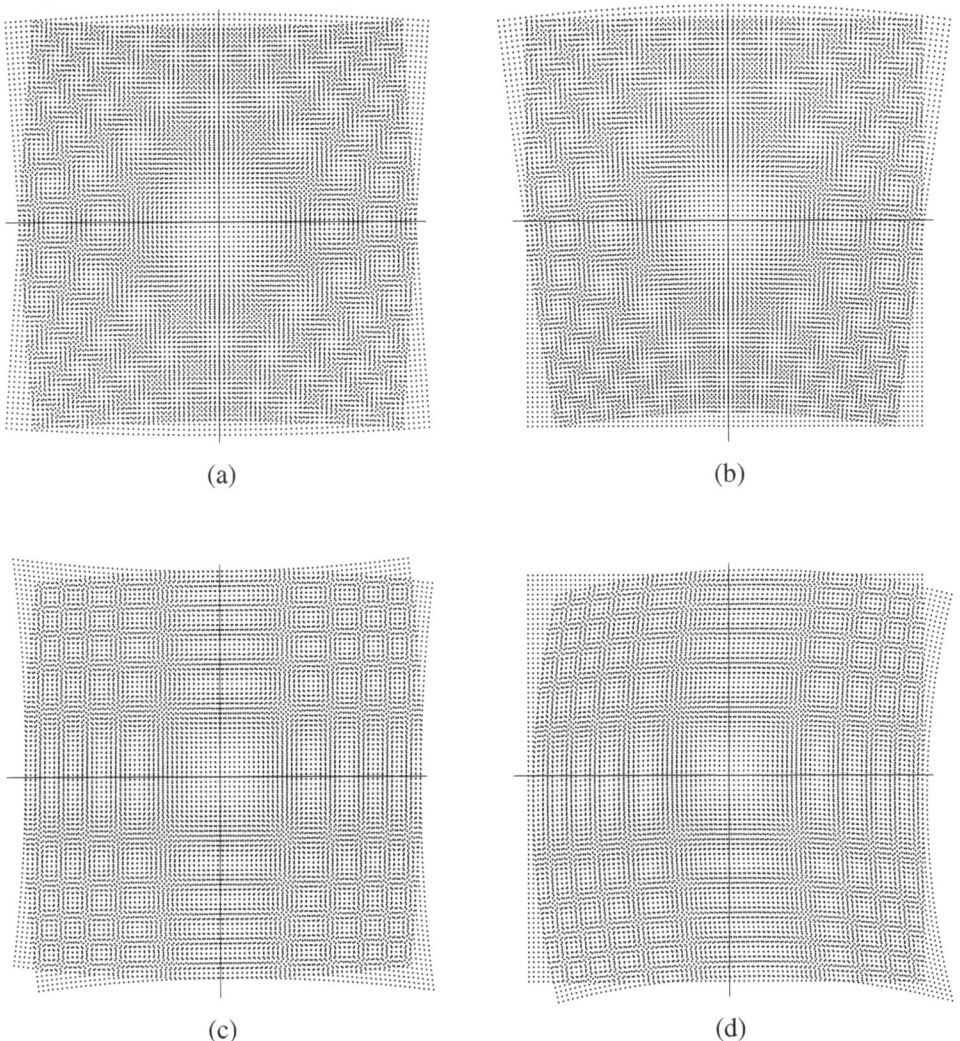

(a) (b)

(c) (d)

Figure D.20: The effects of using forward mapping to transform the original layers in a screen superposition. (a) The moiré effect obtained in Fig. 4.13(b), which is drawn using forward mapping, when the transformation $\mathbf{k}(x,y) = (2xy, y^2 - x^2)$ is equally distributed between the two layers. This moiré effect fully agrees with the results that are obtained when the layers are drawn by backward mapping. (b) When the transformation $\mathbf{k}(x,y)$ is entirely taken care of by the first layer, the resulting moiré is significantly distorted. (c),(d) Same as (a) and (b), but this time using the transformation $\mathbf{k}(x,y) = (ay^2 + x_0, y_0 - ax^2)$, a slightly different variant of Fig. 3.15(b). Note that the same phenomenon occurs also in the aperiodic counterparts of these superpositions, but the distortions in the aperiodic case are less visible.

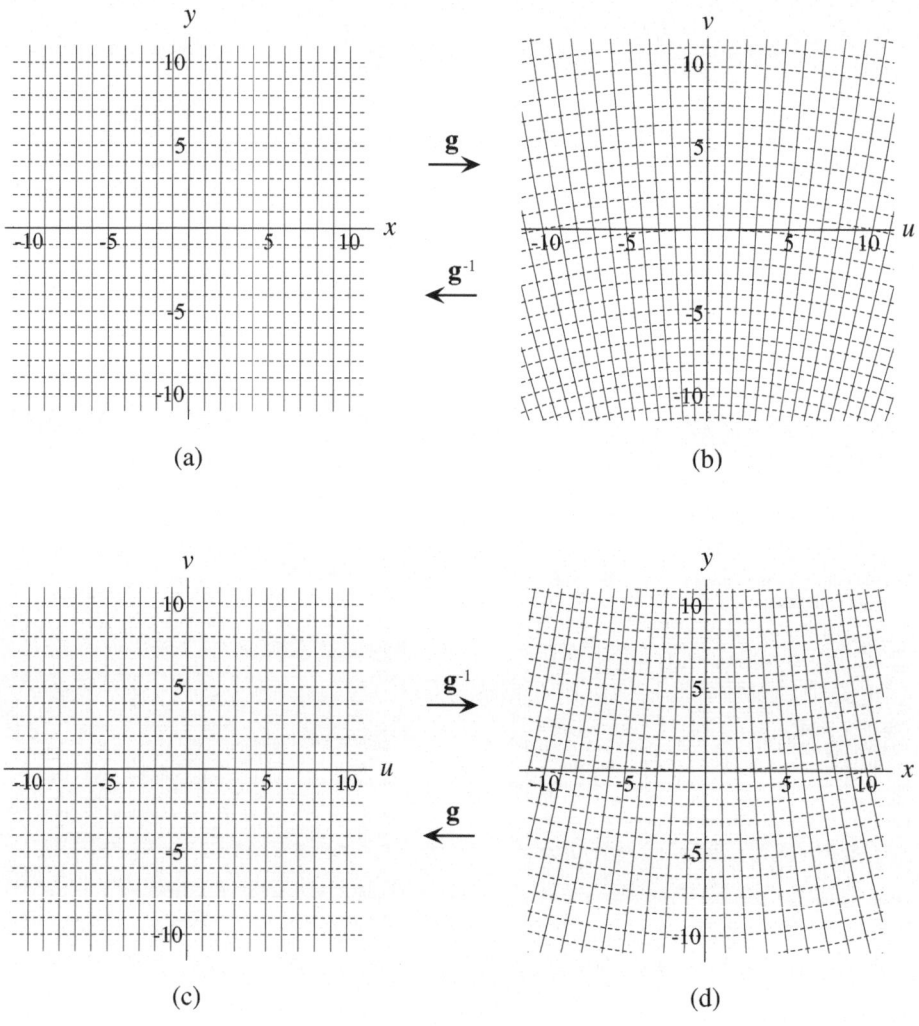

Figure D.21: The effect of the transformation $\mathbf{g}(x,y) = (x,y) + \mathbf{k}(x,y)$ where $\mathbf{k}(x,y)$ is a very weak non-linear transformation, $\mathbf{k}(x,y) = (2xy, y^2 - x^2)/100$. Note that $\mathbf{k}(x,y)$ is in fact a strongly diluted version of the non-linear transformation shown in Fig. D.15, and its influence here on the underlying linear transformation $\mathbf{g}_1(x,y) = (x,y)$ is very moderate. Therefore, the inverse of $\mathbf{g}(x,y)$, shown in (d), is almost identical to $(x,y) - \mathbf{k}(x,y)$: $[(x,y) + \mathbf{k}(x,y)]^{-1} \approx (x,y) - \mathbf{k}(x,y)$.

According to our proposition, these differences express the domain transformation of the resulting moiré in the layer superposition. But clearly, none of these differences can be expected to give the desired result, $\mathbf{k}(x,y)$, as was the case when backward mapping was used and no inversions were needed (see Eqs. (3.41) and (3.42)):

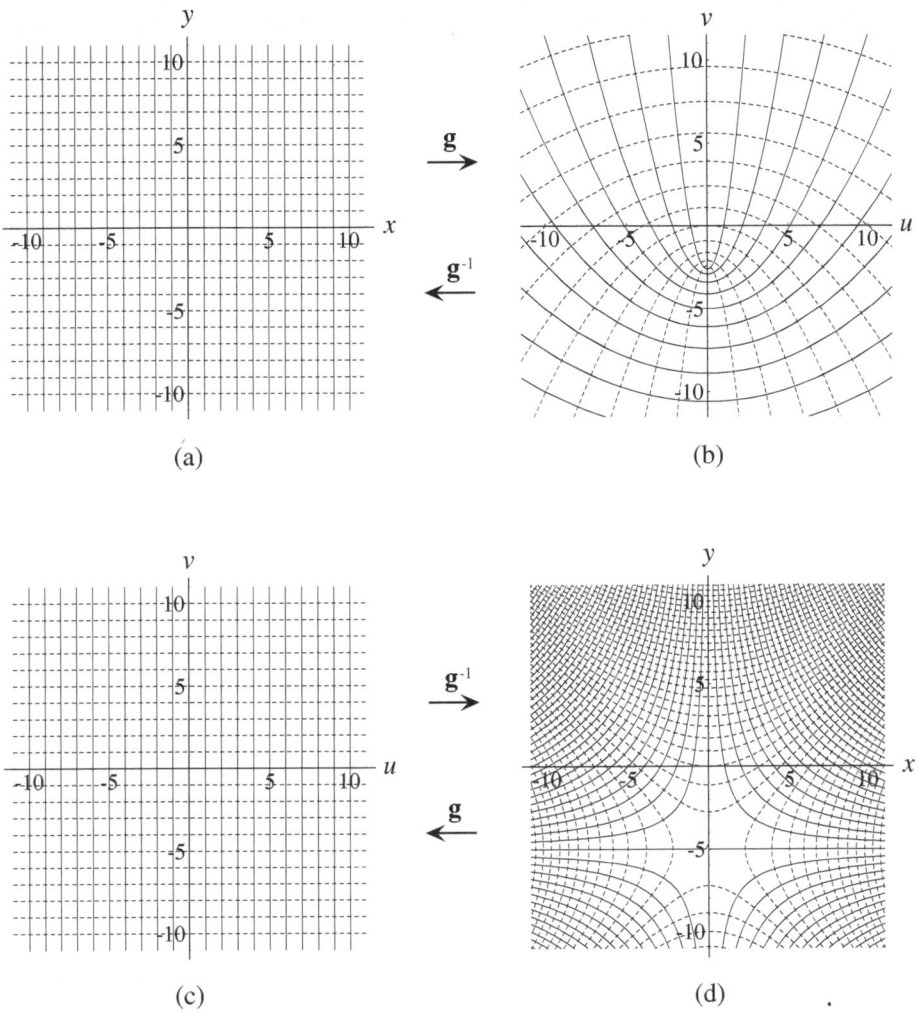

Figure D.22: Same as in Fig. D.21, but here the non-linearity introduced by $\mathbf{k}(x,y)$ is ten times higher: $\mathbf{k}(x,y) = (2xy, y^2 - x^2)/10$. Note that in this case the approximation $[(x,y) + \mathbf{k}(x,y)]^{-1} \approx (x,y) - \mathbf{k}(x,y)$ is no longer valid.

$$[(x,y) + \mathbf{k}(x,y)] - (x,y) = \mathbf{k}(x,y)$$

$$[(x,y) + \tfrac{1}{2}\mathbf{k}(x,y)] - [(x,y) - \tfrac{1}{2}\mathbf{k}(x,y)] = \mathbf{k}(x,y)$$

However, surprisingly, superposition tests with various transformations show that when forward mapping is being used to draw the layers, the difference (D.27), which corresponds to an equal distribution of the distortion between the two layers, *does* give in the layer superposition the same moiré effect $\mathbf{k}(x,y)$ as in the case of backward mapping (up to a sign inversion). Even more surprisingly, it turns out that this occurs *only* if $\mathbf{k}(x,y)$ is equally distributed between the two layers, as in Eq. (D.27). When the distortion is not equally distributed, for instance in the case of Eq. (D.26), the moiré effect obtained with forward mapping is not $-\mathbf{k}(x,y)$, but rather a distorted version of $-\mathbf{k}(x,y)$ (see Figs. D.20(b),(d); note that in this case $-\mathbf{k}(x,y) = \mathbf{k}(x,y)$). In other words: the expected moiré $\mathbf{k}(x,y) = \mathbf{g}_1(x,y) - \mathbf{g}_2(x,y)$, as predicted by Proposition *I*.10.5, *can* be obtained in the superposition (up to a sign inversion) even if we perform the two layer transformations $\mathbf{g}_i(x,y)$ by means of forward mapping, but this happens only if the deformation $\mathbf{k}(x,y)$ is equally distributed between the two original, undistorted layers. This, however, seems to contradict Proposition *I*.10.5, which guarantees that the resulting moiré effect must have the shape of the difference $\mathbf{k}(x,y) = \mathbf{g}_1(x,y) - \mathbf{g}_2(x,y)$ in *all* cases, no matter how $\mathbf{k}(x,y)$ is distributed between $\mathbf{g}_1(x,y)$ and $\mathbf{g}_2(x,y)$. How can it be?

These surprising results are explained as follows: Remember that in our case we only use weak layer transformations $\mathbf{g}(x,y)$, in order not to destroy the correlation between the superposed layers. This means that our layer transformations $\mathbf{g}(x,y)$ are of the form:

$$\mathbf{g}(x,y) = (x,y) - \mathbf{o}(x,y)$$

or:$$\qquad \mathbf{g}(x,y) = (x,y) + \mathbf{o}(x,y)$$

where $\mathbf{o}(x,y)$ is a negligible transformation that differs only slightly, within the zone covered by the layer superposition, from the zero transformation $\mathbf{z}(x,y) = (0,0)$.

And indeed, it turns out (see Proposition D.9 in Sec. D.12 below) that if $\mathbf{g}(x,y)$ is such a weak layer transformation then we have for its inverse, $\mathbf{g}^{-1}(x,y)$, the following close approximations:

$$[(x,y) - \mathbf{o}(x,y)]^{-1} \approx (x,y) + \mathbf{o}(x,y)$$

$$[(x,y) + \mathbf{o}(x,y)]^{-1} \approx (x,y) - \mathbf{o}(x,y)$$

or, in a more compact notation, denoting the identity transformation by $\mathbf{i}$:

$$[\mathbf{i} - \mathbf{o}]^{-1} \approx \mathbf{i} + \mathbf{o} \qquad\qquad\qquad\qquad\qquad (D.28)$$

$$[\mathbf{i} + \mathbf{o}]^{-1} \approx \mathbf{i} - \mathbf{o} \qquad\qquad\qquad\qquad\qquad (D.29)$$

(see also Figs. D.21 and D.22). Furthermore, Proposition D.9 also asserts that the approximation error $\mathbf{e}_1$ in (D.28) and the approximation error $\mathbf{e}_2$ in (D.29) (both of which are functions of x and y, $\mathbf{e}_1(x,y)$, $\mathbf{e}_2(x,y)$) are almost identical. Therefore, denoting this common approximate error by $\mathbf{e}(x,y)$ we obtain for Eq. (D.27):

$$[(x,y) + \tfrac{1}{2}\mathbf{k}(x,y)]^{-1} - [(x,y) - \tfrac{1}{2}\mathbf{k}(x,y)]^{-1} =$$

$$= [(x,y) - \tfrac{1}{2}\mathbf{k}(x,y) + \mathbf{e}(x,y)] - [(x,y) + \tfrac{1}{2}\mathbf{k}(x,y) + \mathbf{e}(x,y)]$$

$$= -\mathbf{k}(x,y) \qquad\qquad\qquad\qquad\qquad (D.30)$$

whereas for Eq. (D.26) we have:

$$[(x,y) + \mathbf{k}(x,y)]^{-1} - (x,y) =$$

$$= [(x,y) - \mathbf{k}(x,y) + \mathbf{e}(x,y)] - (x,y)$$

$$= -\mathbf{k}(x,y) + \mathbf{e}(x,y) \tag{D.31}$$

This explains, indeed, why, when providing to the forward-mapping algorithm the *inverse* layer transformations rather than the *direct* ones, we still obtain the correct moiré effect if the distortion $\mathbf{k}(x,y)$ is equally distributed between the two superposed layers; but if the distortion is only taken care of by one of the two layers, the resulting moiré is rather distorted, as shown in Fig. D.20. Note that Eq. (D.30) also explains why the obtained moiré is identical (up to a sign inversion) to the moiré $\mathbf{k}(x,y)$ which is expected when using a backward-mapping algorithm, and not to its inverse, $[\mathbf{k}(x,y)]^{-1}$.

D.11 Fixed points of a superposition in terms of direct or inverse transformations

In this section we provide several useful results that can help us better understand our discussions on fixed points in layer superpositions. We start with the simpler case which occurs when only one of the superposed layers is transformed, and then we proceed to the more general case where both layers are transformed.

D.11.1 Fixed points when only one layer is transformed

Suppose we are given two identical dot screens $r_1(x,y)$ and $r_2(x,y)$, periodic or not, that are superposed on top of one another in full coincidence, dot on dot. We apply to one of the layers (say, the first one) a transformation $\mathbf{g}(x,y)$; as a result, the original layer $r_1(x,y)$ changes into $r_1(\mathbf{g}(x,y))$.

Let $\mathbf{g}_M(x,y)$ denote the transformation that is undergone by the resulting moiré. We have seen in Chapter 3 that:

$$\mathbf{g}_M(x,y) = \mathbf{g}(x,y) - (x,y) \tag{D.32}$$

(which is exactly the same result as in the case of the (1,-1)-moiré between periodic or repetitive layers; see Propositions 10.2 and 10.5 in Sec. 10.9 of *Vol. I*). Note that $\mathbf{g}(x,y)$ and $\mathbf{g}_M(x,y)$ are understood here as the *domain* transformations undergone by the layer $r_1(x,y)$ and by the resulting moiré, respectively; this means that both $\mathbf{g}(x,y)$ and $\mathbf{g}_M(x,y)$ are used here as *inverse* transformations.

As we have seen in Chapter 3, thanks to the layer transformation $\mathbf{g}(x,y)$ one or more Glass patterns may be generated in the layer superposition $r_1(\mathbf{g}(x,y))r_2(x,y)$. Typically, a Glass pattern occurs around each of the fixed points of the layer superposition, namely, around each point (x,y) that satisfies $\mathbf{g}_M(x,y) = (0,0)$, or equivalently:

$$g(x,y) = (x,y) \tag{D.33}$$

And yet, the same layer transformation $g(x,y)$ can be also expressed in terms of its effect on the locations of the individual screen dots, as we do in Chapter 4. Clearly, each point (x,y) of the original layer $r_1(x,y)$ is moved by our transformation to a new location $\overline{g}(x,y)$:

$$(x,y) \mapsto \overline{g}(x,y)$$

However, in this case the effect of the layer transformation is understood as a *direct* transformation, meaning that $\overline{g}(x,y)$ is, in fact, $g^{-1}(x,y)$, the *inverse* of the domain transformation $g(x,y)$ (the reasons for using the barred notation to denote direct transformations are explained in Sec. 4.4, Remark 4.1). For example, suppose that we are given the original dot screen $r_1(x,y)$. If we apply to it the domain transformation $g(x,y) = (x/2,y/2)$ we obtain the transformed screen $r_1(x/2,y/2)$, which is a two-fold expanded version of the original screen $r_1(x,y)$. But the effect of this two-fold expansion on each individual dot of the screen is expressed by means of the *direct* transformation $\overline{g}(x,y) = (2x,2y)$:

$$(x,y) \mapsto (2x,2y)$$

which is the inverse of the transformation $g(x,y) = (x/2,y/2)$.

Now, as explained in Sec. 4.4, the effect of the direct transformation $\overline{g}(x,y)$ on any individual dot is best described by the vector field:

$$\overline{h}(x,y) = \overline{g}(x,y) - (x,y) \tag{D.34}$$

This vector field assigns to each point (x,y) the vector $\overline{g}(x,y) - (x,y)$ that connects (x,y) to its destination $\overline{g}(x,y)$ under the transformation $\overline{g}$.

As we can see, Eqs. (D.32) and (D.34) represent the same physical reality in two different ways: Eq. (D.32) is expressed in terms of inverse transformations, while Eq. (D.34) is expressed in terms of direct transformations. It is important to note, however, that although $\overline{g}(x,y)$ and $g(x,y)$ are mutually inverse transformations, $\overline{h}(x,y)$ is not the inverse of $g_M(x,y)$.

Now, based on Eq. (D.34), it is clear that we can also define the fixed points of the superposition as those points (x,y) that satisfy $\overline{h}(x,y) = (0,0)$, or equivalently:

$$\overline{g}(x,y) = (x,y) \tag{D.35}$$

We have obtained, therefore, two different ways to express the fixed points in the resulting layer superposition: On the one hand we say in Eq. (D.33), just as we did in Chapter 3, that the fixed points of the superposition are those points (x,y) for which $g(x,y) = (x,y)$; but on the other hand, based on different considerations, we say in Eq. (D.35) that the fixed points of the same superposition are those points for which $\overline{g}(x,y) = (x,y)$. This seems to be incoherent, since obviously $g(x,y)$ and $\overline{g}(x,y)$ (i.e. $g^{-1}(x,y)$)

are different transformations. But fortunately, the following result comes here to the rescue:

Proposition D.6: Let $\mathbf{g}(x,y)$ be a 2D transformation, and let $\mathbf{g}^{-1}(x,y)$ be its inverse (we assume here, as usual, that $\mathbf{g}(x,y)$ is sufficiently well-behaved, and that it satisfies all the conditions required in order to be invertible). Then, any fixed point of $\mathbf{g}$ is also a fixed point of $\mathbf{g}^{-1}$, and any fixed point of $\mathbf{g}^{-1}$ is also a fixed point of $\mathbf{g}$. ■

To see this, suppose that (x_F,y_F) is a fixed point of $\mathbf{g}$. This means that at the point (x_F,y_F) we have:

$$\mathbf{g}(x_F,y_F) = (x_F,y_F)$$

but then, we also have at the same point:

$$\mathbf{g}^{-1}(\mathbf{g}(x_F,y_F)) = \mathbf{g}^{-1}(x_F,y_F)$$

which means:

$$(x_F,y_F) = \mathbf{g}^{-1}(x_F,y_F)$$

so that (x_F,y_F) is, indeed, a fixed point of $\mathbf{g}^{-1}$, too. The other direction can be shown in a similar way.

We see, therefore, by virtue of this proposition, that the two different definitions of fixed points in our layer superposition are fully equivalent: Any point (x_F,y_F) that satisfies $\mathbf{g}(x_F,y_F) = (x_F,y_F)$ satisfies also $\mathbf{g}^{-1}(x_F,y_F) = (x_F,y_F)$, and vice versa. (Incidentally, this implies also that at any fixed point (x_F,y_F) we have $\mathbf{g}(x_F,y_F) = \mathbf{g}^{-1}(x_F,y_F)$). In conclusion:

Proposition D.7: When one of the two superposed layers is transformed by $\mathbf{g}(x,y)$ but the other layer remains unchanged, the fixed points can be expressed by either $\mathbf{g}(x,y) = (x,y)$ or $\overline{\mathbf{g}}(x,y) = (x,y)$. ■

D.11.2 Fixed points when both layers undergo transformations

Unfortunately, this happy situation does not extend to the more general case in which both of the superposed layers undergo transformations. Suppose, for example, that our two original layers $r_1(x,y)$ and $r_2(x,y)$ undergo, respectively, the domain transformations $\mathbf{g}_1(x,y)$ and $\mathbf{g}_2(x,y)$. We have, therefore, the following generalization of Eq. (D.32):

$$\mathbf{g}_M(x,y) = \mathbf{g}_1(x,y) - \mathbf{g}_2(x,y) \tag{D.36}$$

where $\mathbf{g}_M(x,y)$ is the domain transformation undergone by the resulting moiré.

Clearly, the fixed points of this superposition are those points (x,y) that satisfy $\mathbf{g}_M(x,y) = (0,0)$, or equivalently:[38]

$$\mathbf{g}_1(x,y) = \mathbf{g}_2(x,y) \tag{D.37}$$

---

[38] These points are the *mutual fixed points* of the transformations $\mathbf{g}_1$ and $\mathbf{g}_2$ (see Sec. 3.5).

On the other hand, just as we did in the simpler case before, we can also consider the effect of the layer transformations on the locations of the individual screen dots. In this case, each original point (x,y) in the first layer is moved by the transformation $\bar{\mathbf{g}}_1$ to a new location $\bar{\mathbf{g}}_1(x,y)$, while in the other layer the same point (x,y) is moved by the transformation $\bar{\mathbf{g}}_2$ to another destination, $\bar{\mathbf{g}}_2(x,y)$:

$$(x,y) \mapsto \bar{\mathbf{g}}_1(x,y)$$

$$(x,y) \mapsto \bar{\mathbf{g}}_2(x,y)$$

Note that $\bar{\mathbf{g}}_1$ and $\bar{\mathbf{g}}_2$ are the *direct* transformations that express the same physical layer distortions as the domain transformations $\mathbf{g}_1$ and $\mathbf{g}_2$ above, so that we have $\bar{\mathbf{g}}_1 = \mathbf{g}_1^{-1}$ and $\bar{\mathbf{g}}_2 = \mathbf{g}_2^{-1}$. For example, a two-fold expansion of the first layer can be expressed either by $r_1(x/2,y/2)$, using the domain transformation $\mathbf{g}_1(x,y) = (x/2,y/2)$, or by $(x,y) \mapsto (2x,2y)$, using the direct transformation $\bar{\mathbf{g}}_1(x,y) = \mathbf{g}_1^{-1}(x,y) = (2x,2y)$.

As we have seen in Sec. 4.5, the effect of the direct transformations $\bar{\mathbf{g}}_1$ and $\bar{\mathbf{g}}_2$ on any individual dot is described by the vector field:

$$\bar{\mathbf{h}}(x,y) = \bar{\mathbf{g}}_1(x,y) - \bar{\mathbf{g}}_2(x,y) \tag{D.38}$$

This vector field assigns to each point (x,y) the vector $\bar{\mathbf{g}}_1(x,y) - \bar{\mathbf{g}}_2(x,y)$ that connects the destination of the original point (x,y) under the transformation $\bar{\mathbf{g}}_2$ (the point $\bar{\mathbf{g}}_2(x,y)$) to the destination of the same point (x,y) under the transformation $\bar{\mathbf{g}}_1$ (the point $\bar{\mathbf{g}}_1(x,y)$).[39]

As we can see, Eqs. (D.36) and (D.38) represent the same physical reality in two different ways: Eq. (D.36) is expressed in terms of inverse transformations, while Eq. (D.38) is expressed in terms of direct transformations. Note, however, that although $\bar{\mathbf{g}}_1$ and $\mathbf{g}_1$ as well as $\bar{\mathbf{g}}_2$ and $\mathbf{g}_2$ are mutually inverse transformations, $\bar{\mathbf{h}}(x,y)$ is not the inverse of $\mathbf{g}_M(x,y)$.

Now, based on Eq. (D.38) one may also define the fixed points of our superposition as those points that satisfy $\bar{\mathbf{h}}(x,y) = (0,0)$, or equivalently:

$$\bar{\mathbf{g}}_1(x,y) = \bar{\mathbf{g}}_2(x,y) \tag{D.39}$$

Hence, just as in the simpler case with $\bar{\mathbf{g}}_2(x,y) = (x,y)$, we have obtained here two different ways to express the fixed points in the resulting layer superposition: On the one hand we say in Eq. (D.37) that the fixed points of the superposition are those points (x,y) for which $\mathbf{g}_1(x,y) = \mathbf{g}_2(x,y)$; but on the other hand, Eq. (D.39) suggests that the fixed points of the superposition are those points for which $\bar{\mathbf{g}}_1(x,y) = \bar{\mathbf{g}}_2(x,y)$. Based on our experience from the simpler case with $\bar{\mathbf{g}}_2(x,y) = (x,y)$, we might expect that conditions (D.37) and (D.39) give exactly the same fixed points. However, it turns out that this is not always true. As a simple counter-example, consider the case where the first layer undergoes a scaling by

---

[39] Note that for any points (a,b) and (c,d) in the plane, when the tail of the vector $(a,b) - (c,d)$ is attached to the point (c,d), its head is located at the point (a,b); this means that the vector $(a,b) - (c,d)$ connects the point (c,d) to the point (a,b).

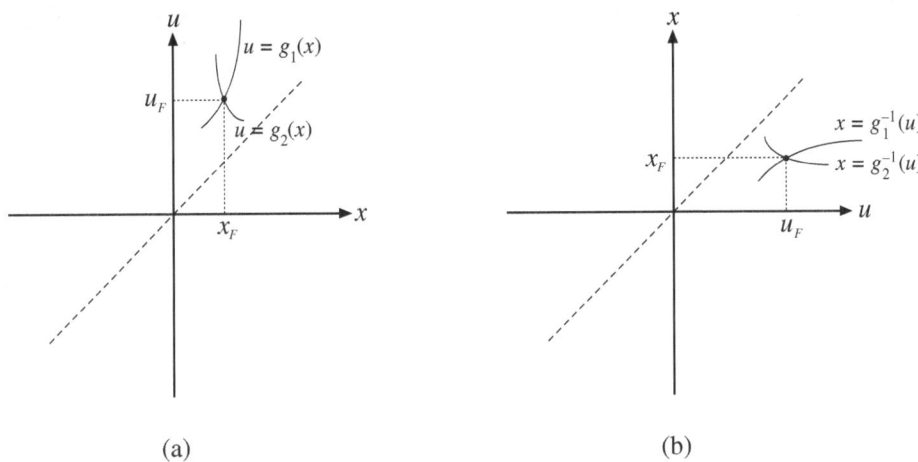

Figure D.23: (a) Two functions $g_1(x)$ and $g_2(x)$ that coincide at a certain point x_F. (b) The inverse functions g_1^{-1} and g_2^{-1} coincide at the point $u_F = g_1(x_F) = g_2(x_F)$. Note that the inverse functions are the reflections of the direct functions with respect to the main diagonal.

factor s and the second layer is shifted by (a,b). It is easy to verify that the fixed point between the direct transformations, i.e. the point which satisfies $(sx,sy) = (x+a, y+b)$, is given by $(x_F,y_F) = (\frac{a}{s-1}, \frac{b}{s-1})$, whereas the fixed point between the inverse transformations, which satisfies $(x/s,y/s) = (x-a, y-b)$, is given by $(x_F,y_F) = (\frac{sa}{s-1}, \frac{sb}{s-1})$. (Incidentally, it is interesting to note that in this case the distance between the two fixed points is exactly $(\frac{sa}{s-1}, \frac{sb}{s-1}) - (\frac{a}{s-1}, \frac{b}{s-1}) = (a,b)$, for any values of a, b and s.)

The reason for this discrepancy between Eqs. (D.37) and (D.39) is that Proposition D.6 cannot be generalized to the case of fixed points between two transformed layers. In other words, the fact that at a given point (x_F,y_F) we have $\mathbf{g}_1(x_F,y_F) = \mathbf{g}_2(x_F,y_F)$ does not guarantee that we have at the same point $\mathbf{g}_1^{-1}(x_F,y_F) = \mathbf{g}_2^{-1}(x_F,y_F)$, too. This is clearly illustrated by the counter-example above. In fact, the reason for this failure is much easier to understand by considering the 1D counterpart of this question: As shown in Fig. D.23, the inverse of any function $g(x)$ is simply its reflection about the main diagonal. And indeed, a glance at Fig. D.23 shows that if $g_1(x)$ and $g_2(x)$ coincide at a certain point x_F, $g_1(x_F) = g_2(x_F)$, then their inverse functions g_1^{-1} and g_2^{-1} do not necessarily coincide at the same point x_F, but rather at the point $u_F = g_1(x_F) = g_2(x_F)$, giving $g_1^{-1}(u_F) = g_2^{-1}(u_F) = x_F$. Only in the particular case where the point of coincidence (x_F,u_F) happens to fall on the main diagonal, so that $u_F = x_F$, we have, indeed, $g_1^{-1}(x_F) = g_2^{-1}(x_F)$. It is obvious, therefore, that this alwasy happens when one of the two functions is the identity function (and hence coincides with the main diagonal). As we have seen above, the same is also true in the case of 2D transformations, but in this case the visualization of the "diagonal" is less obvious since it resides in a 4D space.

But although Proposition D.6 cannot be extended to the more general case, it turns out that under the conditions mentioned below — that are anyway satisfied in our case — we can still obtain the following weaker, yet quite useful result:

Proposition D.8: If $\mathbf{g}_1$ and $\mathbf{g}_2$ are weak transformations, i.e. transformations that only slightly differ from the identity transformation $\mathbf{i}(x,y) = (x,y)$:

$$\mathbf{g}_1(x,y) = (x,y) + \mathbf{o}_1(x,y)$$

$$\mathbf{g}_2(x,y) = (x,y) + \mathbf{o}_2(x,y)$$

then $\mathbf{g}_1(x_F,y_F) = \mathbf{g}_2(x_F,y_F)$ implies that $\mathbf{g}_1^{-1}(x_F,y_F) \approx \mathbf{g}_2^{-1}(x_F,y_F)$, and $\mathbf{g}_1^{-1}(x_F,y_F) = \mathbf{g}_2^{-1}(x_F,y_F)$ implies that $\mathbf{g}_1(x_F,y_F) \approx \mathbf{g}_2(x_F,y_F)$. ∎

The 1D counterpart of this result is, indeed, quite obvious, since here both $g_1(x)$ and $g_2(x)$ are close to the diagonal. In the 2D case this result follows from the fact that any weak transformation $\mathbf{g} = \mathbf{i} + \mathbf{o}$ satisfies the approximation $[\mathbf{i} + \mathbf{o}]^{-1} \approx \mathbf{i} - \mathbf{o}$ (see Proposition D.9 below). Using this result we have, therefore:

$$\mathbf{g}_1^{-1}(x,y) = [(x,y) + \mathbf{o}_1(x,y)]^{-1} \approx (x,y) - \mathbf{o}_1(x,y)$$

$$\mathbf{g}_2^{-1}(x,y) = [(x,y) + \mathbf{o}_2(x,y)]^{-1} \approx (x,y) - \mathbf{o}_2(x,y)$$

But since at the point (x_F,y_F) we have $\mathbf{g}_1(x_F,y_F) = \mathbf{g}_2(x_F,y_F)$ it follows that we also have there $\mathbf{o}_1(x_F,y_F) = \mathbf{o}_2(x_F,y_F)$, and thus we obtain, indeed:

$$\mathbf{g}_1^{-1}(x_F,y_F) \approx (x_F,y_F) - \mathbf{o}_1(x_F,y_F) = (x_F,y_F) - \mathbf{o}_2(x_F,y_F) = \mathbf{g}_2^{-1}(x_F,y_F).$$

Thus, as long as $\mathbf{g}_1$ and $\mathbf{g}_2$ are weak transformations, our two possible definitions of a fixed point between the two superposed layers can be considered as practically equivalent. But since in our case we are anyway forced to use weak layer transformations, in order not to destroy the correlation between the superposed layers, it follows that for our needs we can freely use either of the two definitions of a fixed point, (D.37) or (D.39). Interestingly, in some cases (see Remark D.22) both definitions become fully identical.

Remark D.21: Just for the sake of completeness, it is interesting to note that if we allow both of the superposed layers to be modified, there exist infinitely many different ways to distribute a given additive distortion between the two layers. Consider, for example, the case in which the first layer is scaled by factor s and the second layer remains unchanged, and suppose that the distortion we wish to add consists of a lateral shift of (a,b). If we consider the layer transformations as *domain* transformations, the same effect will be obtained when we apply a shift of $-(a,b)$ to the first layer or when we apply a shift of (a,b) to the second layer, since:

$$g_M(x,y) = (x/s + a, y/s + b) - (x,y) \tag{D.40}$$

is identical to: $$g_M(x,y) = (x/s, y/s) - (x - a, y - b) \tag{D.41}$$

The same effect is also obtained when the shift is equally distributed between the two layers, i.e. when the first layer is shifted by $-\frac{1}{2}(a,b)$ and the second layer is shifted by $\frac{1}{2}(a,b)$:

$$\mathbf{g}_M(x,y) = (x/s + \tfrac{1}{2}a,\ y/s + \tfrac{1}{2}b) - (x - \tfrac{1}{2}a,\ y - \tfrac{1}{2}b)$$

Clearly, there exist infinitely many different ways to distribute the distortion between the two layers, all of which give exactly the same result $\mathbf{g}_M(x,y)$, and hence, the same macroscopic moiré in the superposition (by virtue of Proposition 5.1 in Sec. 5.3, or the second part of Proposition 10.5 in Sec. 10.9.2 of *Vol. I*). However, when we consider the very same distorted layers in terms of *direct* transformations (for example, in order to study the dot trajectories in the layer superposition, as explained in Chapter 4), each of these cases gives a different result. For example, in the case of Eq. (D.40) we have:

$$\overline{\mathbf{h}}_1(x,y) = [(x/s + a,\ y/s + b)]^{-1} - [(x,y)]^{-1}$$

$$= (s(x-a),\ s(y-b)) - (x,y) \tag{D.42}$$

(since the inverse transformation of $u = x/s + a$, $v = y/s + b$ is $x = s(u-a)$, $y = s(v-b)$; see also Remark D.4 on the variable names), while in the case of Eq. (D.41) we obtain:

$$\overline{\mathbf{h}}_2(x,y) = [(x/s,\ y/s)]^{-1} - [(x-a,\ y-b)]^{-1}$$

$$= (sx,\ sy) - (x+a,\ y+b) \tag{D.43}$$

where clearly $\overline{\mathbf{h}}_2(x,y) \neq \overline{\mathbf{h}}_1(x,y)$.

Because all the different distributions of the shift between the two layers are equivalent in terms of domain transformations, it is clear that all of them have the same fixed point in accordance with Eq. (D.37); and indeed, both $(x/s,y/s) = (x-a, y-b)$ and $(x/s+a, y/s+b) = (x,y)$ give the same fixed point, $(x,y) = (\frac{sa}{s-1}, \frac{sb}{s-1})$.

However, using the definition (D.39), which is based on direct transformations, Eq. (D.42) gives the fixed point $(x,y) = (\frac{sa}{s-1}, \frac{sb}{s-1})$ while Eq. (D.43) gives the fixed point $(x,y) = (\frac{a}{s-1}, \frac{b}{s-1})$. But among all of the different possible cases, there exists only a single one that gives the same fixed point as in Eqs. (D.40) and (D.41): By virtue of Proposition D.6, this is precisely the case in which only one of the superposed layers undergoes a transformation, while the other layer remains unchanged (in our example above this is the case given in Eq. (D.42)). Nevertheless, as we have just seen above, if the transformations applied to our layers are weak then the difference between the resulting values is practically negligible. ∎

D.12 Useful approximations

In this appendix we provide some approximations that are often handy to use, because they allow us to formulate our main results in terms of either direct or inverse transformations. We start with the following informal definitions:

Definition D.1: A transformation $(u,v) = \mathbf{o}(x,y)$ is said to be *almost zero* or *almost null* if it differs only slightly, within our zone of interest (i.e. within the area covered by the layer superposition), from the zero transformation $(u,v) = \mathbf{z}(x,y) = (0,0)$. ∎

Definition D.2: A transformation $(u,v) = \mathbf{g}(x,y)$ is said to be *weak* or *almost identity* if it differs only slightly, within our zone of interest (i.e. within the area covered by the layer superposition), from the identity transformation $(u,v) = \mathbf{i}(x,y) = (x,y)$. In other words, $\mathbf{g}(x,y)$ is a weak transformation if it satisfies $\mathbf{g}(x,y) = (x,y) + \mathbf{o}(x,y)$, where $\mathbf{o}(x,y)$ is an almost zero transformation. ∎

We now provide the following useful result:

Proposition D.9: If $\mathbf{o}(x,y)$ is an almost zero transformation, then:

(a) $\qquad [(x,y) - \mathbf{o}(x,y)]^{-1} \approx (x,y) + \mathbf{o}(x,y)$

(b) $\qquad [(x,y) + \mathbf{o}(x,y)]^{-1} \approx (x,y) - \mathbf{o}(x,y)$

or, in a more compact notation, denoting the identity and the zero transformations by $\mathbf{i}$ and $\mathbf{o}$, respectively:

(a) $\qquad [\mathbf{i} - \mathbf{o}]^{-1} \approx \mathbf{i} + \mathbf{o}$

(b) $\qquad [\mathbf{i} + \mathbf{o}]^{-1} \approx \mathbf{i} - \mathbf{o}$

Furthermore, the errors in the two approximations (a) and (b) are almost identical. ∎

Note that this proposition also means that if $\mathbf{g}(x,y)$ is a weak transformation, so that $\mathbf{g}(x,y) = (x,y) + \mathbf{o}(x,y)$, then $\mathbf{g}^{-1}(x,y) \approx (x,y) - \mathbf{o}(x,y)$; see also Figs. D.21 and D.22.

To derive this proposition, note that an almost zero transformation $\mathbf{o}$ is almost linear, because in its 2D Taylor series development (see, for example, [Courant89 pp. 68–70] or [Strogatz94 p. 150]) all the non-linear terms (i.e. the terms of second and higher orders) are negligible and can be dropped out. Therefore we can use the distributive law for the composition of linear transformations [Mansfield76 p. 168], and we obtain, denoting by "∘" the composition of transformations:

$$(\mathbf{i} + \mathbf{o}){\circ}(\mathbf{i} - \mathbf{o}) = \mathbf{i}{\circ}\mathbf{i} + \mathbf{o}{\circ}\mathbf{i} - \mathbf{i}{\circ}\mathbf{o} - \mathbf{o}{\circ}\mathbf{o}$$

$$= \mathbf{i} - \mathbf{o}{\circ}\mathbf{o}$$

$$\approx \mathbf{i}$$

(since the influence of $\mathbf{o}{\circ}\mathbf{o}$ is negligible). This means, indeed, that the transformations $\mathbf{i} + \mathbf{o}$ and $\mathbf{i} - \mathbf{o}$ are approximately the inverse of each other (see [Weisstein99 p. 901] for an equivalent reasoning in terms of matrices).

An alternative approach for deriving this proposition is based on the following well-known identities [Harris89 p. 540]:

$$(1 - x)^{-1} = 1 + x + x^2 + x^3 + x^4 + x^5 + \ldots \qquad (|x| < 1)$$

$$(1 + x)^{-1} = 1 - x + x^2 - x^3 + x^4 - x^5 + \ldots \qquad (|x| < 1)$$

or rather on their matrix counterparts [Horn85 p. 301]:

$$(I - A)^{-1} = I + A + A^2 + A^3 + A^4 + A^5 + \ldots \qquad (\|A\| < 1)$$

$$(I + A)^{-1} = I - A + A^2 - A^3 + A^4 - A^5 + \ldots \qquad (\|A\| < 1)$$

where A is a square matrix (in our case 2×2), $\|A\|$ is its norm, and I denotes the unit matrix.[40]

These matrix identities mean, in terms of the linear transformation $\mathbf{g}(x,y)$ that corresponds to the matrix A (see, for example, [Kreyszig78 p. 375]):[41]

$$(\mathbf{i} - \mathbf{g})^{-1} = \mathbf{i} + \mathbf{g} + \mathbf{g} \circ \mathbf{g} + \mathbf{g} \circ \mathbf{g} \circ \mathbf{g} + \mathbf{g} \circ \mathbf{g} \circ \mathbf{g} \circ \mathbf{g} + \mathbf{g} \circ \mathbf{g} \circ \mathbf{g} \circ \mathbf{g} \circ \mathbf{g} + \ldots \qquad (\|\mathbf{g}\| < 1)$$

$$(\mathbf{i} + \mathbf{g})^{-1} = \mathbf{i} - \mathbf{g} + \mathbf{g} \circ \mathbf{g} - \mathbf{g} \circ \mathbf{g} \circ \mathbf{g} + \mathbf{g} \circ \mathbf{g} \circ \mathbf{g} \circ \mathbf{g} - \mathbf{g} \circ \mathbf{g} \circ \mathbf{g} \circ \mathbf{g} \circ \mathbf{g} + \ldots \qquad (\|\mathbf{g}\| < 1)$$

Consider now the almost-zero transformation $\mathbf{o}(x,y)$. As we have seen above, this transformation is almost linear, and therefore we have:

$$(\mathbf{i} - \mathbf{o})^{-1} \approx \mathbf{i} + \mathbf{o} + \mathbf{o} \circ \mathbf{o} + \mathbf{o} \circ \mathbf{o} \circ \mathbf{o} + \mathbf{o} \circ \mathbf{o} \circ \mathbf{o} \circ \mathbf{o} + \mathbf{o} \circ \mathbf{o} \circ \mathbf{o} \circ \mathbf{o} \circ \mathbf{o} + \ldots \qquad (\|\mathbf{o}\| < 1)$$

$$(\mathbf{i} + \mathbf{o})^{-1} \approx \mathbf{i} - \mathbf{o} + \mathbf{o} \circ \mathbf{o} - \mathbf{o} \circ \mathbf{o} \circ \mathbf{o} + \mathbf{o} \circ \mathbf{o} \circ \mathbf{o} \circ \mathbf{o} - \mathbf{o} \circ \mathbf{o} \circ \mathbf{o} \circ \mathbf{o} \circ \mathbf{o} + \ldots \qquad (\|\mathbf{o}\| < 1)$$

But since we assume that $\mathbf{o}$ is an almost zero transformation, the influence of the mapping compositions $\mathbf{o} \circ \mathbf{o}$, $\mathbf{o} \circ \mathbf{o} \circ \mathbf{o}$, etc. is negligible; and indeed, by omitting them we obtain the two desired approximations:

$$[\mathbf{i} - \mathbf{o}]^{-1} \approx \mathbf{i} + \mathbf{o}$$

$$[\mathbf{i} + \mathbf{o}]^{-1} \approx \mathbf{i} - \mathbf{o}$$

Furthermore, it turns out that the approximation error $\mathbf{e}_1$ in (a) and the approximation error $\mathbf{e}_2$ in (b), both of which are functions of x and y, $\mathbf{e}_1(x,y)$, $\mathbf{e}_2(x,y)$, are almost identical. To see this, note that the approximation errors $\mathbf{e}_1$ and $\mathbf{e}_2$ are given by:[42]

$$\mathbf{e}_1 = \mathbf{o} \circ \mathbf{o} + \mathbf{o} \circ \mathbf{o} \circ \mathbf{o} + \mathbf{o} \circ \mathbf{o} \circ \mathbf{o} \circ \mathbf{o} + \mathbf{o} \circ \mathbf{o} \circ \mathbf{o} \circ \mathbf{o} \circ \mathbf{o} + \ldots$$

$$\mathbf{e}_2 = \mathbf{o} \circ \mathbf{o} - \mathbf{o} \circ \mathbf{o} \circ \mathbf{o} + \mathbf{o} \circ \mathbf{o} \circ \mathbf{o} \circ \mathbf{o} - \mathbf{o} \circ \mathbf{o} \circ \mathbf{o} \circ \mathbf{o} \circ \mathbf{o} + \ldots$$

---

[40] It turns out that these matrix series converge if $\|A\| < 1$ for *any* matrix norm $\|A\|$ [Horn85 pp. 258, 301]. This condition is clearly satisfied by matrices that are very close to the zero matrix O, as in our case.

[41] The norm $\|\mathbf{g}\|$ can be defined, for example, by $\|\mathbf{g}\| = \sup_{|(x,y)| = 1} |\mathbf{g}(x,y)|$ [Kreyszig78 p. 92], where $|(u,v)|$ is the length of the vector (u,v).

[42] Note that $\mathbf{e}_1$ and $\mathbf{e}_2$ express the truncation errors in (a) and (b), and therefore they accurately represent the errors when $\mathbf{o}$ is a linear transformation (that corresponds to the matrix A). But when $\mathbf{o}$ is only *almost* linear, the errors $\mathbf{e}_1$ and $\mathbf{e}_2$ are only *approximate*.

It follows, therefore, because $\mathbf{o}$ is an almost zero transformation, that the difference:

$$\mathbf{e}_1 - \mathbf{e}_2 = 2\mathbf{o} \circ \mathbf{o} \circ \mathbf{o} + 2\mathbf{o} \circ \mathbf{o} \circ \mathbf{o} \circ \mathbf{o} \circ \mathbf{o} + \ldots$$

is negligible with respect to $\mathbf{e}_1$ and $\mathbf{e}_2$, since $\mathbf{o} \circ \mathbf{o} \circ \mathbf{o}$ is of lower order of magnitude than $\mathbf{o} \circ \mathbf{o}$. This means, indeed, that $\mathbf{e}_1 \approx \mathbf{e}_2$.

Using this result we provide now the following useful approximations that connect between Eqs. (D.36) and (D.38):

Proposition D.10: If $\mathbf{g}_1$ and $\mathbf{g}_2$ are weak transformations then:

$$\overline{\mathbf{g}}_1(x,y) - \overline{\mathbf{g}}_2(x,y) \approx \mathbf{g}_2(x,y) - \mathbf{g}_1(x,y) \tag{D.44}$$

This means, using our notations from Eqs. (D.36) and (D.38), that the following approximations:

$$\overline{\mathbf{h}}(x,y) \approx -\mathbf{g}_M(x,y) \tag{D.45}$$

$$\overline{\mathbf{h}}(x,y) \approx \mathbf{g}_2(x,y) - \mathbf{g}_1(x,y) \tag{D.46}$$

$$\mathbf{g}_M(x,y) \approx \overline{\mathbf{g}}_2(x,y) - \overline{\mathbf{g}}_1(x,y) \tag{D.47}$$

also hold.   ∎

To see this, let $\mathbf{g}_1 = \mathbf{i} + \mathbf{o}_1$ and $\mathbf{g}_2 = \mathbf{i} + \mathbf{o}_2$ be weak transformations. We have, therefore:

$$\overline{\mathbf{g}}_1(x,y) - \overline{\mathbf{g}}_2(x,y) = [(x,y) + \mathbf{o}_1(x,y)]^{-1} - [(x,y) + \mathbf{o}_2(x,y)]^{-1}$$

Using Proposition D.9:
$$\approx [(x,y) - \mathbf{o}_1(x,y)] - [(x,y) - \mathbf{o}_2(x,y)]$$

$$= [(x,y) + \mathbf{o}_2(x,y)] - [(x,y) + \mathbf{o}_1(x,y)]$$

$$= \mathbf{g}_2(x,y) - \mathbf{g}_1(x,y)$$

The other approximations, (D.45)–(D.47), are simply different variants of (D.44). This proposition is very useful in cases where we do not have the explicit forms of the inverse (or of the direct) transformations, since it still allows us to obtain a close approximation based on those transformations whose explicit forms *are* known (see, for instance, Example 5.5 in Sec. 5.3).

Remark D.22: It is interesting to note that for some categories of transformations $\mathbf{g}_1(x,y)$ and $\mathbf{g}_2(x,y)$ the approximations given in Proposition D.10 turn into *identities*. For example, it is easy to see that if $(u,v) = \mathbf{g}_1(x,y)$ and $(u,v) = \mathbf{g}_2(x,y)$ are given, respectively, by:

$$\begin{pmatrix} u \\ v \end{pmatrix} = \begin{pmatrix} x + f_1(y) \\ y \end{pmatrix} \qquad\qquad \begin{pmatrix} u \\ v \end{pmatrix} = \begin{pmatrix} x \\ y + f_2(x) \end{pmatrix} \tag{D.48}$$

where $z = f_1(s)$ and $z = f_2(s)$ are arbitrary 1D functions, then their inverse transformations are given, respectively, by:

$$\begin{pmatrix} x \\ y \end{pmatrix} = \begin{pmatrix} u - f_1(v) \\ v \end{pmatrix} \qquad\qquad \begin{pmatrix} x \\ y \end{pmatrix} = \begin{pmatrix} u \\ v - f_2(u) \end{pmatrix}$$

and thus we have:

$$\mathbf{g}_1(x,y) - \mathbf{g}_2(x,y) = (f_1(y), -f_2(x))$$

$$\overline{\mathbf{g}}_2(x,y) - \overline{\mathbf{g}}_1(x,y) = (f_1(y), -f_2(x))$$

meaning that the following identity holds:

$$\mathbf{g}_1(x,y) - \mathbf{g}_2(x,y) = \overline{\mathbf{g}}_2(x,y) - \overline{\mathbf{g}}_1(x,y) \qquad\qquad\qquad (D.49)$$

Another example consists of all the transformations $\mathbf{g}_1(x,y)$, $\mathbf{g}_2(x,y)$ having the form:

$$(u,v) = (f_1(x), f_2(y)) \qquad\qquad (u,v) = (f_1^{-1}(x), f_2^{-1}(y)) \qquad\qquad (D.50)$$

where $z = f_1(s)$ and $z = f_2(s)$ are arbitrary 1D functions. In this case we have: $\overline{\mathbf{g}}_1(x,y) = \mathbf{g}_2(x,y)$ and $\overline{\mathbf{g}}_2(x,y) = \mathbf{g}_1(x,y)$, and therefore identity (D.49) holds, again.

There also exist other families of transformation pairs that satisfy identity (D.49); but for all other pairs $\mathbf{g}_1$ and $\mathbf{g}_2$ that do not satisfy the identity we still have the approximation given by Proposition D.10, provided that $\mathbf{g}_1$ and $\mathbf{g}_2$ are sufficiently weak transformations. ■

Finally, note that whenever the approximations of Proposition D.10 turn into identities, this also implies that the approximations given by Proposition D.8 turn into equalities, so that a point in the plane is a mutual fixed point of $\mathbf{g}_1(x,y)$ and $\mathbf{g}_2(x,y)$ *iff* it is a mutual fixed point of $\mathbf{g}_1^{-1}(x,y)$ and $\mathbf{g}_2^{-1}(x,y)$. The converse, however, is not necessarily true. For example, consider the simple case in which one of the two superposed layers remains unchanged: $\mathbf{g}_2(x,y) = (x,y)$. We already know from Proposition D.6 that in this case any fixed point of $\mathbf{g}$ is also a fixed point of $\overline{\mathbf{g}}$ (and vice versa), so that any point that satisfies $\mathbf{g}_1(x,y) - (x,y) = (0,0)$ also satisfies $(x,y) - \overline{\mathbf{g}}_1(x,y) = (0,0)$. However, this does not yet imply that for *any* point (x,y) we have $\mathbf{g}_1(x,y) - (x,y) = (x,y) - \overline{\mathbf{g}}_1(x,y)$. For example, if one of the layers has undergone a two-fold magnification $\overline{\mathbf{g}}_1(x,y) = (2x,2y)$ and the other layer remains unchanged, then:

$$\overline{\mathbf{g}}_1(x,y) - (x,y) = (2x,2y) - (x,y) = (x,y)$$

while: $\quad \mathbf{g}_1(x,y) - (x,y) = (x/2, y/2) - (x,y) = -(x/2, y/2)$

This clearly shows that in this case identity (D.49) is not satisfied; and yet, the fixed points of $\overline{\mathbf{g}}_1$ and $\mathbf{g}_1$ are, indeed, identical (the point $(0,0)$).

We conclude this appendix with some further approximations that prove to be useful in Chapter 4 (see Remark 4.3):

Proposition D.11: If $\mathbf{g}_1$ and $\mathbf{g}_2$ are weak transformations, i.e. transformations that only slightly differ from the identity transformation $\mathbf{i}(x,y) = (x,y)$ (at least within our zone of interest), then the following approximations:

$$\bar{\mathbf{g}}_1(\mathbf{g}_2(x,y)) - (x,y) \approx \mathbf{g}_2(x,y) - \mathbf{g}_1(x,y) \approx \bar{\mathbf{g}}_1(x,y) - \bar{\mathbf{g}}_2(x,y) \tag{D.51}$$

$$(x,y) - \bar{\mathbf{g}}_2(\mathbf{g}_1(x,y)) \approx \mathbf{g}_2(x,y) - \mathbf{g}_1(x,y) \approx \bar{\mathbf{g}}_1(x,y) - \bar{\mathbf{g}}_2(x,y) \tag{D.52}$$

also hold. ■

This result can be demonstrated as follows:

Denoting $(x',y') = \mathbf{g}_2(x,y)$ we have:

$$\bar{\mathbf{g}}_1(\mathbf{g}_2(x,y)) - (x,y) = \bar{\mathbf{g}}_1(x',y') - (x,y)$$

Now, using $\mathbf{g}_1 = \mathbf{i} + \mathbf{o}_1$ and hence $\bar{\mathbf{g}}_1 = [\mathbf{i} + \mathbf{o}_1]^{-1}$:

$$= [(x',y') + \mathbf{o}_1(x',y')]^{-1} - (x,y)$$

which gives, by virtue of Proposition D.9:

$$\approx (x',y') - \mathbf{o}_1(x',y') - (x,y)$$

and by substituting back $(x',y') = \mathbf{g}_2(x,y)$ and using $\mathbf{g}_2 = \mathbf{i} + \mathbf{o}_2$:

$$= (x,y) + \mathbf{o}_2(x,y) - \mathbf{o}_1((x,y) + \mathbf{o}_2(x,y)) - (x,y)$$

But since within our zone of interest $\mathbf{o}_1$ is almost linear we obtain:

$$\approx (x,y) + \mathbf{o}_2(x,y) - \mathbf{o}_1(x,y) - \mathbf{o}_1(\mathbf{o}_2(x,y)) - (x,y)$$

which gives in turn, because $\mathbf{o}_1(\mathbf{o}_2(x,y))$ is negligible:

$$\approx \mathbf{g}_2(x,y) - \mathbf{g}_1(x,y)$$

or, using (D.44): $\approx \bar{\mathbf{g}}_1(x,y) - \bar{\mathbf{g}}_2(x,y)$

Eq. (D.51) shows that when $\mathbf{g}_1$ and $\mathbf{g}_2$ are weak transformations Eq. (4.7) of Sec. 4.5 is indeed a close approximation to Eq. (4.8). This explains also why vector field (4.7) gives a good approximation to the dot trajectories in cases where both of the original layers are being transformed, provided that both layer transformations are rather weak. Note that we are allowed to use these approximations because in our application we must anyway restrict ourselves to weak layer transformations, in order not to destroy the correlation between the superposed layers within our zone of interest.

Appendix E

Convolution and cross correlation

E.1 Introduction

Convolution and cross correlation are operations on functions: each of these operations takes two functions $f(x)$ and $g(x)$, and produces from them a new function of the same variable x, that is customarily denoted by $h(x) = f(x) * g(x)$ or by $c_{f,g}(x) = f(x) \star g(x)$, respectively.[1] The need for the subscript in the cross correlation $c_{f,g}(x)$ will become clear shortly. In the 2D case the convolution and the cross correlation of the functions $f(x,y)$ and $g(x,y)$ are denoted by $h(x,y) = f(x,y) ** g(x,y)$ and $c_{f,g}(x,y) = f(x,y) \star\star g(x,y)$. A full description of convolution and cross correlation can be found in the literature (see, for example, Chapters 6 and 9 in [Gaskill78], Chapters 5 and 7 in [Cartwright90], or Chapter 5 in [Bracewell95]); in this appendix we only provide a general overview and discuss those properties that may be needed for our application. Because we are basically interested in 2D images this appendix is mostly formulated in terms of 2D functions; but the corresponding 1D results can be deduced from their 2D counterparts without difficulty. Note also that this appendix only deals with the continuous case. The discrete case — where the functions or images being treated are discrete (i.e. composed of pixels) — is basically obtained by replacing integrations by summations, but it remains beyond our scope here. More details on discrete convolution and cross correlation can be found in books on digital image processing or digital signal processing.

E.2 Convolution

The convolution of two real-valued functions $f(x,y)$ and $g(x,y)$ is defined to be:

$$h(x,y) = f(x,y) ** g(x,y) = \int_{-\infty}^{\infty} \int_{-\infty}^{\infty} f(x',y')\, g(x-x',\, y-y')\, dx'dy' \qquad (E.1)$$

Because this integral is clearly a function of the independent variables x and y, the resulting function $h(x,y)$ is again a function of x and y.[2] As we will see soon, the convolution operation may be viewed as one of finding the volume of the product of $f(x',y')$ and $g(x-x',\, y-y')$ as x and y are allowed to vary. In general, the resulting function $h(x,y)$ is smoother than either $f(x,y)$ or $g(x,y)$: The fine structure of the original functions

---

[1] More formally, convolution and cross correlation are *binary operators* that act on functions. Such binary operators are often written in the form $h = L[f,g]$. Simple examples of such operators include $S[f,g] = f + g$, $P[f,g] = fg$, $C[f,g] = f \star g$, etc. Similarly, there exist also *unary operators* that act on a single function, such as $U[f] = f^2$ or the Fourier transform, $\mathcal{F}[f]$ [Cartwright90 p. 135].

[2] Note that inside the integral the x and y axes are renamed x' and y'; thus, after integrating over the plane the dummy integration variables x' and y' disappear, and we are left with a function of x and y.

tends to be washed out, the sharp peaks and valleys tend to be rounded, etc., but the amount of smoothing depends on the exact nature of $f(x,y)$ and $g(x,y)$ (see [Gaskill78 pp. 162–163]).

Definition (E.1) may seem at first sight quite obscure, but in fact it has a quite intuitive graphical interpretation that gives a good visual insight into the nature of the convolution operation — the "move and multiply" interpretation (see, for example, [Rosenfeld82 pp. 13–14] or [Gaskill78 pp. 151–154, 291–292]): In order to obtain graphically the convolution of $f(x,y)$ and $g(x,y)$, we first draw the function $g(-x',-y')$, which is simply a 180° rotation of $g(x',y')$, and then we shift it along the x and y directions on top of $f(x',y')$. For each position x,y of the moving function, the resulting value of $h(x,y)$ is simply the volume under the product of the two functions (the fixed function $f(x',y')$ and the moving function $g(-x',-y')$ when it is shifted by x,y, i.e., $g(-(x'-x), -(y'-y)) = g(x-x', y-y')$).

The convolution operation has several useful properties, such as commutativity:

$$f(x,y) ** g(x,y) = g(x,y) ** f(x,y) \tag{E.2}$$

associativity:

$$[f_1(x,y) ** f_2(x,y)] ** f_3(x,y) = f_1(x,y) ** [f_2(x,y) ** f_3(x,y)] \tag{E.3}$$

homogeneity:

$$[cf(x,y)] ** g(x,y) = c[f(x,y) ** g(x,y)]$$
$$f(x,y) ** [cg(x,y)] = c[f(x,y) ** g(x,y)] \tag{E.4}$$

distributivity over addition:[3]

$$[f_1(x,y) + f_2(x,y)] ** g(x,y) = [f_1(x,y) ** g(x,y)] + [f_2(x,y) ** g(x,y)]$$
$$f(x,y) ** [g_1(x,y) + g_2(x,y)] = [f(x,y) ** g_1(x,y)] + [f(x,y) ** g_2(x,y)] \tag{E.5}$$

and shift preservation:

$$f(x-a, y-b) ** g(x,y) = f(x,y) ** g(x-a, y-b) = h(x-a, y-b) \tag{E.6}$$

Another interesting property is that the volume under the convolution $h(x,y)$ equals the product of the volumes under $f(x,y)$ and $g(x,y)$ [Bracewell95 p. 193]. Other properties of the convolution operation can be found, for example, in [Gaskill78 pp. 159–166, 292–294] and in [Poularikas96 pp. 27–32].

---

[3] More formally, properties (E.4) and (E.5) together imply that the convolution operator is *linear* [Cartwright90 p. 130]. In general, a binary operator $L[f,g]$ is said to be linear if it is homogeneous and additive in each of its two arguments, i.e. if it has the following properties: $L[cf,g] = cL[f,g]$, $L[f_1+f_2, g] = L[f_1,g] + L[f_2,g]$, $L[f,cg] = cL[f,g]$, and $L[f, g_1+g_2] = L[f,g_1] + L[f,g_2]$. Similarly, a unary operator $U[f]$ is said to be linear if $U[cf] = cU[f]$ and $U[f_1+f_2] = U[f_1] + U[f_2]$; an example of such an operator is the Fourier transform $\mathcal{F}[f]$ [Cartwright90 p. 90].

E.3 Cross correlation

Given two real-valued functions $f(x,y)$ and $g(x,y)$, we define the cross correlation of $f(x,y)$ and $g(x,y)$ to be:

$$c_{f,g}(x,y) = f(x,y) \star\star g(x,y) = \int_{-\infty}^{\infty} \int_{-\infty}^{\infty} f(x',y')\, g(x'-x,\, y'-y)\, dx'dy'$$

$$= \int_{-\infty}^{\infty} \int_{-\infty}^{\infty} f(x''+x,\, y''+y)\, g(x'',y'')\, dx''dy'' \qquad (E.7)$$

(where the second integral is obtained from the first one by a simple change of variables, $x''=x'-x$, $y''=y'-y$). The cross-correlation function $c_{f,g}(x,y)$ indicates the relative amount of agreement between the two given functions $f(x,y)$ and $g(x,y)$ for all possible degrees of misalignment (displacements). In other words, the function $c_{f,g}(x,y)$ is a measure of the similarity between $f(x,y)$ and a moving copy of $g(x,y)$, and it gets its maximum at the point (x,y) representing the displacement of g for which its similarity with f is the highest.[4]

As we can see, the cross-correlation operation, too, can be presented graphically using the "move and multiply" interpretation. It is important to note, however, that although this operation is similar to convolution, there is one very significant difference: the function $g(x,y)$ is not rotated as in convolution. This fact implies that convolution and cross correlation behave very differently. An example illustrating graphically the difference between $f(x) * g(x)$ and $f(x) \star g(x)$ for some given functions $f(x)$ and $g(x)$ can be found, for instance, in [Coulon84 pp. 82–83]. In fact, it is easy to see from definitions (E.1) and (E.7) that cross correlation can be expressed in terms of convolution by:

$$f(x,y) \star\star g(x,y) = f(x,y) ** g(-x,-y) \qquad (E.8)$$

It follows, therefore, that unlike convolution, the cross-correlation operation is *not* commutative, meaning that in general $f(x,y) \star\star g(x,y) \neq g(x,y) \star\star f(x,y)$. For this reason we have to clearly distinguish between the two functions $c_{f,g}(x,y) = f(x,y) \star\star g(x,y)$ and $c_{g,f}(x,y) = g(x,y) \star\star f(x,y)$; the relationship between the two is given by:[5]

$$c_{g,f}(x,y) = c_{f,g}(-x,-y) \qquad (E.9)$$

Another consequence of Eq. (E.8) is that if $g(-x,-y) = g(x,y)$ then $f(x,y) \star\star g(x,y)$ is identical to the convolution $f(x,y) ** g(x,y)$. In this particular case the cross correlation *is*, of course, commutative; such exceptional commutativity may also occur in some other special cases (see Problem 5-10 in [Bracewell95 pp. 201 and 643]).

Note that in the case of *complex-valued* functions $f(x,y)$ and $g(x,y)$ one can also define the *complex cross correlation* of $f(x,y)$ and $g(x,y)$:

---

[4] Note that in order to facilitate comparisons of the correlation between different functions it may be useful to normalize the cross correlation by dividing it by the square root of the product of the volume under $[f(x,y)]^2$ and the volume under $[g(x,y)]^2$. This ensures that the values of the normalized cross correlation are always between -1 and 1 [Cartwright90 p. 174].

[5] More formally, this means that the cross correlation operator $C[f,g]$ is anti-symmetric: $C[g(x),f(x)] = C[f(-x),\, g(-x)]$ [Cartwright90 pp. 174, 176].

$$\gamma_{f,g}(x,y) = f(x,y) \star\star g^*(x,y) = \int_{-\infty}^{\infty} \int_{-\infty}^{\infty} f(x',y') \, g^*(x'-x, \, y'-y) \, dx'dy'$$

$$= \int_{-\infty}^{\infty} \int_{-\infty}^{\infty} f(x''+x, \, y''+y) \, g^*(x'',y'') \, dx''dy'' \tag{E.10}$$

where $g^*(x,y)$ is the complex conjugate of the function $g(x,y)$. This definition is largely used in the literature, but because in our case we are only interested in real-valued functions (images), it is clear that $g^*(x,y) = g(x,y)$ and therefore the complex cross correlation (E.10) reduces into the cross correlation as defined in Eq. (E.7).

Remark E.1: There exists in the literature an alternative convention for the definition of convolution and cross correlation, in which the roles of f and g under the integral are inversed (see, for example, [Bracewell95 pp. 176–179]). Although this does not have any effect on the resulting function in the case of convolution (due to the commutativity of this operation), in the case of cross correlation this alternative definition inverses the meanings of $c_{f,g}(x,y)$ and $c_{g,f}(x,y)$, and gives instead of (E.8): $f(x,y) \star\star g(x,y) = f(-x,-y) \ast\ast g(x,y)$. In order to avoid confusion we will stick here to our definitions (E.1) and (E.7), following the notations in [Gaskill78]. ∎

Cross correlation is less generous than convolution in terms of nice mathematical properties. We have already mentioned its lack of commutativity, but in fact, it also lacks associativity (as one can see by rewriting the expressions $[f_1(x,y) \star\star f_2(x,y)] \star\star f_3(x,y)$ and $f_1(x,y) \star\star [f_2(x,y) \star\star f_3(x,y)]$ in terms of convolution, using Eq. (E.8)). And yet, cross correlation still does have some other useful properties, such as homogeneity [Cartwright90 p. 174]:

$$[cf(x,y)] \star\star g(x,y) = c[f(x,y) \star\star g(x,y)]$$

$$f(x,y) \star\star [cg(x,y)] = c[f(x,y) \star\star g(x,y)] \tag{E.11}$$

distributivity over addition [Cartwright90 p. 174]:[6]

$$[f_1(x,y) + f_2(x,y)] \star\star g(x,y) = [f_1(x,y) \star\star g(x,y)] + [f_2(x,y) \star\star g(x,y)]$$

$$f(x,y) \star\star [g_1(x,y) + g_2(x,y)] = [f(x,y) \star\star g_1(x,y)] + [f(x,y) \star\star g_2(x,y)] \tag{E.12}$$

an asymmetric shift-preservation property:

$$f(x+a, \, y+b) \star\star g(x,y) = f(x,y) \star\star g(x-a, \, y-b) = c_{f,g}(x-a, \, y-b) \tag{E.13}$$

and in spite of its non-commutativity it still satisfies the identity [Weisstein99 p. 352]:

$$(f \star\star g) \star\star (f \star\star g) = (f \star\star f) \star\star (g \star\star g) \tag{E.14}$$

Further properties of cross correlation are mentioned at the end of this section.

---

[6] Once again, like in the case of convolution, properties (E.11) and (E.12) together mean that cross correlation is a linear operator. Note that the lack of commutativity, $C[g,f] \neq C[f,g]$, does not contradict the linearity of the operator (i.e., its homogeneity and additivity in each of its two arguments).

An interesting particular case of cross correlation occurs when the functions f and g are identical; in this case the operation (E.7) is called the autocorrelation of $f(x,y)$, and denoted by $c_{f,f}(x,y) = f(x,y) \star\star f(x,y)$. The autocorrelation operation is described in detail in [Bracewell95 pp. 181–193]. It has several interesting properties some of which do not, in general, hold for either cross correlation or convolution, not even for self convolution $f(x,y) \ast\ast f(x,y)$ [Gaskill78 pp. 174–176]:

(1) The autocorrelation $c_{f,f}(x,y)$ has a twofold rotational symmetry about the origin (see a few examples in [Bracewell95 p. 185, Fig. 5-10]).[7]

(2) Its value is maximum at the origin: $c_{f,f}(x,y) \le c_{f,f}(0,0)$.

(3) Its central value $c_{f,f}(0,0)$ is the volume under the function $[f(x,y)]^2$ [Bracewell95 p. 192].

(4) The volume under the autocorrelation function is the square of the volume under $f(x,y)$ [Bracewell95 p. 192]. (Note that this is not the same as the volume under $[f(x,y)]^2$ in the previous property.)

(5) The autocorrelation is invariant under shifts of $f(x,y)$:
$f(x-a, y-b) \star\star f(x-a, y-b) = f(x,y) \star\star f(x,y)$ [Bracewell95 pp. 185–196].

Note that some of these properties of autocorrelation can be *partially* generalized into the case of cross correlation between two different functions $f(x,y)$ and $g(x,y)$. Property (2) can be generalized as follows [Coulon84 p. 82], [Bendat93 p. 48]:

$$[c_{f,g}(x,y)]^2 \le c_{f,f}(0,0)\, c_{g,g}(0,0) \tag{E.15}$$

Note, however, that unlike in autocorrelation this does not mean that the absolute value of cross correlation is maximum at the origin. The counterpart of property (3) says in the case of cross correlation that $c_{f,g}(0,0)$ is the volume under the product $f(x,y)g(x,y)$. The counterpart of property (4) says that the volume under the cross correlation $f(x,y) \star\star g(x,y)$ equals the product of the volumes under $f(x,y)$ and $g(x,y)$; this is obtained from the analogous property of convolution (see at the end of Sec. E.2) by using Eq. (E.8) and noting that the area under $g(-x,-y)$ equals the area under $g(x,y)$. Finally, the counterpart of property (5) for cross correlation is already given in Eq. (E.13) above; note that this shift-preservation property is weaker than the invariance under shifts in autocorrelation, since it only asserts that shifting $f(x,y)$ or $g(x,y)$ gives a shifted version of $c_{f,g}(x,y)$, but it does not give $c_{f,g}(x,y)$ itself.

E.4 Extension to more general cases

Definitions (E.1) and (E.7) are only appropriate in cases where the integrals have finite values. This includes cases where one (or both) of the functions is a *finite-energy signal*

---

[7] This also means that the unary operator of autocorrelation $U[f]$ is symmetric: $U[f(x,y)] = U[f(-x,-y)]$.

(i.e. square integrable) [Champeney73 pp. 59, 68], but it obviously excludes cases with functions of constant character, periodic functions, or stationary random functions. But there exists a way for generalizing the notions of convolution and cross correlation to functions that are *finite-power signals* [Champeney73 pp. 59, 63, 68]; those include also constant functions, step functions, periodic functions and stationary random functions. This generalization is done by replacing the integral by a limiting process (see, for example, [Cartwright90 pp. 174–175], [Gaskill78 p. 158] or [Coulon84 pp. 34, 91–92]):[8]

$$f(x,y) ** g(x,y) = \lim_{a \to \infty} \frac{1}{a^2} \int_{-a/2}^{a/2} \int_{-a/2}^{a/2} f(x',y')\, g(x - x',\, y - y')\, dx'dy' \qquad (E.16)$$

$$f(x,y) \star\star g(x,y) = \lim_{a \to \infty} \frac{1}{a^2} \int_{-a/2}^{a/2} \int_{-a/2}^{a/2} f(x',y')\, g(x' - x,\, y' - y)\, dx'dy' \qquad (E.17)$$

In the periodic case, where both of the given functions f and g are assumed to have the same period T, the limits given in Eqs. (E.16) and (E.17) are identical to the mean values calculated on a single period. Therefore, in this case we have [Coulon84 p. 99]:

$$f(x,y) ** g(x,y) = \frac{1}{T^2} \int_T \int_T f(x',y')\, g(x - x',\, y - y')\, dx'dy' \qquad (E.18)$$

$$f(x,y) \star\star g(x,y) = \frac{1}{T^2} \int_T \int_T f(x',y')\, g(x' - x,\, y' - y)\, dx'dy' \qquad (E.19)$$

This gives, of course, periodic functions having the same period T as f and g. These functions are known, respectively, as *T-convolution* and *T-cross correlation* (or *cyclic* convolution and *cyclic* cross-correlation). An extension to cases where the functions f and g have different periods is also possible, as shown in [Gaskill78 p. 158].

E.5 The Fourier transform of convolution and of cross correlation

The operations of convolution and cross correlation play a major role in the Fourier theory thanks to two fundamental theorems, that are known as the convolution theorem and the cross-correlation theorem. Given two real-valued functions $f(x,y)$ and $g(x,y)$, the convolution theorem asserts that the Fourier transform of the convolution $h(x,y) = f(x,y) ** g(x,y)$ is given by the product of the individual Fourier transforms:

$$H(u,v) = F(u,v)G(u,v) \qquad (E.20)$$

As an immediate result, the autoconvolution theorem says that the Fourier transform of the autoconvolution $h(x,y) = f(x,y) ** f(x,y)$ is given by:

$$H(u,v) = F(u,v)^2 \qquad (E.21)$$

---

[8] In fact, this is a generalization of the normalized versions of Eqs. (E.1) and (E.7) that are divided by the area a^2 in which the integration is taking place (assuming that the functions f and g have a finite spatial extent).

On the other hand, given the same real-valued functions $f(x,y)$ and $g(x,y)$, the cross-correlation theorem says that the Fourier transform of the cross correlation $c_{f,g}(x,y) = f(x,y) \star\star g(x,y)$ is given by [Gaskill78 p. 200]:

$$C_{f,g}(u,v) = F(u,v)G(-u,-v) \tag{E.22}$$

As a particular case, the autocorrelation theorem says that the Fourier transform of the autocorrelation $c_{f,f}(x,y) = f(x,y) \star\star f(x,y)$ is given by:

$$C_{f,f}(u,v) = F(u,v)F(-u,-v) \tag{E.23}$$

but since in our case $f(x,y)$ is real it follows that $F(u,v)$ is Hermitian, and therefore $F(-u,-v)$ is just the complex conjugate $F^*(u,v)$ [Bracewell95 p. 208], and we obtain:

$$C_{f,f}(u,v) = F(u,v)F^*(u,v) = |F(u,v)|^2 \tag{E.24}$$

This means that the Fourier transform of the autocorrelation function $c_{f,f}(x,y)$ is the *power spectrum* of $f(x,y)$.[9] This result is also known as the Wiener-Khintchine theorem.

In the case of *finite-power signals* (see Sec. E.4), the Fourier transform does not always exist; for example, stationary random functions do not have Fourier transforms [Champeney73 p. 59].[10] In such cases Eqs. (E.20) and (E.22) do not hold. And yet, it turns out that Eq. (E.24) does have a valid counterpart: As explained in [Coulon84 p. 93] for the 1D case, if $f(x)$ is a finite-power signal, then $r(x,a) = f(x) \operatorname{rect}(x/a)$ is a finite-energy signal that satisfies $f(x) = \lim_{a \to \infty} r(x,a)$. Let $R(u,a)$ be the Fourier transform of $r(x,a)$; by the 1D analog of Eq. (E.24) we have $C_{r,r}(u,a) = |R(u,a)|^2$; and it turns out that:[11]

$$C_{f,f}(u) = \lim_{a \to \infty} \frac{1}{a} |R(u,a)|^2 \tag{E.25}$$

In the particular case of finite-power signals where the given functions f and g are periodic with the same period T the convolution and cross-correlation theorems do hold, but they must be adapted accordingly, replacing convolution and cross correlation by T-convolution and T-cross correlation. This gives the T-convolution and the T-cross-correlation theorems. More details can be found in Secs. 4.2 and 4.3 of *Vol. I* and in [Champeney87 p. 166].

---

[9] Note the major difference between this result and its counterpart for *autoconvolution* which is given in Eq. (E.21). The difference is that unlike $F(u,v)^2$, the power spectrum $|F(u,v)|^2$ contains no phase information; this also means that the autocorrelation function $c_{f,f}(x,y)$, too, unlike the autoconvolution function $h(x,y)$, contains no information about the phase of $f(x,y)$ [Bracewell86 p. 115]. This difference originates from the fact that for any complex number $z = |z| e^{i\theta}$, the value $|z|^2$ is purely real and has no phase component, whereas the value z^2 equals $|z|^2 e^{i2\theta}$ and has in general a non-zero phase component. Remark that the loss of phase information occurs either while taking the cross correlation ($f(x,y) \Rightarrow c_{f,f}(x,y)$) or while taking the power spectrum ($f(x,y) \Rightarrow |F(u,v)|^2$), but not while taking the Fourier transform ($c_{f,f}(x,y) \Rightarrow C_{f,f}(u,v)$). The Fourier transform does not cause any loss of information.

[10] Note that this fact concerns *theoretical* stationary random functions, that extend throughout the range $-\infty < x < \infty$, but not their *practical* approximations whose spatial extent is finite.

[11] Note that in some references such as [Champeney73, Chapter 4] it is customary to call this limit the *power spectrum* of the finite-power signal $f(x)$, and to use the term *energy spectrum* for $|F(u)|^2$ in cases where $f(x)$ is a finite-energy signal. However, we do not follow this convention, and prefer to use the same term, power spectrum, in all circumstances.

E.6 Methods for quantifying the correlation; similarity measures

It is often desirable to compare the degree of correlation between functions; for example, one may want to know if the functions $f_1(x,y)$ and $f_2(x,y)$ are more correlated (i.e. more similar to each other) than $f_1(x,y)$ and $f_3(x,y)$. How can we formally formulate the degree of correlation between two functions (or in our application, between two images)? As we have seen, the cross correlation operation between two functions $f(x,y)$ and $g(x,y)$ may be a useful starting point. However, the cross correlation is again a 2D function, and the question remains how we can use it to compare the degree of correlation between the two images; in other words, how can we extract from the 2D cross correlation (or even better, from the normalized 2D cross correlation) a single number, known as a matching score, that can be used to evaluate the similarity between $f(x,y)$ and $g(x,y)$? In fact, this question does not have a unique answer, and one could think of several ways of doing so. Let us evaluate here some of the most plausible methods that may come to one's mind.

(1) The volume under the cross correlation. This seems to be a promising method, but in fact it turns out to be useless. Suppose, for example, that $g(x,y)$ is a rotated version of $f(x,y)$. Clearly, the more $g(x,y)$ is rotated, the lower the similarity between $g(x,y)$ and $f(x,y)$. However, as we have seen, the volume under the cross correlation equals the product of the volumes under $f(x,y)$ and under $g(x,y)$; but the volumes under both $f(x,y)$ and $g(x,y)$ are independent of the rotation angle. This means that the area under the cross correlation remains constant when we rotate $g(x,y)$, and therefore it cannot be used to determine the degree of correlation between the two images.

(2) The maximum value or the maximum absolute value of the cross correlation. This method is used, indeed, in *template matching*, where the problem is to find the closest match between a given unknown image and a set of known images.[12] In this approach one computes the cross correlation between the unknown and each of the known images, and the *closest match* is then found by selecting the image that yields the cross-correlation function with the largest value. Since the resultant cross correlations are 2D functions, this involves searching for the largest amplitude of each such function [Gonzalez87 p. 92]. Note that if we only look for the position of $g(x,y)$ in which it is most similar to $f(x,y)$ (for example, if we search the position of a letter "M" within an image consisting of some given text) then the procedure is simpler: In this case the desired location is simply the point (x,y) for which the cross correlation of the two images has the highest value [Gonzalez87 pp. 425–427]. If $g(x,y)$ is identical to $f(x,y)$ the highest value is located at the origin, and it equals the volume under the function $[f(x,y)]^2$ (see properties (2) and (3) of autocorrelation in Sec. E.3).

One may also think of other matching scores based on cross correlation to quantify the similarity between two functions. It should be remembered, however, that the cross-correlation operation (and hence, any similarity score that is based on it) is only capable of detecting similarities between functions $f(x)$ and $g(x)$ that are basically related by a simple

---

[12] Such situations frequently occur in digital image processing when the images in question are discrete, but the principles are the same for both discrete and continuous cases.

relationship of the type: $g(x) = cf(x - b)$, namely, *amplitude scaling* and *lateral shift*. Even simple linear relationships such as $g(x) = f(ax)$, let alone non-linear relationships such as $g(x) = f(x^2)$, cannot be uncovered by cross correlation [Cartwright90 p. 174]. The 2D counterpart of this fact is, indeed, nicely illustrated in our discussion on Glass patterns in Sec. 7.8. In order to overcome this significant restriction, it is customary to *undo* any geometric transformations that were undergone by the signals (or images) to be compared — of course, provided that these transformations are known in advance, or that they can be estimated (for example, there exist methods based on log-polar mapping or on the Mellin transform for recovering scale and rotation transformations that were undergone by an image; see, for instance, [Feitelson88 Sec. 3.2], [Hotta99] and [Casasent76]).

It should be noted that methods based on cross-correlation are not the only ones that can be used to estimate the degree of similarity between two functions $f(x)$ and $g(x)$. Other similarity metrics include the *coherence* [Cartwright90 pp. 179–180; Coulon84 p. 146]; the *scalar product* of f and g, $s[f,g] = \int f(x)g(x)\,dx$ (or its normalized version, that is called in [Cartwright90 pp. 169–171] the *correlation coefficient* of the functions f and g); the Euclidean distance $d_2[f,g] = [\int |f(x) - g(x)|^2\,dx]^{1/2}$; and other distance functions that are based on different norms such as $d_1[f,g] = \int |f(x) - g(x)|\,dx$ or $d_\infty[f,g] = \sup|f(x) - g(x)|$ [Lipschutz01 pp. 252–254].[13] An example showing a graphical comparison between some of these distances can be found in [Coulon84 pp. 44–46]. Among the distance functions the Euclidean distance is the most useful, but the other distances are occasionally used, too, either because they are better adapted for a given context, or simply because they allow easier calculation [Coulon84 p. 43].

Note that unlike cross correlation all of these techniques yield the distances between the given functions as a single number and not as a function, and they do not take into account displacements between the given functions.

Remark E.2: Some of the similarity measures mentioned above are, indeed, interrelated. For example, the Euclidean distance between two functions f and g is related to their scalar product by the relation [Coulon84 p. 48]:

$$(d_2[f,g])^2 = s[f,f] + s[g,g] - 2s[f,g] \qquad (E.26)$$

In particular, if f and g are orthogonal, meaning that $s[f,g] = 0$, this relation reduces into the Pythagoras theorem:

$$(d_2[f,g])^2 = s[f,f] + s[g,g]$$

$$= (d_2[f,0])^2 + (d_2[g,0])^2$$

There also exists a relationship between the cross correlation of f and g and their scalar product: If we denote by g_x the counterpart of g which has been translated by x, namely,

---

[13] Note that the entities we denote here by lower case, such as $s[f,g]$, $d[f,g]$ etc. are not *operators*, like $C[f,g]$, but *functionals*: for any two functions f and g they yield a *number*, and not another function.

$g_x(x') = g(x' - x)$, then the cross-correlation function $c_{f,g}(x)$ expresses the evolution of the scalar product as a function of the translation x:

$$c_{f,g}(x) = s[f, g_x] \tag{E.27}$$

Similarly, there also exists a relationship between the cross correlation of f and g and the Euclidean distance between them as a function of the translation x [Coulon84 p. 81]:

$$(d_2[f, g_x])^2 = c_{f,f}(0) + c_{g,g}(0) - 2c_{f,g}(x) \tag{E.28}$$

(using property (3) of autocorrelation). Note that Eq. (E.26) is simply a particular case of this relationship where the translation is $x = 0$. ∎

Remark E.3: At each value of the translation x where the cross-correlation function $c_{f,g}(x)$ equals zero, the functions f and g are *non-correlated*, and the functions f and g_x are *orthogonal*. ∎

Remark E.4: It may be theoretically possible to define other variants of cross correlation (and convolution) that would account for other misregistrations of $g(x,y)$ than simple translation. For example, a variant of cross correlation that accomodates *scalings* of $g(x,y)$ rather than translations could be defined by:

$$s_{f,g}(a,b) = \int_{-\infty}^{\infty} \int_{-\infty}^{\infty} f(x',y') \, g(ax', by') \, dx' dy'$$

and a variant that accomodates both scalings and translations could be defined by:

$$s_{f,g}(x,y,a,b) = \int_{-\infty}^{\infty} \int_{-\infty}^{\infty} f(x',y') \, g(ax' - x, by' - y) \, dx' dy'$$

This last variant would give peaks in the 4D space spanned by the x, y, a and b axes at those values of (x,y,a,b) for which the best correlation is found between $f(x,y)$ and the scaled and shifted variants of $g(x,y)$. Similar definitions could be also provided to allow for rotations, general linear or affine transformations, or even non-linear transformations. Such an approach has been presented, for the discrete case, in [Pratt91 pp. 669–671].

However, the interest in such generalizations is usually limited, because the same effects can be obtained by first undoing the transformations undergone by $g(x,y)$ (and possibly also by $f(x,y)$), and then applying standard cross correlation to the untransformed $f(x,y)$ and $g(x,y)$. This approach is more economical in terms of computational load since it does not require to run throughout a four- or even higher-dimentional space, and moreover, it allows us to use the large body of already existing theoretical results (including the convolution and the cross-correlation theorems) without having to adapt them (if at all possible) to each particular family of transformations we might wish to consider. ∎

Appendix F

The Fourier treatment of random images and of their superpositions

F.1 Introduction

The Fourier theory is very well adapted to the study of random structures, too (see, for example, [Bracewell86 Chapters 15–16], [Champeney73 Chapter 6] or [Coulon84 Chapters 5–6]). The main difference between the Fourier treatment of deterministic signals and that of random signals is that in the latter case all the phase information is lost, and we no longer have the full spectral representation of the signals but only their power spectra (see, for example, [Castelman79 pp. 199–201], [Bracewell86 p. 381]).

And yet, as we will see below, given a *fully known* random image such as the random dot screen shown in Fig. 2.1(a), we can still consider it as being deterministic, and treat it just like any other image. For example, we can apply to it the Fourier transform and obtain its full spectral representation, including all the phase information. In this appendix we evaluate both the stochastic and the deterministic approaches in the particular context of the moiré theory (the investigation of Glass patterns between random layers). We first provide in Secs. F.2–F.4 a short review of the stochastic approach, and then, in Sec. F.5 we explain why we have chosen to use the deterministic rather than the stochastic approach.

Note that whenever we wish to cover both the 1D and the 2D cases we use the generic term "signal" rather than our usual term "image", which has a strong 2D connotation.

F.2 Stochastic processes and their power spectra

Signals can be classified into two distinct categories: deterministic signals and random (or stochastic) signals. A signal is said to be *deterministic* if its values are fully known throughout its domain of definition, or if its values can be predicted by an appropriate mathematical formula or model. On the other hand, a signal is said to be *random* (or *stochastic*) if its precise values are not fully predictable and they depend, at least partly, on the laws of probability; such a signal does not have an analytic representation and it can only be described using statistical considerations.[1] As we clearly see from this definition, a signal does not need to be completely unknown in order to be random; for example, a signal whose shape is fully known and only its position along the axes is unknown (such

---

[1] Stated in other words, the values of such a signal $f(x)$ do not depend in a completely definite way on the independent variable x, as in a deterministic signal; instead, the evolution of the signal depends also on chance, so one gets in different observations different realizations of $f(x)$.

as a sinusoidal signal whose initial phase is unknown) is already considered as a random signal. Note, however, that although a random signal does not have an analytic representation, it still can be characterized by its *statistical* properties (such as the probability distribution of its values, its average, its standard deviation, etc.) and by its *frequential* properties (its spectral decomposition in terms of its power spectrum).

Any observed random signal should be considered as one particular case among all the similar signals that could be produced by the same phenomenon or random process. Mathematically, a *random process* (or a *stochastic process*) is defined as an ensemble of signals, $\{f_\zeta(\mathbf{x})\}$, where the variable $\mathbf{x}$ represents a point in the signal's domain of definition (for example, along the time axis in the 1D case or within the x,y plane in the 2D case), and the variable ζ represents an element of the ensemble, i.e. one among all the possible signals that may result from the same statistical experiment [Papoulis65 pp. 279–281]. Note that each of the variables ζ and $\mathbf{x}$ may be either continuous or discrete. A stochastic process is said to be *continuous* or *discrete* depending on whether ζ is continuous or discrete; if the variable $\mathbf{x}$ is discrete the process is called a *random sequence*, and if both ζ and $\mathbf{x}$ are discrete the process is called a *point process*. Each member of the ensemble $\{f_\zeta(\mathbf{x})\}$ is called a *particular instance* or a *particular realization* of the random process, and is, in itself, a random signal.[2]

The concept of a stochastic process can be illustrated pictorially as a series of random functions $f_\zeta(\mathbf{x})$ that are stacked along the ζ axis. Assume that we draw the 3D x,y,z space such that the x axis points to the right within the paper's plane, the z axis points upward within the paper's plane, and the y axis is perpendicular to this plane and points toward the observer. If the variable $\mathbf{x}$ is 1D, we can identify the ζ axis with our y direction, so that each of the functions $z = f_\zeta(\mathbf{x})$ of the random process can be drawn within the paper's plane or parallel to it along the ζ axis (see Fig. 5.1 in [Coulon84 p. 112] or Fig. 1.1 in [Bendat93 p. 2]). If the variable $\mathbf{x}$ is 2D, the random process is best represented as a stack of planar gray level plots $f_\zeta(x,y)$ that is viewed from a lateral perspective, where the ζ axis (the stacking direction) coincides with our z direction (see Fig. 8 in [Rosenfeld82 p. 40]).

Clearly, statistical properties of a stochastic process such as its mean value can be approached in two different ways: they can be either computed along the x axis (or the x,y axes in the 2D case), or along the ζ axis. In the first case the mean value we obtain is called a *time average* (or *spatial average*), while in the second case, when the mean is computed over the different realizations of the random process, the value we obtain is called an *ensemble average*. A stochastic process is said to be *ergodic* if (1) the time averages of all its member functions are equal; (2) the ensemble average is constant with time; and (3) the time average and the ensemble average are equal [Castelman79 p. 200].

---

[2] It should be stressed, however, that once a random signal has been observed or recorded, all its values are fully known, and it should be therefore considered as a deterministic signal, albeit of random origin. This fact may cause some terminological confusion, because such a signal is still very often called a random signal. For example, all the random dot screens and line gratings in the figures throughout this volume are, in fact, deterministic signals of random origin, because they are fully known. But in most cases the intended meaning can be understood from the context without difficulty.

(Note that this definition is formulated here for the 1D case, but its 2D counterpart is easily obtained by replacing the term "time average" by "spatial average"). Thus, for ergodic processes, time averages (or spatial averages) and ensemble averages are interchangeable. This is also true for higher-order statistical averages such as the mean quadratic value of the process, its standard deviation, etc. (see Table 5.2 in [Coulon84 p. 117]). Hence, whenever ergodicity can be assumed one may compute the statistical properties of the stochastic process by analyzing the temporal (or spatial) behaviour of a single signal (a single member of the ensemble), which is obviously much easier to do in practice. Fortunately, ergodic processes model commonly encountered random signals quite well [Castelman79 p. 201]. In the following we will assume, indeed, that all our stochastic processes are ergodic.

We now proceed to the spectral properties of a random process; but before doing so, let us first consider its autocorrelation function. The autocorrelation function of a signal is defined as a time (or spatial) average by:

$$c_{f,f}(x,y) = f(x,y) \star\star f(x,y) = \int_{-\infty}^{\infty} \int_{-\infty}^{\infty} f(x',y') \, g(x'-x, y'-y) \, dx'dy'$$

(see Sec. E.3 in Appendix E). In an ergodic stochastic process the autocorrelation function is the same for all member signals, and thus it characterizes the ensemble. Therefore, although the precise values of an ergodic stochastic process are unknown, its autocorrelation function *is fully known*; this function reflects, in fact, our partial knowledge of the stochastic process. Now, since the autocorrelation $c_{f,f}(x,y)$ of the process is known, its Fourier transform $C_{f,f}(u,v)$ is also known; but according to the autocorrelation theorem (see Sec. E.5 in Appendix E), this is precisely the power spectrum of the process. This means that the power spectrum of a stochastic process is also fully known, just like the autocorrelation function; in fact, both of them contain the same information, which is only presented in a different way, in the spectral domain or in the image domain. It is important to note, however, that although we know the *power spectrum* of the stochastic process, and hence its *amplitude spectrum*, too (which is simply the square root of the power spectrum), we do not know the *phase spectrum* of the stochastic process.[3] This means that unlike in deterministic signals, random signals do not have a full Fourier spectrum, and one can only deal with their power (or amplitude) spectra [Castelman79 p. 201; Bracewell86 p. 381]. Another explanation why random signals only have power spectra is provided in [Champeney73 pp. 79–80]: it turns out that any attempt to generalize the Fourier transform to random signals is doomed to be unsuccessful, while the power spectrum *can* be generalized to such cases.

The power spectrum of a stochastic process can be seen as the mean (i.e. expectance) of the power spectra of all the infinitely many individual signals that make up the stochastic process [Coulon84 pp. 135–136]. Under the assumption of *ergodicity*, this is also equal

---

[3] Remember that the Fourier transform $F(u)$ of a function $f(x)$ is a complex-valued entity, so that it can be presented either in terms of its real part $\text{Re}[F(u)]$ and its imaginary part $\text{Im}[F(u)]$, where $F(u) = \text{Re}[F(u)] + i\text{Im}[F(u)]$, or, equivalently, in the polar representation, in terms of its amplitude spectrum $\text{Abs}[F(u)]$ and its phase spectrum $\text{Arg}[F(u)]$, where $F(u) = \text{Abs}[F(u)] \cdot e^{i\text{Arg}[F(u)]}$.

to the mean power spectrum (E.25) of any individual signal (particular instance) of the process [Papoulis65 p. 343]. Due to the averaging effect the power spectrum of a stochastic process is much smoother than the power spectra of the individual signals of the process: Usually, because the structure of each individual signal of the process is rather irregular, its power spectrum may be jumpy or noisy, and it often admits a typical *diffuse* appearance (see [Bracewell95 pp. 586–590, 600–601] and Problem 2-2). However, in the power spectrum of a stochastic process the fluctuations that are inherent to each of the individual spectra are smoothed out, so that we are left with a "clean" representation of the net spectral behaviour of the process. This ideal average power spectrum, in which all random effects have been averaged out, may typically consist of smooth curves and isolated impulses, but it is no longer buried in a diffuse random background noise. Several pictorial illustrations of such power (or rather amplitude) spectra can be found in the central column of the figure in [Coulon84 p. 500].

This clean look of the theoretic, average power spectrum of the stochastic process certainly facilitates the understanding of the underlying spectral information, and is therefore more attractive to use than the noisy spectra of the individual random signals.[4] However, if an individual signal of the process is fully known, we still may prefer to consider it as a *deterministic* signal, even though it originates from a random process. This would allow us to take the Fourier transform of the signal and obtain its full spectral information, i.e. both its amplitude spectrum and its phase spectrum (although both of them would usually contain some diffuse noise, too). This would also allow us to freely pass between the original signal in the image domain and its spectrum in the frequency domain and vice versa, and to benefit from fundamental results such as the convolution theorem (which allows us to consider products in one domain as convolutions in the other domain and vice versa). On the other hand, when we are dealing with a random process we have no longer access to the Fourier spectrum but only to the amplitude (or power) spectrum, and all the phase information (the phase spectrum) is lost. This means that we can no longer freely pass between the image and spectral domains, and furthermore, we can no longer use the convolution theorem (note that power spectra have no equivalent to the convolution theorem; more about the properties and non-properties of power spectra can be found in [Champeney73 pp. 64–65]). We will return to these considerations in more detail in Sec. F.5, where we evaluate the stochastic and the deterministic approaches in the particular context of the moiré theory.

Remark F.1: Note that in a deterministic signal $f(x)$ the power spectrum $\mathcal{P}_f(u)$ can be obtained in two different ways: Either directly from the Fourier transform $F(u)$ of the signal: $\mathcal{P}_f(u) = |F(u)|^2$, or indirectly, using the autocorrelation theorem (see Sec. E.5 in Appendix E). In the indirect way we first have to find the autocorrelation function $c_{f,f}(x)$ of the given signal $f(x)$, and then we apply to it the Fourier transform: $\mathcal{P}_f(u) = C_{f,f}(u)$. However, if $f(x)$ is not a deterministic signal but a random process the direct approach is no longer available, because a random process does not have a Fourier transform. In this

---

[4] A more detailed discussion on the averaged, limit power spectrum and its possible uses can be found in [Gardner88 pp. 5–6, 67–72].

case the power spectrum can only be found in the indirect method, using the autocorrelation theorem; this also explains why the power spectrum of a random process is often defined in the literature as the Fourier transform of the autocorrelation (see, for example, [Papoulis65 p. 338]). ■

F.3 Possible stochastic modelizations of random screens and gratings

Stochastic processes may be used to modelize many different types of random images; a non-exhaustive, illustrated survey of several different types of such random images can be found in Chapter 17 of [Bracewell95]. In the present section we briefly review some of the stochastic processes that can be used for the modelization of random images of the types we mostly use, such as dot screens, line gratings, etc. The different models are presented in increasing order of their suitability to our needs.

F.3.1 Point processes

The simplest and most natural approach for the stochastic modelization of random structures such as random dot screens is based on *spatial point processes*. A spatial point process is any stochastic mechanism which generates a *point pattern*, i.e. a countable set of points x_i in the plane [Diggle83 Chapter 4]. Clearly, such a model only takes into account the random locations of our screen elements, but not the elements themselves (their shapes, sizes, etc.). In fact, we can think of each particular instance of the spatial point process as a nailbed consisting of randomly located impulses; our random dot screen can be then considered as the convolution of this random nailbed with a single dot, as we will see in the following subsection.

The simplest point process is the *Poisson process*. This is, indeed, the cornerstone on which the theory of spatial point processes is built. This model is used in applications as an idealized standard of *complete spatial randomness*; although unattainable in practice, it often provides a useful approximate description of an observed pattern [Diggle83 p. 50]. Informally, a Poisson process can be seen as a 1D or 2D impulse train (i.e. a comb or a nailbed) whose individual impulses are positioned at random:

$$q(x) = \sum_n \delta(x - x_n) \tag{F.1}$$

The formal requirements on (F.1) for being a Poisson process (i.e. the formal requirements for the dot locations x_n to be fully random) are given, for example, in [Diggle83 p. 50].

The statistical properties of a Poisson process are well established, and they can be found in the literature (see, for example, [Diggle83 pp. 50–51] or [Papoulis65 pp. 284–287]; note that the latter uses the term "Poisson impulses" for what we call here a Poisson process). Let us briefly consider now the *spectral* properties of such a process.

The power spectrum of a Poisson process (i.e. the power spectrum of a random nailbed) is given by (see, for example, [Champeney73 pp. 82–83]):

$$P_q(u) = \eta + \eta^2\, \delta(u) \tag{F.2}$$

where η is the average number of dots per unit interval.[5] Upon Fourier inversion this also gives us the autocorrelation function of the Poisson process:

$$c_{q,q}(x) = \eta\delta(x) + \eta^2 \tag{F.3}$$

As we can see from Eq. (F.2), the power spectrum of a random impulse train is evenly distributed over all frequencies, and the Poisson process is therefore an example of "white noise" (with an additional DC impulse at the spectrum origin). Note, however, that this power spectrum is not unique to random combs, and it is shared with infinitely many other random processes, including continuous, non-impulsive ones (remember that unlike the Fourier transform $F(u)$, the power spectrum $P_f(u)$ does not uniquely originate from a single function $f(x)$).

Apart from the Poisson process there exist many other types of point processes, having different statistical distribution rules for the random points and different spatial and spectral properties. These processes include, for example, inhomogeneous Poisson processes, clustered processes, inhibition processes, Markov point processes, lattice-based processes, etc. A short survey on some of the main types of point processes and their properties can be found, for example, in [Diggle83, Chapter 4].

It should be noted, however, that all point processes are based on the simplifying assumption that we are dealing with a random arrangement of *impulses*; as already mentioned above, this means that point processes can only modelize the locations of our screen dots, but they do not take into account the actual geometric shapes and sizes of the individual dots. If we wish to obtain a better approximation by taking also into consideration the shapes of the individual elements of our random layers (dot screens, line gratings, etc.), we may use another type of random process which is known as *shot noise*.

F.3.2 Shot noise

A *shot noise* random process ([Champeney73 p. 82], [Papoulis65 p. 288]) is a random process whose individual signals are produced as a sum of randomly located copies of a given function (for example, a dot or a pulse). In the 1D case, if the individual dot has the profile $d(x)$ then the shot noise process is given by:

$$f(x) = \sum_n d(x - x_n) \tag{F.4}$$

where the values x_n form a random sequence. As we can easily see, this is precisely the convolution of the random nailbed (F.1) with the dot $d(x)$:

---

[5] This last value is usually denoted in the literature by λ, but we prefer to use here η since we have already used λ in a different context, that of colour layers (see Chapter 9 in *Vol. I*).

$$f(x) = d(x) * q(x) \tag{F.5}$$

Remark that we denote here an individual element of the shot noise process by $d(x)$ and call it a *dot* in anticipation of the 2D case where $d(x)$ represents a 2D dot (like in a dot screen). It should be understood, however, that in our application a general random dot screen can only be *approximated* by shot noise, since in shot noise overlapping dots are summed up (as clearly indicated by Eq. (F.4); see also Fig. 9–8 in [Papoulis65 p. 288]), whereas in random screens overlapping black dots remain black and overlapping white dots remain while. This means, indeed, that a multiplicative version of shot noise would be better adapted to our needs; but under the assumption that the random screen dots do not overlap, a random screen made of white dots on black background can be aptly modelized by the shot noise $f(x)$, and its inverse video, consisting of black dots on white background, can be modelized by $1 - f(x)$.

Proceeding now to the spectral domain, it is clear that if the signal $f(x)$ is deterministic, then we immediately have its Fourier transform (using the shift and addition theorems):

$$F(u) = D(u) \sum_n e^{-i2\pi x_n u}$$

On the other hand, if we consider $f(x)$ as a random process we no longer have its Fourier transform, since all the phase information is lost. And yet, we do have a simple expression for the *power spectrum* of the random process $f(x)$: Using the fact that the power spectrum of a convolution of two functions is the product of their individual power spectra [Champeney73 pp. 64–65] we get from Eq. (F.5):

$$\mathcal{P}_f(u) = \mathcal{P}_d(u)\, \mathcal{P}_q(u) \tag{F.6}$$

where $\mathcal{P}_f(u)$, $\mathcal{P}_d(u)$ and $\mathcal{P}_q(u)$ are, respectively, the power spectra of the random process $f(x)$, of the individual dot $d(x)$ and of the Poisson process $q(x)$. Therefore, using Eq. (F.2) we obtain (see also [Papoulis65 p. 358]):

$$\mathcal{P}_f(u) = \eta\, \mathcal{P}_d(u) + \eta^2\, \mathcal{P}_d(u)\, \delta(u)$$

$$= \eta\, \mathcal{P}_d(u) + \eta^2\, [D(0)]^2\, \delta(u)$$

where $\delta(u)$ is an impulse at the origin, and $D(u)$ is the Fourier transform of the dot $d(x)$. There also exist simple expressions for the autocorrelation $c_{f,f}(x)$, the mean value μ and the variance σ^2 of the shot noise process $f(x)$ (see, for example, [Champeney73 pp. 82–83]):

$$c_{f,f}(x) = \eta\, c_{d,d}(x) + \mu^2$$

$$\mu = \eta \int d(x)\, dx = \eta D(0)$$

$$\sigma^2 = \eta \int [d(x)]^2\, dx$$

Using the above expression for μ the power spectrum of the shot noise process becomes (see also [Champeney73 pp. 82, 230–231]):

$$\mathcal{P}_f(u) = \eta\, \mathcal{P}_d(u) + \mu^2\, \delta(u) \tag{F.7}$$

It may be instructive to notice the smoothness of this power spectrum: indeed, it is composed of a smooth curve plus an impulse at the origin, but it includes no diffuse random noise. As explained above in Sec. F.2, this property is common to the power spectra of all random processes due to the averaging effect that is inherent to their definition.

A particular case of Eq. (F.7) occurs when the mean value μ of $f(x)$ is zero. In this case the impulse at the origin disappears and we obtain (see [Champeney73 pp. 82, 230–231]):

$$\mathcal{P}_f(u) = \eta\, \mathcal{P}_d(u) \tag{F.8}$$

It should be remembered, however, that all the above power spectra are based on the power spectrum $\mathcal{P}_q(u)$ of the Poisson process; this means that they are based on the assumption that the dot locations are fully random and hence uncorrelated. If the dot locations (i.e. the impulse locations of the underlying point process) *are* correlated, so that for a dot which has occurred at x_1 the probability of having another dot in the infinitesimal interval between $x_1 + x$ and $x_1 + x + dx$ is, say, $p(x)dx$, then Eq. (F.7) becomes [Champeney73 p. 231]:

$$\mathcal{P}_f(u) = \eta\, \mathcal{P}_d(u)\, (1 + P(u)) \tag{F.9}$$

where $P(u)$ is the Fourier transform of the probability density function $p(x)$. This means that if the dot locations *are* correlated the power spectrum is no longer determined by $\mathcal{P}_d(u)$ alone, and it includes an additional term which depends on $P(u)$, too. Consider, for example, the 2D case of a random dot screen in which the distances between neighbouring dots are highly correlated, with a characteristic nearest-neighbour distance of r. Such dot screens tend to give a ring-like power spectrum where the mean radius of the ring is $1/r$ (see, for example, the figures in [Yellott82] and [Yellott83], Plates 15–16 in [Harburn75], or Fig. 10.25 in [Glassner95 p. 433]; in all of these cases the ring-like envelope of the power spectrum is not explained by the shape of the power spectrum $\mathcal{P}_d(u,v)$ of the individual dot $d(x,y)$, but rather by the shape of $P(u,v)$, the Fourier transform of the probability density function $p(x,y)$).

Further generalizations of Eq. (F.9) are also possible. For example, [Heiden69] derives the power spectrum of shot noise processes whose pulse widths and pulse amplitudes are not constant but rather random entities that are correlated with the pulse locations.

Finally, it should be mentioned that the above power spectra are only valid when the process $f(x)$ is not periodic. If $f(x)$ *is* periodic its power spectrum no longer contains the continuous component $\nu\, \mathcal{P}_d(u)$, and it becomes purely impulsive. This can be easily seen by reconsidering Eqs. (F.5) and (F.6): Suppose, for example, that $f(x)$ is periodic with period 1. If we denote by $\mathrm{III}(x)$ the unit-period comb of impulses, we have in this case instead of Eq. (F.5):

$$f(x) = d(x) * \mathrm{III}(x)$$

and the power spectrum of $f(x)$ becomes:

$$P_f(u) = P_d(u)\, P_{III}(u) \tag{F.10}$$

where the power spectrum $P_{III}(u)$ of $III(x)$ is the unit-frequency impulse comb $III(u)$. This means, indeed, that in this case $P_f(u)$ is a unit-frequency comb whose impulse amplitudes are modulated by $P_d(u)$, the power spectrum of the isolated dot $d(x)$.

A further generalization of Eq. (F.10) is provided in [Williams86] and more recently in [Ridolfi04 p. 67]. It is shown there that if the location of each dot in the periodic process $f(x)$ is slightly randomized (or "jittered") where each dot is perturbed independently of the other dots, and the probability of each of the dots to lie in an infinitesimal area dx is given by $p(x)dx$, then the power spectrum of $f(x)$ becomes:

$$P_f(u) = P_d(u)\, P_p(u)\, P_{III}(u) + \eta\, P_d(u)\, (1 - P_p(u)) \tag{F.11}$$

where $P_p(u)$ is the power spectrum of the probability density function $p(x)$. The first term in (F.11) is a unit-frequency comb whose impulse amplitudes are modulated by the product $P_d(u)\, P_p(u)$, and the second term is a continuous function. This means that the power spectrum (F.11) is no longer purely impulsive: in addition to the impulses of Eq. (F.10) (whose amplitude is modulated here by $P_p(u)$, too), it also contains a new continuous part whose shape depends on both $P_d(u)$ and $P_p(u)$. Note that in the particular case where the dot locations are fully periodic (no random perturbation in the dot locations) we have $p(x) = \delta(x)$, whose Fourier transform is $P(u) = 1$ and whose power spectrum is therefore $P_p(u) = 1$; and indeed, putting this back in Eq. (F.11) gives us again, as expected, the purely impulsive power spectrum of Eq. (F.10).

F.3.3 Random fields

A random field is a stochastic process $\{f_\zeta(\mathbf{x})\}$ whose argument $\mathbf{x}$ varies in a continuous fashion over some subset of $\mathbb{R}^n$, the n-dimensional Euclidean space [Adler81 p. ix]. Of course, in our application we will be most interested in the case of $n = 2$; in this case a random field is a family of 2D functions $\{f_\zeta(x,y)\}$ defined over the x,y plane (or a subset thereof). Typical examples of random fields include, for instance, the infinite family of functions $\{z = f_\zeta(x,y)\}$ that describe the height z of a wavy sea surface, or the surface of any rough plate [Adler81 pp. 1–4]. Unlike point processes and shot noise, which are based on a discrete distribution of points in the plane (the locations of the screen dots), random fields are completely general, and they can take into account all the properties of our original random layers, including the dot shapes, sizes, intensities, etc.

Because the study of random fields is, by definition, the study of random functions over some Euclidean space, this study can cover an extremely wide area, since any question that can be asked about an ordinary non-random function, or class of functions, can just as readily be asked about their random counterparts. Hence, the general theory of random fields is certainly at least as large as the general theory of functions. And indeed, adding a random component to the theory of functions makes it much larger, more interesting, and often more complex subject [Adler81 p. 1].

Due to its vast extent, we will not attempt here to go into details of the random field theory. Interested readers can find further information on random fields and on their spectral representation in references such as [Rosenfeld82 pp. 38–47] or [Adler81]. A general overview on random fields is also provided in [EncStat82, Vol. 7, pp. 508–512].

F.4 Stochastic modelization of layer superpositions

Having understood how random layers (random dot screens, random line gratings, etc.) can be modelized as stochastic processes, we now proceed to the modelization of superpositions of such layers.

As we know from the deterministic case, the superposition of layers is usually best modelized as a *product* of the individual layers (although in some particular cases other models may be more appropriate; see Sec. 2.2). This can be extended to the stochastic case, too. Intuitively, the probability of seeing white at a point (x,y) of the layer superposition is the product of the probabilities of seeing white at the point (x,y) in each of the two original layers. However, this is only true if the two superposed layers are statistically independent of each other. But in our application, when we come to study the moiré or Glass patterns in the superposition of two layers, we cannot make the simplifying assumption that the two layers are independent; on the contrary, we clearly know that they *are* dependent (remember the high correlation that is required between the two layers in order that a Glass pattern be generated).

Although there exist in the literature models that allow the treatment of such layer superpositions (see, for example, the superposition of two Poisson point processes in [Stoyan95 pp. 152–153]), the fact that our original layers are not independent of each other considerably complicates things and may render a detailed investigation intractable. For example (see [Coulon84 p. 152]), if two stochastic signals are independent then the autocorrelation function of their product is the product of the individual autocorrelation functions, and the power spectrum of the product is the convolution of the individual power spectra; but if the signals are correlated these results do not necessarily hold.[6]

F.5 Evaluation of the stochastic vs. deterministic approaches for our application

The need for a statistical treatment may arise in several situations. The most obvious situation occurs when we do not have full information about the phenomenon that is to be

---

[6] This is true for both random and deterministic signals; indeed, the counterpart of the convolution theorem for power spectra only holds for functions that are not correlated. Note, however, that the converse direction of the convolution theorem *does* hold, meaning that the power spectrum of the convolution of two signals is the product of the two power spectra [Champeney73 pp. 64–65].

treated (1D signal, 2D image, etc.), and we only know its statistical properties. This may be the case when we are dealing with a process whose physics is not fully understood, or with a process too complicated to analyze in detail. In other situations a statistical treatment may be needed even though the signal is deterministic (fully known). This may happen, for example, when we are given an ensemble of similar signals, and we simply do not know in advance which of them will occur. In this situation, too, we can only treat the general properties of the ensemble (such as the *average* power spectrum), but not the properties of an individual signal. The need to use a statistical approach may also arise when one is specifically interested in the statistical properties (distribution, mean value, standard deviation, etc.) of the given signals — which may be either deterministic or not — and in the statistical properties of various combinations of these signals (for example, in our case, the layer superposition).

It should be remembered, however, that even if the given signals were originally generated by a random process, once they are *fully known*, it is no longer *needed* to treat them statistically, and we can treat them like any other deterministic signals. In our case, for example, every given dot screen that we fully know can be considered as a deterministic signal and undergo a standard Fourier treatment even if its elements have been positioned in the plane in a random manner (see, for example, Problems 2-1–2-13 in Chapter 2).[7] The advantage of doing so is that this way we have full access to all the information related to the signal, and we do not lose its phase information.

Let us try to see more closely what this means in our particular case of interest, the study of moiré effects in the superposition of dot screens or line gratings whose individual elements, dots or lines, are randomly positioned. For this end, let us consider Fig. 7.7 that illustrates the core of our Fourier-based explanation of the macroscopic moiré or Glass patterns which may occur in a layer superposition. If we choose to treat our given layers statistically, we no longer have access to their Fourier transforms, but only to their power spectra. However, in this case we lose all the Fourier considerations that we have used in Chapter 7: First of all, we can no longer pass freely between the image and spectral domains of Fig. 7.7, since the transition from the power spectrum back to the original images is not possible. But even worse, in this case we can no longer use the convolution theorem, since the power spectrum of the product is not necessarily equal to the convolution of the individual power spectra [Champeney73 pp. 64–65].

As an alternative approach, we may consider altering *both* rows of Fig. 7.7, namely, replacing the original layers shown in the top row of the figure (the image domain) by their respective autocorrelation functions, and the spectra of the original layers shown in the bottom row of the figure (the spectral domain) by the respective power spectra. In this case, the bidirectional Fourier relationship between the entities in the image and in the spectral domains remains fully available, thanks to the autocorrelation theorem which

---

[7] In other words, when we superpose two random dot screens we are not actually superposing two *random processes*, but rather two *specific realizations* of these random processes; and the moiré (or Glass) pattern we obtain is, again, a *specific realization*, not an ensemble of moiré (or Glass) patterns.

states that the autocorrelation function and the power spectrum are a Fourier pair. This guarantees, indeed, that we can freely move from the image domain to the spectral domain and vice versa without any loss of information; furthermore, in this case we can also use the convolution theorem. However, this happy situation should not mislead us to believe that by passing to statistical reasoning we do not lose information: True, the Fourier transform itself does not cause any loss of information when we pass from one domain to the other; but it simply operates on entities in the image and in the spectral domains that *have already lost* all the phase information. Indeed, it is the passage from the original images to their autocorrelation functions (in the image domain), or equivalently, the passage from the Fourier transforms of the original images to the power spectra of the original images (in the spectral domain), that causes the loss of information. Specifically, returning to Fig. 7.7, we would no longer have in its upper row the original layers, their superpositions and their moiré phenomena, but only the corresponding autocorrelations. But this loss of information in the image domain along with the loss of the direct contact with the original layers themselves make our Fourier-based reasoning less effective in the explanation of the moiré phenomenon.

Another drawback of the stochastic approach is that it requires a full, prior mathematical knowledge of the underlying stochastic process, and if the problem at hand does not fall within the framework of a known stochastic process that has already been worked out in the known literature, one will have to invest some efforts in order to find its various properties. In fact, this is precisely the situation that we are facing in our application, since the statistical processes representing our individual layers may be quite complex (remember, for example, that overlapping dots in a random dot screen are not really summed up as in shot noise but rather multiplied). Furthermore, when we wish to study the superposition of two layers we cannot make the simplifying assumption that the two layers are not correlated; on the contrary, we do know that they are correlated, due to the high correlation that is required between the two layers in order that a Glass pattern be generated.

In conclusion, we see that in our application the use of the stochastic approach is not quite appropriate, because it is not easily tractable (at least if we require a good approximating model), and also because it causes the loss of important information. Therefore, since in our case the given random images and their superpositions are fully known, we have all good reasons to treat them as deterministic images instead of using a stochastic approach. This has also the important advantage of providing a unified treatment for all types of images, periodic, aperiodic or random.

Appendix G

Integral transforms

G.1 Introduction

In Chapter 7 we have presented the fundamental Glass pattern theorem for the superposition of geometrically transformed aperiodic layers. But although we mentioned there that this result is valid for both linear and non-linear layer transformations, we only explained it, via spectral domain considerations, for *linear* transformations (see Fig. 7.7). For the more general case involving non-linear transformations we only presented the image-domain results (see Fig. 7.12), without giving their spectral-domain interpretation as we did in Fig. 7.7 for the case of linear transformations. The reason is that in the case of non-linear transformations a general spectral-domain analysis is not possible, since there exists no general expression for the Fourier transform of $f(g(x))$ when $g(x)$ is non-linear. In the present appendix we present a generalized fourier decomposition of $f(g(x))$ that allows us to obtain our main results (including the fundamental Glass pattern theorem and its counterpart for periodic structures, the fundamental moiré theorem) even if $g(x)$ is non-linear. Although in this approach we lose the direct connection with the spectral, frequency domain (since the transformed domain no longer coincides with the Fourier, spectral domain), this approach still proves to be very useful.

G.2 Fourier decomposition of periodic and aperiodic structures

Let $f(x)$ be an aperiodic function whose Fourier transform is $F(u)$. This means, using the definition of the Fourier transform, that:

$$F(u) = \int_{-\infty}^{\infty} f(x)\, e^{-i2\pi ux}\, dx \tag{G.1}$$

Similarly, using the definition of the *inverse* Fourier transform we also have:

$$f(x) = \int_{-\infty}^{\infty} F(u)\, e^{i2\pi ux}\, du \tag{G.2}$$

Note that Eq. (G.2) expresses the decomposition of the image-domain function $f(x)$ into its spectral components. This is, indeed, the continuous-frequency counterpart of the Fourier series decomposition of a *periodic* function $p(x)$, which is given by:

$$p(x) = \sum_{n=-\infty}^{\infty} c_n\, e^{i2\pi f_n x} \tag{G.3}$$

with the Fourier coefficients:

$$c_n = \frac{1}{T} \int_T p(x)\, e^{-i2\pi f_n x}\, dx \tag{G.4}$$

where $f_n = n/T$, $n \in \mathbb{Z}$ (see Eqs. (A.5) and (A.6) in Appendix A of *Vol. I*). As explained in Sec. B.6 of *Vol. I*, $F(u)$ in Eq. (G.2) plays the same role as the Fourier coefficients c_n in Eq. (G.3), namely, it assigns the proper amplitudes (weights) to the various frequencies in the spectral decomposition of $f(x)$. The difference between the two cases is that in Eq. (G.3), where the given image-domain function $p(x)$ is periodic, the spectral decomposition only consists of a denumerable set of frequencies, $f_n = n/T$, $n \in \mathbb{Z}$, and the spectrum is impulsive; while in Eq. (G.2), where the image-domain function $f(x)$ is aperiodic, the spectral decomposition consists of a continuum of frequencies $u \in \mathbb{R}$. In this case the summation over the frequencies is no longer denumerable, and it turns into integration, the continuous counterpart of the discrete summation.

G.3 Generalized Fourier decomposition of geometrically transformed structures

Suppose now that the image-domain function $f(x)$ undergoes a mapping (coordinate transformation) $g(x)$, so that we obtain a new, transformed version of $f(x)$, that we denote by $r(x)$:

$$r(x) = f(g(x))$$

How does the application of the mapping $g(x)$ to $f(x)$ affect the spectrum of $f(x)$? As mentioned in Sec. 10.3 of *Vol. I*, when the transformation $g(x)$ is linear or affine, the spectrum $R(u)$ of the transformed function $r(x)$ can be readily expressed in terms of the original spectrum $F(u)$. However, in the more general case where $g(x)$ is non-linear, no general rule exists which tells us how the spectrum will be influenced. This renders the Fourier approach in the general case intractable unless the non-linear mapping $g(x)$ is particularly simple.

However, in Chapter 10 of *Vol. I*, where we studied geometric transformations of periodic functions $p(x)$, we have found a way to bypass this problem by representing the transformed function $r(x) = p(g(x))$ as a *generalized* Fourier series. As explained in Sec. 10.5 of *Vol. I*, instead of considering the spectrum $R(u)$ of $r(x) = p(g(x))$, whose analytical expression may be unknown or hard to find, we make the following two-step reasoning: We start with the Fourier decomposition (G.3) of the original periodic function $p(x')$, using here the variable x' rather than x:

$$p(x') = \sum_{n=-\infty}^{\infty} c_n \, e^{i 2\pi f_n x'} \tag{G.5}$$

and then we make in this Fourier series the formal substitution $x' = g(x)$, keeping the coefficients c_n unchanged:

$$r(x) = p(g(x)) = \sum_{n=-\infty}^{\infty} c_n \, e^{i 2\pi f_n g(x)} \tag{G.6}$$

Although the resulting generalized Fourier series (G.6) is not the actual spectral decomposition of $r(x) = p(g(x))$, it still turns out to be extremely useful. As we have seen in Sec. 10.9 of *Vol. I*, this approach allowed us to to analyze the superposition of geometrically transformed periodic layers and the resulting moiré effects, without having to resort to the spectra $R(u)$ of these curvilinear structures; and indeed, this approach led us in *Vol. I* to the main results of Chapter 10, the fundamental moiré theorems for the superposition of geometrically-transformed periodic layers.

Based on the success of this approach in the case of periodic functions $p(x)$, it may be asked, therefore, weather a similar reasoning could be also used in our present case of interest, where the original functions $f(x)$ are *aperiodic*.

To answer this question, let us return to the spectral decomposition of the image-domain function $f(x')$ as given by Eq. (G.2), using the variable x' rather than x:

$$f(x') = \int_{-\infty}^{\infty} F(u)\, e^{i2\pi ux'}\, du \tag{G.7}$$

Now, just as we did above in the case of discrete spectral decompositions, let us *formally* replace x' in this expression with $x' = g(x)$, keeping the "coefficients" (i.e. the Fourier transform) $F(u)$ unchanged:

$$r(x) = f(g(x)) = \int_{-\infty}^{\infty} F(u)\, e^{i2\pi ug(x)}\, du \tag{G.8}$$

Obviously, this formal construct is not the spectral decomposition of $r(x)$, but some kind of generalization thereof. The actual spectral decomposition of $r(x)$ would be given, according to Eq. (G.2), by:

$$r(x) = f(g(x)) = \int_{-\infty}^{\infty} R(u)\, e^{i2\pi ux}\, du \tag{G.9}$$

where $R(u)$ is the Fourier transform of $r(x)$; but as we have seen above, the problem here is that when the mapping $g(x)$ is non-linear, we do not always know the Fourier transform $R(u)$, and therefore we do not have the spectral decomposition (G.9), either. But although the generalized decomposition provided by Eq. (G.8) is *not* the spectral decomposition of $r(x)$, it still proves to be very useful for our needs, i.e. for understanding the behaviour of our aperiodic layers, their superpositions and their moiré (or Glass) patterns.

G.4 Integral transforms and their kernels

Before we go any further into these considerations, let us try to better understand the meaning of the construct provided by Eq. (G.8). So far we already know what this construct is *not* — it is not the Fourier spectral decomposition of the transformed function $r(x) = f(g(x))$ — but we would like now to see what exactly it *is*. For this end, we recall that the Fourier transform is, in fact, just one member from a larger class of transforms that are known as *integral transforms*. An integral transform $F_k(u)$ of a function $f(x)$ is defined by the integral:

$$F_k(u) = \int_{a}^{b} f(x)\, k(x,u)\, dx \tag{G.10}$$

where the function $k(x,u)$ is called the *kernel* of the integral transform, and a and b are the integration limits. The class of functions to which $f(x)$ may belong and the range of the variable u are to be prescribed in each case; in particular, they must be so prescribed that the integral (G.10) converges [Churchill72 pp. 2, 317]. An integral transform is uniquely determined by its kernel $k(x,u)$ and by its integration limits a and b.[1] Note that an integral transform is in fact an operator that maps a given function $f(x)$ into another function $F_k(u)$; this is more clearly expressed by the notation $\mathcal{F}_k[f(x)] = F_k(u)$. The resulting function $F_k(u)$ is called the $\mathcal{F}_k$-*transform* of $f(x)$, and its variable u is known as the variable of the *transformed domain*.[2] For example, in the case where $k(x,u) = e^{-i2\pi ux}$, $a = -\infty$ and $b = \infty$ the resulting function $F_k(u)$ is the Fourier transform of $f(x)$, and its variable u represents the frequency in the Fourier, spectral domain.

In principle, any function $k(x,u)$ of two variables gives rise to an integral transform, but in practice, few such kernels yield useful transforms [Cartwright90 p. 195]. Some of the most useful kernels are listed in Table G.1 along with the integral transforms they provide. Note also that for any function $g(x)$, if $k(x,u) = g(u-x)$ then the integral transform $F_k(u)$ is simply the operation of convolution with $g(x)$, since for any given function $f(x)$ it yields the function $f(x) * g(x)$:

$$F_k(u) = \int f(x)\, g(u-x)\, dx$$

In particular, if $g(x) = \delta(x)$ we get the kernel $k(x,u) = \delta(u-x)$, and the resulting integral transform is the so-called "identity transform", usually denoted by I, which maps any given function $f(x)$ into itself: $I[f(x)] = f(x)$.

As we can see, each kernel provides a different integral transform, which maps the given function $f(x)$ into a different function $F_k(u)$. While some of these integral transforms have very specialized uses and are rarely encountered, others have found very important applications in mathematics or in other fields. The most widely known integral transform is the Fourier transform, but in certain applications other integral transforms may prove to be more suitable. For example, because many functions have a Laplace transform but not a Fourier transform,[3] the Laplace transform turns out to be more useful in numerous applications, such as in the solution of certain classes of differential equations [Cartwright90 p. 193]. However, the essential advantage of the Fourier transform over all the other integral transforms is its physical interpretability as a *frequency spectrum*

---

[1] Some references such as [Andrews03 p. 496] use an alternative notation, in which the integration limits are incorporated within the kernel itself, by multiplying $k(x,u)$ with suitable functions that take the value 1 within the integration range and the value 0 everywhere else. In this case the integration is always performed between $-\infty$ and ∞, and the integral transform is uniquely determined by its kernel.

[2] Note the slight terminological ambiguity due to the use of the term "transform" for both the resulting function $F_k(u)$ and the operator $\mathcal{F}_k$ itself. This ambiguity could be avoided, as done in [Churchill72 pp. 2–3], by using the term *transformation* for the operator, while keeping the term *transform* for the resulting function. But this convention would simply shift the ambiguity elsewhere, since we already use the term transformation in the context of coordinate transformations, or more generally, as a synonym for a mapping $g(x)$.

[3] This happens, for instance, in functions with exponential growth such as $f(x) = e^x$ [Cartwright90 pp. 192–193].

[Bracewell86 p. 220]. Although the Laplace and the other transforms can be used as efficient mathematical tools for solving various problems, they do not provide the spectral decomposition (or the frequency content) of the given function $f(x)$. In such integral transforms the "transformed" domain is not the spectral domain, and its variable u does not represent frequencies as it does in the case of the Fourier transform. Therefore, in such cases we no longer speak of the "image domain" and the "frequency domain", but rather of the "original domain" and the "transformed domain".

Integral transform	$k(x,u)$	a	b	Ref.
Fourier transform	$e^{-i2\pi ux}$	$-\infty$	∞	p. 7
Laplace transform	e^{-ux}	$-\infty$	∞	p. 219
Cosine transform	$2\cos(2\pi ux)$	0	∞	p. 17
Sine transform	$2\sin(2\pi ux)$	0	∞	p. 17
Hankel transform	$2\pi x J_0(2\pi ux)$	0	∞	p. 248
Mellin transform	x^{u-1}	0	∞	p. 254
Abel transform	$\dfrac{2x}{\sqrt{x^2-u^2}}$	0	∞	p. 262
Hilbert transform	$\dfrac{1}{\pi(x-u)}$	$-\infty$	∞	p. 267

Table G.1: Some of the most useful integral transforms, their kernels and their integration limits. The page numbers in the last column refer to [Bracewell86]. Note that many of these transforms have in the literature several different variants that differ from each other in their kernel $k(x,u)$ or in the integration limits a, b. For example, different variants of the Fourier transform are given in [Bracewell86 p. 7]. Obviously, each of these variants maps the given function $f(x)$ into a different function $F_k(u)$, and it could be therefore included in the table as a new entry. Note also that many of the transforms in the table have a distinct inverse transform, which is also an integral transform on its own right and could be added to the table. For example, the inverse Fourier transform has the kernel $k(x,u) = e^{i2\pi ux}$, and the other variants of the Fourier transform, too, have their respective inverse transforms with their own kernels [Bracewell86 p. 7].

It is interesting to note that although the properties of the different integral transforms vary widely, they still have some properties in common. For example, every integral transform $\mathcal{F}_k$ is a linear operator, meaning that it satisfies:

$$\mathcal{F}_k[c_1 f_1(x) + c_2 f_2(x)] = c_1 \mathcal{F}_k[f_1(x)] + c_2 \mathcal{F}_k[f_2(x)]$$

for any two functions $f_1(x), f_2(x)$ and constants c_1, c_2; this property is a straightforward consequence of the fact that the integral is a linear operator [Debnath95 p. 4]. In fact, if the kernel is allowed to be a generalized function then the converse is also true, meaning that all linear operators are integral transforms. A properly formulated version of this statement is the Schwartz kernel theorem [Wikipedia05; Ehrenpreis56].[4] Furthermore, it turns out that each integral transform $\mathcal{F}$ has an inverse integral transform $\mathcal{F}^{-1}$ such that $\mathcal{F}^{-1}[F(u)] = f(x)$; accordingly, $\mathcal{F}^{-1}\mathcal{F} = \mathcal{F}\mathcal{F}^{-1} = I$, where I is the identity transform mentioned above [Debnath95 p. 4]. It can be also proved that integral transforms are *unique*, meaning that if $\mathcal{F}_k[f_1(x)] = \mathcal{F}_k[f_2(x)]$ then $f_1(x) = f_2(x)$ under suitable conditions; this is known as the *uniqueness theorem* [Debnath95 p. 4].

The usefulness of integral transforms lies in the simplification that they bring about, for example in dealing with differential equations. A proper choice of the transform often makes it possible to convert an intractable problem in the original domain into a much simpler problem in the transformed domain that can be easily solved. The solution obtained is, of course, the transform of the solution of the original problem, and if the solution is required in the original domain, one still needs to apply the inverse transform to complete the operation. Hence, the typical procedure in such cases is to transform the original problem, solve the transformed problem, and then use the inverse transform to obtain a solution to the original problem. In many situations the type of the integral transform to be used is determined by the nature of the problem at hand; for example, the Fourier transform is the natural choice whenever we wish to make use of spectral considerations in the frequency domain. But in other situations, the art of choosing the best integral transform (or the best kernel) is often the key to a successful solution of the given problem. Some insights into the question of how to construct the best kernel can be found, for example, in [Ferraro88] and [Rubinstein91], in the context of pattern recognition in deformed images. A method for constructing an integral transform that solves a given differential equation can be found in [Churchill72, Chapter 10 and pp. 24–25, 384].

Finally, it should be noted that integral transforms can be easily generalized to functions of several variables. In this case Eq. (G.10) becomes [Debnath95 p. 4]:

$$F_k(\mathbf{u}) = \int_S f(\mathbf{x})\, k(\mathbf{x}, \mathbf{u})\, d\mathbf{x}$$

where $\mathbf{x} = (x_1, ..., x_n)$, $\mathbf{u} = (u_1, ..., u_n)$ and $S \subset \mathbb{R}^n$.

---

[4] A nice explanation showing how the integral transform (G.10) can be seen as the continuous-space counterpart of the linear vector transformation $\mathbf{f}_k = \mathbf{K} \cdot \mathbf{f}$ can be found in [Churchill72 pp. 25–26]. It is based on the interpretation of a function $f(x)$ as the continouos generalization of a vector $\mathbf{f} = (f_1, ..., f_n)$.

Returning now to the question that opened this section, concerning the meaning of the generalized decomposition of $r(x)$ that is provided by Eq. (G.8), we can clearly see now the answer: This expression is simply an integral transform of $F(u)$, whose kernel is given by:

$$k(x,u) = e^{i2\pi u g(x)}$$

In the particular case where $g(x) = x$ (the identity transformation), this gives back the kernel of the inverse Fourier transform, and Eq. (G.8) becomes the spectral decomposition of the function $f(x)$, as in Eq. (G.2). However, when $g(x)$ is not the identity transformation, Eq. (G.8) is no longer the spectral decomposition of $r(x) = f(g(x))$, but simply another integral transform, that we may call the *g-Fourier transform*. But although this mathematical construct no longer has a spectral interpretation (formally, instead of the spectral decomposition of $r(x)$ we obtain here its g-spectral decomposition), it still may be used as a mathematical tool for solving our particular problems, just as the Laplace and the other integral transforms are used for solving various problems without necessarily providing a spectral decomposition of the functions in question.[5]

G.5 The use of generalized Fourier transforms in the moiré theory

Let us see now how the generalized g-Fourier formulation can help us to understand the superposition of two transformed, aperiodic layers and the resulting moiré effects. Note that the following discussion simply extends to the continuous case the generalized Fourier series approach we have already introduced in Chapter 10 of *Vol. I* for the case of transformed periodic layers; the main difference is that in our present case summation is replaced by integration.

Let $f_1(x',y')$ and $f_2(x',y')$ be two 2D aperiodic layers whose Fourier transforms are $F_1(u,v)$ and $F_2(u,v)$, respectively. We therefore have, using the more compact vector notation $\mathbf{x}' = (x',y')$ and $\mathbf{u} = (u,v)$, the following Fourier spectral decompositions, just like in Eq. (G.7):

$$f_1(\mathbf{x}') = \int F_1(\mathbf{u})\, e^{i2\pi \mathbf{u}\cdot\mathbf{x}'}\, d\mathbf{u}$$
$$f_2(\mathbf{x}') = \int F_2(\mathbf{u})\, e^{i2\pi \mathbf{u}\cdot\mathbf{x}'}\, d\mathbf{u} \tag{G.11}$$

Suppose, now, that we apply to the layers $f_1(\mathbf{x}')$ and $f_2(\mathbf{x}')$ the geometric transformations $\mathbf{x}' = \mathbf{g}_1(\mathbf{x})$ and $\mathbf{x}' = \mathbf{g}_2(\mathbf{x})$, respectively. By substituting these transformations in Eqs. (G.11) we obtain the formal expressions:

---

[5] As clearly expressed by Eq. (G.8), g-Fourier transforms decompose any given function $f(x)$ into a continuous set of basis functions $b(u) = e^{i2\pi u g(x)}$, whose proper weights are assigned by $F(u)$, the Fourier transform of $f(x)$. If the mapping $g(x)$ is rather weak, meaning that it differs just slightly from the identity mapping $g(x) = x$, then the resulting g-Fourier transform is still close to the Fourier transform, and its basis functions $b(u)$ are closely sinusoidal, and are therefore strongly localized in the frequency spectrum. In the case of the Fourier transform itself (when $g(x) = x$), each basis function is perfectly sinusoidal and corresponds to a single frequency component in the spectrum.

$$r_1(\mathbf{x}) = f_1(\mathbf{g}_1(\mathbf{x})) = \int F_1(\mathbf{u}) \; e^{i2\pi\mathbf{u}\cdot\mathbf{g}_1(\mathbf{x})} \; d\mathbf{u}$$

$$r_2(\mathbf{x}) = f_2(\mathbf{g}_2(\mathbf{x})) = \int F_2(\mathbf{u}) \; e^{i2\pi\mathbf{u}\cdot\mathbf{g}_2(\mathbf{x})} \; d\mathbf{u}$$

$$\text{(G.12)}$$

which are simply the decompositions of $r_1(\mathbf{x})$ and $r_1(\mathbf{x})$ into their respective $\mathbf{g}_1$- and $\mathbf{g}_2$-Fourier transform components. The superposition of the transformed layers is therefore expressed by:

$$r_1(\mathbf{x}) \, r_2(\mathbf{x}) = \left(\int F_1(\mathbf{u}) \; e^{i2\pi\mathbf{u}\cdot\mathbf{g}_1(\mathbf{x})} \; d\mathbf{u} \right) \left(\int F_2(\mathbf{w}) \; e^{i2\pi\mathbf{w}\cdot\mathbf{g}_2(\mathbf{x})} \; d\mathbf{w} \right)$$

$$= \int\int F_1(\mathbf{u}) F_2(\mathbf{w}) \; e^{i2\pi[\mathbf{u}\cdot\mathbf{g}_1(\mathbf{x}) + \mathbf{w}\cdot\mathbf{g}_2(\mathbf{x})]} \; d\mathbf{u}d\mathbf{w} \qquad \text{(G.13)}$$

Note that in the particular case where $\mathbf{g}_1(\mathbf{x}) = \mathbf{g}_2(\mathbf{x}) = \mathbf{x}$ Eq. (G.13) is simply the inverse Fourier transform of the convolution $F_1(\mathbf{u}) ** F_2(\mathbf{u})$, as indeed predicted by the convolution theorem, since in this case:

$$r_1(\mathbf{x}) \, r_2(\mathbf{x}) = \int\int F_1(\mathbf{u}) F_2(\mathbf{w}) \; e^{i2\pi(\mathbf{u} + \mathbf{w})\cdot\mathbf{x}} \; d\mathbf{u}d\mathbf{w}$$

and by substituting $\mathbf{z} = \mathbf{u} + \mathbf{w}$:

$$= \int \left[\int F_1(\mathbf{u}) F_2(\mathbf{z} - \mathbf{u}) \; d\mathbf{u} \right] e^{i2\pi\mathbf{z}\cdot\mathbf{x}} \; d\mathbf{z}$$

$$= \int [F_1(\mathbf{z}) ** F_2(\mathbf{z})] \; e^{i2\pi\mathbf{z}\cdot\mathbf{x}} \; d\mathbf{z}$$

$$= \mathcal{F}^{-1}[F_1(\mathbf{z}) ** F_2(\mathbf{z})]$$

Returning to Eq. (G.13), let us consider now, just as we did in the discrete case of Fourier series in Sec. 10.7 of *Vol. I*, the partial sum (or rather the partial integral) that consists of all the terms in which $\mathbf{w} = -\mathbf{u}$:

$$m_{1,-1}(\mathbf{x}) = \int F_1(\mathbf{u}) F_2(-\mathbf{u}) \; e^{i2\pi\mathbf{u}\cdot[\mathbf{g}_1(\mathbf{x}) - \mathbf{g}_2(\mathbf{x})]} \; d\mathbf{u} \qquad \text{(G.14)}$$

This partial integral corresponds to a sub-structure that is present in the superposition $r_1(\mathbf{x}) \, r_2(\mathbf{x})$ of Eq. (G.13), but is not present in either of the original layers $r_1(\mathbf{x})$ and $r_2(\mathbf{x})$ themselves. And indeed, just as in the discrete case of Sec. 10.7 in *Vol. I*, this structure is simply the substructive $(1,-1)$-moiré that is generated in the superposition. Note that in Sec. 10.7 of *Vol. I* we defined, more generally, the substructures of the superposition which correspond to its (k_1,k_2)-moirés; but as we already know (see Sec. 7.5), in the case of aperiodic layers no moirés other than the $(1,-1)$-moiré can exist, since in such cases there is no correlation between the superposed layers (for example, the correlation between a random screen $r(x,y)$ and its scaled version $r(2x,2y)$ is practically zero). Therefore, although we *can* technically define $m_{k_1,k_2}(\mathbf{x})$ for any (k_1,k_2) (and in fact, even for non-integer values of k_1 and k_2), the only visible sub-structure in the superposition of aperiodic layers corresponds to the $(1,-1)$-moiré. (As we have seen in Sec. 10.7.1 of *Vol. I*, in the discrete case, too, we can technically define a (k_1,k_2)-moiré $m_{k_1,k_2}(\mathbf{x})$ for any integers (k_1,k_2), but in practice only a few of them are really visible in the given superposition, depending on the case).

Now, if we define the normalized profile of $m_{1,-1}(\mathbf{x})$ by:

$$p_{1,-1}(\mathbf{x}') = \int F_1(\mathbf{u})F_2(-\mathbf{u})\, e^{i2\pi\mathbf{u}\cdot\mathbf{x}'}\, d\mathbf{u}$$

we see that $p_{1,-1}(\mathbf{x}')$ is simply the inverse Fourier transform of the product $F_1(\mathbf{u})F_2(-\mathbf{u})$, and hence, by the convolution theorem we have:

$$p_{1,-1}(\mathbf{x}') = p_1(\mathbf{x}') ** p_2(-\mathbf{x}') = p_1(\mathbf{x}') \star\star p_2(\mathbf{x}')$$

where $p_1(\mathbf{x}')$ and $p_2(\mathbf{x}')$ are the normalized profiles of the original, untransformed layers $f_1(\mathbf{x})$ and $f_2(\mathbf{x})$, and $\mathbf{x}' = \mathbf{g}_1(\mathbf{x}) - \mathbf{g}_2(\mathbf{x})$ is the coordinate transformation which brings $p_{1,-1}(\mathbf{x}')$ into the geometric layout of the $(1,-1)$-moiré. This leads us, indeed, to the fundamental Glass pattern theorem (see Sec. 7.8). Note that this way we obtain, indeed, the generalized version of Proposition 7.7 for any linear or non-linear transformations $\mathbf{g}_i(\mathbf{x})$, whereas in Sec. 7.4.1 the proposition was only justified for the case of linear transformations (see the footnote at the end of Sec. 7.4.1).

Note that in the particular case where $\mathbf{g}_1(\mathbf{x}) = \mathbf{g}_2(\mathbf{x})$ the two superposed layers are identical (up to their intensity profiles) and the moiré (or Glass) pattern becomes singular and hence invisible. On the other hand, when $\mathbf{g}_1$ and $\mathbf{g}_2$ are completely different, the moiré effect is too weak to be visible due to the lack of correlation between the superposed layers. Thus, the most interesting cases occur when $\mathbf{g}_1$ and $\mathbf{g}_2$ are just slightly different, so that the moiré effect is not yet singular, but not too weak, either.

Eq. (G.14) can be considered, in fact, as a $(\mathbf{g}_1-\mathbf{g}_2)$-Fourier transform. This formal construct allows us, indeed, to extract mathematically the moiré (or Glass) pattern from the global structure of the layer superposition — without really knowing their Fourier spectra in the frequency domain.

As we can see, the g-Fourier formalism is simply a working tool that we use here to simplify our problem by considering it in the transformed domain. It is clear that in the particular case where $g(x) = x$ the g-Fourier transform simply reduces into the Fourier transform; but in fact, a similar situation occurs for any linear or affine mapping $g(x) = ax + b$, since in such cases Eq. (G.8) becomes:

$$r(x) = f(ax + b) = \int_{-\infty}^{\infty} F(u)\, e^{i2\pi u(ax + b)}\, du$$

which gives after some short manipulations (see, for example, [Gaskill78 pp. 194–195]):

$$= \int_{-\infty}^{\infty} R(u)\, e^{i2\pi ux}\, du$$

where $R(u)$ is the Fourier transform of $r(x)$. It follows, therefore, that whenever $g(x)$ is linear (or affine) the g-Fourier transform can be seen as a Fourier transform, and therefore in such cases we can also interpret our transformed-domain reasoning in terms of the classical Fourier spectral domain. But even when the mapping g is non-linear and the Fourier spectral interpretation is no longer possible, the g-Fourier formalism still remains extremely useful, as we have seen above.

Appendix H

Miscellaneous issues and derivations

H.1 Classification of the dot trajectories

Suppose we are given two identical aperiodic dot screens that are superposed on top of each other dot on dot, and that we apply to the two layers the direct transformations $\bar{g}_1(x,y)$ and $\bar{g}_2(x,y)$, respectively. As we have seen in Chapter 4, if the layer transformations $\bar{g}_1(x,y)$ and $\bar{g}_2(x,y)$ are not too violent, the microstructure of the resulting superposition may give rise to visible dot trajectories whose shapes are determined, to a close approximation, by the trajectories (field lines) of the vector field:

$$\bar{h}(x,y) = \bar{g}_1(x,y) - \bar{g}_2(x,y)$$

As we already know from Sec. B.6 of Appendix B, the trajectories of a vector field $\bar{h}(x,y)$ are given by the solution curves of the system of differential equations:

$$\frac{d}{dt}x(t) = \bar{h}_1(x(t),y(t))$$
$$\frac{d}{dt}y(t) = \bar{h}_2(x(t),y(t)) \tag{H.1}$$

where $\bar{h}_1(x,y)$ and $\bar{h}_2(x,y)$ are the two cartesian components of $\bar{h}(x,y)$, namely, $\bar{h}(x,y) = (\bar{h}_1(x,y), \bar{h}_2(x,y))$. Although the system of differential equations (H.1) is relatively easy to solve when $\bar{h}(x,y)$ is linear (see, for example, Chapter 4 in [Kreyszig93]), its solution in non-linear cases may present a more difficult challenge. However, it turns out that it is often possible to get a qualitative idea about the shape of these solution curves (trajectories) without even having to solve the system of differential equations (H.1). This follows as a straightforward outcome of the characterization and classification of the critical points of the system of differential equations, since the behaviour of the trajectories surrounding a critical point highly depends on the properties of the critical point itself.

A point (x,y) is called a *critical point* of the vector field $\bar{h}(x,y)$ or of the system of differential equations (H.1) if it satisfies $\bar{h}_1(x,y) = 0$ and $\bar{h}_2(x,y) = 0$ [Birkhoff89 p. 133; Kreyszig93 p. 176].[1] At such a point we have $\frac{d}{dt}x(t) = 0$ and $\frac{d}{dt}y(t) = 0$, and hence the direction of the solution curves of Eq. (H.1) there is indeterminate: $\frac{dy}{dx} = \frac{dy/dt}{dx/dt} = \frac{0}{0}$. (Equivalently, we may say that at a critical point the direction of the trajectories of the vector field $\bar{h}(x,y)$ is indeterminate; another way to see this is that the vector field $\bar{h}(x,y)$ assigns to this point the null vector $(0,0)$, whose direction is obviously undefined.) In the

---

[1] Confusingly, some references use the term "fixed point" for a critical point (see, for example, [Strogatz94 pp. 124, 150; Weisstein99 p. 652]). Note that a point (x,y) for which we have $\bar{h}(x,y) = (0,0)$ is a *zero* of $\bar{h}(x,y)$, and not a fixed point of $\bar{h}(x,y)$; it *is*, however, a mutual fixed point of $\bar{g}_1(x,y)$ and $\bar{g}_2(x,y)$, since it satisfies, of course, $\bar{h}(x,y) = \bar{g}_1(x,y) - \bar{g}_2(x,y) = (0,0)$.

following subsections we will see how the classification of the critical points can give us clues to the qualitative behaviour of the trajectories of the system (H.1), and hence to the shape of the dot trajectories in our layer superpositions. We start, as usual, with the simplest case, in which $\overline{\mathbf{h}}(x,y)$ is linear.

H.1.1 Classification of the dot trajectories in the linear case

Suppose first that the vector field $\overline{\mathbf{h}}(x,y)$ is linear. This obviously occurs when both of the layer transformations $\overline{\mathbf{g}}_1(x,y)$ and $\overline{\mathbf{g}}_2(x,y)$ are linear, but it may also happen when $\overline{\mathbf{g}}_1(x,y)$ and $\overline{\mathbf{g}}_2(x,y)$ are non-linear, and their non-linear components are mutually cancelled out in the difference $\overline{\mathbf{h}}(x,y) = \overline{\mathbf{g}}_1(x,y) - \overline{\mathbf{g}}_2(x,y)$. In both of these cases the linearity of $\overline{\mathbf{h}}(x,y)$ implies, of course, that the set of differential equations (H.1) is linear, too:

$$\frac{d}{dt}x(t) = a_1x(t) + b_1y(t)$$
$$\frac{d}{dt}y(t) = a_2x(t) + b_2y(t) \tag{H.2}$$

Because of its linearity, $\overline{\mathbf{h}}(x,y)$ is clearly zero at the origin, and thus the origin is a critical point of the system of differential equations (H.2). (Note that if the linear transformation $\overline{\mathbf{h}}(x,y)$ is singular it may have a full critical line passing through the origin, i.e. a line all of whose points satisfy $\overline{\mathbf{h}}(x,y) = (0,0)$. However, the origin itself is always a zero of any linear $\overline{\mathbf{h}}(x,y)$, be it singular or regular, and hence Eq. (H.2) always has a critical point at the origin.)

Now, in the theory of differential equations there exists a simple technique that allows us to characterize and classify the critical point of the linear system (H.2), and hence to find qualitatively the behaviour of the trajectories (solution curves) that surround it, without having to solve the system. This technique is based on the two eigenvalues of the matrix of the linear transformation $\overline{\mathbf{h}}(x,y)$ (see, for example, Sec. 16.1 in [Gray97]). For instance, if the matrix of $\overline{\mathbf{h}}(x,y)$ has two real eigenvalues with opposite signs then the critical point is a saddle point and the trajectories surrounding it are hyperbolic, and if the eigenvalues are purely imaginary then the critical point is a *center* with circular or elliptical trajectories surrounding it. The full classification of the critical points is given in Table H.1 below, which is based on [Gray97 p. 566].[2] As we can see from this table, eigenvalues with a positive real part always cause repulsion from the origin, whereas eigenvalues with a negative real part cause attraction to the origin;[3] the imaginary part of the eigenvalues indicates rotation of the trajectories about the origin, and a zero eigenvalue corresponds to degenerate cases with an entire critical line.

---

[2] Note that the nomenclature used to designate the different types of critical points significantly varies from reference to reference. For example, center points, spiral points and degenerate nodes according to the terminology used in [Gray97] are called in [Birkhoff89] vortex points, focal points and star points, respectively; similarly, improper nodes in [Gray97] correspond to degenerate nodes in [Strogatz94 p. 136], while degenerate nodes in [Gray97] refer to stars in [Strogatz94 p. 135]. In Table H.1 we have chosen the names that best suit our needs and our general conventions.

[3] Note that this is immaterial for our needs, since the dot trajectories in the layer superposition do not show the sense of the arrows along the curves.

Remark H.1: This classification can be further simplified thanks to the fact that the eigenvalues of a linear transformation $\overline{\mathbf{h}}(x,y)$ depend only on the trace t and the determinant d of the matrix of that linear transformation [Strogatz94 p. 130; Kreyszig93 p. 176]. This eliminates the need for calculating the eigenvalues of the matrix, and allows an alternative, elegant classification of the different types of critical points and the trajectories surrounding them in terms of the two real numbers t and d. This classification can be represented graphically in the t,d plane, as shown, for example, in figure 5.2.8 in [Strogatz94 p. 137]. Note, however, that this graphical classification still may require the calculation of $t^2 - 4d$ in order to determine in which region of the graph our particular case (t,d) is situated; furthermore, this graphical method may be ambiguous in some particular cases such as stars and improper nodes, which are both located in the t,d plane along the border of the parabola $t^2 - 4d = 0$. For these reasons we prefer to stick here to the original classification of Table H.1, that is based on the eigenvalues of the matrix. ■

Example H.1: Consider the superposition shown in Fig. 2.1(e). In this case the layer transformations are given by:

$$\overline{\mathbf{g}}_1(x,y) = ((1+\varepsilon)x, (1+\varepsilon)y)$$

$$\overline{\mathbf{g}}_2(x,y) = (x,y)$$

where ε is a small positive fraction, and therefore we have:

$$\overline{\mathbf{h}}(x,y) = \overline{\mathbf{g}}_1(x,y) - \overline{\mathbf{g}}_2(x,y) = (\varepsilon x, \varepsilon y)$$

In this case the matrix of the linear transformarion $\overline{\mathbf{h}}(x,y)$ is $\left(\begin{smallmatrix} \varepsilon & 0 \\ 0 & \varepsilon \end{smallmatrix}\right)$, and its two eigenvalues are simply $\lambda_1 = \lambda_2 = \varepsilon$ where $\varepsilon > 0$, meaning that the critical point at the origin is a star (see case 6 in Table H.1). And indeed, this fully agrees with the dot trajectories that surround the origin in the layer superposition that is shown in Fig. 2.1(e). ■

Example H.2: Consider now the superposition shown in Fig. 2.2(a). In this case one layer is obtained from the other by a slight scaling of $s_x = 1 - \varepsilon$ in the x direction, and a slight scaling of $s_y = 1 + \varepsilon$ in the y direction (ε being a small positive fraction). Therefore the layer transformations $\overline{\mathbf{g}}_1(x,y)$ and $\overline{\mathbf{g}}_2(x,y)$ are given by:

$$\overline{\mathbf{g}}_1(x,y) = ((1-\varepsilon)x, (1+\varepsilon)y)$$

$$\overline{\mathbf{g}}_2(x,y) = (x,y)$$

so that:

$$\overline{\mathbf{h}}(x,y) = \overline{\mathbf{g}}_1(x,y) - \overline{\mathbf{g}}_2(x,y) = (-\varepsilon x, \varepsilon y)$$

In this case the matrix of the linear transformarion $\overline{\mathbf{h}}(x,y)$ is $\left(\begin{smallmatrix} -\varepsilon & 0 \\ 0 & \varepsilon \end{smallmatrix}\right)$, and its two eigenvalues are $\lambda_1 = -\varepsilon$ and $\lambda_2 = \varepsilon$. As we can see in case 3 of Table H.1 the critical point in this case is a saddle point, that is surrounded by hyperbolic trajectories. And indeed, this fully agrees with the dot trajectories that surround the origin in the layer superposition that is shown in Fig. 2.2(a). ■

	Eigenvalues	Type of critical points	Fig.	Remarks
1	$0 < \lambda_1 < \lambda_2$	Repelling node		
2	$\lambda_1 < \lambda_2 < 0$	Attracting node		
3	$\lambda_1 < 0 < \lambda_2$	Saddle	2.2(a)	(1)
4	$0 = \lambda_1 < \lambda_2$	Repelling line	2.3(a)	(2)
5	$\lambda_1 < \lambda_2 = 0$	Attracting line		(3)
6	$\lambda_1 = \lambda_2 > 0$ Two eigenvectors	Repelling star	2.1(e)	
7	$\lambda_1 = \lambda_2 < 0$ Two eigenvectors	Attracting star		
8	$\lambda_1 = \lambda_2 > 0$ One eigenvector	Repelling improper node		
9	$\lambda_1 = \lambda_2 < 0$ One eigenvector	Attracting improper node		
10	$\lambda_1 = \lambda_2 = 0$ One eigenvector	Linear center	2.3(c)	(4)
11	$\lambda_1, \lambda_2 = \alpha \pm i\beta$ $\alpha > 0, \beta \neq 0$	Repelling spiral	2.1(g)	
12	$\lambda_1, \lambda_2 = \alpha \pm i\beta$ $\alpha < 0, \beta \neq 0$	Attracting spiral		
13	$\lambda_1, \lambda_2 = \pm i\beta$ $\beta \neq 0$	Center	2.1(c), 2.2(c)	(5)

Table H.1: (*continued on the opposite page*)

Example H.3: Consider the superposition shown in Fig. 2.3(a). In this case the layer transformations are given by:

$$\overline{g}_1(x,y) = (x, (1+\varepsilon)y)$$

$$\overline{g}_2(x,y) = (x,y)$$

Remarks:
(1) The critical point is surrounded by hyperbolic trajectories.
(2) An entire critical line, with parallel straight trajectories emanating from it.
(3) An entire critical line, with parallel straight trajectories pointing to it.
(4) An entire critical line, with parallel straight trajectories parallel to it.
(5) The critical point is surrounded by circular or elliptic trajectories.
Note that for our own needs (the characterization of the dot trajectories in a dot screen superposition) the sense of the trajectories is immaterial, and hence we do not need to distinguish between the repelling and attracting variants in each type of critical points. Nevertheless, we still maintain this distinction in the table for the sake of completeness.

Table H.1: (*continued.*) Summary of all the different types of critical points that may occur in a 2D linear system of differential equations, and the conditions on the eigenvalues of the matrix of the system that give rise to each of these types. Note that in cases 1–3, 6–9, 11–13 the critical point is isolated and located at the origin, while in all the other cases (which are called *degenerate* or *singular* cases) there exists a full line of critical points passing through the origin.

where ε is a small positive fraction, and therefore we have:

$$\overline{\mathbf{h}}(x,y) = \overline{\mathbf{g}}_1(x,y) - \overline{\mathbf{g}}_2(x,y) = (0,\varepsilon y)$$

The matrix of the linear transformarion $\overline{\mathbf{h}}(x,y)$ is $\left(\begin{smallmatrix} 0 & 0 \\ 0 & \varepsilon \end{smallmatrix}\right)$, and its two eigenvalues are $\lambda_1 = 0$ and $\lambda_2 = \varepsilon > 0$, meaning that the critical point at the origin is in this case a repelling line (case 4 in Table H.1). And indeed, this fully agrees with the dot trajectories that surround the origin in the layer superposition that is shown in Fig. 2.3(a). ∎

It should be noted, however, that the classification provided by Table H.1 is only valid for linear cases, and even in affine cases, where a mere constant shift has been added (see, for example, Figs. 2.3(e),(g)), the table can no longer be used to determine the shape of the trajectories.

H.1.2 Classification of the dot trajectories in the non-linear case

What happens now when $\overline{\mathbf{h}}(x,y)$ is not linear? Unlike a linear vector field that always has a critical point at the origin (or, in singular cases, an entire critical line passing through the origin), a non-linear vector field may have, depending on the case, no critical points at all, one or more isolated critical points, or even one or more straight or curved critical lines. But because the behaviour of the trajectories surrounding a critical point highly depends on the properties of the critical point itself, in cases where critical points do exist it should be possible to get a qualitative idea about the shape of the trajectories of $\overline{\mathbf{h}}(x,y)$ (i.e., the solution curves of Eq. (H.1)) simply by identifying the critical points and studying their properties, without having to solve the system (H.1). As said in [Tabor89 p. 20], critical points can be thought of as the "organizing centers" of a system's dynamics; thus, by identifying them and their properties one can build up a fairly global picture of the system's behaviour.

In Sec. H.1.1 we have seen that if $\overline{\mathbf{h}}(x,y)$ is linear it has a critical point at the origin, and that we can characterize and classify this critical point by studying the eigenvalues of the matrix of the linear transformation $\overline{\mathbf{h}}(x,y)$. And indeed, it turns out that this technique can be also extended to the case of non-linear $\overline{\mathbf{h}}(x,y)$, i.e. to cases where the system (H.1) is non-linear.

This extended technique is based on the *linearization* of the non-linear system about each of its critical points separately (see, for example, [Kreyszig93 Sec. 4.5], [Strogatz94 Sec. 6.3], or [Gray97 pp. 588–590]). That is, instead of considering the original non-linear system (H.1) itself, we study its close approximation by the following linear system, which is simply its first-order Taylor approximation about the critical point (x_c, y_c):

$$
\begin{pmatrix} \dfrac{d}{dt} x(t) \\[2mm] \dfrac{d}{dt} y(t) \end{pmatrix} = \begin{pmatrix} \dfrac{\partial \overline{h}_1(x,y)}{\partial x} & \dfrac{\partial \overline{h}_1(x,y)}{\partial y} \\[3mm] \dfrac{\partial \overline{h}_2(x,y)}{\partial x} & \dfrac{\partial \overline{h}_2(x,y)}{\partial y} \end{pmatrix}_{\substack{x = x_c \\ y = y_c}} \begin{pmatrix} x(t) \\[2mm] y(t) \end{pmatrix}
\tag{H.3}
$$

Note that the matrix of the system (H.3) consists of constant coefficients, meaning that (H.3) is, indeed, a linear system of differential equations. But since $\overline{\mathbf{h}}(x,y)$ is non-linear, this linear approximation obviously varies from point to point, and we have to recalculate the constant coefficients of its matrix for each critical point (x_c, y_c) separately. Thus, for each critical point of the non-linear system (H.1) we obtain a separate linear system of the form (H.2) that approximates the non-linear system (H.1) about that critical point. Note that each of these approximating linear systems is, in fact, shifted so as to bring the critical point (x_c, y_c) to the origin; this can be easily seen in the Taylor development that leads to Eq. (H.3), as explained in each of the references mentioned above.

Therefore, all that we have to do in the non-linear case is to solve the system of equations $\overline{h}_1(x,y) = 0$, $\overline{h}_2(x,y) = 0$ in order to identify the *isolated* critical points of the system, and then, for each of these points (x_c, y_c), to calculate the numeric values of the partial derivatives of $\overline{\mathbf{h}}(x,y)$ at this point. This gives us for each of the critical points a linear system (H.2), and thus we can characterize and classify each of the critical points separately using the method of Sec. H.1.1 for linear cases. Note that the matrix we use in the linear approximation (H.3) is simply the Jacobian matrix of $\overline{\mathbf{h}}(x,y)$, evaluated at the critical point (x_c, y_c).

It should be emphasized that for each critical point we have a different linearized system, and the properties of the linearized system in each critical point may be radically different. But when we "piece together" the different linearized systems, we may obtain a fairly accurate picture of the nonlinear system and its trajectories. Note, however, that this technique can only be used for isolated critical points of the non-linear system, but not for critical lines or for non-linear systems having no critical points at all.

Example H.4: As we have seen in Sec. 3.4, when two originally identical aperiodic dot screens undergo non-linear transformations, several Glass patterns may be simultaneously generated in their superposition. Let us consider, as an illustration, Fig. 3.15(a). This figure shows two originally identical aperiodic dot screens that give in their superposition, after each of them has undergone a different non-linear transformation, 4 distinct Glass patterns. The two domain transformations undergone in this case by the original layers are (see Eq. (3.35)):

$$\mathbf{g}_1(x,y) = (x, y + y_0 - ax^2)$$

$$\mathbf{g}_2(x,y) = (x + x_0 - ay^2, y)$$

where $a > 0$, $x_0 > 0$ and $y_0 > 0$. As explained in Examples 3.6 and 5.3, these transformations have 4 different mutual fixed points that are located at:

$$(x,y) = (\pm\sqrt{y_0/a}, \pm\sqrt{x_0/a})$$

(see Eq. (3.39)); and indeed, each of the 4 Glass patterns in the layer superposition is generated about one of these 4 fixed points.

However, as we can clearly see in Fig. 3.15(a), it turns out that the dot trajectories surrounding these Glass patterns are not identical: While the two Glass patterns that are located along the main diagonal are surrounded by *circular* dot trajectories, the two Glass patterns that are located along the other diagonal are surrounded by *hyperbolic* dot trajectories. How can we explain this fact?

According to Proposition 4.3, the dot trajectories that are generated in the superposition due to the application of the direct layer transformations $\bar{\mathbf{g}}_1(x,y)$ and $\bar{\mathbf{g}}_2(x,y)$ are closely approximated by the vector field:

$$\bar{\mathbf{h}}(x,y) = \bar{\mathbf{g}}_1(x,y) - \bar{\mathbf{g}}_2(x,y)$$

In our present case (see Example 5.3) the direct layer transformations are given by:

$$\bar{\mathbf{g}}_1(x,y) = \mathbf{g}_1^{-1}(x,y) = (x, y - y_0 + ax^2)$$

$$\bar{\mathbf{g}}_2(x,y) = \mathbf{g}_2^{-1}(x,y) = (x - x_0 + ay^2, y)$$

where $a > 0$, $x_0 > 0$ and $y_0 > 0$, and therefore we have:

$$\bar{\mathbf{h}}(x,y) = \bar{\mathbf{g}}_1(x,y) - \bar{\mathbf{g}}_2(x,y) = (x_0 - ay^2, ax^2 - y_0)$$

But because $\bar{\mathbf{h}}(x,y)$ is clearly non-linear, we need in order to find the properties of its trajectories to use the linearization technique. As we can see by solving the system of equations $\bar{h}_1(x,y) = 0$, $\bar{h}_2(x,y) = 0$, $\bar{\mathbf{h}}(x,y)$ has 4 critical points that are precisely located at:

$$(x,y) = (\pm\sqrt{y_0/a}, \pm\sqrt{x_0/a})$$

Let us therefore evaluate the matrix of the linearized system (H.3) for each of these 4 critical points separately; this will allow us to characterize each of these critical points using the method of Sec. H.1.1.

It is easy to see that the Jacobian matrix of Eq. (H.3) is in our case:

$$\begin{pmatrix} \dfrac{\partial \bar{h}_1(x,y)}{\partial x} & \dfrac{\partial \bar{h}_1(x,y)}{\partial y} \\[2mm] \dfrac{\partial \bar{h}_2(x,y)}{\partial x} & \dfrac{\partial \bar{h}_2(x,y)}{\partial y} \end{pmatrix} = \begin{pmatrix} 0 & -2ay \\[1mm] 2ax & 0 \end{pmatrix}$$

We now evaluate this matrix and its eigenvalues at each of the 4 critical points of $\bar{h}(x,y)$:

(a) At the critical point $(+\sqrt{y_0/a}, +\sqrt{x_0/a})$ the matrix becomes:

$$\begin{pmatrix} 0 & -2a\sqrt{x_0/a} \\[2mm] 2a\sqrt{y_0/a} & 0 \end{pmatrix}$$

The two eigenvalues of this matrix are purely imaginary:

$$\lambda_1, \lambda_2 = \pm\sqrt{-4a\sqrt{x_0 y_0}}$$

(because x_0, y_0, $a > 0$), and therefore, according to case 13 in Table H.1, this critical point is a center. This explains, indeed, why the dot trajectories about this point in Fig. 3.15(a) are circular.

(b) At the second critical point on the main diagonal, $(-\sqrt{y_0/a}, -\sqrt{x_0/a})$ the matrix is:

$$\begin{pmatrix} 0 & 2a\sqrt{x_0/a} \\[2mm] -2a\sqrt{y_0/a} & 0 \end{pmatrix}$$

and the two eigenvalues are the same as in case (a). And indeed, the dot trajectories about this point in Fig. 3.15(a) are, again, circular.

(c) At the third critical point, $(+\sqrt{y_0/a}, -\sqrt{x_0/a})$, the matrix is:

$$\begin{pmatrix} 0 & 2a\sqrt{x_0/a} \\[2mm] 2a\sqrt{y_0/a} & 0 \end{pmatrix}$$

The two eigenvalues of this matrix are:

$$\lambda_1, \lambda_2 = \pm\sqrt{4a\sqrt{x_0 y_0}}$$

but because x_0, y_0, $a > 0$ these eigenvalues are purely real, with opposite signs. This corresponds to case 3 in Table H.1, meaning that this critical point is a saddle point that is surrounded by hyperbolic trajectories. And indeed, this result fully agrees with the dot trajectories about this point in Fig. 3.15(a).

(d) Finally, at the fourth critical point, $(-\sqrt{y_0/a}, +\sqrt{x_0/a})$, the matrix is:

$$\begin{pmatrix} 0 & -2a\sqrt{x_0/a} \\ -2a\sqrt{y_0/a} & 0 \end{pmatrix}$$

and the two eigenvalues are the same as in case (c). And indeed, the dot trajectories about this point in Fig. 3.15(a) are, again, hyperbolic. ■

Example H.5: Consider the superposition shown in Fig. 2.3(e). This case is identical to that of Example H.3, except that a horizontal shift has been added to one of the layers:

$$\bar{g}_1(x,y) = (x + x_0, (1+\varepsilon)y)$$

$$\bar{g}_2(x,y) = (x,y)$$

and therefore we have:

$$\bar{h}(x,y) = \bar{g}_1(x,y) - \bar{g}_2(x,y) = (x_0, \varepsilon y)$$

Note that because of the shift of x_0 this transformation is no longer linear but rather affine, and therefore the classification of Table H.1 no longer holds for it. And indeed, in this case the vector field $\bar{h}(x,y)$ and its corresponding system of differential equations have no critical points at all, and we cannot derive the behaviour of the trajectories from properties of the critical points. ■

H.2 The connection between the vector fields $\bar{h}_1(x,y)$ and $\bar{h}_2(x,y)$ in Sec. 4.5

We have seen in Sec. 4.5 of Chapter 4 that in cases where both of the superposed layers are distorted by transformations $\bar{g}_1(x,y)$ and $\bar{g}_2(x,y)$ the vector field which accurately represents the dot trajectories is given by Eq. (4.8):

$$\bar{h}_1(x,y) = \bar{g}_1(g_2(x,y)) - (x,y)$$

However, we have seen there that a similar reasoning may lead us to a different vector field, which also represents accurately the dot trajectories in the same superposition, and which is given by Eq. (4.9):

$$\bar{h}_2(x,y) = (x,y) - \bar{g}_2(g_1(x,y))$$

Clearly, vector fields (4.8) and (4.9) are not necessarily identical (see, for example, Figs. 4.11(b)–(d) and 4.12(b)–(d)). In fact, these vector fields happen to be identical only if our transformations satisfy the identity:

$$\bar{\mathbf{g}}_1(\mathbf{g}_2(x,y)) - (x,y) \equiv (x,y) - \bar{\mathbf{g}}_2(\mathbf{g}_1(x,y))$$

which means:

$$\tfrac{1}{2}\left[\,\bar{\mathbf{g}}_1(\mathbf{g}_2(x,y)) + \bar{\mathbf{g}}_2(\mathbf{g}_1(x,y))\right] \equiv (x,y) \tag{H.4}$$

but this identity is certainly not satisfied by all transformation pairs $\bar{\mathbf{g}}_1$, $\bar{\mathbf{g}}_2$. So how is it possible that the two different vector fields (4.8) and (4.9) represent the same dot trajectories in the layer superposition?

The answer to this question is given, indeed, by Proposition 4.2: As we can see from this proposition, the dot trajectories obtained by applying the transformations $\bar{\mathbf{g}}_1(x,y)$ and $\bar{\mathbf{g}}_2(x,y)$ to our original screens are not uniquely obtained by this specific pair of layer transformations, and there exist in fact infinitely many transformation pairs that give the same dot trajectories. Among these transformation pairs there exist precisely two, the pair $\bar{\mathbf{g}}_1(\mathbf{g}_2(x,y))$ and (x,y) and the pair (x,y) and $\bar{\mathbf{g}}_2(\mathbf{g}_1(x,y))$, in which only *one* of the two superposed layers is transformed, and for which we know, therefore, by virtue of Proposition 4.1, the precise vector field representations. These two vector fields have, indeed, different mathematical expressions — but as we have just seen, by virtue of Proposition 4.2, both of them represent the same dot trajectories in the superposition.

Incidentally, it is interesting to note that the two transformations $\bar{\mathbf{g}}_1(\mathbf{g}_2(x,y))$ and $\bar{\mathbf{g}}_2(\mathbf{g}_1(x,y))$ are, in fact, the inverse of each other. This can be seen by applying the general rule $[\mathbf{f}_1 \circ \mathbf{f}_2]^{-1} = \mathbf{f}_2^{-1} \circ \mathbf{f}_1^{-1}$ [Bernstein05 p. 6; Halmos74 p. 40], where "$\circ$" indicates the composition of transformations, to $\bar{\mathbf{g}}_2$ and $\mathbf{g}_1$; this gives us $[\bar{\mathbf{g}}_2 \circ \mathbf{g}_1]^{-1} = \bar{\mathbf{g}}_1 \circ \mathbf{g}_2$, which means, indeed, $[\bar{\mathbf{g}}_2(\mathbf{g}_1(x,y))]^{-1} = \bar{\mathbf{g}}_1(\mathbf{g}_2(x,y))$. And in fact, it can be shown that any two vector fields $\bar{\mathbf{h}}_1(x,y) = \bar{\mathbf{g}}(x,y) - (x,y)$ and $\bar{\mathbf{h}}_2(x,y) = (x,y) - \bar{\mathbf{g}}^{-1}(x,y)$ with an arbitrary $\bar{\mathbf{g}}(x,y)$ represent equivalent dot trajectories: Clearly, the vector field $\bar{\mathbf{h}}_1(x,y)$ connects by an arrow the point (x,y) to its destination under $\bar{\mathbf{g}}$, the point $\bar{\mathbf{g}}(x,y)$, while the second vector field $\bar{\mathbf{h}}_2(x,y)$ connects the point $\bar{\mathbf{g}}^{-1}(x,y)$ to its destination under $\bar{\mathbf{g}}$, the point (x,y).[4] This means that both vector fields correspond to the application of the same transformation $\bar{\mathbf{g}}$, although the departure points to which $\bar{\mathbf{g}}$ is applied are not the same in both cases. But because in both superpositions the same transformation $\bar{\mathbf{g}}$ is applied to one of two identical random screens, the resulting dot trajectories in the superposition are, indeed, equivalent. This can be seen from Proposition 4.2 by taking $\bar{\mathbf{g}}_1(x,y) = \bar{\mathbf{g}}(x,y)$, $\bar{\mathbf{g}}_2(x,y) = (x,y)$ and $\bar{\mathbf{f}}(x,y) = \bar{\mathbf{g}}^{-1}(x,y)$.

H.3 Hybrid (1,-1)-moiré effects whose moiré bands have 2D intensity profiles

As we have seen in Chapters 6 and 7, the transition from periodic line gratings to aperiodic (yet correlated) line gratings results in the loss of repetitivity in the resulting moiré effect: Instead of having infinitely many moiré bands, we are left in the aperiodic

---

[4] Remember that for any points (a,b) and (c,d) in the plane, when the tail of the vector $(a,b) - (c,d)$ is attached to the point (c,d), its head is located at the point (a,b); this means that the vector $(a,b) - (c,d)$ connects the point (c,d) to the point (a,b).

case with a single moiré band, i.e. a Glass pattern (see, for example, Figs. 6.1, 6.2 and 7.13). This typical behaviour is general to both 1D (1,-1)-moirés between line gratings and 2D (1,0,-1,0)-moirés between dot screens, independently of their intensity profiles (see, for example, Figs. 2.1 and 7.1).[5]

It is not surprising, therefore, that the same behaviour subsists in the (1,-1)-moiré even when the intensity of each individual line in the original gratings is modulated by some given 1D or 2D information. Part (b) of Fig. H.1 shows the superposition of two such periodic line gratings: The first grating consists of lines whose individual profiles are modulated by some 2D information (in the present example: a flattened version of the letters "EPFL", as clearly seen in the leftmost part of the figure), while the second grating consists of narrow white lines or "slits" on a black background. In such cases, if both gratings have similar periods and angles, the resulting superposition shows a (1,-1)-moiré effect that is a largely magnified version of the first grating; for example, in the case shown in our figure each of the moiré bands consists of a largely stretched-out (and possibly sheared) version of the letters "EPFL". In other words, in such cases the 2D information that modulates the intensity profile of each of the moiré bands in the superposition is a largely stretched-out (and possibly sheared) version of the 2D information that modulates the intensity of each individual line of the first grating.[6]

Part (a) of Fig. H.1 shows, on its part, the aperiodic counterpart of Fig. H.1(b). Here, the individual lines in each of the two gratings are still the same as in Fig. H.1(b), but their locations have been randomized (using the same random numbers for both of the superposed layers). And indeed, as expected, this aperiodic superposition contains only one moiré band, the Glass pattern, whose intensity profile remains modulated by the same 2D information (the letters "EPFL") as each of the moiré bands in Fig. H.1(b).

Note that replacing in Figs. H.1(a),(b) the modulated lines of both of the superposed gratings by simple black lines on a white background brings us back to the classical moiré (or Glass) patterns of Figs. 6.2(a),(b). The difference between Figs. H.1 and 6.2 in terms of the moiré intensity profile is, in fact, the 1D equivalent of the difference between Figs. 7.1 and 2.1 (see Secs. 7.2.4 and 7.4.2).

A similar phenomenon occurs also in superpositions of curvilinear line gratings. For example, if we replace the simple curved lines in the grating shown in Fig. 6.7 by modulated curved lines having the same profiles as in Fig. H.1, the intensity profile of the resulting curvilinear moiré (or Glass) pattern will be simply modulated by the letters "EPFL". The same result can be also obtained, of course, by applying to the straight gratings of Fig. H.1 the geometric layer transformations that are used in Fig. 6.7. This is, indeed, the 1D equivalent of the transition between Figs. 7.1 and 7.12.

---

[5] Note, however, that this behaviour does not extend to moirés of higher orders, as shown in Sec. 7.5.
[6] This phenomenon, originally discovered in the 1980s by Joe Huck in his artistic work [Huck03], has been investigated in depth in [Hersch04] and [Chosson06]. A Fourier-based explanation of this phenomenon is provided in Sec. C.14 of Appendix C in the second edition of *Vol. I*.

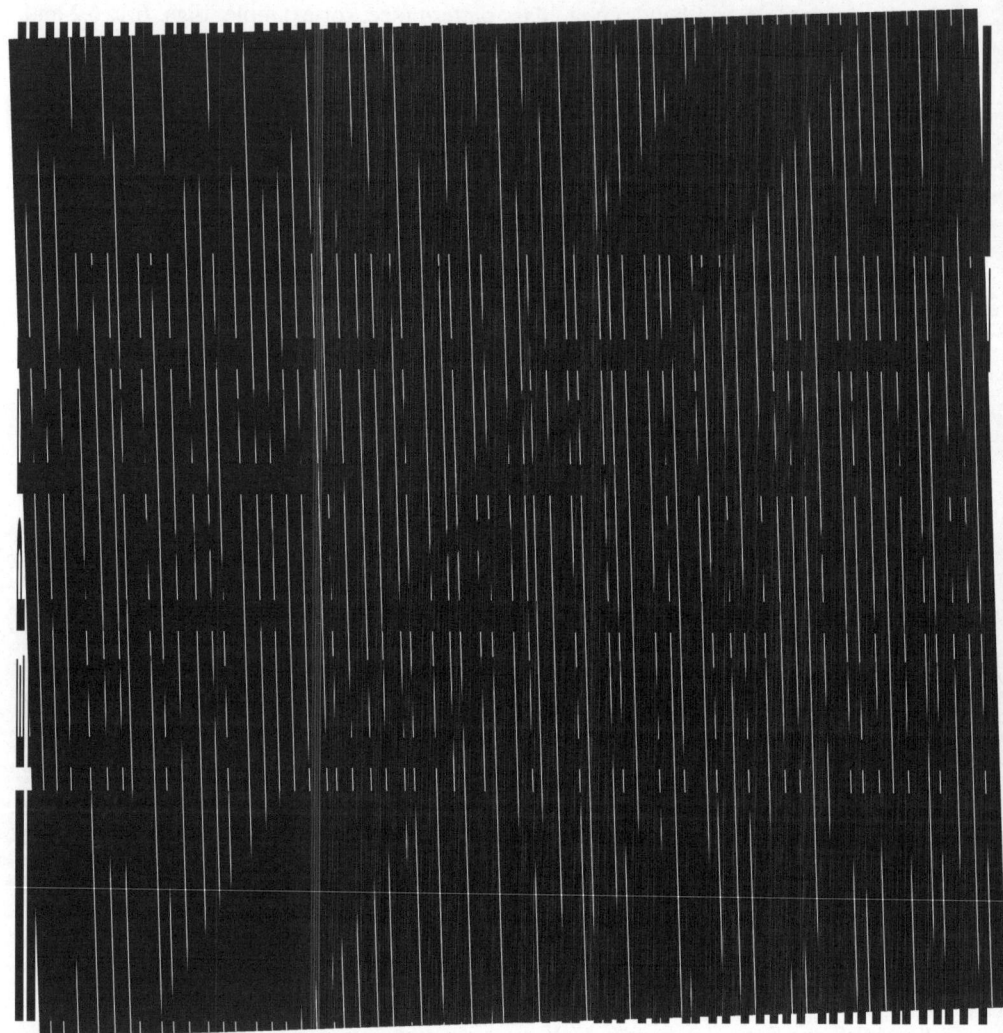

Figure H.1: (a) A superposition of two aperiodic (yet fully correlated) line gratings, one of
which is composed of straight lines that contain some given 2D information
(the flattened letters "EPFL", as clearly seen in the leftmost line), while the
other consists of narrow slits on a black background. As expected, a single
moiré band (Glass pattern) appears in the center of the superposition, and its
intensity profile contains a largely stretched-out (and possibly sheared) version
of the 2D information that appears in each of the individual lines of the first
grating (the letters "EPFL"). Compare with Fig. 6.2(a), in which both of the
superposed gratings consist of simple black lines. The difference between
Figs. H.1(a) and 6.2(a) is similar to the difference between Figs. 7.1(a) and
2.1(c) in the 2D case.

Figure H.1: (*continued.*) (b) The periodic counterpart of (a): The superposition of two *periodic* line gratings, one of which is composed of straight lines that contain some given 2D information (the flattened letters "EPFL", as clearly seen in the leftmost line), while the other consists of narrow slits on a black background. Each of the bands of the resulting (1,-1)-moiré is a duplicate of the central band, and contains the same 2D information. Compare with Fig. 6.2(b), in which both of the superposed gratings consist of simple black lines. The difference between Figs. H.1(b) and 6.2(b) is similar to the difference between Figs. 7.1(b) and 2.1(d) in the 2D case. Note that due to some particularities of the human visual system the effects shown in Figs. H.1(a),(b) are more easily perceived when the figures are rotated by 90° (try and see!).

Finally, it is interesting to see how the fundamental Glass-pattern theorem (Sec. 7.8 in Chapter 7) applies to our present hybrid case, in which the individual lines in one of the superposed gratings have a 2D profile, while the lines in the other grating (the slits) only have a 1D profile. To see this, consider the full componentwise notation of the layer transformations $\mathbf{g}_1(\mathbf{x})$ and $\mathbf{g}_2(\mathbf{x})$ and of the moiré (or Glass pattern) transformation $\mathbf{g}(\mathbf{x})$:

$$\mathbf{g}_1(\mathbf{x}) = \begin{pmatrix} g_{1,1}(x,y) \\ g_{1,2}(x,y) \end{pmatrix}, \quad \mathbf{g}_2(\mathbf{x}) = \begin{pmatrix} g_{2,1}(x,y) \\ g_{2,2}(x,y) \end{pmatrix}, \quad \mathbf{g}(\mathbf{x}) = \begin{pmatrix} g_1(x,y) \\ g_2(x,y) \end{pmatrix}$$

According to Eq. (7.37) of the fundamental Glass-pattern theorem, the transformation $\mathbf{g}(\mathbf{x})$ undergone by the moiré (or Glass pattern) is given by $\mathbf{g}(\mathbf{x}) = \mathbf{g}_1(\mathbf{x}) - \mathbf{g}_2(\mathbf{x})$. In the 2D case, where both of the original, untransformed layers are dot screens, this simply means:

$$\begin{pmatrix} g_1(x,y) \\ g_2(x,y) \end{pmatrix} = \begin{pmatrix} g_{1,1}(x,y) \\ g_{1,2}(x,y) \end{pmatrix} - \begin{pmatrix} g_{2,1}(x,y) \\ g_{2,2}(x,y) \end{pmatrix} \tag{H.5}$$

However, in the 1D case, i.e. when both of the original, undeformed layers consist of lines with a purely 1D profile, the second component in each of the above transformations becomes irrelevant, and we obtain:

$$\mathbf{g}_1(\mathbf{x}) = \begin{pmatrix} g_{1,1}(x,y) \\ 0 \end{pmatrix}, \quad \mathbf{g}_2(\mathbf{x}) = \begin{pmatrix} g_{2,1}(x,y) \\ 0 \end{pmatrix}, \quad \mathbf{g}(\mathbf{x}) = \begin{pmatrix} g_1(x,y) \\ 0 \end{pmatrix}$$

or even, more simply, by dropping the unused components and indices:

$$\mathbf{g}_1(\mathbf{x}) = g_1(x,y), \qquad \mathbf{g}_2(\mathbf{x}) = g_2(x,y), \qquad \mathbf{g}(\mathbf{x}) = g(x,y)$$

In this case Eq. (7.37) reduces into its single-component counterpart:

$$g(x,y) = g_1(x,y) - g_2(x,y) \tag{H.6}$$

which is precisely Eq. (7.35) of the fundamental Glass-pattern theorem for *line gratings*, as we have seen in Sec. 7.8.

We now return to our present hybrid case. In this case, only one of the two original, untransformed gratings (the second one) has a purely 1D profile, and therefore we have:

$$\mathbf{g}_1(\mathbf{x}) = \begin{pmatrix} g_{1,1}(x,y) \\ g_{1,2}(x,y) \end{pmatrix}, \quad \mathbf{g}_2(\mathbf{x}) = \begin{pmatrix} g_{2,1}(x,y) \\ 0 \end{pmatrix}, \quad \mathbf{g}(\mathbf{x}) = \begin{pmatrix} g_1(x,y) \\ g_2(x,y) \end{pmatrix}$$

Hence, Eq. (7.37) of the fundamental Glass-pattern theorem simply becomes here:

$$\begin{pmatrix} g_1(x,y) \\ g_2(x,y) \end{pmatrix} = \begin{pmatrix} g_{1,1}(x,y) \\ g_{1,2}(x,y) \end{pmatrix} - \begin{pmatrix} g_{2,1}(x,y) \\ 0 \end{pmatrix} \tag{H.7}$$

which is, indeed, intermediate between the 2D case of Eq. (H.5) and the 1D case of Eq. (H.6). This suggests that our hybrid superposition could be considered, in fact, as a "$1\frac{1}{2}$D case". A more detailed discussion on this case can be found in the second edition of *Vol. I* in Sec. C.14 of Appendix C.

Appendix I

Glossary of the main terms

I.1 About the glossary

Several thousands of publications on the moiré phenomenon have appeared during the last decades, in many different fields and applications. However, as already mentioned in *Vol. I*, the terminology used in this vast literature is very far from being consistent and uniform. Different authors use different terms for the same entities, and what is even worse, the same terms are often used in different meanings by different authors. To mention just one example, Glass patterns have been also called in literature *moiré fringes* [Glass69], *random-dot interference patterns* [Glass73], *quasi-moiré patterns* [Garavaglia01], and even *flash correlation artifacts*, or in short, *FCAs* [Prokoski99].

Obviously, in such an interdisciplinary domain as the moiré theory it would be quite impossible to adopt a universally acceptable standardization of the terms, because of the different needs and traditions in the various fields involved (optics, mechanics, mathematics, printing, etc.). Nevertheless, even without having any far-reaching pretensions, we were obliged to make our own terminological choices in a systematic and coherent way, in order to prevent confusion and ambiguity in our own work. We tried to be consistent in our terminology throughout this work, even if it forced us to assign to some terms a somewhat different meaning than one would expect (depending on his own background, of course).

We included in the present glossary all the terms for which we felt a clear definition was desirable to avoid any risk of ambiguity. But although this glossary is basically devoted to the terms that are being used in the present volume, we have also included, for the sake of completeness, some entries from the glossary of *Vol. I*, often with some additions or adaptations to our needs in the present volume. Just as in *Vol. I*, this glossary is not ordered alphabetically; rather, we preferred to group the various terms according to subjects. We hope this should help the reader not only to clearly see the meaning of each individual term by itself, but also to put it in relation with other closely related terms (which would be completely dispersed throughout the glossary if an alphabetical order were preferred). Note that terms in the glossary can be found alphabetically through the general index at the end of the book.

I.2 Terms in the image domain

grating (or *line-grating*) —
    A pattern consisting of parallel lines. A grating can be periodic or aperiodic.

curvilinear grating —
> A pattern consisting of parallel curvilinear lines. A curvilinear grating can be seen as a non-linear transformation of an initially uncurved periodic or aperiodic grating of straight lines.

cosinusoidal grating (not to be confused with *cosine-shaped grating*) —
> A grating with a cosinusoidal periodic-profile; for example, a cosinusoidal circular grating is a circular grating with a cosinusoidal periodic-profile. Note, however, that since reflectance and transmittance functions always take values ranging between 0 and 1, the cosinusoidal grating is normally "raised" and rescaled into this range of values. For example, a reflectance function in the form of a vertical straight cosinusoidal grating is expressed by: $r(x,y) = \frac{1}{2}\cos(2\pi fx) + \frac{1}{2}$.

cosine-shaped grating (not to be confused with *cosinusoidal grating*) —
> A grating (with any periodic-profile form) whose corrugations in the x,y plane are bent into a cosinusoidal shape, like in Fig. 6.17.

grid (or *line-grid*; also called in literature *cross-line grating*) —
> A pattern consisting of two superposed line-gratings, periodic or aperiodic, crossing each other at a non-zero angle. Unless otherwise mentioned it will be assumed that a grid consists of two binary straight line gratings. Note that every grid can be also seen as a screen (whose dot-elements are the spaces left between the lines of the grid).

regular grid (or *square grid*) —
> A 2-fold periodic grid composed of two superposed straight line-gratings that are identical but perpendicular to each other.

curved grid —
> A pattern obtained by applying a non-linear transformation to a periodic or to an aperiodic grid.

screen (or *dot-screen*) —
> A pattern consisting of dots. A screen can be periodic (as in Fig. 2.1(b)) or aperiodic (as in Fig. 2.1(a)).

regular screen —
> A 2-fold periodic screen whose dot arrangement is orthogonal and whose periods (or frequencies) to both orthogonal directions are equal.

curved screen —
> A pattern obtained by applying a non-linear transformation to a periodic or to an aperiodic screen (see, for example, Figs. 3.5(a),(b)).

halftone screen —
> A binary screen in which the size (and the shape) of the screen dots may vary (typically, according to the gray level of a given original continuous-tone image). A

halftone screen may be periodic or not. Halftone screens are used in the printing
world for the reproduction of continuous-tone images on bilevel printing devices.

screen gradation (or *wedge*) —
A halftone screen whose screen dots vary gradually (in their size and possibly also
in their shape) across the image, generating a halftoned image with a smooth and
uniform tone gradation.

image (has nothing to do with the image of a transformation) —
The most general term we use to cover "anything" in the image domain. It may be
periodic or not, binary or continuous, etc. In principle, a monochrome (black-and-
white) image has reflectance (or transmittance) values that vary between 0 (black)
and 1 (white); similarly, a colour image has reflectance (or transmittance) values
varying between 0 and 1 for each wavelength λ of its colour spectrum.

period (or *repetition-period* of a function p) —
A number $T \neq 0$ such that for any $x \in \mathbb{R}$, $p(x+T) = p(x)$. Note that the set of all the
periods of $p(x)$ forms a lattice in $\mathbb{R}$. In the case of a 2-fold periodic function $p(x,y)$,
a double period (or period parallelogram) of $p(x,y)$ is any parallelogram A which
tiles the x,y plane so that $p(x,y)$ repeats itself identically on any of these tiles (see
also Sec. A.3.4 in Appendix A of *Vol. I*).

period-vector (of a periodic function $p(x,y)$) —
A non-zero vector $\mathbf{P} = (x_0,y_0)$ such that for any $(x,y) \in \mathbb{R}^2$, $p(x+x_0,y+y_0) = p(x,y)$. If
there exist two non-collinear vectors $\mathbf{P}_1$, $\mathbf{P}_2$ having this property, $p(x,y)$ is said to be
2-fold periodic; in this case, for any point $\mathbf{x} \in \mathbb{R}^2$ the points $\mathbf{x}$, $\mathbf{x}+\mathbf{P}_1$, $\mathbf{x}+\mathbf{P}_2$,
$\mathbf{x}+\mathbf{P}_1+\mathbf{P}_2$ define a period parallelogram of $p(x,y)$.

periodic function —
A function having a period. Note that a 2D function $p(x,y)$ can be 2-fold periodic
(such as $p(x,y) = \cos x + \cos y$) or only 1-fold periodic (such as $p(x,y) = \cos x$).

almost-periodic function —
See Secs. B.3, B.5 in Appendix B of *Vol. I*.

aperiodic function —
A function which is not included in the class of almost-periodic functions. This
also implies that the function in question is not periodic (see Fig. B.3 in Appendix
B of *Vol. I*). Note, however, that although repetitive functions formally fall within
the scope of this definition of aperiodic functions, we prefer to exclude them from
our present definition, because they are still structurally ordered, and they have
already been investigated in Chapters 10 and 11 of *Vol. I*. We therefore adopt the
definition saying that a function is aperiodic if it is neither periodic (or almost-
periodic) nor a geometrically transformed version thereof.

repetitive structure (or *repetitive function*) —

A structure (or a function) which is repetitive according to a certain rule, but which is not necessarily periodic (or almost-periodic). For example: concentric circles; gratings with logarithmic line-distances; screen gradations; etc. Note that such structures are sometimes called in literature *quasi-periodic* (like in [Bryngdahl74 p. 1290]); however, we reserve the term quasi-periodic only to its meaning in the context of the theory of almost-periodic functions (see Sec. B.5 in Appendix B of *Vol. I*).

random structure —

A structure consisting of randomly positioned elements. Note that every random structure is aperiodic, but the converse is not necessarily true. See also Sec. 2.4

coordinate-transformed structure —

A structure $r(x,y)$ which is obtained by the application of a non-linear coordinate transformation $g(x,y)$ to a certain initial periodic or aperiodic structure $p(x,y)$. Note that $g(x,y)$ is applied to the original structure $p(x,y)$ as a *domain transformation*; more formally, using vector notation, $r(\mathbf{x}) = p(\mathbf{g}(\mathbf{x}))$. Curvilinear gratings (such as parabolic or circular gratings) and gratings with a varying frequency (such as a grating with logarithmic line-distances) are coordinate-transformed structures. Examples of coordinate-transformed gratings (both periodic and aperiodic) are shown in the top row of Fig. 6.3, and examples of coordinate-transformed dot screens (both periodic and aperiodic) are shown in the top row of Fig. 3.5.

profile-transformed structure —

A structure $r(x,y)$ which is obtained by the application of a non-linear transformation $t(z)$ to the profile of a certain initial periodic or aperiodic structure $p(x,y)$. Note that $t(z)$ is applied to the original structure $p(x,y)$ as a *range transformation*; more formally, using vector notation, $r(\mathbf{x}) = t(p(\mathbf{x}))$. Screen gradations are an example of profile-transformed structures.

coordinate-and-profile transformed structure —

A structure $r(x,y)$ which is obtained from a certain initial periodic or aperiodic structure $p(x,y)$ by the application of both a non-linear coordinate-transformation $g(x,y)$ and a non-linear profile-transformation $t(z)$. More formally, using vector notation, $r(\mathbf{x}) = t(p(\mathbf{g}(\mathbf{x})))$. An example of a coordinate-and-profile-transformed structure is given in Remark 2 of Sec. 10.2 in *Vol. I*.

intensity profile (of a structure $r(x,y)$) —

A function over the x,y plane whose value at each point (x,y) indicates the intensity (or more precisely, the reflectance or the transmittance) of the structure $r(x,y)$.

periodic profile (of a curvilinear grating, curved grid, etc.) —

The periodic profile of a curvilinear grating or a curved screen $r(x,y)$ is defined as the intensity profile of the original, uncurved periodic grating (or screen), before the non-linear transformation has been applied to it (see Sec. 10.2 in *Vol. I*). Note that in periodic structures the periodic profile coincides with the intensity profile.

normalized periodic profile (of a curvilinear grating, curved grid, etc.) —
   See Sec. 10.2 in *Vol. I*.

geometric layout (of a curvilinear grating, curved grid, etc.) —
   The geometric layout of a curvilinear grating $r(x,y)$ is the locus of the centers of its curvilinear corrugations in the x,y plane; it is defined by the bending transformation of the curvilinear grating (see Sec. 10.2 in *Vol. I*). Similarly, the geometric layout of a curved grid or a curved screen is defined by its two bending functions.

bending transformation (of a curvilinear grating, curved grid, etc.) —
   The bending transformation of a curvilinear grating $r(x,y) = p(g(x,y))$ is the non-linear coordinate transformation $g(x,y)$ which bends the original, uncurved grating $p(x')$ into the curvilinear grating $r(x,y)$. The bending transformation of a curved grid or a curved screen $r(x,y) = p(g_1(x,y),g_2(x,y))$ is the non-linear coordinate transformation $\mathbf{g}(x,y) = (g_1(x,y),g_2(x,y))$ which bends the original, uncurved grid or screen $p(x',y')$ into the curved structure $r(x,y)$. We usually assume that the bending transformation is smooth (a diffeomorphism), so that it has no abrupt jumps or other troublesome singularities.

I.3 Terms in the spectral domain

spectrum (or *frequency spectrum*; not to be confused with *colour spectrum*) —
   The frequency decomposition of a given function, which specifies the contribution of each frequency to the function in question. The frequency spectrum is obtained by taking the Fourier transform of the given function.

visibility circle —
   A circle around the spectrum origin whose radius represents the cutoff frequency, i.e., the threshold frequency beyond which fine detail is no longer detected by the eye. Obviously, its radius depends on several factors such as the viewing distance, the light conditions, etc. It should be noted that the visibility circle is just a first-order approximation. In fact, the sensitivity of the human eye is a continuous 2D bell-shaped function [Daly92 p. 6], with a steep "crater" in its center (representing frequencies which are too small to be perceived), and "notches" in the diagonal directions (owing to the drop in the eye sensibility in the diagonal directions [Ulichney87 pp.79–84]).

frequency vector —
   A vector in the u,v plane of the spectrum which represents the geometric location of an impulse in the spectrum (see Sec. 2.2 and Fig. 2.1 in *Vol. I*).

DC impulse —

The impulse that is located on the spectrum origin. This impulse represents the frequency of zero, which corresponds to the constant component in the Fourier decomposition of the image; the amplitude of the DC impulse corresponds to the intensity of this constant component. This impulse is traditionally called the *DC impulse* because it represents in electrical transmission theory the direct current component, i.e., the constant term in the frequency decomposition of an electric wave; we are following here this naming convention.

comb (or *impulse-comb*, *Dirac-comb*, *impulse-train*) —

An infinite train of equally spaced impulses located on a straight line in the spectrum. Any 1D periodic function is represented in the spectrum by a comb centered on the spectrum origin. The step and the direction of this comb represent the frequency and the orientation of the periodic function; its impulse amplitudes, which are given by the Fourier series development of the periodic function, determine its intensity profile.

nailbed (or *impulse-nailbed*) —

An infinite 2D train of equally spaced impulses located in the spectrum on a dot-lattice (either square-angled or skewed). Any 2D periodic function is represented in the spectrum by an impulse nailbed centered on the spectrum origin. The steps and the two main directions of this nailbed represent the frequency and the orientation of the two main directions of the function's 2D periodicity; the impulse amplitudes, which are given by the 2D Fourier series development of the periodic function, determine its intensity profile.

support (of a *comb*, a *nailbed*, a *spectrum*, etc.) —

The set of the geometric locations on the u,v plane of all the impulses of the specified comb, nailbed, or spectrum.

line-impulse —

A generalized function which is impulsive along a 1D line through the plane, and null everywhere else. A line-impulse can be graphically illustrated as a "blade" whose behaviour is continuous along its 1D line support but impulsive in the perpendicular direction. For example, the spectrum of an aperiodic line grating is a Hermitian line impulse (see Sec. 7.3.1 and Figs. 7.7, 7.8). As another example, the spectrum of a parabolic cosinusoidal grating consists of two parallel line-impulses (see Example 10.5 in Sec. 10.3 of *Vol. I*). Note that the amplitude of a line-impulse does not necessarily die out away from its center, and it may even rapidly oscillate between two constant values.

curvilinear impulse —

A generalized function which is impulsive along a 1D curvilinear path through the plane, and null everywhere else. A curvilinear impulse can be graphically illustrated as a curvilinear "blade" whose behaviour is continuous along its 1D curvilinear support but impulsive in the perpendicular direction.

hump —

A 2D continuous surface, often bell-shaped, elliptic or hyperbolic, which is defined around a given center on the plane. For example, the spectrum of an aperiodic dot screen is a Hermitian hump (see Sec. 7.4.1 and Fig. 2.10(e)). As another example, the convolution of two non-parallel line-impulses gives a hump (see Sec. 7.3.1, or Sec. 10.7.3 and Fig. 10.13 in *Vol. I*). Note that the amplitude of a hump does not necessarily die out away from its center, and in some cases it may even rapidly oscillate between two constant values.

impulsive spectrum —

A spectrum which only consists of impulses, i.e., whose support consists of a finite or at most denumerably infinite number of points. All periodic and almost-periodic functions have impulsive spectra.

line-spectrum —

A spectrum which consists of line impulses (see, for example, Fig. 10.11 in *Vol. I*).

hybrid spectrum —

A spectrum which contains any combination of impulses, line-impulses and continuous humps (as opposed to a purely impulsive spectrum, a purely line-spectrum or a purely continuous spectrum). See, for instance, Example 10.14 of Sec. 10.7.4 and Fig. 10.13 in *Vol. I*.

singular support (of a *spectrum*, etc.; distinguish from a *singular locus of a moiré*) —

The subset of the spectrum support over which the spectrum is impulsive. The singular support of a given spectrum includes the support of all the impulsive elements which are included in the spectrum (impulses, line-impulses, etc.), but not the support of continuous elements such as humps or wakes.

I.4 Terms related to moiré

moiré effect (or *moiré phenomenon*) —

A visible phenomenon which occurs when repetitive structures (such as line-gratings, dot-screens, etc.) are superposed. It consists of a new pattern which is clearly observed in the superposition, although it does not appear in any of the original structures. Moiré effects may occur also in the superposition of correlated (or at least partially correlated) aperiodic layers; such moiré effects are called *Glass patterns*.

Glass pattern —

A moiré pattern that occurs between two aperiodic layers (which are correlated or at least partially correlated; note that uncorrelated aperiodic layers do not generate

Glass patterns). Unlike moiré patterns between periodic or repetitive layers, a Glass pattern typically consists of a single structure (see, for example, Figs. 2.1–2.2). Often (but not always, as clearly shown in Chapter 7) a Glass pattern is brighter in its center (where the correlation between the layers is maximal and their individual elements almost coincide), and farther away it stabilizes at a darker gray level (because the layers are no longer correlated and their elements more often fall between each other, leaving less white area in the superposition). The Glass pattern reflects, therefore, the macroscopic gray level variations in the layer superposition. Usually it is also surrounded by typical geometric dot alignments in the microstructure, which are known as dot trajectories. Glass patterns are called after Leon Glass, who described them in the late 1960s [Glass69, Glass73].

linear Glass pattern —
A Glass pattern that is generated in the layer superposition along a straigt line (a fixed line). See, for example, Figs. 2.3 and 6.1.

curvilinear Glass pattern —
A Glass pattern that is generated in the layer superposition along a curvilinear line (a curvilinear fixed line). See, for example, Figs. 3.16 and 6.7.

$(k_1,...,k_m)$-moiré —
The 1-fold periodic structure in the image domain which corresponds to the $(k_1,...,k_m)$-comb in the spectrum convolution (the spectrum of the superposition); see Sec. 2.8 in *Vol. I*. In other words, this is the moiré which is generated due to the interaction between the k_i harmonic frequencies of the respective layers in the superposition. This moiré may be visible if at least its fundamental impulse, the $(k_1,...,k_m)$-impulse, is located inside the visibility circle. More details on our moiré notational system can be found in Sec. 2.8 of *Vol. I*.

singular moiré (or *singular state, singular superposition*) —
A configuration of the superposed layers in which the period of the moiré in question becomes infinitely large (i.e., its frequency becomes 0), and hence it can no longer be seen in the superposition. Singular moirés are unstable moiré-free states, since the slightest deviation in the angle or scaling of any of the superposed layers may cause the moiré in question to "come back from infinity" and to reappear with a large, visible period. See Sec. 2.3.2.

stable moiré-free state —
A moiré-free configuration of the superposed layers in which no moiré becomes visible even when small deviations occur in the angle or in the scaling (or frequency) of any of the layers (see Sec. 2.3.2).

unstable moiré-free state —
A moiré-free configuration of the superposed layers in which any slight deviation in the angle or in the scaling (or frequency) of any of the layers causes the

reappearance of a moiré with a large, visible period (see Fig. 2.5). Any singular state is an unstable moiré-free state.

moiré profile (or *moiré intensity profile*; *moiré intensity surface*) —
   A function which defines the intensity level (the macroscopic gray level) of the moiré at any point of the image (see Sec. 7.1, or Secs. 2.10 and 4.1 in *Vol. I*).

macrostructure, microstructure (within a superposition) —
   The superposition of two or more layers (gratings, screens, etc.) may generate new structures which appear in the superposition but not in the original layers. These new structures can be classified into two categories: the *macrostructures*, i.e., the moiré effects proper, which are much coarser than the detail of the original layers; and the *microstructures*, i.e., the tiny geometric forms which are almost as small as the periods of the original layers, and are normally visible only from a close distance or through a magnifying glass. In the periodic case the microstructure may consist of *rosettes* (see Chapter 8 in *Vol. I*), while in the aperiodic case the microstructure may consist of *dot trajectories* (see Secs. 2.2 and 2.3.3).

rosettes —
   The various tiny flower-like shapes which are often present in the microstructure of periodic dot-screen or grid superpositions (see Chapter 8 in *Vol. I*).

dot trajectories (not to be confused with *trajectories*) —
   This term is reserved to the typical microstructure dot alignments that appear in the superposition of correlated aperiodic layers. These dot trajectories may have various geometric shapes, depending on the transformations undergone by the superposed layers. In the case of simple linear transformations such as layer rotations, layer scalings, etc. the resulting dot trajectories are rather simple (circular, radial, spiral, elliptic, hyperbolic, linear, etc.); see Figs. 2.1–2.3. But when the layer transformations are more complex, the resulting dot trajectories may have more interesting shapes (see, for example, Fig. 5.1). Dot trajectories are a typical property of the superposition of aperiodic layers, and they are not visible in superpositions of periodic layers.

additive / subtractive moiré (not to be confused with *additive superposition*) —
   Classical terms often used in literature to designate moirés which are generated by frequency sums or frequency differences, respectively, in the spectrum. For example, the (1,-1)-moiré is subtractive, while the (1,1)-moiré is additive. Note, however, that these terms cannot be generalized to more complex cases such as the (1,1,-1)-moiré between three gratings. These terms are mostly useful in the superposition of two curvilinear gratings, where both the additive and the subtractive moiré are often observed simultaneously, each of them having a different shape and location (see, for example, Fig. 10.31 in *Vol. I*). In this case the most convenient way to define them is based on their indicial equations (see Sec. 11.2 in *Vol. I*): the additive moiré is the system of moiré fringes which

corresponds to the indicial equation $m+n = p$, while the subtractive moiré is the system of moiré fringes which is selected by the indicial equation $m-n = p$. Note that additive moirés are not generated between aperiodic layers (see Sec. 7.5).

I.5 Terms related to light and colour

colour spectrum (not to be confused with *frequency spectrum*) —
The wavelength decomposition of a given light, which specifies the contribution of each visible light wavelength λ (approximately between $\lambda = 380$ nm for violet and $\lambda = 750$ nm for red) to the given light. The colour spectrum determines the visible colour of the light in question. See Chapter 9 in *Vol. I* for more details.

monochrome (or *black-and-white*; not to be confused with *monochromatic*) —
Achromatic light, image, etc. involving only black, white and all the intermediate gray levels. The colour spectrum of an ideal monochrome light is flat, i.e., it has a constant value (between 0 and 1) for all wavelengths λ of the visible light.

monochromatic (not to be confused with *monochrome*) —
Chromatic light, image, etc. involving only a single pure wavelength λ of the visible light. The colour spectrum of an ideal monochromatic light consists of a single impulse of intensity 1 at the wavelength λ.

reflectance (or *reflectance function*) —
A function $r(x,y)$ which assigns to any point (x,y) of a monochrome image viewed by reflection a value between 0 and 1 representing its light reflection: 0 for black (or no reflected light), 1 for white (or full light reflection), and intermediate values for in-between shades. More formally, reflectance is defined at any point (x,y) as the ratio of reflected to incident radiant power [Wyszecki82 p. 463].

transmittance (or *transmittance function*) —
A function $r(x,y)$ which assigns to any point (x,y) of a monochrome image viewed by transmission (such as a transparency, a film, etc.) a value between 0 and 1 representing its light transmission: 0 for black (or no transmitted light), 1 for white (or full light transmission), and intermediate values for in-between shades. More formally, transmittance is defined at any point (x,y) as the ratio of transmitted to incident radiant power [Wyszecki82 p. 463].

chromatic reflectance (or *chromatic reflectance function*) —
A function $r(x,y;\lambda)$ which assigns to any point (x,y) of a colour image viewed by reflection its full colour spectrum. In other words, it gives for every wavelength λ of the visible light (approximately between $\lambda = 380$ nm and $\lambda = 750$ nm) a value between 0 and 1, which represents the reflectance of light of wavelength λ at the point (x,y) of the image. This is a straightforward generalization of the reflectance function $r(x,y)$ in the monochrome case (see Chapter 9 in *Vol. I*).

chromatic transmittance (or *chromatic transmittance function*) —

A function $r(x,y;\lambda)$ which assigns to any point (x,y) of a colour image viewed by transmission its full colour spectrum. In other words, it gives for every wavelength λ of the visible light (approximately between $\lambda = 380$ nm and $\lambda = 750$ nm) a value between 0 and 1, which represents the transmittance of light of wavelength λ at the point (x,y) of the image. This is a straightforward generalization of the transmittance function $r(x,y)$ in the monochrome case (see Chapter 9 in *Vol. I*).

I.6 Miscellaneous terms

binary (grating, etc.) —

A structure which contains only two transmittance (or reflectance) levels: 0 and 1.

discrete —

A subset D of $\mathbb{R}^n$ is called *discrete* if there exists a number $d > 0$ so that for any points $a,b \in D$ the distance between a and b is larger than d. Note, however, that the term *discrete* is also used as the opposite of *continuous*. For example: periodic functions have *discrete spectra*, while aperiodic functions have *continuous spectra*.

domain / range transformation —

Any image $r(x,y)$ (or function $r: \mathbb{R}^2 \to \mathbb{R}$) can undergo two types of coordinate transformations: Either a transformation of its *domain*, $r(x,y) \mapsto r(g(x,y))$, or a transformation of its *range*, $r(x,y) \mapsto t(r(x,y))$. As explained in Sec. D.6 of Appendix D, in the first case $g(x,y)$ is applied as an inverse transformation, while in the second case $t(x)$ is used as a direct transformation. Similarly, any mapping $f(x,y)$, $f: \mathbb{R}^2 \to \mathbb{R}^2$, can undergo two types of coordinate transformations: Either a transformation of its *domain*, $f(x,y) \mapsto f(g(x,y))$, or a transformation of its *range*, $f(x,y) \mapsto g(f(x,y))$. Again, in the first case $g(x,y)$ is applied as an inverse transformation, while in the second case it is used as a direct transformation.

direct / inverse mapping (not to be confused with *forward / backward mapping algorithm*) —

The mathematical terms used to designate a mapping (geometric transformation) g and its inverse g^{-1}. We have, therefore, $g \circ g^{-1} = g^{-1} \circ g = i$, where i is the identity transformation $i(x,y) = (x,y)$, and where "$\circ$" is the operation of mapping composition: $(f \circ g)(x,y) = f(g(x,y))$. Note that the designations *direct* and *inverse* are interchangeable, and they depend on our point of view; thus, if we focus our attention to the mapping $h = g^{-1}$, we may consider h as the direct mapping and $h^{-1} = (g^{-1})^{-1} = g$ as its inverse. For example, if $g(x,y) = (2x,2y)$ then $g^{-1}(x,y) = (x/2,y/2)$; and if $g(x,y) = (x/2,y/2)$ then $g^{-1}(x,y) = (2x,2y)$. Note that the terms *direct / inverse transformations* are also used in the same meaning, whereas the terms *direct / inverse transforms* are reserved to operations such as the Fourier transform.

forward / backward mapping algorithm (not to be confused with *direct / inverse mapping*) —

Digital imaging terms used to designate two different types of graphical algorithms or computer programs: A *forward mapping* operates by scanning the original input image pixel by pixel, and copying the value of the image at each input pixel location (x,y) onto the corresponding position (u,v) in the output image. The output position (u,v) is determined by passing the x and y coordinates of each input pixel through the transformation $(u,v) = \mathbf{g}(x,y)$. On the other hand, a *backward mapping* operates by scanning the target output image pixel by pixel, and copying onto each output pixel location (u,v) the value of the input pixel at the corresponding (x,y) position in the input image. The input position (x,y) that corresponds to a given output pixel location is determined by passing the u and v coordinates of each output pixel through the *inverse* transformation, $(x,y) = \mathbf{g}^{-1}(u,v)$. For more details on forward and backward mapping algorithms see Sec. D.9.1 in Appendix D.

trajectories (not to be confused with *dot trajectories*) —

This term is reserved to the *solution curves* of a system of differential equations, or, equivalently, to the *field lines* of the corresponding vector field. A system of differential equations $\frac{d}{dt}x(t) = g_1(x(t),y(t))$, $\frac{d}{dt}y(t) = g_2(x(t),y(t))$ has for solutions a family of curves in the x,y plane, whose parametric representation is $(x(t),y(t))$, and whose members differ from each other by some constants c [Kreyszig93 180–186]. Each of these solution curves is called a *trajectory* since it traces out the evolution of the curve as the parameter t is being varied. These trajectories are also the field lines of the mapping $\mathbf{g}(x,y) = (g_1(x,y),g_2(x,y))$ where this mapping is regarded as a 2D vector field that assigns to each point (x,y) the vector $\mathbf{g}(x,y)$. Note that the trajectories (field lines) of a vector field $\mathbf{g}(x,y)$ are the curves that are tangent to the vectors of the vector field at any point in the plane. For more details on the connections between the vector field $\mathbf{g}(x,y)$, the corresponding system of differential equations and their trajectories see Sec. B.6 in Appendix B.

critical point (not to be confused with a *fixed point*) —

A point (x,y) is called a critical point of the vector field $\mathbf{g}(x,y)$ or of the system of differential equations $\frac{d}{dt}x(t) = g_1(x(t),y(t))$, $\frac{d}{dt}y(t) = g_2(x(t),y(t))$ if it satisfies $g_1(x,y) = 0$, $g_2(x,y) = 0$ [Birkhoff89 p. 133; Kreyszig93 p. 176]. For more details see Sec. H.1 in Appendix H. Note that the term *critical point* is also used in mathematics to designate a point (x,y) where a surface $g(x,y)$ has an extremum or a horizontal inflection point, i.e. where $\frac{\partial}{\partial x}g(x,y) = 0$ and $\frac{\partial}{\partial y}g(x,y) = 0$ (see, for example, Sec. 2.19 in [Kaplan03]).

fixed point (not to be confused with a *critical point*) —

A point (x_F,y_F) is said to be a fixed point of transformation $\mathbf{g}(x,y)$ if it is not affected by the transformation, meaning that $\mathbf{g}(x_F,y_F) = (x_F,y_F)$. Note, however, that some references use the term "fixed point" for a critical point (see, for example, [Strogatz94 pp. 124, 150; Weisstein99 p. 652]).

fixed locus —

Transformation $g(x,y)$ is said to have a fixed locus if it has a locus consisting of fixed points. For example, the transformation shown in Fig. 3.17 has a fixed locus consisting of an isolated point at the origin and a family of equispaced concentric circles surrounding it. Note that each point in a fixed locus is mapped by $g(x,y)$ to itself; it is not sufficient that each point of the locus be mapped by $g(x,y)$ to another point within the locus.

fixed line —

A fixed locus consisting of a straight line. Transformation $g(x,y)$ has a fixed line if it has an entire straight line of fixed points. For example, each of the two transformations shown in Fig. 2.3(a)–(d) has a fixed line along the x axis, while each of the transformations shown in Fig. 3.6 has two fixed lines parallel to the x axis.

mutual fixed point —

A point (x_F,y_F) is said to be a mutual fixed point of transformations g_1 and g_2 if $g_1(x_F,y_F) = g_2(x_F,y_F)$. Note that the term *common fixed point* of g_1 and g_2 is already used in the mathematical literature for a point (x_F,y_F) that satisfies $g_1(x_F,y_F) = (x_F,y_F) = g_2(x_F,y_F)$, but this definition is too restrictive for our needs.

mutual fixed locus —

Transformations g_1 and g_2 are said to have a mutual fixed locus if there exists a locus in the x,y plane that consists of mutual fixed points of g_1 and g_2.

almost fixed point —

A point (x_F,y_F) is said to be an almost fixed point of transformation $g(x,y)$ if it is only very slightly affected by the transformation, meaning that $g(x_F,y_F) \approx (x_F,y_F)$. Similarly, (x_F,y_F) is an almost mutual fixed point of g_1 and g_2 if $g_1(x_F,y_F) \approx g_2(x_F,y_F)$.

correlation (between two signals, images, etc.) —

A general term referring to the agreement or similarity between the two entities in question. Confusingly, this term is routinely used in several different meanings, notably as an abbreviation to the terms *local correlation*, *global correlation* or *cross correlation* (see below).

local correlation (between two signals, images, etc.) —

Two entities (signals, images, functions, etc.) are said to be locally correlated (or to be well correlated in a certain location) if they highly agree with each other in the specified location. As its name indicates, this is a local property, so that two given images can be highly correlated in some areas but not at all correlated in other areas. Local correlation between two superposed layers is the reason for the appearance of a Glass pattern in that area of the superposition. For example, the Glass patterns shown in Fig. 7.12 consist of four dark "2"-shaped areas; within these areas the two superposed layers are well correlated (the pinholes of the second layer coincide with the tiny "2"-shaped dots of the first layer), while in the remaining areas of the figure the correlation between the two layers is low. See also Secs. 2.2, 7.8 and Problem 7-13.

global correlation (between two signals, images, etc.) —
Two entities (signals, images, functions, etc.) are said to be globally correlated if they highly agree with each other throughout their entire domain of definition. For example, two identical copies $r_1(x)$ and $r_2(x)$ of the same aperiodic signal (one or both of which may have undergone some amplitude transformation or have some additive random noise) are globally correlated. However, the signal $r_1(x)$ is no longer globally correlated with the shifted signal $r_2(x - x_0)$ or with the stretched signal $r_2(ax)$. See also Secs. 2.2, 7.8 and Problem 7-13.

cross correlation (between two signals, images, etc.) —
The cross correlation between two functions $f(x)$ and $g(x)$ is a third function $c_{f,g}(x)$ that indicates the relative amount of agreement between $f(x)$ and $g(x)$ for all possible degrees of misalignments (shifts). At each point x the value of the function $c_{f,g}(x)$ is defined as the area under the product of f and g after g has been shifted by x. For example, if $r_1(x)$ and $r_2(x)$ are defined as above (see under "global correlation") then their cross correlation function consists of a high peak about $x = 0$, and low values everywhere else. The reason is that r_1 and r_2 are globally correlated when the shift between them is $x = 0$, but for any other shift they are no longer globally correlated. In the case of 2D images, points (x,y) in which the value of the cross correlation function is high indicate displacements of x,y between the two images in which the volume under the product of the two entire layers is high. Obviously, the cross correlation can detect displacements x,y in which the two *entire* images are *globally* well correlated, but it cannot detect isolated zones of *local* correlation between the two images, since the contribution of such local zones is relatively small, and it may be buried and lost within the global volume under the product of the entire images. The formal definition of cross correlation as well as its main mathematical properties are given in Appendix E. See also Secs. 2.2, 7.8 and Problem 7-13.

finite energy function (or signal) (not to be confused with *finite power function*) —
A function $f(x)$ is called a finite energy function (or a finite energy signal) if it is square integrable, i.e. if it satisfies:

$$\int_{-\infty}^{\infty} |f(x)|^2 \, dx < \infty$$

The class of finite energy functions includes all physically realizable functions, but it does not include many other useful functions such as constant functions, step functions, periodic functions or stationary random functions [Champeney73 p. 59; Coulon84 pp. 33–35].

finite power function (or signal) (not to be confused with *finite energy function*) —
A function $f(x)$ is called a finite power function (or a finite power signal) if it satisfies:

$$\lim_{a \to \infty} \frac{1}{a} \int_{-a/2}^{a/2} |f(x)|^2 \, dx < \infty$$

The class of finite power functions is larger than that of finite energy functions, and it includes, for example, constant functions, step functions, periodic functions and stationary random functions [Champeney73 p. 59; Coulon84 pp. 33–35].

almost-zero transformation (or *almost-null transformation*) —
A transformation $(u,v) = \mathbf{o}(x,y)$ is said to be *almost zero* or *almost null* if it differs only slightly, within our zone of interest (i.e. within the area covered by the layer superposition), from the zero transformation $(u,v) = \mathbf{z}(x,y) = (0,0)$. See Sec. D.12 in Appendix D.

weak transformation (or *almost-identity transformation*) —
A transformation $(u,v) = \mathbf{g}(x,y)$ is said to be *weak* or *almost identity* if it differs only slightly, within our zone of interest (i.e. within the area covered by the layer superposition), from the identity transformation $(u,v) = \mathbf{i}(x,y) = (x,y)$. In other words, $\mathbf{g}(x,y)$ is a weak transformation if it satisfies $\mathbf{g}(x,y) = (x,y) + \mathbf{o}(x,y)$, where $\mathbf{o}(x,y)$ is an almost-zero transformation. See Sec. D.12 in Appendix D.

diffeomorphism —
A diffeomorphism (in our case, on $\mathbb{R}^2$) is a one-to-one continuously differentiable mapping of $\mathbb{R}^2$ onto itself whose inverse mapping is also continuously differentiable.

scaling (of a *comb*, *nailbed*, etc.) —
We distinguish between *amplitude* scalings, and *period* or *frequency* scalings (in which the expansion or contraction occurs along the x,y axes in the image, or the u,v axes in the spectrum).

phase (of a periodic function, a periodic moiré, etc.) —
See Sec. C.4 in *Vol. I* and Chapter 7 Secs. 7.1–7.5 in *Vol. I*.

separable function (of two variables) —
A function $f(x,y)$ is said to be *separable* if it can be presented as (or separated into) a product of a function of x and a function of y: $f(x,y) = g(x) \cdot h(y)$ [Gaskill78 pp. 16–17; Cartwright90 p. 117]. Note, however, that we use this term in a slightly larger sense: A 2D function $f(x,y)$ is separable if it can be presented as a product of two independent 1D functions. Therefore, although $f(x,y) = g(x) \cdot h(y)$ may no longer be separable (in the narrower sense) after it has undergone a rotation or a skewing transformation, we will still consider it as separable (with respect to the rotated or skewed axes x' and y': $f(x',y') = g(x') \cdot h(y')$).

inseparable function (of two variables) —
A function that is not *separable*. For example, the function representing a square white dot is separable: rect(x,y) = rect(x)·rect(y), while the function representing a circular white dot is inseparable.

spatially separable (not to be confused with a *separable function*) —
Two functions $F(u,v)$ and $G(u,v)$ in the spectrum are called spatially separable if their supports in the u,v plane are not overlapping. Spatially separable elements in the spectrum can be separated and extracted by means of filtering, i.e., by multiplying the spectrum with an appropriate 2D low-pass or band-pass filter.

spatially inseparable (not to be confused with an *inseparable function*) —
Two functions $F(u,v)$ and $G(u,v)$ in the spectrum are called spatially inseparable if their supports in the u,v plane are at least partially overlapping. Spatially inseparable elements in the spectrum cannot be separated or extracted by multiplying the spectrum with 2D low-pass or band-pass filters.

dots per inch (dpi) (or *dots per centimeter*; not to be confused with *lines per inch*) —
A term used to specify the resolution of a digital device such as a printer, a scanner, etc. For example, a device whose resolution is 300 dpi can only address points on an underlying pixel-grid whose period is 1/300 of an inch, and no in-between points or pixel-fractions can be addressed. Some devices have different resolutions in the horizontal and in the vertical directions.

lines per inch (lpi) (or *lines per centimeter*; not to be confused with *dots per inch*) —
A term used to specify the frequency of gratings, dot-screens, etc. This term specifies the number of periods per inch. For example: the finest grating that can be produced on a 300 dpi device, namely: a sequence of alternating one-pixel wide black and white lines, is a grating of 150 lpi (since one period consists here of two device pixels).

List of notations and symbols

This list consists of the main symbols used in the present volume. They appear with a very brief description and a reference to the page in which they are first used or defined. Obvious symbols such as '+', '−', etc. have not been included.

Symbol	Short description	Page
x, y	The coordinates (axes) of the image plane	15
u, v	The coordinates (axes) of the spectral plane	15
x', y'	Rotated coordinates (axes) in the image plane	56
$r(x,y)$	A 2D reflectance (or transmittance) function	15
$R(u,v)$	The spectrum of $r(x,y)$	15
$d(x,y)$	A single dot (of a dot-screen)	32
$D(u,v)$	The spectrum of $d(x,y)$	32
$*$	1D convolution (or T-convolution)	245, 436
$**$	2D convolution (or T-convolution)	15, 251, 411
$\star$	1D cross correlation	246, 268
$\star\star$	2D cross correlation	258, 269, 413
$c_{f,g}$	The cross correlation between the functions f and g	17, 413
$\alpha, \theta, ...$	Angles	56, 58
φ_M	Angle of a moiré effect	231
$\alpha \to 0$	The angle α tends to 0	76
$f, f_1, ...$	Frequencies of 1D periodic functions $p(x), p_1(x), ...$	228, 231
f_M	Frequency of a moiré effect	231
$T, T_1, ...$	Periods of 1D periodic functions $p(x), p_1(x), ...$	59, 231
T_M	Period of a moiré effect	59, 231

F, P	Matrices	62
F^{-1}	The inverse of matrix F	63, 322
F^T	The transpose of matrix F	390
$\lvert F \rvert$	The determinant of matrix F	255, 282
$J(x,y)$	The Jacobian determinant of a transformation	255, 309
a, b, r, s, t	Real numbers (sometimes also used as integer numbers)	58
i, j, m, n, p, q	Integer numbers	159, 189
i	(In complex numbers): the imaginary unit, $\sqrt{-1}$	32, 38
a, b, x, u	Vectors	62, 231, 438
$\mathbf{f}_1,...,\mathbf{f}_m$	Frequency vectors in the u,v plane of the spectrum	230
$\mathbb{Z}$	The set of all integer numbers (positive, negative, and 0)	159
$\mathbb{R}$	The set of all real numbers	16, 48
$\mathbb{R}^n$	The n-dimensional Euclidean space	47, 429, 463
dim V	Dimension of vector space V	294
Im **g**	The image of linear transformation **g**	294
Ker **g**	The kernel of linear transformation **g**	294
Re[ ]	The real-valued part of a complex entity	34, 38
Im[ ]	The imaginary-valued part of a complex entity	34, 38
Abs[ ]	The magnitude of a complex entity	36, 38
Arg[ ]	The phase of a complex entity	36, 38
$\mathbf{g}(x,y)$	A 2D transformation	48
$\overline{\mathbf{g}}(x,y)$	A 2D transformation (used as direct mapping)	107, 110
$(x,y) \mapsto \mathbf{g}(x,y)$	The transformation **g** maps the point (x,y) to $\mathbf{g}(x,y)$	110
$\mathbf{g}: \mathbb{R}^2 \to \mathbb{R}^2$	A transformation **g** from $\mathbb{R}^2$ to $\mathbb{R}^2$	289, 327
$a \cdot b$	Multiplication	15
v·w	Scalar product of two vectors	246, 439

$\mathbf{f}_1 \circ \mathbf{f}_2$	The composition of transformations $\mathbf{f}_1$ and $\mathbf{f}_2$, i.e. $\mathbf{f}_1(\mathbf{f}_2(\mathbf{x}))$	337, 406, 452
ε, δ	Arbitrarily small, positive real numbers	83, 88
$|a|$	The absolute value of the number a (real or complex)	42, 179, 417
$F(u,v) = \mathcal{F}[f(x,y)]$	$F(u,v)$ is the Fourier transform of $f(x,y)$	32
$\approx$	Approximately equal	131, 406
$\ll$	Much smaller than	76
m	The number of superposed layers	15
$(k_1,...,k_m)$	An index-vector: an m-tuple of integers	231
$\mathbf{f}_{k_1,...,k_m}$	The frequency-vector of the $(k_1,...,k_m)$-impulse	230
$a_{k_1,...,k_m}$	The amplitude of the $(k_1,...,k_m)$-impulse	230, 235
$(k_1,...,k_m)$-moiré	The 1D moiré corresponding to the $(k_1,...,k_m)$-comb	231, 235
$\delta(u)$	The impulse symbol	426
$\delta(u,v)$	The 2D impulse symbol	32
$\text{rect}(x)$	A square pulse: 1 in the range $-0.5 \leq x \leq 0.5$, and 0 elsewhere	214
$\text{rect}(x,y)$	A 2D square pulse: 1 in the range $-0.5 \leq x,y \leq 0.5$, and 0 elsewhere	32
$\text{sinc}(x)$	$\frac{\sin(\pi x)}{\pi x}$ for $x \neq 0$, and 1 for $x = 0$	32
■	End of example, proof, etc.	16

List of abbreviations

Symbol	Short description	Page
1D	1-dimensional	48, 187
2D	2-dimensional	48, 187
DC	The impulse at the spectrum origin (i.e., at frequency zero)	438

CMYK	The four process ink colours: Cyan, Magenta, Yellow, blacK	45
DFT	Discrete Fourier transform	34
FFT	Fast Fourier transform	46
dpi	Dots per inch (printer resolution)	448
lpi	Lines per inch (frequency of a grating or a screen)	448
iff	If and only if	22

References

[Adler81]   R. J. Adler, *The Geometry of Random Fields*. John Wiley & Sons, Chichester, 1981.

[Ahumada83]   A. J. Ahumada, Jr., D. C. Nagel and A. B. Watson, "Reduction of display artifacts by random sampling," in *Applications of Digital Image Processing VI*, Proceedings of the SPIE, Vol. 432, 1983, pp. 216–221.

[Allebach76]   J. P. Allebach and B. Liu, "Random quasiperiodic halftone process," *Journal of the Optical Society of America*, Vol. 66, No. 9, September 1976, pp. 909–917.

[Amidror00]   I. Amidror, *The Theory of the Moiré Phenomenon*. Kluwer Academic Publishers, Dordrecht, 2000.

[Amidror02]   I. Amidror, "A new print-based security strategy for the protection of valuable documents and products using moiré intensity profiles," in *Optical Security and Counterfeit Deterrence Techniques IV*, R. L. Van Renesse (Ed.), Proceedings of SPIE Vol. 4677, SPIE, Bellingham, 2002, pp. 89–100.

[Amidror02a]   I. Amidror, "Scattered data interpolation methods for electronic imaging systems: a survey", *Journal of Electronic Imaging*, Vol. 11, No. 2, April 2002, pp. 157–176.

[Amidror03]   I. Amidror, "Glass patterns as moiré effects: new surprising results," *Optics Letters*, Vol. 28, No. 1, January 2003, pp. 7–9.

[Amidror03a]   I. Amidror, "Unified approach for the explanation of stochastic and periodic moirés", *Journal of Electronic Imaging*, Vol. 12, No. 4, October 2003, pp. 669–681.

[Amidror03b]   I. Amidror, "Glass patterns in the superposition of random line gratings," *Journal of Optics A: Pure and Applied Optics*, Vol. 5, No. 3, May 2003, pp. 205–215.

[Amidror03c]   I. Amidror, "Moiré patterns between aperiodic layers: quantitative analysis and synthesis," *Journal of the Optical Society of America A*, Vol. 20, No. 10, October 2003, pp. 1900–1919.

[Amidror04]   I. Amidror, "Dot trajectories in the superposition of random screens: analysis and synthesis," *Journal of the Optical Society of America A*, Vol. 21, No. 8, August 2004, pp. 1472–1487.

[Amidror04a]   I. Amidror, "Authentication with built-in encryption by using moire intensity profiles between random layers," Published US Patent Application No. 20040001604, 2004; issued as US Patent No. 7,058,202, 2006.

[Andrews03]   L. C. Andrews and R. L. Phillips, *Mathematical Techniques for Engineers and Scientists*. SPIE Press, Bellingham, Washington, 2003.

[Anstis70]   S. M. Anstis, "Phi movement as a subtraction process," *Vision Research*, Vol. 10, 1970, pp. 1411–1430.

[Arnold73]   V. I. Arnold, *Ordinary Differential Equations*. MIT Press, Cambridge, Massachusetts, 1973.

[Bahuguna88]   R. D. Bahuguna, A. B. Western and S. Lee, "Young's double-slit experiment using speckle photography", *American Journal of Physics*, Vol. 56, No. 6, June 1988, pp. 531–533.

[Barlow04]   H. B. Barlow and B. A. Olshausen, "Convergent evidence for the visual analysis of optic flow through anisotropic attenuation of high spatial frequencies," *Journal of Vision*, Vol. 4, No. 6, 2004, pp. 415–426.

[Barrett97]   K. E. Barrett, "Intersection of conics and a rounding error problem," *International Journal of Mathematical Education in Science and Technology*, Vol. 28, No. 1, 1997, pp. 141–145.

[Bendat93]   J. S. Bendat and A. G. Piersol, *Engineering Applications of Correlation and Spectral Analysis*. John Wiley & Sons, NY, 1993 (second edition).

[Bernstein05]   D. S. Bernstein, *Matrix Mathematics*. Princeton University Press, Princeton, 2005.

[Birkhoff89]   G. Birkhoff and G-C. Rota, *Ordinary Differential Equations*. John Wiley & Sons, NY, 1989 (fourth edition).

[Bracewell86]   R. N. Bracewell, *The Fourier Transform and its Applications*. McGraw-Hill Publishing Company, Reading, NY, 1986 (second edition).

[Bracewell95]   R. N. Bracewell, *Two Dimensional Imaging*. Prentice Hall, NJ, 1995.

[Bronshtein97]   I. N. Bronshtein and K. A. Semendyayev, *Handbook of Mathematics*. Springer, Berlin, 1997 (third edition).

[Bronstein90]   I. N. Bronstein and K. A. Semendiaev, *Aide-Mémoire de Mathématiques*. Eyrolles, Paris, 1990 (9th edition; French translation).

[Bryngdahl74]   O. Bryngdahl, "Moiré: formation and interpretation," *Jour. of the Optical Society of America,* Vol. 64, No. 10, 1974, pp. 1287–1294.

[Callahan74]   J. Callahan, "Singularities and plane maps," *The American Mathematical Monthly*, Vol. 81, 1974, pp. 211–240.

[Cantwell02]   B. J. Cantwell, *Introduction to Symmetry Analysis*. Cambridge University Press, Cambridge, 2002.

[Cardinal03]   K. S. Cardinal and D. C. Kiper, "The detection of colored Glass patterns," *Journal of Vision*, Vol. 3, 2003, pp. 199–208.

[Cartwright90]   M. Cartwright, *Fourier Methods for Mathematicians, Scientists and Engineers*. Ellis Horwood, UK, 1990.

[Casasent76]   D. Casasent and D. Psaltis, "Position, rotation, and scale invariant optical correlation," *Applied Optics*, Vol. 15, No. 7, July 1976, pp. 1795–1799.

[Casselman04]   W. Casselman, *Mathematical Illustrations: A Manual of Geometry and PostScript*. Cambridge University Press, Cambridge, 2004. Full text available on the Internet at: `http://www.math.ubc.ca/people/faculty/cass/graphics/text/www/index.html`

[Castelman79]   K. R. Castelman, *Digital Image Processing*. Prentice-Hall, New Jersey, 1979.

[Champeney73]   D. C. Champeney, *Fourier Transforms and their Physical Applications*. Academic Press, London, 1973.

[Champeney87] D. C. Champeney, *A Handbook of Fourier Theorems*. Cambridge University Press, Cambridge, 1987.

[Chosson06] S. Chosson, *Synthèse d'Images Moiré*. Ph.D. Thesis No. 3434, EPFL, Lausanne, 2006 (in French).

[Churchill72] R. V. Churchill, *Operational Mathematics*. McGraw-Hill Kogakusha, Tokyo, 1972 (third edition).

[Cloud95] G. L. Cloud, *Optical Methods of Engineering Analysis*. Cambridge University Press, Cambridge, 1995.

[Colley98] S. J. Colley, *Vector Calculus*. Prentice Hall, New Jersey, 1998.

[Coulon84] F. de Coulon, *Theorie et Traitement des Signaux*. Presses Polytechniques Romandes, Lausanne, 1984 (in French).

[Courant88] R. Courant, *Differential and Integral Calculus (Vol. II)*. Wiley-Interscience, USA, 1988.

[Courant89] R. Courant and F. John, *Introduction to Calculus and Analysis (Vol. II)*. Springer, NY, 1989.

[Dakin97] S. C. Dakin, "The detection of structure in Glass patterns: psychophysics and computational models," *Vision Research*, Vol. 37, 1997, pp. 2227–2246.

[Daly92] S. Daly, "The visible differences predictor: an algorithm for the assessment of image fidelity," *Human Vision, Visual Processing and Digital Display III*, SPIE proceedings, Vol. 1666, USA, 1992, pp. 2–15.

[Debnath95] L. Debnath, *Integral Transforms and Their Applications*. CRC Press, Boca Raton, 1995.

[Dey91] T. W. Dey, "Optical alignment method using arbitrary geometric figures," US Patent No. 5,054,929, 1991.

[Diggle83] P. J. Diggle, *Statistical Analysis of Spatial Point Processes*. Academic Press, London, 1983.

[Ehrenpreis56] L. Ehrenpreis, "On the theory of kernels of Schwartz," Proceedings of the American Mathematical Society, Vol. 7, 1956, pp. 713–718.

[Einstein02] *The Collected Papers of Albert Einstein (Vol. 7)*. Princeton University Press, NJ, 2002.

[EncMath88] *Encyclopaedia of Mathematics,* Vols. 1–10. Kluwer Academic Publishers, Dordrecht, 1988–1994.

[EncStat82] *Encyclopedia of Statistical Sciences,* Vols. 1–9. John Wiley & sons, NY, 1982–1988.

[Erf78] R. K. Erf (Ed.), *Speckle Metrology*. Academic Press, NY, 1978.

[Feitelson88] D. G. Feitelson, *Optical Computing: A Survey for Computer Scientists*. MIT Press, Cambridge, 1988.

[Ferraro88] M Ferraro and T. M. Caelli, "Relationship between integral transform invariances and Lie group theory," *Journal of the Optical Society of America A*, Vol. 5, No. 5, May 1988, pp. 738–742.

[Foley90] J. D. Foley, A. van Dam, S. K. Feiner and J. F. Hughes, *Computer Graphics: Principles and Practice*. Addison-Wesely, Reading, Massachusetts, 1990 (second edition).

[Garavaglia01] M. Garavaglia and P. F. Melián, "Making operations with superposed black and white and colors transparencies: Does quasi-moiré patterns observation suggest two forms to correlate retinal visual signals?" in Proceedings of SPIE Vol. 4419, SPIE, Bellingham, 2001, pp. 581–584.

[Gardner88] W. A. Gardner, *Statistical Spectral Analysis: A Nonprobabilistic Theory.* Prentice Hall, Englewood Cliffs, 1988.

[Gaskill78] J. D. Gaskill, *Linear Systems, Fourier Transforms, and Optics.* John Wiley & Sons, NY, 1978.

[Gåsvik95] K. J. Gåsvik, *Optical Metrology.* John Wiley & Sons, NY, 1995 (second edition).

[Glass69] L. Glass, "Moiré effect from random dots," *Nature*, Vol. 223, August 1969, pp. 578–580.

[Glass73] L. Glass and R. Pérez, "Perception of random dot interference patterns," *Nature*, Vol. 246, December 1973, pp. 360–362.

[Glass76] L. Glass and E. Switkes, "Pattern recognition in humans: correlations which cannot be perceived," *Perception*, Vol. 5, 1976, pp. 67–72.

[Glass02] L. Glass, "Looking at dots," *The Mathematical Intelligencer*, Vol. 24, 2002, pp. 37–43.

[Glassner95] A. S. Glassner, *Principles of Digital Image Synthesis (Vol. 1).* Morgan Kaufmann, San Francisco, 1995.

[Gonzalez87] R. C. Gonzalez and P. Wintz, *Digital image processing.* Addison-Wesley, Reading, Massachusetts, 1987 (second edition).

[Gray97] A. Gray, M. Mezzino and M. A. Pinsky, *Introduction to Ordinary Differential Equations with Mathematica.* Springer, NY, 1997.

[Halmos74] P. R. Halmos, *Naive Set Theory.* Springer, NY, 1974.

[Harburn75] G. Harburn, C. A. Taylor and T. R. Welberry, *Atlas of Optical Transforms.* G. Bell & Sons, London, 1975.

[Hardy48] A. C. Hardy and F. L. Wurzburg, Jr., "A photoelectric method for preparing printing plates," *Journal of the Optical Society of America*, Vol. 38, No. 4, pp. 295–300, 1948.

[Harris98] J. W. Harris and H. Stocker, *Handbook of Mathematics and Computational Science.* Springer, NY, 1998.

[Heiden69] C. Heiden, "Power spectrum of stochastic pulse sequences with correlation between the pulse parameters," *Physical Review*, Vol. 188, No. 1, pp. 319–326, 1969.

[Hersch04] R. D. Hersch and S. Chosson, "Band moiré images," Proceedings of SIGGRAPH 2004, *ACM Transactions on Graphics*, Vol. 23, No. 3, 2004, pp. 239–248.

[Hines75] M. E. Hines, "Image scrambling technique," US Patent No. 3,914,877, 1975.

[Horn85] R. A. Horn and C. R. Johnson, *Matrix Analysis.* Cambridge University Press, Cambridge, 1985.

[Hotta99] K. Hotta, T. Kurita and T. Mishima, "Scale invariant face recognition method using spectral features of log-polar image," in Proceedings of the SPIE Conference on Applications of Digital Image Processing XXII, Denver, Colorado, July 1999, SPIE Vol. 3808, pp. 33–43.

[Howse95] J. W. Howse IV, *Gradient and Hamiltonian Dynamics: Some Applications to Neural Network Analysis and System Identification*. Ph.D. thesis, The University of New Mexico, USA, 1995. http://www.eece.unm.edu/controls/theses/Howse.pdf

[Huck03] J. Huck, *Mastering Moirés: Investigating Some of the Fascinating Properties of Interference Patterns*. Private publication by J. Huck, 2003. http://pages.sbcglobal.net/joehuck

[Indebetouw92] G. Indebetouw and R. Czarnek (Eds.), *Selected Papers on Optical Moiré and Applications*. SPIE Milestone Series, Vol. MS64, SPIE Optical Engineering Press, Washington, 1992.

[Ivanov95] V. I. Ivanov and M. K. Trubetskov, *Handbook of Conformal Mapping with Computer Aided Visualization*. CRC Press, Boca Raton, 1995.

[Kang99] H. R. Kang, *Digital Color Halftoning*. SPIE, Bellingham, Washington, 1999.

[Kaplan03] W. Kaplan, *Advanced Calculus*. Addison-Weseley, Boston, 2003 (5th edition).

[Kipphan01] H. Kipphan, *Handbook of Print Media*. Springer, Berlin, 2001.

[Knopp74] P. J. Knopp, *Linear Algebra: an Introduction*. Hamilton Publishing Company, California, 1974.

[Koenderink90] J. J. Koenderink, *Solid Shape*. The MIT Press, Cambridge, Massachusetts, 1990.

[Kreyszig78] E. Kreyszig, *Introductory Functional Analysis with Applications*. John Wiley & Sons, NY, 1978.

[Kreyszig93] E. Kreyszig, *Advanced Engineering Mathematics*. John Wiley & Sons, NY, 1993 (7th edition).

[Lang87] S. Lang, *Calculus of Several Variables*. Springer, New York, 1987 (3rd edition).

[Lau98] D. L. Lau, G. R. Arce and N. C. Gallagher, "Green-noise digital halftoning," *Proc. of the IEEE*, Vol. 86, pp. 2424–2444, 1998.

[Lau01] D. L. Lau, A. M. Khan and G. R. Arce, "Stochastic moiré," in Proceedings of the IS&T's PICS Conference 2001, Montréal, April 22–25 2001, pp. 96–100.

[Lau02] D. L. Lau, "Stochastic moiré II," in Proceedings of the IS&T's PICS Conference 2002, Portland, April 7–10 2002, pp. 250–255.

[Lau02a] D. L. Lau, A. M. Khan and G. R. Arce, "Minimizing stochastic moiré in frequency-modulated halftones by means of green-noise masks," *Journal of the Optical Society of America A*, Vol. 19, November 2002, pp. 2203–2217.

[Lau03] D. L. Lau, R. Ulichney and G. R. Arce, "Blue- and green-noise halftoning models: a review of the spatial and spectral characteristics of halftone textures," *IEEE Signal Processing Magazine*, Vol. 20, July 2003, pp. 28–38.

[Lay03] D. C. Lay, *Linear Algebra and its Applications*. Addison-Wesley, Boston, 2003 (third edition).

[Lipschutz68] S. Lipschutz, *Theory and Problems of Linear Algebra*. Schaum's Outline Series, McGraw-Hill, New York, 1968.

[Lipschutz01] S. Lipschutz and M. Lipson, *Theory and Problems of Linear Algebra*. Schaum's Outline Series, McGraw-Hill, New York, 2001 (third edition).

[Mansfield76]   L. E. Mansfield, *Linear Algebra with Geometric Applications.* Marcel Dekker, NY, 1976.

[McGrew95]   S. P. McGrew, "Anticounterfeiting method and device using holograms and pseudo-random dot patterns," US Patent No. 5,396,559, 1995.

[Needham97]   T. Needham, *Visual Complex Analysis.* Oxford University Press, Oxford 1997.

[Nishijima64] Y. Nishijima and G. Oster, "Moiré patterns: their application to refractive index and refractive index gradient measurements," *Journal of the Optical Society of America,* Vol. 54, No. 1, January 1964, pp. 1–5.

[Olver86]   P. J. Olver, *Applications of Lie Groups to Differential Equations.* Springer, NY, 1986.

[Oster63]   G. Oster and Y. Nishijima, "Moiré patterns," *Scientific American,* Vol. 208, May 1963, pp. 54–63.

[Papoulis65]   A. Papoulis, *Probability, Random Variables and Stochastic Processes.* McGraw-Hill Kogakusha, Tokyo, 1965.

[Patorski93]   K. Patorski, *Handbook of the Moiré Fringe Technique.* Elsevier, Amsterdam, 1993.

[Pocheć95]   P. Pocheć, "Moiré based stereo matching technique," in Proceedings of the IEEE International Conference on Image Processing (Vol. 2), Washington DC, October 23–26 1995, pp. 370–373.

[Poston78]   T. Poston and I. Stewart, *Catastrophe Theory and its Applications.* Dover, NY, 1978.

[Poularikas96] A. D. Poularikas (Ed.), *The Transforms and Applications Handbook.* CRC Press, Boca Raton, 1996.

[Pratt91]   W. K. Pratt, *Digital Image Processing.* John Wiley & Sons, NY, 1991 (second edition).

[Prokoski99]   F. J. Prokoski, "Method and apparatus for flash correlation," US Patent No. 5,982,932, 1999.

[Renesse98]   R. L. van Renesse (Ed.), *Optical Document Security.* Artech House, Boston, 1998 (second edition).

[Renesse05]   R. L. van Renesse, *Optical Document Security.* Artech House, Boston, 2005 (third edition).

[Ridolfi04]   A. Ridolfi, *Power Spectra of Random Spikes and Related Complex Signals with Application to Communications.* Ph.D. Thesis No. 3157, EPFL, Lausanne, 2004. http://lcavwww.epfl.ch/~ridolfi/research/thesis_andrea.pdf

[Rodriguez94] M. Rodriguez, "Promises and pitfalls of stochastic screening in the graphics arts industry," in Proceedings of the IS&T 47th Annual Conference, Rochester, NY, May 15–20 1994; also reprinted in *Recent Progress in Digital Halftoning,* R. Eschbach (Ed.), IS&T, 1994, pp. 34–37.

[Rosenfeld82] A. Rosenfeld and A. C. Kak, *Digital Picture Processing (Vol.1).* Academic Press, Florida, USA, 1982 (second edition).

[Rubinstein91] J. Rubinstein, J. Segman and Y. Zeevi, "Recognition of distorted patterns by invariance kernels," *Pattern Recognition,* Vol. 24, No. 10, 1991, pp. 959–967.

[Schläpfer94] K. Schläpfer, "Are fine screens an alternative to frequency modulation screening?" Proceedings of the TAGA Conference, Baltimore, May 1994, pp. 34–41.

[Schuette97]   W. Schuette, "Glass patterns in image alignment and analysis," US Patent No. 5,613,013, 1997.

[Schwalbe97]   D. Schwalbe and S. Wagon, *VisualDSolve: Visualizing Differential Equations with Mathematica*. Springer, NY, 1997.

[Sharma03]   G. Sharma (Ed.), *Digital Color Imaging Handbook*. CRC Press, Boca Raton, 2003.

[Spiegel63]   M. R. Spiegel, *Theory and Problems of Advanced Calculus*. Schaum's Outline Series, McGraw-Hill, New York, 1963.

[Spiegel68]   M. R. Spiegel, *Handbook of Formulas and Tables*. Schaum's Outline Series, McGraw-Hill, New York, 1968.

[Steeb96]   W-H. Steeb, *Continuous Symmetries, Lie Algebras, Differential Equations and Computer Algebra*. World Scientific, Singapore, 1996.

[Stoyan95]   D. Stoyan, W. S. Kendall and J. Mecke, *Stochastic Geometry and its Applications*. John Wiley & Sons, Chichester, 1995 (second edition).

[Strogatz94]   S. H. Strogatz, *Nonlinear Dynamics and Chaos*. Addison-Wesely, Reading, Massachusetts, 1994.

[Svanbro04]   A. Svanbro, *Speckle Interferometry and Correlation Applied to Large-Displacement Fields*. Ph.D. thesis, Luleå University of Technology, Sweden, 2004. http://www.sirius.luth.se/expmek/Angelica/Thesis.pdf

[Tabor89]   M. Tabor, *Chaos and Integrability in Nonlinear Dynamics: an Introduction*. John Wiley & Sons, NY, 1989.

[Ulichney87]   R. Ulichney, *Digital Halftoning*. MIT Press, USA, 1987.

[Ulichney88]   R. Ulichney, "Dithering with blue noise," *Proc. of the IEEE*, Vol. 76, pp. 56–79, 1988.

[Vargady64]   L. O. Vargady, "Moiré fringes as visual position indicators," *Applied Optics*, Vol. 3, No. 5, May 1964, pp. 631–636; reprinted in [Indebetouw92, pp. 604–609].

[Walker80]   J. Walker, "The amateur scientist: Visual illusions in random-dot patterns and television 'snow'," *Scientific American*, Vol. 242, No. 4, April 1980, pp. 136–140.

[Walker80a]   J. Walker, "The amateur scientist: More about random-dot displays, plus computer programs to generate them," *Scientific American*, Vol. 243, No. 5, November 1980, pp. 158–168.

[Weisstein99]   E. W. Weisstein, *CRC Concise Encyclopedia of Mathematics*. CRC, Boca Raton, 1999.

[Wesner74]   J. W. Wesner, "Screen patterns used in reproduction of continuous-tone graphics," *Applied Optics*, Vol. 13, No. 7, July 1974, pp. 1703–1710.

[Widmer92]   E. Widmer, K. Schläpfer, V. Humbel and S. Persiev, "The benefits of frequency modulation screening," Proceedings of the TAGA Conference, Vancouver, April 1992, pp. 28–43.

[Wikipedia05]   http://en.wikipedia.org/wiki/Integral_transform, April 2005.

[Williams86]   P. P. Williams, D. H. Davies and G. Harburn, "On randomization techniques for the suppression of unwanted moiré patterns in images generated by a scanning system with a periodic amplitude defect," *Optica Acta*, Vol. 33, No. 10, 1986, pp. 1311–1319.

[Williams92]   D. R. Williams, "Photoreceptor sampling and aliasing in human vision," Chapter 2 in *Tutorials in Optics*, D. T. Moore (Ed.), OSA Annual Meeting, Rochester, NY, 1992.

[Wilson98]   H. R. Wilson and F. Wilkinson, "Detection of global structure in Glass patterns: implications for form vision," *Vision Research*, Vol. 38, 1998, pp. 2933–2947.

[Wolberg90]   G. Wolberg, *Digital Image Warping*. IEEE Press, California, 1990.

[Wolf79]   K. B. Wolf, *Integral Transforms in Science and Engineering*. Plenum Press, NY, 1979.

[Wyszecki82]   G. Wyszecki and W. S. Stiles, *Color Science: Concepts and Methods, Quantitative Data and Formulae*. John Wiley & Sons, NY, 1982 (second edition).

[Yellott82]   J. I. Yellott Jr., "Spectral analysis of spatial sampling by photoreceptors: topological disorder prevents aliasing," *Vision Research*, Vol. 22, 1982, pp. 1205–1210.

[Yellott83]   J. I. Yellott Jr., "Spectral consequences of photoreceptor sampling in the rhesus retina," *Science*, Vol. 221, July 1983, pp. 382–385.

Index

Page numbers followed by the letter "g" indicate entries in the glossary (Appendix I).

A

additive moiré, 260, 465g
additive superposition, *see under* superposition rules
affine transformation, 51–63, 281–284
almost fixed point, *see under* fixed point
almost-identity transformation, *see under* transformation
almost-null transformation, *see under* transformation
almost-periodic function, 459g
almost-zero transformation, *see under* transformation
amplitude spectrum, *see under* spectrum
anti-counterfeiting, *see* applications of Glass patterns: document security
aperiodic function, 16, 459g
aperiodic layer, *see under* layer
applications of Glass patterns, 4–5
  art, 4
  comparison of patterns, 4, 46
  comparing the scale of two copies of an image, 96
  detection and measurement of slight displacements or deformations, 4, 42, 97, 221, 259
  detection of periodic noise (or periodic residues) in an aperiodic structure, 275
  determination of the axis of rotation, 4, 96
  determination of the similarity or the degree of correlation between patterns, 4, 19
  document security (authentication, anti-counterfeiting), 4, 97, 259, 270
  high-precision registration, 4, 96, 221, 259
  human visual system, 4, 259
  identification of a given pattern within an image, 4, 46
  latent images, 97, 224
  optical alignment, 4, 96, 221, 259
  speckle interferometry, 43–44
  speckle metrology, 4, 42–43
  stereo matching, 44, 48
  visualizing the curve family of an indefinite integral of a function, 4, 151
  visualizing the flow lines of a vector filed, 4, 105, 148
  visualizing the solutions of differential equations, 105, 148
  visualizing the solutions of equations, 4, 98
  visually locating (or illustrating) the fixed points of a transformation, 4, 63, 97
approaches for investigating Glass patterns:
  indicial equations, 159, 193–196
  probabilistic (or stochastic) approach, 279–280, 422–432
  spectral, Fourier-based approach, 15, 225, 238–280
authentication, *see* applications of Glass patterns: document security
autocorrelation, *see under* correlation
autocorrelation theorem, 417

B

backward mapping algorithm (in digital imaging), 386–389, 468g
bending rate, 65
bending transformation, 215, 461g
binary, 467g
blade, *see* line-impulse

C

cardioid, 100
Cartesian coordinates, *see under* coordinates
charecteristic function, 355

cluster, 228, 235
colour printing, 4, 24, 45–46
colour spectrum, 466g
comb, 228, 462g
conformal transformation (or mapping), 296–298, 305, 324
congruent layers, 239, 243, 249–250
constraint, 191–193, 208–214
contour lines, *see* level lines
convolution, 15, 271, 411–420
of combs, 230
of line-impulses (blades), 240–242
of nailbeds, 234–235
convolution theorem, 15, 226, 230, 235, 240, 244, 249–250, 416–417
coordinate-and-profile transformed structure, 460g
coordinate transformation, 295–298
see also transformation
coordinate-transformed structure, 460g
coordinates:
Cartesian, 85–86, 100–102, 230–231, 329, 332–336, 341, 354, 365–374
curvilinear, 295–298, 332–336
polar, 85–86, 100–102, 228, 341, 354, 365–374
correlation, 17, 418–420, 469g
autocorrelation, 415, 423
cross correlation, 17, 246, 258, 264–269, 271–273, 411–420, 470g
global correlation, 17, 264–269, 272–273, 470g
local correlation, 17, 264–269, 272–273, 469g
cosinusoidal grating, *see under* grating
counterfeit deterrents, *see* applications of Glass patterns: document security
critical point:
of a system of differential equations, 443, 468g
of a transformation, 331, 468g
of a vector field, 443, 468g
cross correlation, *see under* correlation
cross-correlation theorem, 416–417
curved grid, 458g
curved screen, 65–82, 254–258, 458g
curvilinear grating, 196–208, 453, 458g
curvilinear impulse, 462g
cutoff frequency, 228
cyclic convolution, *see* T-convolution

D

DC impulse, 462g
decorrelation, 96

deterministic signal or image, 32, 421, 430–432
diffeomorphism, 471g
difference moiré, *see* subtractive moiré
diffraction, 6
direct transformation, 109–110, 133, 151–152, 160–161, 183–186, 292, 295, 327–410, 467g
discrete, 467g
displacement, *see* shift
document security, *see under* applications of Glass patterns
domain, 327, 332–336, 347
domain transformation, 56, 65, 82, 109–110, 133, 151–152, 158–160, 183–186, 218, 273, 293–295, 347–356, 460, 467g
dot-screen, *see* screen
dot shape, 156, 236–238
dot trajectories, 8, 12–14, 19–29, 105–156, 158–161, 465g
classification, 443–451
curve equations, 114–123, 152
invariance properties, 175–181
morphology, 106–107
synthesis, 134–137, 149–151, 259, 280
under different superposition rules, 139–140, 156
visual interpretation, 140–148
dots per inch (dpi), 472g

E

eigenvalues of a matrix, 444–454
element distribution, *see under* layer
extraction of a moiré, *see* moiré extraction

F

field line, *see* trajectory
finite-energy signal (or function), 415–416, 470g
finite-power signal (or function), 416–417, 471g
first order moiré, 158, 180, 222
fixed dot shape, 240, 252
fixed element shape, 240, 248
fixed line, 60–62, 65, 254–258, 469g
almost fixed line, 90–95, 104
fixed line shape, 240
fixed locus, 47–72, 469g
almost fixed locus, 64, 82, 87–96, 104
mutual fixed locus, 72–82, 469g
synthesis, 83–87, 100–102, 280
fixed point, 47–72, 189–193, 399–405, 443, 468g
almost fixed point, 64, 82, 87–96, 104, 469g
mutual fixed point, 72–82, 154–156, 258, 288, 401–405, 469g

multiple fixed points, 65, 80–82, 254–258, 287

fixed point theorem, 47–49, 281–288
  fixed point theorem for affine mappings, 48, 281–284
  fixed point theorem for second-order polynomial mappings, 284–288

flash correlation artifacts, 46, 457

flow line, *see* trajectory

forward mapping algorithm (in digital imaging), 386–389, 468g

Fourier-based approach, *see also under* approaches for investigating Glass patterns
  generalized, 252, 269, 433–442
  in the aperiodic case, 238–269
  in the periodic case, 226–238

Fourier decomposition, 433–434
  generalized, 434–435

Fourier series, 228, 433
  coefficients, 433
  generalized, 434

Fourier spectrum, *see* spectrum

Fourier transform, 15, 32, 421, 423, 433
  g-Fourier transform, 439
  generalized, 433–442
  inverse, 232, 236, 244, 250, 433

frequency domain, *see* spectral domain; spectrum

frequency vector, 228, 230, 233, 461g

functional, 419

fundamental Glass-pattern theorem:
  for the superposition of aperiodic gratings, 268, 456
  for the superposition of aperiodic screens, 269, 456

G

generalized Fourier series, *see under* Fourier series

generalized Fourier transform, *see under* Fourier transform

geometric layout, 159, 174, 280, 461g

geometric location of an impulse, *see* impulse: location

Glass pattern, 1–5, 11–14, 19, 51–53, 64, 82, 158–161, 180, 222, 457g, 463g
  applications, *see* applications of Glass patterns
  artificial, 275–277
  behaviour under affine layer transformations, 59–63
  behaviour under layer rotations, 51–53
  behaviour under layer scalings, 53–54
  behaviour under layer shifts, 54–59, 96
  behaviour under non-linear layer transformations, 63–72
  between aperiodic line gratings, 187–224, 238–248
  curvilinear, 464g
  historical background, 4–5
  hybrid (1,-1)-Glass pattern whose band carries 2D information, 452–456
  in multilayer superpositions, 30–31, 104
  intensity profile, 214–220, 224, 225, 238–269
  invariance properties, 175–181, 222, 224
  linear, 66–67, 70, 149, 240, 244, 288, 464g
  radius of, 277–278
  singular, 272
  synthesis, 47, 83–87, 100–102, 222–223, 259, 270, 280
  theorem, *see* fundamental Glass pattern theorem

global correlation, *see under* correlation

gradient lines (or curves), 302

grating, 187–224, 457g
  aperiodic, 196
  cosine shaped, 458g
  cosinusoidal, 458g
  curvilinear, *see* curvilinear grating
  periodic, 196
  repetitive, 196

gray-level surface, 215

grid, 458g
  curved, *see* curved grid
  regular (or square), 458g

H

halftone screen, 458g
  see also halftoning

halftoning, 4, 24, 45

Hermitian, 240, 249

higher order moiré, 54, 158

human visual system, 4, 7, 228, 259

hump, 242, 249, 463g

hybrid between Glass and moiré pattern, *see* superposition: of partly periodic layers

hybrid (1,-1)-Glass pattern whose band carries 2D information, 452–456

hybrid (1,-1)-moiré whose bands carry 2D information, 452–456

hybrid spectrum:
  continuous and impulsive, 463g

I

image, 15, 459g

image domain, 15

image of a linear transformation, 294

impulse, 228
   amplitude, 228
   DC, *see* DC impulse
   line, *see* line-impulse
   location, 228
impulsive spectrum, *see under* spectrum
indicial equations, *see under* approaches for
   investigating Glass patterns
inseparable, 472g
integral transform, 433–442
intensity profile, 214–220, 224, 225, 238–259,
   452–456, 460g
   see also under moiré, Glass pattern
inverse Fourier transform, *see* Fourier transform:
   inverse
inverse transformation, 109–110, 327–410, 467g
isometric layers, 238, 244–246, 250, 258
isotropic mapping, 324

J
Jacobian, 255, 284, 286, 291, 309
   geometric interpretation, 309–311
Jacobian matrix, 309, 322–326, 448–451

K
kernel of a linear transformation, 294
kernel of an integral transform, 435–439

L
laser speckle, *see* speckle
latent images, *see under* applications of Glass
   patterns
layer:
   aperiodic, 1–3, 11, 15–19, 63–64
   constrained, *see* constraint
   deterministic, 421, 430–432
   element distribution within a layer, 28–30
   intermediate between periodic and aperiodic,
      17, 50–51, 98–101, 172–174, 260–262
   periodic, 1–3, 11, 15–19, 63–64
   pseudo-random, 28, 275–276
   random, 1–3, 11, 15–19, 28–29, 421–432
   repetitive, 1–3, 11, 15–19, 63–64
   stochastic, 16
level lines, 100, 159, 162, 164,–167, 170, 174,
   289–291, 295, 303–305, 330, 338–339, 358
limaçon, 100
line-grating, *see* grating
line-grid, *see* grid
line impulse, 240–248, 462g
line of coincidence, 189–193
linear algebra, 345–346
lines per inch (lpi), 472g
local correlation, *see under* correlation

locally reflecting transformation, *see* reflecting
   transformation
lpi (lines per inch), 472g

M
macrostructure, 18, 26–28, 180, 214, 222, 465g
   and microstructure, 18, 26–28
magnitude spectrum, *see* spectrum: amplitude
   spectrum
mapping, *see* transformation
measuring (displacements, deformations, etc.),
   see various entries under applications of
   Glass patterns
metrology, *see under* applications of Glass
   patterns
microstructure, 18, 19, 26–28, 105–156, 180,
   465g
   and macrostructure, 18, 26–28
   artifacts, 28, 142–143
   morphology, 106–107
   under different superposition rules, 139–140,
      156
   invariance properties, 175–181
   visual interpretation, 140–148
moiré, 1–3, 11, 24, 64, 158–161, 463g
   additive, *see* additive moiré
   between aperiodic layers, *see* Glass pattern
   between periodic and aperiodic layers, 46
   between periodic layers, 1, 11, 64
   between random layers, *see* Glass patterns
   hybrid (1,-1)-moiré whose bands carry 2D
      information, 452–456
   indexing, *see* moiré notational system
   intensity profile, 214–220, 225–238, 465g
   invariance properties, 175–181
   notational system for, *see* moiré notational
      system
   of higher order, *see* higher order moiré
   of the first order, *see* first order moiré
   period, 230–231
   profile, *see* moiré: intensity profile
   profile extraction:
      in superposed gratings, 232–233, 244–
         245
      in superposed screens, 235–236, 249–250
   singular, 464g
   subtractive, *see* subtractive moiré
   temporal, 6
   theorem, *see* fundamental moiré theorem
   unwanted, *see* unwanted moirés
moiré analysis:
   qualitative, 225
   quantitative, 225
moiré extraction, 232, 236, 244, 249–250

moiré-free superposition, 24–26
    stable (non-singular), 24–28, 464g
    unstable (singular), 24–26, 464g
moiré notational system, 231
    $(1,-1)$-moiré, 231–233, 242–243, 259–260, 440
    $(1,1)$-moiré, 260
    $(1,0,-1,0)$-moiré, 235, 249, 254, 259–260
    $(1,0,1,0)$-moiré, 260
    (k_1,k_2)-moiré, 231, 260, 440
    (k_1,k_2,k_3,k_4)-moiré, 235, 260
    $(k_1,...,k_m)$-moiré, 464g
monochromatic, 466g
monochrome, 15, 259, 466g
multichromatic, *see* polychromatic
multilayer superposition, *see under* superposition
multiplicative superposition, *see under* superposition rules
mutual fixed locus, *see under* fixed locus
mutual fixed point, *see under* fixed point

N

nailbed, 38, 228, 230, 233, 462g
non-reflecting transformation, 255, 306–308, 312–313
normalization, 232, 236, 244–245, 250–251
notational system for moirés, *see* moiré notational system

O

operator:
    binary, 411
    linear, 412
    unary, 411
optical alignment, *see under* applications of Glass patterns

P

parallax, 6
period, 459g
period-vector, 459g
periodic function, 16, 459g
    1-fold periodic function, 459g
    2-fold periodic function, 459g
periodic layer, *see under* layer
periodic profile, 460g
    normalized, 461g
phase, 471g
phase spectrum, *see under* spectrum
pinhole screen, *see under* screen
point of coincidence, 52, 70, 77
point process, 422, 425–426
Poisson process, 425–426

polar coordinates, *see under* coordinates
polychromatic, 15, 259
power spectrum, *see under* spectrum
precision alignment, *see* applications of Glass patterns: optical alignment
precision measurement, *see various entries under* applications of Glass patterns
profile, *see* intensity profile; periodic profile
profile-transformed structure, 460g
pseudo-random, *see under* layer

Q

qualitative moiré analysis, *see under* moiré analysis
quantitative moiré analysis, *see under* moiré analysis

R

random scanning, 45
random field, 429–430
random image, *see* layer: random
random layer, *see* layer: random
random process, 32, 422
random sampling, 44–45
random structure, 460g
range, 327, 332–336, 347
range transformation, 218, 347–356, 460, 467g
reflectance function, 15, 466g
    chromatic, 466g
reflecting transformation, 255, 306–308, 312–313
repetitive layer, *see under* layer
repetitive, non-periodic structure, 460g
rosettes, 465g

S

sampling moiré, 44–45
scaling, 411, 471g
scanning moiré, *see* sampling moiré
screen, 458g
    curved, *see* curved screen
    halftone, *see* halftone screen; halftoning
    pinhole, 227, 236, 252–253
    regular, 458g
screen gradation, 459g
separable, 471g
    see also spatially separable
shift, 54–59
shot noise, 426–429
similar layers, 239, 249–250
similarity matrix (or transformation), 58, 324
singular:
    Glass pattern, *see* Glass pattern: singular
    linear transformation, 282

moiré, *see* moiré: singular

state, *see* superposition: singular

superposition, *see* superposition: singular

support, 463g

slits, 453

spatially separable, 472g

speckle, 42–44

interferometry, *see under* applications of
Glass patterns

metrology, *see under* applications of
Glass patterns

spectral approach, *see under* approaches for
investigating Glass patterns

in the periodic case, 226–238

in the aperiodic case, 238–269

spectral domain, 15, 226

see also spectrum

spectrum, 15, 461g

amplitude spectrum, 32–42, 423

colour, *see* colour spectrum

continuous, 16, 38, 226, 230, 248

diffuse, 16, 32–33, 38, 226, 230, 424

hybrid, *see* hybrid spectrum

imaginary part, 32–42, 423

impulsive, 16, 39, 226, 230, 248, 463g

magnitude spectrum, *see* spectrum: amplitude
spectrum

of a (k_1,k_2)-moiré between two periodic
gratings, 231

of a (k_1,k_2,k_3,k_4)-moiré between two periodic
screens, 235

of a partially periodic function (intermediate
between periodic and aperiodic), 17, 38

of a periodic function, 16, 32–42, 226, 228

of a random screen, 32–42

of a superposition, 15

of an aperiodic function, 16, 32–42, 226,
228, 230

phase spectrum, 32–42, 423

power spectrum, 32, 417, 421–432, 423–425

real part, 32–42, 423

stable moiré-free state, *see under* moiré-free
superposition

stochastic process, *see* random process

subtractive moiré, 260, 465g

superposition, 1, 15

macrostructure of, *see* macrostructure

microstructure of, *see* microstructure

of aperiodic layers, 1–5, 11, 19–29, 50, 157

of correlated aperiodic gratings, 240–246

of correlated aperiodic screens, 249–252

of line-gratings, 187–224, 230–233

of partly correlated layers, 21, 262–264

of partly periodic (or partly random) layers,
50–51, 98–101, 172–174, 260–262

of periodic layers, 1–2, 11, 50, 157

of random layers, 1–5, 11

of repetitive, non-periodic layers, 1–2, 11, 63

of screens, 12–31, 233–238, 248–259

of several layers, 30–31, 104

of uncorrelated aperiodic gratings, 246–248

of uncorrelated aperiodic screens, 258–259

singular, 24–26

superposition moiré in colour printing, 45–46

superposition rules, 15–16, 139–140, 156, 280

additive, 15–16

inverse additive, 16

multiplicative, 15

support (of a comb, a nailbed, a spectrum etc.),
462g

singular, 463g

synthesis of dot trajectories, *see* dot trajectories:
synthesis

synthesis of Glass patterns, *see* Glass pattern:
synthesis

T

T-convolution, 233, 236, 416

T-convolution theorem:

in one dimension, 233

in two dimensions, 236

T-cross-correlation, 416

Talbot effect, 6

temporal moiré, 6

trajectory, 105, 107, 148, 298, 331, 468g

transform, 436, 467

see also Fourier transform, integral transform

transformation, 289, 327, 347, 436

see also direct transformation, inverse
transformation, domain transformation,
range transformation

active and passive interpretations, 343–347

affine, *see* affine transformation

almost identity, 406, 471g

almost null, 406, 471g

almost zero, 406–410, 471g

Cartesian to polar coordinate transformation,
340–341, 354, 365–374

conformal, *see* conformal transformation

direct, *see* direct transformation

identity, 337, 404, 406

interpretations of a 2D transformation, 289–
308, 329–331

inverse, *see* inverse transformation

linear, 281–282

parabolic, 64–72

polar to Cartesian coordinate transformation, 340–341, 354, 365–374
  second-order polynomial, 65, 284–288
  weak, 98, 123, 132–133, 155, 161, 167, 174–175, 182, 406–410, 471g
  zero, 406
transformation surface, 215
translation, *see* shift
transmittance function, 15, 466g
  chromatic, 467g

U

unstable moiré-free state, *see under* moiré-free superposition

unwanted moirés, 4–5
  avoiding, 4–5, 44–45

V

vector field, 102, 109–111, 298–301, 330
visibility circle, 228, 461g

W

weak transformation, *see under* transformation
wedge, *see* screen gradation

Z

zero transformation, *see under* transformation

Computational Imaging and Vision

1. B.M. ter Haar Romeny (ed.): *Geometry-Driven Diffusion in Computer Vision.* 1994
   ISBN 0-7923-3087-0
2. J. Serra and P. Soille (eds.): *Mathematical Morphology and Its Applications to Image Processing.* 1994 ISBN 0-7923-3093-5
3. Y. Bizais, C. Barillot, and R. Di Paola (eds.): *Information Processing in Medical Imaging.* 1995 ISBN 0-7923-3593-7
4. P. Grangeat and J.-L. Amans (eds.): *Three-Dimensional Image Reconstruction in Radiology and Nuclear Medicine.* 1996 ISBN 0-7923-4129-5
5. P. Maragos, R.W. Schafer and M.A. Butt (eds.): *Mathematical Morphology and Its Applications to Image and Signal Processing.* 1996 ISBN 0-7923-9733-9
6. G. Xu and Z. Zhang: *Epipolar Geometry in Stereo, Motion and Object Recognition. A Unified Approach.* 1996 ISBN 0-7923-4199-6
7. D. Eberly: *Ridges in Image and Data Analysis.* 1996 ISBN 0-7923-4268-2
8. J. Sporring, M. Nielsen, L. Florack and P. Johansen (eds.): *Gaussian Scale-Space Theory.* 1997 ISBN 0-7923-4561-4
9. M. Shah and R. Jain (eds.): *Motion-Based Recognition.* 1997 ISBN 0-7923-4618-1
10. L. Florack: *Image Structure.* 1997 ISBN 0-7923-4808-7
11. L.J. Latecki: *Discrete Representation of Spatial Objects in Computer Vision.* 1998
    ISBN 0-7923-4912-1
12. H.J.A.M. Heijmans and J.B.T.M. Roerdink (eds.): *Mathematical Morphology and its Applications to Image and Signal Processing.* 1998 ISBN 0-7923-5133-9
13. N. Karssemeijer, M. Thijssen, J. Hendriks and L. van Erning (eds.): *Digital Mammography.* 1998 ISBN 0-7923-5274-2
14. R. Highnam and M. Brady: *Mammographic Image Analysis.* 1999
    ISBN 0-7923-5620-9
15. I. Amidror: *The Theory of the Moiré Phenomenon.* 2000 ISBN 0-7923-5949-6;
    Pb: ISBN 0-7923-5950-x
16. G.L. Gimel'farb: *Image Textures and Gibbs Random Fields.* 1999 ISBN 0-7923-5961
17. R. Klette, H.S. Stiehl, M.A. Viergever and K.L. Vincken (eds.): *Performance Characterization in Computer Vision.* 2000 ISBN 0-7923-6374-4
18. J. Goutsias, L. Vincent and D.S. Bloomberg (eds.): *Mathematical Morphology and Its Applications to Image and Signal Processing.* 2000 ISBN 0-7923-7862-8
19. A.A. Petrosian and F.G. Meyer (eds.): *Wavelets in Signal and Image Analysis. From Theory to Practice.* 2001 ISBN 1-4020-0053-7
20. A. Jaklič, A. Leonardis and F. Solina: *Segmentation and Recovery of Superquadrics.* 2000 ISBN 0-7923-6601-8
21. K. Rohr: *Landmark-Based Image Analysis. Using Geometric and Intensity Models.* 2001 ISBN 0-7923-6751-0
22. R.C. Veltkamp, H. Burkhardt and H.-P. Kriegel (eds.): *State-of-the-Art in Content-Based Image and Video Retrieval.* 2001 ISBN 1-4020-0109-6
23. A.A. Amini and J.L. Prince (eds.): *Measurement of Cardiac Deformations from MRI: Physical and Mathematical Models.* 2001 ISBN 1-4020-0222-X

Computational Imaging and Vision

24. M.I. Schlesinger and V. Hlaváč: *Ten Lectures on Statistical and Structural Pattern Recognition.* 2002   ISBN 1-4020-0642-X
25. F. Mokhtarian and M. Bober: *Curvature Scale Space Representation: Theory, Applications, and MPEG-7 Standardization.* 2003   ISBN 1-4020-1233-0
26. N. Sebe and M.S. Lew: *Robust Computer Vision.* Theory and Applications. 2003
                                                                        ISBN 1-4020-1293-4
27. B.M.T.H. Romeny: *Front-End Vision and Multi-Scale Image Analysis.* Multi-scale Computer Vision Theory and Applications, written in Mathematica. 2003
                                                                        ISBN 1-4020-1503-8
28. J.E. Hilliard and L.R. Lawson: *Stereology and Stochastic Geometry.* 2003
                                                                        ISBN 1-4020-1687-5
29. N. Sebe, I. Cohen, A. Garg and S.T. Huang: *Machine Learning in Computer Vision.* 2005   ISBN 1-4020-3274-9
30. C. Ronse, L. Najman and E. Decencière (eds.): *Mathematical Morphology: 40 Years On.* Proceedings of the 7th International Symposium on Mathematical Morphology, April 18–20, 2005. 2005   ISBN 1-4020-3442-3
31. R. Klette, R. Kozera, L. Noakes and J. Weickert (eds.): *Geometric Properties for Incomplete Data.* 2006   ISBN 1-4020-3857-7
32. K. Wojciechowski, B. Smolka, H. Palus, R.S. Kozera, W. Skarbek and L. Noakes (eds.): *Computer Vision and Graphics.* International Conference, ICCVG 2004, Warsaw, Poland, September 2004, Proceedings. 2006   ISBN 1-4020-4178-0
33. K. Daniilidis and R. Klette (eds.): *Imaging Beyond the Pinhole Camera: Proceedings of the Twelfth Workshop Theoretical Foundations of Computer Vision.* 2006
                                                                        ISBN 1-4020-4893-9
34. I. Amidror: *The Theory of the Moiré Phenomenon.* Volume 2. 2007
                                                                        ISBN 1-4020-5457-2